PRO/ENGINEER 2001 INSTRUCTOR

THE MCGRAW-HILL GRAPHICS SERIES

Providing you with the highest quality textbooks that meet your changing needs requires feedback, improvement, and revision. The team of authors and McGraw-Hill are committed to this effort. We invite you to become part of our team by offering your wishes, suggestions, and comments for future editions and new products and texts.

Please mail or fax your comments to: David Kelley
c/o McGraw-Hill
1333 Burr Ridge Parkway
Burr Ridge, IL 60521
fax 630-789-6946

TITLES IN THE MCGRAW-HILL GRAPHICS SERIES INCLUDE:

Graphic Drawing Workbook
Gary Bertoline, 1999

Technical Graphics Communications, 2e
Bertoline, Wiebe, Miller, and Mohler, 1997

Fundamentals of Graphics Communication, 3e
Bertoline, Wiebe, and Mohler, 2001

Engineering Graphics Communication
Bertoline, Wiebe, Miller, and Nasman, 1995

I-DEAS Student Edition SDRC, 2000

AutoCAD Instructor Release 12
James A. Leach, 1995

AutoCAD Companion Release 12
James A. Leach, 1995

AutoCAD 13 Instructor
James A. Leach, 1996

AutoCAD 13 Companion
James A. Leach, 1996

AutoCAD 14 Instructor
James A. Leach, 1998

AutoCAD 14 Companion
James A. Leach, 1998

AutoCAD 2000 Instructor
James A. Leach, 2000

AutoCAD 2000 Companion
James A. Leach, 2000

The Companion for CADKEY 97
John Cherng, 1998

Hands-on CADKEY: A Guide to Versions 5, 6, and 7
Timothy J. Sexton, 1996

Modeling with AutoCAD Designer
Dobek and Ranschaert, 1996

Engineering Design and Visualization Workbook
Dennis Stevenson, 1995

Graphics Interactive CD-ROM
Dennis Lieu, 1997

PRO/ENGINEER 2001 INSTRUCTOR

David S. Kelley
Purdue University

Boston Burr Ridge, IL Dubuque, IA Madison, WI New York San Francisco St. Louis
Bangkok Bogotá Caracas Kuala Lumpur Lisbon London Madrid Mexico City
Milan Montreal New Delhi Santiago Seoul Singapore Sydney Taipei Toronto

This book is dedicated to my wife, Dixie, for her love and support.

McGraw-Hill Higher Education 

A Division of The ***McGraw-Hill*** *Companies*

PRO/ENGINEER 2001 INSTRUCTOR

Published by McGraw-Hill, a business unit of The McGraw-Hill Companies, Inc., 1221 Avenue of the Americas, New York, NY 10020.

Some ancillaries, including electronic and print components, may not be available to customers outside the United States.

This book is printed on recycled, acid-free paper containing 10% postconsumer waste.

1 2 3 4 5 6 7 8 9 0 QPD/QPD 0 9 8 7 6 5 4 3 2 1

ISBN 0-07-249940-0

General manager: *Thomas E. Casson*
Publisher: *Elizabeth A. Jones*
Senior developmental editor: *Kelley Butcher*
Executive marketing manager: *John Wannemacher*
Project manager: *Joyce M. Berendes*
Senior production supervisor: *Sandy Ludovissy*
Coordinator of freelance design: *Rick D. Noel*
Cover illustration: *©Vehicle Research Institute/Dr. Michael Seal/William Connelly*
Cover designer: *Rick D. Noel*
Supplement producer: *Brenda A. Ernzen*
Media technology senior producer: *Phillip Meek*
Compositor: *Interactive Composition Corporation*
Typeface: *10/12 Times Roman*
Printer: *Quebecor World Dubuque, IA*

Library of Congress Cataloging-in-Publication Data

Kelley, David S.
Pro/Engineer 2001 instructor / David S. Kelley. — 1st ed.
cm. — (The McGraw-Hill graphics series)
ISBN 0-07-249940-0
Includes index.
1. Pro/ENGINEER. 2. Computer-aided design. 3. Mechanical drawing. I. Title.
II. Series.

TA174 .K445 2002
620'.0042—dc21 2001034101
CIP

www.mhhe.com

PREFACE

PURPOSE

My decision to write a Pro/ENGINEER textbook was based on the lack of a comprehensive textbook on this popular computer-aided design package. I focused on several objectives and ideas when I started to develop this project:

1. To write a textbook for an introductory course in engineering graphics.
2. To meet the needs of institutions teaching a course on parametric design and constraint-based modeling.
3. To create a book that would serve as a self-paced, independent study guide for the learning of Pro/ENGINEER for those who do not have the opportunity to take a formal course.
4. To incorporate a tutorial approach to the learning of Pro/ENGINEER in conjunction with detailed reference material.
5. To include topics that make the text a suitable supplement for an upper division course in mechanical design.

APPROACHES TO USING THE TEXTBOOK

This textbook is designed to serve as a tutorial, reference, and lecture guide. Chapters start by covering selected topics in moderate detail. Following the reference portion of each chapter are one or more tutorials covering the chapter's objectives and topics. At the end of each chapter are practice problems used to reinforce concepts covered in the chapter and previously in the book.

I had several ideas in mind when developing this approach to the book:

1. Since Pro/ENGINEER is a menu-intensive, computer-aided design application, the most practical pedagogical method to cover Pro/ENGINEER's capabilities (that would be the most beneficial both to students and instructors) would be to approach this book as a tutorial.
2. The book provides detailed reference material. A typical approach to teaching Pro/ENGINEER would be to provide a tutorial exercise followed by a nontutorial practice or practical problem. Usually students can complete the tutorial, but they may run into problems on the practice exercise. One of the problems that Pro/ ENGINEER students have is digging back through the tutorial to find the steps for performing specific modeling tasks. The reference portion of each chapter in this text provides step-by-step guides for performing specific Pro/ENGINEER modeling tasks outside of a tutorial environment.
3. A supplemental CD with existing Pro/ENGINEER part and configuration files is available to enhance reference material. Each chapter contains reference guides for performing specific Pro/ENGINEER tasks. When appropriate in the reference guide, Pro/ENGINEER part files have been provided. This serves two purposes. First, the provided model provides a good starting point for instructors lecturing on specific topics of Pro/ENGINEER. Second, a student of Pro/ENGINEER can use a reference guide and part file to practice specific tasks.
4. The book is flexible in the order that topics are covered. Chapters 1 and 2 are primarily reference material. The first in-depth modeling tutorial starts with Chapter 3 (sketching fundamentals), and the first three-dimensional modeling begins in Chapter 4 (extruding features). Using this textbook in a course, an instructor may decide to start with any of Chapters 1, 2, 3, or 4. Additionally, many of the chapters after Chapter 4 can be reordered to meet the needs of an individual instructor.

STUDENTS OF PRO/ENGINEER

One of the objectives of this book is to serve as a stand-alone text for independent learners of Pro/ENGINEER. This book is approached as a tutorial to help meet this objective. Since Pro/ENGINEER is menu intensive, tutorials in this book use numbered steps to guide the selection of menu options. The following is an example of a tutorial step:

STEP 6: **Place Dimensions according to Design Intent.**

Use the Dimension icon to match the dimensioning scheme shown in Figure 4–24. Placement of dimensions on a part should match design intent. With Intent Manager activated (Sketch >> Intent Manager), dimensions and constraints are provided automatically that fully define the section. Pro/ENGINEER does not know what dimensioning scheme will match design intent, though. Due to this, it is usually necessary to change some dimension placements.

> **MODELING POINT** If possible, a good rule of thumb to follow is to avoid modifying the section's dimension values until your dimension placement scheme matches design intent.

The primary menu selection is shown in bold. In this example, you are instructed to use the dimension option (portrayed by the Dimension icon) to create dimensions that match the part's design intent. Following the specific menu selection, when appropriate, is the rationale for the menu selection. In addition, Modeling Points are used throughout the book to highlight specific modeling strategies.

CHAPTERS

The following is a description and rationale for each chapter in the book:

CHAPTER 1 INTRODUCTION TO PARAMETRIC DESIGN This chapter covers the basic principles behind parametric modeling, parametric design, and constraint-based modeling. Discussed is how Pro/ENGINEER can be used to capture design intent, and how it can be an integral component within a concurrent engineering environment.

CHAPTER 2 PRO/ENGINEER'S USER INTERFACE This chapter covers basic principles behind Pro/ENGINEER's interface and menu structure. The purpose is to serve as a guide and reference for later modeling activities. A tutorial is provided to reinforce the chapter's objectives.

CHAPTER 3 CREATING A SKETCH Parametric modeling packages such as Pro/ENGINEER rely upon a sketching environment to create most features. This chapter covers the fundamentals behind sketching in Pro/ENGINEER's sketcher mode. Two tutorials are provided along with practice problems.

CHAPTER 4 EXTRUDING, MODIFYING, AND REDEFINING FEATURES Chapter 4 is the first chapter covering Pro/ENGINEER's solid modeling capabilities. Pro/ENGINEER's protrusion and cut commands are introduced, and the extrude option is covered in detail. In addition, modification and datum plane options are introduced. Two tutorials are provided.

CHAPTER 5 FEATURE CONSTRUCTION TOOLS While the protrusion and cut commands are Pro/ENGINEER's basic tools for creating features, this chapter covers additional feature creation tools. Covered in detail are the hole, round, rib, and chamfer commands; creating draft surfaces; shelling a part; cosmetic features; and creating linear patterns. Two tutorials are provided.

CHAPTER 6 REVOLVED FEATURES Many Pro/ENGINEER features are created by revolving around a center axis. Examples include the revolve option found under the protrusion and cut commands, the sketched hole option, thc shaft command, the flange command, and the neck command. These options, along with creating rotational patterns and datum axes, are covered in this chapter. Two tutorials are provided.

CHAPTER 7 FEATURE MANIPULATION TOOLS Pro/ENGINEER provides tools for manipulating existing features. Manipulation tools covered include the group option, copying features, user-defined features, creating relations, family tables, and cross sections. In addition, using the model tree to manipulate features is covered.

CHAPTER 8 CREATING A PRO/ENGINEER DRAWING Since Pro/ENGINEER is primarily a modeling and design application, the creation of engineering drawings is considered a downstream task. Despite this, there is a need to cover the capabilities of Pro/ ENGINEER's drawing mode. This chapter covers the creation of general and projection views. Other topics covered include sheet formatting, annotating drawings, and creating draft entities. Two tutorials are provided, one of which covers the creation of geometric and dimensional tolerances.

CHAPTER 9 SECTIONS AND ADVANCED DRAWING VIEWS Due to the length and depth of Chapter 8, the creation of section and auxiliary views is covered in a separate chapter.

CHAPTER 10 SWEPT AND BLENDED FEATURES The protrusion and cut commands have options for creating extruded, revolved, swept, and blended features. Extruded and revolved features are covered in previous chapters. This chapter covers the fundamentals behind creating sweeps and blends. Three tutorials are provided.

CHAPTER 11 ADVANCED MODELING TECHNIQUES The protrusion and cut commands have options for creating advanced features. Covered in this chapter are helical sweeps, variable section sweeps, and swept blends. Three tutorials are provided.

CHAPTER 12 ASSEMBLY MODELING This chapter covers the basics of Pro/ENGINEER's assembly mode. Other topics covered include the creation of assembly drawings using report mode, the control of assemblies through layout mode, top-down assembly design, and mechanism design. Three tutorials are provided.

CHAPTER 13 SURFACE MODELING This chapter covers the basics behind creating surface features within Pro/ENGINEER. Two tutorials are provided.

www.mhhe.com Please visit our web page. Ancillary materials are available for reading and download. For instructors, solutions to end-of-chapter Questions and Discussion are available. Also available are solutions to chapter problems and additional problems.

ACKNOWLEDGMENTS

I would like to thank many individuals for contributions provided during the development of this text. I would like to thank faculty, friends, and students at Purdue University and Western Washington University. I am grateful to McGraw-Hill for producing the beta edition of this book. I would especially like to thank the following individuals for testing and reviewing this edition:

Dan Beller, *University of Wisconsin—Milwaukee*
Vojin R. Nikolic, *Indiana Institute of Technology*
Tao Yang, *California Polytechnic State University—San Luis Obispo*

I would also like to thank the following reviewers for their excellent comments and critiques:

Holly K. Ault, *Worcester Polytechnic Institute*
John R. Baker, *University of Kentucky—Paducah*
Malcolm Cooke, *Case Western Reserve University*
Quentin Guzek, *Michigan Technological University*
Thomas Malmgren, *University of Pittsburgh at Johnstown*
Marie Planchard, *Massachusetts Bay Community College*
John Renuad, *University of Notre Dame*
Donald Wright, *Duke University*
Douglas H. Baxter, *Rensselaer Polytechnic Institute*
Patrick Connolly, *Purdue University*
Rollin C. Dix, *Illinois Institute of Technology*
Lawrence K. Hill, *Iowa State University*
Mike Pierce, *Oklahoma State University—Okmulgee*
Sally Prakash, *University of Missouri—Rolla*
C. Steve Suh, *Texas A&M University*

The editorial and production groups of McGraw-Hill have been wonderful to work with. I would especially like to thank Betsy Jones, Kelley Butcher, Joyce Berendes, and Melinda Dougharty. Thanks also to Gary Bertoline and Craig Miller at Purdue University for their support, help, and encouragement.

TRADEMARK ACKNOWLEDGMENTS

ABOUT THE AUTHOR

David S. Kelley is Assistant Professor of Manufacturing Graphics in the Department of Computer Graphics Technology at Purdue University. Prior to joining Purdue's faculty, David served as an assistant professor in the Engineering Technology Department at Western Washington University. He has also taught engineering graphics at Itawamba Community College in Fulton, Mississippi; drafting and design technology at Northwest Mississippi Community College; engineering design at Northeastern State University in Tahlequah, Oklahoma; and engineering graphics technology at Oklahoma State University—Okmulgee. He earned his A.A. degree from Meridian Community College in Meridian, Mississippi, his B.S. and M.S. degrees from the University of Southern Mississippi, and his Ph.D. degree from Mississippi State University. He may be reached at dskelley@tech.purdue.edu.

CONTENTS

Preface v

Chapter 1

INTRODUCTION TO PARAMETRIC DESIGN 1

Introduction 1
Definitions 1
Introduction to Computer-Aided Design 1
Engineering Graphics 2
Parametric Modeling Concepts 3
- Feature-Based Modeling 5
- Sketching 6
- Constraint Modeling 6
- Dimensional Relationships 7
- Feature References 7
- Model Tree 8
- Integration 8
- Datum Features 9

Concurrent Engineering 10
Design Intent 11
Pro/ENGINEER Modes 13
Summary 15

Chapter 2

PRO/ENGINEER'S USER INTERFACE 17

Introduction 17
Definitions 17
Menu Bar 17
Toolbar 19
- File Management 20
- View Display 20
- Model Display 20
- Datum Display 21
- Context-Sensitive Help 21

File Management 21
- Filenames 21
- Memory 22
- Working Directory 22
- Opening an Object 22
- Creating a New Object 24
- Saving an Object 24
- Activating an Object 25

Viewing Models 25
- Dynamic Viewing 25
- Model Display 26
- View Orientation 27
- Naming and Saving Views 28

Setting Up a Model 28
Units 29
- Setting a System of Units 29
- Creating a System of Units 29

Materials 30
- Defining a Material 30
- Writing a Material to Disk 30
- Assigning Materials 31

Dimensional Tolerance Set Up 31
- ANSI Tolerance Standard 32
- ISO Tolerance Standard 32
- Modifying a Tolerance Table 33
- Geometric Dimensioning and Tolerancing 34
- Creating a Geometric Tolerance 35

Naming Features 38
Obtaining Model Properties 38
- Parent/Child Relationships 38
- Model Analysis 38

Printing in Pro/ENGINEER 39
- Configuring the Printer 39

Pro/ENGINEER's Environment 40
Configuration File 42
Mapkeys 43
- Defining Mapkeys 44

Layers 45
- Creating a Layer 46
- Setting Items to a Layer 46
- Default Layers 47

Selecting Features and Entities 47
Summary 48
Pro/ENGINEER Interface Tutorial 49
- Opening an Object 49
- Viewing the Object 50
- Setting an Object's Units 51
- Establishing Layers 52

Chapter 3

CREATING A SKETCH 55

Introduction 55
Definitions 55
Fundamentals of Sketching 56
- Capturing Design Intent 56
- Sketching Elements 57
- Sketch Plane 57

Section Tools 58
- Grid Options 58
- Placing Sections 58
- Section Information 60

Constraints 60
- Constraints with Intent Manager 60
- Constraints without Intent Manager 61
- Constraint Options 61

Sketcher Display Options 61
Sketching with Intent Manager 63
- Order of Operations 64

Sketching Entities 64
- Sketching Lines 64
- Sketching Arcs 64
- Sketching Circles 66
- Sketching a Rectangle 67
- Splines 67
- Sketched Text 67
- Axis Point 67
- Elliptic Fillet 68
- Construction Entities 68

Sketching without Intent Manager 68
- Alignments 69
- Order of Operations 69

Dimensioning 70
- Linear Dimensions 70
- Radial Dimensions 71
- Angular Dimensions 72
- Perimeter Dimensions 73
- Ordinate Dimensions 73
- Reference Dimensions 73
- Modifying Dimensions 74

Sketcher Relations 75
Geometric Tools 76
Summary 78
Sketcher Tutorial 1 79
- Creating a New Object in Sketch Mode 79
- Sketch Entities 80

Sketcher Tutorial 2 83
- Creating a New Section 83
- Creating the Sketch 83

Chapter 4

EXTRUDING, MODIFYING, AND REDEFINING FEATURES 91

Introduction 91
Definitions 91
Feature-Based Modeling 91
Parent-Child Relationships 92
The First Feature 93
Steps for Creating a New Part 93
Protrusions and Cuts 95
- Solid versus Thin Features 96

Extruded Features 97
- Extrude Direction 97
- Depth Options 98
- Open and Closed Sections in Extrusions 99
- Material Side 100

Creating Extruded Features 100
Datum Planes 102
- Creating Pro/ENGINEER's Default Datum Planes 103
- Creating Datum Planes 103
- Stand-Alone Constraint Options 103
- Paired Constraint Options 104
- On-the-Fly Datum Planes 104

Modifying Features 105
- Dimension Modification 105
- Cosmetic Dimension Modification 106

Redefining Features 107
Summary 108
Extrude Tutorial 109
- Starting a New Model 109
- Creating an Extruded Protrusion 110
- Sketching the Section 111
- Finishing the Feature 115
- Creating an Extruded Cut 116
- Redefining the Feature 119
- Creating an Extruded Cut 120
- Sketching the Section 121
- Creating an Extruded Protrusion 123
- Dimension Modification 125
- Redefining a Feature's Depth 128
- Redefining a Feature's Section 129

Datum Tutorial 131
- Creating the Part 131
- Creating Datum Planes 132
- Creating a Datum Axis 135
- Creating a Coordinate System 136

Chapter 5

FEATURE CONSTRUCTION TOOLS 139

Introduction 139
Definitions 139
Hole Features 139
- Hole Placement Options 140
- Hole Types 140
- Hole Depth Options 141
- Creating a Straight Linear Hole 142
- Creating a Straight Coaxial Hole 143

Rounds 143
- Round Radii Options 143
- Round Reference Options 144
- Creating a Simple Round 144

Chamfer 145
Draft 146
- Neutral Planes and Curves 146
- Neutral Plane Drafts 146
- Creating a Neutral Plane No Split Draft 147
- Creating a Neutral Plane Split Draft 149
- Neutral Curve Drafts 149
- Creating a Neutral Curve Draft 150

Shelled Parts 151
Ribs 152
- Creating a Rib 153

Cosmetic Features 154
- Creating a Cosmetic Thread 155

Patterned Features 156
- Pattern Options 156

Dimensions Variation 157
Creating a Linear Pattern 157
Summary 160
Feature Construction Tutorial 1 161
Starting a New Object 161
Creating the Base Geometric Feature 161
Adding Extruded Features 163
Creating Rounds 164
Creating a Chamfer 165
Creating a Standard Coaxial Hole 167
Creating a Linear Hole 169
Creating an Advanced Round 171
Inserting a Shell 173
Feature Construction Tutorial 2 175
Starting a New Part 175
Creating the Base Geometric Feature 175
Creating an Offset Datum Plane 175
Creating an Extruded Protrusion 177
Creating a Coaxial Hole 179
Creating a Linear Hole 180
Creating a Linear Pattern 180
Creating a Chamfer 182
Creating a Cut 182
Creating a Rib 183
Creating a Draft 185
Creating a Round 187

Chapter 6

REVOLVED FEATURES 191
Introduction 191
Definitions 191
Revolved Feature Fundamentals 191
Sketching and Dimensioning 191
Revolved Protrusions and Cuts 192
Revolved Feature Parameters 192
Creating a Revolved Protrusion 193
Revolved Hole Options 195
Sketched Holes 195
Creating a Sketched Hole 196
Radial Hole Placements 197
Creating a Straight-Diameter (Radial) Hole 198
The Shaft Option 199
The Flange and Neck Options 199
Rotational Patterns 201
Creating a Radial Pattern 202
Datum Axes 203
Creating Datum Axes 203
Summary 204
Revolved Features Tutorial 205
Creating a Revolved Protrusion 205
Creating a Radial Sketched Hole 209
Creating a Radial Hole Pattern 211
Creating a Revolved Cut 212
Modifying the Number of Holes 214
Shaft Tutorial 215
Setting Configuration Options 215
Creating a Base Protrusion 216
Creating a Shaft 217
Creating a Cut 219
Creating a Pattern of the Cut 221

Chapter 7

FEATURE MANIPULATION TOOLS 225
Introduction 225
Definitions 225
Grouping Features 225
The Group Menu 226
Group Types 227
Patterning a Group 227
Copying Features 229
Copy Options 229
Independent Versus Dependent 230
Selecting a Model 230
Mirroring Features 230
Rotating Features 231
Translated Features 232
Copying with New References 233
The Mirror Geometry Command 234
User-Defined Features 234
The UDF Menu 234
Creating a User-Defined Feature 235
Placing a User-Defined Feature 236
Relations 238
Conditional Statements 238
Adding and Editing Relations 240
Family Tables 240
Adding Items to a Family Table 241
Creating a Family Table 241
Cross Sections 245
Modifying Cross Sections 246
Creating a Planar Cross Section 246
Creating Offset Cross Sections 247
Model Tree 248
Suppressing Features 248
Inserting Features 248
Reordering Features 249
Rerouting Features 249
Regenerating Features 250
Regeneration Failures 250
Feature Simplified Representations 251
Summary 252
Manipulating Tutorial 1 253
Creating the Base Protrusion 253
Creating a Coaxial Hole 255
Mirroring the Extruded Feature 256
Rotating the Extruded Feature 258
Adding Relations to a Part 259
Manipulating Tutorial 2 261
Creating the Base Protrusion 261
Creating a Through >> Axis Datum Plane 261
Creating the Boss Feature 262
Creating a Coaxial Hole and Round 264

Grouping Features 264
Patterning the Boss Group 265
Establishing a Conditional Relationship 266

Chapter 8
CREATING A PRO/ENGINEER DRAWING 272
Introduction 272
Definitions 272
Drawing Fundamentals 272
Drawing Setup File 273
Sheet Formats 274
Modifying Formats 275
Creating Formats 276
Creating a New Drawing 276
Drawing Views 278
The Views Menu 278
View Types 279
View Visibilities 281
Multiple Models 282
Creating a General View 282
Setting a Display Mode 283
Detailed Views 283
Showing and Erasing Items 285
Showing All Item Types 286
Showing/Erasing Limited Item Types 287
Pop-Up Menu 287
Dimensioning and Tolerancing 288
Manipulating Dimensions 289
Dimension Tolerances and Modification 290
Geometric Tolerances 292
Creating Notes 292
Note without Leader 292
Note with a Standard Leader 293
Creating Drawing Tables 293
Two-Dimensional Drafting 295
Draft Geometry 295
Construction Geometry 296
Line Styles and Fonts 297
Draft Dimensions 298
Draft Cross Sections 298
Manipulating Draft Geometry 299
Summary 301
Drawing Tutorial 1 303
Creating the Part 304
Starting a Drawing 304
Adding a Drawing Format 305
Creating a General View 306
Creating Projection Views 307
Creating a Detailed View 308
Establishing Drawing Setup Values 310
Creating Dimensions 310
Creating Notes 317
Setting Display Modes 318
Drawing Tutorial 2 321
Creating the Part 321
Starting a Drawing with a Template 322
Establishing Drawing Setup Values 323
Creating the General View 324
Creating the Right-Side View 326
Setting and Renaming Datum Planes 326
Creating Dimensions 328
Setting Geometric Tolerances 330
Setting Dimensional Tolerances 335
Creating the Title Block 338

Chapter 9
SECTIONS AND ADVANCED DRAWING VIEWS 341
Introduction 341
Definitions 341
Section View Fundamentals 341
Section View Types 342
Full Sections 343
Half Sections 345
Offset Sections 346
Broken Out Sections 348
Aligned Section Views 350
Revolved Sections 351
Auxiliary Views 352
Summary 354
Advanced Drawing Tutorial 1 355
Creating the Part 355
Starting a Drawing 355
Establishing Drawing Setup Values 357
Creating the General View 357
Creating an Aligned Section View 358
Creating a Partial Broken Out Section View 361
Centerlines and Dimensions 362
Title Block Notes 366
Advanced Drawing Tutorial 2 367
Creating the Part 367
Starting a Drawing 367
Establishing Drawing Setup Values 368
Creating the Broken Front View 369
Creating a Partial Auxiliary View and Left-Side View 370
Adding Dimensions and Centerlines 371
Modifying Dimension Values 373
Title Block Information 374

Chapter 10
SWEPT AND BLENDED FEATURES 377
Introduction 377
Definitions 377
Sweep and Blend Fundamentals 377
Swept Features 378
Creating a Sweep with a Sketched Section 379
Blended Features 381
Creating a Parallel Blend 382

Datum Curves 384
Creating Datum Curves 384
Projected and Formed Datum Curves 385
Creating Datum Points 386
Coordinate Systems 387
Types of Coordinate Systems 388
Creating a Cartesian Coordinate System 388
Summary 390
Blend Tutorial 391
Creating the Base Feature 391
Creating a Blend 391
Creating a Second Blend 394
Creating a Cut Feature 395
Sweep Tutorial 1 397
Creating the Base Feature 397
Creating Flange Features 399
Sweep Tutorial 2 401
Creating the Base Feature 401
Creating the Wheel Handle 401
Creating a Datum Curve 402
Sweep Creation 404
Creating a Round 405
Grouping Features 405
Copying the Spoke Group 406

Chapter 11
ADVANCED MODELING TECHNIQUES 410
Introduction 410
Definitions 410
Swept Blend Option 410
Creating a Swept Blend 412
Variable Section Sweep 414
Creating a Variable Section Sweep 416
Helical Sweeps 419
Creating a Constant Pitch Helical Sweep Feature 420
Summary 422
Swept Blend Tutorial 423
The First Feature 423
The Second Feature 424
Swept Blend Feature 424
Spring Tutorial 429
Bolt Tutorial 433
Creating the Bolt's Shaft 433
Bolt Threads 433
Extrusion Extension 435
Bolt Head Creation 436
Bolt Head Cut 437
Variable Section Sweep Tutorial 439
Creating the Origin Trajectory 439
X-Trajectory Creation 440
Trajectory Creation 441
Variable Section Sweep Feature 442
Swept Feature 444
Extruded Protrusion 446

Chapter 12
ASSEMBLY MODELING 450
Introduction 450
Definitions 450
Introduction to Assembly Mode 450
Assembly Mode's Menu 451
Placing Components 451
Assembly Constraints 451
Moving Components 455
Packaged Components 455
Placing a Parametric Component 455
Mechanism Design 456
Modifying Assemblies and Parts 458
Modifying Dimensions 459
Creating New Part Features 459
Redefining a Component Feature 459
Creating Assembly Features 460
Top-Down Assembly Design 460
Creating Parts in Assembly Mode 461
Skeleton Models 464
Assembly Relations 465
Layout Mode 466
Simplified Representation 467
Creating a Simplified Representation 468
Exploded Assemblies 468
Creating an Explode State 468
Summary 470
Assembly Tutorial 471
Creating Components for an Assembly 471
Placing Components into an Assembly 472
Creating an Exploded Assembly 478
Creating an Assembly Drawing (Report) 480
Top-Down Assembly Tutorial 485
Creating a Layout 485
Creating a Start Part 488
Creating the First Component (base.prt) 488
Creating the Second Component (shaft.prt) 492
Creating the Third Component (pulley.prt) 495
Declaring and Using a Layout 498
Mechanism Tutorial 501
Modeling Assembly Parts 501
Assembling a Mechanism 503
Manipulating a Mechanism 507
Running a Mechanism's Motion 508
Animating a Mechanism 511

Chapter 13
SURFACE MODELING 518
Introduction 518
Definitions 518
Introduction to Surfaces 518
Surface Options 520
Surface Operations 521
Advanced Surface Options 523
Merging Quilts 524
Boundaries Option 526
Creating a Blended Surface from Boundaries 527

Use Quilt 529
Summary 530
Surface Tutorial 1 531
- Create the Base Extruded Surface Feature 531
- Creating Datum Curves 532
- Trimming Surface Features 533
- Creating an Approximate Boundaries Surface 534
- Merging Quilts 535
- Creating a Solid from a Quilt 537

Surface Tutorial 2 539
- Creating Datum Curves 539
- Creating Surfaces from Boundaries 543
- Merging Surfaces 548
- Creating Additional Surfaces 549
- Creating a Draft Offset 551
- Trimming a Surface 552
- Converting a Surface to a Solid 554

Appendix A
SUPPLEMENTAL CD FILES 558

Appendix B
CONFIGURATION FILE OPTIONS 560

Index 562

CHAPTER

1

INTRODUCTION TO PARAMETRIC DESIGN

Introduction

This chapter introduces the basic concepts behind parametric design and modeling. **Parametric design** is a powerful tool for incorporating design intent into computer-aided design models. Parametric models, often referred to as feature-based models, can be intuitively created and modified. Within this chapter, engineering graphics and three-dimensional modeling concepts will be explored. Additionally, parametric modeling principles will be covered. Upon finishing this chapter, you will be able to

- Describe the utilization of computer-aided design within engineering graphics.
- Compare three-dimensional modeling techniques.
- Describe concepts associated with parametric modeling and design.
- Describe the use of parametric design within a concurrent manufacturing environment.

DEFINITIONS

Associativity The sharing of a component's database between application modes.

Design intent The intellectual arrangement of assemblies, parts, features, and dimensions to meet a design requirement.

Parametric design The incorporation of component design intent in a graphical model by means of parameters, relationships, and references.

Parametric modeling A computer model that incorporates design parameters.

INTRODUCTION TO COMPUTER-AIDED DESIGN

Engineering design graphics has made significant changes since the early 1980s. For the most part, these changes have occurred due to the evolution of computer-aided design (CAD). Before CAD, design was accomplished by traditional board drafting utilizing paper, pencil, straightedges, and various other manual drafting devices. Concurrently with manual drafting were sketching techniques, which allowed a designer to explore ideas freely without being constrained within the boundaries of drafting standards.

Many of the drafting and design standards and techniques that existed primarily due to the limitations of manual drafting still exist today. Popular midrange CAD packages still emphasize two-dimensional orthographic projection techniques. For example, these techniques allow a design to be portrayed on a computer screen by means once accomplished on a drafting table. Drafting standards have changed little since the beginning of CAD. These standards still place an emphasis on the two-dimensional representation of designs.

Many engineering fields continue to rely on orthographic projection to represent design intent. Some fields, such as manufacturing and mechanical engineering, foster a paperless environment that does not require designs to be displayed orthographically. In this theoretically paperless environment, products are designed, engineered, and produced without a hard-copy drawing. Designs are modeled within a CAD system, and the electronic data is utilized concurrently in various departments, such as manufacturing, marketing, quality control, and production control. Additionally, CAD systems are becoming the heart of many product data management systems. Utilizing a computer network, CAD designs can be displayed throughout a corporation's intranet. With Internet capabilities, a design can be displayed using the World Wide Web.

ENGINEERING GRAPHICS

The fundamentals of engineering graphics and the displaying of three-dimensional (3D) designs on a two-dimensional surface has changed little since the advent of CAD. Despite the explosion of advanced 3D modeling packages, many design standards and techniques that once dominated manual drafting remain relevant today.

Sketching is an important tool in the design process. Design modeling techniques using two-dimensional CAD, three-dimensional CAD, or manual drafting can restrict an individual's ability to work out a design problem. It takes time to place lines on a CAD system or to construct a solid model. Sketching allows a designer to work through a problem without being constrained by the standards associated with orthographic projection or by the time required to model on a CAD system.

There are two types of sketching techniques: artistic and technical. Many individuals believe that artistic sketching is a natural inborn ability. This is not always the case. There are techniques and exercises that engineering students can perform that will improve their ability to think in three dimensions and solve problems utilizing artistic sketching skills. Despite this, few engineering students receive this type of training.

If engineering or technology students receive training in sketching, it is often of the technical variety. Technical sketching is similar to traditional drafting and two-dimensional computer-aided drafting. This form of sketching allows a design to be displayed orthographically or pictorially through sketching techniques.

The design process requires artistic sketching and technical sketching to be utilized together. Conceptual designs are often developed through artistic sketching methods. Once a design concept is developed, technical sketches of the design can be developed that will allow the designer to display meaningful design intent information. This information can then be used to develop orthographic drawings, prototypes, and/or computer models.

The traditional way to display engineering designs is through orthographic projection. Any object has six primary views (Figure 1–1). These views display the three primary dimensions of any feature: height, width, and depth. By selectively choosing a combination of the primary views, a detailer can graphically display the design form of an object. Often, three or fewer views are all that are necessary to represent design intent (Figure 1–2). A combination of views such as the front, top, and right side will display all three primary dimensions of any feature. By incorporating dimensions and notes, design intent for an object can be displayed.

Orthographic projection is not a natural way to display a design. The intent of orthographic drawings is to show a design in such a way that it can be constructed or manufactured. Pictorial drawings are often used to represent designs in a way that nontechnically trained individuals can understand. Pictorial drawings display all three primary dimensions

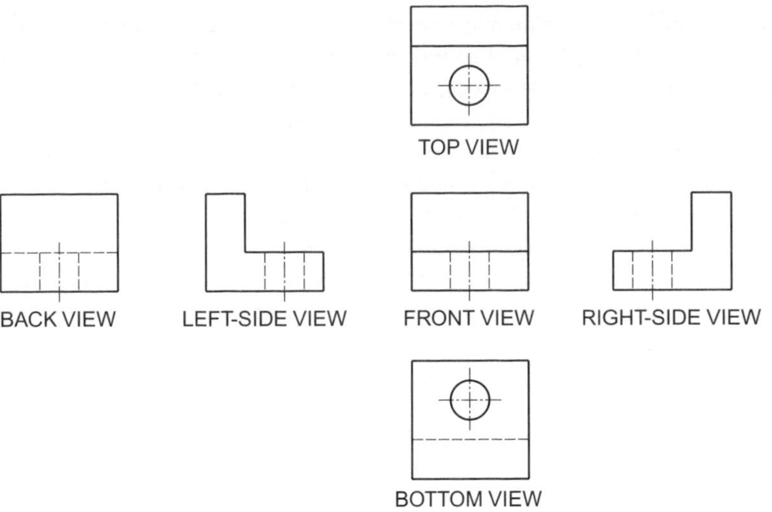

Figure 1-1 Six primary views

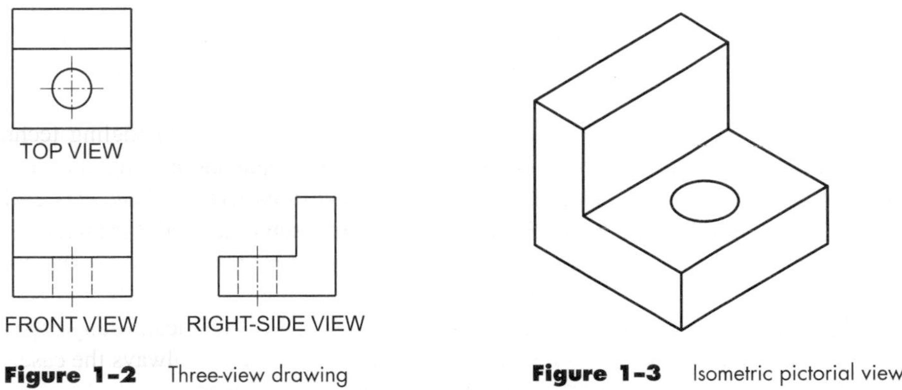

Figure 1-2 Three-view drawing

Figure 1-3 Isometric pictorial view

(height, width, and depth) in one view (Figure 1–3). There are many forms of pictorial drawings. The most common are isometric, diametric, and trimetric. Naturally, objects appear to get smaller as one moves farther away. This is known as *perspective.* Perspective is another form of pictorial drawing. Orthographic, isometric, diametric, and trimetric projections do not incorporate perspective. Perspective drawings are often used to display a final design concept that can be easily understood by individuals with no technological training.

Before an object can be manufactured or constructed, technical drawings are often produced. Technical drawings display all information necessary to properly build a product. These drawings consist of orthographic views, dimensions, notes, and details. Details are governed by standards that allow for ease of communications between individuals and organizations.

Parametric Modeling Concepts

Parametric modeling is an approach to computer-aided design that gained prominence in the late 1980s. A commonly held assumption among CAD users is that similar modeling techniques exist for all CAD systems. To users that follow this assumption, the key to learning a different CAD system is to adapt to similar CAD commands. This is not entirely true when a two-dimensional CAD user tries to learn, for the first time, a parametric modeling application. Within parametric modeling systems, though, you can find commands that resemble 2D CAD commands. Often, these commands are used in a parametric modeling system just as they would be used in a 2D CAD package. The following is a partial list of commands that crossover from 2D CAD to Pro/ENGINEER.

LINE

The line option is used within Pro/ENGINEER's sketcher mode (or environment) as a tool to create sections. Within a 2D CAD package, precise line distances and angles may be entered using coordinate methods, such as absolute, relative, and polar. Pro/ENGINEER does not require an entity to be entered with a precise size. Feature size definitions are established after finishing the geometric layout of a feature's shape.

CIRCLE

As with the line command, the circle option is used within Pro/ENGINEER's sketcher environment. Precise circle size is not important when sketching the geometry.

ARC

As with the line and circle options, the arc command is used within Pro/ENGINEER's sketching environment. Pro/ENGINEER's arc command also includes a fillet command for creating rounds at the intersection of two geometric entities.

DELETE

The delete command is used within a variety of Pro/ENGINEER modes. Within the sketcher environment, delete removes geometric entities such as lines, arcs, and circles; within Part mode, delete removes features from a part. For assembly models, the delete command deletes features from parts and to delete parts from assemblies.

OFFSET

Offset options can be found within various Pro/ENGINEER modes. Within the sketcher environment, existing part features can be offset to form sketching geometry. Additionally, planes within Part and Assembly modes can be offset to form new datum planes.

TRIM

The trim command is used within Pro/ENGINEER's sketching environment. Geometric entities that intersect can be trimmed at their intersection point.

MIRROR

The mirror option is used within Pro/ENGINEER's Sketch and Part modes. Geometry created as a sketch can be mirrored across a centerline. Also, part features can be mirrored across a plane by executing the copy option.

COPY

The copy option is used within Part mode to copy existing features. Features can be copied linearly, mirrored over a plane, or rotated around an axis. Within Assembly mode, parts can be copied to create new parts.

ARRAY

Polar and rectangular array commands are common components among 2D CAD packages. Pro/ENGINEER's Pattern command serves a similar function. Features may be patterned using existing dimensions. Selecting an angular dimension will create a circular pattern.

Parametric modeling represents a different approach to CAD, when compared to 2D drafting and to Boolean-based 3D modeling. Often, an experienced CAD user will have trouble learning a parametric modeling package. This is especially true when a user tries to approach three-dimensional parametric modeling as he or she would approach Boolean solid modeling. There are similar concepts, but the approaches are different.

Feature-Based Modeling

Parametric modeling systems are often referred to as *feature-based modelers.* In a parametric modeling environment, parts are composed of features (Figure 1–4). Features can comprise either positive or negative space. Positive space features are composed of actual mass. An example of a positive space feature is an extruded boss. A negative space feature is where a part has a segment cut away or subtracted. An example of a negative space feature is a hole.

Parametric modeling systems such as Pro/ENGINEER incorporate an intuitive way of constructing features. Often, the feature is first sketched in two dimensions then either extruded, revolved, or swept to form the three-dimensional object. When sketching the feature, design intent is developed in the model by dimensioning and constraining the sketch.

Features can be predefined or sketched. Examples of predefined features include holes, rounds, and chamfers. Many parametric modeling packages incorporate advanced ways of modeling holes. Within a parametric modeling package, predefined holes can be simple, counterbored, countersunk or drilled. Pro/ENGINEER's hole command allows users the opportunity to sketch unique hole profiles, such as may be required for a counterbore. Sketched features are created by sketching a section that incorporates design intent. Sections may be extruded, revolved, or swept to add positive or negative space features.

Compared to Boolean modeling, feature-based modeling is a more intuitive approach. In Boolean modeling, a common way to construct a hole is to model a solid cylinder and then subtract it from the parent feature. In a parametric design environment, a user can simply place a hole by either using a predefined hole command or by cutting a circle through the part. With most Boolean-based modelers, if the user has to change a parameter of the hole, such as location or size, he or she has to plug the original hole, then subtract a second solid cylinder. With a feature-based hole, the user can change any parameter associated with the hole by modifying a dimension or parameter. Additionally, a feature's sketch can be redefined or modified.

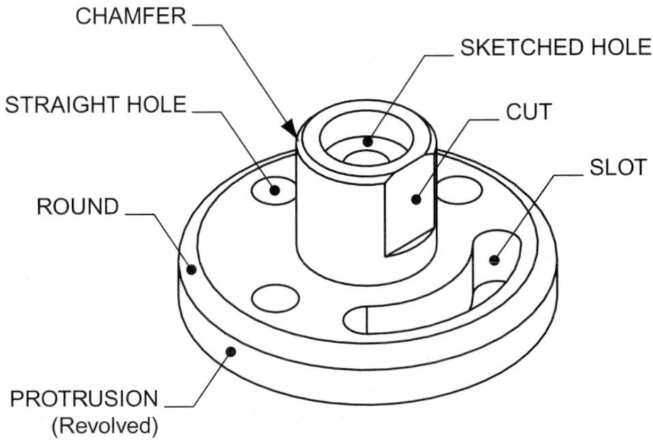

Figure 1–4 Features in a model

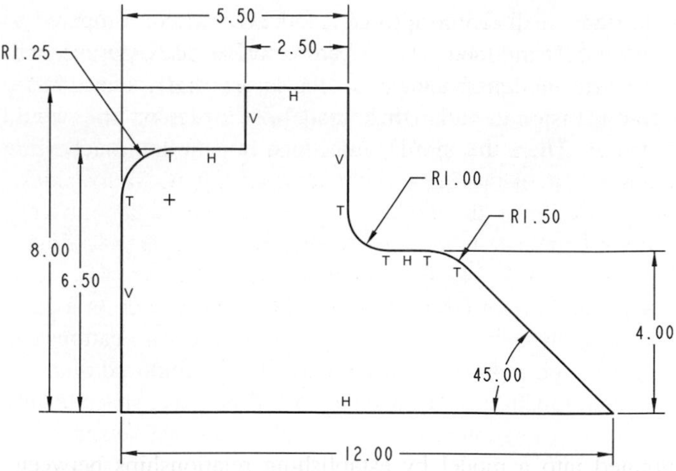

Figure 1–5 Section created in the sketcher environment

SKETCHING

As previously mentioned, parts consist of features. As shown in Figure 1–5, features are normally created by first sketching a section of the feature's profile. Sketch construction techniques are similar to 2D CAD drawing methods. In 2D CAD, a user has to use precision CAD techniques to draw a design. Parametric modeling sketcher environments do not require this. The sketching component within parametric design systems was developed to allow a user to quickly construct a design feature without having to be concerned with time-consuming precision. As an example, if a user wanted to draw a 4-inch square polygon in 2D CAD, he or she might draw four lines each exactly 4 inches long and each perpendicular to one another. In a parametric design system, the user would sketch four lines forming roughly the shape of a square. If it is necessary to have a precise 4-inch square object, the user could dimension the object, constrain each line perpendicular to one another, and then modify the dimension values to equal 4 inches. Modifying the size of a feature (or sketch) in a parametric design system is as simple as modifying the dimension value and then updating the model.

CONSTRAINT MODELING

Design intent is incorporated in a feature by applying dimensions and constraints that meet the intent of the design. If the intent for a hole is to be located 2 inches from a datum edge, the user would locate the hole within the modeling or sketching environment by dimensioning from the desired edge. Constraints are elements that further enhance design intent. Table 1–1 is a list of common parametric constraints. Once elements are constrained, they typically remain constrained until the user deletes them or until a dimension overrides them.

Table 1–1 Common constraints used in the sketcher environment

Constraint	Meaning
Perpendicular	Two lines are perpendicular to each other.
Parallel	Two lines are parallel to each other.
Tangent	Two elements are tangent to each other.
Coincident	Two elements lie at the same location.
Vertical	A line is vertical.
Horizontal	A line is horizontal.

Some parametric modeling packages, such as Pro/ENGINEER, require a sketch to be fully constrained before it can be protruded into a feature. Other packages do not require this. There are advantages and disadvantages to packages that require fully constrained sketches. One advantage is that a fully constrained sketch requires all necessary design intent to be incorporated into the model. Additionally, a fully constrained sketch does not present as many surprises later in the modeling of a part. Many times, when a sketch is not completely constrained, the model will solve regenerations in a way not expected by the CAD user. Under-constrained sketches can have advantages though. Complicated sketches are often hard to fully constrain. Additionally, fully constraining a sketch may inhibit the design process. Often, designers do not want to fully define a feature.

DIMENSIONAL RELATIONSHIPS

Design intent can be incorporated into a model by establishing relationships between dimensions. A dimension can be constrained to be equal to another dimension (i.e., length = width). Additionally, mathematical relationships can be set between dimensions. As an example, the length of a feature can be set to twice its width (length = width × 2). Or the width could be set to equal half the length (width = length/2). Most forms of algebraic equations can be used to establish relationships between dimensions. An example of this would be to set the length of a feature equal to half the sum of the width and depth of a second feature (length = (width + depth)/2).

FEATURE REFERENCES

Within a part, features are related to each other in a hierarchical relationship. The first feature of a part is the base feature, which is the parent of all features that follow. Once a base feature is constructed, child features can be added to it. Child features may comprise positive or negative space. A part can have an elaborate and complicated family tree. This tree is graphically displayed within Pro/ENGINEER as a Model Tree (Figure 1–6). It is important for the user to understand the relationship between a parent and a child. For any

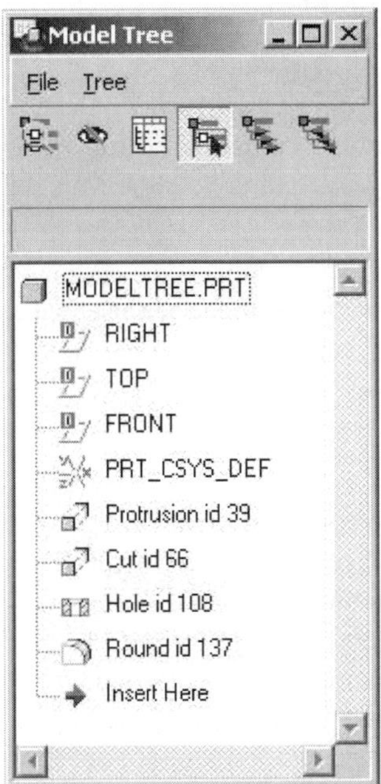

Figure 1–6 Pro/ENGINEER's Model Tree

feature, if a parent is modified, it can have a devastating effect on its child features. Also, if a parent is deleted, all child features will be deleted. The user should be aware of this when constructing features.

Model Tree

Parametric modeling packages utilize a graphical model (or history) tree to list in chronological order all features that make up a part or assembly (Figure 1–6). The Model Tree does more than just list features, though. The following is a list of some of the uses of Pro/ENGINEER's Model Tree.

Redefining a Feature

A feature's definition can by modified by selecting it from the Model Tree. Some definitions that can be modified using the Redefine command include:

- **Section** A feature's section can be modified.
- **Placement Refs** A feature's placement plane and references can be modified.
- **Depth** A feature's extrusion depth can be modified.
- **Hole Attributes** Attributes associated with a hole, such as diameter and direction, may be modified.

Deleting a Feature

A feature can be deleted or redefined by selecting the feature in the work screen. Sometimes a part can become so complicated that selecting a feature in this manner is difficult. To ensure that the correct feature is deleted, the feature can be selected on the Model Tree.

Reordering Features

Features that rely on other features in the modeling process are referred to as *child features.* A child feature must follow a parent feature in the order of regeneration. Usually, this means it should come after it on the Model Tree. Sometimes it is necessary to reorder a feature on the Model Tree to place it after a potential parent feature.

Inserting Features

Normally, new features are placed chronologically at the end of the Model Tree. At times it may be necessary to place a new feature before an existing feature. Using the insert feature command, Pro/ENGINEER allows features to be inserted before or after existing features in the Model Tree.

Suppressing Features

Being able to remove a feature from the Model Tree can be a useful tool. The Suppress option is used to remove a feature temporarily from the order of regeneration.

Integration

Modules of Pro/ENGINEER are fully integrated. These modules share a common database. Often referred to as **associative,** this database sharing allows modifications and redefinitions made in one module, such as Part mode, to be reflected in other modules, such as Drawing and Manufacturing. As an example, an object can be created in Part mode. This model can be used directly to create an orthographic drawing in Drawing mode. Dimensions used to create the model can be displayed in Drawing mode. If a dimension value is changed in Part mode, the same dimension and feature is changed in Drawing mode. Additionally, dimension values can be changed in Drawing mode, and part mode will reflect these changes.

DATUM FEATURES

A datum is not a new idea to engineering design. A datum plane is a theoretically perfectly flat surface. It is used often in quality control for the inspection of parts. Geometric dimensioning and tolerancing practices utilize datum surfaces to control the size, location, and orientation of features. Many (but not all) parametric modeling systems use datum surfaces (or datum planes) to model features. Within Pro/ENGINEER, datum planes are often the parent features of all geometric features of a part. Datum planes are used as sketching surfaces, especially when sketching the first feature of a part. Datum planes are also used in parametric packages to locate features or to create new features. As an example, within Pro/ENGINEER features can be mirrored about a datum plane to create new features (see Figure 1–7). In this case, the new features would be children of both the original features and the datum plane. The following is a description of the types of datums available within Pro/ENGINEER.

DATUM PLANE

A datum plane is a theoretically perfectly flat surface. Within Pro/ENGINEER, datum planes are used as sketching surfaces and references. Features may be sketched on any plane. Often, a suitable part plane does not exist. Datum planes can be created that will serve this purpose.

Many feature creation and construction processes require the use of a plane. A datum plane can be created for use as a reference when creating or constructing a feature. As an example, the *Radial-Hole* option requires a hole to be located at an angle from an existing plane. A datum plane can be created for this purpose. Another example would be patterning a feature around an axis. This construction technique requires that an angular dimension exist. A datum plane can be created that forms an angle with an existing plane. This new datum can then be combined with the feature to be patterned using the Group option. By grouping a datum plane with a feature, the angle parameter used to create the datum plane can be used within the Group >> Pattern option.

DATUM AXIS

A datum axis is similar to the centerline required at the center of holes and cylinders on orthographic drawings. Pro/ENGINEER provides an option for creating a datum axis. When the Create >> Datum >> Axis option sequence is used to create an axis,

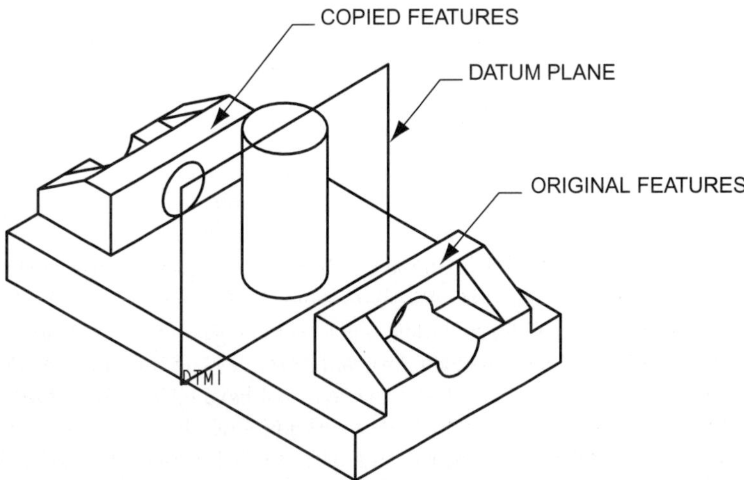

Figure 1–7 Feature mirrored about a datum plane

the axis is represented as a feature on the Model Tree. Pro/ENGINEER will also create a datum axis at the center of revolved features, holes, and extruded circles. These datum axes are not considered part features.

Datum axes are useful modeling tools. As an example, the Radial option under the Hole command requires the user to select an axis to reference the hole location. Another example is the Move >> Rotate option under Copy. A feature can be rotated around an axis or coordinate system.

Datum Curve

Datum curves are useful for the creation of advanced solid and surface features. Datum curves are considered part features and can be referenced in the sketching environment. They can be used as normal or advanced sweep trajectories or as edges for the creation of surface models. There are several techniques available for creating datum curves. One method is to sketch the curve using normal sketching techniques. Another construction methodology is to select the intersection of two surfaces.

Datum Point

Datum points are used in the construction of surface models, to locate holes, and to attach datum target symbols and notes. They are required for the creation of Pipe features. Datum points are considered features of a part, and Pro/ENGINEER labels the first point created PNT0. Each additional point is sequentially increased one numeric value.

Coordinate System

Pro/ENGINEER does not utilize a Cartesian coordinate system as most mid-ranged computer-aided design packages do. Mid-ranged two-dimensional drafting and three-dimensional Boolean-based modeling applications are based on a Cartesian coordinate world. Most parametric modeling packages, including Pro/ENGINEER, do not model parts based on this system. Because of this, many users fail to understand the importance of establishing a datum coordinate system. Coordinate systems are used for a variety of purposes within Pro/ENGINEER. Many analysis tasks such as Mass Properties and Finite-Element analysis utilize a coordinate system. Coordinate systems are also used in modeling applications. As an example, the Copy-Rotate command provides an option to select a coordinate system to revolve around. Coordinate systems are also used frequently in Assembly mode and in Manufacturing mode.

Concurrent Engineering

The engineering design process was once linear and decentralized. Modern engineering philosophies are integrating team approaches into the design of products. As shown in Figure 1–8, team members may come from a variety of fields. Teaming stimulates a non-linear approach to design, with the CAD model being the central means of communicating design intent.

Concurrent engineering has many advantages over a traditional design process. Individuals and groups invest significant resources and time into the development of products. Each individual and group has needs that have to be met by the final design solution. As an example, a service technician wants a product that is easy to maintain while the marketing department wants a design that is easy to sell. Concurrent engineering allows everyone with an interest in a design to provide input.

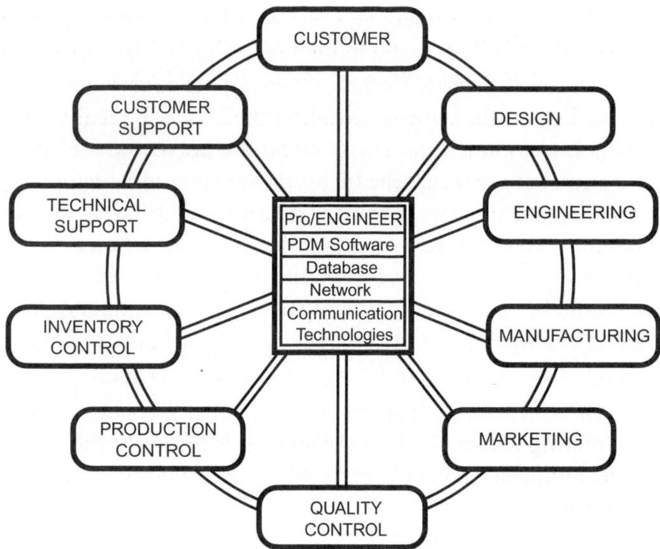

Figure 1-8 Concurrent engineering members

Modern engineering and communication technologies allow the easy sharing of designs among team members. CAD three-dimensional models graphically display designs that can be interpreted by individuals not trained in blueprint reading fundamentals. Because of this accessibility, the CAD system has become the heart of many product-data-management systems. Internet capabilities allow designs to be shared over long distances. Most CAD applications have Internet tools that facilitate the sharing of design data. For example, Parametric Technology Corporation's Pro/Web-Publish allows for the publishing of CAD data over the Internet. PTC's Pro/Fly-Through application displays VRML (Virtual-Reality-Modeling-Language) models and allows for the markup and animation of these models over an Internet or intranet.

Design Intent

A capability unique to parametric modeling packages, when compared to other forms of CAD, is the ability to incorporate design intent into a model. Most computer-aided design packages have the ability to display a design, but the model or geometry does not hold design information beyond the actual vector data required for construction. Two-dimensional packages display objects in a form that graphically communicate the design, but the modeled geometry is not a virtual image of the actual shape of the design. Traditional three-dimensional models, especially solid models, display designs that prototype the actual shape of the design. The problem with solid-based Boolean models is that parameters associated with design intent are not incorporated. Within Boolean operations, when a sketch is protruded into a shape or when a cylinder is subtracted from existing geometry to form a hole, data associated with the construction of the part or feature is not readily available.

Parameters associated with a feature in Pro/ENGINEER exist after the feature has been constructed. An example of this would be a hole. A typical method used within Pro/ENGINEER to construct a straight hole is to locate the hole from two edges. After locating the hole, the hole diameter and depth are provided. The dimensional values used to define the hole can be retrieved and modified at a later time. Additionally, parametric values associated with a feature, such as a hole diameter, can be used to control parameters associated with other dimensions.

With most Boolean operations, of primary importance is the final outcome of the construction of a model. When modeling a hole, the importance lies not in parameters used to locate a hole but where the hole eventually is constructed. When the subtraction process is accomplished, the cylinder location method is typically lost. With parametric hole construction techniques, these parameters are preserved for later use.

The dimensioning scheme for the creation of a feature, such as a hole, is important for capturing design intent. Figure 1–9 shows two different ways to locate a pair of holes. Both examples are valid ways to dimension and locate holes. Which technique is the best? The answer depends upon the design intent of the part and feature. Does the design require that each hole be located a specific distance from a common datum plane? If it does, then the second example might be the dimensioning scheme that meets design intent. If the design requires that the distance between the two holes be carefully controlled, the first example might prove to be the best dimensioning scheme.

Designs are created for a purpose. **Design intent** is the intellectual arrangement of assemblies, parts, features, and dimensions to solve a design problem. Most designs are composed of an assembly of parts. Each part within a design is made up of various features. Design intent governs the relationship between parts in an assembly and the relationship between features in a part. As shown in Figure 1–10, a hierarchical ordering of intent can be created for a design. At the top of the design intent tree is the overall intent of the design.

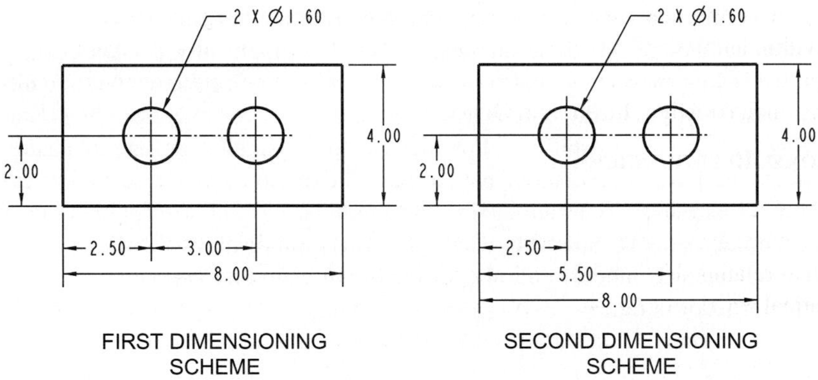

Figure 1–9 Dimensioning scheme differences

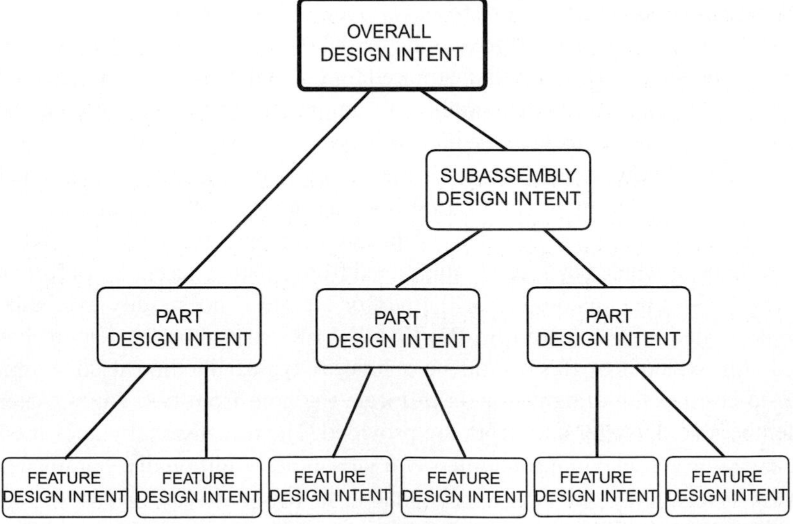

Figure 1–10 Hierarchical order of design intent

Below the overall design intent is the component design intent. Components are composed of parts and subassemblies. The intent of each component of a design is to work concurrently with other components as a solution to the design problem. Features comprise parts. Features must meet the design intent of the parts of which they are composed.

Parametric modeling packages provide a variety of tools for incorporating design intent. The following is a list of these tools.

DIMENSIONING SCHEME

The placement of dimensions is extremely important for the incorporation of design intent into a model. During sketching within Pro/ENGINEER, dimensions that will fully define a feature are placed automatically (when intent manager is activated). These dimensions may not match design intent, though. Dimensions within a section or within the creation of a feature should match the intent of a design.

FEATURE CONSTRAINTS

Constraints are powerful tools for incorporating design intent. If a design requires a feature's element to be constrained perpendicular to another element, a perpendicular constraint should be used. Likewise, design intent can be incorporated with other constraints, such as parallel, tangent, and equal length.

ASSEMBLY CONSTRAINTS

Assembly constraints are used to form relationships between components of a design. Within an assembly, if a part's surface should mesh with the surface of another part, a Mate constraint should be used. Examples of other common assembly constraints include Align, Insert, and Orient.

DIMENSIONAL RELATIONSHIPS

Dimensional relationships allow the capture of design intent between and within features while in Part mode, and between parts while in Assembly mode. A dimensional relationship is an explicit way to relate features in a design. Mathematical equations are used to relate dimensions. An example of a dimensional relationship would be to make two dimensions equal in value. Within Pro/ENGINEER, for this example, the first dimension would drive the second. Most algebraic and trigonometric formulas can be included in a relationship. In addition, simple conditional statements can be incorporated.

REFERENCES

Feature references can be created within Part and Assembly modes of Pro/ENGINEER. An example of a reference within Part mode is to use existing feature edges to create new geometry within the sketcher environment. A Parent-Child relationship is established between the two features. If the reference edge is modified, the child feature is modified correspondingly.

Within Assembly mode, an external reference can be established between a feature on one part and a feature on a second. Pro/ENGINEER allows for the creation of parts and subassemblies within Assembly mode. By creating a component using this technique, relationships can be established between two parts. Modification of the parent part reference will modify the child part.

PRO/ENGINEER MODES

Pro/ENGINEER is an integrated, fully associative package. Integrated parametric design packages such as Pro/ENGINEER share data with its various other operating modules. Pro/ENGINEER is the fundamental application in a powerful suite of tools capable of an integrated and concurrent environment. Objects created within Pro/ENGINEER can be shared with other applications. Due to a part's parametric associativity, changes made to an object in one mode will be reflected in other modes. As an example, a part can be modeled in Part mode. Following the modeling process, an orthographic drawing can be created in Drawing mode. Additionally, the part can be assembled with other components in Assembly mode. While in Part, Drawing, or Assembly mode, a change can be made to a parameter of the part. Upon regeneration, this change will be reflected in the other modes in which the part resides.

The following is a description of the basic modules found with Pro/ENGINEER; they are included in Pro/ENGINEER's foundation package. Additionally, modules such as Pro/Designer, Pro/NC, Pro/Mechanica, Pro/Fly-Through, and Pro/Layout are available as well.

Sketch Mode

A fundamental technique within most parametric modeling packages is to sketch feature entities and then to invoke a three-dimensional construction operation, such as Extrude, Revolve, or Sweep. Most features created utilizing a sketching technique are constructed within Part or Assembly mode. Sketches can be created separate of the part and assembly environment utilizing Sketch mode. By doing this, the sketch can be saved for use in later modeling situations.

Part Mode

Part mode is the primary environment for the creation of solid and surface models. For many manufacturing enterprises, Part mode is the center of the design and production environment. Objects created in Part mode can be utilized in downstream applications, such as Pro/ENGINEER's Drawing and Manufacturing modes. Additionally, part design intentions can be shared concurrently over Internet or intranet networks using Pro/Fly-Through and/or Pro/Web-Publish.

Drawing Mode

Drawing mode is the primary means within Pro/ENGINEER for constructing documentation drawings. While technical drawings were once considered the primary tool in engineering graphics, a drawing is now a downstream application in the parametric modeling design process. In a true "paperless" manufacturing environment, an orthographic drawing is no longer required. A design can be developed in Part and Assembly mode, analyzed in Pro/Mechanica, and have its manufacturing code generated in Pro/NC. Despite this integrated philosophy, companies still need documentation.

Drawing mode can take an existing part or assembly and produce an orthographic drawing. It can produce detailed drawings with a variety of section and auxiliary view capabilities. Dimensioning tools, including geometric tolerancing, are available. Additionally, through Pro/Detail, drawing and construction tools are available. Figure 1–11 shows an example of a detailed drawing produced in Drawing mode.

Assembly Mode

Assembly mode allows for the combining of design components into a final design solution. A variety of tools exist for the integration of design intent. When parts are placed within an assembly, relationships can be established with existing parts, features, and subassemblies. Parts can be created within Assembly mode or placed from preexisting parts.

Pro/ENGINEER allows for bottom-up or top-down assembly design. Bottom-up design requires components to be modeled in Part mode and then assembled in Assembly mode. In top-down design, a skeleton model creates an assembly

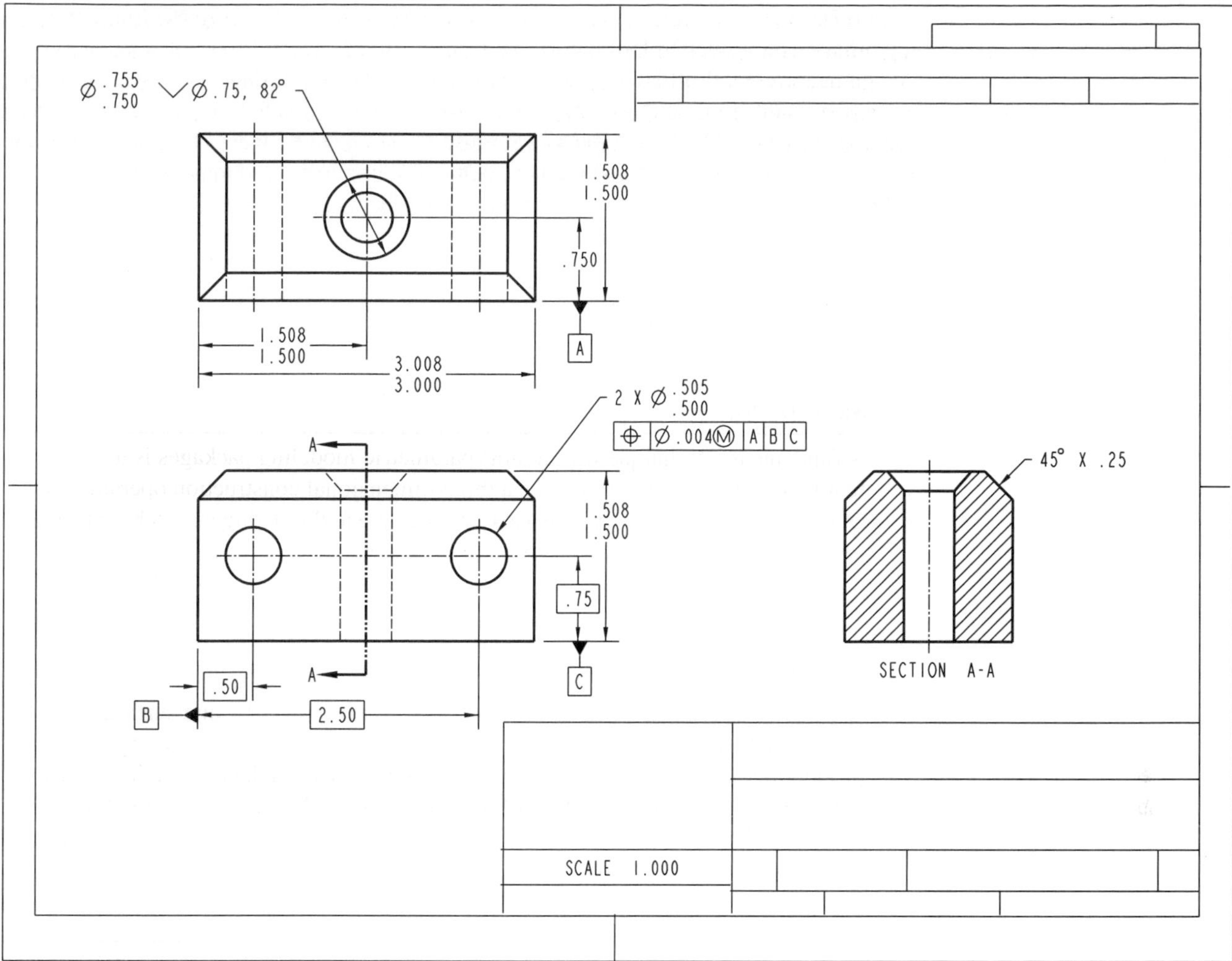

Figure 1-11 Pro/ENGINEER drawing

that starts at the overall design intent level and works down to the individual part level. Parts can be created within top-down design by modeling within Assembly mode.

Summary

Parametric design packages such as Pro/ENGINEER are revolutionizing the engineering design process. Early CAD systems were capable of producing technical drawings in an electronic format but added little to the actual design process beyond what could be accomplished with paper and pencil. Three-dimensional CAD applications, especially solid modeling systems, have design tools that allow a designer to model a design as a virtual prototype.

Parametric design fundamentals have increased the design capabilities of three-dimensional modeling CAD systems. As with solid modeling applications, parametric modeling systems can construct a design as an electronic prototype. Parametric modeling objects have intelligence that not only display a design as a graphic image but also incorporates parameters that can describe the intent of a design.

Integrated design applications such as Pro/ENGINEER are being used in companies not simply as a modeling tool but as the center of the product data management system. Design data from a Pro/ENGINEER modeling file can be viewed and retrieved throughout a company's intranet. Additionally, Internet tools such as Pro/Web-Publish allow data to be viewed by individuals external to a company's localized network. This powerful capability has increased the collaborative tools of computer-aided design and enhanced Pro/ENGINEER's concurrent engineering capabilities.

Questions and Discussion

1. Describe two types of sketching done in engineering graphics.
2. How many primary views are there possible in orthographic projection?
3. Define and describe the term *feature* within a parametric design package.
4. List and describe five types of geometric constraints used during the sketch construction process of a parametric design package.
5. Explain what is meant by the Parent-Child relationship that exists between parametric features.
6. Describe uses of Pro/ENGINEER's Model Tree.
7. Explain what is meant when a parametric modeling package is fully associative.
8. What is a datum surface? How might datums be used within Pro/ENGINEER?
9. Describe how Pro/ENGINEER can be used within a concurrent manufacturing environment.
10. List and describe the uses of the various modes found within Pro/ENGINEER's foundation package.

CHAPTER

2

PRO/ENGINEER'S USER INTERFACE

Introduction

Pro/ENGINEER has both a UNIX and Windows version (NT, 95, and 98). Starting with Release 20, Pro/ENGINEER introduced a new interface. This interface resembles, on its face, a typical Windows application. When manipulating Pro/ENGINEER, it is important to remember that it does not function like a true Windows application. This chapter will introduce the fundamentals of Pro/ENGINEER's interface. Upon finishing this chapter, you will be able to

- Describe the purpose behind each menu on Pro/ENGINEER's menu bar.
- Use Pro/ENGINEER's file management capabilities to save object files.
- Set up a Pro/ENGINEER object to include units, tolerances, and materials.
- Customize Pro/ENGINEER through the use of the Configuration file.
- Customize Pro/ENGINEER commands using mapkeys.
- Organize items using the Layers option.

DEFINITIONS

Configuration file A Pro/ENGINEER file used to customize environmental and global settings. Configuration options can be set through the Utilities >> Preferences option.

Mapkeys Keyboard macros used to define frequently used command sequences.

Model An object that represents the actual sculptured part, assembly, or work piece.

Nominal dimension A dimension with no tolerance.

Object A file representing an item, part, assembly, drawing, layout, or diagram created in Pro/ENGINEER.

Tolerance The allowable amount that a feature's size or location may vary.

MENU BAR

The following describes many of the options available on Pro/ENGINEER's Part mode menu bar (see Figure 2–1). While this interface may appear to make Pro/ENGINEER a true Windows application, many typical Windows functions are not available (copy, paste, etc.).

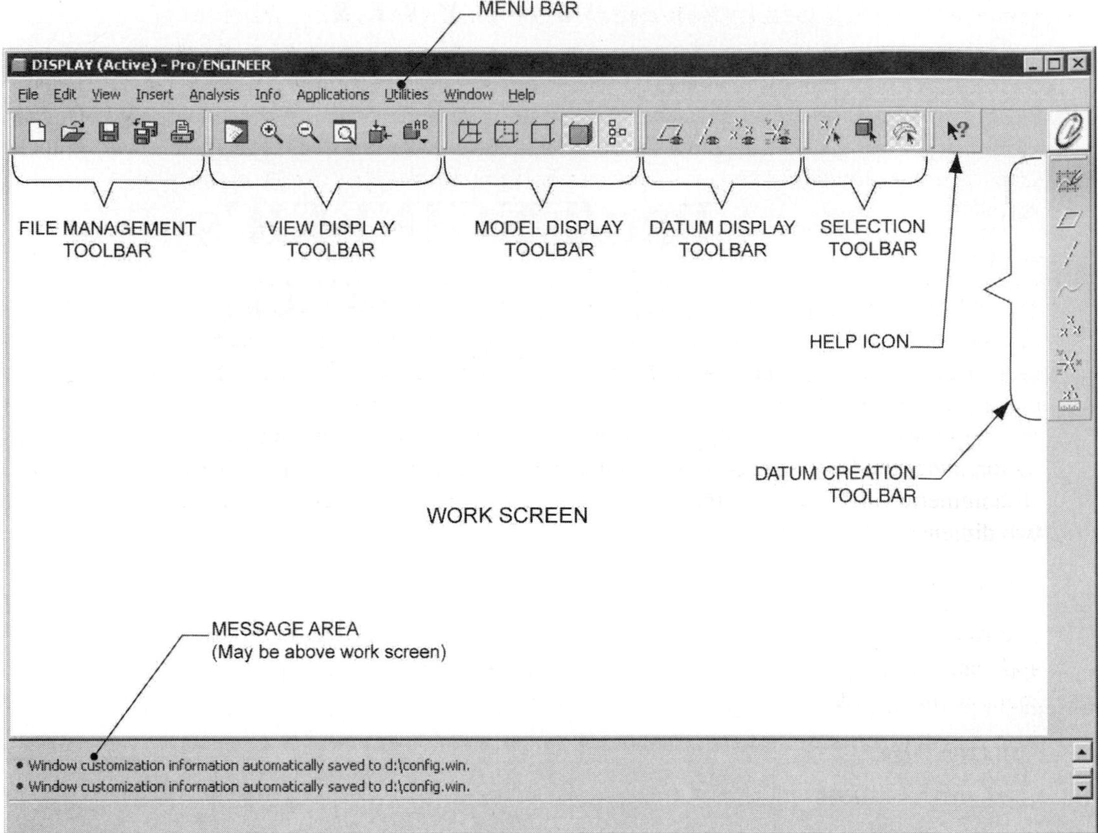

Figure 2-1 Pro/ENGINEER's work screen

File Menu

The File menu is Pro/ENGINEER's interface for the manipulation of files and objects. Found under the File menu are options for saving, opening, printing, and exporting objects.

Edit Menu

The edit menu provides option for the modification of geometric elements. Within Part mode, options are available for performing feature manipulation and modification techniques such as redefine, reroute, suppress, and delete. Within sketch mode, options are available for moving, copying, and trimming sketched entities.

View Menu

The View menu is used to change the appearance of models and Pro/ENGINEER's work screen. Many of the view options available exist as shortcut keys or can be found on the toolbar. Commonly used view manipulation options under the View menu include zooming, repainting the view, and retrieving the default view. Options also are available for orienting a model, saving a view, modifying a model's color and appearance, and for changing the lighting of Pro/ENGINEER's work screen.

Insert Menu

The Insert menu provides selections for the creation of traditional Pro/ENGINEER features (e.g., protrusion, hole, datum plane, cosmetic thread, etc.). Unlike previous versions of Pro/ENGINEER, datums can be created at anytime during object modeling, including while sketching.

Analysis

Options for finding assembly and part properties can be found under the Analysis menu. As an example, the mass of a part can be obtained through the Model Analysis option.

Info Menu

The Info menu provides information about Pro/ENGINEER objects. Information can be found on Parent-Child relationships, features, references, and geometry. Messages, such as error messages created during regeneration failures, can be displayed using the Message Log option. Additional information about failed regenerations can be found using the Geometry Check option. A commonly used option under the Info menu is Switch Dims, in which dimensions can be displayed with numeric values or with dimension symbols. This option toggles between the two dimension display modes.

Applications Menu

The Applications menu allows a user to switch between Pro/ENGINEER modes and applications. As an example, a user can switch between Part mode and Manufacturing mode.

Utilities Menu

The Utilities menu allows for the customization of Pro/ENGINEER's interface. Before Release 20, the Environment option was a common tool for changing the work screen appearance. Many of the features found under the Environment menu, such as datum and model display, can now be found on the toolbar. The Utilities menu provides access to Pro/ENGINEER's configuration file. Additional options are available for customizing colors and for creating **mapkeys.**

Window Menu

The Window menu is used to manipulate Pro/ENGINEER windows. Windows can be activated, opened, or closed. Within Pro/ENGINEER, multiple windows of multiple parts can be open at once. To work in one menu, a user has to first activate it. Opened windows are displayed under the Window menu, thus allowing a user to easily switch from one object to another.

Help Menu

Pro/ENGINEER utilizes a web browser to access help information. The Pro/Help CD must be loaded before the full help option can be utilized. While Pro/Help provides search capabilities for Pro/ENGINEER options, a context-sensitive help option is also available. Use context-sensitive help to find information on individual Pro/ENGINEER menus and options.

Toolbar

As shown previously in Figure 2–1, Pro/ENGINEER provides a toolbar for easy access to frequently used options through the use of icons. By default, Pro/ENGINEER's initial toolbar is divided into five groups. Additional options can be added to the toolbar under the Utilities menu.

File Management

The file management group of icons is available for manipulating files.

Figure 2-2 File management icons

- **New** The New icon is used to start a new Pro/ENGINEER file.
- **Open** The Open icon is used to open a Pro/ENGINEER file.
- **Save** The Save icon is used to save a Pro/ENGINEER file.
- **Save A Copy** The Save A Copy icon is used to save a Pro/ENGINEER file as a different name and to a different location.
- **Print** The Print icon is used to print or plot a Pro/ENGINEER object.

View Display

The view display group of icons is available for modifying the display of Pro/ENGINEER objects on the work screen.

Figure 2-3 View display icons

- **Repaint** The Repaint icon is used to redraw the work screen.
- **Zoom In** The Zoom In icon is used to zoom in to a user-defined window.
- **Zoom Out** The Zoom Out icon is used to zoom out from the work screen.
- **Refit** The Refit icon is used to fit the extent of a Pro/ENGINEER object into the work screen.
- **Orient** The Orient icon is used to orient a Pro/ENGINEER object on the work screen.
- **Saved Views** The Saved Views icon is used to access saved views.

Model Display

The model display group of icons is available for changing the display of Pro/ENGINEER objects. Only one of the four available icons under this group may be activated at a time.

Figure 2-4 Model display icons

- **Wireframe** The Wireframe icon displays a Pro/ENGINEER object as a wireframe.
- **Hidden Line** The Hidden Line icon displays a Pro/ENGINEER object with hidden lines.
- **No Hidden** The No Hidden icon displays a Pro/ENGINEER object without hidden lines.
- **Shade** The Shade icon shades a Pro/ENGINEER object.
- **Model Tree** The Model Tree icon turns the display of the model tree on or off.

Datum Display

The datum display group of icons is used to control the display of datums.

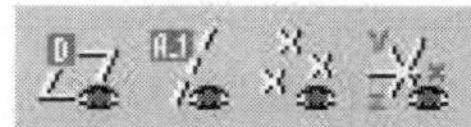

Figure 2-5 Datum display icons

- **Datum Planes** The Datum Plane icon is used to turn on or off the display of Pro/ENGINEER datum planes.
- **Datum Axes** The Datum Axes icon is used to turn on or off the display of Pro/ENGINEER datum axes.
- **Point Symbols** The Point Symbols icon is used to turn on or off the display of Pro/ENGINEER datum points.
- **Coordinate Systems** The Coordinate Systems icon is used to turn on or off the display of Pro/ENGINEER coordinate systems.

Context-Sensitive Help

The Context-Sensitive Help icon is used to display help information on individual menu or dialog box options. To use this help function, select the context-sensitive help icon, then select the menu item. Pro/Help will launch a web browser and display information about the selected item.

Figure 2-6 Context-sensitive help

File Management

Various options are available for manipulating Pro/ENGINEER files. Pro/ENGINEER's file management capabilities provide a wide range of functions for managing projects and models. On first appearance, Pro/ENGINEER's file opening and saving commands resemble a Windows application. However, there are some significant differences between Pro/ENGINEER's file management and a Windows application.

- Pro/ENGINEER filename requirements are more restricted than Windows application filenames.
- Saving a Pro/ENGINEER object creates a new version of the object each time the object is saved. It does not override older versions.
- Pro/ENGINEER will not allow an object to be saved to a specific filename if that filename already exists. Pro/ENGINEER will not save on top of an existing file.

Filenames

Pro/ENGINEER has different file extensions according to the mode being utilized. Table 2–1 shows file extensions based on six common Pro/ENGINEER modes.

Notice the extra asterisk at the end of each file extension. This asterisk represents the version of the file. The first time Pro/ENGINEER saves a file, this extra extension has a value of 1. The second time a file is saved, a new file is created with a 2 as this value. The

Table 2-1 File extensions for Pro/ENGINEER modes

Mode	Extension
Sketch	*.sec.*
Part	*.prt.*
Assembly	*.asm.*
Manufacturing	*.mfg.*
Drawing	*.drw.*
Format	*.frm.*

Table 2-2 Invalid and valid filenames

Invalid Filename	Problem	Valid Filename
part one	Space in filename	part_one
part@11	Nonalphanumeric character	part_11
Part[1_10]	Brackets used in filename	Part_1_10

third time a file is saved, a new file is created with a 3 as this value. Pro/ENGINEER creates a new object file each time a file is saved. If an object file is saved 10 times, 10 Pro/ENGINEER files will be created. To delete the previous Pro/ENGINEER files, select File >> Delete >> Old Versions.

Pro/ENGINEER file and directory names cannot be longer than 31 characters. Brackets, parentheses, periods, nonalphanumeric characters, and spaces cannot be used in a filename. An underscore (_) may be used in a file name, though. Table 2–2 shows examples of invalid and valid filenames.

Memory

When an object is opened, referenced, or created in Pro/ENGINEER, it is placed in memory. It remains there until it is erased or until Pro/ENGINEER is exited. Also, when opening an assembly, every part referenced by the assembly is placed in memory. Parts in active memory are displayed in a window. Multiple parts, assemblies, and drawings can be in active memory at once. This allows for ease of access between objects. Objects may also be in session memory. Session memory is the condition where the object is in memory, but not displayed in a graphics window.

Working Directory

Pro/ENGINEER utilizes a *working directory* to help manage files. The working directory is usually the modeling point for all Pro/ENGINEER objects. When a new file is saved, it is saved in the current working directory unless a new directory is specified. To change the current working directory, from the File menu select Working Directory. Then select the desired directory as the working directory.

Opening an Object

As seen in Figure 2–7, Pro/ENGINEER objects are retrieved using the File Open dialog box. This dialog box may be retrieved by selecting Open from the File menu or by selecting the Open icon on the Toolbar menu.

Utilizing the File Open dialog box, any type of Pro/ENGINEER object can be retrieved. When opening an object, Pro/ENGINEER defaults to one of the following directories:

- Directory associated with the active object (first alternative).
- Working directory (second alternative).
- Directories contained in the search path (third alternative).

Perform the following steps to open a Pro/ENGINEER object:

Step 1: Select FILE >> OPEN to reveal the File Open dialog box.

The File Open dialog box can be revealed using the Open icon also.

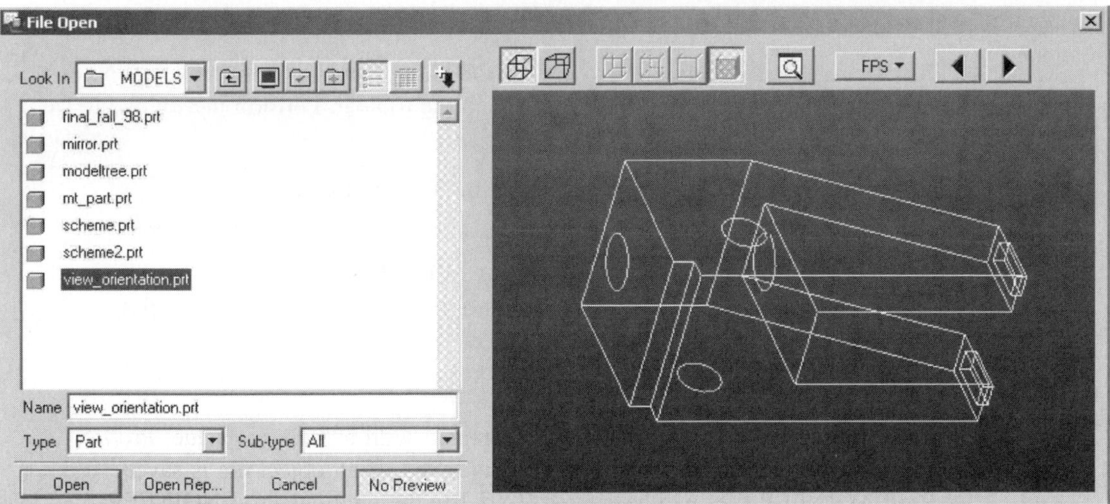

Figure 2-7 File open dialog box

STEP 2: Select the directory in which the object is located.

As shown in Figure 2–7, options are available for manipulating the Look In directory. The following is a list of the icons found on the File Open dialog box that are available for this purpose.

- This icon option moves the Look In directory up one level.
- This icon option list files currently in memory.
- This icon option returns the Look In directory to the current working directory.
- This icon option changes the way that files are viewed in the Look In directory. An option is available for showing all versions of an object, which is useful for opening old versions of an object.

STEP 3: Select the object to open.

Objects can be selected by picking the object from the file list or by typing in the file name. Older versions of an object can also be opened. If a version number is not known, you can type in the complete file name with an extension number relative to the older version. As an example, if you want to open a part file with the name *part_one,* but open the file three versions earlier, you would enter the filename *part_one.prt.-3.*

Wildcards are available for opening objects. An asterisk can replace multiple characters (e.g., *part*.prt*), while a question mark can replace a single character (e.g., *part?a.prt*).

> **MODELING POINT** Finding an object in a directory with many files can be difficult. There are several options available to lower the number of objects shown on the file list. One option is to select the type of file to open by using the Type option. Another option is to sort files alphabetically using the Sort icon.

STEP 4: Select OPEN on the dialog box.

Creating a New Object

Files for most basic modes of Pro/ENGINEER are created using the File >> New option from the menu bar or using the New icon on the toolbar. Perform the following steps to create a new object:

Step 1: Select FILE >> NEW to open the New dialog box.

New Pro/ENGINEER files can also be created using the File icon.

Step 2: Select a mode type and subtype of Pro/ENGINEER.

Available mode types are shown in Figure 2–8. Pro/ENGINEER defaults to Part mode with Solid as the subtype.

Step 3: Enter a filename.

You can enter a filename or take the default name.

> **MODELING POINT** If you forget to enter a filename or if you want to change a filename, use the Rename option found under the File menu. This option will rename all versions of a filename.

Step 4: Select OK from the dialog box.

Saving an Object

Various options are available for saving objects. New objects are saved by default in the current working directory. If an object is retrieved from a directory other than the working directory, the object is saved in its original directory. Additionally, selecting Save while in a sketcher environment will save the section (*.sec.*) and not the object being modeled. The following options are available for saving objects.

Save

This option saves an object to disk. When saving an assembly, all individual parts that comprise the assembly are saved. When saving a drawing, the model used to

Figure 2–8 New dialog box

create the drawing is saved only when changes have been made to the object. While sketching in a sketcher environment, the section under modification or creation is saved, but not the object file. Pro/ENGINEER objects can be saved to a computer hard drive, floppy disk, or zip disk.

Save A Copy

The Save As option either saves an object as a new filename or saves an object to a new directory. When an object is saved using Save As, the original filename is not deleted and is still the active model. Save As practically creates a copy of the object being modeled. Any changes made to the original object are not reflected in the copied object.

Backup

The Backup option creates a copy of the object being modeled. The name of the object cannot be changed with this option. Any saves of an object conducted after a backup will be to the directory of the backup.

Rename

The Rename option changes the name of a Pro/ENGINEER object. A suboption is available for renaming the object on disk and in memory, or just in memory. When renaming an object that already exists, all previous versions of the object are saved.

Delete

Saving an object multiple times can create many versions of the object on disk. The Delete option is available to purge old versions. Options are available for deleting old versions or all versions of an existing object.

Erase

Closing a window that contains a Pro/ENGINEER object does not remove it from memory. The Erase command must be used to remove an object from memory. An object that is referenced by another opened object cannot be erased. The Erase dialog box shows all objects referenced by a selected object. Options are available for erasing referenced objects from memory or for keeping them in memory.

Activating an Object

Multiple objects can be open at once within Pro/ENGINEER. Additionally, multiple windows can be opened. To modify an object, its associated window must be activated. To make a window active, use the Activate option found under the Window menu.

Viewing Models

There are many different ways to view a Pro/ENGINEER object and to view Pro/ENGINEER's work screen. Options are available for panning, rotating, and zooming an object dynamically. Other options are available for changing the display of a model.

Dynamic Viewing

A useful feature of Pro/ENGINEER that enhances its model building capabilities is its dynamic viewing functions. A model can be dynamically zoomed, rotated, and panned using a combination of the control key and a mouse button (see Figure 2–9).

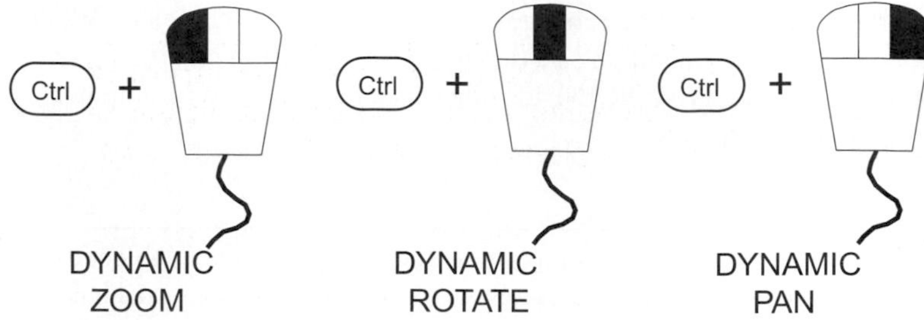

Figure 2-9 Dynamic viewing options

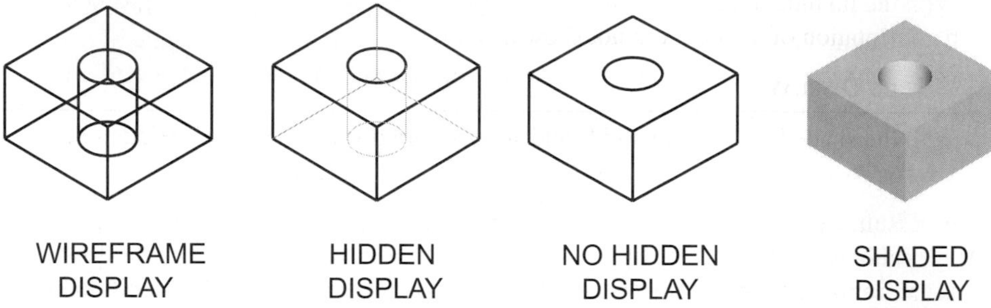

Figure 2-10 Model display options

Dynamic Zoom

A user can dynamically zoom in or out on a model by using Pro/ENGINEER's Dynamic Zoom option. Dynamic zoom is activated when the keyboard's control key is selected at the same time as the left mouse button. While simultaneously holding down the control key and the left mouse button, move the cursor from the top of the work screen to the bottom to zoom in on a model. Correspondingly, move the cursor from the bottom of the work screen to the top to zoom out on a model.

Dynamic Rotate

A user may dynamically rotate a model by using Pro/ENGINEER's Dynamic Rotate option. Dynamic rotate is activated when the keyboard's control key is selected at the same time as the middle mouse button. While simultaneously holding down the control key and the middle mouse button, move the cursor around the work screen to rotate the model around a specified spin center.

Dynamic Pan

A user may dynamically pan a model by using Pro/ENGINEER's Dynamic Pan option. Dynamic pan is activated when the keyboard's control key is selected at the same time as the right mouse button. While simultaneously holding down the Control key and the right mouse button, move the cursor around the work screen to pan the model.

Model Display

As shown in Figure 2–10, four styles are used to display a model in part, assembly, and manufacturing modes. Similarly, three styles are used within drawing mode. There are situations when each display style is the most practical. The Model Display dialog box is located under the View menu and contains other display options. Each display style can be selected dynamically from the Toolbar menu.

Wireframe Display

Within all relevant modes of Pro/ENGINEER, the wireframe style displays all edges of a model as a wireframe. Edges that would be hidden from view during a true representation of the model are displayed, as are edges that would not be hidden.

Hidden Display

With the hidden display style, lines that would be hidden from view during a true representation of a model are shown in gray. Within Drawing mode, these gray lines represent hidden lines and will be printed as hidden lines.

No Hidden Display

With the no hidden display style, lines that would be hidden from view during a true representation of a model are not shown.

Shaded Display

With the shaded display, all solids and surfaces are displayed shaded. Hidden lines are not shown. This option is not available in drawing mode.

Three default views are possible within Pro/ENGINEER: trimetric, isometric, and user-defined. When selecting *default view* from the View menu, the model returns to this viewpoint. The initial setting within Pro/ENGINEER is trimetric. This setting can be permanently changed by the configuration file option *orientation,* or temporarily changed in the Environment menu or in the Orientation dialog box.

View Orientation

When modeling in Part or Assembly mode, it can be advantageous to orient the model to one of the six primary orthographic views. Also, within Drawing mode, when a view is first established the model is initially placed as the default view. A correct view orientation can be created with the Orientation dialog box.

The Orientation option is available under the View menu. This option opens the Orientation dialog box (Figure 2–11). Three orientation types are available under this dialog box:

Angles

The Angles option orients the model by defining axis rotation angles.

Dynamic Orientation

The Dynamic Orient option is similar to Pro/ENGINEER's dynamic view options. The Dynamic Orient option provides a dialog box for zooming, rotating, and panning a model.

Orientation by Reference

The Orient by Reference option is used to create an orthographic view of a model. This option is also available on the toolbar. Figure 2–11 shows the Orientation dialog box with this option selected. Two references are required with this option. These references correspond to a selected primary orthographic view of the model (e.g., front, top, right, etc.). The initial references are *front* for the first reference and *top* for the second. These references can be changed. When selecting a reference with a selected orthographic view, select a planar surface to orient the model toward that direction. As an example, as shown in Figure 2–12, when using *top* as the first reference and *front* as the second, selecting as shown will orient the model in the direction specified.

Figure 2-11 Orientation dialog box

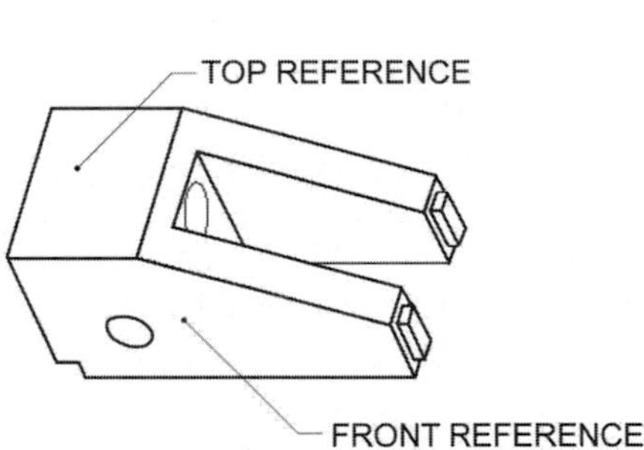

Figure 2-12 Orienting a model

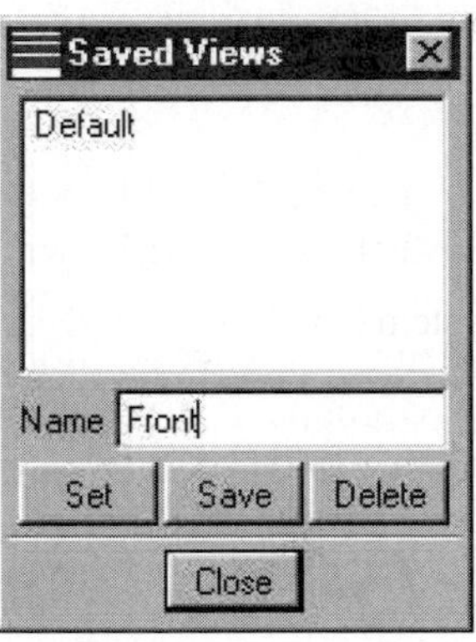

Figure 2-13 Saved views dialog box

Naming and Saving Views

Often, it is necessary to return to a user-defined orientation of a model. View orientations can be obtained using dynamic viewing or using the Orientation option. As shown in Figure 2–13, the Saved Views dialog box is used to save and retrieve views. To access the Saved Views dialog box, select Saved Views from the View menu. On the dialog box, enter a name for the view, and then select the Save button. To retrieve a saved view, either use the Saved Views option under the View menu or select the Saved Views icon on the toolbar.

Setting Up a Model

Pro/ENGINEER provides several options for setting up a model. Properly establishing an object's parameters is an often overlooked, but important step in the modeling process.

This section will describe how to establish units for a model and how to define and assign materials. Additionally, setting tolerances and renaming features will be introduced.

UNITS

Within Pro/ENGINEER, there exist four principle unit categories: length, mass or force, time, and temperature. Each category has a full range of possible units. As an example, available within the length category are inches, feet, millimeters, centimeters, and meters.

SETTING A SYSTEM OF UNITS

Pro/ENGINEER utilizes a system of units to group the four principle categories. Listed below are six predefined systems of units available within Pro/ENGINEER. These systems may be accessed through the System of Units tab found on the Units Manager dialog box.

- Meter kilogram second (MKS)
- Centimeter gram second (CGS)
- Millimeter newton second (mmNs)
- Foot pound second (FPS)
- Inch pound second (IPS)
- Inch lbm second (Pro/E default)

From the preceding list of predefined systems of units, *inch-lbm-second* is the default system. To set a different system of units, select the specific system on the System of Units tab, then select the Set button.

CREATING A SYSTEM OF UNITS

A user-defined system of units can be created to allows a user to establish units that meet design intent for a given product. Perform the following steps to create a new system of units.

STEP 1: Select SET UP >> UNITS on Pro/ENGINEER's Menu Manager.

When selecting the Units option, the Units Manager dialog box will appear (Figure 2–14). Options available under this dialog box include: (*a*) Creating a new system, (*b*) copying an existing system to create a new system, (*c*) reviewing a system's units, (*d*) deleting a user-defined system, and (*e*) setting a system of units.

STEP 2: Select NEW from the Units Manager dialog box.

STEP 3: In the System of Units Definitions dialog box, enter a NAME for the user-defined system (Figure 2–14).

STEP 4: Select UNITS from each principle category (Figure 2–14).

Notice under the dialog box that an option exists for choosing between mass and force. Each subcategory has its own available units.

STEP 5: Select OK, then close the Units Manager dialog box.

MODELING POINT It is good modeling practice to set your units before creating a feature. If you forget to set them before creating the first feature, units can be converted from one system to another. Pro/ENGINEER will interpret your existing dimensions as another defined unit (e.g., 1" will be interpreted as 1 mm). To allow for this interpretation, use the Interpret Existing Numbers option after setting a new unit.

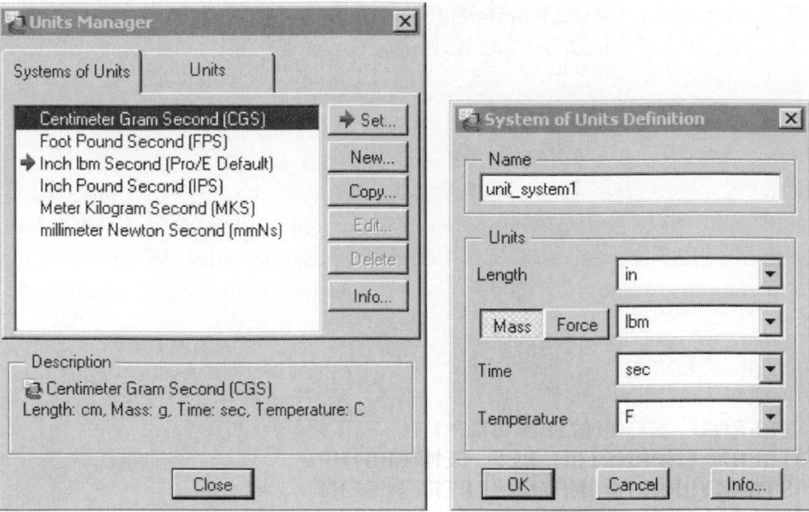

Figure 2-14 Units menus

MATERIALS

Material parameters can be defined for a part. Additionally, a defined material can be written to a file and used with other parts. A recommended procedure is to create a library of materials. Figure 2–15 shows parameters that can be established for any given material. Not all parameters have to be defined.

DEFINING A MATERIAL

When defining a material, the information is stored in the database of the part being modeled. To allow established material parameters to be used by other parts, the information has to be written to a file. To define a material, perform the following steps.

STEP 1: Select SET UP >> MATERIALS >> DEFINE.

STEP 2: Enter a material name.

Any name can be used. It is recommended that the name be descriptive of the material being defined.

STEP 3: Enter parameters associated with the material.

Pro/ENGINEER will display a default Materials file. Figure 2–15 shows an example of parameters that can be created for a material. Upon finishing the material parameters, exit out of the editing environment by selecting File >> Save >> Exit.

WRITING A MATERIAL TO DISK

Materials created for a part are not automatically saved to disk. To save a material specification so other parts can use it, the material has to be saved to a permanent location. It is recommended that all materials be stored in a common Materials Library directory. Perform the following steps to write a material to disk:

STEP 1: Select SET UP >> MATERIALS >> WRITE.

STEP 2: Select a material specification to write to disk.

On the menu, Pro/ENGINEER will display materials that have been defined within the current part. Select a material specification then accept it.

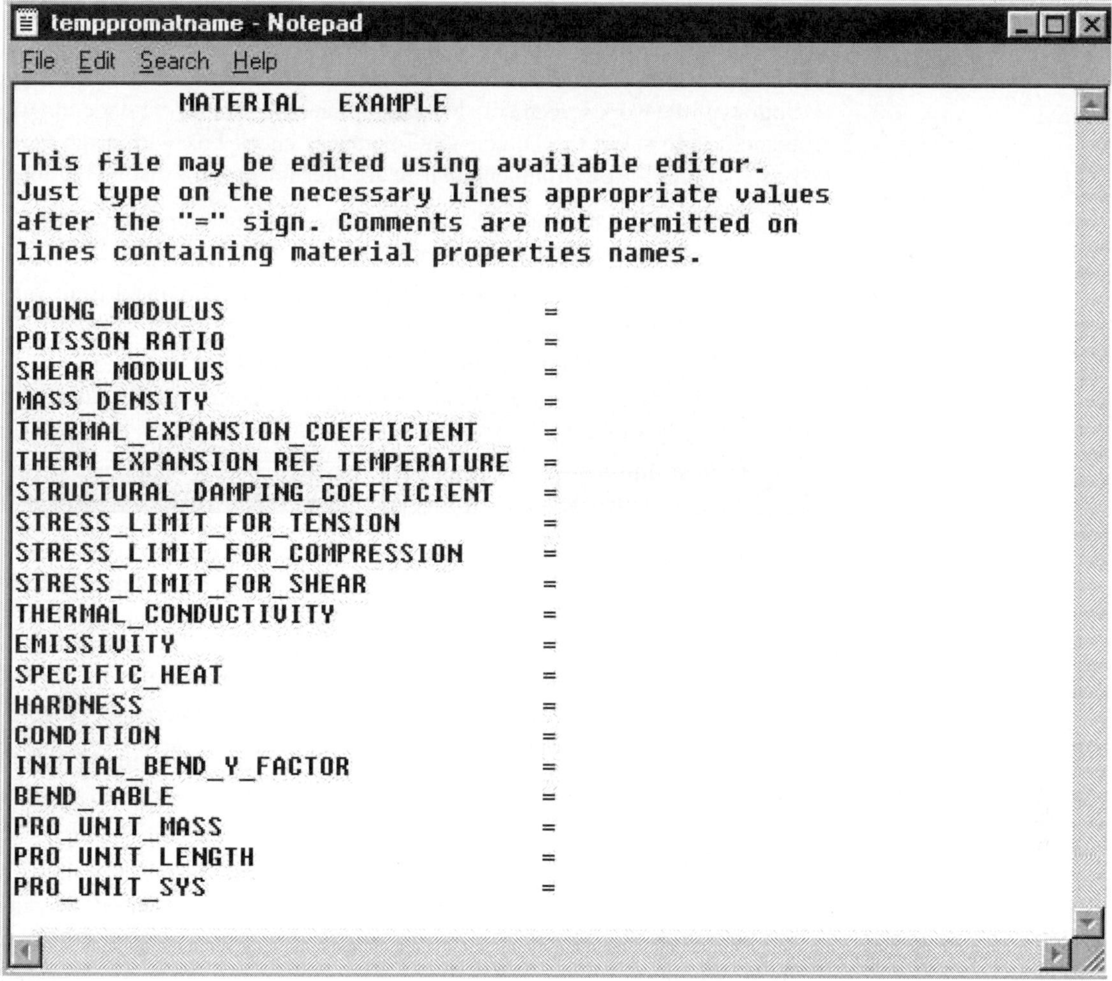

Figure 2-15 Available material parameters

STEP 3: Enter a name for the material file.

Material files are in ASCII format and have an **.mat* extension.

ASSIGNING MATERIALS

When a material is defined within a part, it is not automatically assigned to a part. Perform the following steps to assign a material to a part.

STEP 1: Select SET UP >> MATERIALS >> ASSIGN.

STEP 2: Select FROM PART or FROM FILE as the source of the material specification.

Material may be selected from within the part or from disk.

DIMENSIONAL TOLERANCE SET UP

Pro/ENGINEER provides various methods for the display of dimensional tolerances. A **tolerance** is the allowable amount that a feature's shape or size may vary from the exact specification. Tolerances are a necessity for most precision parts.

By default, Pro/ENGINEER displays tolerances as nominal values. There are two ways to display tolerance values. The first way is to set the configuration file option *tol_display* to yes. Additionally, when setting this option to yes, a table is displayed at the bottom of the work screen with the default tolerance values. The second way to display tolerances is to select the Dimension Tolerances check box within the Environment dialog box (Utilities >> Environment). There are four formats for displaying tolerances:

- **Nominal** The Nominal option does not display tolerance values. This is Pro/ENGINEER's initial format.
- **Limits** The Limits option displays tolerances as an upper limit value and a lower limit value.
- **Plusminus** The Plusminus option displays tolerance values as a positive and negative deviation from the nominal value. The positive value and negative value may be different.
- **Plusminussym** The Plusminussym option displays tolerance values as a positive and negative deviation from the nominal value. The positive value and negative value are the same.

Tolerance formats can be changed using the Modify Dimension dialog box. This dialog box can be accessed through the Dimension option located under the Modify menu. The default format may be changed in the configuration file option *tol_mode.*

Within Pro/ENGINEER, a user can select ANSI or ISO for the tolerance display standard. The ANSI standard is selected initially. A specific tolerance standard can be selected through the Standard option found under the Tol Setup menu. Use the configuration file option *tolerance_standard* to change the default value.

ANSI Tolerance Standard

Tolerance values using the ANSI standard are initially set based on values found in the configuration file. The *linear_tol* and *angular_tol* options are used to set linear and angular tolerance values, respectively. The tolerance values are assigned to dimensions based on the number of decimal places. To change variational values and number of decimal places for dimensions, use the Modify >> Dimension option.

ISO Tolerance Standard

Tolerance tables drive dimensional tolerances with the ISO tolerance standard. Tables are loaded when a model is created with an ISO standard or when the standard is changed to ISO from ANSI. Figure 2–16 shows an example of a tolerance table. The TABLE_TYPE range is used to specify the type of tolerance to use.

Within the ISO standard, tolerances have an additional variation referred to as the *tolerance class.* The tolerance class is the relative looseness or tightness of a tolerance. There are four general classes of fit: fine, medium, coarse, and very coarse. The coarser the fit, the more variation there is in the tolerance. Medium is Pro/ENGINEER's default class. Tolerance classes are specified in each tolerance table. To change classes of fit for individual dimensions, select Model Class from the TOL SETUP menu.

MODELING POINT Tables (referred to as *Pro/TABLE*) are used throughout Pro/ENGINEER to input information. A tolerance table is just one example of a Pro/TABLE. Family tables and the configuration file are other examples of functions that utilize tables.

Many Pro/TABLE cells have established values. As an example, the possible values for the Table_Unit cell (Figure 2–16) within a tolerance table are inch, foot, micrometer, millimeter, centimeter, and meter. Instead of typing one of the possible units into the cell, selecting the F4 key on the keyboard will display all of the available options.

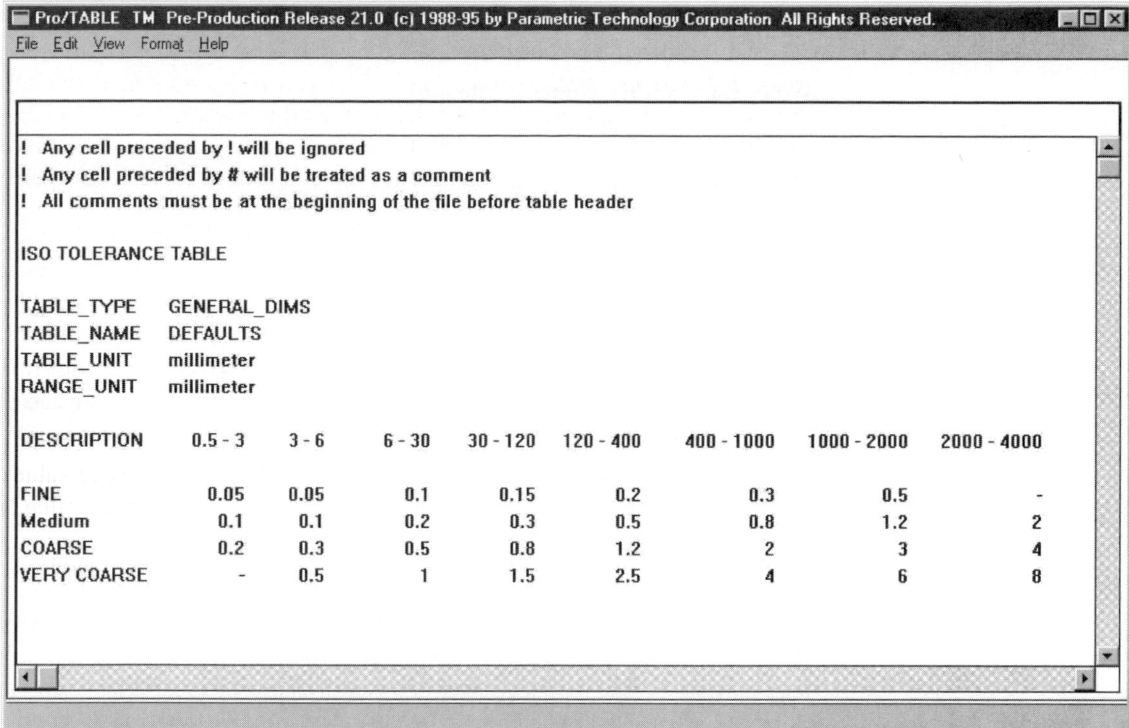
Pro/TABLE TM Pre-Production Release 21.0 (c) 1988-95 by Parametric Technology Corporation All Rights Reserved.

File Edit View Format Help

! Any cell preceded by ! will be ignored
! Any cell preceded by # will be treated as a comment
! All comments must be at the beginning of the file before table header

ISO TOLERANCE TABLE

TABLE_TYPE	GENERAL_DIMS
TABLE_NAME	DEFAULTS
TABLE_UNIT	millimeter
RANGE_UNIT	millimeter

DESCRIPTION	0.5 - 3	3 - 6	6 - 30	30 - 120	120 - 400	400 - 1000	1000 - 2000	2000 - 4000
FINE	0.05	0.05	0.1	0.15	0.2	0.3	0.5	-
Medium	0.1	0.1	0.2	0.3	0.5	0.8	1.2	2
COARSE	0.2	0.3	0.5	0.8	1.2	2	3	4
VERY COARSE	-	0.5	1	1.5	2.5	4	6	8

Figure 2–16 ISO tolerance standard table

Tolerances are based on the nominal size of the dimension. Figure 2–16 shows an example of a tolerance table with metric units. The following list describes information found in a tolerance table.

- **Table_Type** Available options include general, broken edge, holes, and shafts. Use the F4 key to select between available options.
- **Table_Name** This cell represents the name of the table.
- **Table_Unit** Examples of tolerance units include inches, millimeters, and meters. Use the F4 key to select between available options.
- **Range_Unit** Examples of tolerance ranges include inches, millimeters, mci, and meters. Use the F4 key to select between available options.
- **Description** Description represents the nominal sizes of the table. The full range of available sizes of a unit should be covered. The table shown in Figure 2–16 ranges from less than 0.5 to 4,000 millimeters.
- **Tolerance class** Tolerances are provided for one or more of the available classes of fit. Providing all four classes will ensure that a change in a class will not be invalid.
- **Tolerance value** Each tolerance value represents a symmetrical plus/minus variation from the nominal value. As an example, a value entered as 0.2 would provide a tolerance value of plus and minus 0.2 mm from a dimension's nominal value.

Modifying a Tolerance Table

Perform the following steps to modify an ISO tolerance table. The tolerance standard first has to be set to ISO/DIN.

Step 1: **Select SET UP >> TOL SETUP option.**

Step 2: **Select the TOL TABLES option on the Tolerance Setup menu.**

STEP 3: Select the MODIFY VALUE >> GENERAL DIMS options from the Tolerance Tables menu.

STEP 4: Modify values in the table.

GEOMETRIC DIMENSIONING AND TOLERANCING

Geometric dimensioning and tolerancing (GD&T) controls the shape of geometric features of a part. Unlike dimensional tolerances, the application of a geometric tolerance to a Pro/ENGINEER model does not affect the actual shape of the model.

Traditional tolerancing controls the location and size of part features, but does not control the geometric shape. As an example, a cylinder can have a tolerance applied to its diameter dimension. This dimension controls the size of the cylinder, but not variations in the cylinder's shape. If extreme variations in the shape, such as a lack of cylindricity, will have negative affects on the design, GD&T can be used to control this form variation.

Five classes of features are controlled: form, profile, runout, orientation, and location. Cylindricity is one type of form control; other form types include straightness, flatness, and circularity.

Another use of GD&T is to control the location of features. Traditional location tolerances create a square tolerance zone. A square tolerance zone may inadvertently lead to design problems. A geometric position tolerance creates a circular tolerance zone. Position is one type of location control; other location types include concentricity and symmetry.

Many geometric tolerance values can be modified based on a material condition. There are three possible *material condition modifiers:* maximum material condition (MMC), least material condition (LMC), and regardless of feature size (RFS). The MMC of a feature exists when a feature has the most material possible. The MMC for a shaft is the largest shaft possible within its established limits. Since a part will have more material if it has a smaller hole, the MMC for a hole is the smallest hole possible. The GD&T symbol for MMC is a circled letter *M*. The LMC for a part is the condition with the least amount of material is present. The GD&T symbol for LMC is a circled letter *L*. *Regardless of feature size* means that a tolerance value is not modified by a material condition. The symbol for RFS is a circled letter *S*. Having no material condition modifier, by definition, interprets to mean RFS.

Datum references may be datum planes, axes, or points. Some types of feature controls do not require a datum reference. As an example, no Form type control (e.g., flatness or straightness) requires a datum reference. With many other types of controls, a datum reference is an absolute necessity. As an example, a position tolerance might be used to control the location of a hole. It is common to locate a hole with references to two cdges. When applying a position tolerance to a hole, the hole has to be located with respect to two datum references. Datum references are also common with orientation controls.

When a geometric tolerance controls a feature's size or location, the size or location dimension becomes a *basic dimension*. Basic dimensions are theoretically perfect and are represented on a drawing as a nominal value enclosed in a rectangle. Since the geometric tolerance is controlling the dimension, there is no need to apply a traditional tolerance.

Figure 2–17 shows an example of a feature control frame. Feature control frames are used in GD&T to control geometric features. The following is a description of each compartment.

TYPE

There are 14 types of GD&T controls available. Examples include position, perpendicular, parallel, and cylindricity.

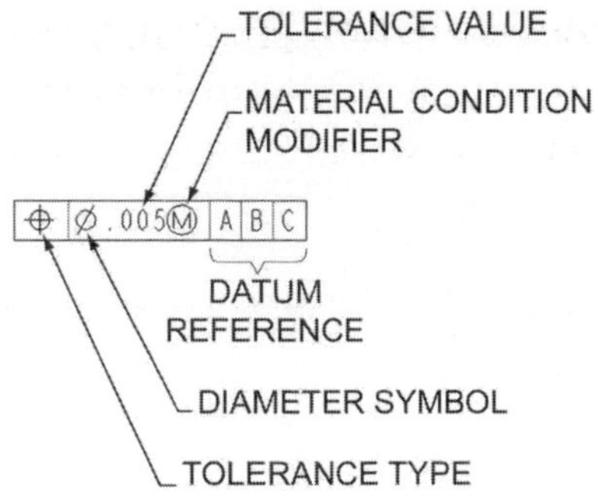

Figure 2-17 Feature control frame

TOLERANCE VALUE

The tolerance value determines the amount of variation allowed in a feature. Many GD&T types produce a cylindrical tolerance zone. If this occurs, a phi (ϕ) symbol representing diameter is placed before the tolerance value.

TOLERANCE MATERIAL CONDITION MODIFIER

A feature's tolerance value can be modified based on a material modifier. Three possible conditions exist: maximum material (MMC), least material (LMC), and regardless of feature size (RFS).

DATUM REFERENCE

Datum references are used within a feature control frame only when necessary. As an example, a flat form control does not need a datum reference, but a perpendicular orientation control does. In the case of the latter example, a feature will be controlled, within a tolerance, to be perpendicular to a datum reference. With perpendicularity, only one datum reference is needed.

CREATING A GEOMETRIC TOLERANCE

Geometric tolerances can be created within Part, Assembly, or Drawing mode. Since Pro/ENGINEER models have high accuracy (default value of 0.0012), a geometric tolerance applied to a feature does not affect the actual shape or form of the feature. Geometric dimensions and tolerances serve to provide graphical information only. Perform the following steps to create a geometric tolerance within Part mode.

> **MODELING POINT** Geometric dimension and tolerances are important considerations for the intent of a design. Care should be taken when applying geometric tolerances to a part. Of considerable importance is the location of datum references. Too often, datum planes are placed without consideration for the design of a part. Many geometric tolerances require one or more datum references. These datums should be created before applying a geometric tolerance. Also, each datum reference must be set using the Set Datum option before it can be utilized as a reference.

STEP 1: Select SET UP >> GEOM TOL on Pro/ENGINEER's Menu Manager.

STEP 2: Select SPECIFY TOL on the Geometric Tolerance menu.

As shown in Figure 2–18, the Geometric Tolerance dialog box will appear. This dialog box is used to create and place feature control frames.

STEP 3: Select the type of feature control.

Fourteen feature control types are used for geometric tolerancing. Each is available at the left end of the Geometric Tolerance dialog box.

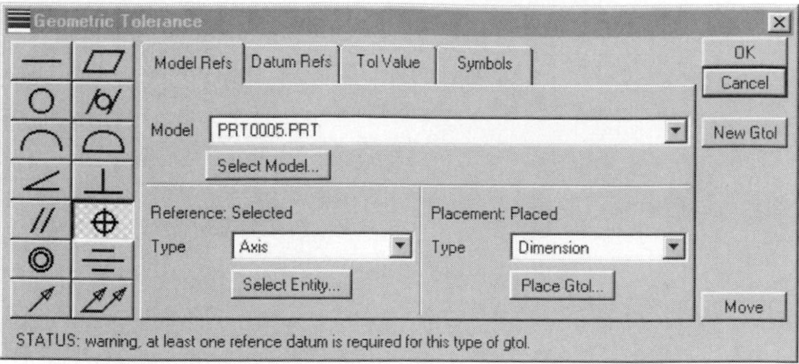

Figure 2–18 Geometric tolerance dialog box

STEP 4: Select the model to apply feature control.

As shown in Figure 2–18, four tabs are available under the Geometric Tolerance dialog box. The first tab specifies model references. The first model reference is the model to which the feature control is applied. By default, the active model is selected.

STEP 5: Select the reference to apply feature control.

An understanding of GD&T principles is needed to select the proper reference for applying a feature control. As an example, when applying a position tolerance to a hole, the tolerance is actually controlling the axis of the hole. The axis of the hole should be selected as the feature reference. With a circularity tolerance, the tolerance is controlling the surface of the feature. This surface should be selected as the reference.

STEP 6: Select a placement for feature control frame.

Several methods are available for attaching a feature control frame to a model. An understanding of GD&T drafting standards is needed to properly apply a frame. As an example, when attaching a feature control frame to a diameter dimension, the axis of the hole or cylinder is the element being controlled by the geometric tolerance.

When selecting Leader as the type of placement, a leader menu will appear. Under the leader menu is the option of attaching the leader to an entity or to a surface. Entity is selected by default. Often it is necessity to change this to surface.

After selecting the placement for the feature control frame, notice how the frame appears on the model. During the addition of references and symbols throughout the Geometric Tolerance dialog box, notice how this symbol is updated.

STEP 7: Under the DATUM REFS tab, select datum references to add to the feature control frame.

As shown in Figure 2–19, it is possible to utilize a primary, secondary, and tertiary datum reference. Only use the datum references necessary for a feature control. Datums have to exist on the model before selecting a datum reference. Additionally, datums have to be set using the Set Datum option found under the Geom Tol menu before they can be utilized as a reference. Datum material condition modifiers also exist under each sub-tab.

STEP 8: Under the TOL VALUE tab, enter the tolerance value to apply.

As shown in Figure 2–20, tolerance values and tolerance material condition modifiers are entered through the Tol Value tab.

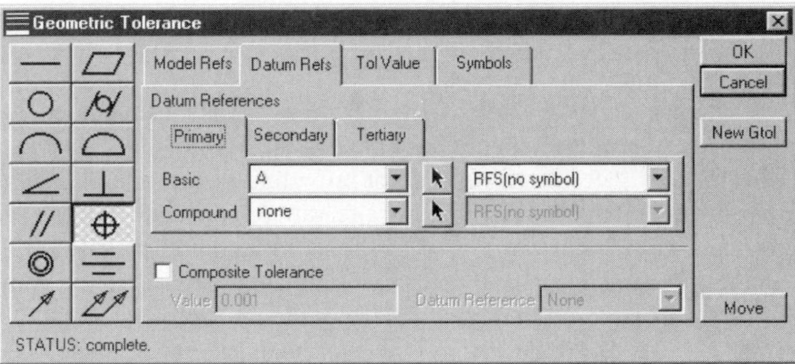

Figure 2-19 Datum refs tab

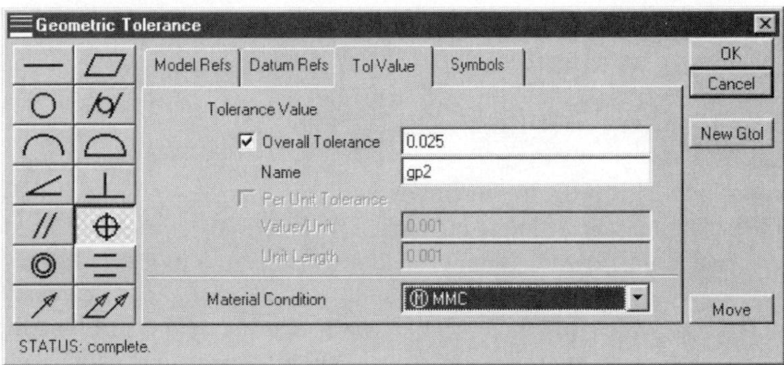

Figure 2-20 Tolerance value tab

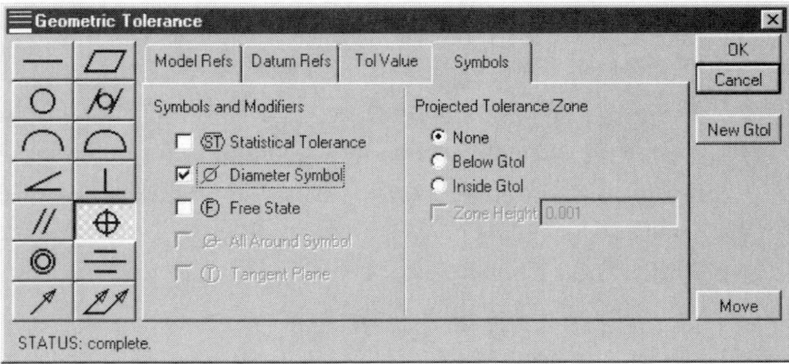

Figure 2-21 Symbols tab

STEP 9: Under the SYMBOLS tab, select the symbols and modifiers to apply.

As shown in Figure 2–21, symbols and/or modifiers such as statistical tolerance, diameter symbol, and free state are added under this tab.

> **MODELING POINT** Before selecting OK to apply the feature control frame, check the model to observe that the frame is being applied correctly. Values for the feature control frame are added dynamically to the model.

STEP 10: Select OK to apply the feature control frame.

Naming Features

Features created within Pro/ENGINEER are assigned a name based on the feature type. As an example, a protrusion might be named on the Model Tree *Protrusion ID 82*. One power of Pro/ENGINEER is its capability for providing a user with the ability to select a feature or part from the Model Tree. Finding a feature or part on the Model Tree is relatively easy for simple models but can become difficult with complicated objects. It is recommended that a user rename features to better discriminate them from other components. To rename a feature (or part in Assembly mode), from the Set Up menu, select Name, then select the feature to be renamed. Follow this by entering a descriptive name.

Datums can be renamed using the Name option. They can also be named using the Set Datum option. Before a datum can be referenced using geometric tolerancing, it must first be set. Options are available under Set Datum for renaming a datum, setting a datum, and unsetting a datum.

Obtaining Model Properties

Pro/ENGINEER provides several options for obtaining information about an object. Most of these options can be found under the Info and Analysis menus. Functions available range from obtaining Parent-Child information to analyzing model properties.

Parent/Child Relationships

Due to its nature as a parametric, feature-based modeling package, Pro/ENGINEER's features are related to other features through references. When a feature references another feature during the modeling process, the referenced feature becomes a parent of the part being modeled.

Information about parent/child relationships can be found utilizing the Parent/Child option from the Info menu. The following options are available.

- **Parents** The Parents option shows parent features of the selected feature.
- **Children** The Children option shows child features of the selected feature.
- **References** The Reference option shows all references used in the construction of a feature.
- **Child Ref** The Child References option shows how a child feature references a selected parent feature.

Model Analysis

A variety of model analysis types exist for obtaining model information. The Model Analysis dialog box is available under the Analysis menu.

Mass Properties Mass properties can be obtained on parts and assemblies. The following properties are available:

- Volume
- Surface area
- Density
- Mass
- Center of gravity
- Inertia tensor
- Principle moments of inertia

- Rotation matrix and rotation angles
- Radii of gyration

X-Section Mass Properties Mass properties can be obtained on a predefined X-section. The following properties are available:

- Area
- Inertia tensor
- Principle area moments of inertia
- Polar moment of inertia
- Rotation matrix and rotation angle
- Radii of gyration
- Section moduli

One-Side Volume The one-sided volume type is used to obtain the volume of a model on one side of a specified plane.

Pairs Clearance The pairs clearance type is used to determine the clearance between two entities of a part or assembly. This option is available in Part, Assembly, and Drawing modes.

Global Clearance The global clearance type is used to determine the clearance between two components of an assembly. This option is available in Assembly and Drawing modes.

Global Interference The global interference type is used to calculate the amount of interference that exists between two components in an assembly.

Thickness The thickness type is used to determine the minimum and maximum thickness of a part. This type is available in Part and Assembly modes.

Printing in Pro/ENGINEER

Pro/ENGINEER has the capability to print to a variety of printers and plotters. Additionally, objects can be printed to a file. The Print dialog box, as shown in Figure 2–22, may be opened by selecting Print from the File menu or from the toolbar.

The Print dialog box Destination is used to select a specific printer. To change printers, select the Add Printer icon.

> **MODELING POINT** The Windows Printer Manager can be used to print objects. To use the default Windows printer, in the configuration file enter *Windows Printer Manager* as the value for the Plotter option.

Configuring the Printer

To configure a printer, select the Configure button located on the Printer dialog box. Multiple print configurations can be set. To save a configuration, select the Save button. The following tab options are available under Configure.

Page Tab

As shown in Figure 2–23, the Page tab is used to configure the sheet size. Standard sheet sizes are available (A, B, C, D, E, F, A0, A1, A2, A3, and A4), or a user can specify a variable sheet size.

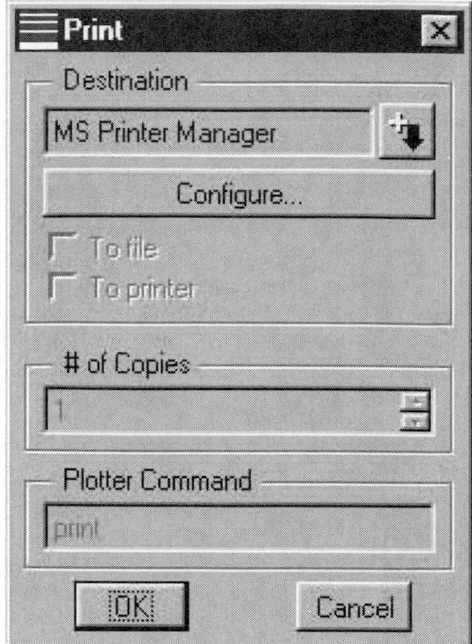

Figure 2–22 Print dialog box

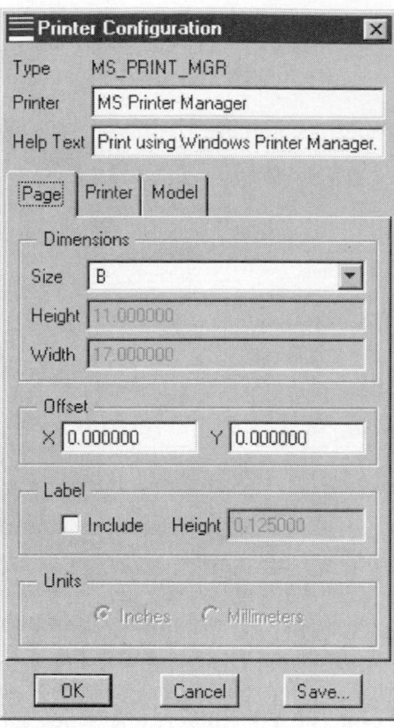

Figure 2–23 Page tab

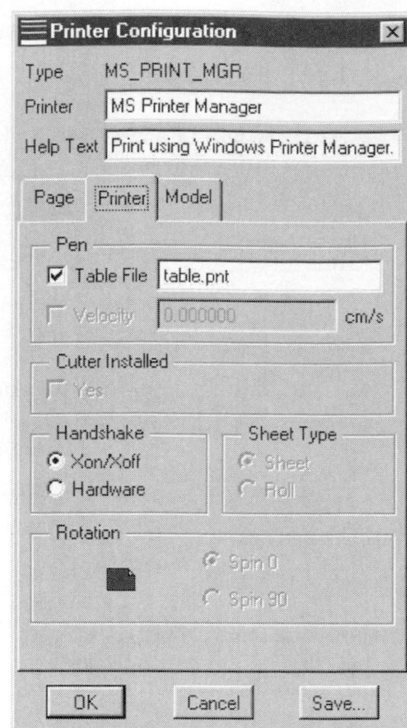

Figure 2–24 Printer tab

PRINTER TAB

As shown in Figure 2–24, the Printer tab is used to specify printer options that might be available. Not all options are available with every printer.

MODEL TAB

The Model tab is used to adjust the way an object is printed on a sheet. The Plot field is used to adjust the area to be plotted. Options are available for creating a full plot, a clipped plot, a plot based on zoom, a plot of an area, and a plot based on the model size. The Base-on-Zoom option is the default value. For full-sized plots, this should be changed to Full Plot.

PRO/ENGINEER'S ENVIRONMENT

Various selections are available for controlling Pro/ENGINEER's working environment. Figure 2–25 shows Pro/ENGINEER's Environment dialog box. This dialog box is available under the Utilities menu. Many of the Environment selection options are also available through other avenues, such as the Toolbar menu. The following is a description of each available selection.

- **Dimension Tolerances** This selection specifies whether an object's dimensions are displayed as a tolerance or as a nominal value. This option is also controllable with the configuration file option *Tolerance Display.* A dimension's tolerance mode can be set with the Modify >> Dimension option, or by default with the configuration file option *Tolerance Mode.*
- **Datum Planes** This selection option controls the display of datum planes. This option is readily accessible through Pro/ENGINEER's Toolbar menu.
- **Datum Axes** This selection option controls the display of datum axes. This option is readily accessible through Pro/ENGINEER's Toolbar menu.

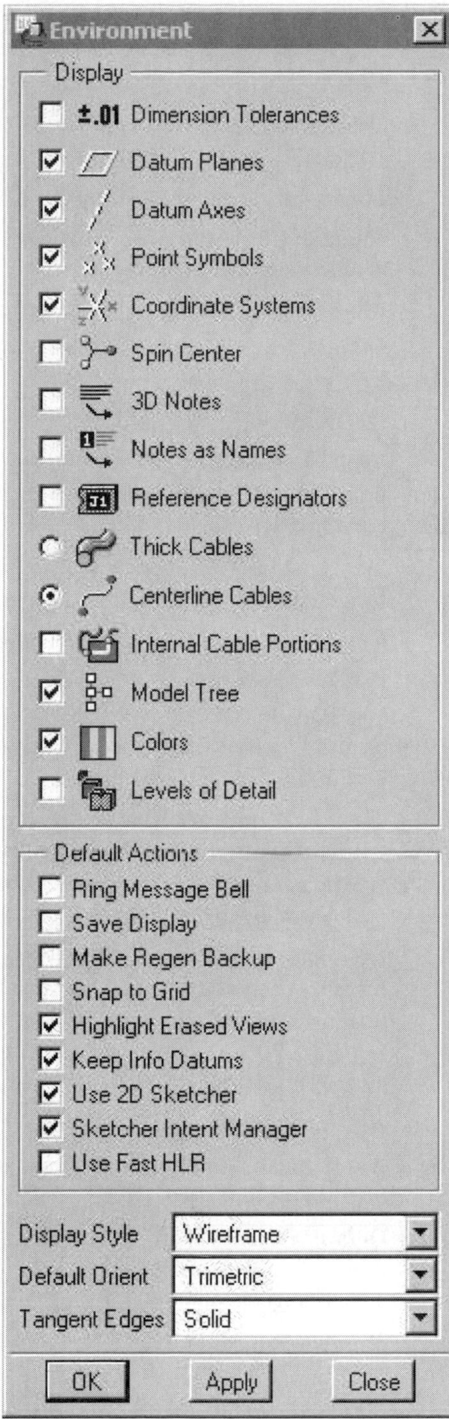

Figure 2-25 Environment dialog box

- **Point Symbols** This selection option controls the display of datum points. This option is readily accessible through Pro/ENGINEER's Toolbar menu.
- **Coordinate Systems** This selection option controls the display of coordinate systems. This option is readily accessible through Pro/ENGINEER's Toolbar menu.
- **Spin Center** The Spin Center is used to show the center of a spin when an object is rotated. This option is activated by default. This selection can be changed permanently using the configuration file option *spin_center_display.*

- **3D Notes** This option controls the display of 3D notes while in part or assembly mode.
- **Thick Cables** This selection controls the three-dimensional display of cables. The Centerline Cables option and this option cannot be selected at the same time.
- **Centerline Cables** This selection displays the centerlines of cables. The Thick Cables option and this option cannot be selected at the same time.
- **Internal Cable Portions** This selection option allows for the display of cables that are hidden by other geometry.
- **Model Tree** This option controls the display of Pro/ENGINEER's Model Tree.
- **Colors** This option will display a model in color.
- **Level of Detail** This selection will allow for the use of levels-of-detail while dynamically viewing a shaded model.
- **Ring Message Bell** When selected, a bell will sound when Pro/ENGINEER provides a message. This selection can be changed permanently with the configuration file *Bell* option.
- **Save Display** This option will save the display of an object, reducing the recalculations needed when reopening the object at a future time.
- **Make Regen Backup** This selection creates a backup copy of the object before regenerations. This allows for the retrieval of a previously valid model. Backups are deleted when ending the object's session.
- **Snap to Grid** When Grid is activated, elements will snap to them, particularly within the sketcher environment.
- **Keep Info Datums** This selection option controls the display of datums created on-the-fly with the Info functionality. When this option is selected, these datums will be considered as features within the model.
- **Use 2D Sketcher** By default, sketching is set up within a two-dimensional environment. Using this setting, the sketching plane is oriented parallel to the screen. When not selected, the sketcher environment remains in a three-dimensional orientation.
- **Use Fast HLR** This option allows for the faster acceleration of hardware while dynamically viewing a model.
- **Display Style** There are four display styles available for viewing a model: wireframe, hidden line, no hidden, and shading. These options are also available on Pro/ENGINEER's Toolbar menu.
- **Default Orient** There are three settings available for Pro/ENGINEER's default orientation: isomeric, trimetric, and user defined. Trimetric is the initial setting. This can be changed with the configuration file option *Orientation.*
- **Tangent Edges** This selection option controls the display of tangent edges. There are five options available: solid, no display, phantom, centerline, and dimmed (dimmed menu color).

CONFIGURATION FILE

Pro/ENGINEER's **configuration file** (Figure 2–26) is used to customize a variety of environmental and global settings. It is accessed through the Utilities >> Options command. Options such as model orientation, system geometry color, background color, tolerance mode, and sketcher grid display can be set by default. As shown in the figure, the left-most column of the table is used to establish a specific option, and the next column is used to define the value for the option. Available options can be found through Pro/ENGINEER's

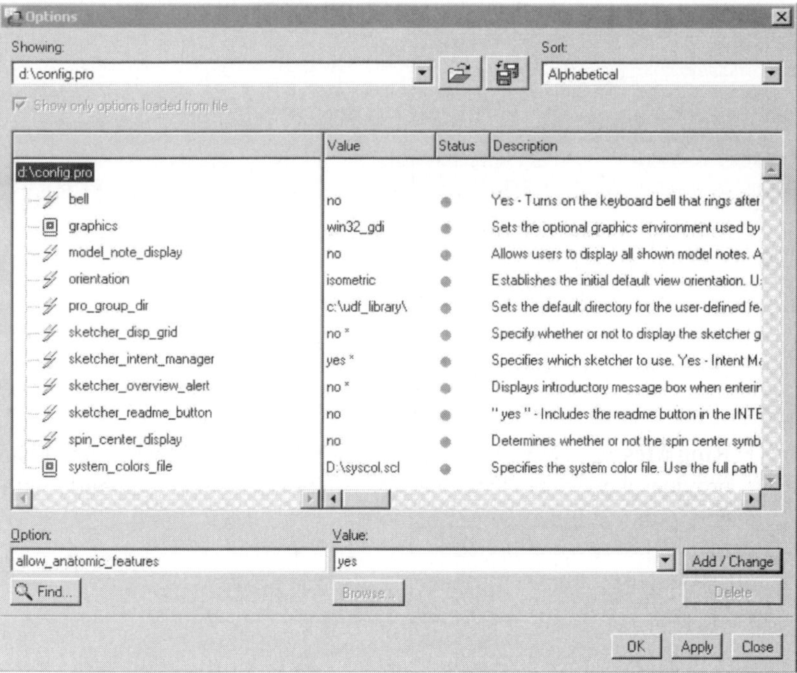

Figure 2-26 Configuration file

online help. A user can set an option by entering the option's name and value (e.g., *orientation* and *isometric* as shown in the illustration), followed by selecting the Add/Change icon.

> **MODELING POINT** The configuration file is used to permanently set environmental and global settings. Most settings can be changed temporarily using other options, such as under the Environment dialog box.

Configuration files may be defined in a variety of directories. When Pro/ENGINEER is first launched, it reads configuration files in order from the locations listed below. The last settings read from a configuration file are the ones that Pro/ENGINEER utilizes:

1. **Pro/ENGINEER LOADPOINT** The loadpoint directory is the location where Pro/ENGINEER is installed. A configuration file saved here is loaded first.
2. **LOGIN DIRECTORY** The login directory is the home directory for a login ID. This configuration file is read after the loadpoint directory and is used to save individualized configuration options.
3. **STARTUP DIRECTORY** The startup directory is the working directory for the current object. This is the last configuration file read by Pro/ENGINEER and any settings that are read from this file will override settings from previous configuration files.

MAPKEYS

Mapkeys are keyboard macros of frequently used command sequences. A possible use of a mapkey would be defining a macro for the command sequence to establish Pro/ENGINEER's default datum planes. The following command steps are required to set the default datum planes: Datum Plane Icon >> Default. A mapkey can be used to create the default datum planes in one keyboard selection.

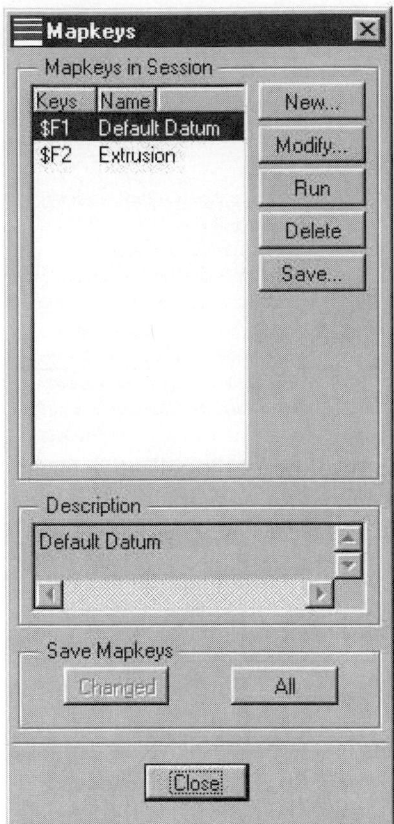

Figure 2–27 Mapkeys dialog box

The Mapkeys dialog box (Figure 2–27) is used to create and edit mapkeys. The following options are found under this dialog box:

- **New** The New option is used to define a new mapkey.
- **Modify** The Modify option is used to modify an existing mapkey.
- **Run** Run is used to execute an existing mapkey. Mapkeys can be run outside of the Mapkey dialog box.
- **Delete** The Delete option is used to delete an existing mapkey.
- **Save** The Save option is used to save the mapkey to the Configuration file. Saving a mapkey will allow it to be used with other Pro/ENGINEER sessions.

Defining Mapkeys

Mapkeys are created by recording keyboard command sequences. Pauses to allow for the selection of objects on the work screen are included automatically in the recording process. Perform the following steps to create a mapkey.

Step 1: Select UTILITIES >> MAPKEYS on Pro/ENGINEER's menu bar.

The Mapkeys dialog box will appear (Figure 2–27).

Step 2: Select NEW from the Mapkeys dialog box.

The Record Mapkey dialog box will appear.

Step 3: Enter a Key Sequence, Mapkey name, and Mapkey description.

The key sequence is the keyboard entry that will execute the mapkey. For function keys, enter a dollar sign in front of the function key sign (e.g., $F2).

STEP 4: Select RECORD and select command sequence from the keyboard and from Pro/ENGINEER's menu.

After selecting Record, enter command sequence just as you would if you were performing the function. You can enter pauses to allow for the entry of sequences outside of the macro.

STEP 5: Select STOP to end the recording of a mapkey.

STEP 6: Select OK on the Record Mapkey dialog box.

STEP 7: Select SAVE on the Mapkey dialog box to permanently store the mapkey.

Mapkeys are saved in the current configuration file (see Figure 2–26).

LAYERS

Layers are used to organize features and parts together to allow for the collective manipulation of all included items. Features and parts can be included on more than one layer. The Layers dialog box is used to create and manipulate layers (see Figure 2–28). It is accessed through the Layers option found under the View menu. The New Layer icon, found on the Layers dialog box, is used to create new layers. The dialog box has options for setting features and parts to selected layers or to remove items from a layer. It is used to control how layers are displayed in a model. The following options are available.

- **Show** Used to display a layer.
- **Blank** Used to blank the display of a layer. Items will not be displayed if their associated layers are blanked.
- **Isolate** Layers that are selected for isolation will be shown while unselected layers and items not on a layer will be blanked.

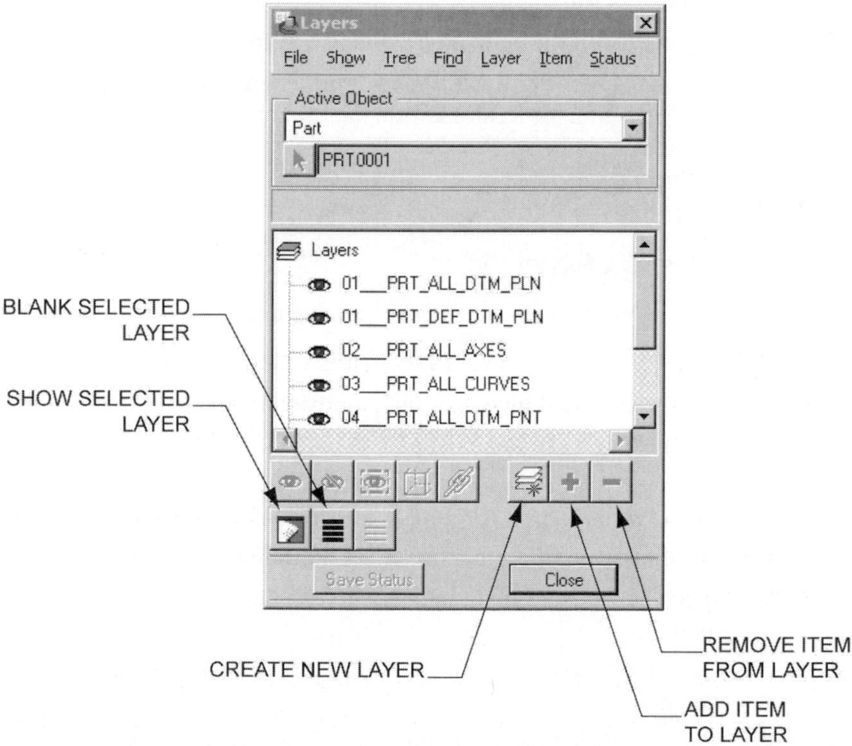

Figure 2-28 Layers dialog box

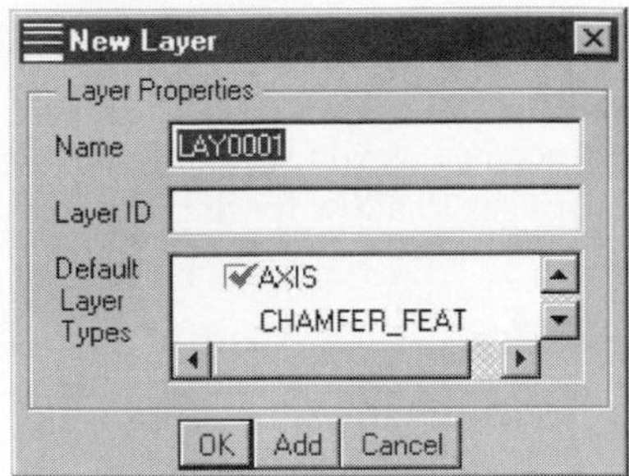

Figure 2–29 New layer dialog box

- **Hidden** The Hidden option is available in Assembly mode only. Parts in hidden layers will not be displayed according to environmental settings for hidden-line display.
- **Add Items** The Add Items option is used to add selected items to a layer.
- **Remove Items** The Remove Items option is used to remove items from layers.

Creating a Layer

Perform the following steps to create a new layer:

Step 1: Select VIEW >> LAYERS on Pro/ENGINEER's Menu Bar.

The Layers dialog box will appear after selecting the **Layers** option (Figure 2–28).

Step 2: On the Layers dialog box, select the CREATE NEW LAYER icon (Figure 2–28).

Step 3: On the New Layer dialog box, enter a Name for the new layer (Figure 2–29).

Step 4: Select OK on the New Layer dialog box to create the new layer.

Selecting OK will create the new layer and return you to the Layers dialog box. Selecting the Return key on the keyboard will create the layer and allow you to remain in the New Layer dialog box. This allows for the creation of multiple layers.

Setting Items to a Layer

Perform the following steps to set an item or feature to a layer:

Step 1: Select VIEW >> LAYERS on Pro/ENGINEER's menu bar.

Step 2: On the Layers dialog box, pick an existing layer (Figure 2–28).

Step 3: On the Layers dialog box, pick the ADD ITEM icon (Figure 2–28).

After selecting the Add Item icon, Pro/ENGINEER will launch the Layer Object menu. This menu is used to select an item type to add to the layer.

Step 4: On the Layer Object menu, select an item type to add to the layer.

Examples of items that can be added to a layer include features, curves, and quilts.

Step 5: On the work screen or on the Model Tree, select items to add to the layer.

Step 6: On the Layer Feature menu, select DONE/RETURN.

Step 7: On the Layer Object menu, select DONE/RETURN.

Table 2-3 Item types available to include on a layer

Item Type	Items Included
FEATURE	All features
AXIS	Datum axes and cosmetic threads
GEOM_FEAT	Geometric features
DATUM_PLANE	Datum planes
CSYS	Coordinate systems
DIM	Dimensions
GTOL	Geometric Tolerances
POINT	Datum points
NOTE	Drawing notes

DEFAULT LAYERS

Default layers can be created for all new objects. Created items can be automatically placed on predetermined default layers. The following configuration file option is used to create a default layer and to automatically add an item type to the default layer: **Def_layer *item-type layername.***

Def_layer is a configuration file option. This option may be used multiple times to correspond to as many default layers as might be needed in an object. The *item-type* value is used to specify a type of item to include on a default layer, while the *layername* value is the name of the layer that will be created. Table 2–3 shows a partial list of item name values that can be used as an *item-type.*

SELECTING FEATURES AND ENTITIES

Parts with many features and assemblies with many parts can make selecting entities, features, and objects difficult. Entities, features, and objects can be selected on the Model Tree or on the work screen. Often, it is necessary to select from the work screen. The Get Select menu provides options that can make selections easier. An example of the Get Select menu is shown in Figure 2–30. Available options on this menu vary depending upon the Pro/ENGINEER routine being executed. The following are some of the options that might be found on the Get Select menu:

Figure 2-30 Get select menu

PICK

The Pick option allows the user to select visible individual entities and datums on the work screen or on the Model Tree.

PICK MANY

The Pick Many option allows a user to construct a rectangular selection window around items to be selected. The user must input with the mouse the diagonal corners of the rectangle. All selectable entities that lie within the rectangle will be selected. The Pick Many option is not available with every Get Select menu.

QUERY SELECT

The Query Sel option allows a user to toggle through available selectable entities. After picking the Query Sel option, the user must select on the work screen where the desired entity, feature, or object is located. The user is then provided a menu for advancing through items that are possible selections. While toggling through items, when the desired item is highlighted an option is available for accepting it.

Figure 2–31 Query bin dialog box

A useful tool with the Query Sel option is the Query Bin. When using the Query Sel option, the Query Bin will display all selectable items in a dialog box (Figure 2–31). Query Bin is available only when it is selected in the Environment dialog box. This is Pro/ENGINEER's default setting.

SEL BY MENU

The Select by Menu option allows for the selecting of items based on an item's feature name, feature ID, or regeneration number.

DONE SEL

The Done Select option will accept all picked items and exit the Get Select menu.

SUMMARY

Pro/ENGINEER provides a variety of tools for interfacing with the modeling environment. Most of these tools are readily available on the toolbar. Other options can be found under the menu bar.

Pro/ENGINEER's data-based management system manipulates files differently than standard Windows applications. Options are available for saving and backing up object files. A unique feature of Pro/ENGINEER is that when it saves, a new version of the object file is created.

A variety of view manipulation options are available within Pro/ENGINEER. Objects can be dynamically zoomed, rotated, and panned. Views can be oriented and saved for later use. Additionally, the display of an object can be represented in one of four possible ways: wireframe, hidden line, no hidden, and shaded. Also, features and entities may be placed on layers and hidden from display.

Pro/ENGINEER's interface provides customization tools. The Configuration file is a powerful tool for personalizing the work environment. Mapkeys can be created that provide a shortcut to commonly used menu pick sequences.

TUTORIAL

PRO/ENGINEER INTERFACE TUTORIAL

This tutorial will provide instruction for the establishment of a Pro/ENGINEER object. The part (*interface.prt*) shown in Figure 2–32 can be found on the supplemental CD and will serve as the part to be manipulated in this tutorial. This tutorial will cover

- Manipulating an existing object file.
- Viewing a Pro/ENGINEER object.
- Setting up an object's units and materials.
- Changing an object's tolerance display.
- Creating a geometric tolerance.
- Creating layers and setting items.

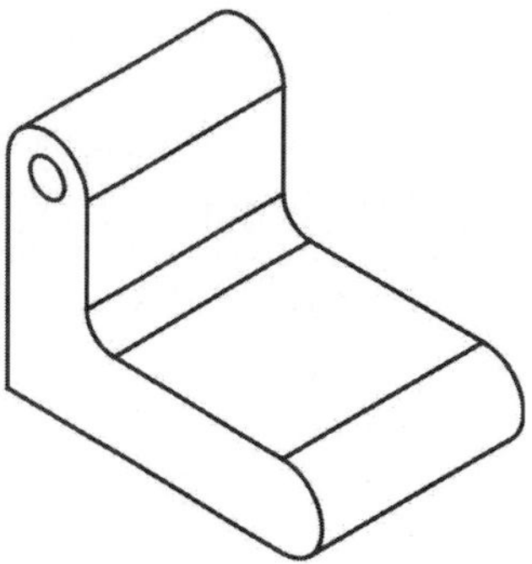

Figure 2-32 Interface part

OPENING AN OBJECT

This segment of the tutorial will cover the opening and saving of a Pro/ENGINEER part file. Highlighted within this segment will be the saving of a part as a different name using the Save A Copy option.

STEP 1: Start Pro/ENGINEER.

STEP 2: Use the FILE >> SET WORKING DIRECTORY option to change Pro/ENGINEER's Working Directory to a location where you have read/write privileges.

From the File menu, select Working Directory then change the working directory to the directory where your part file will be saved.

STEP 3: Use the FILE >> OPEN option to open the part *interface*.

This part (*interface.prt*), shown in Figure 2–32, can be found on the supplemental CD. Select the File >> Open option, then manipulate the Look In box to find the location of this part file.

STEP 4: Select FILE >> SAVE A COPY and save the part with the name *interface_part*.

If your working directory is different from the directory where the original *interface.prt* file is located, notice how the Save A Copy option defaults to save to the currently working directory. The Save A Copy option will allow you to change the directory and object name.

Notice that after performing the Save A Copy option the original *interface* part file is still active. The Save A Copy option will allow for a name change, but will keep the original object active.

STEP 5: Use FILE >> ERASE >> CURRENT to erase the current object from memory.

The Erase >> Current object will erase the *interface* part from session memory.

STEP 6: Use the FILE >> OPEN option to open the newly created part *interface_part.*

Open the part file created in step 4. This part will now be the active object in session memory.

VIEWING THE OBJECT

This segment of the tutorial will introduce Pro/ENGINEER's viewing options. Covered under this segment will be shading a model, dynamically rotating and zooming a model, and setting the default orientation.

STEP 1: Select the SHADE icon on the toolbar.

As shown in Figure 2–33, select the Shade icon from the toolbar.

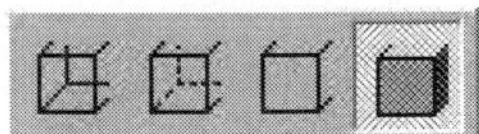

Figure 2–33 Shade icon

STEP 2: Dynamically rotate the model.

Simultaneously selecting the keyboard's control key and the middle mouse button, dynamically rotate the object by moving the cursor on the work screen.

STEP 3: Dynamically zoom the model.

Simultaneously selecting the keyboard's control key and the left mouse button, dynamically zoom the object by moving the cursor on the work screen. Moving the cursor from the bottom of the screen toward the top will zoom out. Moving the cursor from the top of the screen toward the bottom will zoom in on the model.

STEP 4: Using the UTILITIES >> ENVIRONMENT option, set Isometric as the object's default display setting (Figure 2–34).

Using the Environment dialog box found under the Utilities option on the menu bar, select Isometric as the default orientation.

STEP 5: Use VIEW >> DEFAULT to change to the default orientation.

The default orientation is shown in Figure 2–35.

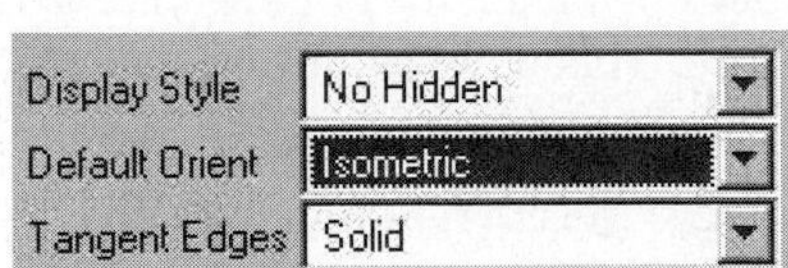

Figure 2–34 Model display

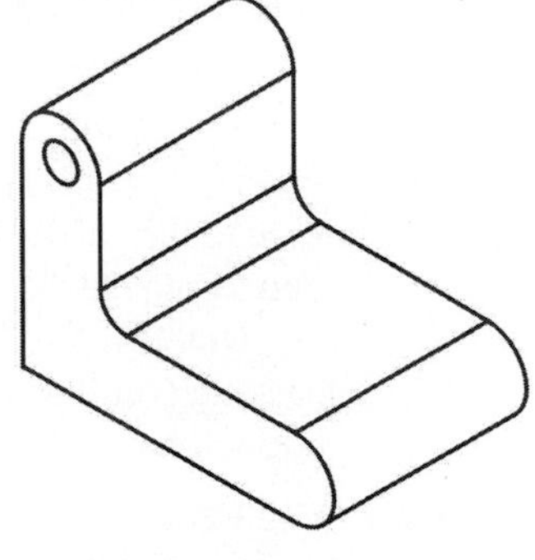

Figure 2–35 Isometric view of model

Setting an Object's Units

This segment of the tutorial will set the units for the part. A part's units can be set at any time during the modeling process. This segment will cover creating a new system of units and change the units for your part.

Step 1: Select SET UP >> UNITS on Pro/ENGINEER's Menu Manager.

Utilizing Pro/ENGINEER's Menu Manager, from the Set Up menu select the Units option.

Step 2: Select the NEW option on the Units Manager dialog box.

As shown in Figure 2–36, select New from the dialog box. New will allow for the creation of a user-defined system of units.

Step 3: Enter *UNITS* as the system name.

Step 4: Enter the unit values shown in Figure 2–37.

Figure 2–37 displays the units to be set for this system. Enter these values as shown.

Step 5: Select OK to close the Systems of Units Definition dialog box.

Step 6: Select the newly created Units system on the Units Manager dialog box.

Step 7: Select the SET option on the Units Manager dialog box.

Step 8: Select INTERPRET EXISTING NUMBERS on the Warning dialog box.

The ***Interpret Existing Numbers (Same Dims)*** option will interpret the existing unit numbers as the new unit numbers. As an example, a value of 2 inches will be converted to a value of 2 millimeters.

Step 9: Close the Units Manager dialog box.

Step 10: SAVE your object.

From the File menu, select Save. Enter the object to save.

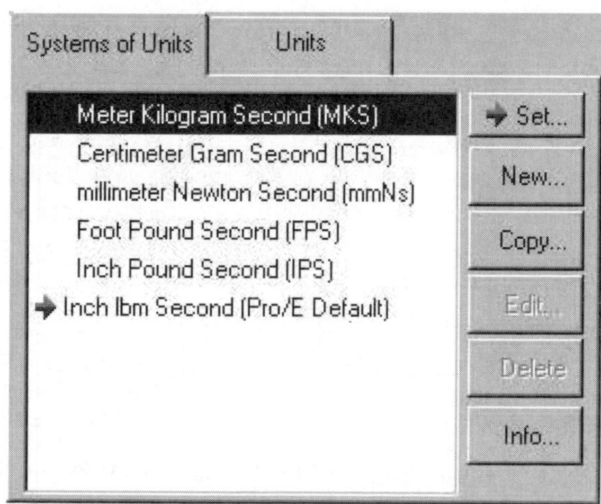

Figure 2–36 Units manager dialog box

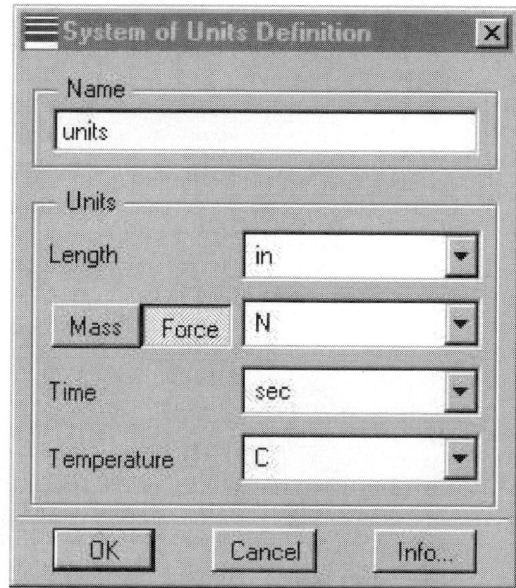

Figure 2–37 New system of units

Establishing Layers

This segment of the tutorial will establish layers. The model used in this tutorial is composed of datums and geometric features. You will create a layer for all geometric features and a layer for all datums.

STEP 1: Select the VIEW >> LAYERS option on the Menu Bar.

After selecting the Layers option, the Layers dialog box will appear (Figure 2–38).

STEP 2: Select the NEW LAYER icon on the Layers Dialog box (Figure 2–38).

After selecting the New Layer icon, the New Layer dialog box will appear on the work screen (Figure 2–39).

STEP 3: On the New Layer dialog box, enter *GEOMETRIC_FEATURE* as a layer name (Figure 2–39).

STEP 4: In the Default Layer Types box, scroll down and select the *GEOM_FEAT* type (Figure 2–39).

STEP 5: Select the ADD option on the Create New Layer dialog box.

STEP 6: In the Name textbox, enter *DATUMS* as a layer name.

STEP 7: In the Default Layer Types box, scroll down and select the *DATUM* type.

STEP 8: Select the OK option on the New Layer dialog box.

Your Layers dialog box should appear as shown in Figure 2–38.

STEP 9: On the Layers dialog box, select the *DATUMS* layer then select the ADD ITEM icon (Figure 2–38).

After selecting the Add Item icon, the LAYER OBJ (Layer Objects) menu will appear on the work screen.

STEP 10: Select the DATUM PLANE option on the Layer Objects menu.

STEP 11: On the Model Tree, select the first three datum planes (Figure 2–40), then select DONE/RETURN on the Layer Objects menu.

STEP 12: On the Layers dialog box, select the *GEOMETRIC_FEATURE* layer then select the ADD ITEM icon (Figure 2–38).

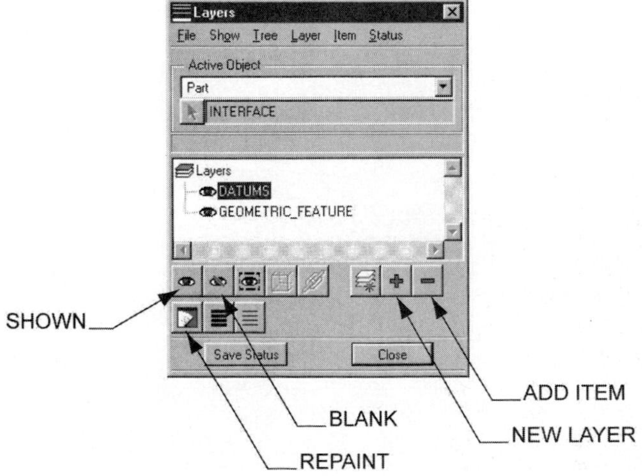

Figure 2-38 Layers menu

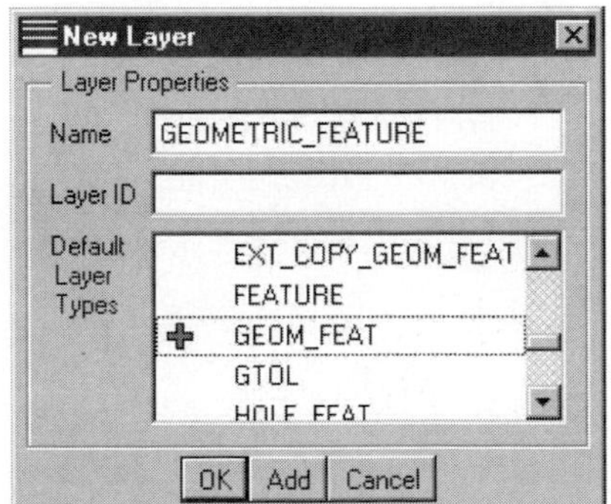

Figure 2-39 New layer dialog box

STEP 13: Select the FEATURE option on the Layer Object menu (Figure 2–41).

STEP 14: Select the ALL OF TYPE option on the Layer Feature menu (Figure 2–41).

STEP 15: Check all available geometric features on the All Features menu (do not select datum planes or datum axes).

This selection will place all the features on the layer *GEOMETRIC_FEATURE.*

STEP 16: Select DONE on the All Features menu, then select DONE/RETURN on the Layer Feature menu (Figure 2–41).

STEP 17: Select DONE/RETURN on the Layer Object menu (Figure 2–41).

STEP 18: On the Layer dialog box, select the *DATUMS* layer then select the BLANKED icon (Figure 2–38).

STEP 19: Select the SAVE STATUS icon, then REPAINT the screen (see Figure 2–38).

What happens to your datums after repainting the screen?

STEP 20: Select the *DATUMS* layer then select the SHOWN icon.

The Shown icon will unblank a layer.

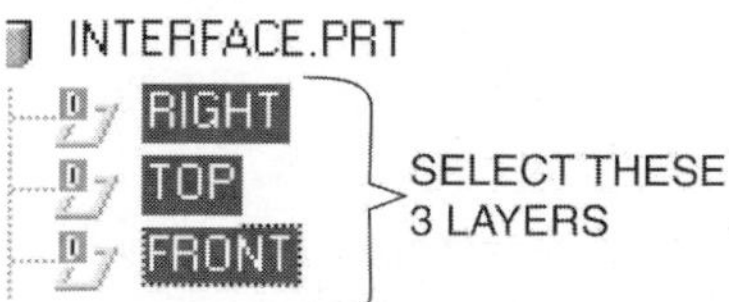

Figure 2–40 Default datum planes displayed on model tree

Figure 2–41 Layer object menu

STEP 21: Select the SAVE STATUS icon, then REPAINT the screen.

What effect does this have on the display of your datum planes?

STEP 22: Select the CLOSE icon.

STEP 23: Save your model.

From the File menu, select the Save option. Enter the object to save.

PROBLEMS

1. Create a material defining 1040 steel. Save this material to disk as a material file.
2. Create a material defining aluminum. Save this material to disk as a material file.
3. Create a system of units with the following configuration:
   - Length = cm
   - Mass = kg
   - Time = micro-sec
   - Temperature = K
4. Create a configuration file with the following options and settings:
   - ***BELL*** option set to NO
   - ***TOL_DISPLAY*** option set to YES
   - ***TOL_MODE*** option set to NOMINAL
   - ***SKETCHER_INTENT_MANAGER*** option set to YES
   - ***ALLOW_ANATOMIC_FEATURES*** option set to YES

QUESTIONS AND DISCUSSION

1. Describe the difference between the Backup option and the Save As option. Give some examples of when each option would be appropriate.
2. In the object filename *revolve.sec.2,* what does the number *2* represent? What does *sec* represent?
3. What is a Pro/ENGINEER working directory?
4. What file management option is used to close an object from Pro/ENGINEER's memory?
5. Describe the process for creating a material file that can be utilized by other Pro/ENGINEER parts.
6. How does Pro/ENGINEER's ANSI tolerance standard differ from its ISO tolerance standard?
7. What is a mapkey and how is it used in Pro/ENGINEER?
8. What is the purpose of Pro/ENGINEER's configuration file?

CHAPTER

3

CREATING A SKETCH

Introduction

Sketching is a fundamental skill within Pro/ENGINEER. Geometric features such as Protrusions or Cuts require the use of a sketch to define the features' section. Other features such as datum curves, swept trajectories, and sketched holes also require the use of sketching to define elements. A sketched Pro/ENGINEER section can be created within a feature creation option (e.g., Cut or Protrusion), or it can be created as a separate object, then used within later construction techniques. This chapter will cover the fundamental sketching techniques required to construct a Pro/ENGINEER feature. Upon finishing this chapter, you will be able to

- Start a new Pro/ENGINEER object utilizing sketcher mode.
- Establish Pro/ENGINEER's sketcher environment.
- Sketch a section using Intent Manager.
- Apply dimensions to a Pro/ENGINEER sketch.
- Apply relations to section dimensions.
- Use geometric construction tools to create a sketch.
- Apply constraints to section entities.

DEFINITIONS

Constraint An explicit relationship that exists between two sketched entities.

Entity An element within the sketcher environment, such as a line, arc, or circle.

Parametric dimension A dimension used as a parameter to define a feature.

Reference An existing feature entity, such as a part edge or datum, used within a sketcher environment to construct a new feature. The referenced entity becomes a parent of the feature being constructed.

Section A combination of sketched entities, dimensions, and constraints that defines a feature's basic geometry and intent.

Sketch The entities of a section that define the basic shape of a feature.

Sketcher environment The environment within a feature creation process where sections are defined.

Sketch mode A Pro/ENGINEER mode for the creation of sections. Sections may be created in Sketcher mode and placed in a sketcher environment at a later time.

FUNDAMENTALS OF SKETCHING

Sketches combined with dimensions, constraints, and references form a **section** (Figure 3–1). There are two categories of sections: those used to directly create a feature and those created in **Sketcher mode.** Protrusions, cuts, shafts, and ribs are examples of features that require a sketched section. When creating a new Pro/ENGINEER object, Sketcher is one of the modes available. Elements created in Sketcher mode are saved with an **.sec* file extension. When creating a feature in Part mode, selecting Save while sketching will save the section, not the part being created. A saved section can be inserted in a sketcher environment by using the Sketch >> Data from File option.

CAPTURING DESIGN INTENT

Pro/ENGINEER requires a fully defined section before completing a feature. The sketcher environment provides a variety of tools that will fully define a sketch and capture design intent. When sketching a section, care should be taken to ensure that the intent of a design is met through the definition of the feature. The following list describes sketcher tools that can be used to capture a design's intent.

DIMENSIONS

Dimensions are the primary tool for capturing the intent of a design. Within a section, dimensions are used to describe the size and location of entities.

CONSTRAINTS

Constraints are used to define the relationship of section entities to other entities. As an example, a constraint might make two lines equal in length, or it might confine two lines parallel to each other.

REFERENCES

When constructing a feature in a sketcher environment, a section can reference existing features of a part or assembly. References can consist of part surfaces, datums, edges, or axes. An example of a reference would be aligning the end point of a sketched entity with the edge of an existing feature. Another example would be creating sketch entities from existing edges by utilizing the Use Edge option. Sections created in Sketcher mode cannot utilize external feature references.

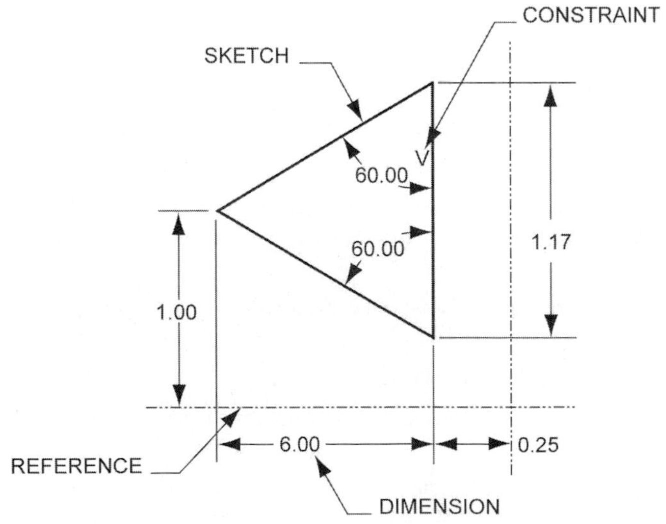

Figure 3-1 Section elements

RELATIONS

Relationships can be established between two dimensions. Most algebraic and trigonometric equations can be utilized to form a mathematical relationship. Also, conditional statements may be applied to create a relationship.

SKETCHING ELEMENTS

Users of two-dimensional computer-aided design applications are accustomed to entering precise values for geometric elements. In low- and mid-range CAD packages, a line 2 inches in length is drawn exactly 2 inches in length. Many of the sketching tools found within Pro/ENGINEER resemble two-dimensional drafting options. Despite this, within Pro/ENGINEER it is not important to draw a section precisely. Instead of creating exact elements, geometry is sketched, similar to how one might sketch freehand. The following guidelines are important when sketching a section.

- When sketching a section, the shape of the sketch is important, not the size.
- When creating a section, the dimensioning scheme should match design intent.
- When creating a section, geometric constraints should match design intent.

After sketching a section and applying dimensions and constraints, it is unlikely that the section will be the correct size. The sketcher environment has dimension modification tools. Within Pro/ENGINEER, when a dimension is modified and the section is regenerated, if no conflicts exist the sketch adjusts to the size of the dimension values.

SKETCH PLANE

When constructing a section for the direct creation of a feature (e.g., the Protrusion command), the section has to be sketched on a plane. The only exception to this rule is when creating a geometric feature that will be the first feature of a part. The sketch for this situation is not placed on a plane.

Two types of planes can be sketched upon. The first type is a feature surface or plane (Figure 3–2). Any flat surface of a part can be used to sketch upon. The second type is a datum plane. Datum planes can provide a suitable surface for sketching when one does not exist on a part. Datum planes can be completed on-the-fly using the Make Datum option. On-the-fly datum planes are not considered features and will not be displayed after the creation of the feature.

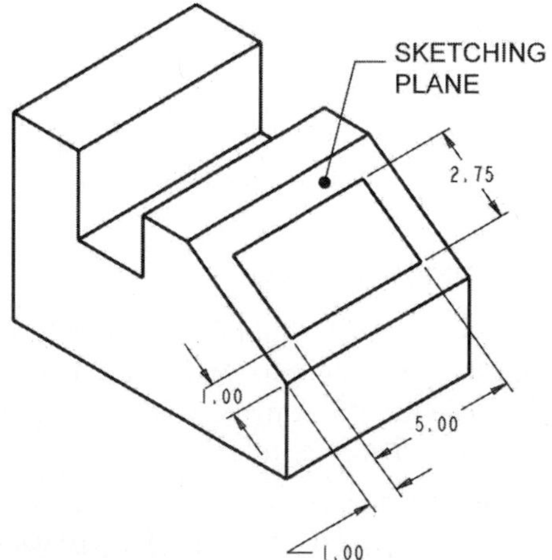

Figure 3–2 Sketching on a part surface

Section Tools

The **sketcher environment** is a customizable mode used to create Pro/ENGINEER sections. Initially, the sketcher environment is set to allow for two-dimensional sketching with grids on and Intent Manager activated. Each of these environmental options can be temporarily or permanently set to different values. As an example, a user can opt to sketch in three-dimensional mode by toggling off the Use 2D Sketcher option found under the Environment dialog box. Other section tools available include placing sections and toggling the display of vertices, constraints, and dimensions.

Figure 3-3 Sketcher preferences dialog box

Grid Options

Grid parameters can be modified on the Sketcher Preferences dialog box (Figure 3–3). This dialog box is available by selecting Utilities >> Sketcher Preferences on Pro/ENGINEER's menu bar. The following options are available.

Grid Display

Grids can be toggled on or off using the Grid On/Off toggle located on the Sketcher Preferences dialog box. Also, as shown in Figure 3–4, grids can be toggled on or off using the Section Tools icon located on Pro/ENGINEER's main Toolbar menu.

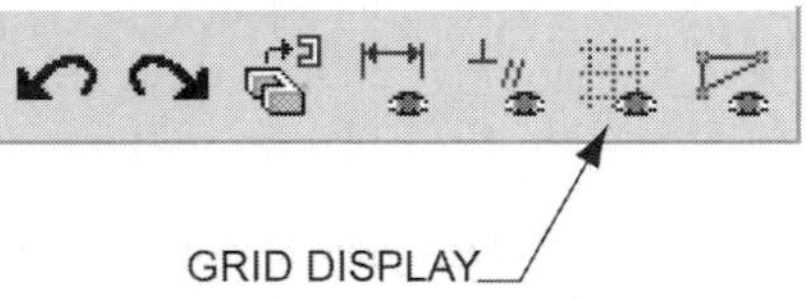

Figure 3-4 Grid display option

Grid Type

The Parameters tab of the Sketcher Preferences dialog box allows for the selection of a Cartesian grid or a polar grid. The Cartesian grid is the default.

Grid Origin

The Origin option under the Parameters tab of the Sketcher Preferences dialog box allows the grid absolute origin to be moved. It can be set at the center or end of an entity, at a point and coordinate system, or at an entity vertex.

Grid Parameters

The Grid Spacing option under the Parameters tab of the Sketcher Preferences dialog box allows for the modification of Cartesian X and Y direction grid spacing or polar grid angle and distance spacing.

Placing Sections

Sections created within Part, Assembly, or Sketcher mode can be saved and used in other sketcher environments. When sketching a section, the File >> Save option is used to save the section. Perform the following steps to place a predefined section into a current sketcher environment.

Step 1: In a sketcher environment, from Pro/ENGINEER's menu bar select SKETCH >> DATA FROM FILE.

Upon selecting the Data From File option, the Open dialog box will appear.

STEP 2: Select a section utilizing the OPEN dialog box.

The Open dialog box will default to the current working directory to allow for the selection of Pro/ENGINEER section files. You can move up or down on the directory tree to locate a section, or you can select a new drive.

STEP 3: If necessary, adjust the Handle Points for Moving, Rotating, or Scaling (see Figure 3–5) by clicking with the right mouse button the desired handle, then dragging the handle to a new location.

STEP 4: Place the section by dragging the moving handle point to the desired location.

The selected section can be placed dynamically into the current sketcher environment. Place the section in a way that will allow for proper alignments when the section is regenerated.

STEP 5: If necessary, rotate the section by either dragging the rotation handle point or by adjusting the Rotate value in the Scale Rotate dialog box (Figure 3–5).

STEP 6: If necessary, enter a scale value in the Scale Rotate dialog box.

STEP 7: When the section has been placed, select the Accept Changes checkmark on the Scale Rotate dialog box.

> **MODELING POINT** For a new object, the display can be zoomed out to such an extent that the section cannot be readily recognizable. Use dynamic viewing (control key and left mouse button) to zoom in on the section being placed.

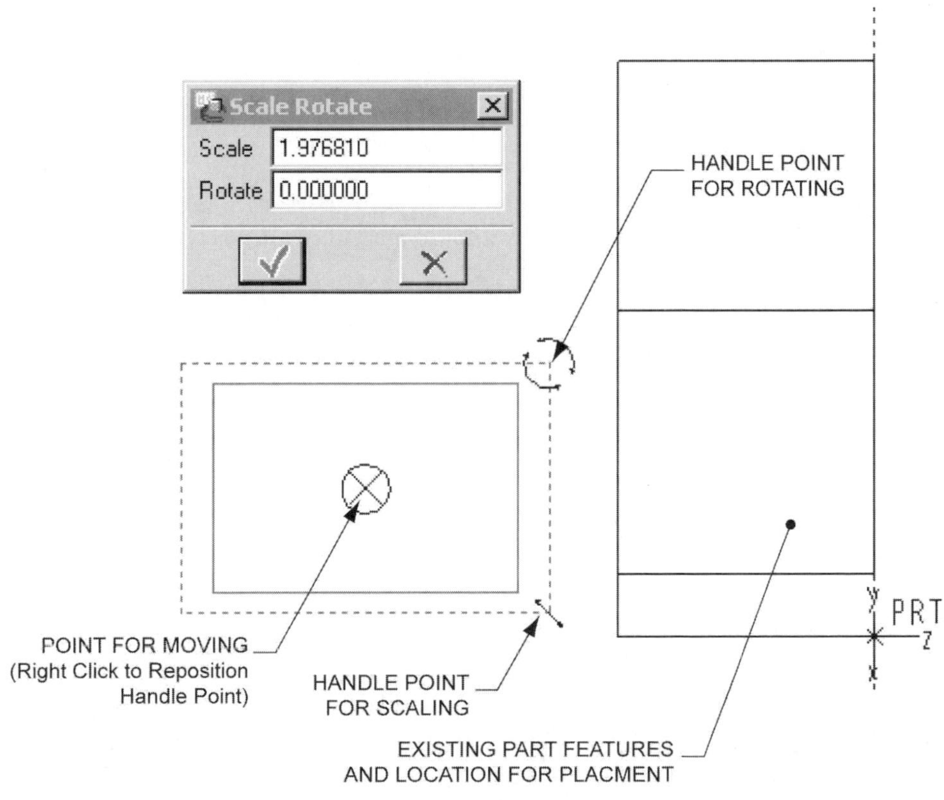

Figure 3-5 Placing a section

SECTION INFORMATION

The Sketch menu on Pro/ENGINEER'S menu bar provides options for obtaining information about entities within the current sketcher environment. The following options are available.

- **Distance** The Distance option measures the distance between two parallel lines, between two points, or between a point and a line.
- **Angle** The Angle option measures the angle between two entities.
- **Entity** The Entity option obtains information such as geometry type, endpoint tangencies, and endpoint coordinates on a selected entity. To obtain endpoint coordinate values, a coordinate system must exist.
- **Intersection point** The Intersection Point option locates the intersection point of two entities. For entities that do not physically intersect, an interpolated intersection point is obtained.
- **Tangency point** The Tangency Point option provides the location where two touching entities are tangent. For entities that do not touch, it shows the location on each entity that has the same slope.
- **Curvature** The Curvature option graphically displays the relative curvature of an entity, especially curved entities such as arcs and splines.

CONSTRAINTS

Constraints are used within Pro/ENGINEER to capture design intent. A **constraint** is a defined relationship that exists between two geometric entities. An example would be constraining two lines parallel two each other. Figure 3–6 shows examples of several types of constraints. Pro/ENGINEER approaches constraints differently depending upon whether Intent Manager is activated or not.

CONSTRAINTS WITH INTENT MANAGER

With Intent Manager activated, Pro/ENGINEER creates constraints on-the-fly while sketching. When an entity being sketched comes within a predefined tolerance value of meeting a

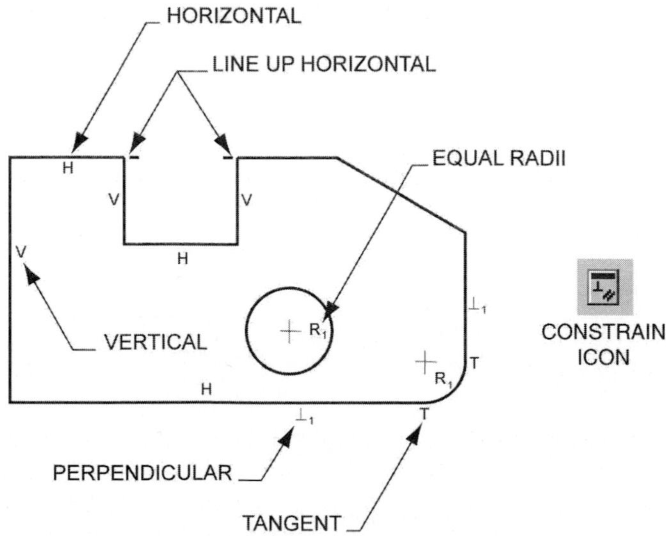

Figure 3–6 Constraint examples

constraint, such as being close to parallel to another entity, the entity is snapped to that constraint. Constraints can be applied manually using the Constrain icon on the sketcher toolbar or the Sketch >> Constrain option on the menu bar.

Constraints are considered weak or strong. Constraints created by Pro/ENGINEER during the sketching process are weak and can be overridden by a user-placed constraint or by a user-placed dimension. Constraints created using the Constrain option are strong and remain on the sketch unless removed using the Delete option. Weak constraints can be converted to strong constraints by using the Edit >> Convert To >> Strengthen option.

Constraints without Intent Manager

Intent Manager can be toggled off by unselecting the Sketch >> Intent Manager option. With Intent Manager deactivated, constraints are applied to sections after a successful regeneration based on the following assumptions:

- When arcs and circles are sketched with approximately the same radius, Pro/ENGINEER assumes that the radii are equal.
- When a line is sketched approximately horizontal or vertical, Pro/ENGINEER assumes that it is horizontal or vertical.
- When a line is sketched approximately parallel to another line, Pro/ENGINEER assumes that the two entities are parallel.
- When a line is sketched approximately perpendicular to another line, Pro/ENGINEER assumes that the two entities are perpendicular.
- When two entities are sketched approximately tangent to each other, Pro/ENGINEER assumes that the two entities are tangent.
- When lines are sketched approximately the same length, Pro/ENGINEER assumes that the lines are of equal length.
- When entities are sketched approximately symmetrical around a centerline, Pro/ENGINEER assumes that the entities are symmetrical.
- When a point entity (such as the endpoint of a line) is sketched near an arc, circle, or line, Pro/ENGINEER assumes that the point entity lies on the arc, circle, or line.
- When the centers or endpoints of arcs are sketched at approximately the same location, Pro/ENGINEER assumes that they have the same X and/or Y coordinate values.

With Intent Manager deactivated, there is no option for a user to apply a constraint after the regeneration of a section.

Constraint Options

Intent Manager allows for the dynamic application of constraints to sketched entities. Constraints applied through the Constrain option are considered strong and will override weak constraints and/or dimensions. Table 3–1 shows constraints that can be applied to sketcher elements.

Sketcher Display Options

As shown in Figure 3–7, when entering the sketcher environment, seven additional icons are added to the Toolbar menu. Five of these icons control the display of items exclusively within the sketcher environment. Each display option is also available under the Sketcher Preferences dialog box (Utilities >> Sketcher Preferences). The following list describes each icon option.

Table 3-1 Constraint options

Constraint Type	Symbol	Use
Same points	⌖	The Same Points option is used to make two points coincident.
Horizontal	H	The Horizontal option is used to constrain a line horizontal.
Vertical	V	The Vertical option is used to constrain a line vertical.
Point on entity	⤥	The Point on Entity option is used to constrain a point on a selected entity.
Tangent	T,	The Tangent option is used to constrain two entities tangent to each other.
Perpendicular	⊥	The Perpendicular option is used to constrain two entities perpendicular to each other.
Parallel	//,	The Parallel option is used to constrain two entities parallel to each other.
Equal radii	R,	The Equal Radii option is used to constrain the radii of two arcs or curves equal to each other.
Equal lengths	L,	The Equal Lengths option is used to constrain two entities to have equal lengths.
Symmetric	→←	The Symmetric option is used to make entities symmetrical around a centerline.
Line up horizontal	--	The Line Up Horizontal option is used to line up two vertices horizontally.
Line up vertical	¦	The Line Up Vertical option is used to line up two vertices vertically.
Collinear	–	The Collinear option is used to make two lines collinear.
Alignment	¦	The Alignment option is used to align two entities.

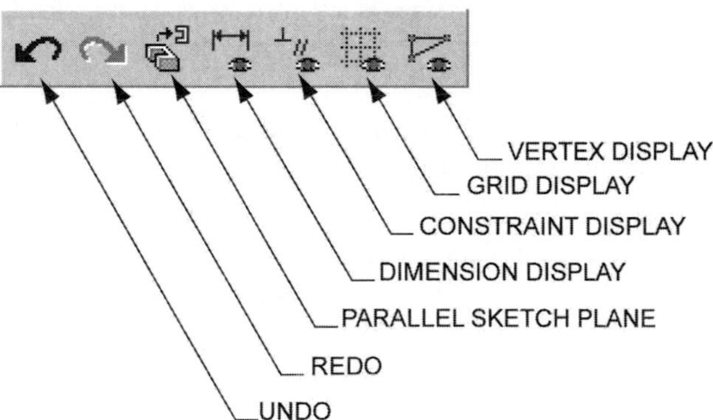

Figure 3-7 Sketcher display options

Undo and Redo

The Undo icon will undo the last executed sketching function (e.g., line, delete, constrain), while the Redo icon will redo a function that was undone with the Undo option.

Parallel Sketch Plane

The default sketching environment is oriented parallel to the display screen. Since a sketcher environment can be dynamically rotated with the control-key middle-mouse button combination, the Parallel Sketch Plane icon will return the sketch plane to its default two-dimensional orientation.

Dimension Display

The Dimension Display icon controls the display of dimensions within the sketcher environment. This option allows dimensions to be toggled on or off.

Constraint Display

Constraint symbols are shown directly on the sketch. The Constraint Display icon controls the display of constraint symbols. Using this option, constraint symbols can be toggled on or off.

Grid Display

The Grid Display icon controls the display of grids within the sketcher environment. Often, grids can cause visualization problems while sketching a section.

Vertex Display

Vertices are displayed at the end of sketcher entities. The Vertex Display icon controls the display of these vertices. Selecting this icon will turn off all vertices.

Sketching with Intent Manager

Intent Manager is used by Pro/ENGINEER to facilitate the capturing of a part's design intent. It is activated by default. If desired, it can be deactivated with the Sketch >> Intent Manager option or it can be deactivated permanently by setting the configuration file option *sketcher_intent_manager* to a value of No. Intent Manager provides several sketching enhancements.

Fully Defined Section

Pro/ENGINEER is a parametric modeling application. Parametric packages require sections to be fully dimensioned and constrained. Within conventional two-dimensional drafting standards, if two lines appear parallel or perpendicular, then they are assumed to be so. Within parametric modeling, a parallel or perpendicular constraint has to be applied. Intent Manager attempts to keep a sketch fully defined by applying dimensions and constraints during the sketching process. Additionally, Intent Manager will not allow a section to be over defined.

Constraints

Constraints are applied during the sketching process. Additionally, constraints can be applied with the Constrain icon and removed with the delete key.

Alignments and References

With Intent Manager deactivated, Pro/ENGINEER requires the user to align entities to existing features using the Alignment option. Intent Manager requires that the user specify existing features as references before sketching. Normally, the Specify Reference option is used for this purpose.

Automatic Dimensioning

Intent Manager applies dimensions automatically upon ending a sketch option. Dimensions initially created by Pro/ENGINEER are weak and can be overridden by user-placed dimensions and/or constraints.

Pro/ENGINEER does not know the intent for a design. Thus, it cannot know the correct dimensioning scheme. Since Pro/ENGINEER objects have stored parameters that can be accessed by other modules (e.g., manufacturing and assembly), proper placement of dimensions is important for capturing design intent. With the exception of simple geometric shapes, it is likely that dimensions will have to be changed on a sketch. When using the Dimension option to change the dimensioning scheme, weak dimensions and constraints are subject to being overridden.

ORDER OF OPERATIONS

There is an established procedure for sketching a section with Intent Manager. For most sketching situations, these steps should be followed:

STEP 1: Specify References.

The position of a sketched section must be located with respect to existing part features. The Reference dialog box is used to identify references to use within the sketcher environment. Part edges, datum planes, vertices, and axes can be selected as references.

> **MODELING POINT** The References dialog box is used when creating a section within a feature creation option (e.g., protrusion or cut). Since sketches created within Sketcher mode are stand-alone sections, no references will be available, and this step will be skipped. Additionally, when creating a protrusion as the first feature of a part, references will not be available. When specifying references, select references that correspond to design intent. Pro/ENGINEER will then apply location dimensions from references that you specify.

STEP 2: Sketch the Section.

Sketch and/or construct the section using appropriate sketching tools.

STEP 3: Apply dimensions and constraints that match design intent.

Intent Manager will apply dimensions and constraints automatically. These dimensions and constraints will be considered weak and can be overridden by the creation of a new constraint or by adding a new dimension.

STEP 4: Modify dimension values.

Use the Modify option to individually change dimension values. Do not regenerate the section until all dimension values are modified.

STEP 5: Add dimension relationships (optional step).

If necessary, use the Sketch >> Relation >> Add option on the menu bar to create dimension relationships.

SKETCHING ENTITIES

Pro/ENGINEER's sketcher environment provides a variety of options for sketching two-dimensional entities. For experienced two-dimensional computer-aided design users, these tools will resemble basic geometry creation techniques.

SKETCHING LINES

Lines are created with the Line icon or with the Sketch >> Line option. Two entity types are available: Line and Centerline (Figure 3–8). Line is the default selection and is used to create feature entities. Centerline entities are used for construction techniques only (e.g., the axis of revolution for a revolved feature).

SKETCHING ARCS

Multiple arc creation tools are available within a sketcher environment. The following list and Figure 3–9 describe each of them.

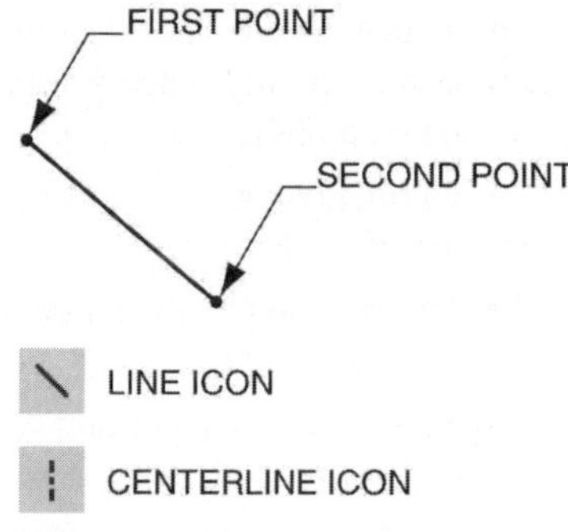

Figure 3-8 Line construction

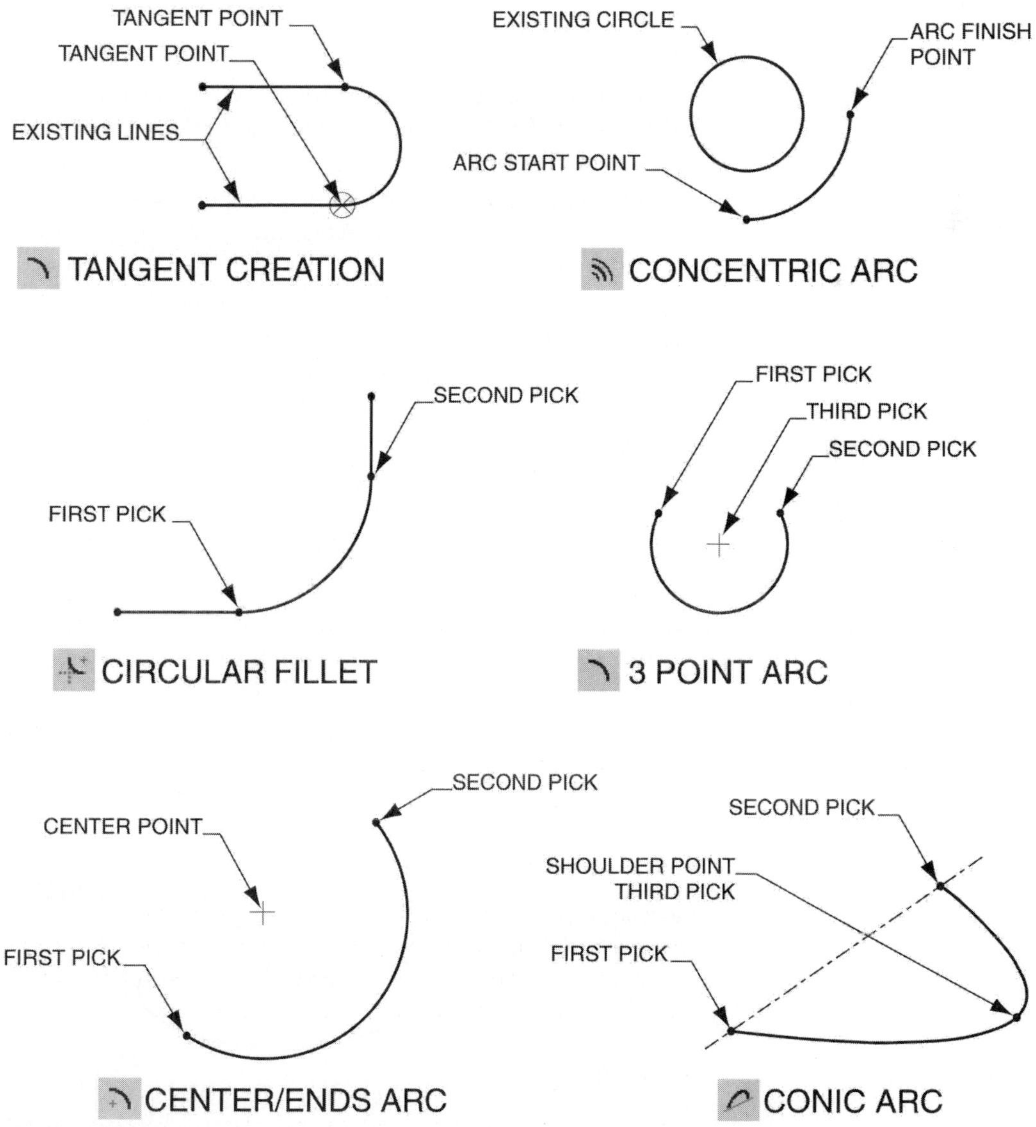

Figure 3-9 Arc creation options

- **Tangent end** The Tangent End option creates an arc tangent to the endpoint of a selected entity. The first endpoint of the arc or both endpoints can be selected as tangent points.
- **Concentric arc** The Concentric Arc option places an arc concentric to an existing arc or circle.
- **Circular fillet** The Circular Fillet option is used to create a fillet between two selected entities.
- **3 point arc** The 3 Point Arc option is used to create an arc by selecting the endpoints of the arc and then a third point on the arc.
- **Center/ends arc** The Center/Ends Arc option is used to create an arc by selecting the center point of the arc and the two endpoints of the arc.
- **Conic arc** The Conic Arc option creates a conical shaped entity through the selection of the conic's endpoints and shoulder point.

SKETCHING CIRCLES

Circles are created with available circle icons on the sketcher toolbar or though the Sketch menu on the menu bar. As shown in Figure 3–10, three sketching options are available.

- **Center/Point Arc** The Center/Point Arc option is used to create a circle by first selecting the circle's center point, then selecting a second point on the perimeter of the circle.
- **Concentric Arc** The Concentric Arc option is used to construct a circle with a center point common to a second circle. The referenced circle can be a sketched entity or an entity referenced from an existing part feature. When selecting an existing feature, this feature will become a section reference.
- **Full Ellipse** The Full Ellipse option creates an ellipse by first selecting the ellipse's center point, then selecting a point on the perimeter of the ellipse.

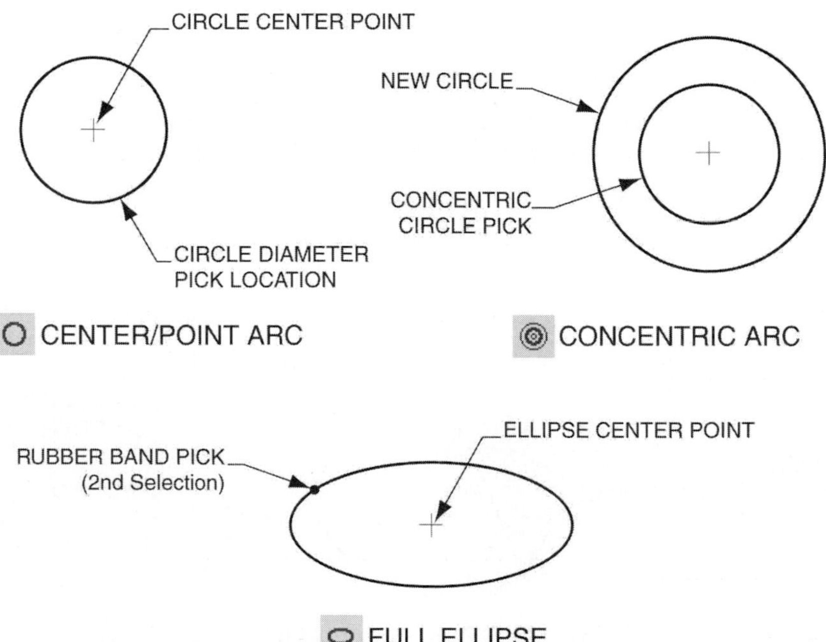

Figure 3-10 Circle creation options

Sketching a Rectangle

A rectangle can be created using the Rectangle icon or the Sketch >> Rectangle option. As shown in Figure 3–11, a rectangle is created by, first, selecting one vertex of the rectangle and, second, by selecting the opposite vertex.

Splines

A spline is a variable radius curve that passes through multiple control points (see Figure 3–12). A spline is created using either the Spline icon on the sketcher toolbar or the Sketch >> Spline option. To draw a spline, on the work screen pick control point locations that lie on the spline. A spline can be constructed tangent to sketched geometry or existing part features. If required, Intent Manager creates dimensions to define endpoints of a spline; the Dimension option can be used to manually create dimensions to additional control points.

Sketched Text

Text can be used in solid extruded Protrusion, Cut, and Cosmetic features. Sketched text is created using the Sketch >> Text option on the menu bar or the Text icon on the sketcher toolbar. Perform the following steps to create text.

Step 1: **Select the TEXT icon on the sketcher toolbar.**

Step 2: **On the work screen, pick the start point for the string of text (Figure 3–13).**

Step 3: **On the work screen, pick the point on the screen that will define the height of your text font.**

This selection point will also define the orientation of your string of text.

Step 4: **Enter text string.**

Step 5: **If necessary, on the Text dialog box, adjust the text font, aspect ratio, and slant angle (see Figure 3–13).**

Axis Point

Axis points created within a sketcher environment are extruded as axes. Axis points are created using the Sketch >> Axis Point option on Pro/ENGINEER's menu bar.

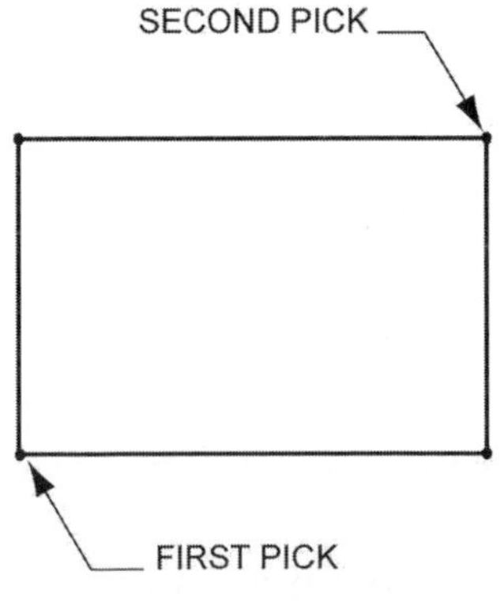

Figure 3-11 Creating a rectangle

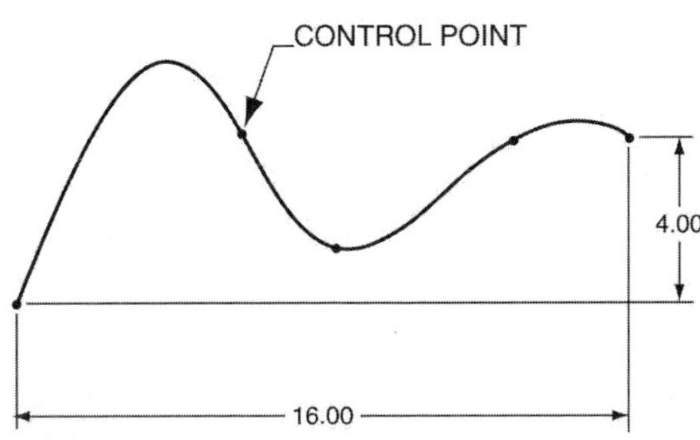

Figure 3-12 Spline creation

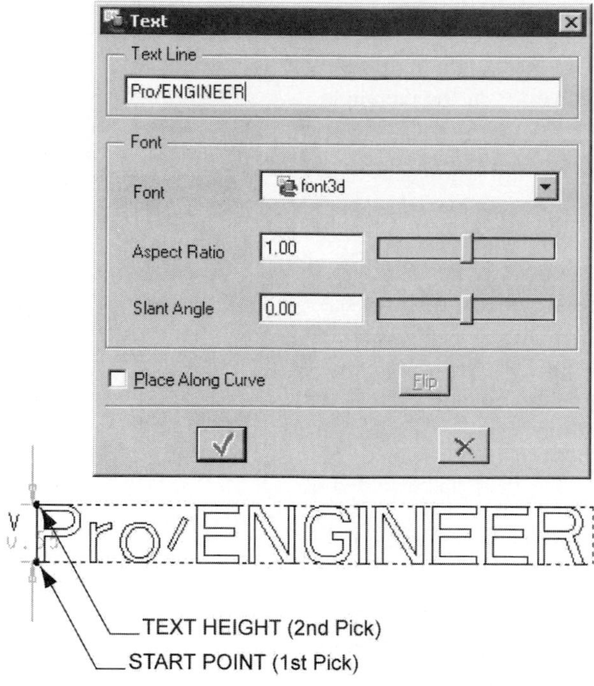

Figure 3–13 Selecting the text box

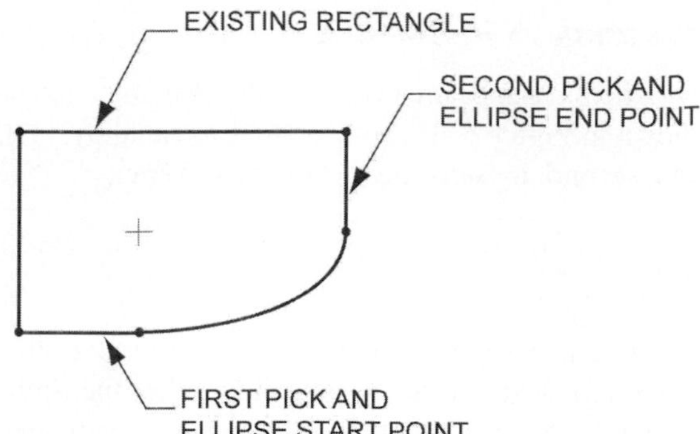

Figure 3–14 Elliptical fillet

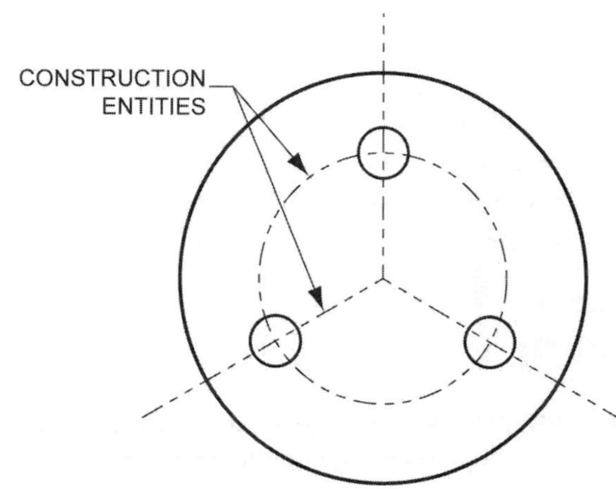

Figure 3–15 Construction entities

Elliptical Fillet

An Elliptical Fillet is a sketch entity created between two selected entities. As shown in Figure 3–14, the fillet is created in an elliptical shape. To create an elliptical fillet, either use the Elliptical Fillet icon or the Sketch >> Fillet >> Elliptical option.

Construction Entities

Normal sketch entities such as lines, arcs, circles, and splines can be converted to construction entities through the Edit >> Toggle Construction option (Figure 3–15). This is an extremely useful technique for creating complex shapes and features. Construction entities are used as references only and will not protrude with feature creation processes.

Sketching without Intent Manager

Intent Manager was first introduced with Release 20 of Pro/ENGINEER. The following are some of the differences between sketching without Intent Manager and sketching with Intent Manager:

- Without Intent Manager, constraints are applied after regeneration and are based on assumptions applied by Pro/ENGINEER.
- Without Intent Manager, dimensions have to be added manually by the user, though the Automatic Dimensioning option is available for fully defining a sketch.

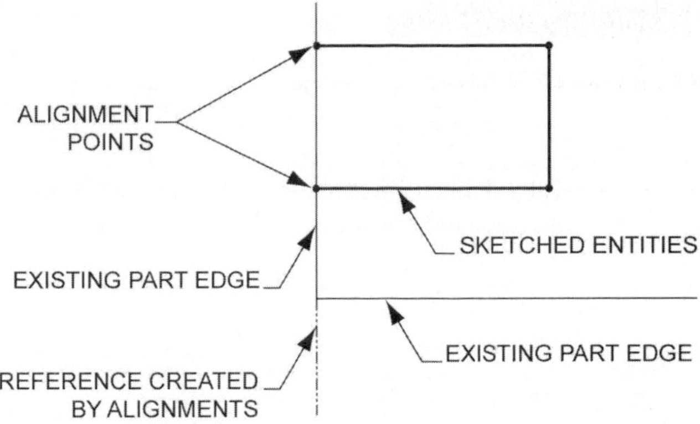

Figure 3-16 Aligning an entity

- Without Intent Manager, a section does not remain fully defined during section modification.
- Without Intent Manager, section references are created from existing part features by using the Alignment option or by dimensioning to the edge of an existing feature.
- Without Intent Manager, mouse functions are available for sketching lines, circles, and arcs.

Alignments

Pro/ENGINEER uses alignments to reference existing part geometry. As shown in Figure 3–16, the design intent for the endpoints of sketched entities might require them to lie on the edge of an existing feature. Without Intent Manager, the Alignment option is used to create section references. There are two ways that Alignment can be used to create a reference:

- By picking the part geometry, such as an edge, with the right mouse button, then entering the element as a known entity with the middle mouse button.
- By picking the part geometry, then picking the sketched entity.

Order of Operations

Just as when sketching with Intent Manager activated, sketching without Intent Manager requires that an established procedure be followed. The following is the recommended order of operations for sketching without Intent Manager.

Step 1: Sketch section.

Sketch the section with tools found under the Sketcher menu. Additionally, geometric construction tools such as Use Edge and Offset Edge are available under the Geom Tools menu.

Step 2: Align section entities to existing geometry (optional step).

Use the Alignment option to align sketched entities to existing feature edges. This is an optional step and may not be used for all sections. As an example, a sketched cut may not intersect an existing feature edge.

Step 3: Dimension the section using appropriate dimensioning techniques.

Use the Dimension option to dimension the sketch.

Step 4: REGENERATE the section.

Use the Regenerate option to regenerate the section. In the message area, the valued words *Section Regenerated Successfully* will identify when a section is

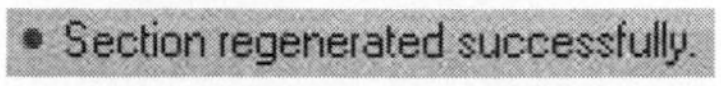

• Section regenerated successfully.

Figure 3–17 The valued message

fully defined (Figure 3–17). If a section is not fully regenerated, add additional dimensions and/or alignments.

MODELING POINT If the automatic dimension option is used to fully define a section, ensure that the section's dimensioning scheme meets design intent before modifying any dimension values. This will help to ensure that the model's dimensions meet design requirements.

STEP 5: MODIFY dimension values.

Use the Modify option to change any dimension values.

STEP 6: REGENERATE the section.

Use the Regenerate option to update the section to the new dimension values.

STEP 7: Add a dimensional RELATION to the section (optional step).

If necessary, create relationships between dimensions.

STEP 8: Regenerate the section.

DIMENSIONING

Pro/ENGINEER provides a variety of dimensioning types for defining a sketch. Typical computer-aided design dimensions such as linear, radial, and angular are available. Additionally, dimensioning options such as perimeter and ordinate exist. All dimensioning types are available under the sketcher environment's Dimension option.

LINEAR DIMENSIONS

Linear dimensions are used to indicate the length of line segments and to measure the distance between two entities, such as two parallel lines or the distance from the center of a circle to a line. The following procedures describe how to dimension entities within the sketcher environment. Refer to Figure 3–18 for a graphical representation.

LINE SEGMENT

To dimension a line segment, select Dimension, then with the left mouse button pick the line to dimension. Next, with the middle mouse button, pick the location on the work screen for the dimension placement.

PARALLEL LINES

To dimension the distance between two parallel lines, select Dimension, then with the left mouse button, pick the parallel line segments. Next, with the middle mouse button, pick the location on the work screen for the dimension placement.

LINE TO POINT

To dimension the distance between a line segment and a point, select Dimension, then with the left mouse button, pick the line and the point. Next, with the middle mouse button, pick the location on the work screen for the dimension placement.

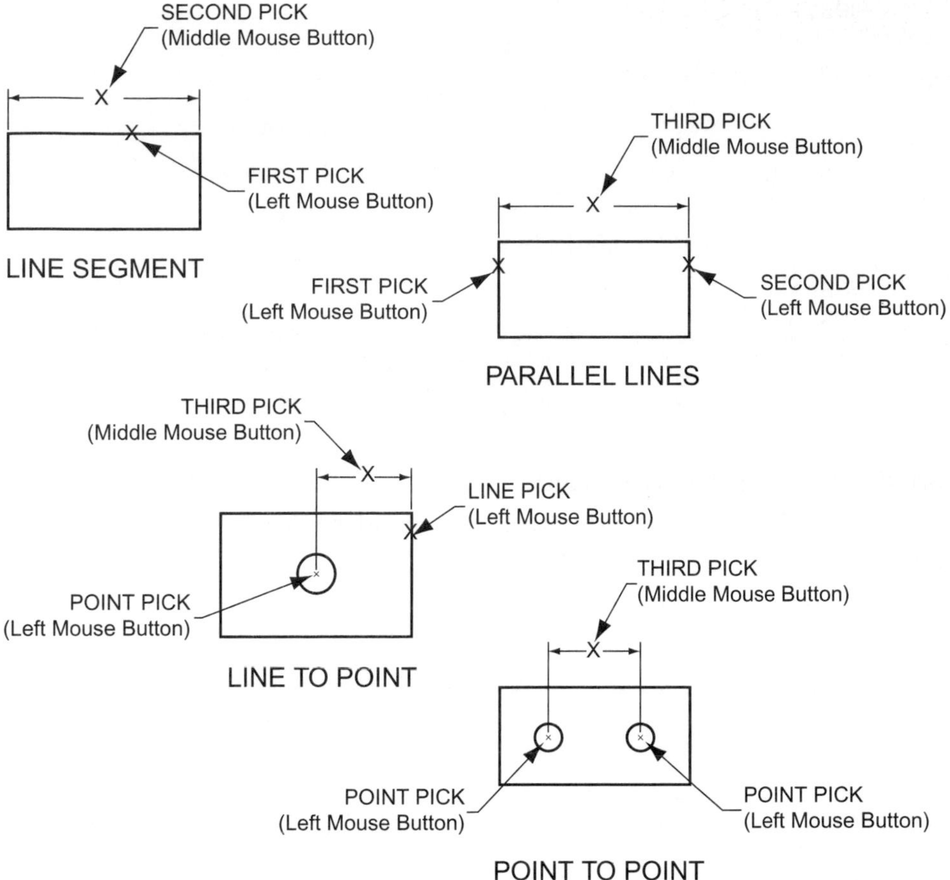

Figure 3–18 Dimensioning options

Point to Point

To dimension the distance between two points, select Dimension; then with the left mouse button, pick the two points. Any dimension to a line entity will assume that the dimension will be oriented perpendicular to the line. With a point-to-point dimension, a menu provides the option for creating a horizontal, vertical, or slanted dimension. After selecting the dimension orientation, with the middle mouse button pick the location on the work screen for the dimension placement.

Radial Dimensions

Arcs and circles are considered radial entities. The Dimension option is available for creating radius and diameter dimensions. Refer to Figure 3–19 for a graphical representation on dimensioning radial entities.

Radius Dimensions

A radius is the distance from the center of an arc or circle to the perimeter of the entity. To dimension a radial entity as a radius, select Dimension, then with the left mouse button, pick the radial entity. Next, with the middle mouse button, pick the location on the work screen for the dimension placement.

Diameter Dimensions

A diameter is the maximum distance across a circle. To dimension a radial entity as a diameter, select Dimension, then with the left mouse button, pick the radial entity

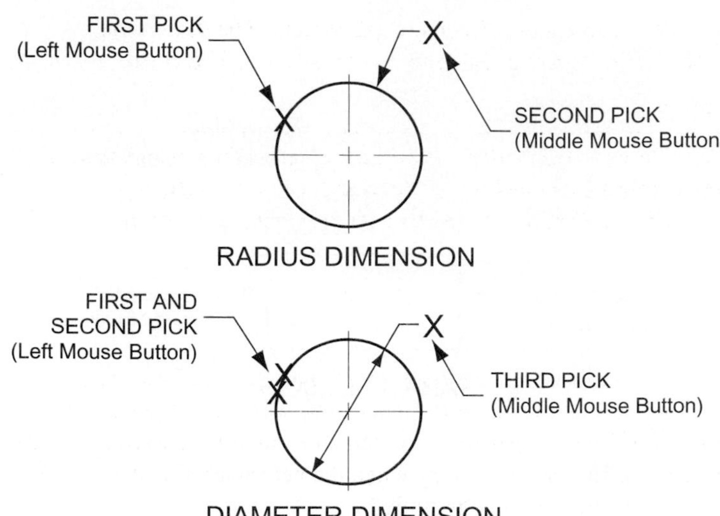

Figure 3-19 Radial dimensioning options

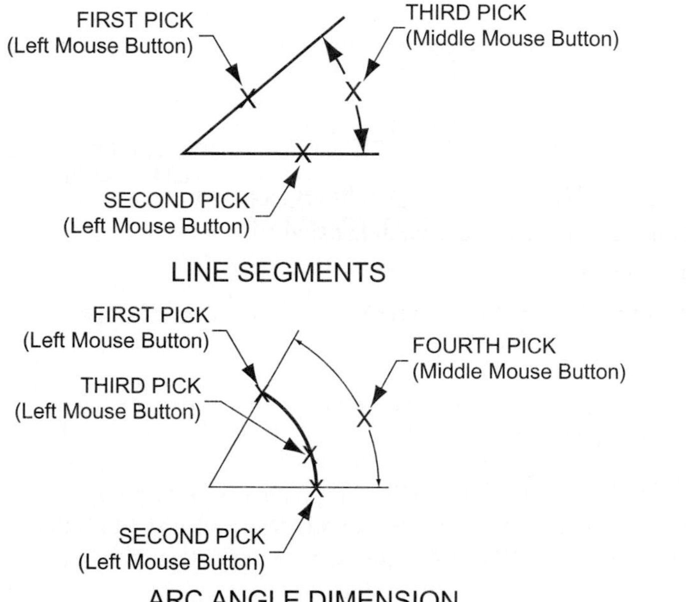

Figure 3-20 Angular dimensioning options

two times. Next, with the middle mouse button, pick the location on the work screen for the dimension placement.

Angular Dimensions

The angle formed between two line segments can be dimensioned within the sketcher environment. Additionally, the angle formed by the endpoints of an arc can be dimensioned. Refer to Figure 3–20 for a graphical representation on dimensioning angular entities.

Line Segments

To dimension the angle between two line segments, select Dimension, then with the left mouse button, pick the two line segments. Next, with the left mouse button, pick the location on the work screen for the dimension placement. The dimension placement location determines whether an acute or obtuse angle is formed.

Arc Angle Dimension

To dimension the angle formed by the endpoints of an arc, select Dimension, then with the left mouse button, pick the endpoints of the arc. Next, select the arc being dimensioned. Finally, pick the location on the work screen for the dimension placement.

Perimeter Dimensions

The Perimeter option under the dimension menu measures the perimeter of a sketched loop or chain of entities. Since a perimeter's dimension value can be modified, it requires the selection of a variable dimension. Due to a perimeter dimension's unique characteristic, this variable dimension is used as the dimension to vary when the perimeter dimension's value is modified. The variable dimension cannot be modified.

Ordinate Dimensions

Pro/ENGINEER provides an option for creating ordinate dimensions (Figure 3–21). Before an ordinate dimension can be created, a baseline is required. The Dimension option uses the baseline as a reference for creating the ordinate dimension. Perform the following steps to create ordinate dimensions.

Step 1: Select SKETCH >> DIMENSION >> BASELINE from Pro/ENGINEER's menu bar.

Before an ordinate dimension can be created, Pro/ENGINEER has to know the baseline from which to measure each dimension.

Step 2: Pick the baseline entity (Figure 3–21).

Step 3: Select the location of the baseline dimension with the middle mouse button.

Step 4: Select the Dimension icon on the sketcher toolbar (or select Sketch >> Dimension >> Normal).

Step 5: Pick the baseline dimension's numeric value.

Step 6: Pick the entity to dimension (4th pick in Figure 3–21), then select the placement location for the dimension.

Reference Dimensions

Reference dimensions are used to show the size or location of an entity, but their values cannot be modified. Reference dimensions do not play a role in the definition of a feature.

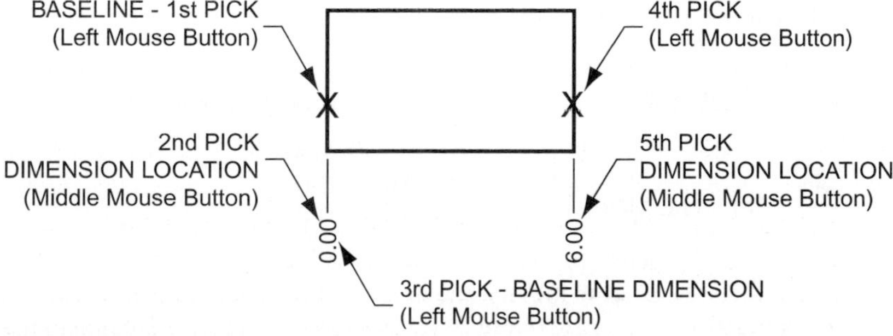

Figure 3–21 Ordinate dimensioning

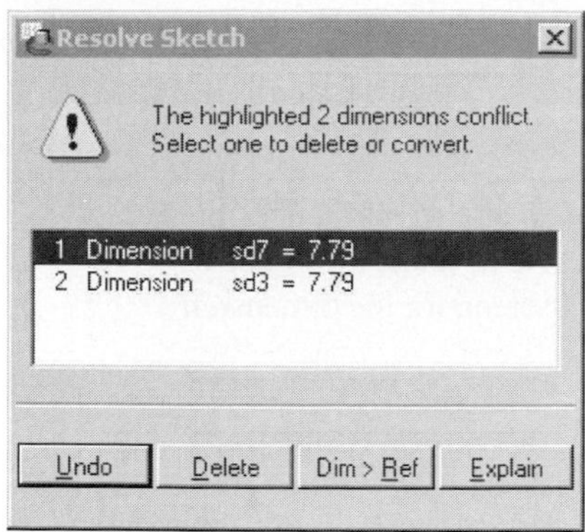

Figure 3–22 Resolve sketch dialog box

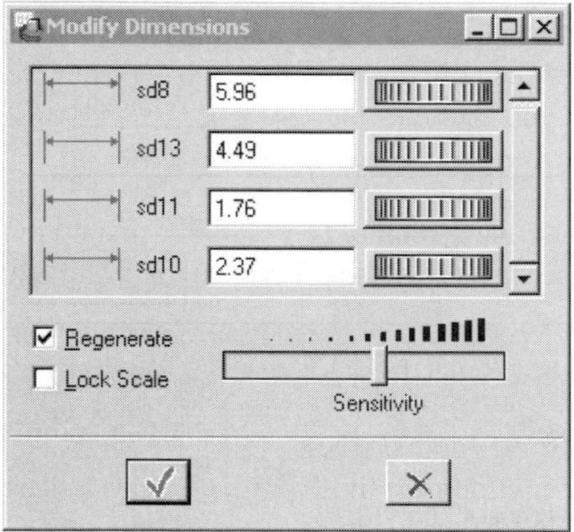

Figure 3–23 Modify dimension dialog box

They can be used, however, for defining dimensional relationships. Reference dimensions are created in a manner similar to linear, angular, and radial dimensions. When a dimension is placed that creates an over-constrained situation, Pro/ENGINEER recognizes this and requires you (through the Resolve Sketch dialog box) to either undo the creation, delete the dimension, or convert the dimension to a reference (see Figure 3–22).

Modifying Dimensions

When sketching a feature in Pro/ENGINEER, the first priority is to create a section that meets design intent through the dimensioning scheme and through the proper utilization of constraints. Once a section is sketched and all dimensions and constraints applied, dimension values can be modified to their correct value.

Individual Modification

An individual dimension can be modified by double picking the dimension's value with the pick icon. The section will be regenerated automatically upon entering the new value.

Multiple Modification

The Modify Dimensions dialog box is accessible through the Modify Dimensions icon and can be used to modify one or more dimension values (Figure 3–23). There are two techniques for selecting dimensions. With the first technique, the user can preselect dimensions with the Pick icon by using the shift key or by drawing a pick box around all necessary dimension values. The user can then individual modify each dimension value through the dialog box.

Scale Modification

Often, it is difficult to modify the values of a complicated section. A complicated section can lead to dimensions that conflict. A conflict in dimensions will lead to a failed regeneration. A solution to this problem is using the Lock Scale option on the Modify Dimensions dialog box (Figure 3–23). The Lock Scale option allows for the modification of one dimension. After entering the new dimensions value, the remaining dimensions are scaled to the same factor.

Sketcher Relations

The Relation option is a powerful tool for capturing design intent. Within the sketcher environment, this option creates relationships between dimension values. As an example, if the design intent is to have the width of a prism two times the length, a mathematical equation can be entered through the Relation option that makes the width value always twice the length.

When a feature is created, dimensions are used to define size and location. Each dimension has a value and a dimension symbol associated with it. Figure 3–24 shows an example of a section with normal dimension values provided in one view and respective dimension symbols provided in the second. The dimension symbol is used within a relation equation to establish a dimensional relationship.

Notice in Figure 3–24 that the format of a dimension symbol while in the sketcher environment is sd*. The *s* represents the sketcher environment (it is not used in part mode), and the *d* represents a dimension type. Pro/ENGINEER sequentially provides the dimension symbol number.

Two types of relationships can be created: equality and comparison. An equality relationship requires an algebraic equation. The following are examples of equality relationships.

sd1 = sd2

sd3 = sd2 + sd1

sd4 = sd1*(sd2 + sd3)

sd5 = sd4/(sd1 + sd2 + sd3)

sd5 = sd4*SQRTsd2

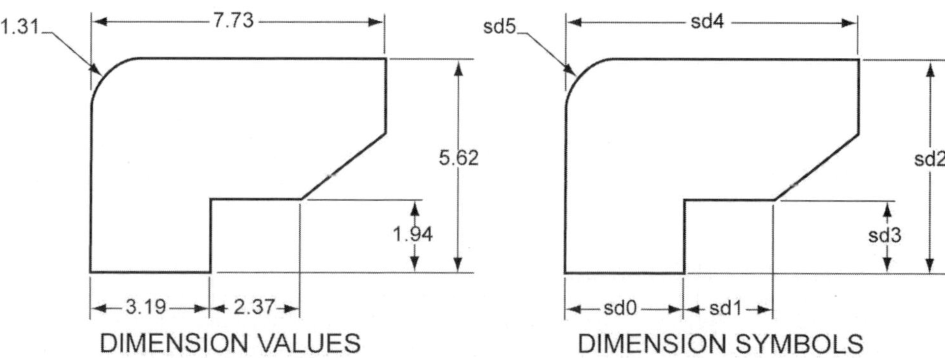

Figure 3-24 Dimension symbols

A comparison relationship can take the form of an equation or a conditional statement. The following are examples of comparison relationships:

$$sd1 < sd2$$

$$(sd2 + sd1) < (sd3 + sd4)$$

$$IF\ (sd1 + sd3) > (sd3 + sd4)$$

The Sketch >> Relation >> Add option is used to create a relationship between two dimensions. When this option is selected, the section's dimensions are converted from the value format to the symbol format (Figure 3–24). This allows for the selection of appropriate dimension symbols. Perform the following steps to add a relation to a section.

STEP 1: Determine dimensional relationships that will satisfy design intent.

Relations are used to intelligently incorporate design intent into a model. Relations should be used only when appropriate.

STEP 2: Select SKETCH >> RELATIONS >> ADD from the menu bar.

The Add option is used to create a dimensional relationship.

> **MODELING POINT** The Relation menu provides options in addition to adding dimensional relationships. The Edit Rel option allows relations to be created and edited through a text editor. The Show Rel option shows created relations without opening a text editor.

STEP 3: Enter a relational equation in Pro/ENGINEER's text box.

Add a valid relational equation then enter the value.

STEP 4: Select ENTER a second time to leave the text box.

Selecting Enter with a blank Text Box will leave the Add option. The section will regenerate automatically with Intent Manager activated.

GEOMETRIC TOOLS

A variety of geometric construction tools are available for manipulating sketched entities. These options are available as icon on the sketcher toolbar.

 DYNAMIC TRIM

The Dynamic Trim option trims a selected entity up to its nearest vertex point or points (Figure 3–25). Pick the segment of the entity to delete.

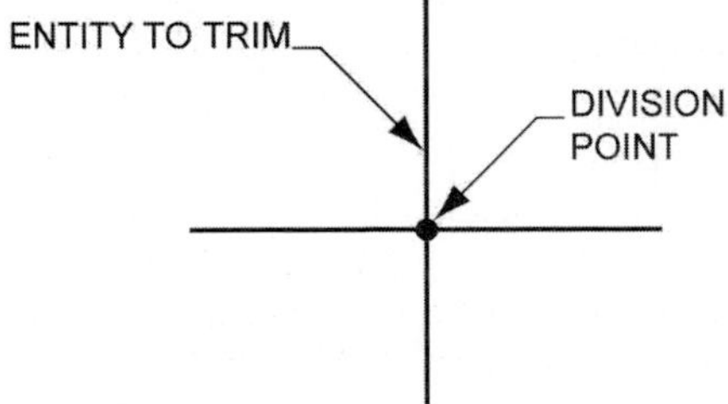

Figure 3–25 Dynamic trim option

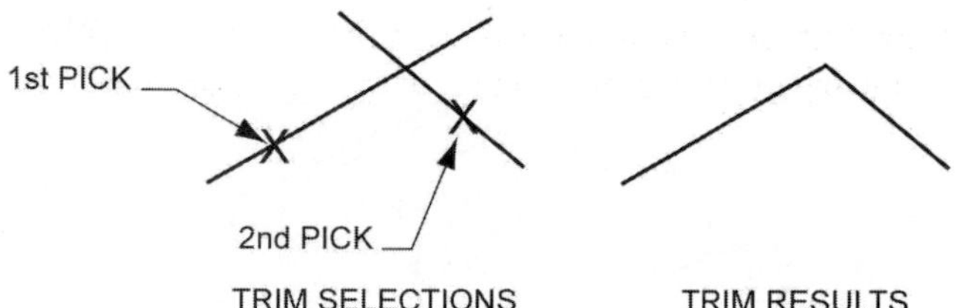

Figure 3-26 Trim options

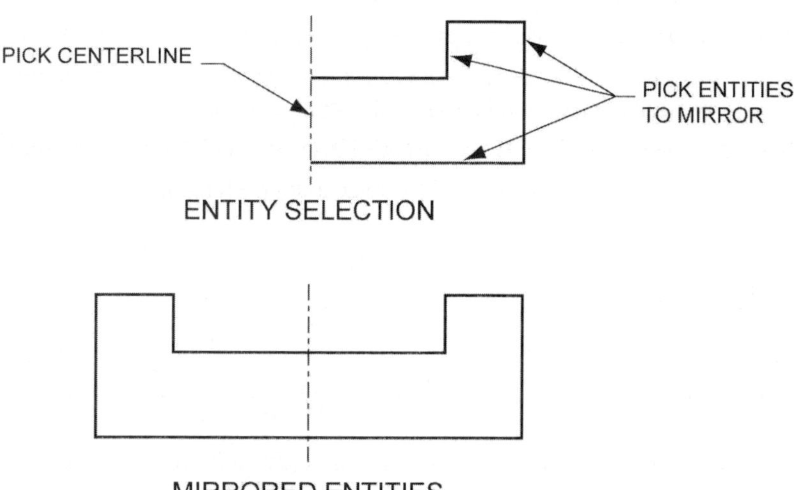

Figure 3-27 Mirroring entities

Trim

The Trim option trims two selected entities at their intersection point. As shown in Figure 3–26, this option deletes the selected entities on the opposite side of the intersection point from where each entity was selected.

Divide

The Divide option divides an entity into two segments. The entity is divided at the selection point.

Mirror

The Mirror option mirrors selected entities over a picked centerline (Figure 3–27). The first step in the mirror process is to pick the entities to mirror. Multiple entities can be picked through the shift key and Pick icon combination. Follow entity selection by executing the Mirror option, then picking the centerline to mirror about.

Use Edge

The Use Edge option creates sketcher geometry from existing feature edges. Selected feature edges are projected onto the sketching plane as sketcher entities. As shown in Figure 3–28, the selected edges do not have to lie on or parallel to the sketching plane. Once entities are projected onto the sketching plane, they can be trimmed, divided, and filleted.

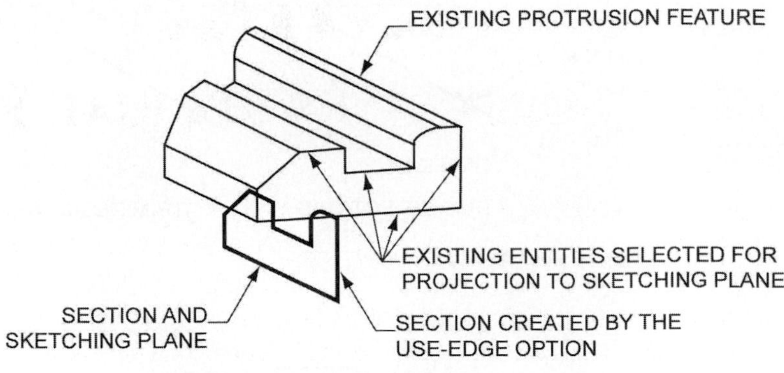

FEATURE CREATION

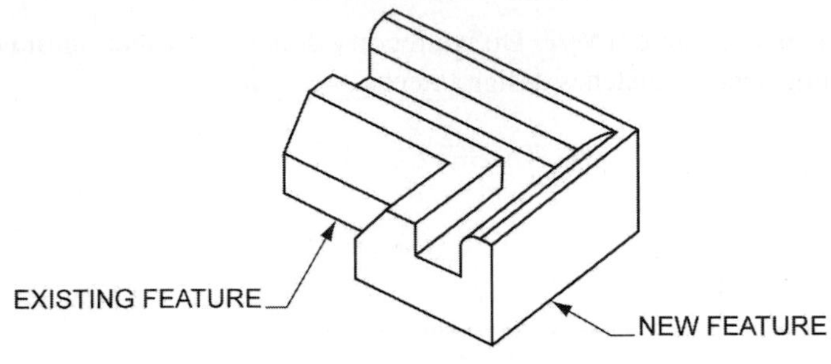

FINISHED FEATURE

Figure 3-28 Using existing feature edges as sketch geometry

 OFFSET EDGE

The Offset Edge option creates sketcher geometry offset from existing feature edges. This option is similar to the Use Edge option except an offset value is required. The projected edges will be offset the specified distance.

SUMMARY

Many of Pro/ENGINEER's feature construction tools require the sketching of a section. Sections are comprised of geometric entities, dimensions, constraints, and references. Sections can be sketched with or without Intent Manager. Intent Manager applies constraints and dimensions during the sketch construction process. Without Intent Manager, constraints and dimensions are applied during regeneration.

TUTORIAL

SKETCHER TUTORIAL 1

Tutorial in this chapter well explore sketching with Intent Manager. Figure 3–29 shows the section to be sketched. When creating a section with Intent Manager activated, adhere to the following order of operations:

1. Sketch section (*Note:* When creating a section in Part or Assembly mode, the first step is to specify references. Since in Sketcher mode no existing features exist to reference, this step is skipped.)
2. Apply dimensions that match design intent.
3. Modify dimension values (*Note:* Do not modify dimension values until the dimensioning scheme matches design intent.)
4. Regenerate.
5. Apply relations (if required).
6. Regenerate (if necessary).

This tutorial will cover the following topics:

- Starting a new object file in Sketch mode.
- Sketching with Intent Manager.
- Creating entities with the Line, Arc, and Circle options.
- Saving a section.

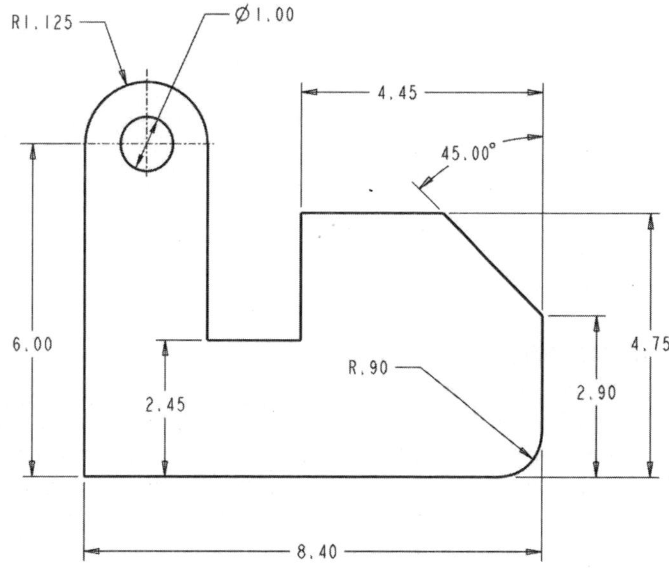

Figure 3–29 Section sketch

CREATING A NEW OBJECT IN SKETCH MODE

This segment of the tutorial will start a section file in Sketch mode.

STEP 1: Start Pro/ENGINEER.

STEP 2: Select FILE >> SET WORKING DIRECTORY, then select an appropriate working directory.

The working directory is the default directory where Pro/ENGINEER will save model files and where Pro/ENGINEER will look when the File >> Open

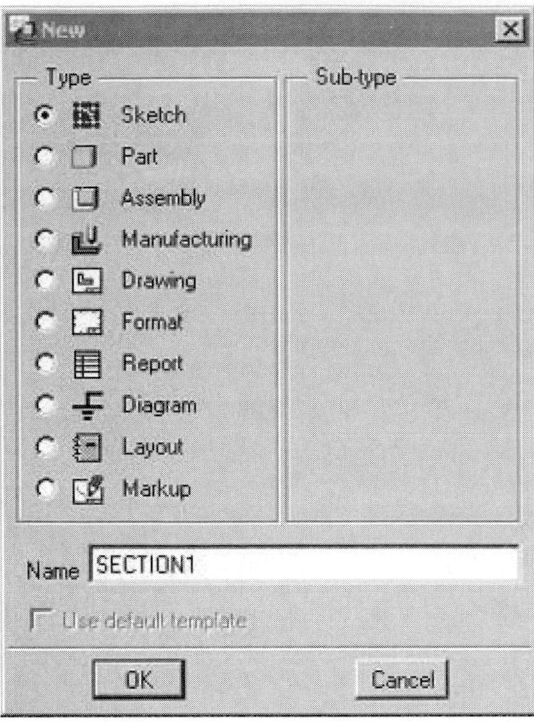

Figure 3–30 New dialog box

option is selected. You should select a directory where you have read and write privileges.

STEP 3: Select FILE >> NEW (or select the New icon).

STEP 4: Select SKETCH as the model type, then enter SECTION 1 as the object's name (Figure 3–30).

STEP 5: Select the OK option to create the section file.

SKETCH ENTITIES

This sketching tutorial will utilize Intent Manager. When creating new features on an existing part, Pro/ENGINEER requires the establishment of feature references. With Intent Manager activated, references are established using the References dialog box. Without Intent Manager, references are established with the Alignment option and with dimensioning tools. Since the object of Pro/ENGINEER's Sketch mode is to create a stand-alone section, no existing part features will be present in the sketching environment. Due to this, the References dialog box is not available.

STEP 1: From the Start Point shown in Figure 3–31, use the Line option to sketch the entities shown.

The left mouse button is used to pick line end points, and the middle mouse button is used to cancel the line command.

When sketching entities, sketch the approximate shape. Do not worry about the size of the entities being sketched. Appropriate dimensions will be modified in a later step. The Start Point and the Endpoint of the sketch should lie approximately horizontal to each other.

Note: Disregard any dimensions that will be created when you cancel the line command.

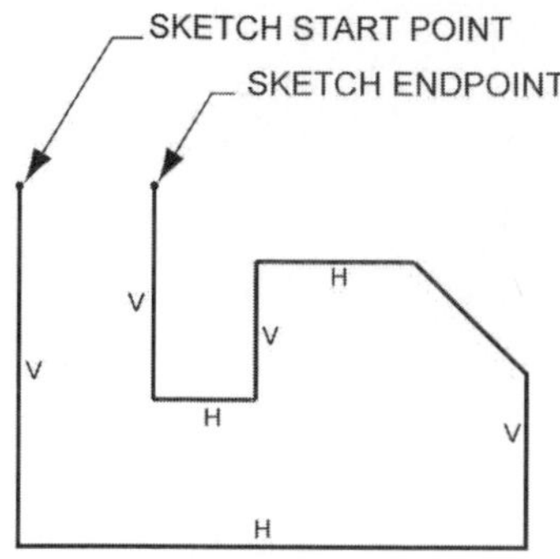

Figure 3–31 Sketching the section

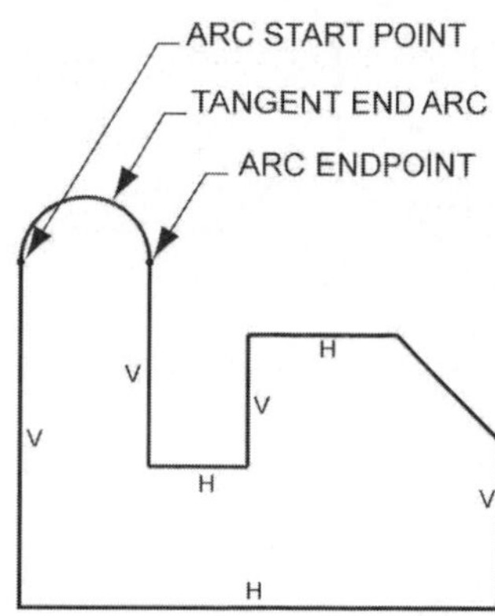

Figure 3–32 Creating a tangent end arc

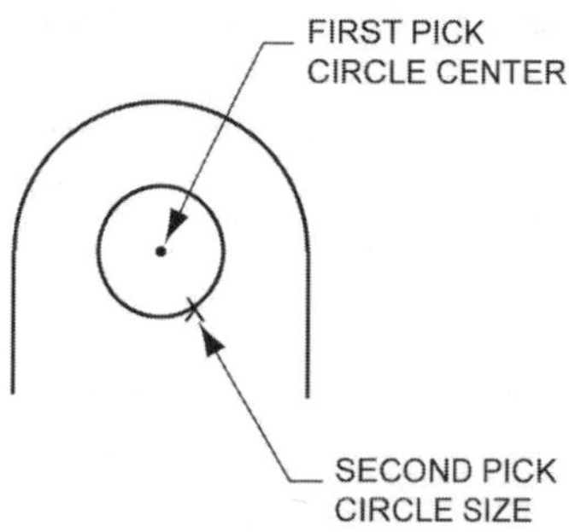

Figure 3–33 Creating a center/point circle

> **MODELING POINT** Lines that appear vertical or horizontal should be sketched accordingly. Intent Manager will apply constraints that will snap each line horizontal or vertical. In Figure 3–31, the lines labeled H and V represent horizontally and vertically constrained lines respectively.
>
> After terminating the Line option, Intent Manager will fully define the sketch with appropriate dimensions. The dimensioning scheme for your sketch will probably not match design intent.

STEP 2: **Use the ARC option to create the arc shown in Figure 3–32.**

STEP 3: **Use the CIRCLE option to create the circle shown in Figure 3–33.**

As shown in Figure 3–33, select the center-point of the circle, making it coincident with the center vertex of the existing arc. Next, define the size of the circle by dragging the perimeter of the entity. The diameter value will be modified in a later step.

STEP 4: **Use the CIRCULAR FILLET option to create the fillet shown in Figure 3–34.**

Use the Circular Fillet option to create the fillet. This option requires the selection of two entities to fillet between. The fillet's radius will be defined in a later step.

STEP 5: **Use the DIMENSION option to create the dimensioning scheme shown in Figure 3–35.**

Use the Dimension option to place dimensions according to Figure 3–35. When defining the dimensioning scheme for a section, it is helpful to disregard

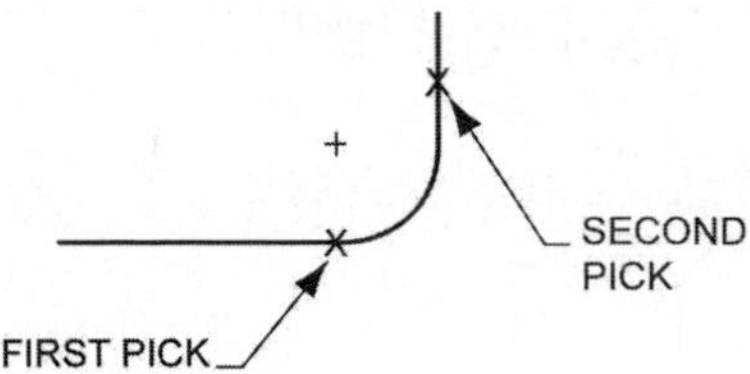

Figure 3–34 Creating a filleted arc

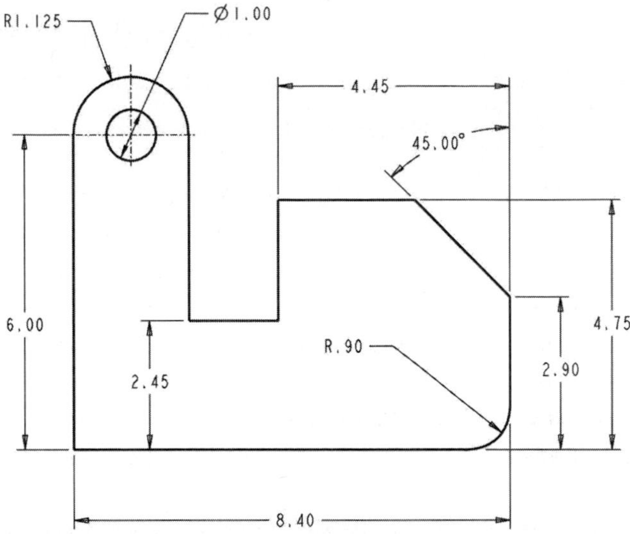

Figure 3–35 Dimensioning scheme

the current weak dimensions on the sketch. Weak dimensions and weak constraints will be overridden through the placement of strong dimensions. Use the following dimensioning techniques when placing your dimensions.

- The left mouse button is used to select entities to dimension and the middle mouse button is used to place the dimension.
- To dimension the distance between parallel lines, select each line with the left mouse button, then place the dimension with the middle mouse button.
- Radii dimensions are created with a single selection of the entity, while diameter dimensions are created with two selections of the entity.

STEP 6: **Use the MODIFY option to modify the dimension values to match Figure 3–35 (modify smaller dimensions first).**

Use the Modify option to change dimension values. After selecting the Modify icon, you can select all available dimensions. It is advisable to start with smaller dimensions first, followed by larger dimensions last. This helps to avoid unusual regeneration problems. Use the Undo option to undo any modification errors.

STEP 8: **SAVE the section.**

From the File menu select the Save option. Your object will be saved as a section file (**.sec*).

STEP 9: **Select the Continue icon to exit sketch mode.**

Selecting the continue icon will close the current sketch window. The object will remain in session memory.

TUTORIAL

Sketcher Tutorial 2

This tutorial will create the section shown in Figure 3–36. Within this tutorial, the following options will be used.

- Trim
- Delete
- Mirror
- Divide
- Adding constraints
- Sketch >> Centerline
- Sketch >> Rectangle

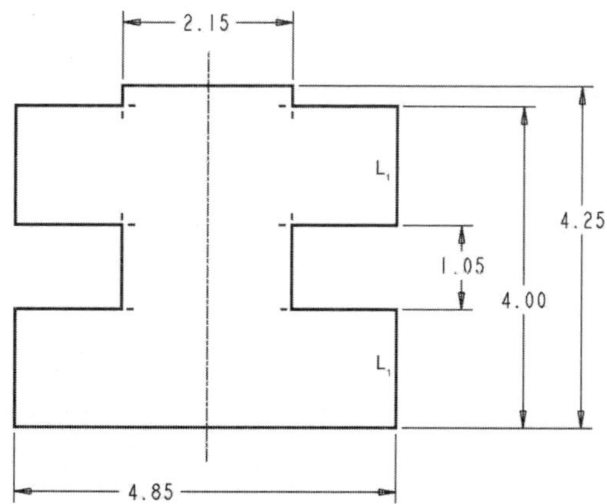

Figure 3–36 Finished sketch

Creating a New Section

This segment of the tutorial will describe the creation of a new Sketch mode object file.

Step 1: Start Pro/ENGINEER.

Step 2: Select an appropriate working directory.

Use the File >> Set Working Directory option to select a working directory.

Step 3: Create a new section file.

Using the File >> New option, select sketch as the mode and enter *Section 2* as the object name.

Creating the Sketch

This segment of the tutorial will describe the process for creating the sketch shown in Figure 3–36.

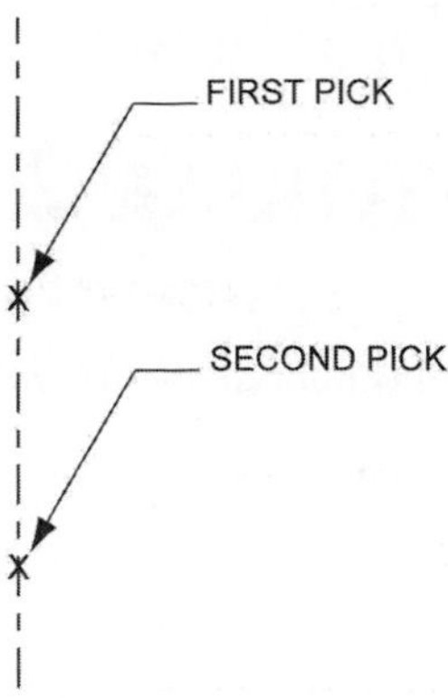

Figure 3–37 A vertical centerline

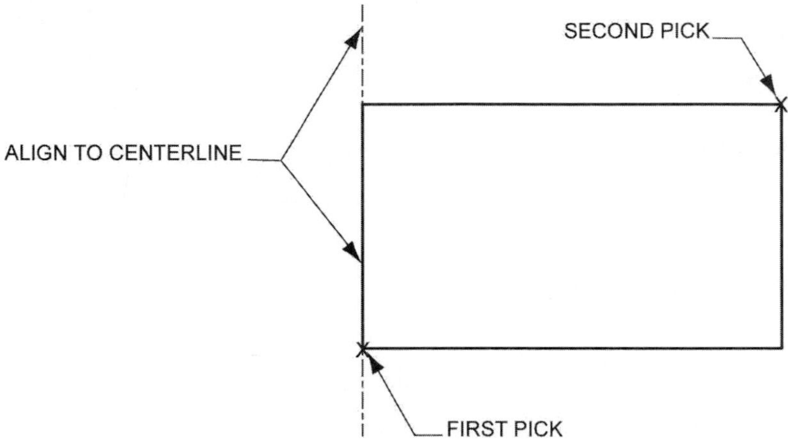

Figure 3–38 Creating a rectangle

STEP 1: Use the CENTERLINE icon to construct a vertical centerline (Figure 3–37).

The Centerline icon is located behind the Line icon on the sketcher toolbar. Since centerlines do not extrude into features, they are useful as construction lines. Additionally, many geometric tools, such as mirrored entities and revolved features, require a centerline. It is important to note that your centerline may not actually look like a centerline on the work screen.

STEP 2: Use the RECTANGLE option to create the rectangular entity shown in Figure 3–38.

As shown in Figure 3–38, the Rectangle option requires the selection of opposite corners of the rectangle. At this time, do not worry about entering a precise size for the rectangle. When sketching, be sure to align the left edge of the rectangle with the centerline.

STEP 3: Create a second rectangle as shown in Figure 3–39.

Sketch the second rectangle the same size as the first rectangle. When sketching, use Intent Manager to ensure that the length of the horizontal side of the new rectangle is the same length as the first rectangle. If necessary, dynamically zoom out from the sketch (control key and middle mouse button).

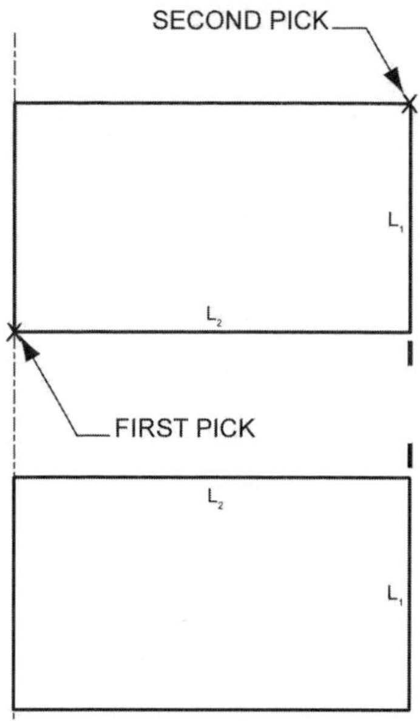

Figure 3–39 The second rectangle

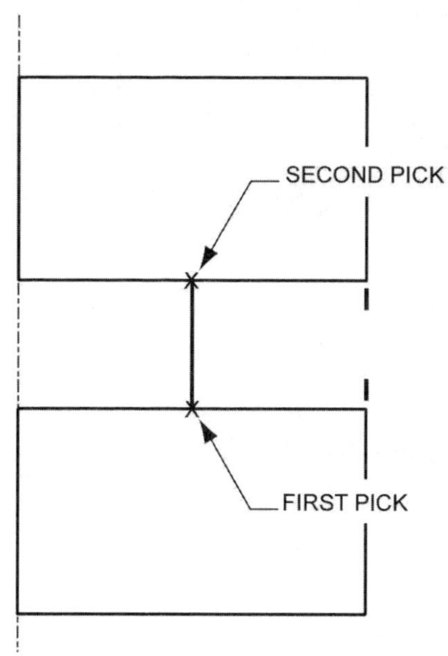

Figure 3–40 Line construction

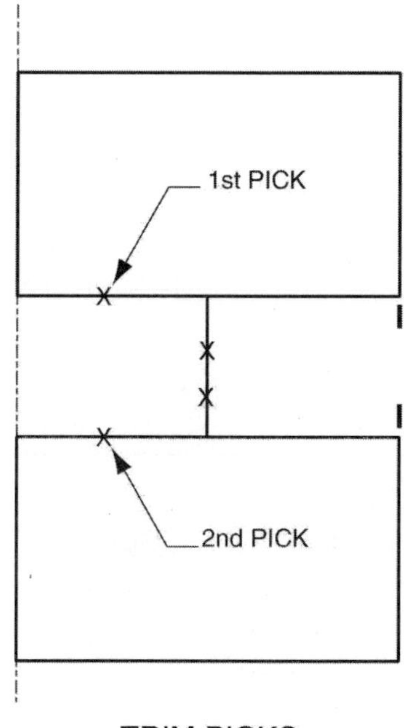

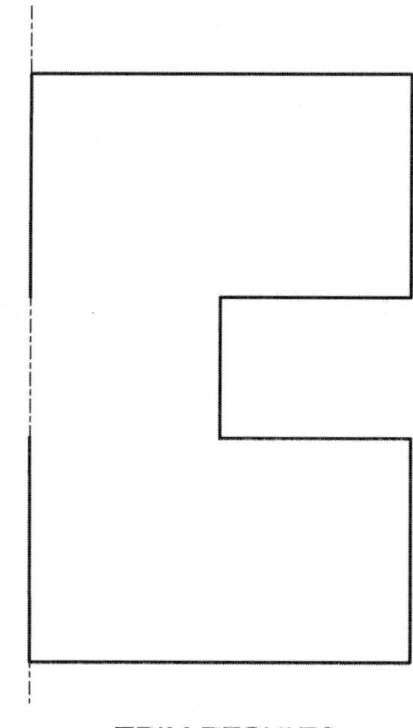

Figure 3–41 Trimming entities

STEP 4: Sketch the line shown in Figure 3–40.

Use the Line option to sketch the vertical line between the points shown in Figure 3–40. Do not worry about the exact location of the line.

STEP 5: Use the DYNAMIC TRIM option to trim the entities as shown in Figure 3–41.

The Dynamic Trim option will trim an entity up to its nearest vertex point(s). Select the segment of each entity that you want to delete.

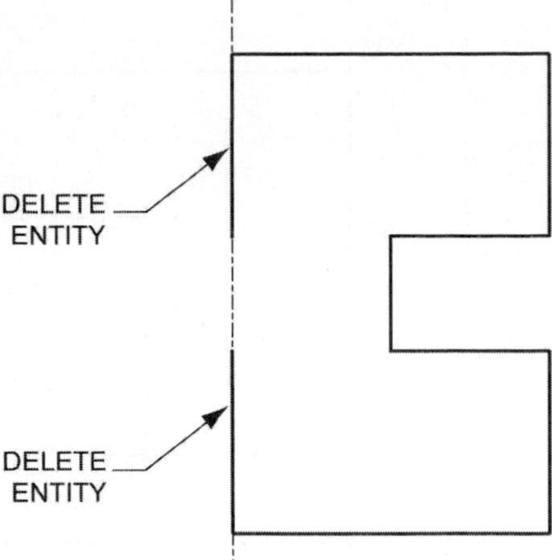

Figure 3–42 Deleting entities

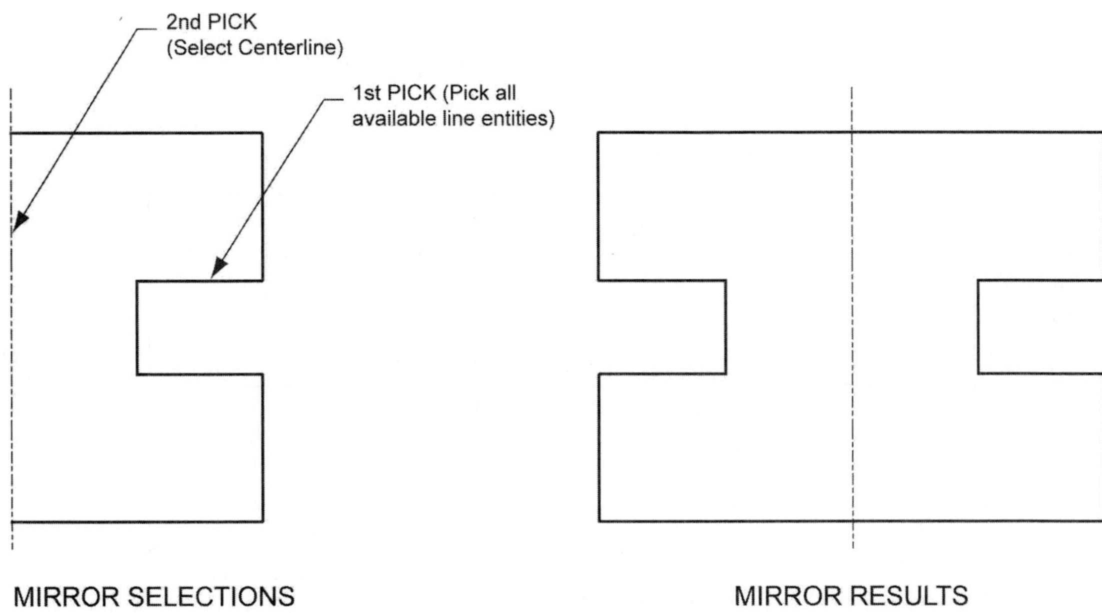

Figure 3–43 Mirroring entities

STEP 6: **Use the Pick icon to select the entities shown in Figure 3–42, then use the keyboard's delete key to delete each entity.**

STEP 7: **Using the Pick option, drag a selection box around all available entities within your sketch.**

The next step of this tutorial will mirror your lines entities about the vertical centerline. To use the Mirror option, entities to mirror must be preselected.

STEP 8: **Select the MIRROR option, then pick the centerline to mirror the entities about (Figure 3–43).**

All line entities must be preselected before you can select the Mirror option. If your results do not match Figure 3–43, undo the last command then start over at step 7.

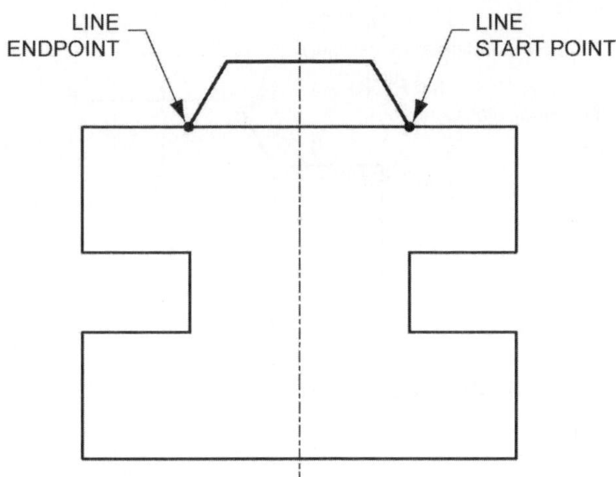

Figure 3–44 Sketching a line

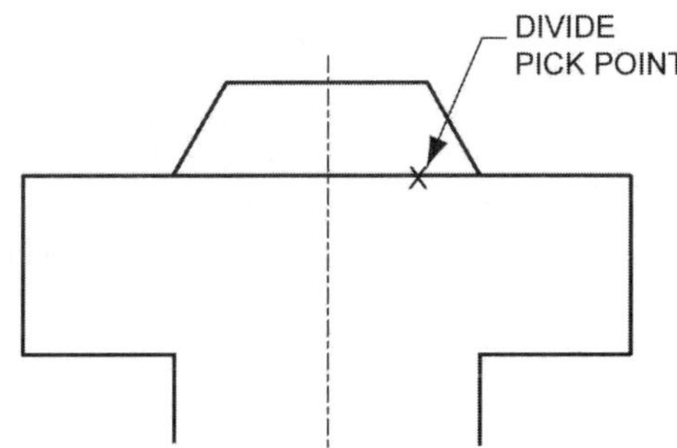

Figure 3–45 Divide selection

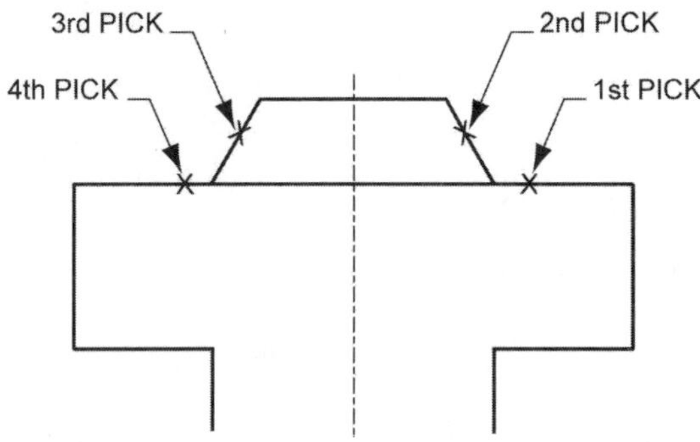

Figure 3–46 Trim selection

STEP 9: **Sketch the line entities shown in Figure 3–44.**

STEP 10: **Use the DIVIDE option to break the single entity at the point shown in Figure 3–45.**

The Divide option will break an entity at the point selected. Its icon is located behind the Dynamic Trim option.

STEP 11: **Use the TRIM option to trim the four lines as shown in Figure 3–46. (Ensure you select Trim, not Dynamic Trim)**

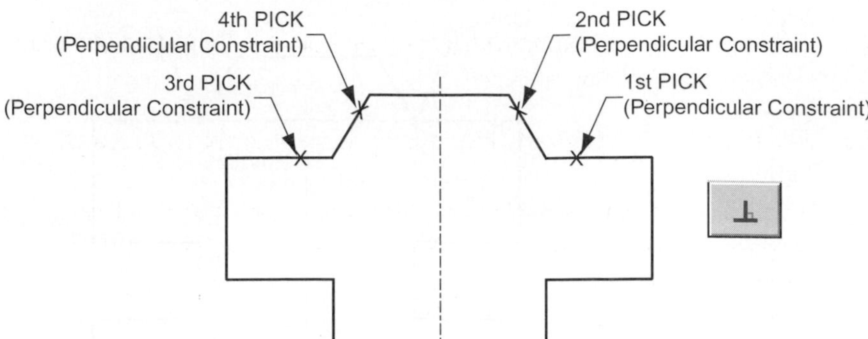

Figure 3–47 Perpendicular constraints

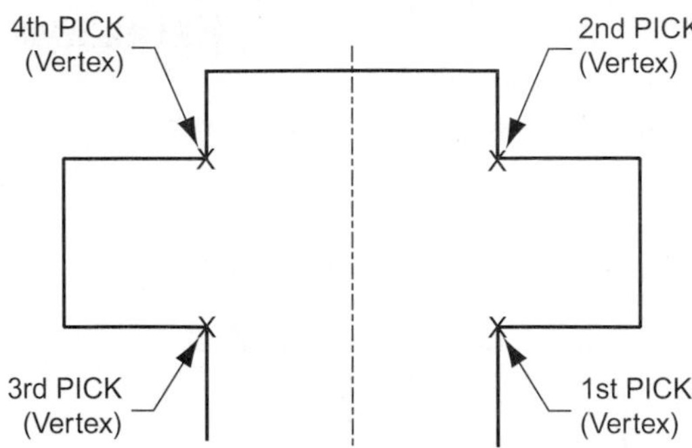

Figure 3–48 Line-up-vertical constraints

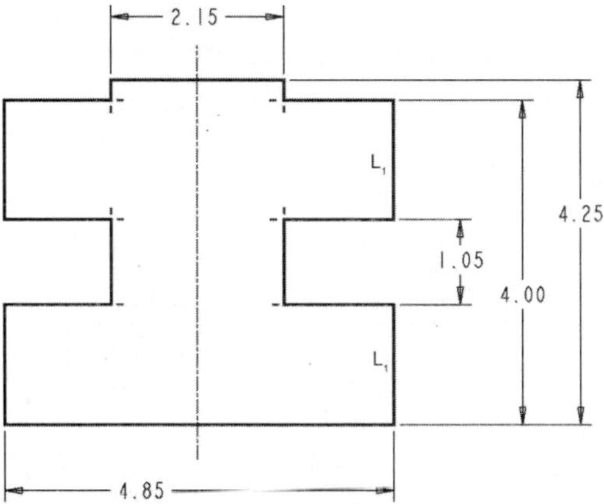

Figure 3–49 Final sketch

STEP 12: **Add perpendicular constraints to the entities shown in Figure 3–47.**

To add a constraint, select the Constraints icon on the sketcher toolbar, then select the Perpendicular constraint icon. Select the entities at the points shown in Figure 3–47.

STEP 13: **Add LINE-UP-VERTICAL constraints to the vertices shown in Figure 3–48.**

The Line-Up-Vertical constraint option will line up entity endpoints in a vertical orientation. Select each vertex as shown in the figure.

STEP 14: **Use the DIMENSION option to create the dimensioning scheme shown in Figure 3–49.**

STEP 15: **Use the MODIFY option to change dimension values to match Figure 3–49.**

The Modify option is used to change dimension values. You could also double select each dimension with the Pick icon.

STEP 16: **Save your section.**

STEP 17: **Select the Continue icon to exit the sketcher environment.**

PROBLEMS

Use Pro/ENGINEER's Sketch mode to create the following sections. The constraints and dimensions shown match design intent.

1. Problem 1

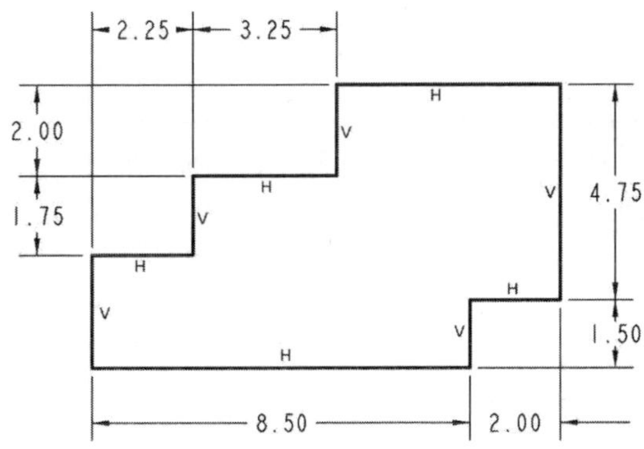

Figure 3-50 Problem one

2. Problem 2

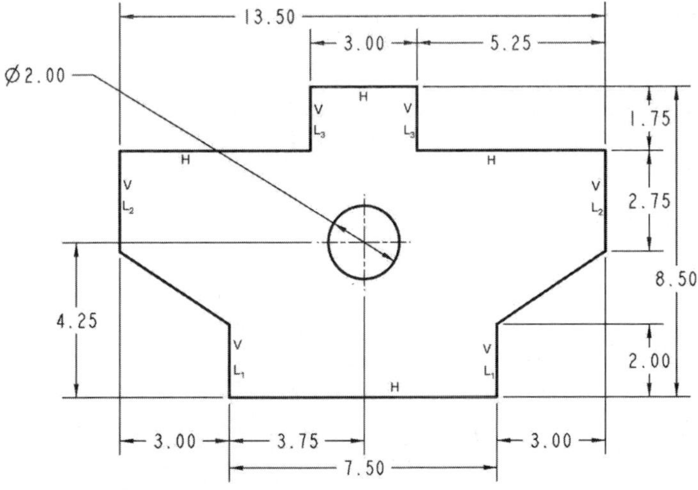

Figure 3-51 Problem two

3. From Problem 2, add a dimensional relationship that will make the circle centered horizontally within the base geometry. Within your relationship, your hole's location should be controlled by the 13.50 dimension.

4. Problem 4.

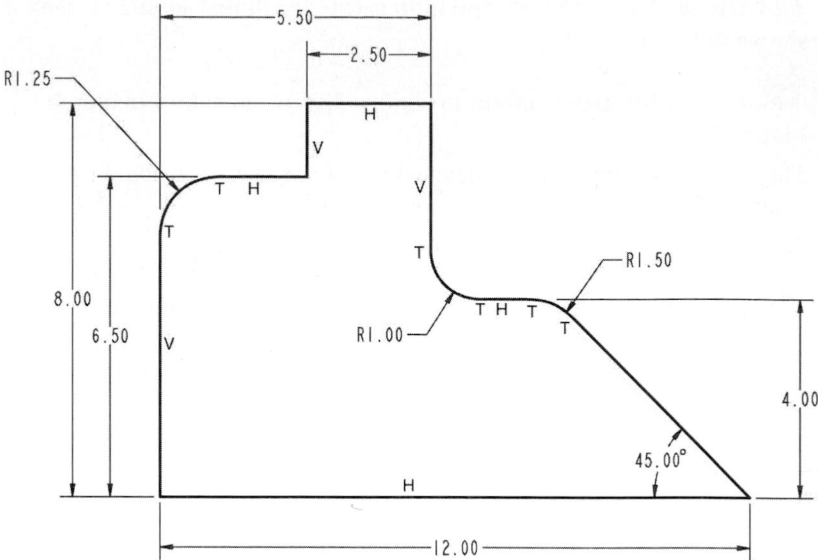

Figure 3-52 Problem four

Questions and Discussion

1. Describe ways of capturing the intent of a design within a parametric model.
2. What makes a suitable sketching surface within Pro/ENGINEER?
3. How does the application of constraints differ when Intent Manager is activated compared to when Intent Manager is not activated?
4. List possible constraints that can be applied within a Pro/ENGINEER section.
5. Describe the order of operations for sketching without Intent Manager.
6. Describe the order of operations for sketching with Intent Manager.
7. Write a valid relations equation for making dimension *sd1* equal to twice dimension *sd2*.
8. Within a sketcher environment, what is a feature reference?

CHAPTER

4

EXTRUDING, MODIFYING, AND REDEFINING FEATURES

Introduction

This chapter introduces extruded features, a concept associated with basic modeling fundamentals. Within Pro/ENGINEER, the Extrude option is common among the Protrusion and Cut commands. Additionally, this chapter will introduce the Redefine command, feature modification techniques, and datum construction. Upon finishing this chapter, you will be able to

- Model solid features as extruded protrusions.
- Remove material from features using extruded cuts.
- Modify feature dimension values using the Modify command.
- Modify feature definitions using the Redefine command.
- Create datum planes.

DEFINITIONS

Base feature The first geometric feature created in a part. It is the parent feature for all other features.

Child feature A feature whose definition is partially or completely referenced to other part features. A feature referenced by a child feature becomes a parent of this feature.

Cut A negative space feature created from a sketched section.

Definition A parameter of a part. An example of a definition of a hole feature would be the depth of the hole.

Protrusion A positive space feature created from a sketched section.

Negative space feature A feature created by removing material from a model. Examples of negative space features include holes, cuts, and slots.

Parent feature A feature referenced by another feature.

Positive space feature A feature created by adding material to a model. Examples of positive space features include protrusions, shafts, and ribs.

FEATURE-BASED MODELING

Parametric design packages are often referred to as feature-based modelers (Figure 4–1). A *feature* is a subcomponent of a part that has its own parameters, references, and geometry. *Geometry* is the graphic description of a feature. Geometry can be sketch-defined or

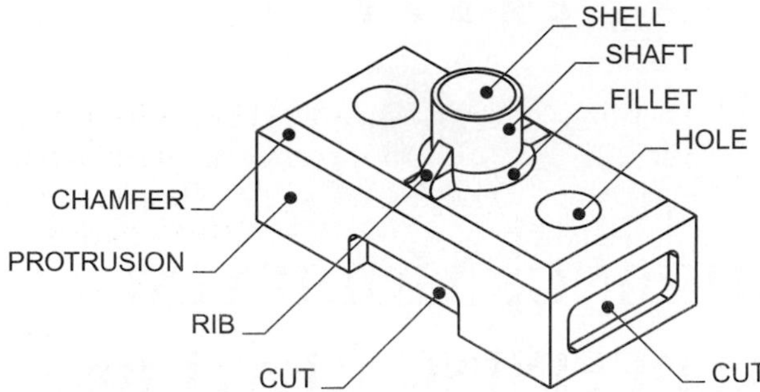

Figure 4-1 Features of a part

predefined. Sketch-defined features consist of sketched sections that are protruded or cut to form either positive or negative space. Predefined geometry has a common section such as a hole, round, or chamfer. *Parameters* are the dimensional values and definitions that define a feature. A hole may have a diameter of 1 inch and can be extruded completely through all existing features. The diameter is a parameter, as is the through-all definition. Parametric modeling packages allow users to modify parameters after the feature has been modeled. This is one of the unique properties that separate parametric modelers from boolean-based modelers. *References* are ways that features are related to other features in a part or assembly. Examples of references include axes, sketch planes, placement planes, reference planes, and reference edges. The surface of one feature may serve as the sketch plane for a second feature. The edges of the first feature may also serve as reference lines for parameters defining the second feature. In both examples, the first feature is a parent of the second feature.

PARENT-CHILD RELATIONSHIPS

Parametric models are composed of features that have established relationships. Features build upon other features in a way that resembles a family tree, hence the phrase *parent-child relationship*. Actually, a history tree of the relationships between features in a Pro/ENGINEER model resembles a web. The first feature created in a part is the center of the web and is the **parent feature** for all features. **Child features** branch off from the base feature and themselves become parent features. Unlike a typical family tree, a child feature may have several parent features.

Parent-child relationships can be established between features implicitly or explicitly. Implicit relationships can be established through the adding of a numeric equation using the Relations option. An example of this would be making two dimensions of equal value. In this process, one dimension governs the value of another. The feature with the governing dimension is the parent feature of the feature with the governed dimension. Care should be taken when modifying a feature that has a dimension that governs another. If a parent feature is selected for deletion, Pro/ENGINEER will provide an error message requesting an action to be accomplished to satisfy the void relationship. The user has the option of deleting, modifying, redefining, or rerouting the relationship.

Explicit relationships are created when one feature is used to construct another. An example would be selecting a plane of one feature as the sketch plane for a second feature. The new feature will become a child of the feature being sketched upon. Another similar example of an explicit relationship would be using existing feature edges within the sketcher environment to create a new feature. By specifying references while sketching, these selected references will create a relationship between the feature being sketched and the existing feature being referenced. The new sketched feature becomes a child of any referenced feature.

THE FIRST FEATURE

Pro/ENGINEER parts are composed of features. Determining what will be the first feature, or **base feature,** of a part is an important decision. Due to the parent-child relationships that exist between features, the first feature created will likely become the parent feature for all the features of a part. There are three possible first features of a part. The following is a discussion of each one.

DATUM PLANES

Datum planes are the recommended first features created for a new part. Datum planes can be used for future feature creation techniques. They are valuable within Assembly mode, and they make good reference features. Pro/ENGINEER has an option for creating a set of default datum planes. These three datum planes are created orthogonal to each other and are listed as the first three features on the model tree. By default, these default datum planes are named DTM1, DTM2, and DTM3.

PROTRUSION

The Protrusion command creates a geometric feature. Examples of protrusions include extruded, revolved, and swept features. When creating a Protrusion as the first feature, this feature will become the parent feature for all subsequent features in the part.

USER-DEFINED FEATURE

A User-Defined feature has been created by a user and saved to disk for use in later modeling applications. Several features can be combined in the creation process to form one grouped user-defined feature. When creating a part, a user-defined feature can be retrieved as the first feature.

STEPS FOR CREATING A NEW PART

Pro/ENGINEER follows an established procedure for starting a new part. This procedure is valid for most solid and thin protrusions. Listed below is a recommended sequence for starting a new part. Keep in mind these three assumptions when following these steps.

- It is assumed that Pro/ENGINEER's default part template file will be used. A part template is an existing file that has preexisting parameters, settings, and features. Upon first installation, the default part template includes Pro/ENGINEER's three default orthogonal datum planes (Front, Top, and Right) and default coordinate system. The default part template also comes with "inch pounds per second" as the model file's units. During the creation of a new part file, the default template can be changed by deselecting the Use Default Template option on the New dialog box. Additional template files come preexisting with Pro/ENGINEER, including an empty template file. The configuration file setting *template_solidpart* can be used to permanently set a default part template file.
- It is assumed that the first features created will be Pro/ENGINEER's default datum planes. This assumption affects two steps in the process. First, one of the steps in this sequence is the creation of the default datum planes. Second, establishing the sketcher environment is not necessary if the default datum planes are not available.
- It is assumed that a user wants to establish model parameters such as part material and units. A part can be created without entering the Set Up menu. As noted in the first assumption, Pro/ENGINEER's default template comes with units already established.

Perform the following steps to create a new protruded part.

STEP 1: Establish the correct working directory.

You can change Pro/ENGINEER's default working directory at anytime during the modeling process. When searching for a file, Pro/ENGINEER, by default, first looks in the working directory. Additionally, when saving a new part, Pro/ENGINEER saves the object in the current working directory. The Working Directory option can be found under the File menu.

INSTRUCTIONAL NOTE For more information on steps 1–3, see appropriate sections in Chapter 2.

STEP 2: Create a NEW object with Pro/ENGINEER's default template.

The File >> New option or the New icon on the menu bar will allow you to create a new object file. The New dialog box gives options for selecting a Pro/ENGINEER mode, with Part mode being the default. A file name can be entered, or the default name can be selected. File names must be limited to 32 characters (usually alphanumeric) or less and have no spaces.

STEP 3: Set up the Model.

Several options are available for setting up a part. Two common selections are part material and modeling units. Both part material and units can be set later in the modeling process if desired.

STEP 4: Select the Feature Creation Method.

For the first geometric feature of a part, creation tools are limited. For solid parts, protrusion is the primary tool available. Examples of protrusion options available include extruded features, revolved features, swept features, and blended features.

STEP 5: Establish the sketch plane.

There are several requirements for establishing a sketch plane. First, a plane has to be selected. Pro/ENGINEER's default datum planes provide three possible choices. Additionally, the Make Datum option can be used to create an on-the-fly datum plane. Other requirements might include, depending upon the feature creation method selected, extrusion direction and sketch orientation.

STEP 6: Sketch the feature's section.

Features created under the Protrusion and Cut commands require sketched geometric entities. Sketching can occur with or without Intent Manager activated. It is recommended that you take advantage of Intent Manager's sketching enhancements.

After sketching a feature, most Protrusion and Cut options require one or more additional definitions. As an example, extrusions require a depth definition.

STEP 7: Finish the feature.

Before finishing a feature, it is recommend that you Preview the results. Any definition set during the feature creation process can be changed through the Feature Definition dialog box (Figure 4–2).

MODELING POINT Often, when a mistake is made in the modeling process, it is tempting to cancel the creation process and start over. When making a mistake, or when skipping a modeling step, continue with the modeling process and modify the definition later using the Feature Definition dialog box.

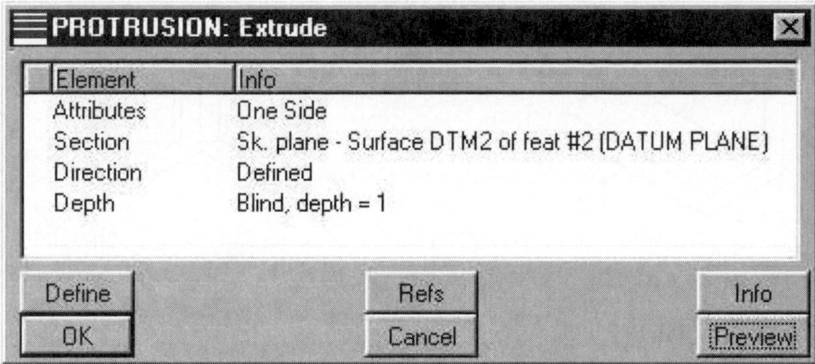

Figure 4-2 Feature definition dialog box

STEP 8: Perform File Management Requirements.

It is recommended that you save an object file after finishing a feature.
Other file management options available include Backup and Save As.
If a part is complete, purging old part files can be accomplished using the File >> Delete >> Old Versions option.

PROTRUSIONS AND CUTS

The procedures for performing a Protrusion and Cut in Pro/ENGINEER are virtually identical. The primary difference between the Protrusion command and the Cut command is that a **protrusion** is a **positive space feature,** while a **cut** is a **negative space feature.** When you protrude a feature, you actually create a solid object. With the Cut command, an extruded feature removes material from existing features.

The menu structure for both commands is similar. For both commands the following options exist:

> **INSTRUCTION POINT** Starting with Release 2000i, the Shaft, Slot, Neck, and Flange commands have been removed from Pro/ENGINEER's Solid menu by default. The Configuration File option *allow_anatomic_features* set to Yes will return these commands to the menu. See the Configuration File section in Chapter 2 for information on modifying this file.

EXTRUDE

The Extrude option sweeps a sketched section along a straight trajectory. The user draws the section in the sketcher environment and then provides an extrude depth. The section is protruded the depth entered by the user.

REVOLVE

The Revolve option sweeps a section around a centerline. The user sketches a profile of the revolved feature and a centerline from which to revolve. The user then inputs the degrees of revolution.

SWEEP

The Sweep option protrudes a section along a user-sketched trajectory. The user sketches both the trajectory and the section.

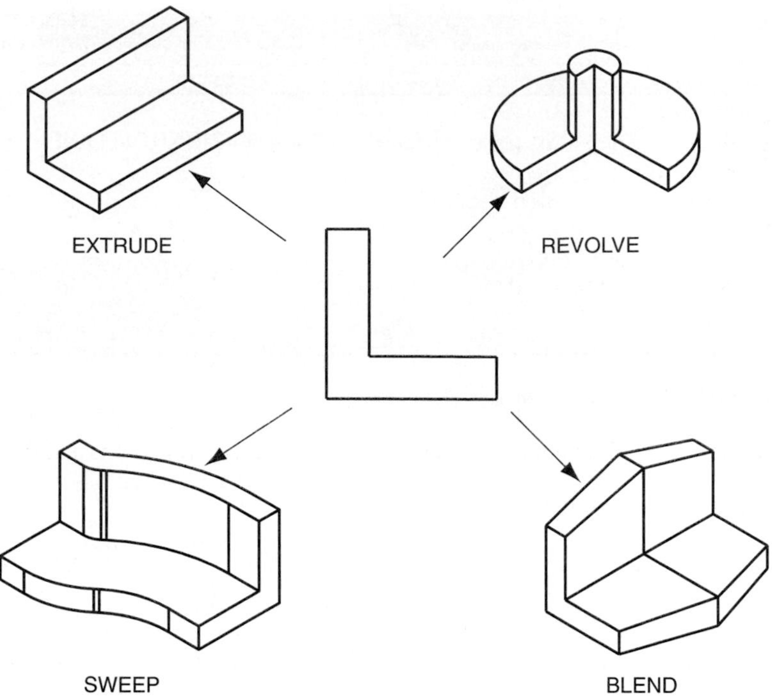

Figure 4–3 Variations in menu option features

BLEND

The Blend option joins two or more sketched sections. The trajectory may be straight or revolved.

USE QUILT

Quilts are patchworks of surfaces. The Use Quilt option turns a quilt into a solid feature.

ADVANCED

Common modeling options under the Advanced menu include Variable Section Sweep, Swept Blend, and Helical Sweep.

Shown in Figure 4–3 is an illustration of how one section can be used to create an Extrude, Revolve, Sweep, or Blend feature.

SOLID VERSUS THIN FEATURES

When creating a Protrusion or Cut, Pro/ENGINEER gives the option of choosing either a solid feature or a thin feature. Solid features are objects that are completely enclosed with material. Thin features are often confused with surfaces. In Pro/ENGINEER, surfaces are quilts with no defined thickness, whereas thin features are actually solids with a user-defined thickness. As shown in Figure 4–4, when the section is extruded as a solid, the section's feature is completely enclosed with material. When the section is extruded as a thin feature, the walls of the section are protruded with the provided wall thickness only.

Thin features can be used with all forms of the Extrude, Revolve, Sweep, and Blend options under the Protrusion and Cut commands. An example of an extruded thin cut is shown in Figure 4–5. The Thin option may be used with the Protrusion command for the base or secondary feature of a part or with the Cut command for secondary features.

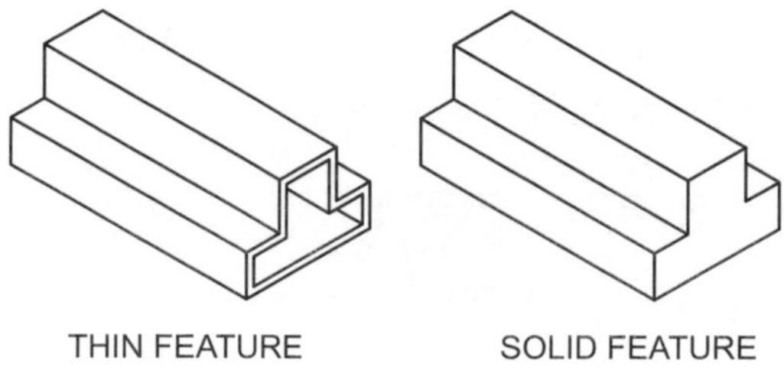

Figure 4-4 Thin versus solid features

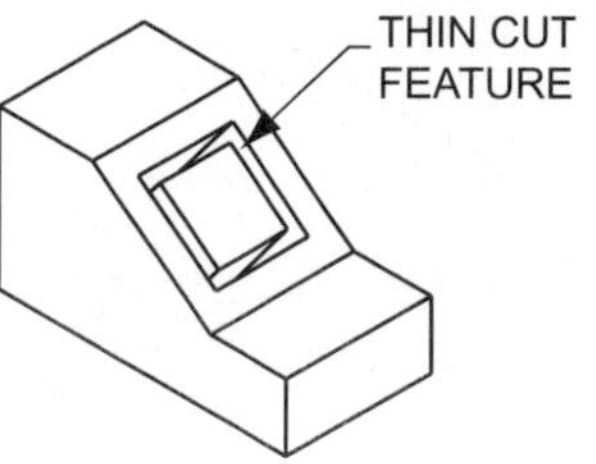

Figure 4-5 Thin cut feature

EXTRUDED FEATURES

The following section will explore options available within extruded Protrusions and Cuts.

EXTRUDE DIRECTION

When sketching on a plane, Pro/ENGINEER, by default, specifies an extrude direction. When sketching on a datum plane, this direction is in the positive direction. When sketching on an existing feature, a Protrusion, as shown in Figure 4–6, will be extruded away from the feature. Since the objective of a Cut is to remove material, a Cut will be extruded toward the feature.

The Extrude option gives the user the option of flipping the direction of extrusion or specifying an extrusion in both directions. The Both Sides selection under the Extrude option protrudes a section outward from the sketch plane in both directions. If the extrude depth is input to be 1 inch, the total extrusion will be 1 inch, not 1 inch in both directions. A typical Both Sides extrusion will divide the specified depth and extrude equally on both sides of the sketching plane. The 2 Side Blind depth option, though, allows the user to input unequal extrusion distances on both sides of the sketch plane.

> **MODELING POINT** Features are composed of definitions established by the user. A definition is not permanently set. It can be changed before finishing a feature or with the Redefine command. If a definition such as the extrude direction, depth option, or material removal side is incorrectly set, do not cancel or delete the feature. Redefine it later.

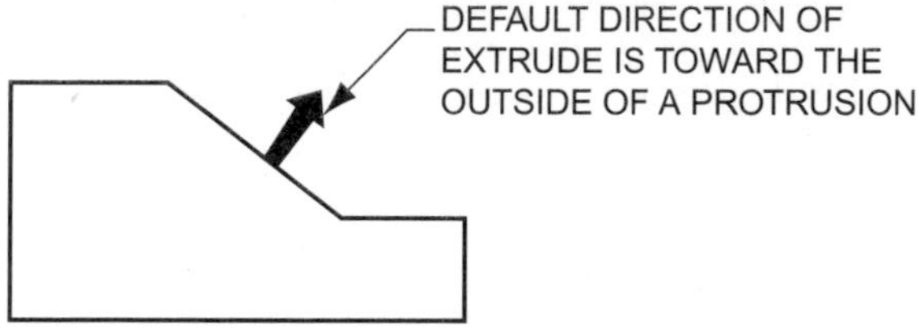

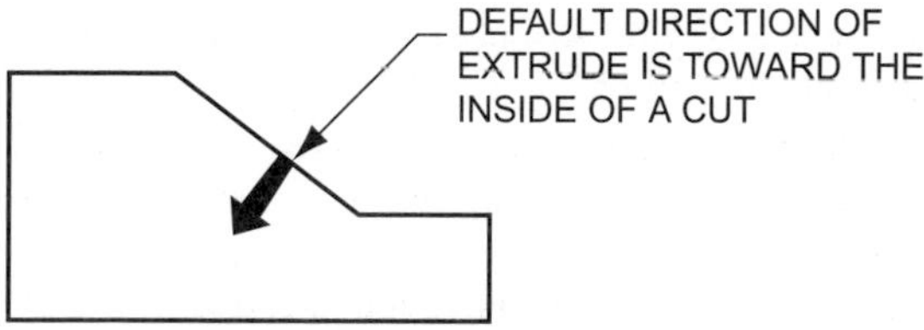

Figure 4-6 Default extrude direction

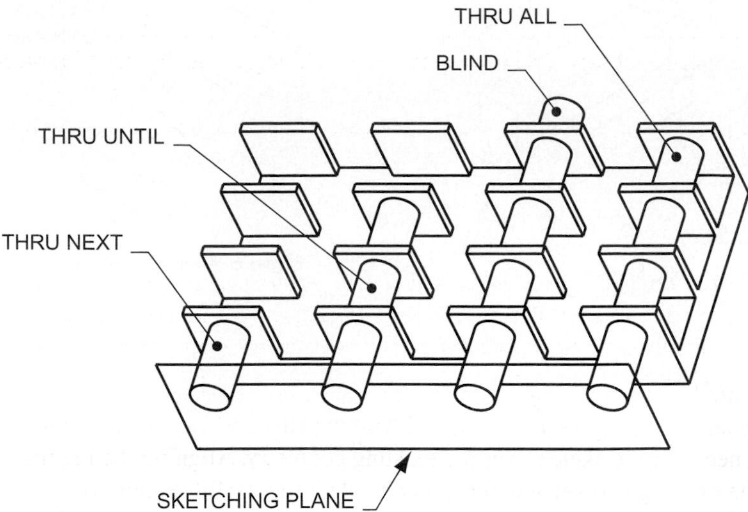

Figure 4–7 Depth options available for extruded features

Depth Options

For Extruded Protrusion and Cuts, an important parameter is the distance of extrusion. Pro/ENGINEER provides eight basic ways to specify an extrusion's depth. Four common depth options are shown in Figure 4–7. The depth for an extrusion is entered for a feature after exiting the sketching environment.

Blind

Blind is the simplest and most basic of the depth options. The Blind option allows a user to input an extrusion distance. It is the most common option for extruded base features.

2 Side Blind

A 2 Side Blind is used with a Both Sides direction option only. This depth option will allow the user to enter separate extrude depths for both sides of the sketch plane.

Thru Next

The Thru Next option extrudes a feature to the next part surface. Part geometry must exist prior to using this option.

Thru All

Thru All is one of the most common depth options for cut features. It extrudes a feature through the entirety of a part. The design intent of many material removal features (such as a Cut, Slot, or Hole) is to cut completely through a part. Entering a blind depth that will extrude through the part may not be adequate if the part thickness changes. The Thru All option adjusts for changing part dimensions. This option is available for parts with existing features and is not available for surface features.

Thru Until

The Thru Until option extrudes a feature until a user-selected surface. The surface can be any geometry, but cannot be a datum.

PNT/VTX

The Pnt/Vtx option extrudes a feature up to a selected datum point or vertex.

UpTo Curve

The UpTo Curve option extrudes a feature up to a selected edge, axis, or datum curve.

UpTo Surface

The UpTo Surface option extrudes a feature up to a selected surface.

Open and Closed Sections in Extrusions

Extruded sections can be sketched opened or closed. With the obvious exception of a base feature, many sections for an extruded Protrusion or Cut will suffice with an open section. The following are guidelines to follow when considering an open or closed section.

- Sections may not branch, and they can have only one loop. As shown in Figure 4–8, when sketching a section aligned with the edges of an existing feature, it often is not necessary to sketch over the existing geometry. Aligning the required sketch with the existing geometry will usually create a successful section. If Pro/ENGINEER is not sure which side of the section to protrude or cut, it will require the user to select a side (see Figure 4–9).
- Thin feature sections may be open or closed.
- For thin features, sections can be open when not aligned with existing geometry.
- Multiple closed sections can be included in a sketch. As shown in Figure 4–10, when a section is included within another, the inside section creates negative space.

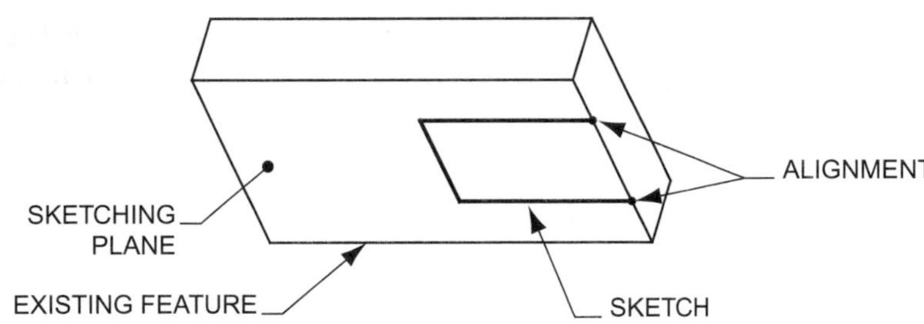

Figure 4–8 Aligning a sketch with existing part geometry

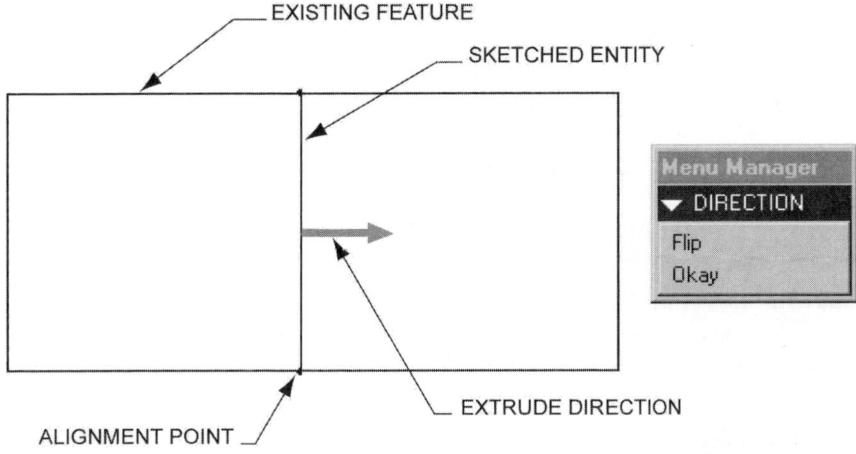

Figure 4–9 Selecting an extrude direction

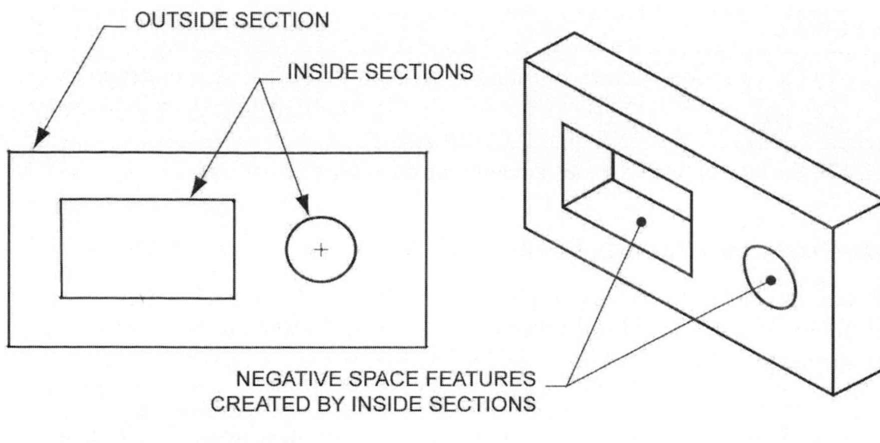

Figure 4-10 Creating negative space with included sections

MATERIAL SIDE

Two definitions are associated with Material Side. The most common is the Material Removal Side. Within the Extrude option, the Material Removal Side definition is relevant only to the Cut command. The Material Removal Side definition is used to specify what side of a section material will be removed from. By default, the removal side is toward the inside of a section. The user has the option of flipping the direction. Material Side definitions can be used with Protrusions when a sketch is an open section. Often, Pro/ENGINEER cannot determine which side of the sketch should be extruded. When this situation occurs, the user has to input the material side.

CREATING EXTRUDED FEATURES

Extrude is a common modeling technique for creating features within protrusions and cuts. An extrude is a section that is protruded along a straight line. It is a common option for creating primary and secondary features and is one of the fundamental skills needed to model in Pro/ENGINEER.

> **MODELING POINT** Pro/ENGINEER is a menu intensive modeling package. It is this menu structure that can seem to make Pro/ENGINEER difficult to learn. Actually, many feature creation techniques within Pro/ENGINEER follow similar procedures. Once these common procedures are understood, modeling becomes easier.

The following steps will outline the process for creating an Extruded Protrusion. Extruded Cuts and Slots are created with similar steps and options.

STEP 1: Select FEATURE >> CREATE >> SOLID >> PROTRUSION (or CUT).

Protrusions and Cuts are features. With Pro/ENGINEER, in most cases, when cutting or adding material, a feature is being created.

STEP 2: Select EXTRUDE as the solid creation option.

Extrude is a common option under the Protrusions and Cuts commands.

> **MODELING POINT** In Pro/ENGINEER, most extrusions can be performed by other options under the Protrusion command. As previously mentioned, an Extrude is a section swept along a straight line. The Sweep option could be used instead of Extrude. When modeling, it is best to use the simplest command available. The Extrude option is easier to perform than the Sweep command.

STEP 3: Select between a SOLID and THIN feature, then select DONE.

By default, the Solid option is selected. Unlike the solid option, thin features protrude or cut with a user-defined wall thickness. This wall thickness is defined after exiting the sketching environment.

STEP 4: Choose between ONE SIDE and BOTH SIDES, then select DONE.

This option will determine how a section will be extruded from the sketching plane. One side is the default choice. The Both Sides option will extrude the feature both directions from the defined sketching plane. The combined total depth extruded will be the blind depth entered by the user.

If you are creating the first feature of a part and you are not using Pro/ENGINEER's default datum planes, you will skip this step. Without a plane to sketch on, Pro/ENGINEER assumes that you will be extruding in one direction.

> **MODELING POINT** Often, design intent can be met better by extruding both directions from the sketching plane. Additionally, having a datum located in the middle of the part can enhance future modeling techniques, such as the Copy >> Mirror option.

STEP 5: Select a sketching plane and orient the sketcher environment.

Except for the base feature of a part, sections have to be sketched on a plane. A plane can be an existing planar feature surface, an existing feature plane, an existing datum plane, or an on-the-fly datum plane.

If you are creating the first feature of a part and you are not using Pro/ENGINEER's default datum planes, you will skip this step. Using a protrusion as a part's base feature is the only time that Pro/ENGINEER does not require you to sketch on a plane.

> **MODELING POINT** The first feature of a part can be a Protrusion, Datum Plane, or a User-Defined Feature. It is recommended that you create Pro/ENGINEER's default datum planes when starting a new model and use them as the base feature. Datum planes can be useful with many modeling and assembly techniques. When creating a Protrusion as the base feature, the user is not given the option of selecting a sketching plane.

STEP 6: Sketch the section.

Sketch the section according to design intent. The following is the order of operations for sketching a section with Intent Manager activated.

1. Specify references.
2. Sketch section.
3. Place dimensions according to design intent.
4. Modify dimensional values.
5. Add relations.

See Chapter 3 for more information on sketching.

STEP 7: Select the Continue icon when the section is complete.

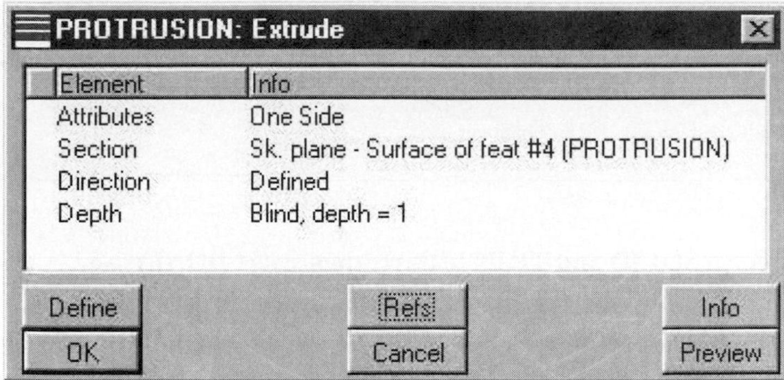

Figure 4–11 Protrusion feature definition dialog box

STEP 8: Specify a DEPTH definition.

For a base protrusion, the most practical depth option is a blind depth. Remember design intent when selecting a depth option.

STEP 9: Preview the feature.

The Preview option is located on the Feature Definition dialog box (Figure 4–11).

MODELING POINT At this point, you can still modify any definitions established for the feature. Definition parameters are shown on the Feature Definition dialog box. Use the Define option to modify any definitions, including the sketch.

STEP 10: Finish the feature.

Select OKAY on the Feature Definition dialog box to finish the feature.

DATUM PLANES

Datum planes are used as references to construct features. Datum planes are considered features, but they are not considered model geometry. When a datum plane is created, the datum plane will show as a feature on the model tree. Datum planes can be created and used as a sketch plane where no suitable one currently exists. As an example, a datum plane can be constructed tangent to a cylinder. This will provide a sketching environment that can be used to construct an extruded feature through the cylinder (Figure 4–12). When a feature is sketched on a datum plane, the datum plane is considered a parent feature.

A datum plane continues to infinity in all directions, with one side yellow and the opposite side red. Protrusions and orientations occur initially toward the yellow side of the datum plane. By default, datum planes are named in sequential order starting with DTM1. As a note, template files may have datum planes that have been renamed (e.g., Front, Top, and Right) with the Set up >> Name option.

Many times, a datum plane would be useful, but cluttering the model with additional features could be detrimental. One solution to this problem is to make a datum plane that is used for single feature creation only. To do this, datum planes can be created on-the-fly using the Make Datum option. Datum planes created on-the-fly belong to the feature undergoing creation. These datum planes do not show on the model tree and become invisible after the feature is created.

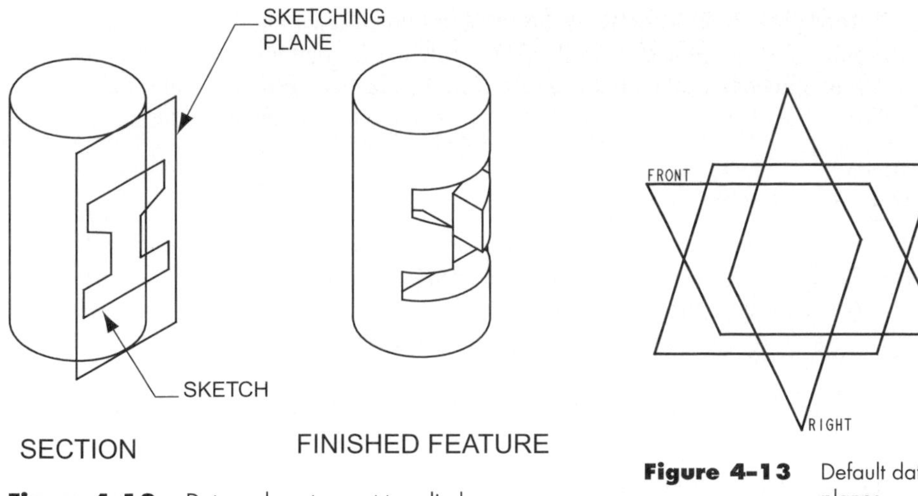

Figure 4–12 Datum plane tangent to cylinder

Figure 4–13 Default datum planes

Creating Pro/ENGINEER's Default Datum Planes

Datum planes are commonly used as the first feature of a part or assembly. Pro/ENGINEER provides a way to create three orthogonal datum planes, referred to as the *default datum planes* (Figure 4–13). When utilizing an empty template file, selecting the datum plane icon will automatically create Pro/ENGINEER's default datum planes.

Creating Datum Planes

A datum plane can be created at any point in the modeling process. One of the primary uses of a datum plane is as a sketching surface. Datum planes can be used as a mirror plane within the Copy command, or they can be used as references when sketching a feature. Datum planes are powerful features for aligning and mating parts within Assembly mode. Additionally, datum planes can be combined with other features using the Group option and then patterned.

Creating datum planes is a vital skill needed by all Pro/ENGINEER users. Several constraint options exist for datum plane definition. Some constraint options are stand-alone, while some are not. Paired constraint options are not stand-alone. They require two or more constraint definitions during the datum plane construction process. As an example, the Angle option requires the selection of an existing plane from which to reference the angle, then a Through constraint option to pass the datum plane through an axis or edge. Stand-alone constraint options require one option only.

Stand-Alone Constraint Options

Through >> Plane

The Through >> Plane constraint option creates a datum plane that passes through an existing part plane.

Offset Plane

The Offset Plane constraint option creates a datum plane that is offset from an existing plane. The user selects the plane from which to offset. Two offset options are available:

- **Point** Select a point to pass the datum plane through. The plane will be created through the point and parallel to the existing part plane.

- **Enter value** Enter an offset distance value. When selecting this option, the user is prompted to enter the offset value. An arrow in the graphics screen shows the default direction of the offset. To offset in the opposite direction, enter a negative value. The plane will be created offset from the reference plane at the value entered.

Offset/Coord Sys

The Offset Coordinate System constraint option creates a datum plane offset from the coordinate origin and normal to a selected coordinate axis. A coordinate system has to exist prior to the use of this option.

Blend Section

The Blend Section constraint option creates a datum plane through a section used to create a feature.

Paired Constraint Options

Through >> AxisEdgeCurv

This option is similar to the Through >> Plane option, except this option places a plane through an axis, edge, or curve. An axis, edge, or curve selected with the Through option will not fully constrain a datum plane; hence, a second constraint option is required.

Through >> Point/Vertex

This option places a datum plane through a point or vertex. Similar to the Through >> AxisEdgeCurv option, this option will not fully define a datum plane and needs an additional constraint option.

Normal >> AxisEdgeCurv

This option places a datum plane perpendicular to an axis, edge, or curve. As with the Through >> AxisEdgeCurv option, an additional constraint is needed.

Tangent >> Cylinder

This constraint option places a datum plane tangent to a hole or cylindrical surface. This is an extremely useful option since it allows a feature to be constructed on the surface of a cylinder. This constraint option is often paired with the Normal >> Plane option or the Angle >> Plane option.

Angle >> Plane

This option places a datum at an angle to an existing plane. It is often paired with the Through option, or the Tangent >> Cylinder option.

On-the-Fly Datum Planes

Datums created with an option on the datum toolbar are considered part features. Datums, especially datum planes, can clutter a part and make viewing and selecting features difficult. Datums intended for use as a modeling tool for individual features can be created on-the-fly with the Make Datum command. On-the-fly datum planes are available only for the feature in which they are intended. They are not considered part features and will not appear on the model tree. Additionally, when the feature creation modeling process is finished, the datum will disappear from the work screen.

Modifying Features

What separates parametric modeling packages, such as Pro/ENGINEER, from boolean-based modeling packages are their feature modification capabilities. Features created within Pro/ENGINEER are composed of parameters. Examples of parameters include parametric dimensions, extrude depth, and material side. Parameters such as these are established during feature construction. These feature parameters and other feature definitions such as a section's sketch and sketch plane, can be modified later in the part modeling process.

As shown in Figure 4–14, a variety of feature modification options can be found under the Modify menu.

Figure 4–14 Modify options

Dimension Modification

Parametric dimensions are used to define a feature. They can be modified at any time. Modifying a dimension value is the most common dimension modification function, but other modification tools exist. The number of decimal places in a dimension can be modified along with the tolerance format. The following are dimension modification techniques that are available within Pro/ENGINEER.

Modifying a Dimension Value

The value of a parametric dimension value can be modified. To modify a dimension, select the Value option from the Modify menu, then select the feature associated with the dimension to be modified. Select the dimension to modify and enter a new value. Modifying a dimension value requires the regeneration of the part. Select Regeneration from the Part menu.

MODELING POINT To display tolerances for all objects created within Pro/ENGINEER, the configuration file option *tolerance_mode* must be set to a value (such as limit) and the option *tolerance_display* must be set to Yes. To display tolerances for individual objects, select Tolerance Display on the Environment dialog box.

Modifying a Tolerance Mode

Dimension tolerances can be displayed in a variety of modes. To modify the tolerance mode for individual dimensions, access the Dimension Modification dialog box by selecting Dimension under the Modify menu. As shown in Figure 4–15, an option exists on the Dimension Properties dialog box that allows for the changing of a tolerance mode.

Modifying Tolerance Values

The Value option under the Modify menu allows for the modification of tolerance values. As an example, if a dimension is set to Limits as the tolerance mode, either the upper or the lower dimension value can be changed. A problem with this approach is that the nominal value of the limit dimension cannot be modified. Another approach is to modify the nominal value and/or the tolerance values with the Dimension Properties dialog box (Figure 4–15). To access this dialog box, select Dimension from the Modify menu.

Dimension Decimal Places

Initially, dimension decimal places are set to two. This value can be changed permanently with the configuration file option *default_dec_places*. To change decimal places for individual dimensions, access the Dimension Properties dialog box by selecting Dimension from the Modify menu. As shown in Figure 4–15, an option exists for changing the decimal places of a dimension.

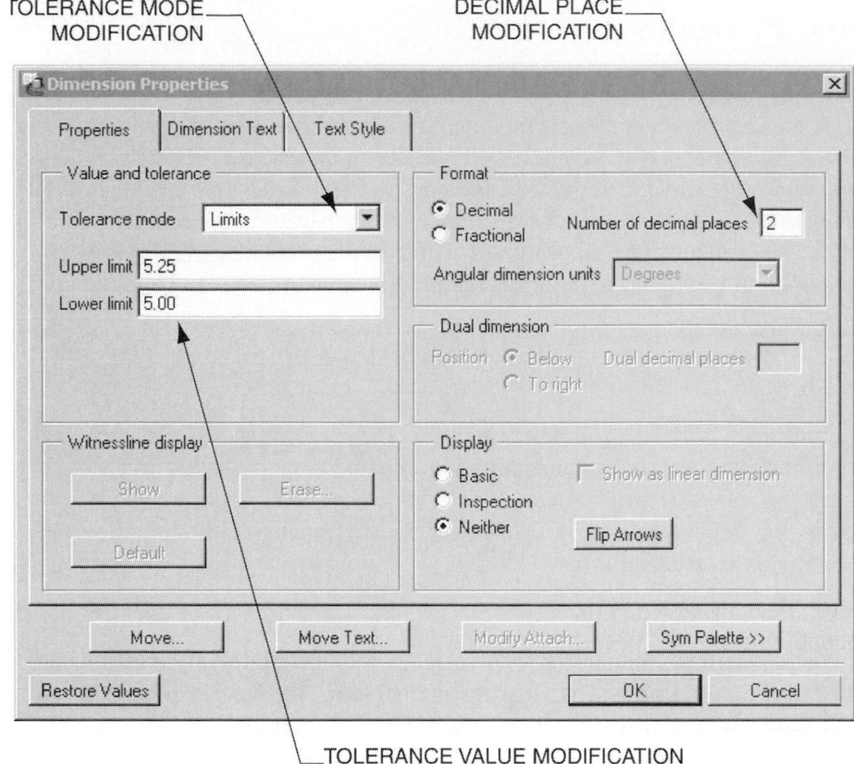

Figure 4–15 Modify dimension dialog box

Cosmetic Dimension Modification

The DimCosmetics (Dimension Cosmetics) option located under the Modify menu provides tools for modifying dimensions. A dimension's cosmetics are the way that the dimension appears on the object. Modifying a dimension's cosmetics does not modify the value of the dimension. The following are techniques that the DimCosmetics option provides for modifying dimensions.

Tolerance Format

A tolerance's format mode can be changed using the Format option found under the DimCosmetics menu. Select the dimensions for format change, then select the desired format to use.

Number of Digits of a Dimension

The number of decimal places displayed by a dimension can be changed using the Num Digits option found under the DimCosmetics menu. To change the number of decimal places for dimensions, select Num Digits, then enter the number of significant digits. Follow this by selecting the dimensions to change.

Adding Text Around a Dimension

Text can be added around a dimension value. A dimension's value is shown in a dimension note with the symbol @D. This symbol must remain in the dimension note. To add text around a dimension value, select Text from the DimCosmetics option then select the dimension to modify. Follow this by entering lines of text.

Changing Dimension Symbols

Dimensions within Pro/ENGINEER can be displayed in two ways. The first way is by showing the actual dimension or tolerance value. The second way is by showing the dimension symbol. As shown in Figure 4–16, when a dimension is created, it is

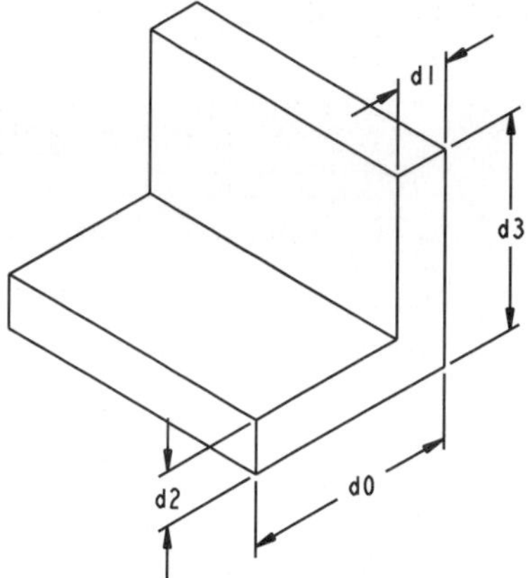

Figure 4–16 Dimension symbols

provided with a dimension symbol. For the first dimension created on a part, the default symbol is d0. This number increases in sequential order for every new dimension. To change a dimension's symbol to make it more descriptive, utilize the Symbol option from the DimCosmetrics menu.

REDEFINING FEATURES

Features are composed of parameters. Parameters can be modified and changed through the Redefine command or through the Modify menu. Many different varieties of parameters exist. The following is a partial list.

- A feature's section.
- The sketch plane for a feature.
- The depth option for a feature.
- The material removal side.
- The extrude direction of a feature.
- The trajectory of a swept feature.
- The value of a dimension.

Figure 4–17 shows an example of a Feature Definition dialog box for an extruded protrusion. This dialog box is accessible at the end of the construction of a part feature or it can be accessed through the Feature menu's Redefine command. Perform the following steps to redefine a feature parameter.

STEP 1: Select FEATURE >> REDEFINE.

STEP 2: From the work screen or on the Model Tree, select the feature to be redefined.

After selecting the feature to redefine, the Feature Definition dialog box associated with the feature will appear (Figure 4–17). This dialog box is composed of two columns. The first column displays the definition's name, while the second column displays the definition's current value.

STEP 3: On the Feature Definition dialog box, select the definition to redefine.

Select the definition's name displayed in the first column of the dialog box.

STEP 4: Select DEFINE on the dialog box.

After selecting Define, Pro/ENGINEER will step you through the remodeling of the parameter.

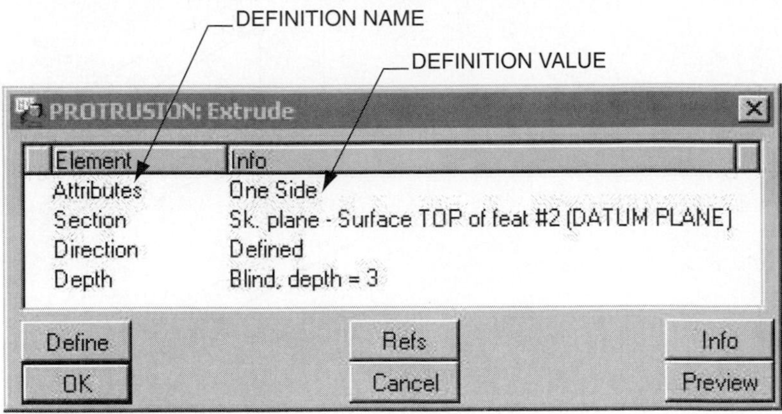

Figure 4-17 Feature definition dialog box

STEP 5: Redefine the parameter according to Pro/ENGINEER modeling procedures.

STEP 6: View the feature's new parameters by selecting PREVIEW on the dialog box.

STEP 7: Select OKAY on the dialog box.

SUMMARY

Pro/ENGINEER is often found difficult to use. Part of the reason is the menu intensity required to create a feature. It is true that several menu selections are required to create a feature, but many of these selections are common throughout different types of features. As an example, the Protrusion and Cut options have basically the same menu structure and suboptions. One of the keys to learning Pro/ENGINEER is to become familiar with these options and to adapt them to new modeling situations.

TUTORIAL

EXTRUDE TUTORIAL

This tutorial exercise provides step-by-step instruction on how to model the part shown in Figure 4–18.

This tutorial will cover

- Starting a new model.
- Setting default datum planes.
- Creating an extruded protrusion.
- Creating an extruded cut.
- Dimension modification.
- Redefining a feature.
- Saving a part.

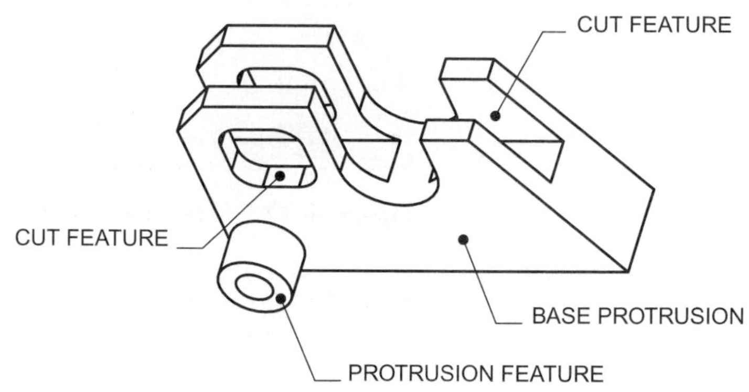

Figure 4–18 Finished part

STARTING A NEW MODEL

This segment of the tutorial explores the starting of a new part file.

STEP 1: Start Pro/ENGINEER.

STEP 2: Use the FILE >> SET WORKING DIRECTION option to establish a working directory for your part file.

Pro/ENGINEER utilizes a working directory to help manage files. When a new file is saved, it is saved in the current working directory, unless a new directory is specified. When the Open command is executed, the default directory is the current working directory.

STEP 3: Use the FILE >> NEW option to create a new part file.

From the File menu, select the New option. Part is one of the modes of Pro/ENGINEER, and is the default mode. Enter a name for the new part. Part names must be less than 31 characters and cannot include spaces.

Notice on your New dialog box how the Use Default Template option is checked (see Figure 4–19). This setting will use a specific part file as a seed file for the new object. Pro/ENGINEER's initial default template file for a new part includes a set of three default datum planes and the default coordinate system. The part file's default units is set at *Inch lbm Second.* This book assumes you will use this template file. The default template file for a part can be changed with the configuration file option *template_solidpart.*

STEP 4: Enter a name for the new part file, then select OKAY on the New dialog box.

STEP 5: Select the SET UP >> UNITS option on the main menu.

STEP 6: On the Units Manager dialog box, select *Inch Pound Second (IPS),* then select the SET icon.

STEP 7: On the Warning dialog box, select the CONVERT EXISTING NUMBERS option, then select OKAY.

STEP 8: Close the Units Manager dialog box.

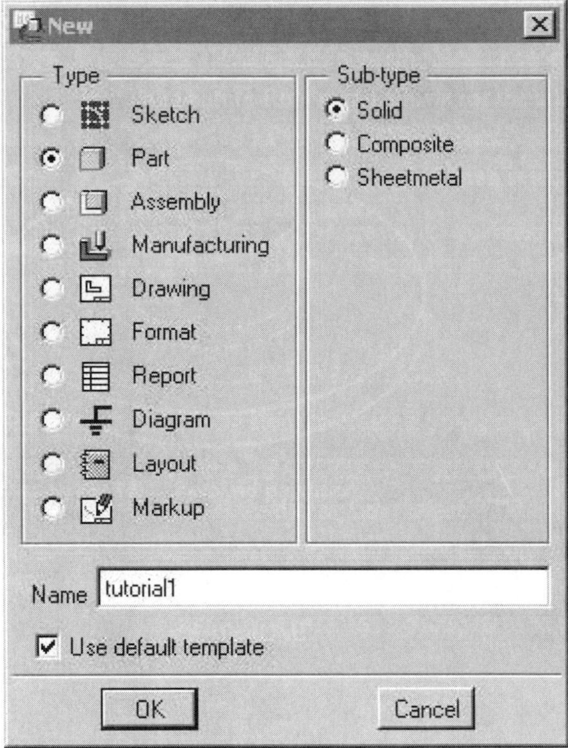

Figure 4–19 New dialog box

Creating an Extruded Protrusion

Protrusion features are the most common positive space feature found in a Pro/ENGINEER solid model. Protrusions can consist of extruded, revolved, swept, or blended features. This segment of the tutorial will extrude a sketched section.

Step 1: Select the FEATURE >> CREATE >> PROTRUSION command option.

Protrusions are the most common type of geometry feature first created in a solid part.

Step 2: Select the EXTRUDE >> SOLID >> DONE option.

Extrude is one of the types of protrusions that can be created (see Figure 4–20). Other options available include Revolve, Sweep, Blend, Helical Sweep, and Swept/Blend. You will be creating a solid feature in this exercise. Solid is selected by default in Pro/ENGINEER.

Step 3: Select ONE SIDE >> DONE as an Extrude option (Figure 4–20).

This exercise will extrude the section in one direction from the sketching plane. Features may be extruded both directions by selecting Both Sides.

Step 4: On the work screen, select Datum Plane FRONT as the sketching plane.

This extruded section will be sketched on datum plane FRONT. You can select the label associated with this datum (FRONT), you can select any portion of the boundary of the datum, or you can select the datum on the model tree.

Step 5: Select OKAY to accept the feature creation direction.

After selecting the sketching plane, the red arrow shown on the work screen points from the datum plane in the positive direction. For one-sided extrusions, this is the direction of feature creation.

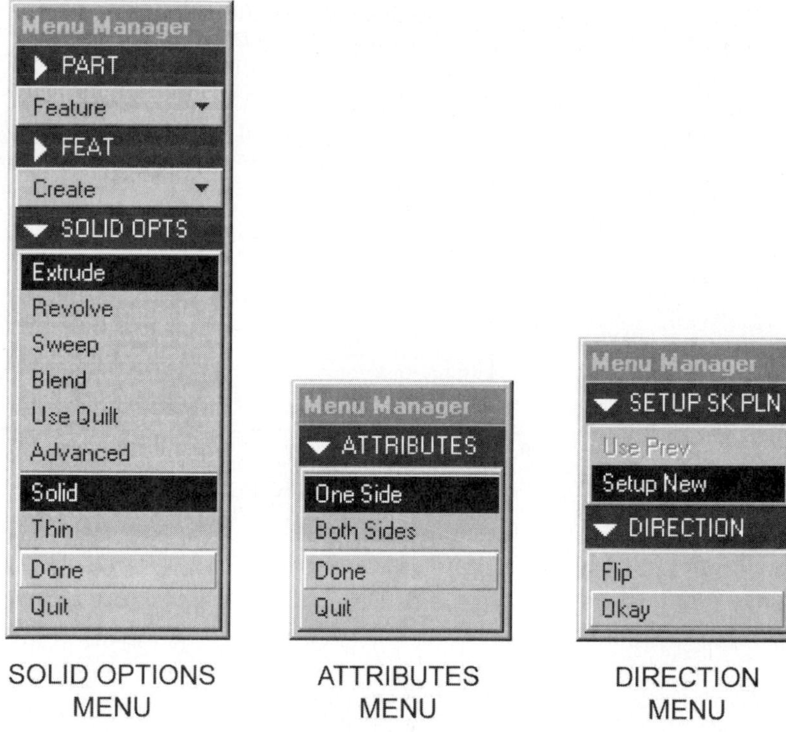

Figure 4–20 Menu options

STEP 6: To orient the sketcher environment, select TOP on the Sketch View menu, then select datum plane TOP.

You can orient your model in two different ways. The first method is by selecting Default. With default, Pro/ENGINEER determines the direction to orient the sketcher environment. The second method is to select a direction of orientation followed by a plane to face in the direction of orientation. If you select to use the latter method, select Top from the sketch view menu, then select datum plane TOP. This will orient datum plane TOP toward the top of the work screen.

SKETCHING THE SECTION

Extruded Protrusion features require a sketched section. Most extruded Protrusion features have a closed section. This segment of the tutorial will provide instruction on the sketching of the section for the extruded feature.

STEP 1: Close the References dialog box.

References are used by Pro/ENGINEER to pass design intent from parent features to child features. Pro/ENGINEER will automatically select a minimum number of references to allow for the sketching of a feature. You can use the References dialog box to add additional references or to delete references. In this example, datum planes RIGHT and TOP have been selected as references.

STEP 2: Use the LINE icon to sketch the section shown in Figure 4–21.

In Figure 4–21, the dimensioning scheme defining the size of the sketch has been purposely hidden. When sketching, you should not worry about the size of your sketch. What is important is the sketching of geometry that

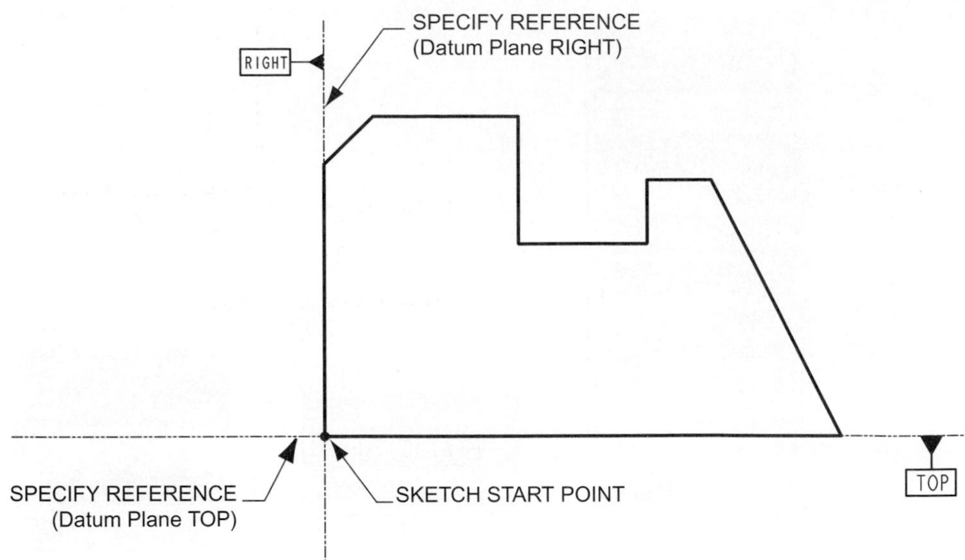

Figure 4–21 Sketched section of feature

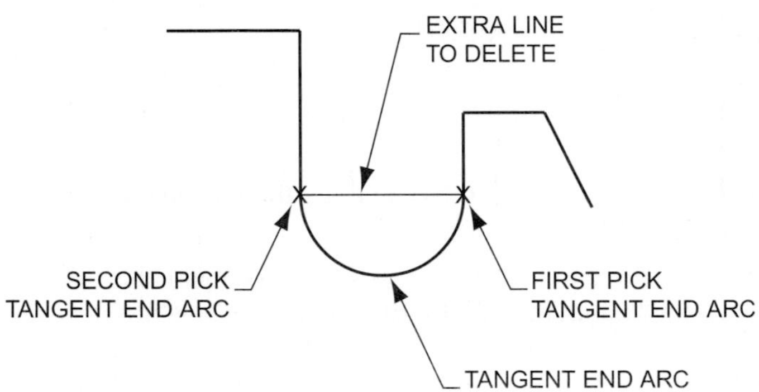

Figure 4–22 Tangent end arc

matches the shape of your design intent. Start your sketch at the intersection of datum planes RIGHT and TOP, sketching in a counterclockwise direction. The left mouse button is used to pick entity locations, and the middle mouse button is used to cancel a command. Align the bottom edge of the feature with datum plane TOP, and align the vertical edge of the sketch with datum plane RIGHT.

STEP 3: **Use the ARC icon option to add a Tangent Arc to the sketch (see Figure 4–22).**

STEP 4: **Delete the extra line shown in Figure 4–22 by first selecting the entity with the Pick icon, followed by selecting the delete key on your keyboard.**

STEP 5: **Use the CIRCULAR FILLET icon to add a Filleted Arc to the sketch.**

Add an additional arc as shown in Figure 4–23. The Circular Fillet option will require you to select two nonparallel entities. A fillet will be created between the selected entities.

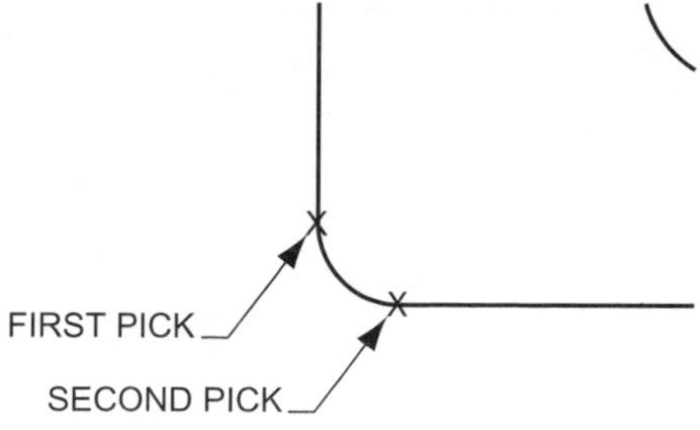

Figure 4–23 Filleted arc

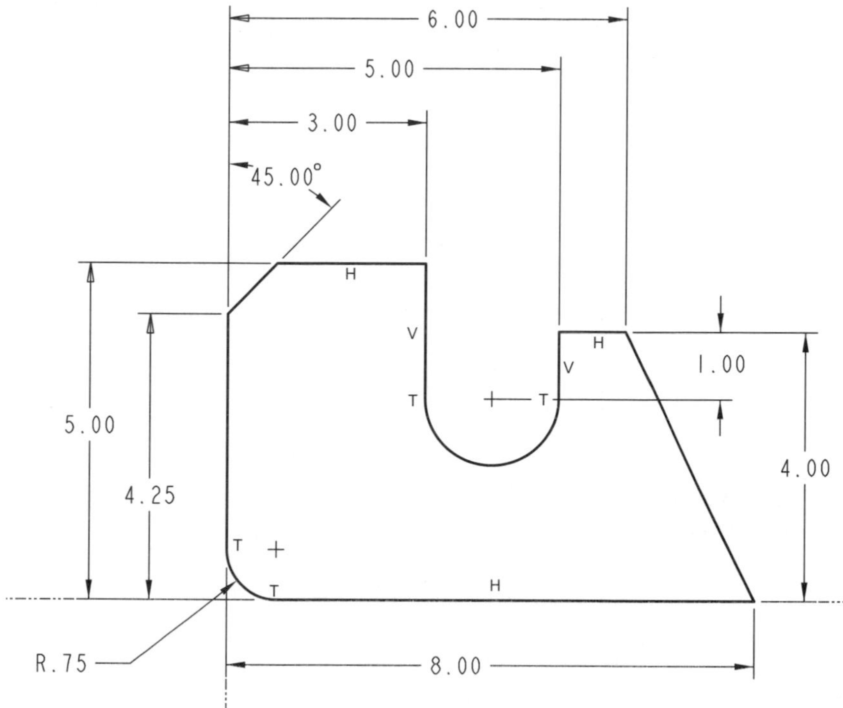

Figure 4–24 Dimensioning scheme

STEP 6: **Place Dimensions according to Design Intent.**

Use the Dimension icon to match the dimensioning scheme shown in Figure 4–24. Placement of dimensions on a part should match design intent. With Intent Manager activated (Sketch >> Intent Manager), dimensions and constraints that fully define the section are provided automatically. Pro/ENGINEER does not know what dimensioning scheme will match design intent, though. Due to this, it is usually necessary to change some dimension placements.

MODELING POINT If possible, a good rule of thumb to follow is to avoid modifying the section's dimension values until your dimension placement scheme matches design intent.

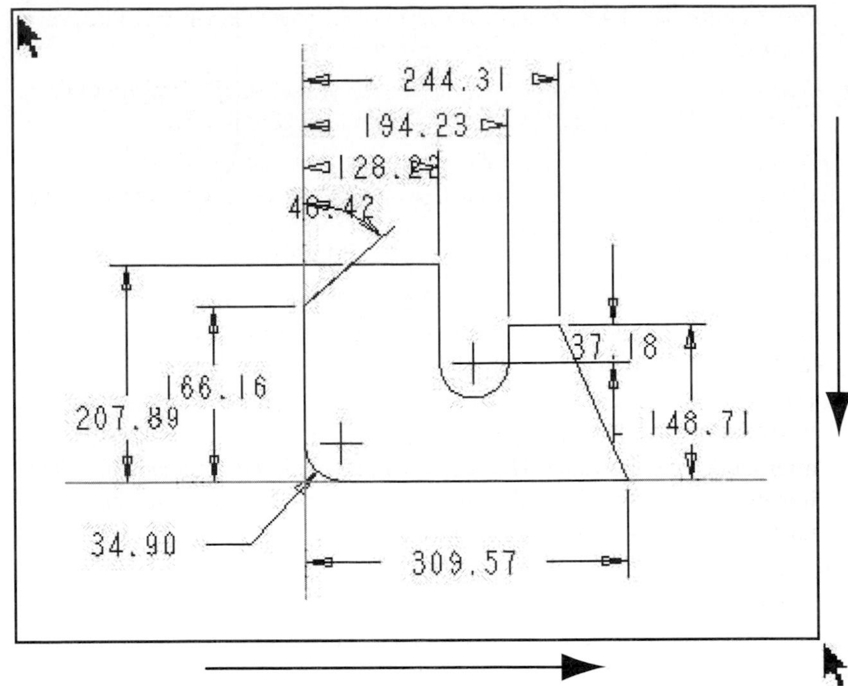

Figure 4–25 Select dimension using the pick icon

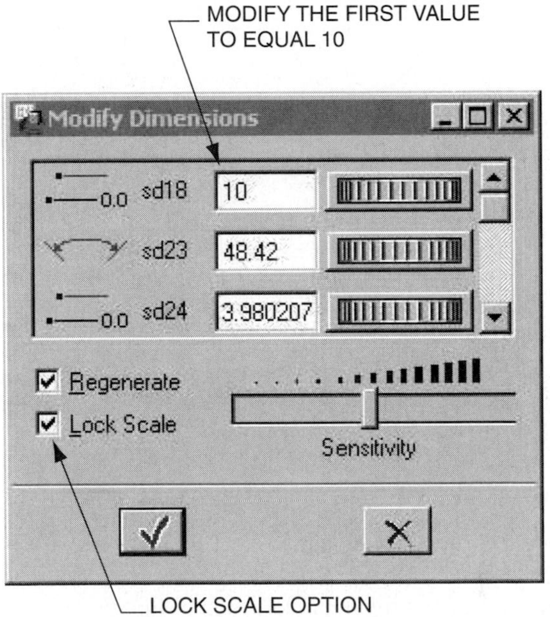

Figure 4–26 Dimension scale modification

STEP 7: **Using the Pick icon, drag a box around the entirety of the sketch making sure to include all dimensions (see Figure 4–25).**

Your next task is to modify the sketch's dimension values. By preselecting all available dimensions, you will be able to simultaneously modify each dimension.

STEP 8: **Select the Modify icon, then check the Lock Scale option (Figure 4–26).**

The Lock Scale option will allow you to modify one dimension, with the remaining dimensions scaling the same factor.

STEP 9: Modify the first dimension value to equal 10, select ENTER on your keyboard, then select the CHECK icon on the dialog box.

At this point, it does not matter which dimension you modify to 10 units, though it is best to select the largest dimension. Your objective is to scale the model down to a workable size. The next step of this tutorial will require you to modify each dimension individually.

STEP 10: Using the Pick icon, double select the arc's radius dimension value then modify its value to equal .750.

Double picking a dimension value will allow it to be modified through Pro/ENGINEER's textbox. Remember to select Enter on your keyboard after modifying the value. The sketch will be modified automatically. Often, it is best to modify smaller dimension values first. If your sketch has unexpected results, select the Undo icon on the toolbar.

STEP 11: Starting with the smallest dimension values first, modify the remaining dimensions to match Figure 4–24.

Use the Undo icon on the toolbar to undo unexpected results.

> **MODELING POINT** Sketched protrusion sections have to be completely enclosed, with no intersections. If a section has intersecting entities or open geometry, an open loop error will occur.

STEP 12: Select the Continue icon to exit the sketching environment.

FINISHING THE FEATURE

The following steps define the process for finishing the creation of a feature.

STEP 1: Select BLIND >> DONE as the Depth option then enter 2.00 as the Depth value.

STEP 2: Preview the feature.

On the Feature Definition Dialog Box (Figure 4–27), select Preview to observe your feature.

STEP 3: Dynamically view your part.

Shade your part by selecting the Shade icon from the toolbar menu. Dynamically view your part using the following options:

- **Dynamic rotation** Control key and middle mouse button.
- **Dynamic pan** Control key and right mouse button.
- **Dynamic zoom** Control key and left mouse button.

STEP 4: Select OKAY to finish the feature.

Selecting OKAY will create the feature. Observe the Model Tree to see the new feature (Figure 4–28).

STEP 5: Save your part file.

Use the File >> Save option to save your part file.

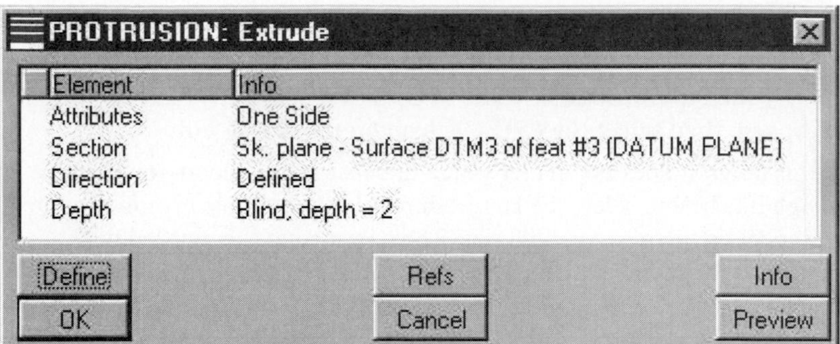

Figure 4–27 Feature definition dialog box

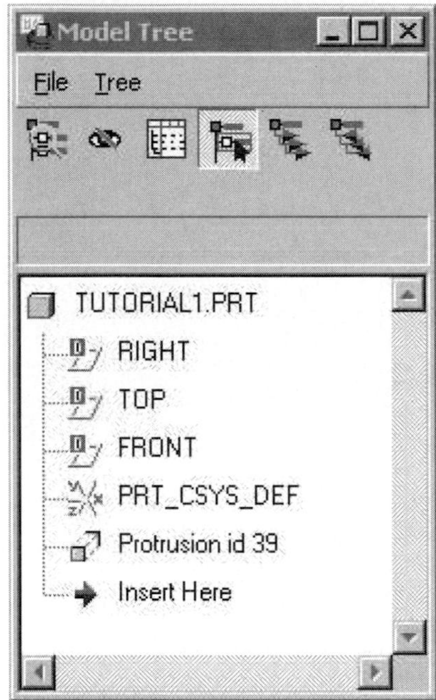

Figure 4–28 New feature added to model tree

Creating an Extruded Cut

This segment of the tutorial will create the extruded Cut feature shown in Figure 4–29. Additionally, the Define option will be used to modify the sketch.

Step 1: Select FEATURE >> CREATE >> CUT.

The Cut option is a common way to remove material from existing features.

Step 2: Select the EXTRUDE >> SOLID >> DONE option.

Extrude >> Solid is the default setting for Pro/ENGINEER. Select Done to continue with the command sequence.

Step 3: Select ONE SIDE >> DONE as the extrude attribute for the feature.

The One Side option will extrude the section one direction from the sketching plane.

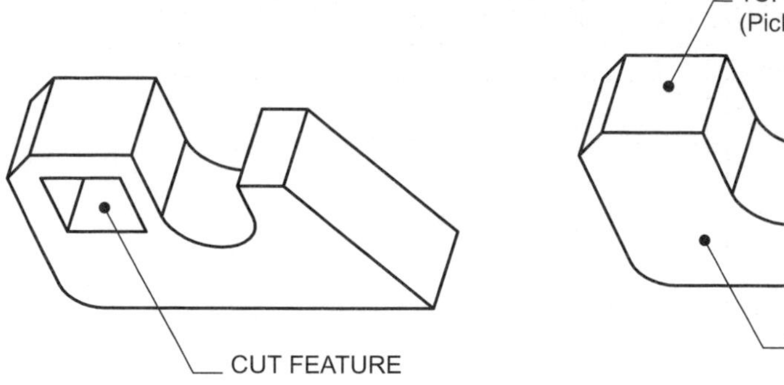

Figure 4–29 Cut feature

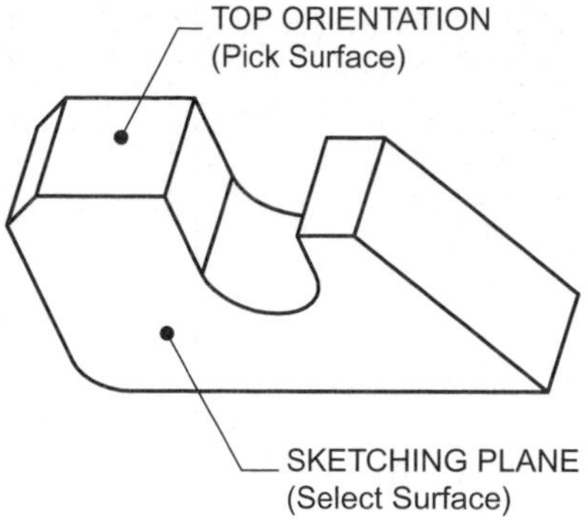

TOP ORIENTATION
(Pick Surface)

SKETCHING PLANE
(Select Surface)

Figure 4–30 Orienting the sketch

STEP 4: Pick the front of the part (Figure 4–30) as the sketching plane.

You will be sketching on the front of the part. Any planar surface can be used as a sketch surface.

> **MODELING POINT** By selecting the first protruded feature as your sketching plane, this new cut feature will become a child feature of the protruded feature. Any changes to a parent feature can affect its children.

STEP 5: Select OKAY to accept the direction of feature creation.

When using the Extrude Cut command, Pro/ENGINEER will attempt to determine the correct direction of feature creation. You should see a red arrow that points toward the interior of the part. This arrow points in the direction of extrusion. The Flip option can be used, if necessary, to change the direction.

STEP 6: Select TOP then pick the top of the part (Figure 4–30).

To orient the sketching environment, from the Sketch View menu, select Top, then pick the top of the part. When selecting a surface for orientation, the selected surface becomes a parent feature of the feature under construction. Try to select an orienting surface that is already a parent feature.

STEP 7: Utilizing the References dialog box, pick the two references shown in Figure 4–31, then close the dialog box.

The design intent for this part requires this cut feature to be located from the top and left edges of the first part feature. As shown in Figure 4–31, select these two edges as references.

STEP 8: Use the RECTANGLE icon to sketch the section.

With the Rectangle option, you select diagonal corners of the rectangular feature.

STEP 9: Apply the correct dimensioning scheme.

Use the Dimension icon to apply the dimensioning scheme that matches the design intent. In this example the dimensioning scheme shown in Figure 4–32 matches the intent of the design.

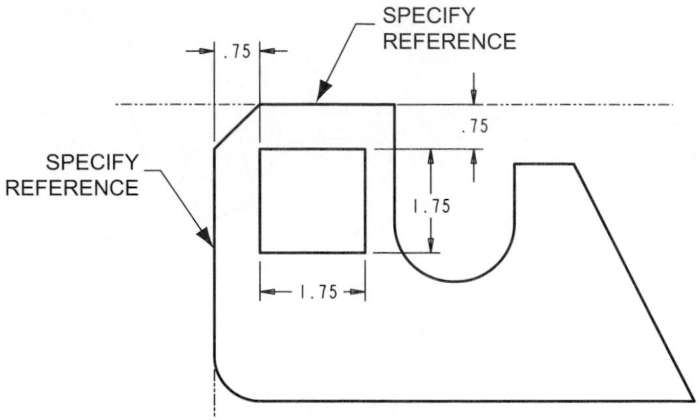

Figure 4–31 Sketching the cut

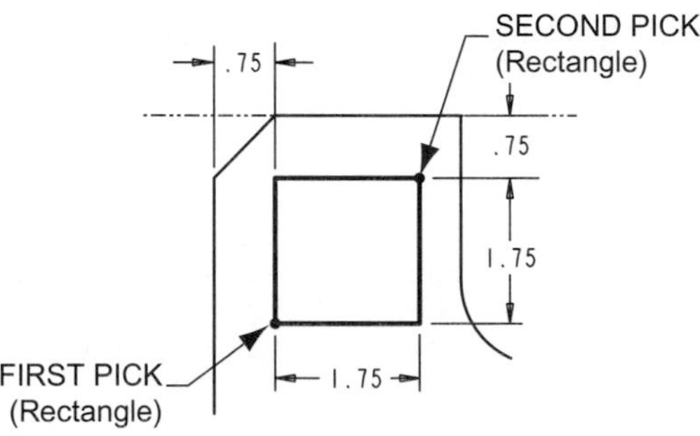

Figure 4–32 Sketching a rectangle

STEP 10: **MODIFY dimension values.**

Use the Modify icon to change the dimension values to match Figure 4–32. Each dimension can be preselected with the Pick icon and the shift key before selecting the Modify option. Another valid option would be to modify each dimension individually by double selecting it with the pick option.

STEP 11: **On the menu bar, use the SKETCH >> RELATIONS option to add a dimensional relationship.**

The design intent for the part requires that this cut feature remains square. Use the Add option on the Relations menu to add a relationship that will make the horizontal dimension equal to the vertical dimension. As shown in Figure 4–33, the dimension symbol for the vertical dimension is *sd2*, while the dimension symbol for the horizontal dimension is *sd3*. Your dimension symbols may be different. Select Add from the Relation menu. For the dimension symbols shown in Figure 4–33, enter the equation sd3 = sd2 in Pro/ENGINEER's textbox. Again, your dimension symbols may be different from those in the figure. Select Enter on the keyboard to exit the Add menu.

STEP 12: **Select the Continue icon to exit the sketcher environment.**

STEP 13: **Select OKAY to accept the material removal side.**

STEP 14: **Dynamically rotate the object.**

Before selecting an extrude depth, dynamically rotate your object using the control key and the middle mouse button. Observe the red arrow. This arrow points in the direction of extrusion.

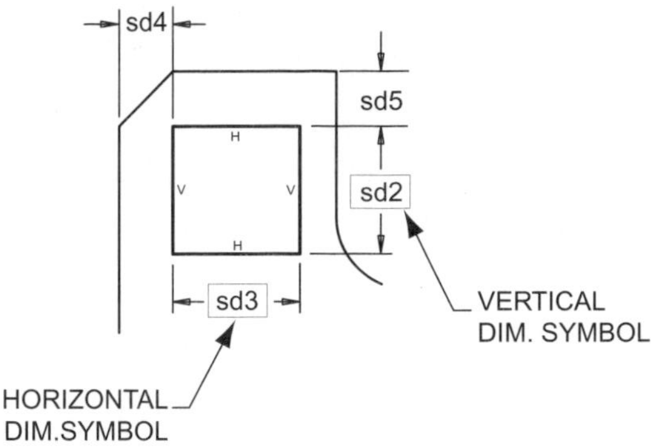

Figure 4–33 Dimension symbols

STEP 15: Select THRU ALL >> DONE as the depth option.

The Thru All depth option will create a cut whose depth always extrudes completely through any previously created features.

STEP 16: On the Feature Definition dialog box, select the PREVIEW option (Do not select OKAY).

Select Preview on the Feature Definition dialog box. Dynamically rotate your part to observe changes. At this time, **DO NOT SELECT OKAY** on the dialog box.

> **MODELING POINT** Previewing a feature is an important step during part modeling. Any conflicts that might exist between the new feature and existing features will normally be revealed during the preview process. If an error does occur, definitions associated with the feature can be redefined by using the Define option on the Feature definition dialog box.

REDEFINING THE FEATURE

In this section of the tutorial, you will modify the sketch of the Cut feature. If you inadvertently finished the feature by selecting OKAY in the previous step, you can access the Feature Definition dialog box by selecting the Feature >> Redefine option. Follow this by selecting the Cut feature on the model tree.

STEP 1: Select the SECTION element on the Feature Definition dialog box.

Select the Section element on the dialog box (Figure 4–34). You will redefine the sketch of the Cut feature. The redefinition of the sketch is a suboption under Section.

STEP 2: Select DEFINE on the Feature Definition dialog box.

With the Section option highlighted (Figure 4–34), select Define.

STEP 3: Select SKETCH on the Section menu.

As shown on the menu, other options available in addition to Sketch include Scheme and Sketching Plane. The Scheme option is used to modify the dimensioning scheme of a sketch while the Sketching Plane option is used to change the planar surface upon which the section was sketched.

STEP 4: Modify the dimension value shown in Figure 4–35.

As shown in Figure 4–35, use the Modify option to change the 0.75 dimension value to equal 0.70.

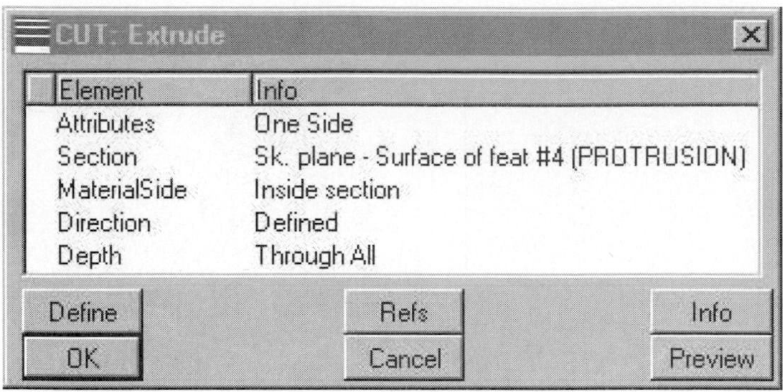

Figure 4–34 Cut feature definition dialog box

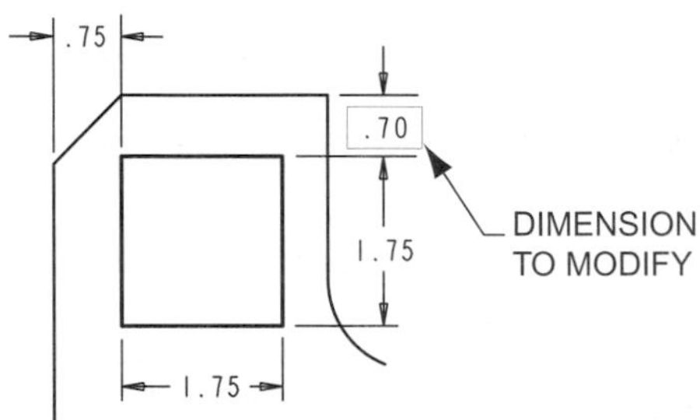

Figure 4–35 Modifying a dimension

STEP 5: **Select the Continue icon to exit the sketching environment.**

STEP 6: **Preview your part.**

Select Preview on the Feature Definition dialog box. Dynamically rotate your model.

STEP 7: **Select OKAY on the dialog box to finish the feature.**

STEP 8: **Save the part.**

CREATING AN EXTRUDED CUT

This segment of the tutorial will create the Cut feature shown in Figure 4–36.

STEP 1: **Select FEATURE >> CREATE >> CUT.**

STEP 2: **Select EXTRUDE >> SOLID >> DONE.**

STEP 3: **Select BOTH SIDES >> DONE as an attribute of the feature.**

This section of the tutorial will require you to extrude a cut in two directions from the sketching plane.

STEP 4: **Select the sketching plane upon which to sketch the Cut.**

Select the planar surface shown in Figure 4–37 as the sketching plane.

STEP 5: **Observe the direction of Cut.**

Notice the red arrow on your part protruding from the sketching plane. Since you will be cutting in both directions, this arrow represents the first direction of cut. This arrow is also important for the orientation of the sketcher environment. Within a sketcher environment, this arrow shows the direction that your sketch plane will be facing.

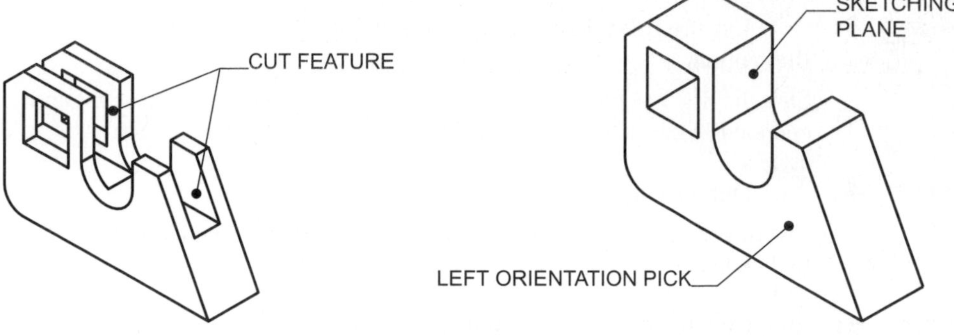

Figure 4–36 Cut feature

Figure 4–37 Sketching plane and left orientation

STEP 6: Select OKAY to accept the first direction of Cut.

STEP 7: Select LEFT then pick the plane shown in Figure 4–37 to orient your sketching environment

By selecting Left and by selecting the front of the part, you will be orienting the front of the part toward the left of the sketcher environment.

SKETCHING THE SECTION

This segment of the tutorial will sketch the Cut section shown in Figure 4–38.

STEP 1: On Pro/ENGINEER's toolbar, turn off the display or datum planes.

STEP 2: Use the References dialog box to specify the two references shown in Figure 4–38.

If necessary, you can also use the References dialog box to delete unwanted references.

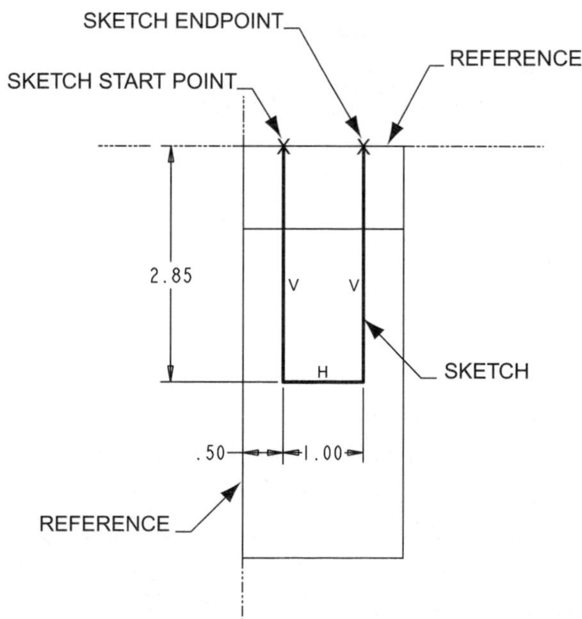

Figure 4–38 Sketching the cut

STEP 3: Use the LINE icon options to sketch the three lines representing the section.

Start the first line by using Line option. Make sure that the start point and endpoint of the sketch are aligned with existing geometry.

STEP 4: Dimension geometry according to design intent.

Figure 4–38 portrays design intent for the feature. Use the Dimension icon to dimension the feature.

STEP 5: MODIFY dimension values.

Use the Modify option to modify your dimension values, or double pick each dimension with the Pick icon.

STEP 6: Select the Continue icon to exit the sketcher environment.

If you do not have a fully defined section, Pro/ENGINEER will give you an error message stating this fact. If this occurs, try to realign your elements or to add dimensions that will fully define the sketch.

STEP 7: Select OKAY on the Material Direction menu.

Pro/ENGINEER tries to determine the side of the section to remove material. As shown in Figure 4–39, an arrow points in the direction of material removal. If this is not the correct side, you can flip the arrow.

STEP 8: Dynamically rotate your model.

Dynamically rotate your model to observe the direction of cut. Notice the direction that the arrow points. Since you are performing a Both Sides extrusion, this will be your first cut direction.

STEP 9: Select THRU ALL >> DONE as the first Cut depth.

STEP 10: Select THRU ALL >> DONE as the second Cut depth.

STEP 11: Preview your part.

Your part should look as shown in Figure 4–40.

MODELING POINT If you missed the step to perform a Both Sides extrusion, it is not too late to change this attribute. You can select the Attributes element in the Feature Definition dialog box and use the Define option to change from One Side to Both Sides.

STEP 12: Select OKAY to finish the part.

STEP 13: SAVE your part.

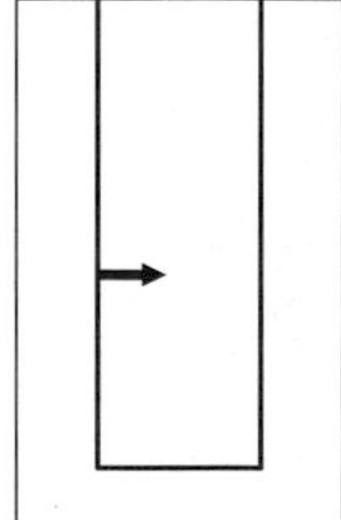

Figure 4–39 Material removal side

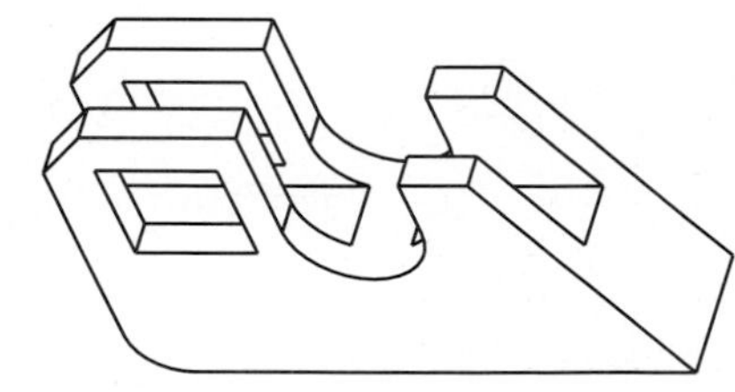

Figure 4–40 Finished cut

CREATING AN EXTRUDED PROTRUSION

This segment of the tutorial will create the extruded Protrusion feature shown in - Figure 4–41. Within this tutorial, the Use Edge option will be introduced. The Use Edge option turns existing part features and edges into sketch geometry.

STEP 1: Select the FEATURE >> CREATE >> PROTRUSION option.

This section of the tutorial will create a Protrusion on the front surface of the part.

STEP 2: Select the EXTRUDE >> SOLID >> DONE option.

STEP 3: Select ONE SIDE >> DONE as an attribute of the feature.

Your protrusions will extrude in one direction from the sketch plane.

STEP 4: Select the front of the part as the sketching plane (Figure 4–42).

STEP 5: Observe the direction of Protrusion.

The arrow shown on the work screen points in the direction that your sketch plane will be facing while sketching. It also represents the direction of feature creation.

STEP 6: Select OKAY to accept the extrude direction.

STEP 7: Select TOP, then pick the top of the part (Figure 4–42) to orient the sketching environment.

The Top option will orient the selected surface toward the top of the sketcher environment. After selecting the orienting surface, Pro/ENGINEER will launch the sketching environment.

STEP 8: On the toolbar, turn off the display of datum planes.

STEP 9: Utilizing the USE EDGE icon, select the edge shown in Figure 4–43. (Make sure you only select the edge once.)

During this portion of the tutorial, you will select an existing edge to project as an entity in this feature's section. The Use Edge option will turn existing feature edges into entities of the current sketch. When selecting existing feature edges, the feature from which the edge is obtained becomes a parent feature of the feature under construction.

Only select the edge once. Since the newly formed sketch entity will lie coincident with its parent edge, it may not be easily identifiable. Selecting a

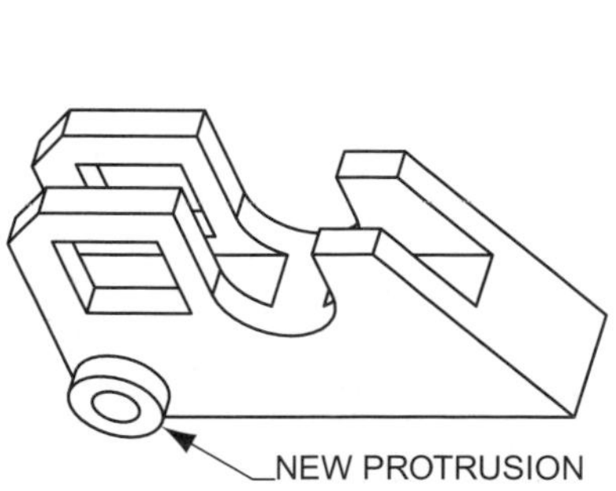

Figure 4–41 Extruded feature

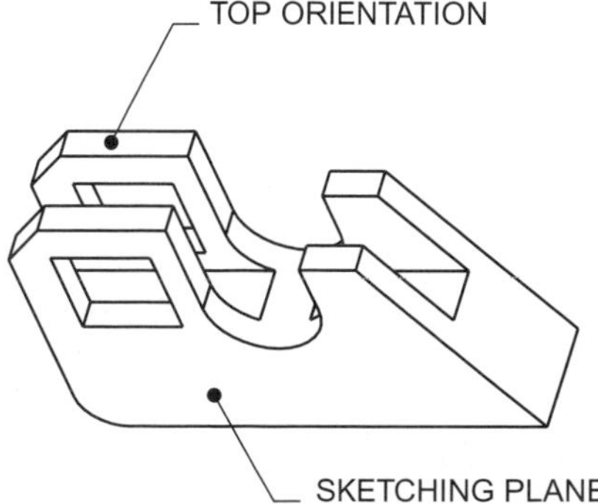

Figure 4–42 Sketching plane and orientation

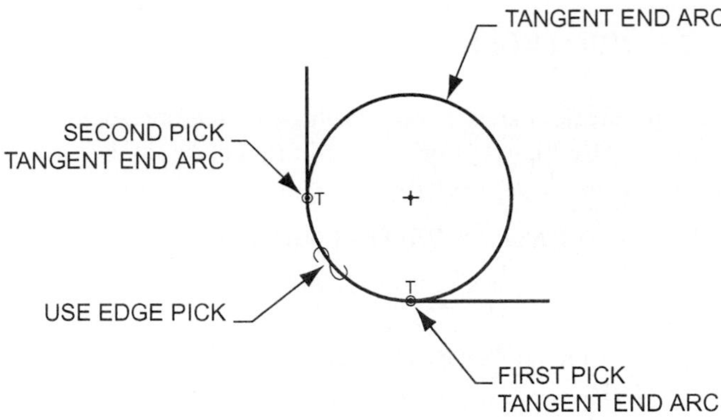

Figure 4–43 Tangent end arc

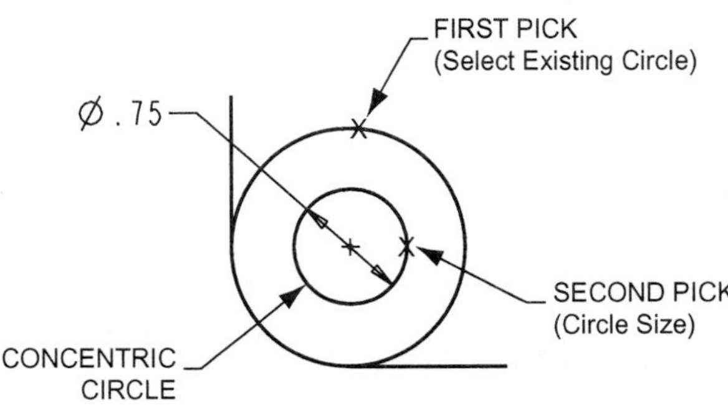

Figure 4–44 Concentric circle

second time will create a second entity. This will cause an error when you attempt to exit the sketcher environment.

STEP 10: Use the ARC icon to create the tangent end arc shown in Figure 4–43.

You will create an arc that joins the arc created in the previous step. In combination with the arc from the previous step, this newly created arc will form a circle. As shown in Figure 4–43, select at the first tangent point, then the second tangent point. When moving the cursor from Point 1 to Point 2, ensure that the arc is created toward the inside of the part.

STEP 11: Use the CONCENTRIC circle icon to create the circle shown in Figure 4–44.

The Concentric Circle icon can be found under the normal Circle icon on the sketch toolbar. Perform the two picks shown in the figure, then select the middle mouse button to end the Concentric Circle creation process.

STEP 12: MODIFY the circle's diameter dimension value.

Modify the circle's dimension value to equal 0.75.

STEP 13: Select the Continue icon to exit the sketcher environment.

After selecting continue, if you get the error message *Cannot have mixture of open and closed sections,* you probably picked the edge more than once while executing the Use Edge option. If necessary, use the Pick option and the delete key to delete any extra entities.

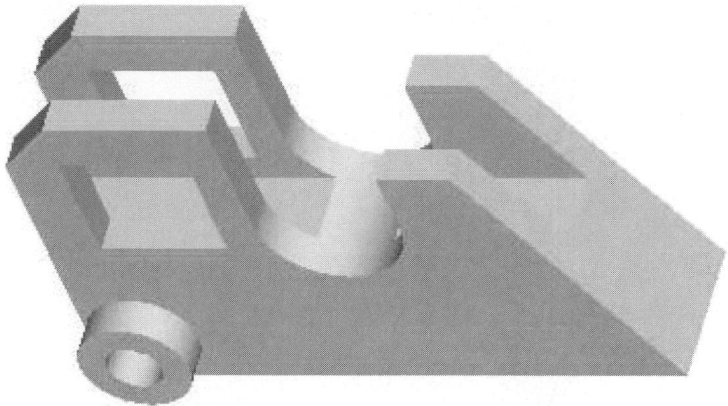

Figure 4–45 Preview of model

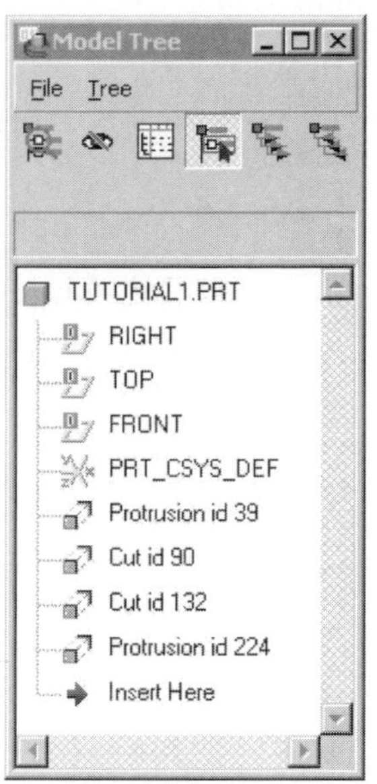

Figure 4–46 Model tree

STEP 14: Observe the direction of protrusion.

Notice the two concentric circles formed on the inside of the sketched circle. These circles represent the pointing end of the extrusion direction arrow. Dynamically rotate the part to get a better view of the extrusion direction.

STEP 15: Select BLIND >> DONE as the extrude depth.

The Blind option will allow you to enter an extrude depth.

STEP 16: Enter 0.500 as the Blind extrude depth.

In Pro/ENGINEER's Textbox, enter 0.500 as the Blind extrude depth.

STEP 17: Preview your model.

Select the Preview option on the Feature Definition dialog box.

STEP 18: Shade your model and dynamically rotate to observe the protrusion.

Your model should appear as shown in Figure 4–45.

STEP 19: Select OKAY on the Feature Definition dialog box.

STEP 20: Observe the Model Tree.

The Model Tree for the part is shown in Figure 4–46. This part contains, in order of creation, a Protrusion, a Cut, a second Cut, and a second Protrusion. This is reflected on the Model Tree.

STEP 21: SAVE your part.

DIMENSION MODIFICATION

This segment of the tutorial will modify the dimensions used to define the cut feature. Covered will be options for modifying a dimension's value and tolerance format.

STEP 1: Select MODIFY >> VALUE on Pro/ENGINEER's Menu Manager.

STEP 2: On the Model Tree select the second Cut feature (Figure 4–46).

This step of the exercise requires you to select the second cut feature. You may select this feature directly on the work screen, or you may select the feature from the model tree. When you select the feature, parametric dimensions used to construct the feature will appear (Figure 4–47).

MODELING POINT Often, it is difficult to select a feature directly on the work screen. When it is necessary to select a feature or element directly on the work screen, consider using the Query Sel option from the Get Select menu. Query select will give you an option to toggle through available features.

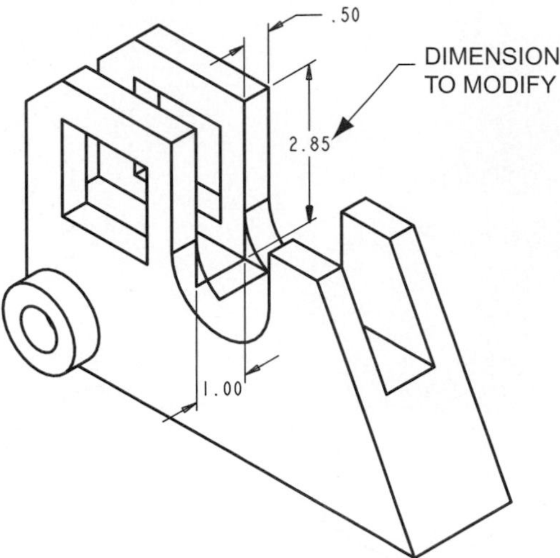

Figure 4–47 Dimension modification

STEP 3: Pick the 2.85 dimension and modify its value to equal 2.90.

After picking the dimension, Pro/ENGINEER requires you to enter the new dimension value in the textbox.

STEP 4: REGENERATE the part.

After modifying a dimension value, Pro/ENGINEER requires a regeneration of the part. Select Regenerate from the Part menu.

STEP 5: On the menu bar, select the UTILITIES >> ENVIRONMENT option.

STEP 6: On the Environment dialog box, check the DIMENSION TOLERANCES setting, then select OKAY.

The Environment dialog box is located under the Utilities menu bar. Check the Dimension Tolerance setting. This will display dimensions with tolerances.

STEP 7: Select MODIFY >> DIMCOSMETICS from Pro/ENGINEER's Menu Manager.

STEP 8: Select the FORMAT >> PLUS-MINUS format type.

STEP 9: On the Model Tree, select the last cut feature.

STEP 10: To set a Plus-Minus Tolerance Mode, select each dimension defining the Cut feature (Figure 4–48).

Selected dimensions will be displayed in a Plus-Minus tolerance format. Be sure to only select the text. While selecting dimension values, you might inadvertently pick a geometric feature on the work screen. This will cause dimensions associated with this feature to be displayed also.

STEP 11: Select DONE SEL on the Get Select menu.

Done Sel will allow you to finish selecting dimensions to modify. The middle mouse button serves the same purpose.

STEP 12: On the Modify menu, select the DIMENSION option.

STEP 13: On the Model Tree or work screen, select the second Cut feature.

STEP 14: Select the three dimensions defining the Cut feature, then select the DONE SEL option on the Get Select menu.

When selecting dimensions, you must select the nominal value of each displayed dimension since tolerances are displayed.

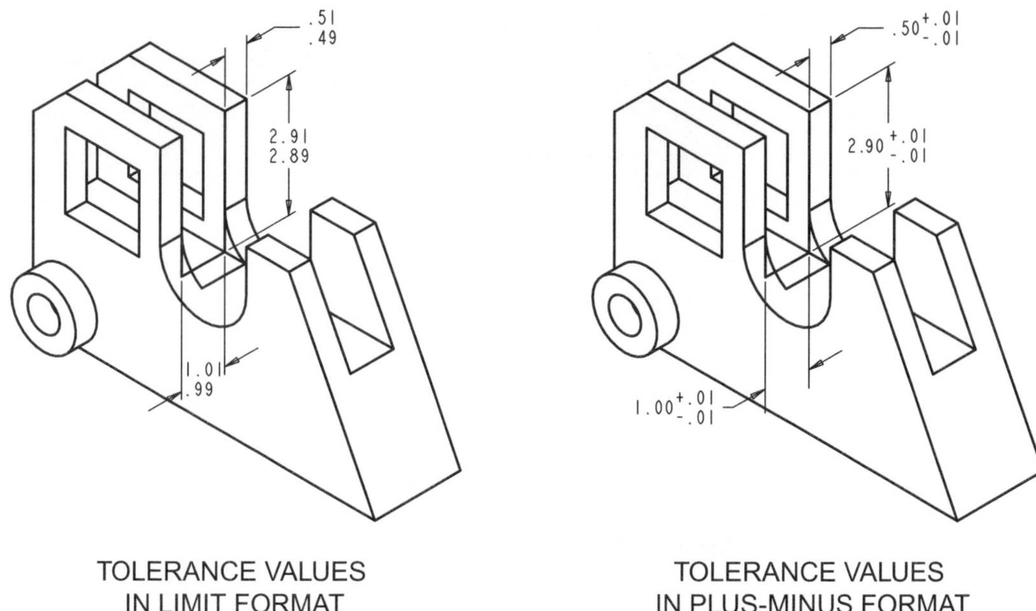

Figure 4–48 Dimension tolerance values

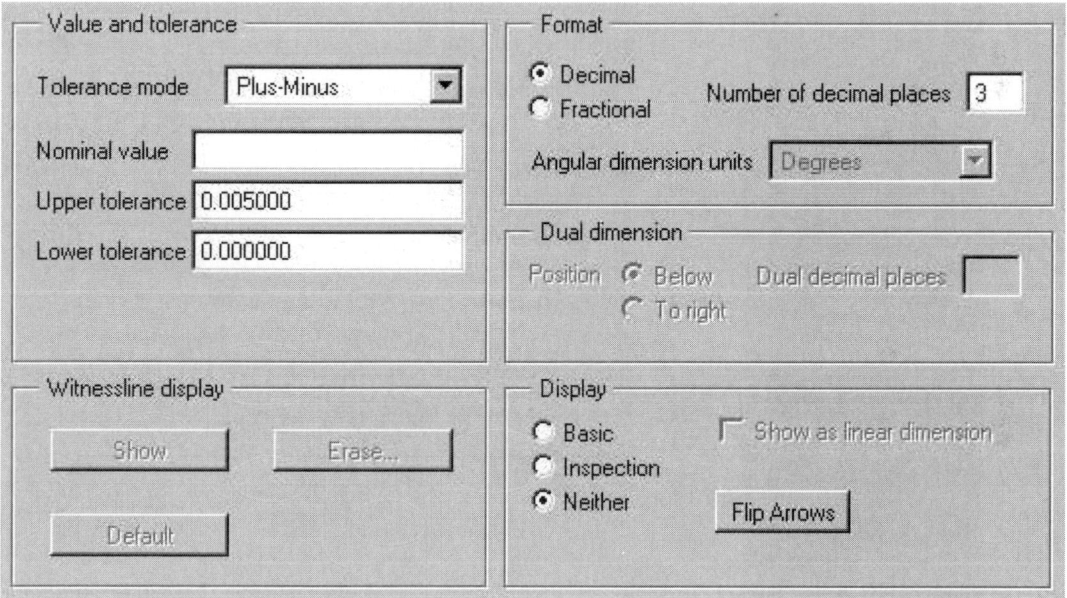

Figure 4–49 Modify dimension dialog box

STEP 15: On the DIMENSION PROPERTIES dialog box, enter 3 as the number of decimal places.

As shown in Figure 4–49, under the Dim Format option, enter 3 as the number of decimal places.

STEP 16: Enter Upper and Lower Tolerance values.

As shown in Figure 4–49, enter 0.005 as the Upper Tolerance value and 0.000 as the Lower Tolerance value. Notice, on the dialog box, how the Tolerance Mode is set to Plus-Minus.

MODELING POINT Other options are available on the Modify Dimension dialog box. The DimText tab is used for entering additional text, notes, and symbols around an existing dimension value. A Sym Pallette (Symbol Pallette) option is used to select predefined graphical symbols that can be used under the Dim Text tab.

STEP 17: Select the OKAY option on the dialog box.

STEP 18: Save your model.

REDEFINING A FEATURE'S DEPTH

This segment of the tutorial will redefine the extrusion depth of the last Protrusion feature. You will change the depth from a value of 0.500 to a value of 1.00. The Redefine command is one of Pro/ENGINEER's most useful and powerful options. It is used to redefine attributes and parameters associated with features. Feature parameters such as Feature Creation Direction, Material Removal Side, and Depth value can be modified.

STEP 1: Select the FEATURE >> REDEFINE command.

STEP 2: On the Model Tree, select the last Protrusion feature (Figure 4–50).

As shown on the model tree, this tutorial has required the creation of two protrusion features and two cut features. This section of the tutorial will require you to modify the last protrusion feature.

STEP 3: On the Protrusion Feature Definition dialog box, select the DEPTH definition (Figure 4–50).

STEP 4: Select the DEFINE option on the dialog box.

After selecting the Define option, Pro/ENGINEER will display the depth option menu. This will allow you to redefine the feature's depth parameter.

STEP 5: Select BLIND >> DONE on the depth option menu.

Notice on the Feature Definition dialog box that the attribute associated with the Depth option is now showing a value equal to Changing.

STEP 6: Enter 1.00 as the new Depth value.

STEP 7: Select PREVIEW on the Feature Definition dialog box.

STEP 8: Select OKAY to accept the changes.

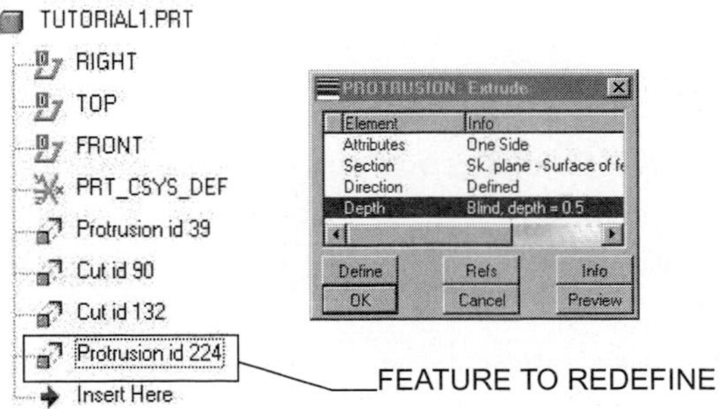

Figure 4–50 Redefining a feature

Redefining a Feature's Section

This section of the tutorial will use the Redefine option to add filleted corners to the first Cut feature.

Step 1: On the Model Tree, select the first Cut feature with your right mouse button then select the REDEFINE command.

Within this segment, you will access the Redefine command through the model tree. Optionally, you could select the Redefine command on the menu, followed by selecting the protrusion feature.

Step 2: As shown in Figure 4–51, select the SECTION definition on the Feature Definition dialog box.

Step 3: Select the DEFINE option on the dialog box.

Step 4: Select SKETCH on the Section menu.

The Section menu gives you the option of modifying the sketch of the feature, the dimensioning scheme of the sketch, or the sketch plane for the section. After selecting Sketch, Pro/ENGINEER will take you into the sketcher environment that defined the original feature.

Step 5: Using the CIRCULAR FILLET icon, construct the filleted arcs shown in Figure 4–52.

The Circular Fillet option requires you to pick the two entities bordering the arc. The first pick will define the initial radius of the fillet.

Step 6: Modify the radius dimensions of each fillet to equal 0.500.

When selecting a dimension value to modify, Pro/ENGINEER requires that you select the text of the dimension. Your section should appear as shown in Figure 4–53.

Step 7: On the menu bar, select the SKETCH >> RELATION option.

Step 8: On the Relations menu, select the EDIT REL option to adjust dimension symbols.

The original sketch was created with the Rectangle sketch option. The default scheme dimensions the length of each line of the rectangle. A previous step of this tutorial required the establishment of a dimensional relationship between the two defining dimensions. The Circular Fillet option will create new dimensions with new dimension symbols. Use the Edit Rel option to adjust the symbols defining the dimensional relationship.

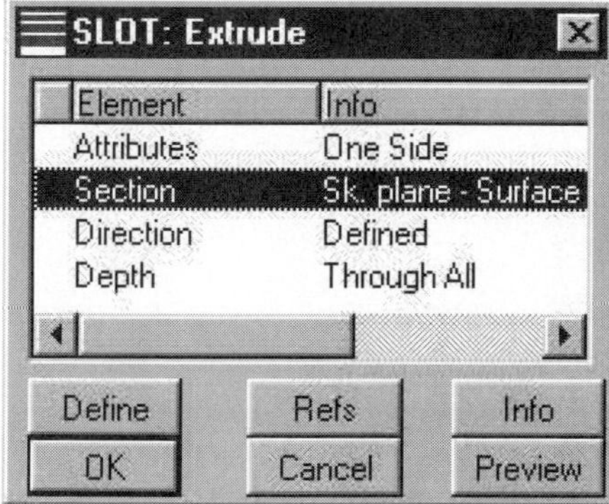

Figure 4–51 Cut feature definition dialog box

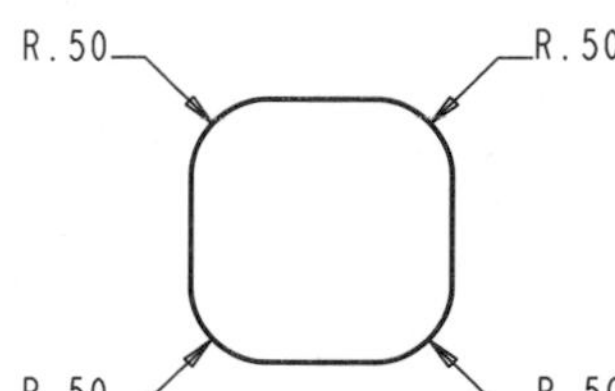

Figure 4–52 New section

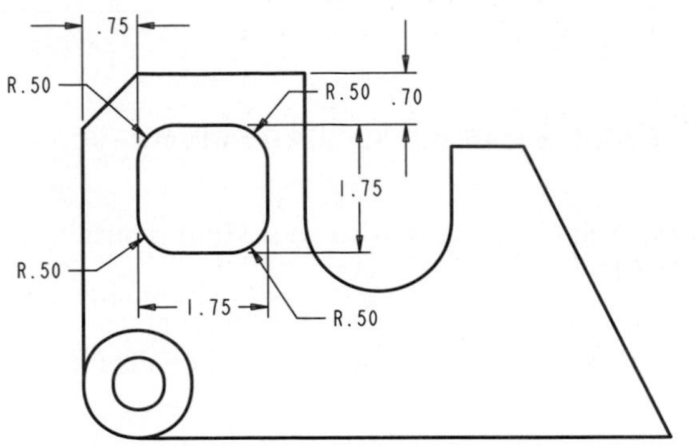

Figure 4–53 Final section

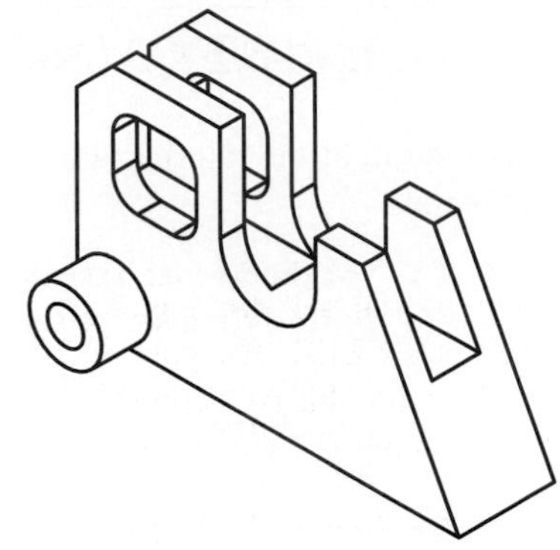

Figure 4–54 Completed part

STEP 9: Select the Continue icon to exit the sketching environment.

STEP 10: Select PREVIEW on the Feature Definition dialog box.

Your final part should look as shown in Figure 4–54.

STEP 11: SHADE and dynamically rotate the part.

STEP 12: Select OKAY on the Feature Definition dialog box.

STEP 13: SAVE your part.

STEP 14: Use the FILE >> DELETE >> OLD VERSIONS option to purge old versions of the part.

Every save of the part creates a new version. Use the File >> Delete >> Old Versions option to delete every version except for the last saved.

TUTORIAL

Datum Tutorial

This tutorial will provide step-by-step instruction on how to create various datums within Pro/ENGINEER. The part shown in Figures 4–55 and 4–56 will be the base for this tutorial. Creating this part comprises the first segment of this tutorial.

This tutorial will cover

- Creating a part.
- Creating datum planes.
- Creating datum points.
- Creating datum axes.
- Creating a coordinate system.

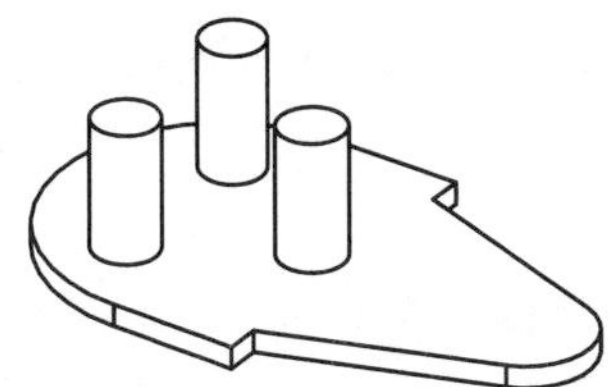

Figure 4–55 Base part

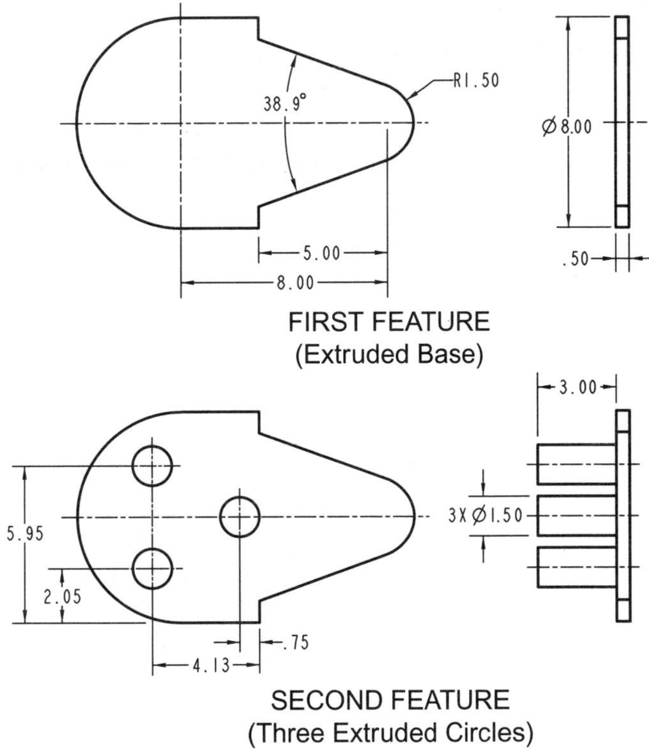

Figure 4–56 Drawing of part

Creating the Part

The first step in this tutorial is to model the part shown in Figure 4–55. This part (*datum1.prt*) is also available on the supplemental CD. Due to the nature of this tutorial, do not use Pro/ENGINEER's default datum planes. This requirement can be accomplished by deselecting the Use Default Template option on Pro/ENGINEER's New dialog box, followed by selecting Empty as the template file.

Create this part with two extruded Protrusions. Figure 4–56 shows details for each feature. Consider the following when creating this part:

- Name the part file *datum1*.
- Do not use Pro/ENGINEER's default datum planes.
- Sketch the Flat base feature first as an extruded Protrusion (Figure 4–56). Use the Centerline icon (located behind the Line icon) to create the two centerlines shown.
- Use the Edit >> Toggle Construction option to create construction entities as needed.
- Remember to utilize available entity constraints.
- Create the three shaft features as one extruded Protrusion utilizing the top of the first feature as the sketching plane.
- When creating the part, use the dimensioning schemes shown in Figure 4–56. These schemes match the design intent for the part.

Creating Datum Planes

This section of the tutorial will explore Pro/ENGINEER's datum plane creation capabilities. Do not start this section of the tutorial until you have created the part shown in Figures 4–55 and 4–56. The following datum planes will be created:

- As shown in Figure 4–57, a datum plane is placed through datum axes A_2 and A_3. This datum plane will be used to construct the remaining datum planes.
- A datum plane will be offset from the first datum plane.
- A datum plane will be made tangent to two selected surfaces.
- A datum plane will be constructed through an edge and at an angle to the first datum plane.

Step 1: Select the Datum Plane icon on the Datum toolbar.

Step 2: Select the THROUGH constraint option, then pick the first axis shown in Figure 4–57.

The Through constraint option allows datum planes to be constructed through an axis, edge, curve, point, vertex, cylinder, or plane. On the work screen, select axis A_2 (your axis symbol might be different). With the exception of Through >> Plane, all Through options require a second paired constraint option. In this example, the second constraint option will place the datum through the second axis.

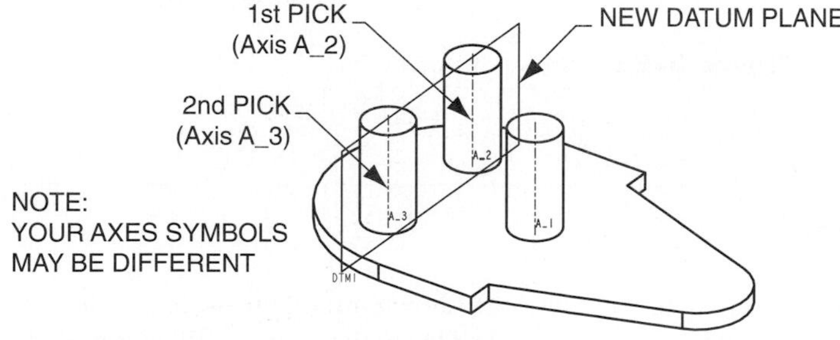

Figure 4–57 Through axes datum plane

STEP 3: Select the THROUGH constraint option, then pick the second axis shown in Figure 4–57.

STEP 4: Select DONE on the Datum Plane menu.

Your datum should appear as shown in Figure 4–57.

STEP 5: Select the Datum Plane icon on the Datum toolbar.

The next several steps of this tutorial will require you to create a datum plane offset from an existing plane (Figure 4–58).

STEP 6: Select OFFSET on the Datum Plane menu.

The Offset option will create a datum plane offset from an existing plane. The existing plane may be any planar surface.

STEP 7: As shown in Figure 4–58, pick the previously created datum plane.

When selecting a datum plane, the edge of the datum plane may be selected or the datum plane label can be selected.

STEP 8: On the Offset menu, select the ENTER VALUE option.

Pro/ENGINEER defaults to the placement of an offset datum plane through a datum point. The Enter Value option will allow you to enter an offset value. After selecting the Enter Value option, notice on the work screen that an arrow displays the direction of offset. If the required direction of offset is in the opposite direction, you can enter a negative value.

STEP 9: Offset the new datum plane 2 inches the direction shown in Figure 4–58.

Enter a negative value if necessary to place the required datum plane.

STEP 10: Select DONE on the Datum Plane menu.

The datum plane should appear similar to Figure 4–58. If the offset side is incorrect, use the Modify >> Value option to enter a −2.00 value.

STEP 11: Select the Datum Plane icon on the Datum toolbar.

STEP 12: Select the TANGENT constraint option.

The Tangent constraint option places a datum tangent to a selected cylinder or round.

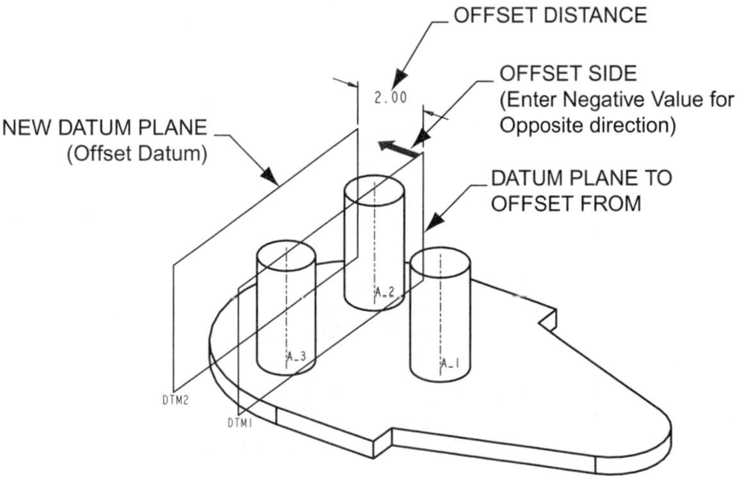

Figure 4–58 Offset datum plane

STEP 13: Select the First Tangent Pick location shown in Figure 4–59.

Select the cylinder surface shown in Figure 4–59. The Tangent option requires a second constraint option to define the datum plane. In this example, the second constraint option will be a second Tangent option.

STEP 14: Select the TANGENT Constraint option then select the second surface of tangency (Figure 4–59).

STEP 15: Select DONE on the Datum Plane menu.

Your model should appear as shown in Figure 4–59.

STEP 16: Select the Datum Plane icon on the Datum toolbar.

STEP 17: Select the THROUGH constraint option; then select the edge shown in Figure 4–60.

This datum plane will be constructed through the edge shown. The Through >> Edge option requires a second constraint option.

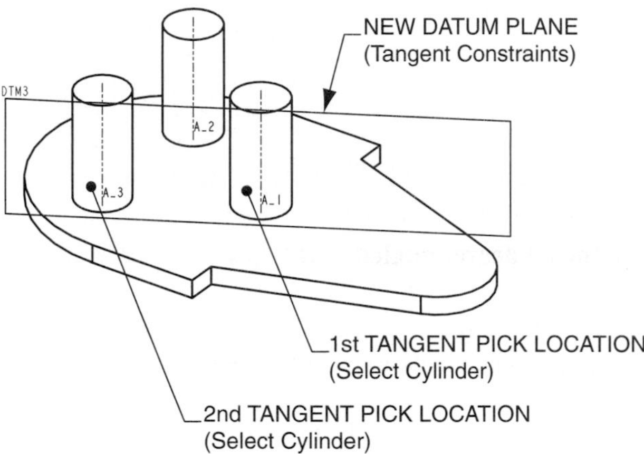

Figure 4–59 Tangent datum plane

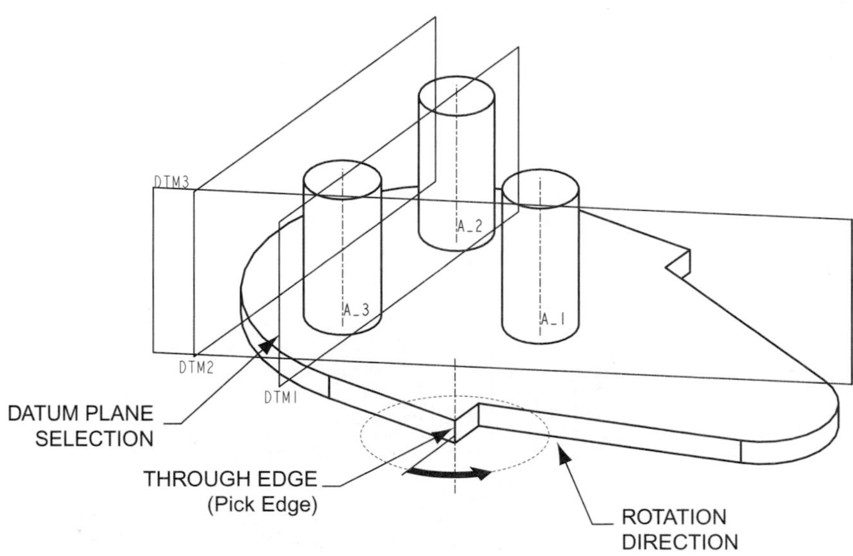

Figure 4–60 Through edge datum plane

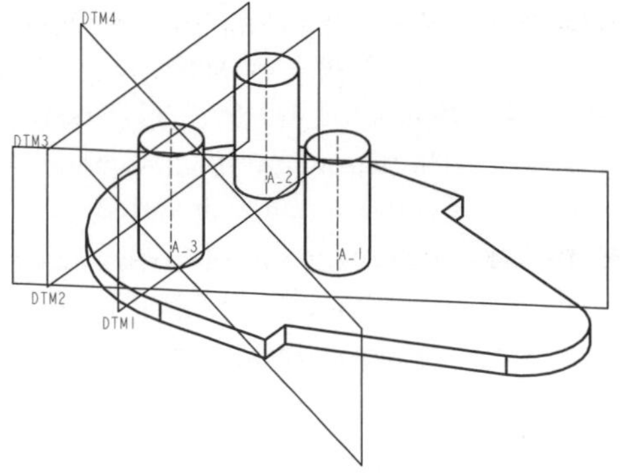

Figure 4–61 Finished datum planes

STEP 18: Select the ANGLE constraint option, then select datum plane DTM1 (Figure 4–60).

On the work screen, select datum plane DTM1. The datum plane being constructed will be placed at an angle to this datum.

STEP 19: Select DONE on the Datum Plane menu.

STEP 20: Select the ENTER VALUE option from the Offset menu.

After selecting the Enter Value option, you will enter an angular value. Notice on the work screen that Pro/ENGINEER shows you the direction that the angle will form. The arrow points in the positive direction, and the base of the arrow is the zero point from which to measure the degrees of rotation. In this example, you will enter a 60-degree angular value.

STEP 21: Enter a value to create the 60-degree angled datum plane as shown in Figures 4–60 and 4–61.

If your rotation direction matches Figure 4–60, enter a positive 60-degree value. If your direction of rotation is opposite the figure, enter a negative value.

STEP 22: Save your model.

CREATING A DATUM AXIS

This section of the tutorial will explore the creation of a datum axis. The Datum axis shown in Figure 4–62 will be created.

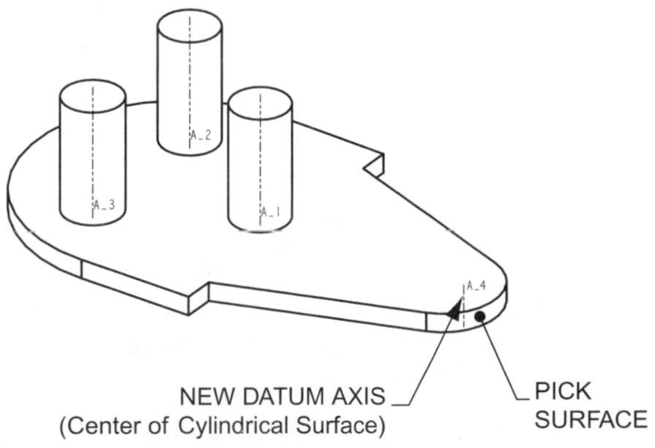

Figure 4–62 Datum axis

STEP 1: **Select the Datum Axis icon on the Datum toolbar.**

STEP 2: **Select the THRU CYL constraint option from the Datum Axis menu.**

The Through Cylinder option will place a datum axis through the center of a cylinder or circular surface.

STEP 3: **Select the curved surface shown in Figure 4–62.**

The Datum Axis will be created through the center of the selected surface. Notice on the model tree the addition of an axis feature.

CREATING A COORDINATE SYSTEM

This section of the tutorial will create the Coordinate System shown in Figure 4–63.

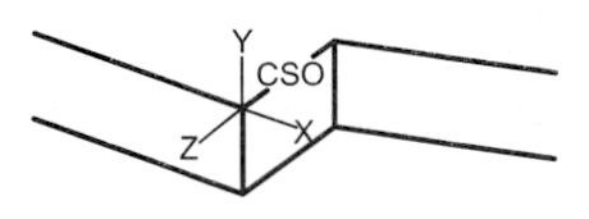

Figure 4–63 Coordinate system

STEP 1: **Select the Coordinate System icon on the Datum toolbar.**

Coordinate systems are considered by Pro/ENGINEER to be a type of datum feature.

STEP 2: **Select the 3 PLANES >> DONE option.**

The 3 Planes option will create a coordinate system at the intersection of three planes.

STEP 3: **Pick in order the three surfaces shown in Figure 4–64.**

Select the surfaces in the order shown in Figure 4–64. Upon selecting the surfaces, Pro/ENGINEER will require each axis to be defined (Figure 4–65). At the origin of the coordinate system, Pro/ENGINEER reveals three axes. One axis is red while the remaining two are yellow (with default color settings). On the Coordinate System menu, Pro/ENGINEER provides an option for defining the red axis as either the X axis, the Y axis, or the Z axis.

STEP 4: **Select the Y Axis option from the Coordinate System menu.**

Select the Y axis option to define the red axis as the Y axis of the coordinate system. Defining two axes will define a coordinate system. After defining the first axis, Pro/ENGINEER will require you to define a second red axis.

STEP 5: **Select the X Axis option from the Coordinate System menu.**

Select the X axis option to define the red axis as the X axis of the coordinate system.

STEP 6: **Save your object.**

From the File menu, select the Save option.

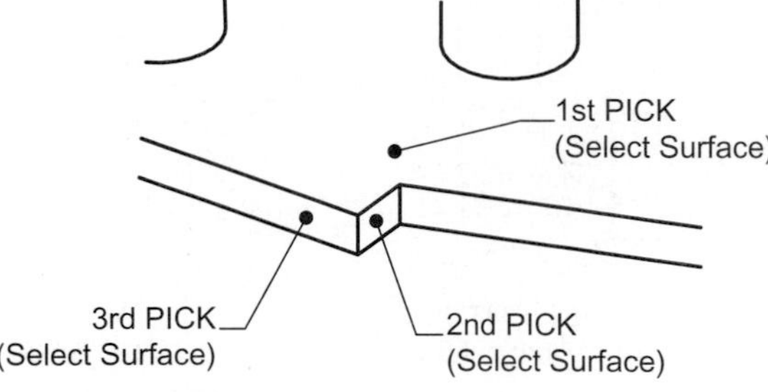

Figure 4–64 Plane selection

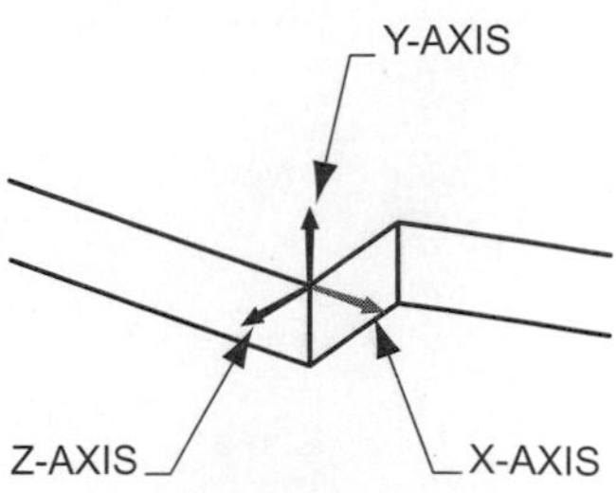

Figure 4–65 Axes definitions

PROBLEMS

1. Use Pro/ENGINEER to model the parts shown in Figures 4–66 through 4–70.

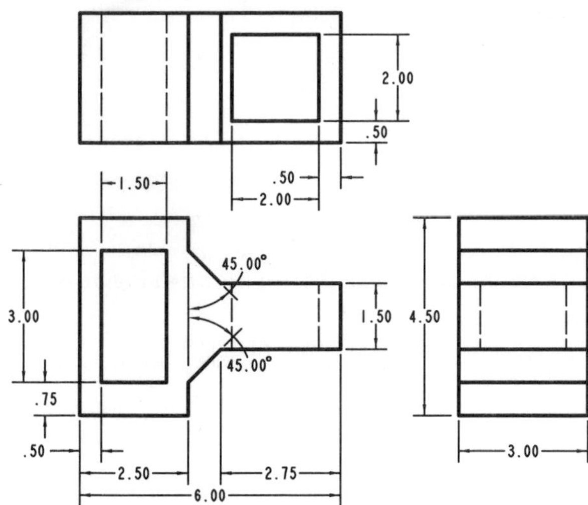

Figure 4–66 Problem 1a

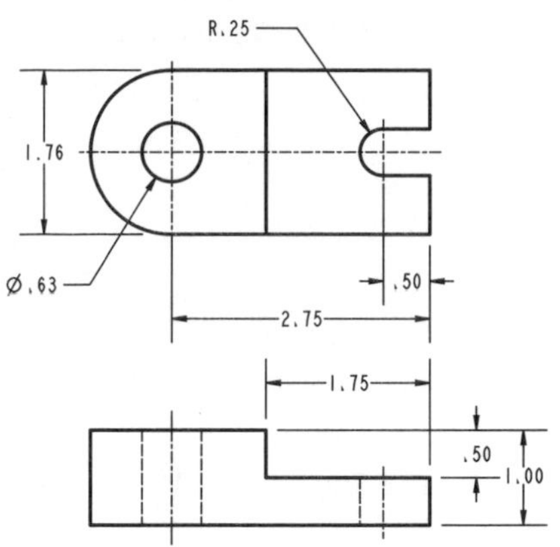

Figure 4–67 Problem 1b

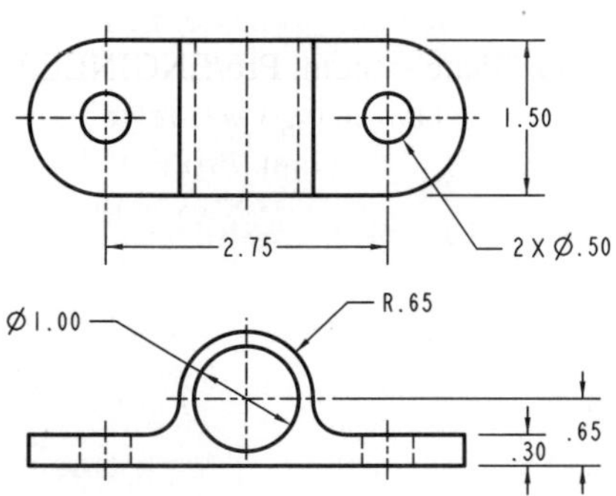

Figure 4–68 Problem 1c

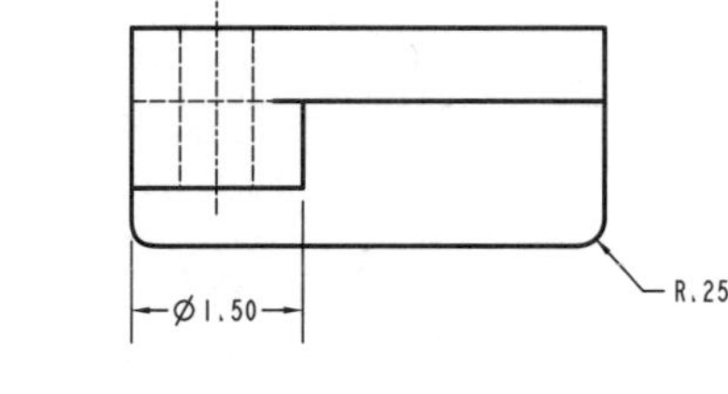

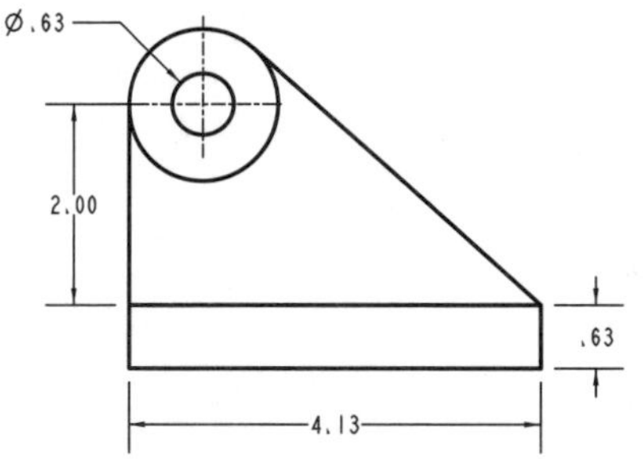

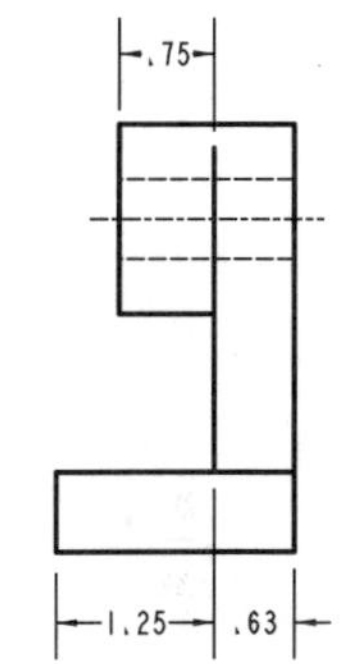

Figure 4–69 Problem 1d

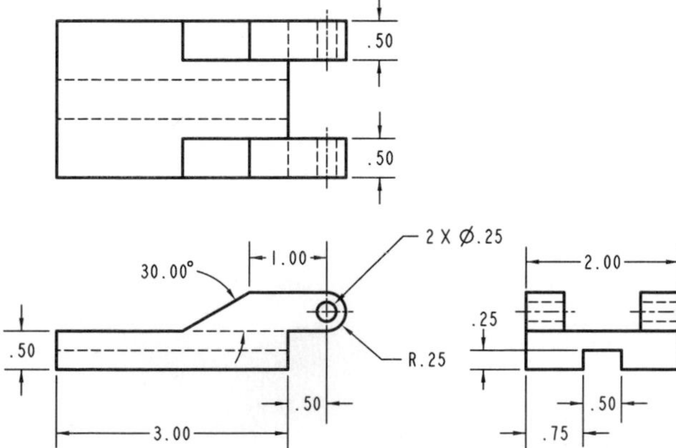

Figure 4–70 Problem 1e

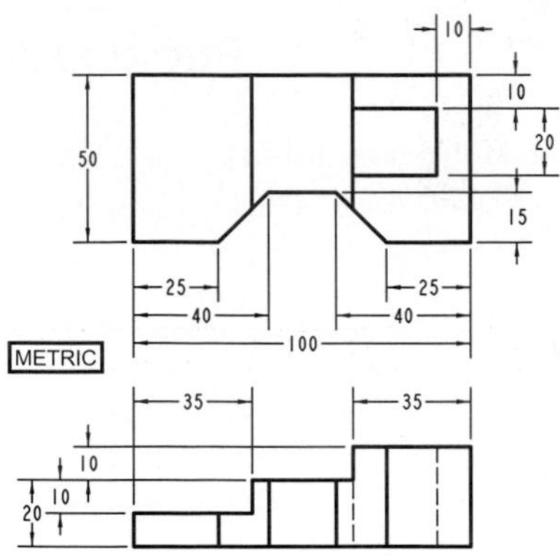

Figure 4–71 Problem 2

2. Use Pro/ENGINEER to model the part shown in Figure 4–71. Use millimeter-newton-seconds as the units for the part.

QUESTIONS AND DISCUSSION

1. Describe each of the feature creation options found under the Protrusion and Cut commands.
2. What is the difference between a solid feature and a thin feature?
3. How does a Both Sides extrusion differ from a One Side extrusion?
4. Compare the Thru Until depth option with the UpToSurface depth option.
5. Describe the methods available within Part mode to modify a dimension's value.
6. How does a cosmetic feature differ from a geometric feature?
7. What option within Pro/ENGINEER is available for changing the depth definition for an extruded feature?

CHAPTER

5

FEATURE CONSTRUCTION TOOLS

Introduction

Pro/ENGINEER provides many feature construction tools that do not rely upon a sketcher environment. Examples include holes, rounds, and chamfers. This chapter will cover the basics of these and other common feature construction commands. Upon finishing this chapter, you will be able to

- Construct a straight-linear hole.
- Construct a straight-coaxial hole.
- Construct a standard hole.
- Create fillets and rounds on a part.
- Create a chamfer on a part.
- Tweak features using the draft option.
- Create a part shell.
- Create a cosmetic thread.
- Create a linear pattern.

DEFINITIONS

Draft surface A part surface angled to allow for easy removal of the part from a mold or cavity.

Cosmetic features A part feature that allows for cosmetic details that do not require complicated regenerations. An example of a cosmetic feature would be a cosmetic thread or a company logo sketched on a surface.

Sketched hole A hole, such as a counterbored or countersunk hole, that has a sketched profile.

HOLE FEATURES

The Hole command is used to create either straight holes, holes with varying profiles, or holes defined by standard fastener tables. A hole is considered a negative space feature. Straight holes can be created with the Cut command also. Why use the Hole command instead of Cut? The Hole command has a predefined section that does not require sketching. Additionally, the Hole command's algorithms are not as complicated as the Cut command's, thus requiring less computer power to regenerate.

Hole Placement Options

Four hole placement options are available (Figure 5–1). A description of each follows.

Linear

The Linear option locates a hole from two feature edges. These edges may be part surfaces or datum planes. For each location edge, the user must pick an edge and then enter a dimension value.

Radial

The Radial option locates a hole at a distance from an axis and at an angle from a reference plane. This option is used often for radial patterns such as might be found on a bolt-circle pattern. This option is covered in detail in Chapter 6.

Coaxial

The Coaxial option locates the center of a hole coincident with an existing axis. The user must provide the axis, placement surface, and hole diameter.

On Point

The On Point option locates the center of a hole on a datum point. The user must provide the point, placement surface, and hole diameter.

Hole Types

Pro/ENGINEER provides three options for defining the profile of a hole: Straight, Sketched, and Standard. The Straight option produces a hole that has a constant diameter throughout the length of the hole. The Sketched option requires the user to sketch the profile of the hole within a sketcher environment. Figure 5–2 shows an example of a sketched profile of a counterbored hole. The Sketched Hole option is covered in Chapter 6.

A standard hole is defined through a fastener table. An example would be a 1.00 unified national course fastener. It has 8 threads per inch with a tap drill of 7/8 inch. As shown in Figure 5–3, a Standard hole is specified along with a tapped screw size equal to *1-8*. This will produce an interior thread minor diameter equal to 0.875 (or 7/8 inch). Pro/ENGINEER derives this information from three default text files: *ISO.hol, UNC.hol.* and *UNF.hol.* Additional user-defined hole charts can be created and specified with the configuration file option *hole_parameter_file_path*.

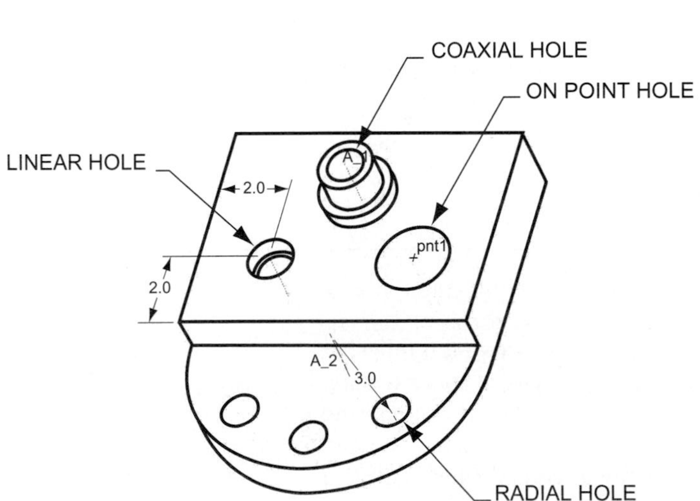

Figure 5–1 Hole placement options

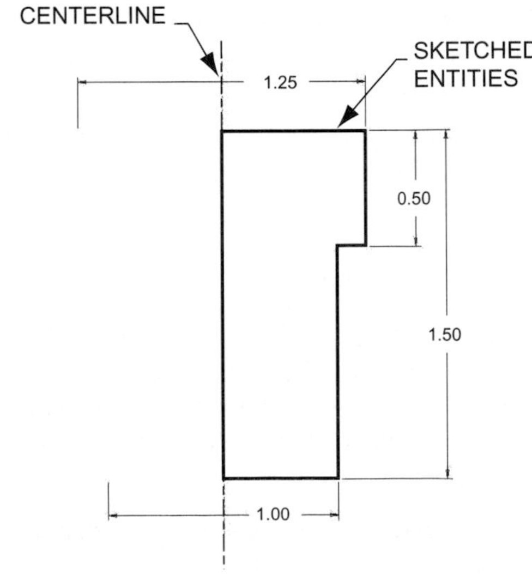

Figure 5–2 Sketched hole

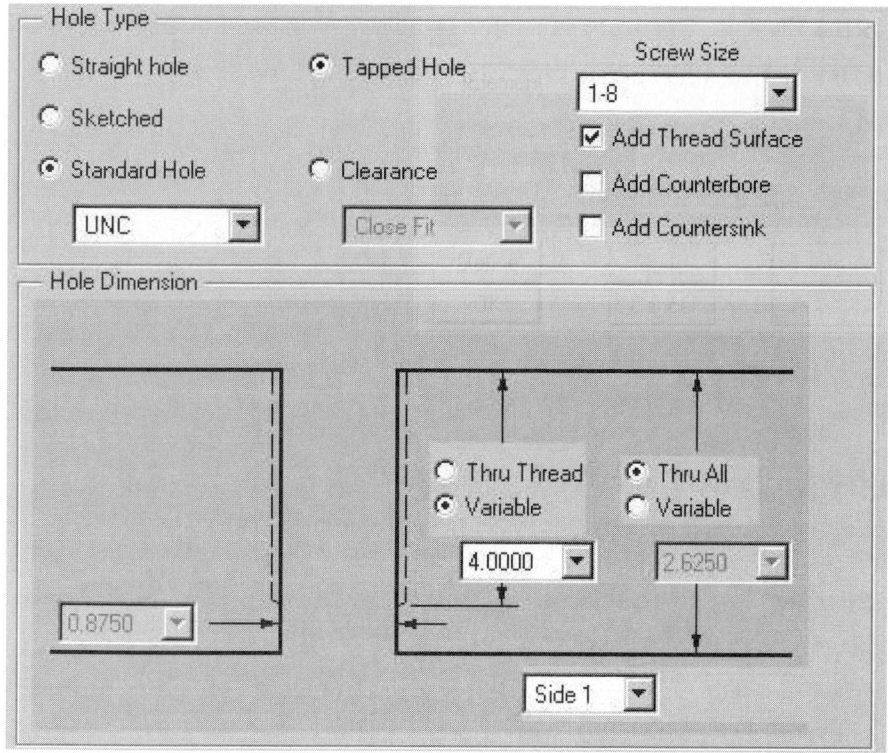

Figure 5-3 Standard hole options

The following options are available for standard holes.

- **Tapped hole** The Tapped Hole option is selected when a fastener's threads are engaged within the hole. Setting this option will provide hole parameters for an internal thread (e.g., minor diameter and thread length).
- **Clearance** The Clearance option is selected when a fastener's threads are not engaged within the hole. Setting this option provides a clearance fit for the fastener.
- **Add thread surface** This option is only available when the Tapped Hole option is selected. It provides a cosmetic representation of the thread. A Clearance hole has Thru All as its thread depth.
- **Add counterbore** As the name implies, the Add Counterbore option creates a counterbored hole. The user must enter the counterbore's diameter and depth.
- **Add countersink** Similar to the Add Counterbore option, the Add Countersink option creates a countersunk hole. The user must enter parameters for the hole.

Hole Depth Options

Similar to the Extrude option, straight holes have several depth options available. As with extruded features, the selection of a hole's depth option is important for incorporating design intent. As an example, it is common for holes to cut completely through a part. If this is the intent of a design then the Thru All option should be used. The following are available depth options.

- **Variable** The Variable option cuts a hole to a user-defined depth. The constructed hole has a flat bottom.
- **Thru next** This option cuts the hole to the next part surface.

- **Thru all** This option cuts a hole completely through a part.
- **Thru until** This option cuts a hole to a user-selected surface. The constructed hole has a flat bottom.
- **To reference** This option cuts a hole to a user-selected point, vertex, curve, or surface. The reference has to exist before executing this option. The hole will have a flat bottom after construction.
- **Symmetric** This option creates a two-sided hole with equal depths on both sides of the placement plane. This option is only selectable under the Depth Two parameter.

Creating a Straight Linear Hole

Perform the following steps to create a Straight-Linear Hole.

Step 1: Select the FEATURE >> CREATE >> HOLE command.

Step 2: On the Hole dialog box, select STRAIGHT as the Hole Type.

Step 3: Enter a Diameter value for the hole.

Step 4: Select a Depth One parameter (e.g., Thru All, Variable, etc.), then perform the operation appropriate to the option.

> **INSTRUCTIONAL NOTE** If you select a depth parameter such as Variable, Thru Until, and so forth, you will have to perform an additional step. As an example, if you select the To Reference option, you will have to pick a point, curve, or surface to cut the hole to.

Step 5: If required, select a Depth Two parameter, then perform the operation appropriate to the option.

The Depth Two parameter creates a Both Sides extruded hole.

Step 6: Select a Primary Reference for the placement of the hole (Figure 5–4).

The Primary Reference parameter defines the hole's placement plane.

Step 7: On the work screen, select a reference edge or surface for locating the hole.

The Linear option requires the selection of two edges or planes to locate the hole. Pick a part edge or surface to position the hole in the first direction.

Step 8: On the Hole dialog box, enter a distance to locate the hole from the first linear reference.

Step 9: On the work screen, select a reference edge or surface for locating the hole.

Enter a value for the distance that the hole will be located from the selected edge.

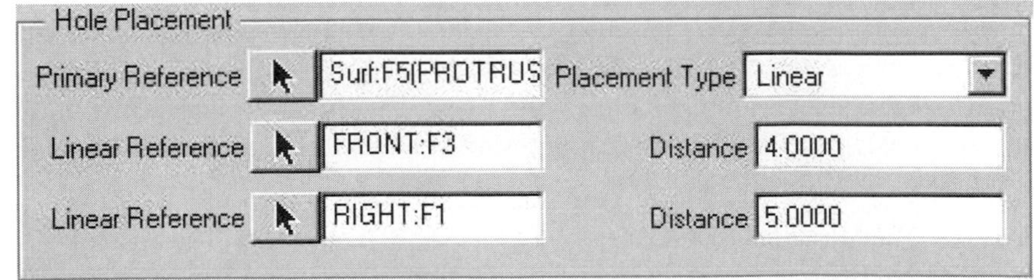

Figure 5–4 Linear hole placement references

STEP 10: On the Hole dialog box, enter a distance to locate the hole from the second linear reference.

STEP 11: Select the Build Feature checkmark.

CREATING A STRAIGHT COAXIAL HOLE

The Coaxial option locates the center of a hole coincident with an existing axis. Perform the following steps to create a straight-coaxial hole.

STEP 1: Select FEATURE >> CREATE >> HOLE.

STEP 2: Select STRAIGHT hole type.

A Coaxial hole can be either a Straight Hole, a Sketched Hole, or a Standard Hole.

STEP 3: Enter a Diameter value for the hole.

STEP 4: Select a Depth One parameter (e.g. Thru All, Variable, etc.), then perform the operation appropriate to the option.

> **INSTRUCTIONAL NOTE** If you select a depth parameter such as Variable, Thru Until, and so forth, you will have to perform an additional step. As an example, if you select the To Reference option, you will have to pick a point, curve, or surface to cut the hole to.

STEP 5: If required, select a Depth Two parameter, then perform the operation appropriate to the option.

The Depth Two parameter creates a Both Sides extruded hole.

STEP 6: On the work screen, select the axis along which to place the hole's centerline.

The axis selection for a coaxial hole is the primary reference.

STEP 7: Select the Hole's placement plane.

STEP 8: Select the Build Feature checkmark.

ROUNDS

Pro/ENGINEER utilizes the Round command to create both fillets and rounds. Despite its apparent simplicity, the Round command can be one of Pro/ENGINEER's most difficult tools to master. Rounds constructed on complex features often result in failures. The following modeling techniques should be used to help avoid round conflicts.

- Create rounds toward the end of the modeling process.
- Create smaller radii rounds before larger.
- Avoid using round geometry as references for the creation of features.
- If a surface is to be drafted, draft the surface first, then create any necessary rounds.

ROUND RADII OPTIONS

Several different Round options are available within Pro/ENGINEER (Figure 5–5).

- **Constant** The Constant option creates a round with a constant radius.
- **Variable** The Variable option creates a round with a variable radius. Radii values are defined from the end of chained segments.

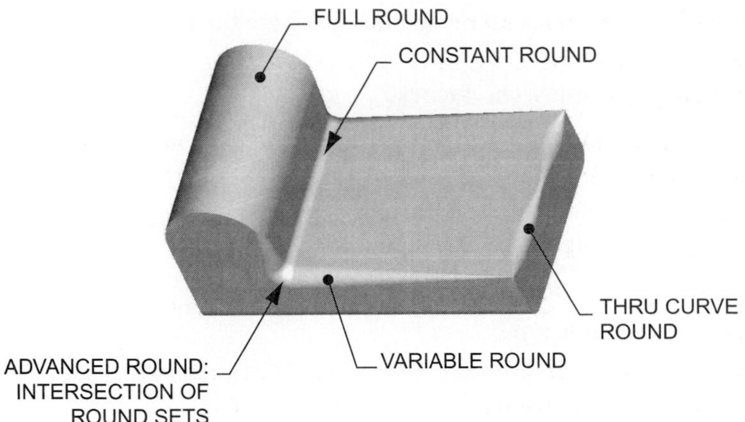

Figure 5-5 Round options

- **Thru Curve** The Thru Curve option defines a round's radius based on a selected curve.
- **Full Round** The Full Round option creates a round in place of a selected surface.

ROUND REFERENCE OPTIONS

Rounds are normally created on the edge of a feature. Pro/ENGINEER provides the Edge Chain option to perform this task. Other options are available to allow for flexibility in the round creation process.

- **Edge Chain** This option allows for the selection of edges to place rounds. Suboptions exist for selecting edges one at a time or for selecting tangent edges.
- **Surf-Surf** This option allows for the selection of two surfaces to place a round. The round will be formed between the two surfaces.
- **Edge-Surf** This option places a round between a selected surface and a selected edge.
- **Edge Pair** This option is similar to the Full Round radius option. With this option, the surface between two selected edges will be replaced with a round.

CREATING A SIMPLE ROUND

Perform the following steps to create a simple Round while utilizing the Edge Chain option.

STEP 1: Select FEATURE >> CREATE >> ROUND.

STEP 2: Select SIMPLE >> DONE on the Round Type menu.

An advanced round type is also available. The Advanced option allows for more complex rounds. As an example, advanced rounds often have varying cross sections and multiple round sets.

STEP 3: Select the CONSTANT >> EDGE CHAIN >> DONE options on the Round Set Attribute menu.

Round Radius options available include Variable, Thru Curve, and Full Round. Other Round Reference options include Surf-Surf, Edge-Surf, and Edge Pair.

STEP 4: Select ONE-BY-ONE on the Chain menu.

The following chain options exist:

- **One By One** Individual edges are selected.
- **Tangent Chain** Edges that lie tangent are selected (default selection).
- **Surf Chain** Edges are defined by selected surfaces.
- **Unselect** This option allows for the unselection of a reference.

STEP 5: Select feature edges to round.

Due to the **One-By-One** option selection, each edge to round must be selected.

STEP 6: Select DONE from the Chain menu.

When you are through selecting edges to round, select Done.

STEP 7: Enter a Radius value for the edges.

STEP 8: Select PREVIEW on the Round Feature definition dialog box.

Previewing a feature, especially rounds, is an important step. Most failed features will be revealed with the Preview option. Additional options are available for resolving any conflicts.

STEP 9: Select OKAY on the Round Feature definition dialog box.

CHAMFER

Pro/ENGINEER provides a construction option for creating edge and corner chamfers (Figure 5–6). An Edge Chamfer creates a beveled surface along a selected edge. The Corner Chamfer option creates a beveled surface at the intersection of three edges. Perform the following steps to create an Edge Chamfer.

STEP 1: Select FEATURE >> CREATE >> CHAMFER.

The Chamfer command creates a beveled surface at a selected solid edge. Creating a chamfer does not require entering a sketching environment.

STEP 2: Select EDGE as the chamfer method.

You have the option of selecting either Edge or Corner (Figure 5–6).

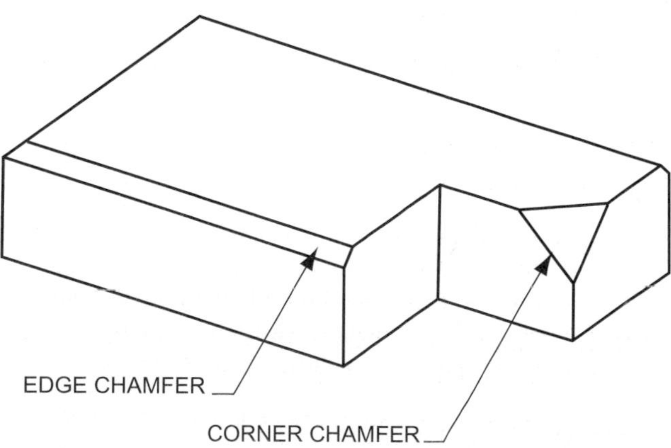

Figure 5–6 Chamfer types

STEP 3: Select an edge chamfer dimensioning scheme.

The following dimensioning schemes are available.

- **45 × d** This option creates a chamfer with a 45-degree angle and with a user-specified distance.
- **d × d** This option creates a chamfer at a user-specified distance from an edge.
- **d1 × d2** This option creates a chamfer at two specified distances from an edge.
- **Ang × d** This option creates a chamfer at a user-specified distance from an edge and at a user-defined angle.

STEP 4: In Pro/ENGINEER's textbox, enter appropriate dimension values.

STEP 5: Select Edges to chamfer.

A single edge or multiple edges can be selected. Use the **Query Sel** option for edge selection if necessary.

STEP 6: Select DONE SEL on the Get Select menu.

Select Done Select after selecting edges to chamfer.

STEP 7: Select DONE REFS on the Feature Refs menu.

The Feature Refs menu allows for the selection of additional edges to chamfer.

STEP 8: Select PREVIEW on the Feature Definition dialog box.

MODELING POINT Chamfer parameters such as selected edges and dimensioning scheme can be modified within the Chamfer Feature definition dialog box. Select the parameter to modify then select the Define button.

STEP 9: Select OKAY on the Feature Definition dialog box.

DRAFT

Cast and molded parts often require a **drafted surface** for ease of removal from the mold. Under the Tweak menu option (Feature >> Create >> Tweak), Pro/ENGINEER provides a command for creating drafted surfaces. The maximum angle that may be created is plus-or-minus 30 degrees.

MODELING POINT The Tweak menu provides a variety of options for modifying an existing part surface. Draft is just one option available. Other options include Offset for offsetting a surface and Radius Dome for creating a dome from a selected surface.

NEUTRAL PLANES AND CURVES

Selected surfaces are drafted by pivoting around a neutral plane or curve. Planes may be surface planes or datum planes, while curves may be datum curves or edges. Additionally, a surface can be split at the neutral plane or curve.

NEUTRAL PLANE DRAFTS

With a Neutral Plane Draft, picked surfaces are pivoted around a selected neutral plane. The plane can be a part surface or a datum. Three split options are available (Figure 5–7).

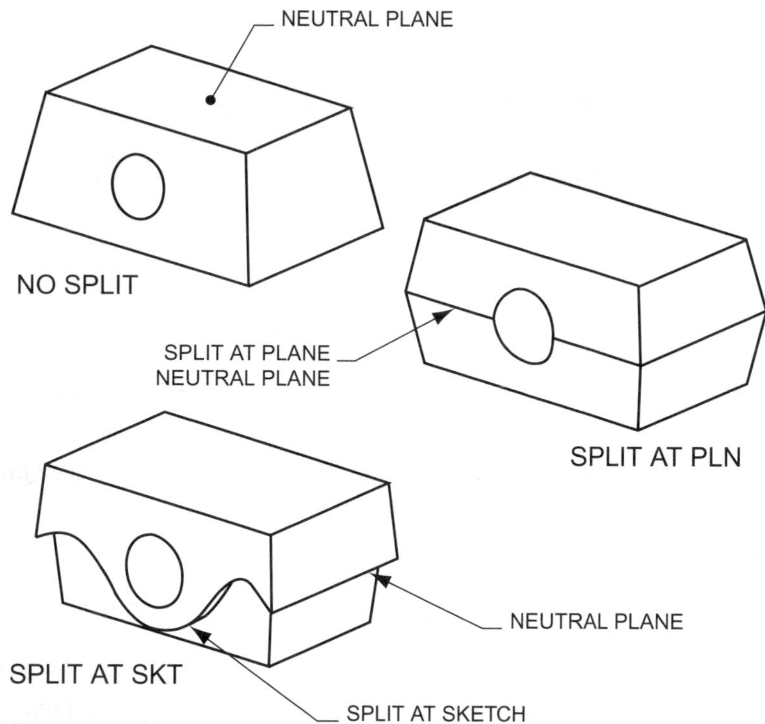

Figure 5–7 Neutral plane drafts

No Split

The No Split option creates a draft without a split in the drafted surface (Figure 5–7). The user selects the neutral plane and the draft surface, then enters a draft angle. The angle can be positive or negative.

Split at Plane

The Split at Plane option creates a draft with the drafted surface split at the neutral plane (Figure 5–7). The portion of the surface selected for drafting is the surface that the draft angle is applied to. The user selects the neutral plane and the draft surface, then enters a draft angle.

Split at Sketch

The Split at Sketch option creates at drafted surface out of a user-sketched section (Figure 5–7). The sketched section is pivoted around the neutral curve. The user selects the neutral plane and a surface to sketch upon. A sketcher environment provides the user with the requirement of sketching the surface to be drafted.

Creating a Neutral Plane No Split Draft

Perform the following steps to create a neutral plane draft. The part (*draft1.prt*) shown in Figure 5–8 is available on the supplemental CD.

Step 1: Select the FEATURE >> CREATE >> TWEAK menu option.

Step 2: Select the DRAFT command on the Tweak menu.

Step 3: Select NEUTRAL PLN >> DONE on the Draft Options menu.

Draft options available include Neutral Plane for pivoting around a plane or surface and Neutral Curve for pivoting around a datum curve or edge.

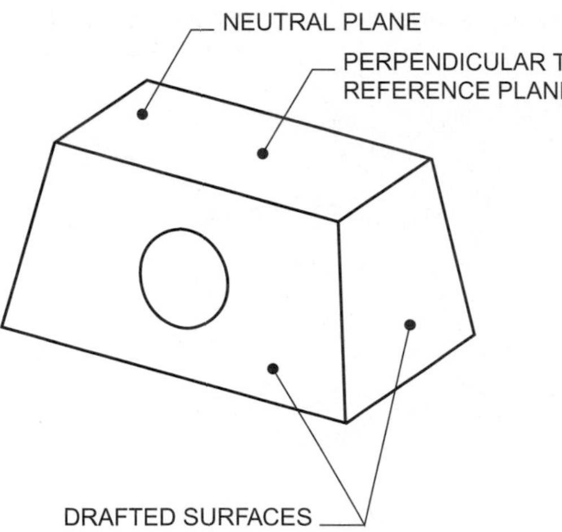

Figure 5–8 Neutral plane no split draft

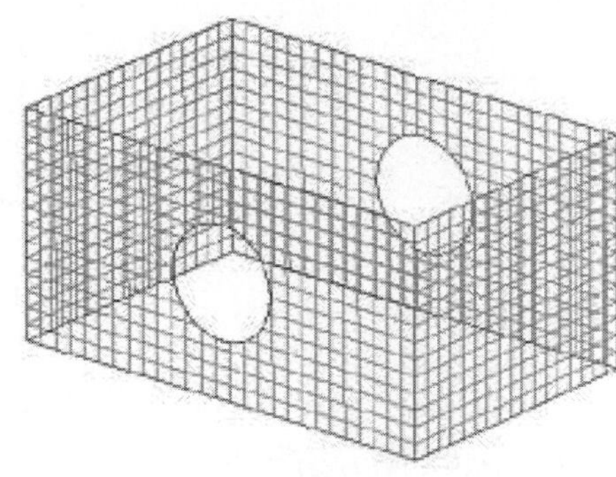

Figure 5–9 Meshed surface

STEP 4: Select TWEAK >> NO SPLIT >> CONSTANT >> DONE on the Draft Attributes menu.

Using the Intersect option in place of Tweak would allow the drafted surfaces to overhang the edge of an adjacent surface. No Split will create a draft on the complete side of a selected surface. The Split at Plane option will create a split surface at the neutral plane, while the Split at Sketch option will create a draft out of a user-sketched surface.

STEP 5: Select surfaces to draft.

The default option is to select individual surfaces.

STEP 6: Select DONE SEL on the Get Select menu.

Selecting Done Sel will return you to the Surface Select menu. The Surface Select menu provides you with options for adding surfaces, excluding surfaces, and showing existing selected surfaces.

STEP 7: Select SHOW >> MESH on the Surface Select menu.

The Show >> Mesh option will show selected surfaces as a mesh (Figure 5–9).

STEP 8: Select DONE on the Surface Select menu.

STEP 9: Select a NEUTRAL PLANE (Figure 5–8).

The Neutral Plane is the part surface that will remain intact during the drafting procedure. Since the drafted surfaces will pivot from the neutral plane, this is a critical selection. The Make Datum option exists for creating an on-the-fly datum plane.

STEP 10: Select a reference perpendicular plane (Figure 5–8).

The reference plane must lie perpendicular to the draft surfaces. Often, this reference plane is the same as the neutral plane.

STEP 11: Enter a draft angle.

As shown in Figure 5–10, Pro/ENGINEER will graphically display the positive direction of rotation. A negative value can be entered.

STEP 12: PREVIEW the draft on the Draft Feature Definition dialog box.

STEP 13: Select OKAY on the dialog box to complete the draft.

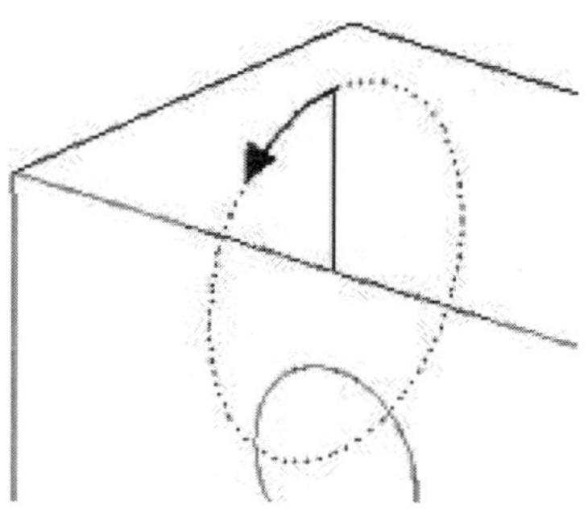

Figure 5-10 Draft angle direction

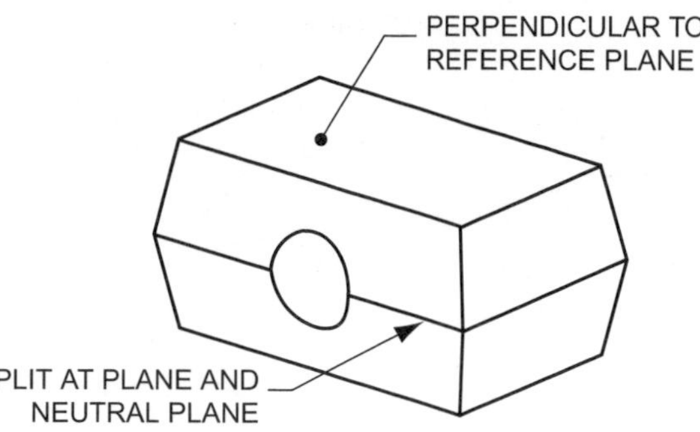

Figure 5-11 Neutral plane split draft

Creating a Neutral Plane Split Draft

Perform the following steps to create a neutral plane draft with a split surface (see Figure 5–11). The part for this guide (*draft2.prt*) is available on the supplemental CD.

Step 1: Select the FEATURE >> CREATE >> TWEAK menu option.

Step 2: Select the DRAFT command on the Tweak menu.

Step 3: Select NEUTRAL PLN >> DONE on the Draft Options menu.

Draft options available include Neutral Plane for pivoting around a plane or surface and Neutral Curve for pivoting around a datum curve or edge.

Step 4: Select TWEAK >> SPLIT AT PLN >> CONSTANT >> DONE on the Draft Attributes menu.

Using the Intersect option in place of Tweak would allow the drafted surfaces to overhang the edge of an adjacent surface. The Split at Plane option will create a split surface at the neutral plane. The No Split option will create a draft of the complete side of a selected surface, while the Split at Sketch option will create a draft out of a user-sketched surface.

Step 5: Select surfaces to draft.

The default selection option is to select individual surfaces.

Step 6: Select DONE on the Surface Select menu.

Step 7: Select the Neutral Plane.

The drafted surfaces will be pivoted around the neutral plane. The Make Datum option is available.

Step 8: Select a perpendicular plane as a reference.

Step 9: Enter a draft angle.

Pro/ENGINEER will graphically display the positive direction of rotation. Negative values can be entered.

Step 10: Select PREVIEW on the Draft Feature Definition dialog box.

Step 11: Select OKAY on the dialog box.

Neutral Curve Drafts

The Neutral Curve option allows for the creation of a draft surface by pivoting around a datum curve or edge. Neutral curve drafts can be used in two situations. First, Neutral Plane drafts require the selection of a reference plane that is perpendicular to the drafted surface. Neutral Curve drafts do not require this. Second, Neutral Curve drafts allow for nonplanar surfaces as the neutral selection. There are three split options, available (Figure 5–12).

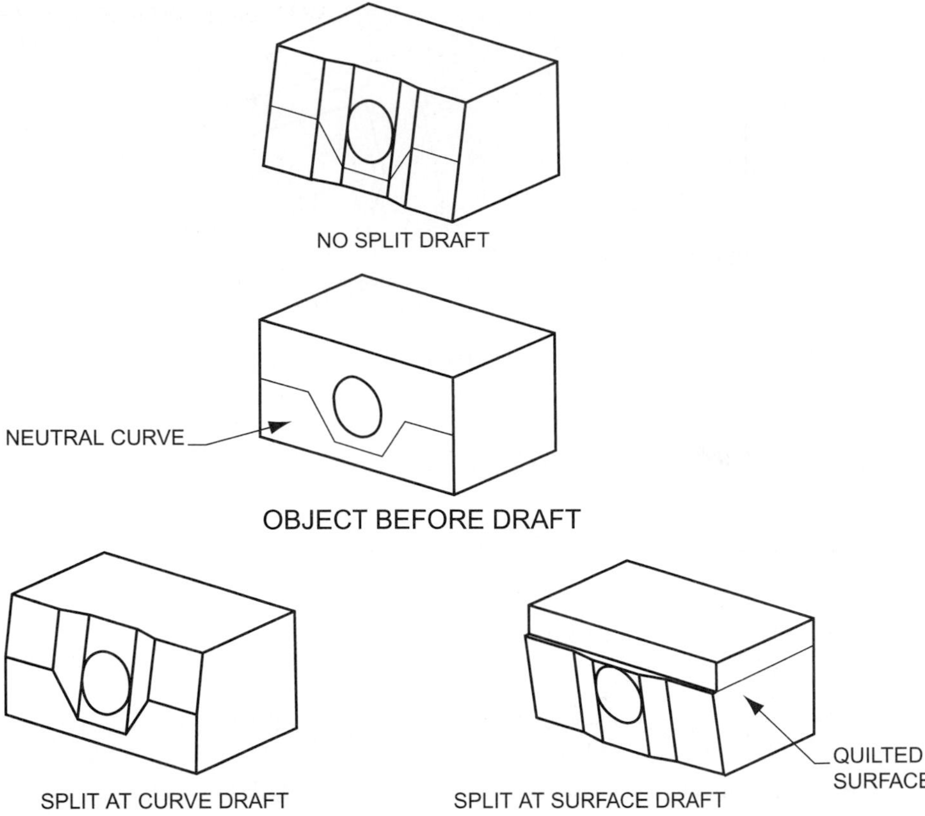

Figure 5–12 Neutral curve drafts

No Split

The No Split option does not allow for a split of the drafted surface at the Neutral Curve (Figure 5–12).

Split at Curve

The Split at Curve option splits a drafted surface at a Neutral Curve (Figure 5–12). Drafted surfaces can be created on both sides or just one side of the Neutral Curve.

Split at Surface

The Split at Surface option splits a drafted surface at a quilted surface. As shown in Figure 5–12, this option requires the selection of Neutral Curves on both sides of the quilted surface. As with the Split at Curve option, the drafted surfaces can be created on both sides or just one side of the splitting surface.

Neutral Curve draft surfaces created with the Split at Curve and Split at Surface options allow for draft surfaces to be created on Both Sides or One Side of the splitting surface. With the Both Sides option, the surfaces on both sides of the splitting curve or surface can have the same draft angle or different draft angles. The Independent option allows for the specification of different draft angles, while the Dependent option allows for one shared draft angle.

Creating a Neutral Curve Draft

This step-by-step guide will detail the construction of a Neutral Curve Draft. The Split at Curve >> Both Sides >> Independent suboptions will be highlighted. The part for this guide (*draft3.prt*) is available on the supplemental CD (Figure 5–13).

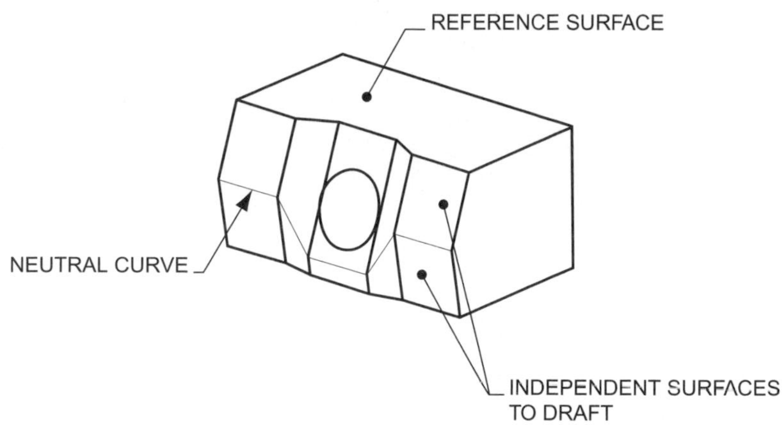

Figure 5–13 Neutral curve split at curve draft

STEP 1: Select FEATURE >> CREATE >> TWEAK >> DRAFT.

STEP 2: Select NEUTRAL CRV >> DONE on the Draft Options menu.

STEP 3: Select TWEAK >> SPLIT AT CRV >> BOTH SIDES >> INDEPENDENT >> CONSTANT >> DONE on the Attributes menu.

Available split options on the Attributes menu include No Split, Split at Curve, and Split at Surface. The Both Sides option will allow drafts to be created on both sides of the neutral curve, with the Independent option allowing the draft angles on each side of the splitting curve to be different. Using the Intersect option in place of Tweak would allow the drafted surfaces to overhang the edge of an adjacent surface.

STEP 4: Select surfaces to add to the draft (Figure 5–13).

Select surfaces to include in the draft operation. The default option is selecting individual surfaces.

STEP 5: Select DONE on the Surface Select menu when all appropriate surfaces are selected.

STEP 6: On the model, select appropriate entities that will compose the neutral curve.

Use the mouse to pick a neutral curve. Selection options include One-by-One, Tangent Chain, Curve Chain, Boundary Chain, and Surface Chain.

STEP 7: Select DONE from the Chain menu when all Neutral Curves are selected.

STEP 8: Select a reference plane perpendicular to the drafted surfaces.

Select a reference to determine the angle of draft.

STEP 9: Observe the graphic angle of rotate symbol provided by Pro/ENGINEER and enter a draft angle for the first surface.

STEP 10: Enter a draft angle for the second surface.

STEP 11: Preview the feature on the Draft Definition dialog box.

STEP 12: Select OKAY on the dialog box to create the draft.

SHELLED PARTS

The Shell command removes a selected surface from a part and hollows the part with a user-defined wall thickness. The wall is created around the outside surfaces of the part, with any features created before the shell feature being included (Figure 5–14). Perform the following steps to create a shelled part.

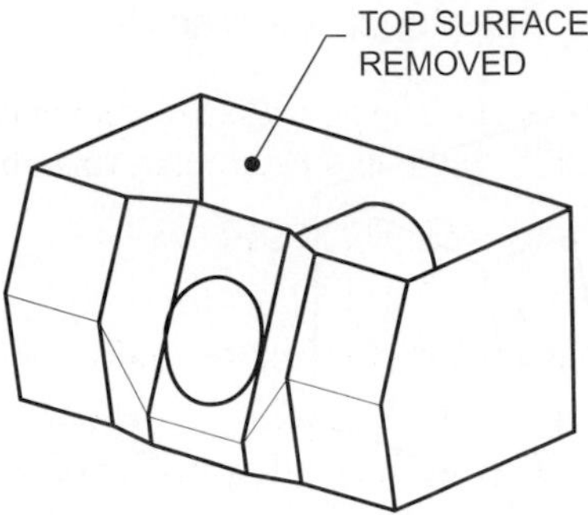

Figure 5-14 Shelled feature

STEP 1: Select FEATURE >> CREATE >> SHELL.

STEP 2: Select surfaces to remove.

The Shell command functions by removing selected surfaces from the part. Select a surface or surfaces to remove.

STEP 3: Select DONE SEL on the Get Select menu.

The Done Sel option ends the selection of surfaces to be removed by the Shell command.

STEP 4: Select DONE REFS from the Feature Refs menu.

The Feature References menu allows for the addition or removal of selected shell surfaces.

STEP 5: Enter a shell wall thickness.

STEP 6: Select PREVIEW on the feature definition dialog box.

STEP 7: Select OKAY on the dialog box.

RIBS

A Rib is a thin web feature created between features on a part (Figure 5–15). It can be revolved or straight. Ribs are similar to protrusions that are extruded both sides from a sketching plane, with the exception that the section of a rib must be open. Additionally, the ends of a section of a rib must be aligned with existing part surfaces. Ribs sketched on a Through >> Axis datum plane and referenced to a surface of revolution form a conical surface on the top of the rib (Figure 5–16).

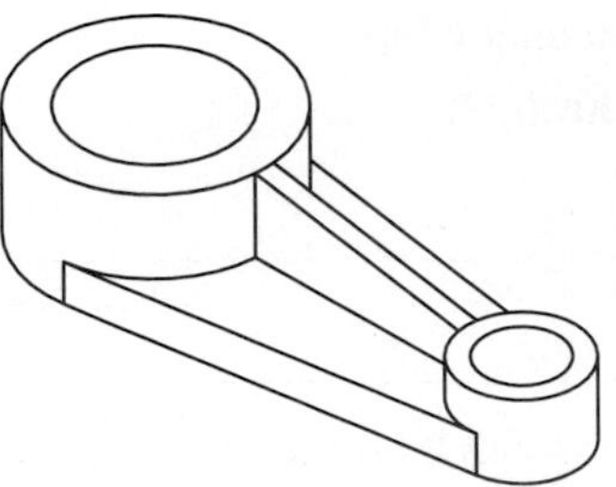

Figure 5-15 Rib on a part

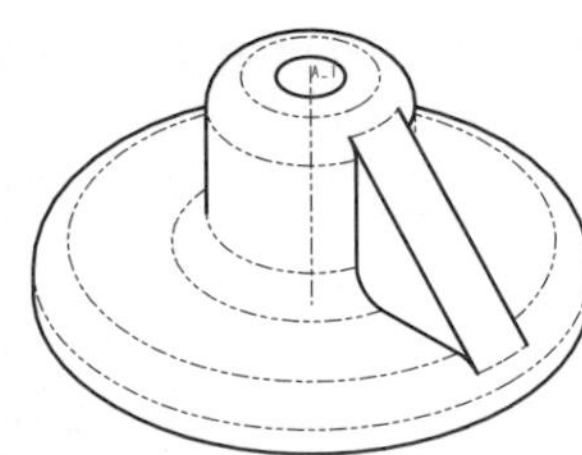

Figure 5-16 Rib on a revolved feature

Creating a Rib

Figure 5–17 shows a part before and after a rib has been created. The part (*rib.prt*) shown in the figure is available on the supplemental CD. Perform the following steps to create this rib.

Step 1: Starting with the part shown in Figure 5–17, select FEATURE >> CREATE >> RIB.

Step 2: Select a datum plane as the sketching plane for the rib (Figure 5–18).

A rib must be sketched on a datum plane. In Figure 5–18, datum plane FRONT intersects the middle of the two cylindrical features and is used in this example.

Step 3: Orient the sketching environment.

Step 4: Within the sketcher environment, specify references to meet design intent (Figure 5–19).

The sketched entity defining the rib will be aligned with each selected edge.

Step 5: Sketch the outline of the rib feature (Figure 5–20).

Sketch a line from the endpoints shown in Figure 5–20. This will be the only entity required to define the section.

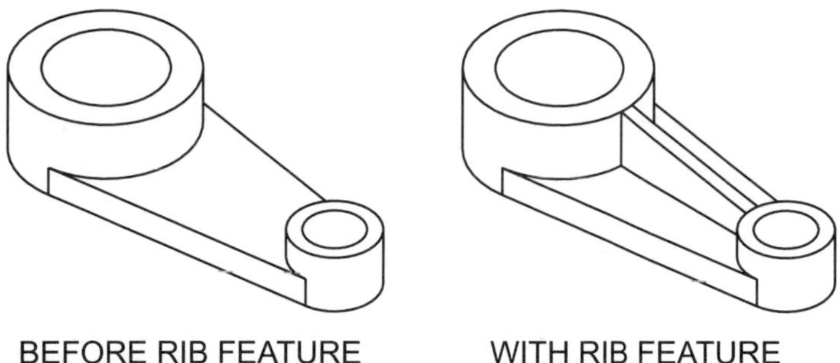

Figure 5-17 Creating a rib

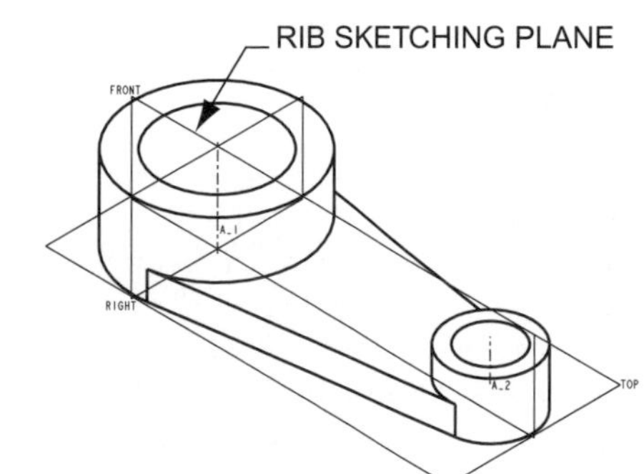

Figure 5-18 Rib sketching plane

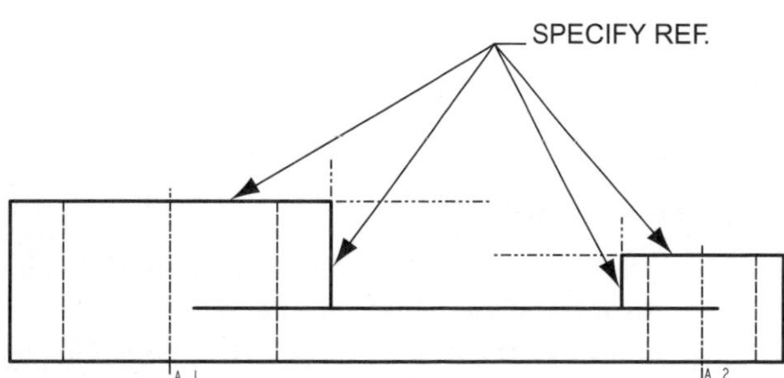

Figure 5-19 Specifying references for a rib

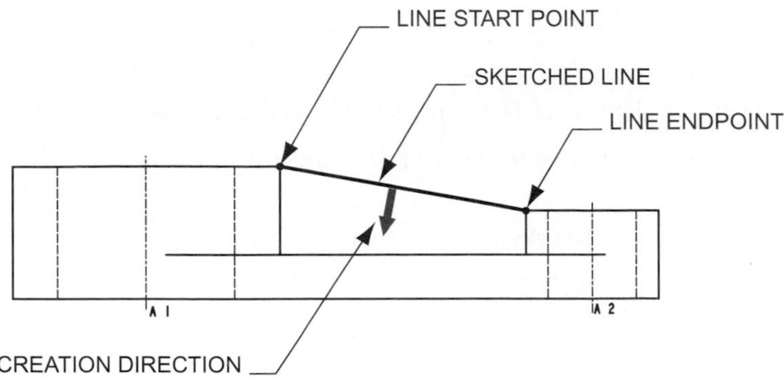

Figure 5-20 Sketching the rib

STEP 6: Select the Continue option to exit the sketcher environment.

STEP 7: The direction of extrusion should point toward the part. If necessary, select FLIP on the direction menu, then select OKAY.

STEP 8: Enter an Extrusion thickness.

This extrusion thickness functions similarly to the Both Sides option. The thickness provided will extrude both directions from the sketching plane.

COSMETIC FEATURES

Cosmetic features enhance the display of parts without complicated geometric features that require regenerations. Since cosmetic features are considered part features, they can be redefined and modified. Unlike geometric features, the line style defining a cosmetic feature (with the exception of cosmetic threads) can be modified using the Modify >> Line Style option. This option allows for the changing of line color, font, and style. Four cosmetic features are available: sketched, threads, grooves, user-defined, and ECAD Areas.

SKETCHED COSMETIC FEATURES

Sketched cosmetic features are useful for including names and logos on a part. These cosmetic features are sketched on a part surface using normal sketching environment techniques (Figure 5–21). They can be constructed with or without feature parameters. To create a nonparametric sketched cosmetic feature, delete the dimensions defining the feature before exiting the sketcher environment.

COSMETIC GROOVES

Cosmetic Grooves are sketched sections projected onto a part surface. The section is sketched with normal sketcher tools. The projected cosmetic feature has no defined depth. An example of a Cosmetic Groove is shown in Figure 5–21. In this example, the Cosmetic Groove was sketched on the same plane as the Cosmetic Sketch and projected on to the receiving surfaces.

COSMETIC THREADS

Threads can be created within Pro/ENGINEER using the Helical option under the Protrusion and Cut commands. Despite this, often it is satisfactory to only symbolically display a thread. The Cosmetic Thread option allows for the creation of thread symbols that correspond to a simplified thread representation (Figure 5–22).

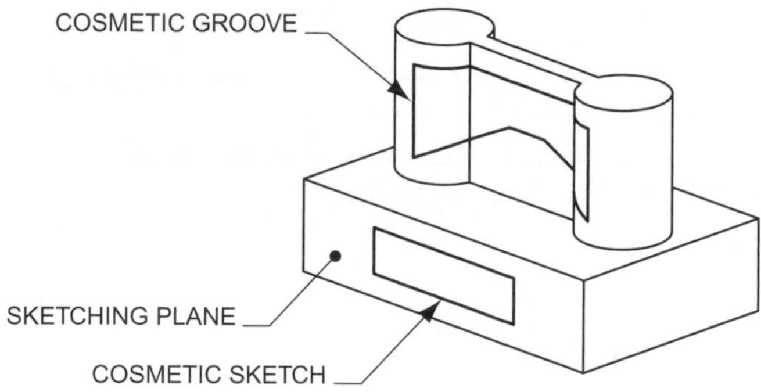

Figure 5-21 Sketched cosmetic feature

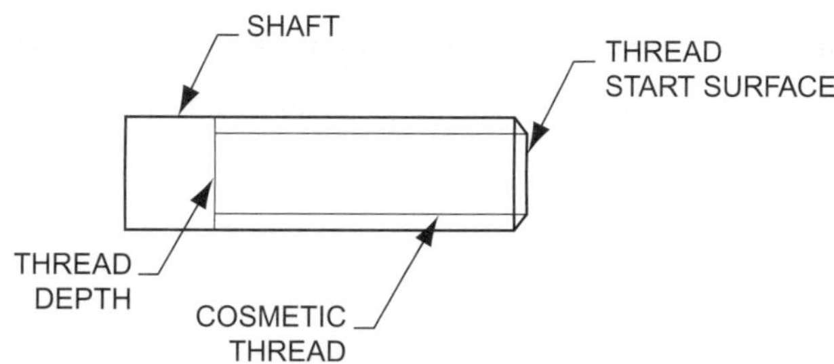

Figure 5-22 Cosmetic threads

The parameters that define a cosmetic thread resemble parameters associated with true threads. As an example, internal and external threads can be defined as well as the major and minor diameters. The following are definable parameters for a cosmetic thread.

- **Major diameter** This parameter allows for the defining of the thread's major diameter. For external threads, the default value is 10 percent smaller than the thread surface diameter; for internal threads, the default value is 10 percent larger.
- **Threads per inch** This parameter defines the number of threads per inch.
- **Thread form** This parameter allows for the selection of a thread form. An example would be the attribute UN (Unified National).
- **Thread class** The class of a thread is defined with this parameter. Examples include Fine, Extra Fine, and Coarse.
- **Placement** This parameter defines an external or internal thread.
- **Metric** This parameter sets the thread to metric (true) or not to metric (false).

Creating a Cosmetic Thread

Perform the following steps to create a cosmetic thread.

Step 1: Select FEATURE >> CREATE >> COSMETIC.

Notice that the feature creation options include solid features, datum features, user-defined features, and cosmetic features.

Step 2: Select the THREAD option from the Cosmetic menu.

Selecting the Thread option will display the Cosmetic Thread dialog box. This dialog box displays the parameters required for the creation of a cosmetic thread.

STEP 3: Select the Thread Surface.

On the work screen, select the cylindrical surface upon which to place the threads. Only one surface can be selected.

STEP 4: Select the starting surface for the thread (Figure 5–22).

The starting surface may be a part surface, quilt, or datum plane.

STEP 5: Choose the direction of thread creation.

A red arrow displays the default thread direction. Select OKAY to accept the direction or select Flip to change the direction.

STEP 6: Select a thread depth specification, then provide appropriate depth information.

Depth specifications include Blind, UpTo PntVtx, UpTo Curve, and UpToSurface. After selecting a specification, provide the necessary information that corresponds to the selected specification.

STEP 7: Enter a thread diameter.

The default diameter is displayed. For external threads, the diameter is 10 percent smaller than the thread surface diameter. For internal threads, the diameter is 10 percent larger than the thread surface diameter.

STEP 8: Select one or more options from the Feature Parameters menu.

The following options are available on the Feature Parameters menu:

- **Retrieve** This option allows for the retrieval of an existing thread parameter.
- **Save** This option allows a defined thread parameter to be saved for later use.
- **Mod Params** This option allows for the modification of a thread's parameters.
- **Show** This option shows parameters set for a thread.

STEP 9: Select DONE/RETURN on the Feature Parameters menu.

STEP 10: Select PREVIEW on the Cosmetic Thread dialog box then select OKAY.

PATTERNED FEATURES

The Pattern command is used to create multiple instances of a feature. A feature is patterned by varying and duplicating one or more of its parametric dimensions. Instances of a pattern are copies of the feature. An instance is an exact duplicate of the parent feature. Two pattern configurations are available: linear and angular (Figure 5–23). One example of an angular pattern would be copying a hole around an axis, as with a bolt-circle pattern. Patterned features can be used effectively with dimensional relations to enhance the design intent of a model.

Two types of patterns are available: dimension patterns and reference patterns. With dimension patterns, varying one of the dimensions that defines the feature creates the new instances. For reference patterns, referencing a previously created pattern creates the new instances. The varying dimensions from the reference pattern govern the new pattern.

PATTERN OPTIONS

Three pattern options are available within Pro/ENGINEER. The decision of which option to use is based on the complexity of the part. Less complex patterns allow for more

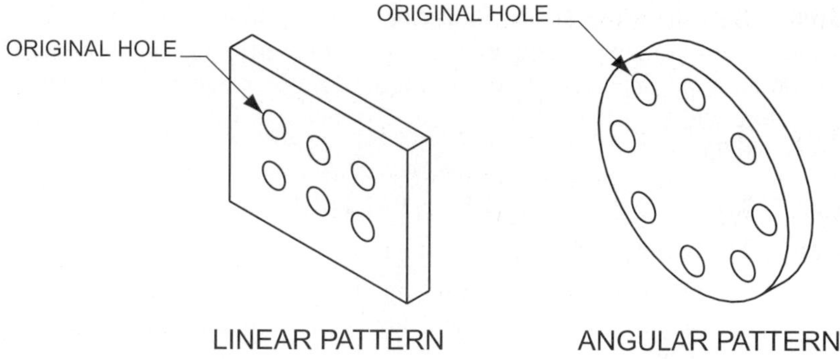

Figure 5-23 Pattern configurations

assumptions during the pattern construction. This allows Pro/ENGINEER to regenerate the pattern quicker. The following pattern options are available.

Identical Patterns

Identical patterns are the least complex and allow for the most assumptions. Identical pattern instances must be of the same size and must lie on the same placement surface. Identical pattern instances cannot intersect other features, instances, or the edge of the placement plane.

Varying Patterns

Varying patterns are more complex than identical patterns. Varying pattern instances can vary in size and can lie on different placement surfaces. As with identical patterns, varying pattern instances cannot intersect other instances.

General Patterns

General patterns are the most complex and require longer to regenerate. With general patterns, no assumptions are made during the construction process. Instances can intersect other instances and placement plane edges. Additionally, instances can vary in size and lie on different surfaces.

Dimensions Variation

Dimensions can be varied using three options: Value, Relation, and Table. The following is a description of each option:

- **Value** With the value option, dimension values are incremented.
- **Relation** With the relation option, relations are used to drive dimension variations.
- **Table** With the table option, dimension variations are controlled by a table.

MODELING POINT When creating a feature that will be patterned, placement of dimensions defining the feature is critical. As an example, when creating a linear pattern, the dimensioning scheme defining the location of the feature will be used to create the pattern. For angular patterns, an angular dimension must exist.

Creating a Linear Pattern

Linear patterns are created by varying linear dimensions. A feature can be patterned unidirectionally or bidirectionally. Additionally, more than one dimension of a feature can be

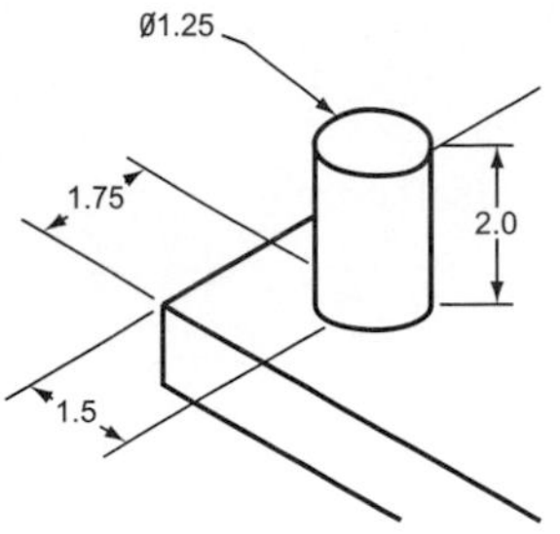

Figure 5–24 Feature to be patterned

varied during the process. The following is a step-by-step approach for patterning the cylindrical feature shown in Figure 5–24. The part used in this guide (*pattern1.prt*) is available on the supplemental CD. As shown, the 1.75-inch and 1.5-inch dimensions locating the feature will be patterned, as will the 2.0-inch and 1.25-inch dimensions defining the height and the diameter.

STEP 1: Select the FEATURE >> PATTERN command.

STEP 2: Select the feature to pattern.

On the work screen or on the Model Tree, select the cylindrical feature (Figure 5–24). With the Pattern command, only one feature can be pattern at a time.

STEP 3: Select VARYING >> DONE as the pattern option.

Other pattern options include Identical and General. Unlike the Identical option, the Varying option allows dimensional values on each pattern instance to vary. Also, unlike Identical, instances can lie on different placement surfaces.

STEP 4: Select VALUE on the Pattern Dimension Increment menu.

Value is selected by default. Other options include Relation, Table, and Redraw Dims.

STEP 5: Pick the 1.5-inch leader dimension for patterning in the first direction (Figure 5–25).

On the work screen, select the first dimension to vary. This pattern will copy the feature in the first direction.

STEP 6: Enter 2.75 as the dimension increment value.

STEP 7: As shown in Figure 5–25, pick the height dimension for varying.

The height value will be varied in the first direction.

STEP 8: Enter .25 as the dimension increment value.

STEP 9: Select DONE from the Exit menu.

Selecting Done will end the definition of varying dimensions in the first direction.

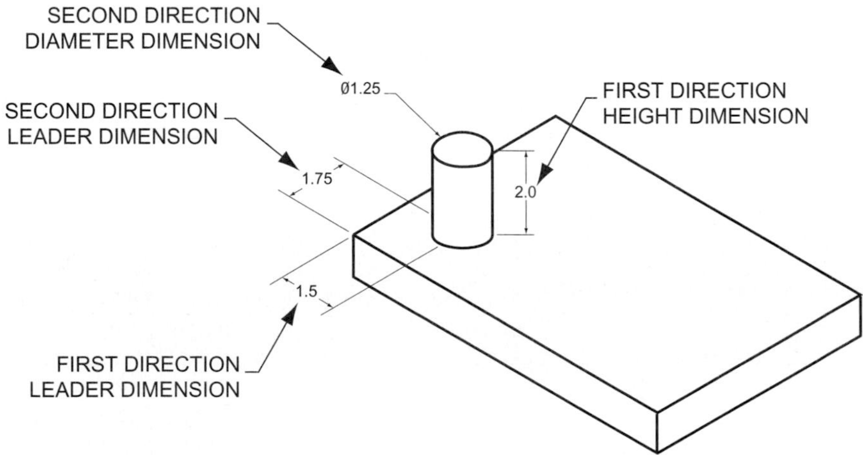

Figure 5–25 Dimension selection

STEP 10: Enter 3 as the number of instances in the first direction.

Three instances of the feature will be created in the first direction.

STEP 11: Select VALUE from the Pattern Dimension Increment menu.

The selection of a Dimension Variation option at this point will create a pattern in two directions. Selecting Done on the Exit menu will create a one direction pattern.

STEP 12: As shown in Figure 5–25, pick the 1.75 inch leader dimension for varying in the second direction.

STEP 13: Enter 2.50 as the dimension increment value.

STEP 14: As shown in Figure 5–25, pick the cylinder diameter dimension for varying.

You will also vary the diameter dimension in the second direction of the pattern.

STEP 15: Enter .500 as the dimension increment value.

STEP 16: Select DONE from the Exit menu.

Selecting Done will end the selection of dimensions for varying.

STEP 17: Enter 2 as the number of instances in the second direction.

Two instances in the second direction in combination with three instances in the first direction will create a total of six instances. Due to the variations selected for the height and the diameter dimensions of the cylinder, each instance of the pattern will be a different size.

STEP 18: Select DONE to create the pattern.

After selecting Done, the pattern will appear as shown in Figure 5–26.

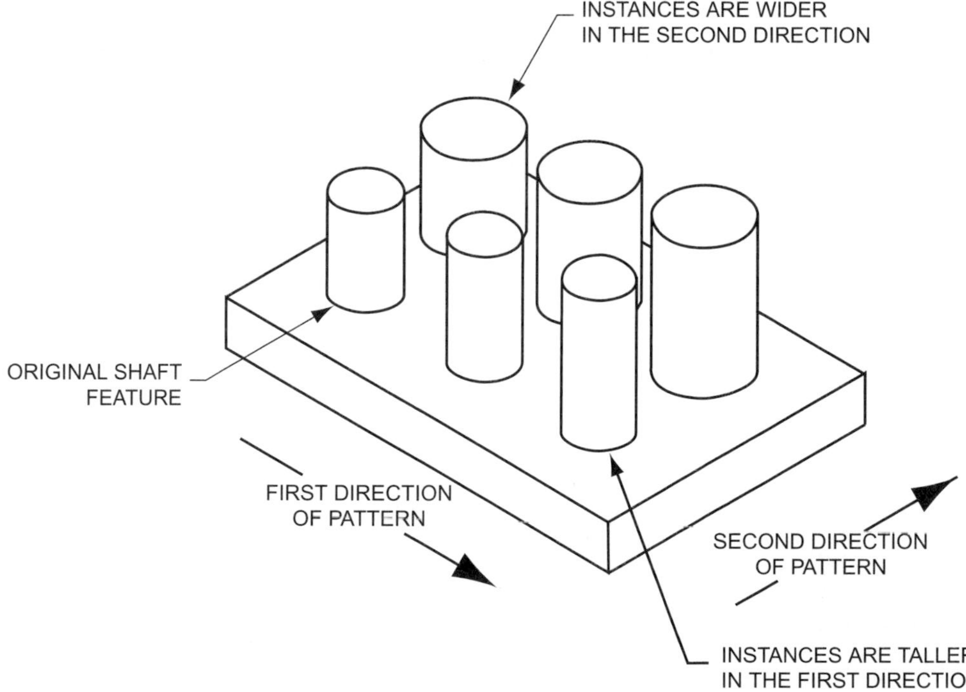

Figure 5–26 Pattern feature

SUMMARY

While the Protrusion and Cuts commands are the primary tools to create features, other options are available that enhance the power of Pro/ENGINEER's modeling capabilities. As an example, if a part has multiple instances of a feature, it would be time consuming to individually model each feature. In this case, Pro/ENGINEER provides the Pattern command to create a rectangular or polar array of the feature. Feature creation tools such as Round, Chamfer, Shell, and Draft are available for unique modeling situations. It is important to understand the capabilities of Pro/ENGINEER's various construction tools and to know when one is appropriate over another.

TUTORIAL

FEATURE CONSTRUCTION TUTORIAL 1

This tutorial exercise will provide instruction on how to model the part shown in Figure 5–27.

Within this tutorial, the following topics will be covered:

- Creating a new object.
- Creating an extruded protrusion.
- Creating a simple round.
- Creating a straight chamfer.
- Creating a standard coaxial hole.
- Creating an advanced round.

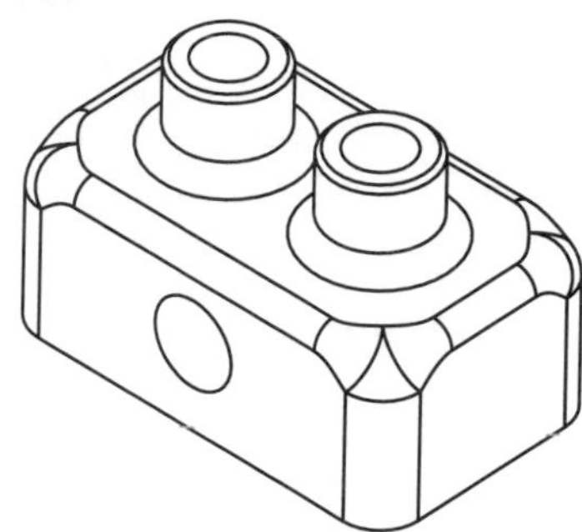

Figure 5-27 Finished model

STARTING A NEW OBJECT

This segment of the tutorial will establish Pro/ENGINEER's working environment.

STEP 1: Start Pro/ENGINEER.

STEP 2: Set Pro/ENGINEER's Working Directory.

From the File menu, select the Set Working Directory option then establish an appropriate working directory for the part.

STEP 3: Start a NEW Pro/ENGINEER part and name it *construct1.*

From the File menu, select New, then create a new part file named ***construct1.*** Use the default template file as supplied by Pro/ENGINEER. Your part model should start with a set of default datum planes, the default coordinate system, and *inch_lbm_second* as the units.

CREATING THE BASE GEOMETRIC FEATURE

This segment of the tutorial will create the extruded feature shown in Figure 5–28.

STEP 1: Select FEATURE >> CREATE >> PROTRUSION.

This step will allow for the creation of a solid protrusion.

STEP 2: Select EXTRUDE >> SOLID >> DONE on the Solid Opts menu.

STEP 3: Select ONE SIDE >> DONE on the Attributes menu.

This feature will be extruded one direction from the sketching plane.

STEP 4: Select datum plane RIGHT as the sketching plane.

STEP 5: Select OKAY on the Direction menu.

Selecting Okay will accept the default feature creation direction.

STEP 6: Select TOP on the Sketch View menu, then pick datum plane TOP on the work screen.

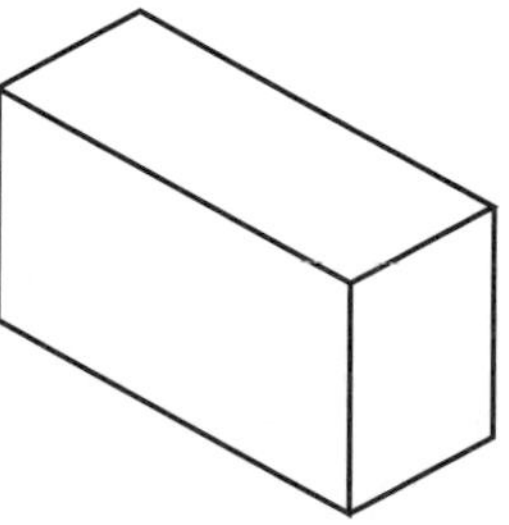

Figure 5-28 Extruded base feature

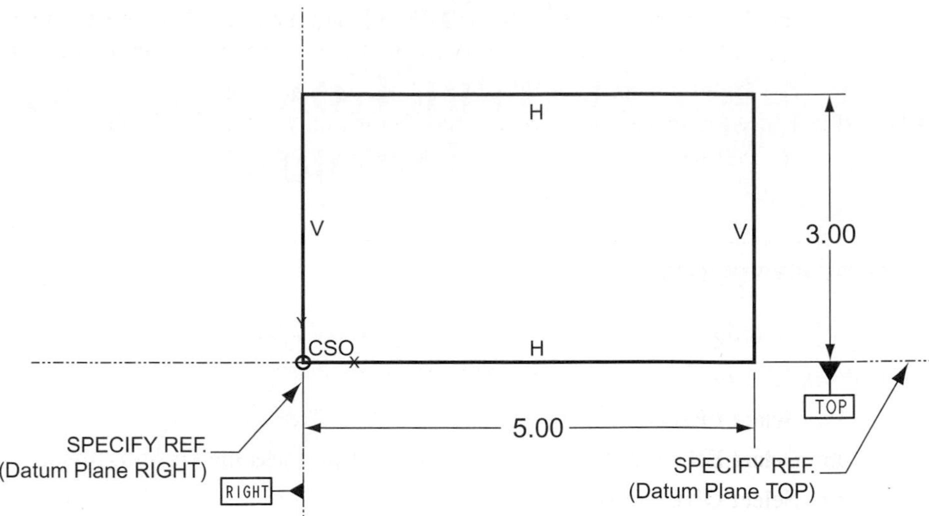

Figure 5-29 Regenerated section

This option will allow datum plane TOP to be oriented toward the top of the sketcher environment. The Default option would perform the same function.

STEP 7: Close the References dialog box.

When no existing geometric features exist for a part, Pro/ENGINEER will automatically specify datum planes RIGHT and FRONT as references. These are the only required references for this model.

STEP 8: Using the LINE icon, sketch the section shown in Figure 5–29.

As shown in Figure 5–29, sketch the Line entities aligned with datum planes RIGHT and TOP. When sketching, do not worry about the size of the entities being sketched. Entity size parameters will be modified in another step. Allow Intent Manager to apply the vertical and horizontal constraints as shown.

> **MODELING POINT** When sketching, your left mouse button is used to pick entity points on the work screen and your middle mouse button is used to cancel commands.

STEP 9: Use the MODIFY icon to match the dimension values shown in Figure 5–29.

Select Modify, then select the dimension value to change. Follow this by entering a new value for the dimension.

STEP 10: Select the Continue icon when the section is complete.

STEP 11: Select BLIND >> DONE as the depth option.

STEP 12: Enter 2.00 as the extrude depth.

STEP 13: Preview the feature.

After selecting the Preview option, shade and dynamically rotate the object. It is not too late to change a parameter defining the feature. Selecting the parameter on the Feature Definition dialog box followed by the Define option will allow the selected parameter to be modified. As an example, the depth of the extrusion can be changed by redefining the depth parameter.

STEP 14: If the Protrusion is correct, select OKAY on the Feature Definition dialog box.

ADDING EXTRUDED FEATURES

An Extruded Protrusion can be sketched on an existing part surface. This section of the tutorial will create the Protrusion shown in Figure 5–30.

STEP 1: Select FEATURE >> CREATE >> PROTRUSION.

STEP 2: Select EXTRUDE >> SOLID >> DONE on the Solid Options menu.

STEP 3: Select ONE SIDE >> DONE.

STEP 4: Select the top surface of the part as the sketching plane (Figure 5–30).

As shown in Figure 5–30, select the top surface of the existing part to use as a sketching plane. Observe the dimensions of the part; care should be taken not to select one of the side surfaces.

STEP 5: Select OKAY to accept the default feature creation direction.

STEP 6: Select DEFAULT to accept the default sketch orientation.

STEP 7: Close the References dialog box.

STEP 8: On the toolbar, turn off the display of datum planes.

STEP 9: On the toolbar, select No Hidden as the model display.

Sketching a new feature on an existing part surface is usually easier with No Hidden or Hidden set as the model display style.

STEP 10: Using the CIRCLE icon, sketch the two circles shown in Figure 5–31.

The Circle option defines a circle by selecting the circle's center point, then dragging the size of the circle. Allow Intent Manager to create a Horizontal Alignment constraint between the centers of the two circles.

STEP 11: Use the DIMENSION icon to match the dimensioning scheme shown in Figure 5–31.

To dimension the location of a circle, with the left mouse button, select the center of the circle, then the locating edge. Place the dimension with the middle mouse button. Diameter dimensions are created by double picking

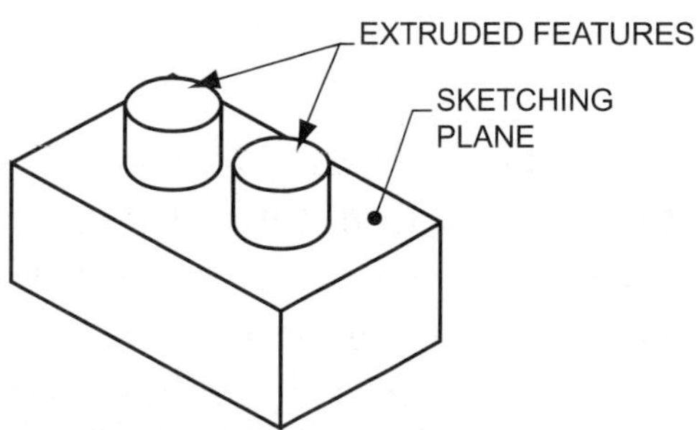

Figure 5–30 Extruded features

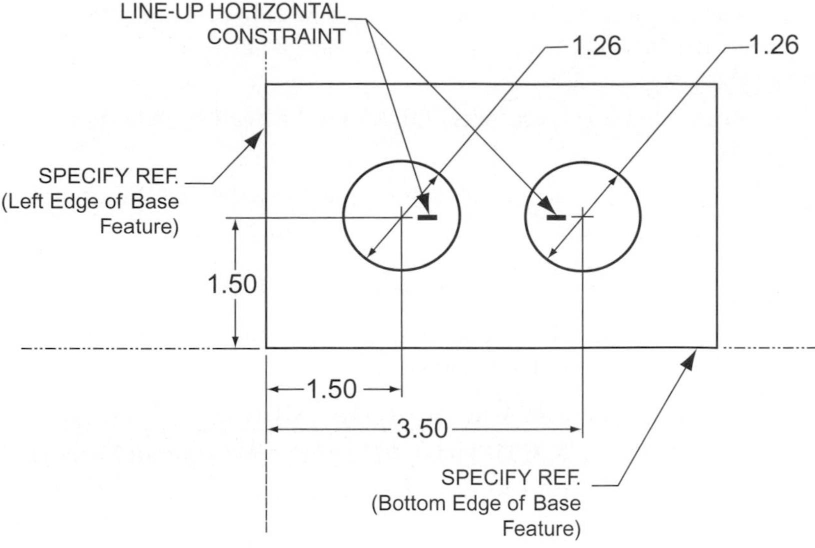

Figure 5–31 Sketched regenerated section

the circle with the left mouse button, followed by placing the dimension with the middle mouse button.

> **MODELING POINT** An alternative to the two separate diameter dimensions shown in Figure 5–31 would be to create an equal radii constraint between the two circles. With an equal radii constraint, one dimension will control the size of both circles. This is a common way to capture design intent. An equal radii constraint can be created through the Constraint icon and the Equal option.

STEP 12: **Using the Select icon, double pick dimensions and modify their values to match Figure 5–31.**

STEP 13: **Select the Continue icon on the Sketch menu when the section is complete.**

STEP 14: **Enter a BLIND depth of 1.00.**

Select Blind as the depth option, then enter a value of 1.00.

STEP 15: **Preview the feature.**

STEP 16: **Select OKAY from the dialog box to create the feature.**

CREATING ROUNDS

This segment of the tutorial will create the simple rounds shown in Figure 5–32.

STEP 1: **Select FEATURE >> CREATE >> ROUND.**

The Round command is available to create rounds and fillets. A round is considered a part feature.

STEP 2: **Select SIMPLE >> DONE on the Round Type menu.**

STEP 3: **Select CONSTANT >> EDGE CHAIN >> DONE on the Round Set Attributes menu.**

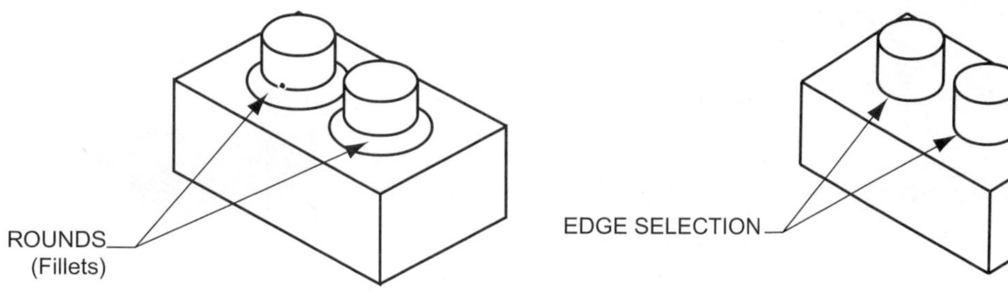

Figure 5–32 Part with simple rounds

Figure 5–33 Round edge selection

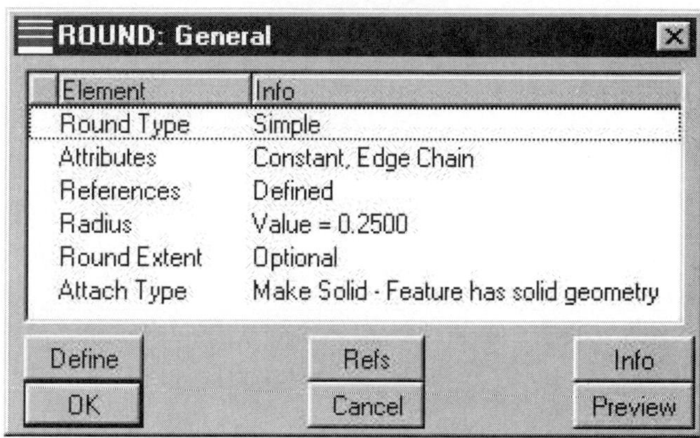

Figure 5–34 Round feature definition dialog box

The Constant option will create a round with a constant radius while the Edge chain option will allow for the selection of edges to round.

STEP 4: Select ONE BY ONE >> DONE on the Chain menu.

The One-By-One option requires the individual selection of each edge.

STEP 5: As shown in Figure 5–33, select the two edges at the base of the circular protrusion.

STEP 6: Select DONE on the Chain menu.

Select Done on the chain menu when you are through selecting edges to round.

STEP 7: Enter 0.25 as the radius of the rounds.

STEP 8: Preview the rounds, then select OKAY on the Round Feature Definition dialog box (Figure 5–34).

CREATING A CHAMFER

Pro/ENGINEER provides the option to create either a corner chamfer or an edge chamfer. This segment of the tutorial will create the edge chamfer shown in Figure 5–35.

STEP 1: Select the FEATURE >> CREATE >> CHAMFER command.

STEP 2: Select EDGE as the type of chamfer to create.

The Edge option will create an angled surface between two planes. The Corner option will create an angled surface between three planes.

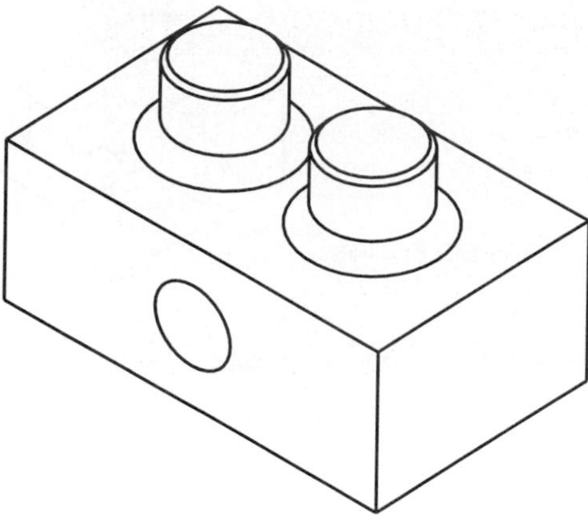

Figure 5-35 Edge chamfer

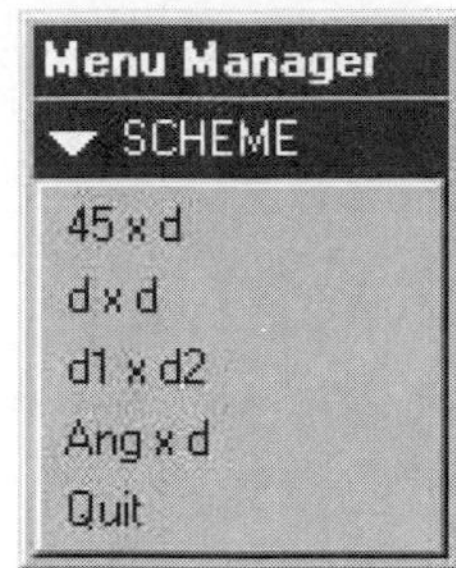

Figure 5-36 Chamfer schemes

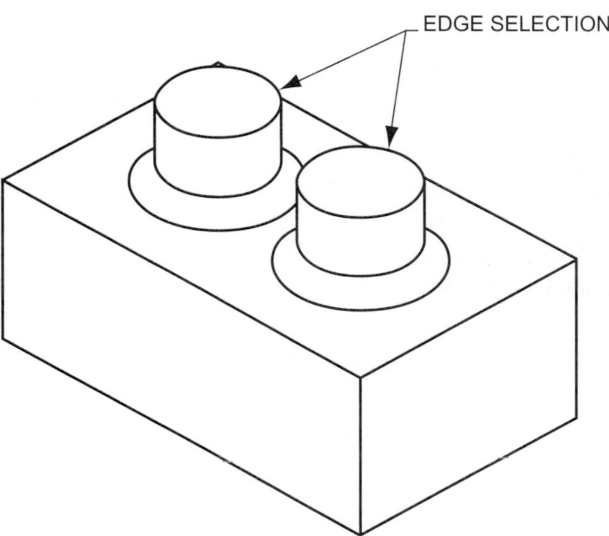

Figure 5-37 Edge selection

STEP 3: On the Scheme menu, select 45 × d (Figure 5–36).

The 45 × d option will create a 45-degree angle with a user-defined distance.

STEP 4: Enter .0625 as the Chamfer Dimension.

STEP 5: As shown in Figure 5–37, select the edges on the top of the two circular protrusions.

STEP 6: Select DONE SEL on the Get Select menu.

When you are through selecting edges to round, use the Done Sel option to exit the Get Select menu.

STEP 7: Select DONE REFS on the Feature Refs menu.

The Feature Refs menu allows for the addition or removal of edges from the edge selection set.

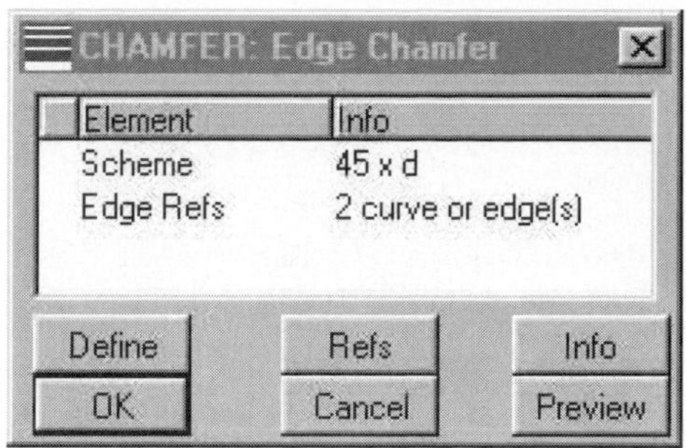

Figure 5–38 Chamfer feature definition dialog box

STEP 8: From the Feature Definition dialog box, select Preview then select OKAY (Figure 5–38).

CREATING A STANDARD COAXIAL HOLE

This segment of the tutorial will create the two coaxial threaded holes shown in Figure 5–39. The specification will be a standard 0.75 inch unified nation course internal thread.

STEP 1: Select FEATURE >> CREATE >> HOLE.

STEP 2: On the Hole dialog box, select STANDARD as the Hole Type.

Standardized thread specification tables are used to define standard holes. As an example, a 3/4 inch nominal size hole with unified national course threads has 10 treads per inch. Pro/ENGINEER has three preexisting hole tables: UNC, UNF, and ISO. Each table provides parameters needed to detail a threaded hole.

STEP 3: On the Hole dialog box, set the options shown in Figure 5–40.

Set the following options: Tapped Hole, 3/4–10 Screw Size, and Add Thread Surface. Ensure that the Add Counterbore and the Add Countersink options are unchecked.

STEP 4: Set THRU THREAD and THRU ALL as shown in Figure 5–41.

The Thru All option will construct the hole through the entire part. Similarly, the Thru Thread option will construct the thread through the entire length of the hole. Notice in the figure the dimmed out 0.6562 value.

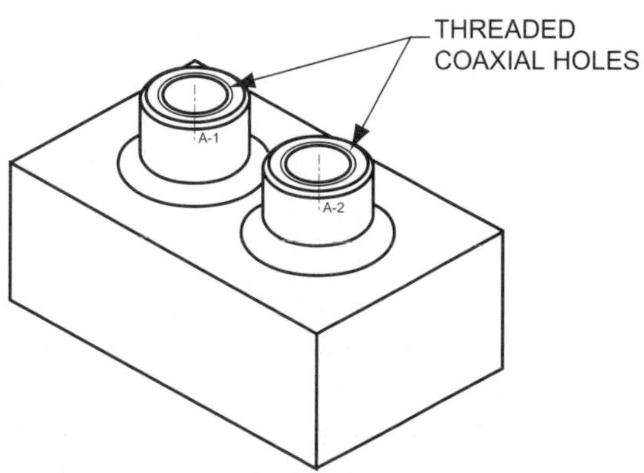

Figure 5–39 Coaxial holes

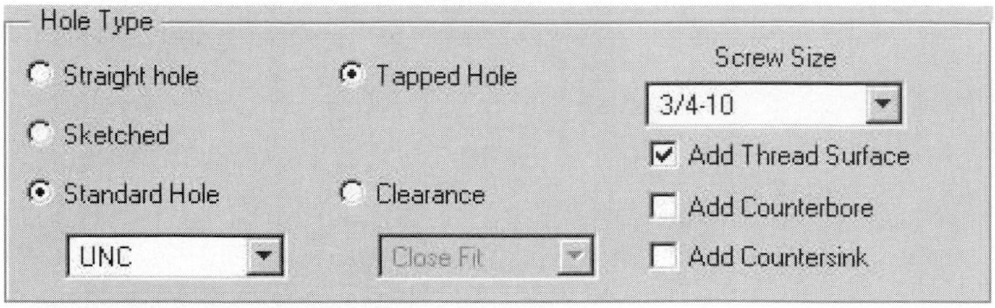

Hole Type
Straight hole
Tapped Hole
Screw Size
3/4-10
Add Thread Surface
Sketched
Standard Hole
Clearance
Add Counterbore
UNC
Close Fit
Add Countersink

Figure 5–40 Hole type

Hole Dimension
Thru Thread
Variable
Thru All
Variable
1.5749
1.9686
0.6562
Side 1

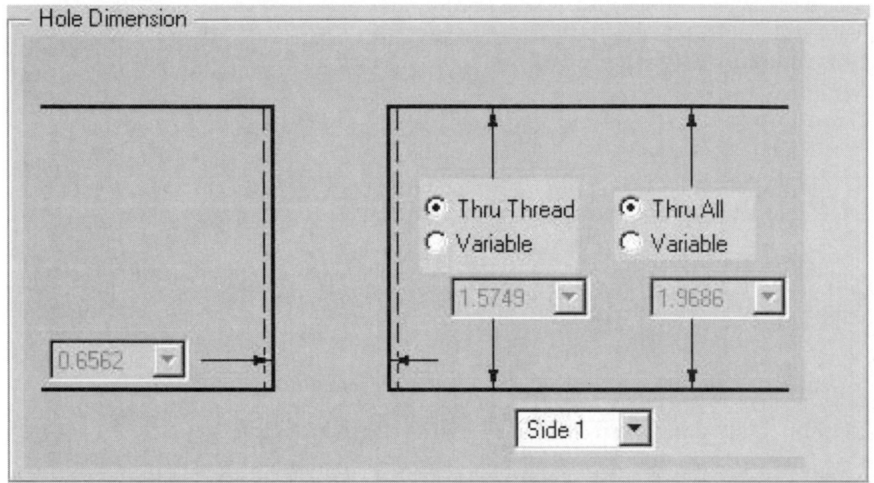

Figure 5–41 Hole depth options

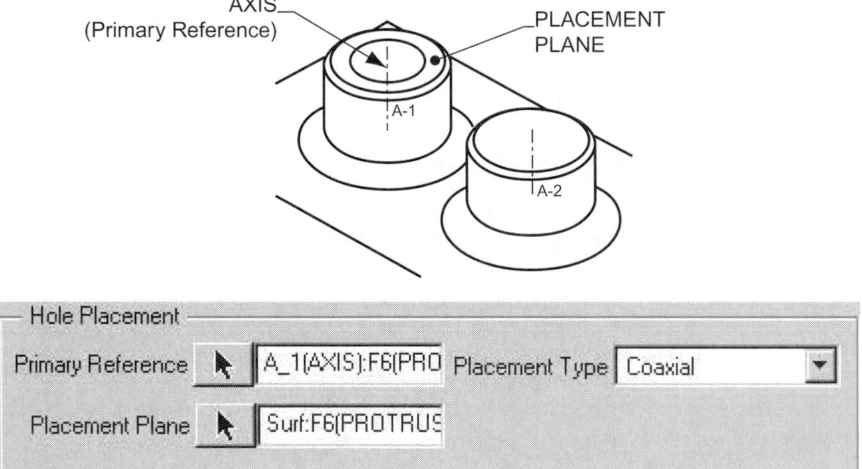

AXIS
(Primary Reference)
PLACEMENT
PLANE
A-1
A-2
Hole Placement
Primary Reference
A_1(AXIS):F6(PRO
Placement Type
Coaxial
Placement Plane
Surf:F6(PROTRUS

Figure 5–42 Hole placement references

This value represents the tap drill size. The hole within your part model will be constructed with this diameter value. This value is obtained from the UNC hole chart.

STEP 5: As shown in Figure 5–42, select the first axis as the Primary Reference.

STEP 6: As shown in Figure 5–42, select the holes placement plane.

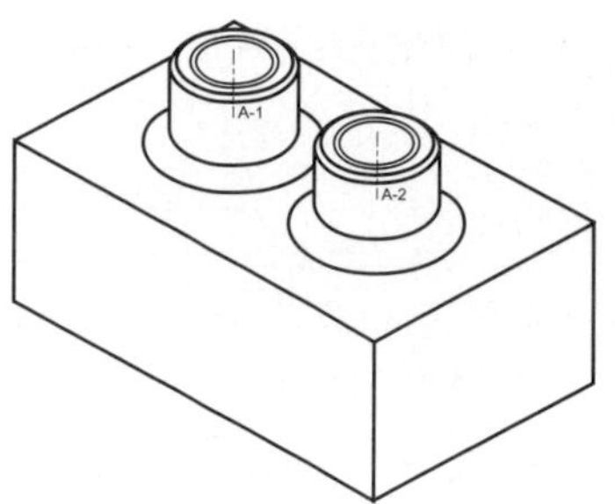

Figure 5-43 Two coaxial holes

STEP 7: **On the dialog box, select the Preview icon.**

STEP 8: **Select the Build Feature checkmark.**

STEP 9: **On the menu bar, select UTILITIES >> ENVIRONMENT and uncheck the 3D NOTES option.**

Deselecting the 3D Notes environmental option will turn off the display of the standard hole's thread representation note. This option can also be set with the configuration file option *model_note_display*.

STEP 10: **Repeat steps 1 through 8 to create the second standard coaxial hole (Figure 5–43).**

STEP 11: **Save your part.**

CREATING A LINEAR HOLE

The Linear Hole placement option will locate a hole from two reference edges. This tutorial will create the linear hole shown in Figure 5–44.

STEP 1: **Select FEATURE >> CREATE >> HOLE.**

STEP 2: **Select STRAIGHT as the Hole Type.**

STEP 3: **Enter 1.000 as the hole's Diameter (see Figure 5–45).**

STEP 4: **Select THRU ALL as the Depth One parameter. (Leave Depth Two set at None)**

STEP 5: **Select the front surface of the part as the Primary Reference (see Figure 5–46).**

STEP 6: **Pick the first Linear Reference edge for locating the hole (Figure 5–46).**

The reference edges shown in Figure 5–46 are used to locate the hole. Locating the hole from these edges matches the intent of the design.

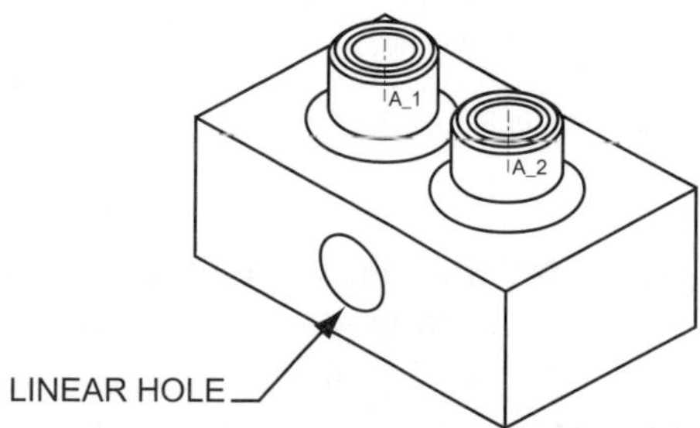

Figure 5-44 Linear hole placement

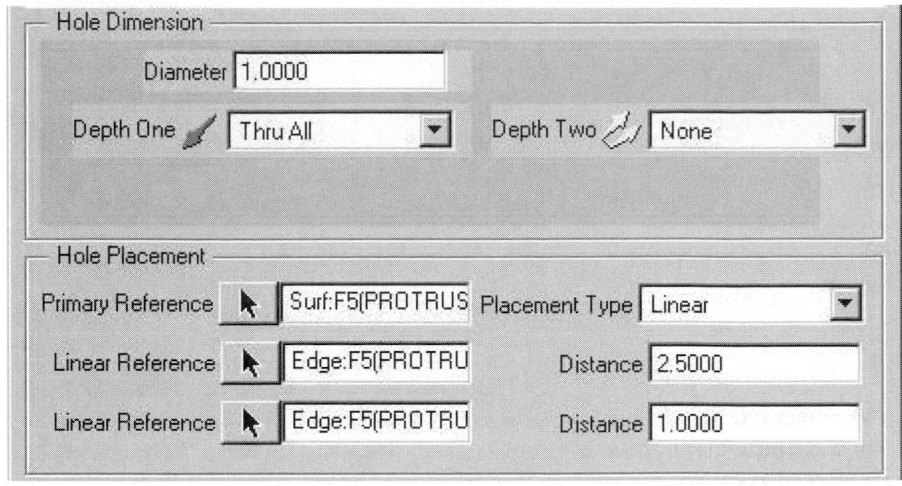

Figure 5–45 Straight hole parameters

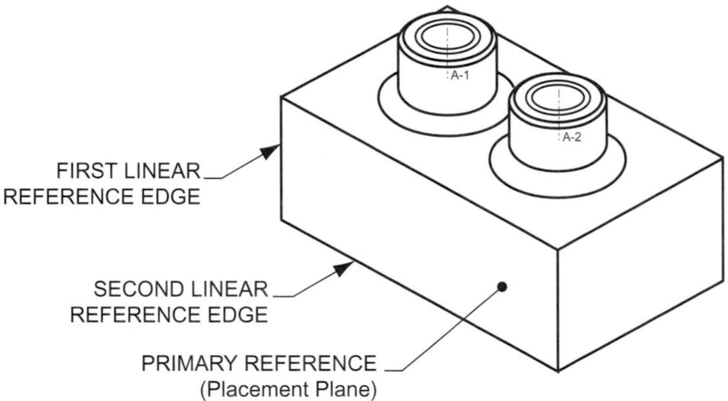

Figure 5–46 Hole references

STEP 7: In the Hole dialog box, enter a distance of 2.50 for the first Linear Reference (Figure 5–45).

> **DESIGN INTENT** When utilizing the Linear placement option and the placement plane shown in Figure 5–46, any edge of the placement plane could be used to locate the hole. Since the design intent requires the hole to be located from the edges shown in the figure, these references should be the edges selected during part modeling. Pro/ENGINEER is a Feature-Based/Parametric modeling system. These types of systems allow for the incorporation of design intent into a model. Placing a hole is one example of where design intent can be incorporated. In this example, placing the hole from these edges incorporates design intent.

STEP 8: Pick the second Linear Reference edge for locating the hole, then enter a distance of 1.00.

STEP 9: Select the Build Feature checkmark to create the hole.

STEP 10: Save your part.

Creating an Advanced Round

The Advanced Round created in this segment of the tutorial will consist of two Round Sets. The first round set will have a radius value of 0.500, while the second round set will have a radius value of 0.235. The result of this round feature is shown in Figure 5–47.

Step 1: Select FEATURE >> CREATE >> ROUND.

Step 2: Select ADVANCED >> DONE on the Round Type menu.

Advanced rounds allow for varied cross-sections and multiple round sets.

Step 3: Select ADD on the Round Sets menu.

The Add option will add a Round set to the round definition.

Step 4: Select CONSTANT >> EDGE CHAIN >> DONE on the Round Set Attribute menu.

Step 5: Select ONE BY ONE on the Chain menu.

The **One-by-One** option will require you to select each individual edge.

Step 6: As shown in Figure 5–48, pick the four vertical corners of the part.

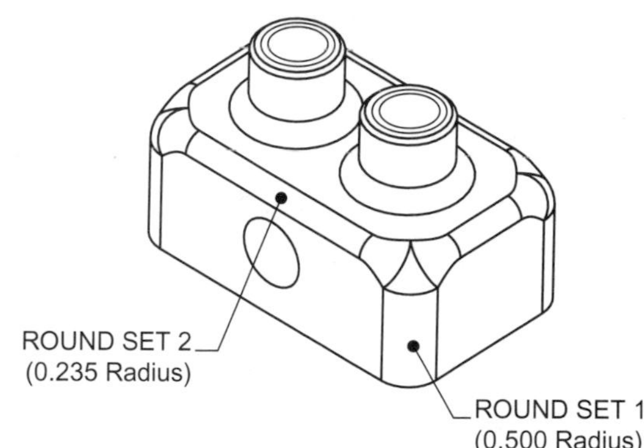

Figure 5–47 Advanced round feature

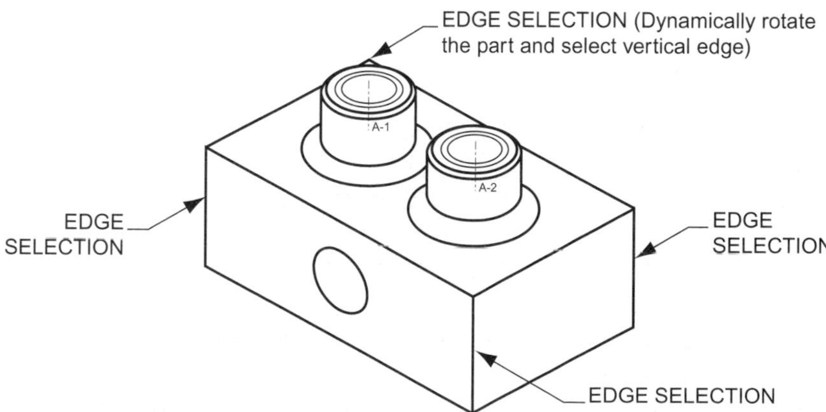

Figure 5–48 Edge selection

> **INSTRUCTIONAL NOTE** The fourth corner is hidden in Figure 5–48. Use the Query Sel option to pick this edge or dynamically rotate the part.

STEP 7: After the four corners are selected, choose DONE on the Chain menu.

STEP 8: Enter .500 as this round set's radius value.

After entering the radius value, observe the Round Set 1 dialog box (Figure 5–49).

STEP 9: Select OKAY on the ROUND SET 1 Feature Definition dialog box.

Selecting OKAY will accept the parameters defined for Round Set 1. Before selecting OKAY, each parameter can be redefined with the Define option.

STEP 10: Select ADD on the Round Sets menu.

Selecting the Add option will create another round set.

STEP 11: Select CONSTANT >> EDGE CHAIN >> DONE.

STEP 12: Select ONE BY ONE on the Chain menu.

STEP 13: As shown in Figure 5–50, pick the four edges that define the top of the part.

STEP 14: Select DONE on the Chain menu.

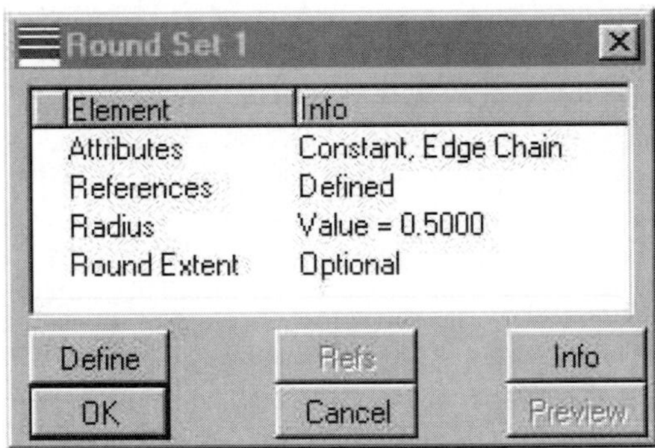

Figure 5–49 Round set one

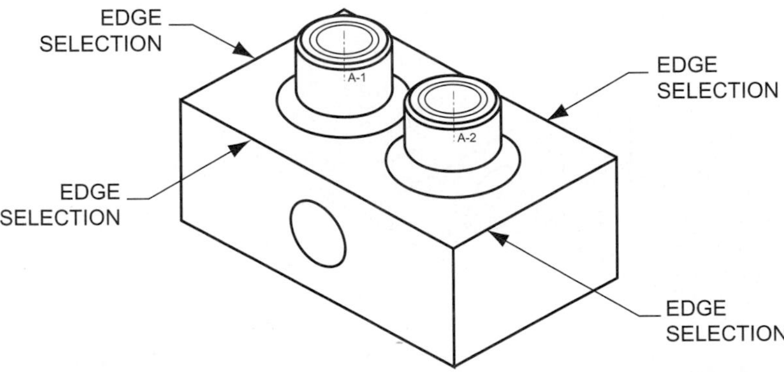

Figure 5–50 Edge selection

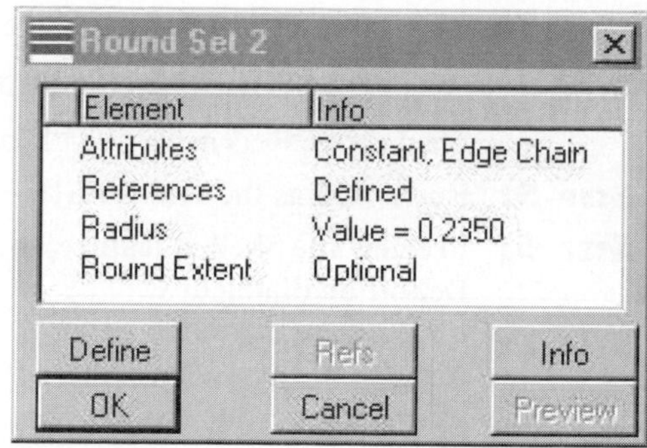

Figure 5–51 Round set two

STEP 15: Enter 0.235 as this round set's radius value.

STEP 16: Select OKAY on the ROUND SET 2 Feature Definition dialog box (Figure 5–51).

STEP 17: Select DONE SETS on the Round Sets menu.

Select Done Sets when you are through creating round sets.

STEP 18: Select PREVIEW on the Round Feature definition dialog box.

STEP 19: Finish the round by selecting OKAY on the dialog box.

INSERTING A SHELL

This segment of the tutorial will create the shelled feature shown in Figure 5–52. The Shell command removes selected surfaces from a part and provides the remaining surfaces with a user-defined wall thickness.

STEP 1: Select FEATURE >> CREATE >> SHELL.

STEP 2: Select the Surface Shown in Figure 5–53.

The selected surface will be removed from the part.

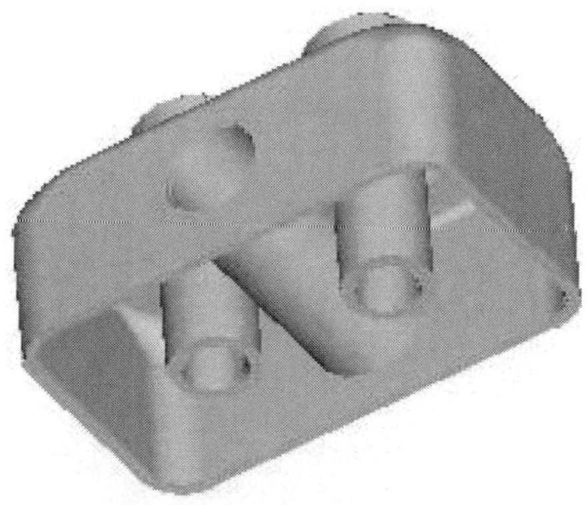

Figure 5–52 Shelled feature

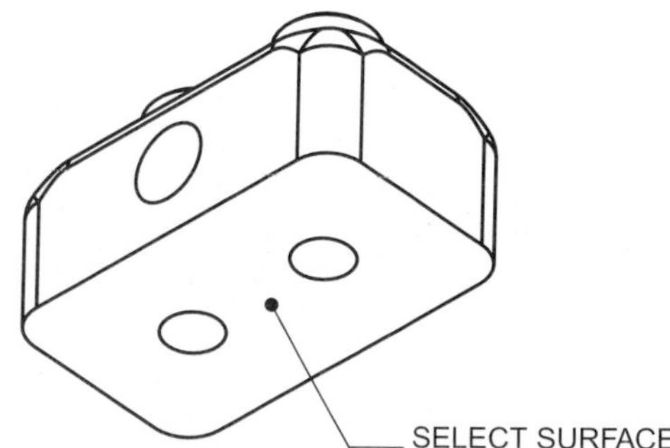

Figure 5–53 Select shell surface

STEP 3: Select DONE SEL on the Get Select menu.

STEP 4: Select DONE REFS on the Feature Refs menu.

The Feature Refs menu allows for the addition or removal of part surfaces.

STEP 5: Enter .125 as the Wall Thickness.

STEP 6: Preview the Shelled feature, then select OKAY on the Shell Feature Definition dialog box.

STEP 7: Save your part.

TUTORIAL

FEATURE CONSTRUCTION TUTORIAL 2

This tutorial exercise will provide instruction on how to model the part shown in Figure 5–54.

Within this tutorial, the following topics will be covered:

- Creating a new object.
- Creating an offset datum plane.
- Creating an extruded protrusion.
- Creating a straight coaxial hole.
- Creating a linear-straight hole.
- Creating a linear pattern.
- Creating a straight chamfer.
- Creating a cut feature.
- Creating a rib.
- Creating a simple round.

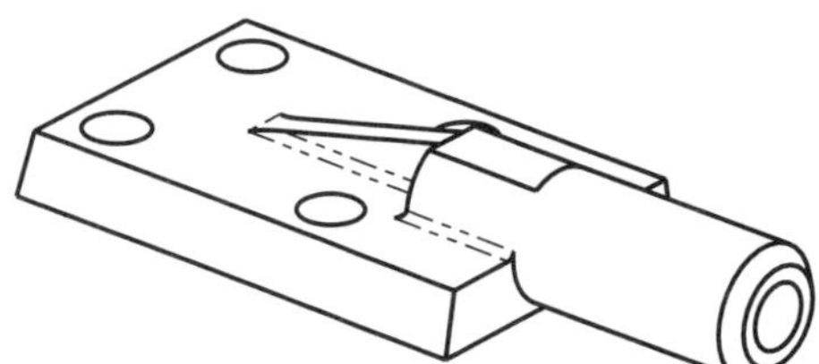

Figure 5–54 Finished model

STARTING A NEW PART

This segment of the tutorial will establish Pro/ENGINEER's working environment.

STEP 1: Start Pro/ENGINEER.

STEP 2: Establish an appropriate working directory.

STEP 3: Create a new part file with the name *CONSTRUCTION2* (Use Pro/ENGINEER's Default Template).

CREATING THE BASE GEOMETRIC FEATURE

Model the Base Geometric Feature. The Base Feature is constructed from an 8 by 5-inch rectangular section extruded one direction a blind distance of 1 inch. Sketch the section on datum plane FRONT. Align the middle of the part with datum plane TOP and the left edge of the part with datum plane RIGHT. Figure 5–55 shows the finished feature and the sketched section.

CREATING AN OFFSET DATUM PLANE

This section of the tutorial will create an offset datum plane. As shown in Figure 5–56, the datum plane will be offset from datum plane RIGHT a distance of 6 inches. A datum plane may be offset from an existing part planar surface or an existing datum plane.

STEP 1: Select the CREATE DATUM PLANE icon on the Datum Toolbar.

STEP 2: Select OFFSET on the Datum Plane menu.

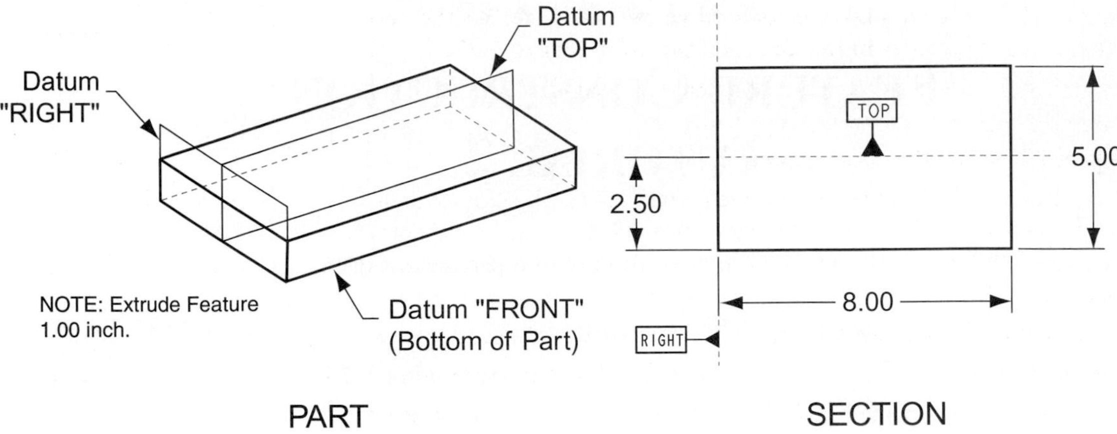

Figure 5-55 Base feature

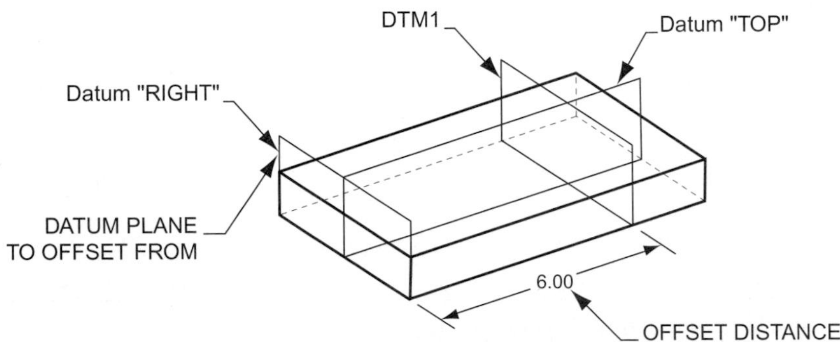

Figure 5-56 Datum offset

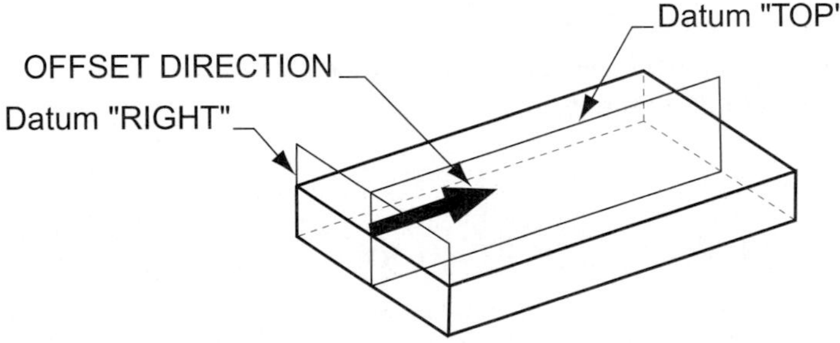

Figure 5-57 Datum offset direction

STEP 3: As shown in Figure 5–56, select datum plane RIGHT as the surface to offset from.

On the work screen or on the model tree select datum plane RIGHT.

STEP 4: On the Offset Menu, select the ENTER VALUE option.

Pro/ENGINEER provides two options for setting the offset of a datum plane. The default option, Thru Point, will create the offset datum plane through an existing datum point. The Enter Value option requires you to enter an offset distance.

After selecting the Enter Value option, notice on the work screen how Pro/ENGINEER graphically displays the direction of offset (Figure 5–57). To offset in the opposite direction, enter a negative value.

STEP 5: In Pro/ENGINEER's textbox, enter a value that will offset the new datum plane 6 inches in the direction shown in Figure 5–57.

A positive or negative value can be entered. If your offset direction is in the opposite direction, enter a negative value.

STEP 6: Select DONE on the Datum Plane menu.

Notice on the work screen and on the model tree the addition of datum plane DTM1. Pro/ENGINEER names datum planes in sequential order starting with DTM1. The next two steps in the tutorial will rename datum plane DTM1.

STEP 7: Select SET UP >> NAME from Pro/ENGINEER's Menu Manager.

STEP 8: On the Model Tree, select the datum plane DTM1 feature (Figure 5–58).

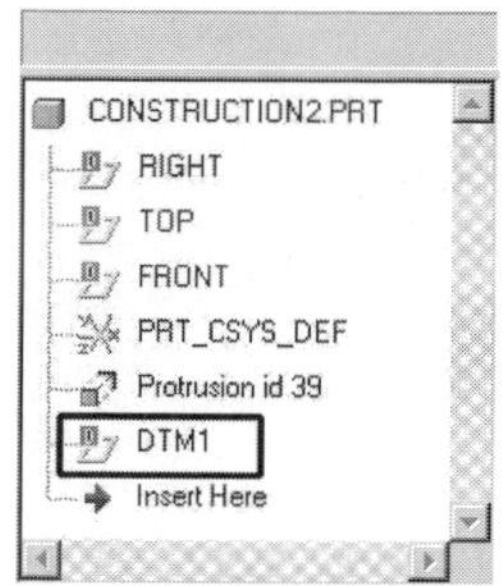

Figure 5–58 Datum plane selection

STEP 9: In the textbox enter *OFFSET_DATUM* as the new name for datum plane DTM1.

Notice the underscore between *Offset* and *Datum*. On the Model Tree, new feature names cannot have spaces.

> **MODELING POINT** The default names that Pro/ENGINEER gives features are not descriptive. It is helpful to rename them to allow for ease of identification and selection.

STEP 10: Save your part.

CREATING AN EXTRUDED PROTRUSION

This section of the tutorial will create the cylindrical protrusion shown in Figure 5–59. This protrusion will consist of a circle entity sketched on the offset datum plane (*Offset_Datum*). The section will be extruded a depth of 6 inches.

STEP 1: Set up a ONE SIDE Extruded Protrusion with datum plane *OFFSET_DATUM* as the sketching plane and the extrude direction set as shown in Figure 5–60.

Set up this feature as a One-Sided extruded Protrusion. Select datum plane *Offset_Datum* as the sketching plane (Figure 5–60). Within Pro/ ENGINEER, there are two sides to a datum plane: the positive side and the negative side. Using Pro/ENGINEER's default color scheme (see the Color option under the Utilities menu), the positive side of a datum plane is yellow, and the negative side is red. When selecting a datum as the sketching plane for an extruded protrusion, by default the feature will be extruded in the positive direction. To extrude in the negative direction, used the Direction menu's Flip option.

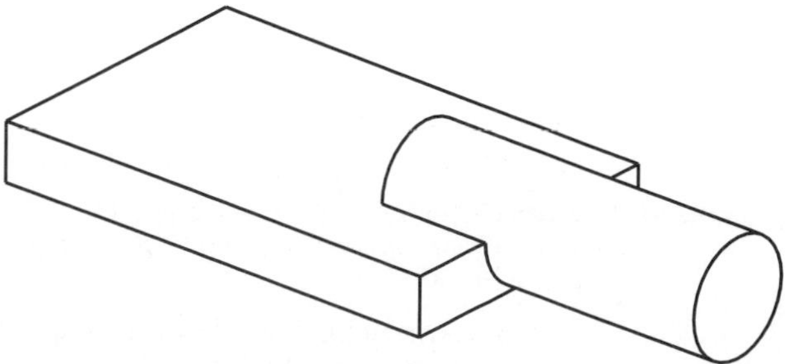

Figure 5–59 Extruded protrusion

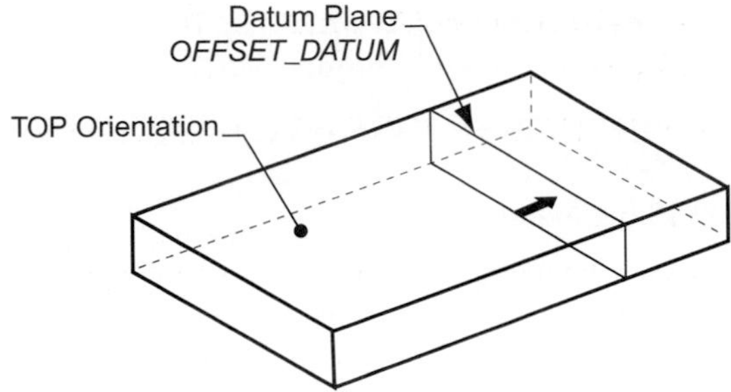

Figure 5–60 Extrude direction

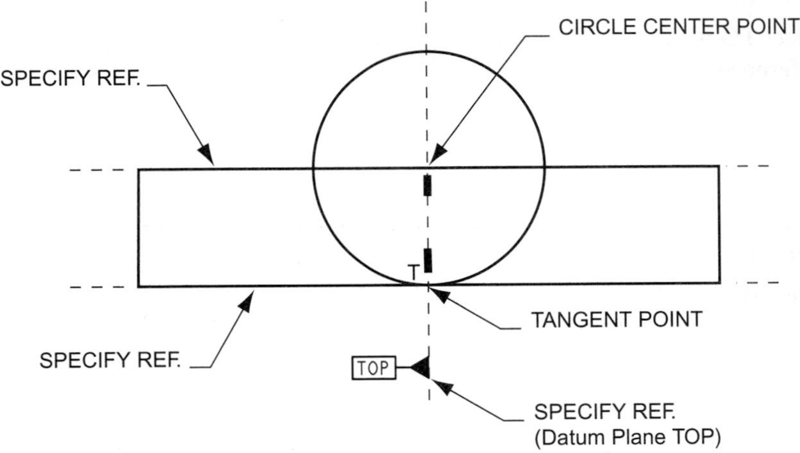

Figure 5–61 Feature's section

STEP 2: Orient the sketching environment as shown in Figure 5–61.

Use the Top option and select the top of the part.

STEP 3: On the Model Display toolbar, select HIDDEN as the model's display.

STEP 4: On the Datum Display toolbar, turn off the display of datum planes.

STEP 5: Using the References dialog box, specify the three references shown in Figure 5–61.

STEP 6: Sketch a Circle entity as shown in Figure 5–61.

The center of the circle is aligned with the top of the part and with datum plane TOP. The bottom edge of the circle is aligned with the bottom of the part (creating a Tangent constraint).

DESIGN INTENT The design intent for this extruded feature requires the diameter of the cylindrical feature to be twice the thickness of the base feature. Also, the design intent requires the center of the feature to be aligned with the top of the part. By sketching the feature in the manner shown in Figure 5–61, the intent of the design will be captured. For this part, if the thickness of the base feature is changed (it is currently set at 1 inch), the size of the cylindrical feature will change accordingly. This meets the intent for this design.

STEP 7: Select the Continue icon to exit the sketcher environment.

STEP 8: Construct a BLIND depth of 6 inches.

STEP 9: Preview the feature, then select OKAY on the Extrude Feature Definition dialog box.

CREATING A COAXIAL HOLE

The segment of the tutorial will create the Straight-Coaxial Hole shown in Figure 5–62.

STEP 1: Select FEATURE >> CREATE >> HOLE.

STEP 2: On the Hole dialog box, select STRAIGHT as the Hole Type.

STEP 3: Enter 1.00 as the diameter for the hole.

STEP 4: Select TO REFERENCE as the Depth One parameter, then select the To Reference Surface shown in Figure 5–63.

STEP 5: Pick the axis of the previously created protrusion feature (Figure 5–63) as the Primary Reference for the hole.

STEP 6: Pick the end of the protrusion feature as the Placement Plane.

STEP 7: Preview the hole.

STEP 8: Select the Build Feature checkmark.

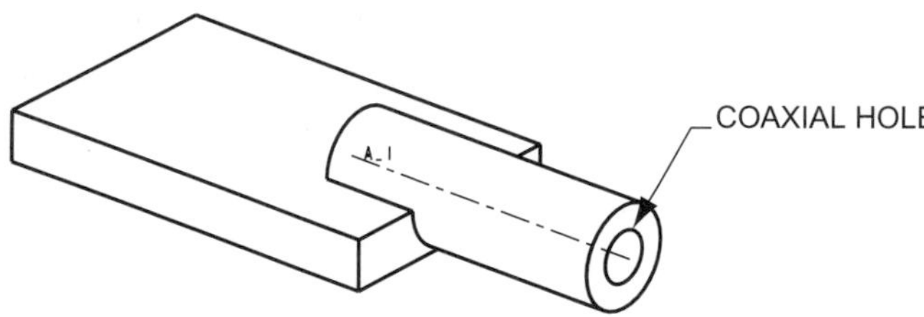

Figure 5–62 Coaxial hole

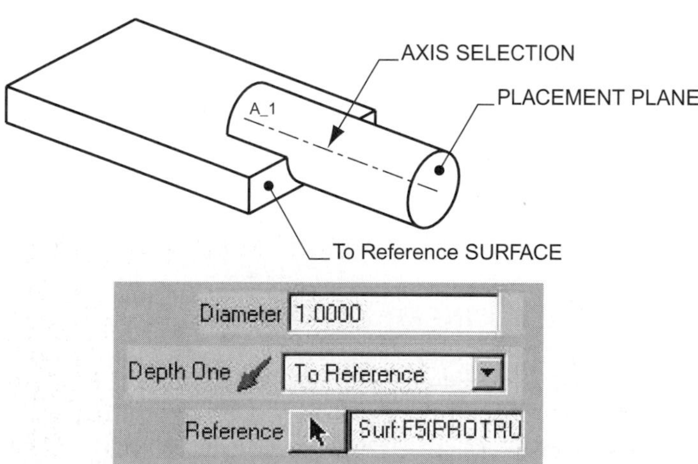

Figure 5–63 Coaxial hole references

Creating a Linear Hole

This section of the tutorial will create the Straight-Linear Hole shown in Figure 5–64.

Step 1: **Select FEATURE >> CREATE >> HOLE.**

Step 2: **Select STRAIGHT as the Hole Type.**

Step 3: **Enter 1.00 as the Hole's Diameter.**

Step 4: **Select THRU ALL as the Depth One parameter.**

Step 5: **Pick Placement Plane, shown in Figure 5–65, as the PRIMARY REFERENCE.**

Step 6: **Select the First LINEAR REFERENCE (see Figure 5–65), then enter a value of 1.00 for the reference's distance.**

Step 7: **Select the Second LINEAR REFERENCE, then enter a value of 1.00 for the reference's distance.**

Step 8: **Select the Build Feature checkmark.**

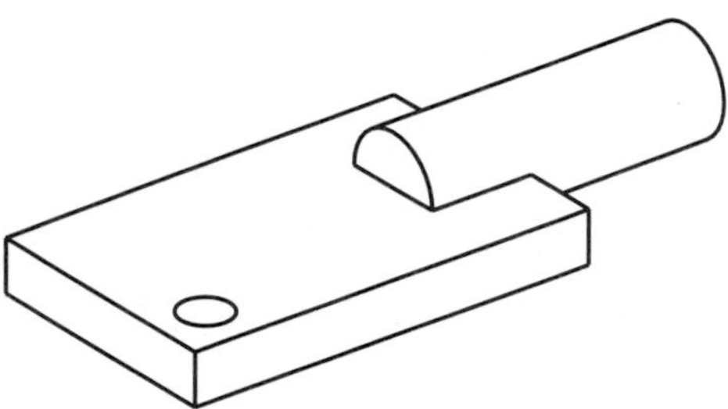

Figure 5–64 Straight-Linear Hole

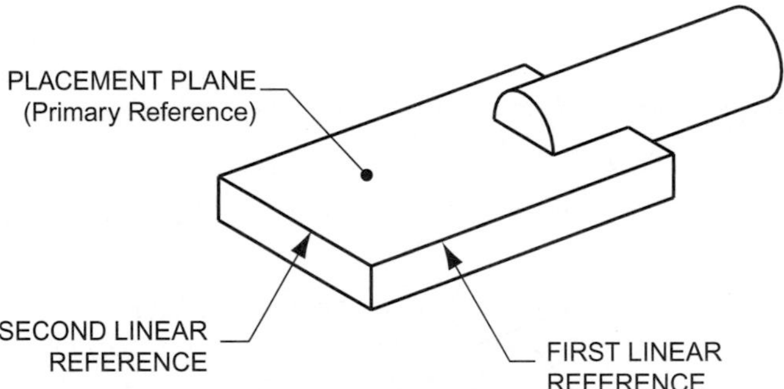

Figure 5–65 Linear hole references

Creating a Linear Pattern

This segment of the tutorial will create a linear pattern of the hole created in the previous section of this tutorial. The Pattern command allows for the copying of a feature in one or two directions. To pattern a feature, a leader dimension is required. In this example, the two linear dimensions used to locate the parent hole will serve as the leader dimensions used. Figure 5–66 shows the finished pattern.

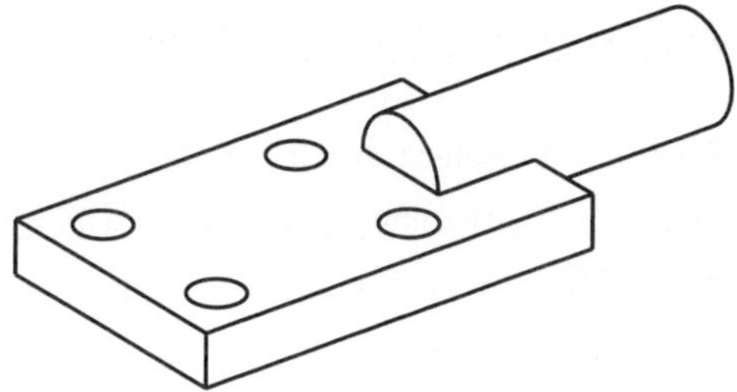

Figure 5–66 Patterned hole

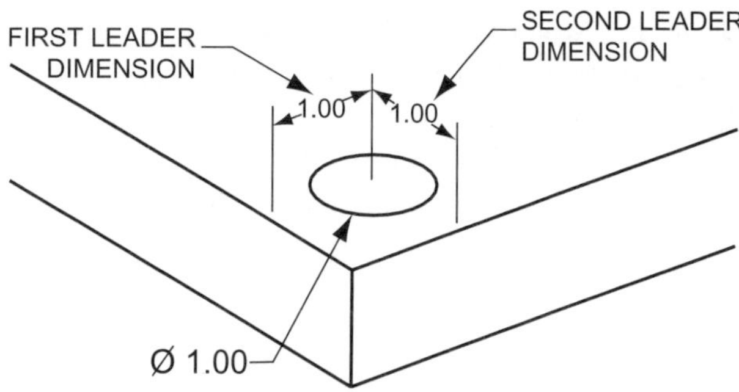

Figure 5–67 Dimension selection

STEP 1: Select the FEATURE >> PATTERN command.

STEP 2: Select the Straight-Linear hole created in the previous section of this tutorial.

On the work screen or on the model tree, select the previously created Hole feature.

STEP 3: Select IDENTICAL as the Pattern option.

The Identical option requires the most assumptions. The following assumptions must be met:

- The pattern must be on one placement surface.
- The pattern cannot intersect an edge.
- The pattern's instances cannot intersect.

STEP 4: As shown in Figure 5–67, pick the First Leader dimension for patterning in the First Direction.

Features can be patterned in two directions. In this step, you are selecting a leader dimension that will be used to determine the first direction of pattern. Additionally, this leader dimension will be used to determine the distance between each instance of the pattern.

STEP 5: Enter 4.00 as the dimension increment value.

Each instance of the pattern in the first direction will be 4 inches apart.

STEP 6: Select DONE on the Exit menu, then enter 2 as the number of instances.

In addition to the leader dimension, multiple dimensions can be varied in a direction of pattern. As an example, the hole's diameter can be varied with each instance of the pattern. In this tutorial, no additional dimensions will be varied.

STEP 7: As shown in Figure 5–67, select the Second Leader dimension for patterning in the Second Direction.

This tutorial will create a pattern in two directions. Select the second hole placement dimension as shown in Figure 5–67.

STEP 8: Enter 3.00 as the dimension increment value.

Each instance of the pattern in the second direction will be 3 inches apart.

STEP 9: Select DONE on the Exit menu then enter 2 as the number of instances.

The pattern will be created after entering the number of instances.

STEP 10: Save the part.

CREATING A CHAMFER

This segment of the tutorial will create the Straight Chamfer shown in Figure 5–68.

STEP 1: Select FEATURE >> CREATE >> CHAMFER.

STEP 2: Select EDGE as the chamfer type.

STEP 3: Select 45 × d as the chamfer's dimensioning scheme.

The 45 × d option will create a chamfer that consists of a 45-degree angle with a user-specified distance.

STEP 4: Enter 0.25 as the chamfer's dimension value.

STEP 5: Pick the feature edge as shown in Figure 5–69, then select DONE SEL on the Get Select menu.

STEP 6: Select DONE REFS on the Feature Reference menu.

The Feature Reference menu allows for the selection or removal of edges to chamfer.

STEP 7: Preview the chamfer, then select OKAY on the Feature Definition dialog box.

STEP 8: Save the part.

Figure 5-68 Chamfer feature

Figure 5-69 Chamfer edge selection

CREATING A CUT

This section of the tutorial will create the Cut feature shown in Figure 5–70.

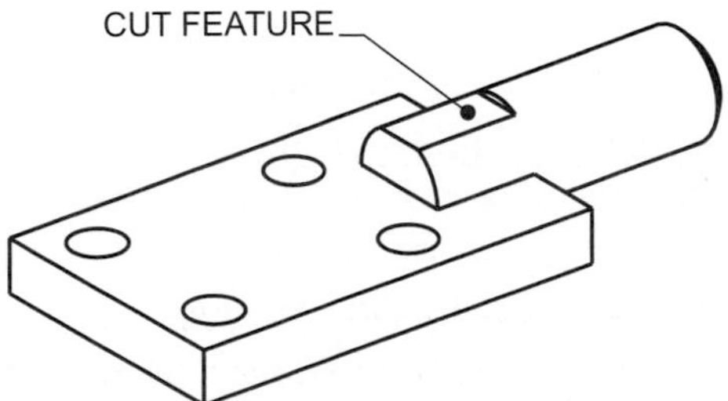

Figure 5-70 Cut feature

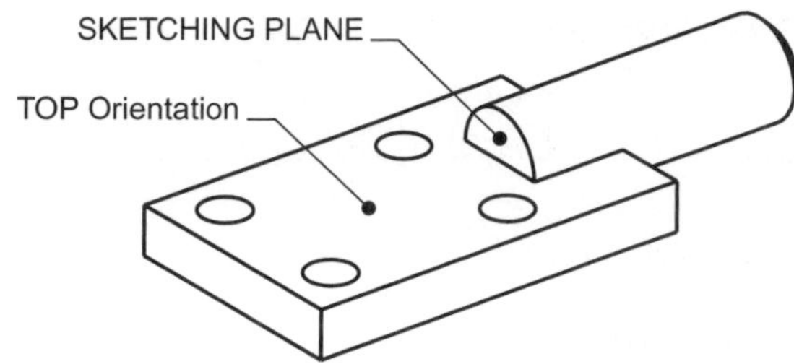

Figure 5-71 Cut references

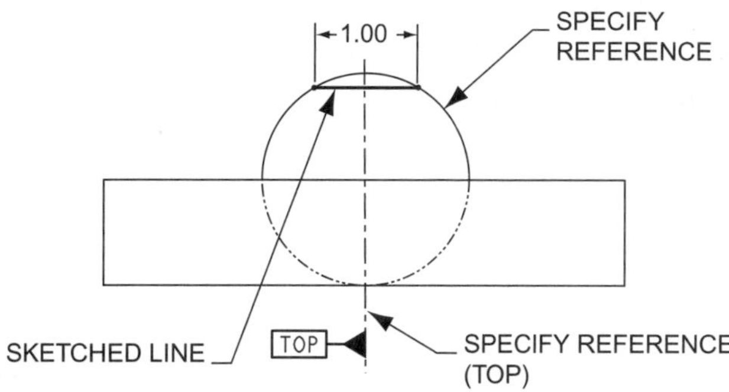

Figure 5-72 Cut section

STEP 1: **Select FEATURE >> CREATE >> CUT.**

STEP 2: **Set up a One-Sided Extruded Cut with the sketching plane and orientation shown in Figure 5–71.**

STEP 3: **In the sketching environment, on the Model Display toolbar select HIDDEN as the model's display style.**

STEP 4: **On the Datum Display toolbar, turn off the display of datum planes.**

STEP 5: **Using the References dialog box, pick the edge of the circular protrusion as a reference for this section (see Figure 5–72).**

STEP 6: **Sketch the Line entity shown in Figure 5–72, then use the DIMENSION and MODIFY options to create the dimensioning scheme shown.**

STEP 7: **Select the Continue icon to exit the Sketcher menu, then accept the default material removal side.**

STEP 8: **Create a BLIND depth of 2.00.**

STEP 9: **Preview the feature, then select OKAY on the Feature Definition dialog box.**

CREATING A RIB

This segment of the tutorial will create the Rib feature shown in Figure 5–73. A Rib feature is similar to an Extruded Protrusion. With Ribs, the section has to be opened and the extrude direction is both-sides by default. In this tutorial, the Rib will be sketched on Datum Plane TOP.

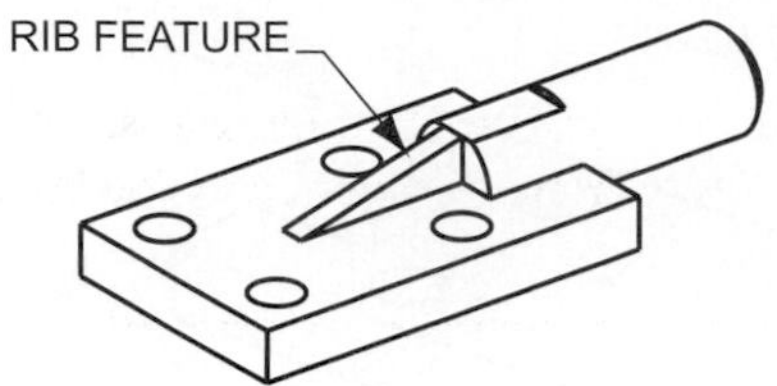

Figure 5–73 Rib feature

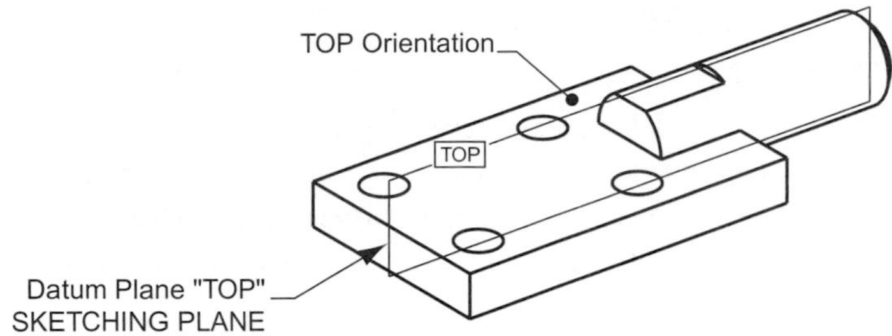

Figure 5–74 Orientation selection

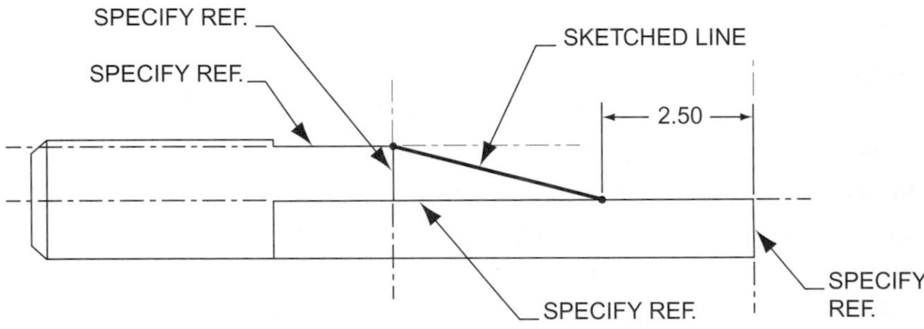

Figure 5–75 Sketched section

STEP 1: Select FEATURE >> CREATE >> RIB.

STEP 2: Select Datum Plane TOP as the sketching plane (Figure 5–74).

Datum plane TOP can be picked on the work screen or on the Model Tree. For parts with multiple features, it is often easier to use the Model Tree to select a specific feature.

STEP 3: Orient the top of the part toward the top of the work screen (Figure 5–74).

STEP 4: Using the References dialog box, specify the four references shown in Figure 5–75.

STEP 5: Sketch the single LINE shown in Figure 5–75.

STEP 6: Use the DIMENSION and MODIFY options to create the dimensioning scheme shown in Figure 5–75.

With the correct references specified, only one dimension is needed to fully define the section. In this example, the design intent requires the end of the rib to be measured from the end of the part.

STEP 7: Select the Continue icon to exit the sketcher environment.

STEP 8: If necessary, select FLIP >> OKAY to change the direction of material creation.

Due to the open section, you have to specify the side of the sketch on which to create material. Your arrow should point toward the direction where material will be added.

STEP 9: Enter .500 as the Rib's thickness.

Pro/ENGINEER will create the feature after entering the Rib's thickness. Notice how the Rib command does not utilize a Feature Definition dialog box.

CREATING A DRAFT

This section of the tutorial will draft the three surfaces shown in Figure 5–76.

STEP 1: Select FEATURE >> CREATE >> TWEAK.

STEP 2: Select DRAFT on the Tweak menu.

STEP 3: Select NEUTRAL PLN >> DONE as the Draft option.

Draft options available include Neutral Plane for pivoting around a plane or surface and Neutral Curve for pivoting around a datum curve or edge.

STEP 4: Select TWEAK >> NO SPLIT >> CONSTANT >> DONE on the Draft Attributes menu.

No Split will create a draft on the complete side of a selected surface. The Split at Plane option will create a split surface at the neutral plane while the Split at Sketch option will create a draft out of a user-sketched surface.

STEP 5: Select surfaces to draft (Figure 5–77).

Select the three surfaces shown in Figure 5–77. You will have to rotate the model dynamically or use the Query Sel option to select the third surface.

STEP 6: Select DONE SEL on the Get Select menu.

Selecting Done Sel will return you to the Surface Select menu. The Surface Select menu provides you with options for adding surfaces, excluding surfaces, and showing existing selected surfaces.

STEP 7: Select SHOW >> MESH on the Surface Select menu.

The Show >> Mesh option will show selected surfaces as a mesh (Figure 5–78).

STEP 8: Select DONE on the Surface Select menu.

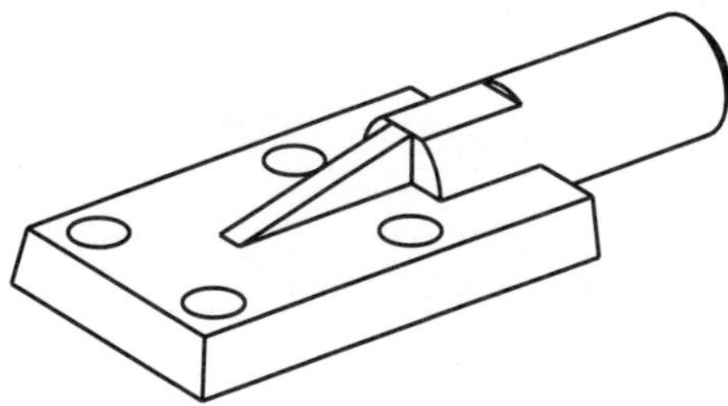

Figure 5-76 Drafted surfaces

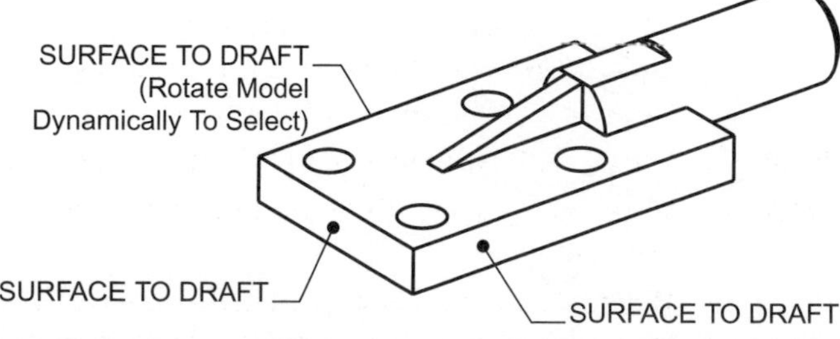

Figure 5-77 Surface selection

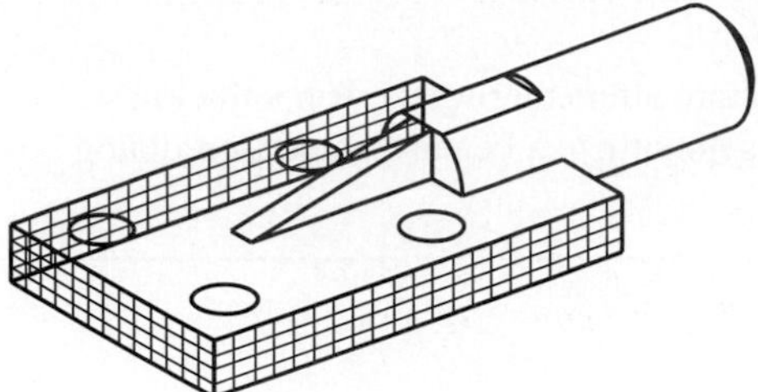

Figure 5–78 Show draft surfaces

STEP 9: Select the bottom of the part as the Neutral Plane (Figure 5–79).

The Neutral Plane is the part surface that will remain intact during the drafting procedure. Since the drafted surfaces will pivot from the neutral plane, this is a critical selection.

STEP 10: Select the bottom of the part as the perpendicular reference plane (Figure 5–79).

The reference plane has to lie perpendicular to the draft surfaces. Often, this reference plane is the same as the neutral plane.

STEP 11: Enter a draft angle that will create a 10-degree draft in the direction shown in Figure 5–80.

As shown in Figure 5–80, Pro/ENGINEER graphically displays the draft's rotation direction. The rotation direction should be as shown in the figure. If necessary, a negative value can be entered to reverse this direction.

STEP 12: Preview the Draft, then select OKAY on the Feature Definition dialog box.

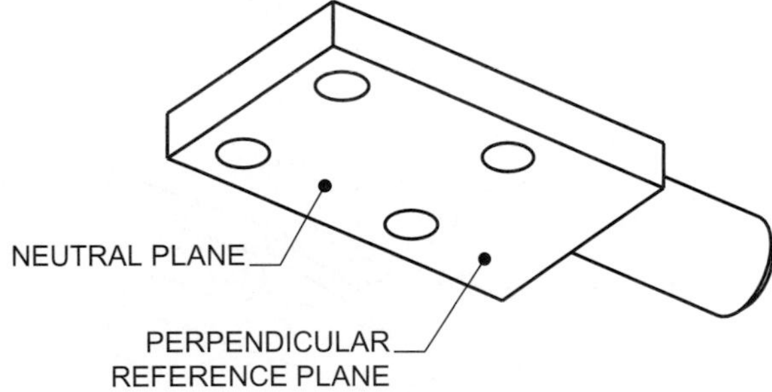

Figure 5–79 Neutral plane and reference plane selection

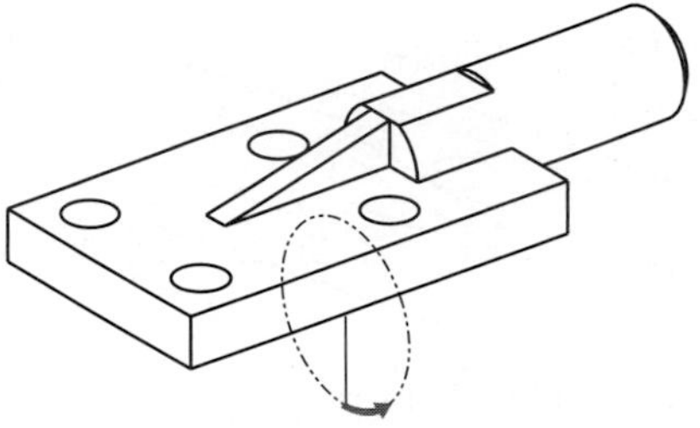

Figure 5–80 Draft rotation reference

Creating a Round

This segment of the tutorial will create the rounds shown in Figure 5–81. Despite their apparent simplicity, rounds can be one of the most troublesome commands in Pro/ENGINEER. Since it is advisable not to use a round as a reference for the creation of a feature, it is recommend that rounds be created as the last features of a part.

Step 1: Select FEATURE >> CREATE >> ROUND.

Step 2: Select SIMPLE >> DONE as the Round type to create.

Step 3: Select CONSTANT >> EDGE CHAIN >> DONE as the Round Set Attributes.

Step 4: Select ONE BY ONE as the Chain selection option.

Step 5: Select the Edges shown in Figure 5–82 and the corresponding edges on the opposite side of the Rib and Cylinder.

On the work screen, select the edges shown in Figure 5–82. Use the Query Sel option to select the corresponding edges on the opposite side of the rib and cylinder features.

Step 6: Select DONE on the Chain menu, then enter a round radius value of 0.125.

Step 7: Preview the rounds on the Feature Definition dialog box, then select OKAY.

Step 8: Save your part.

Step 9: Purge the old part file by using the FILE >> DELETE >> DELETE OLD option.

Every time a file is saved in Pro/ENGINEER, a new version of the file is created. The Delete Old option will delete old versions.

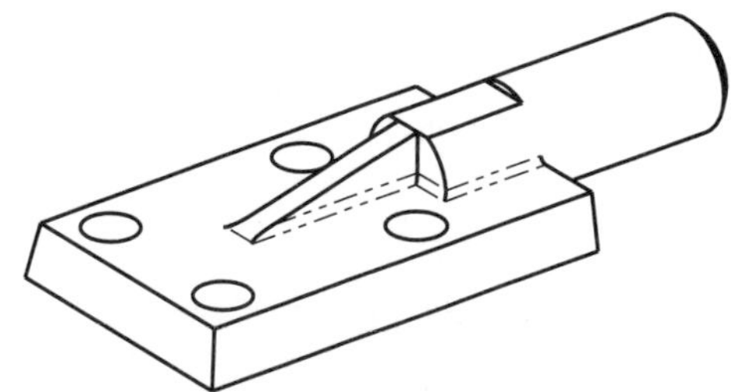

Figure 5–81 Rounded features

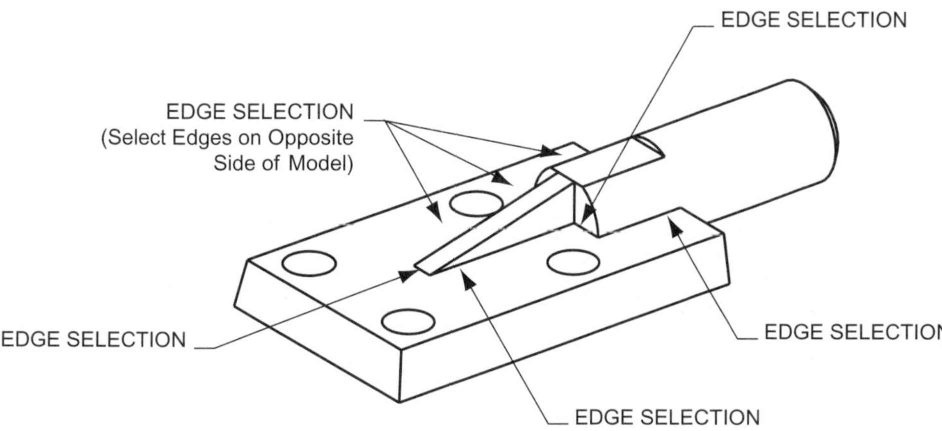

Figure 5–82 Round selection

PROBLEMS

1. Using Pro/ENGINEER's Part mode, model the part shown in Figure 5–83. Construct the part using the following order of operation:
   a. Use Pro/ENGINEER's default template file.
   b. Sketch the base feature on Datum Plane FRONT.
   c. Model the Base Protrusion as a Both-Sides extrusion.
   d. Construct the Rounded features.
   e. Model the first hole as a Linear-Straight Hole feature.
   f. Pattern the Hole feature.
   g. Construct the Rib feature by sketching on Datum Plane FRONT.
2. Using Part mode, model the part shown in Figure 5–84.

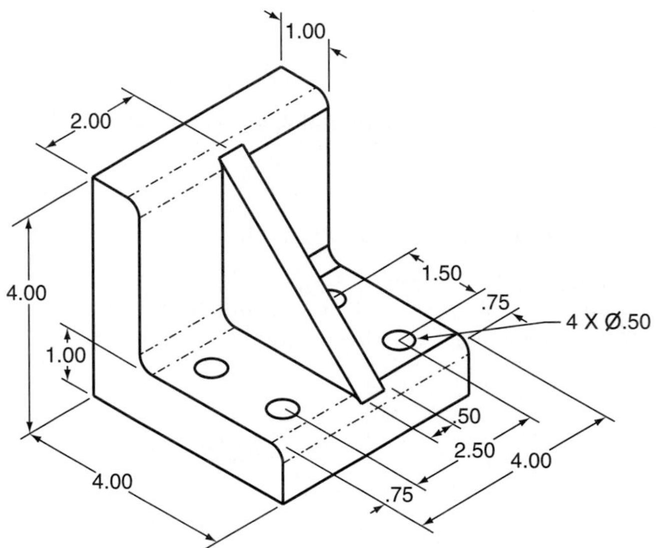

Figure 5–83 Problem one

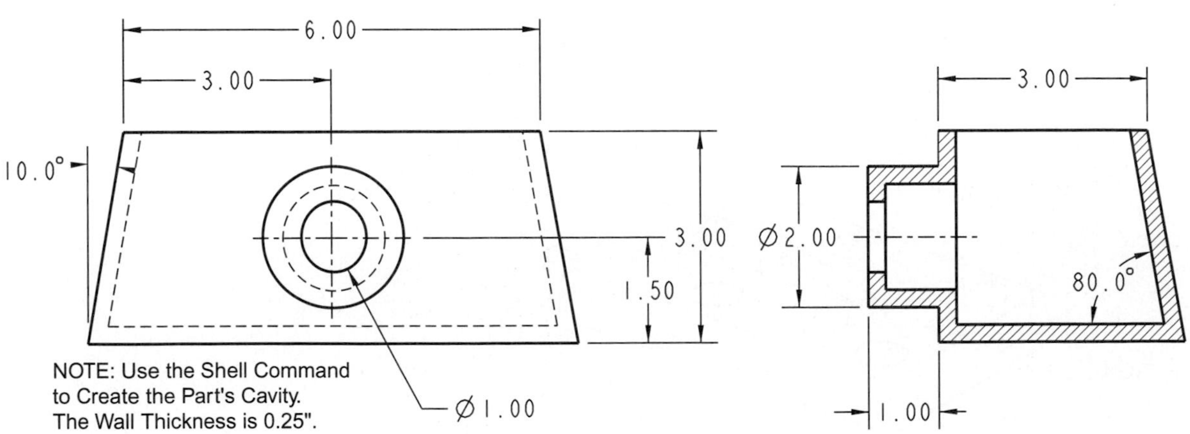

Figure 5–84 Problem two

3. Use Pro/ENGINEER to model the part shown in Figure 5–85. This part will be used within an assembly model problem in Chapter 12.
4. Model the part shown in Figure 5–86.

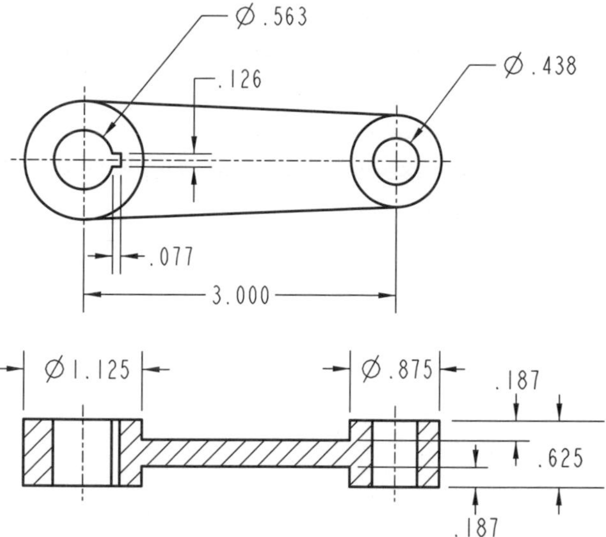

Figure 5-85 Problem three (arm part)

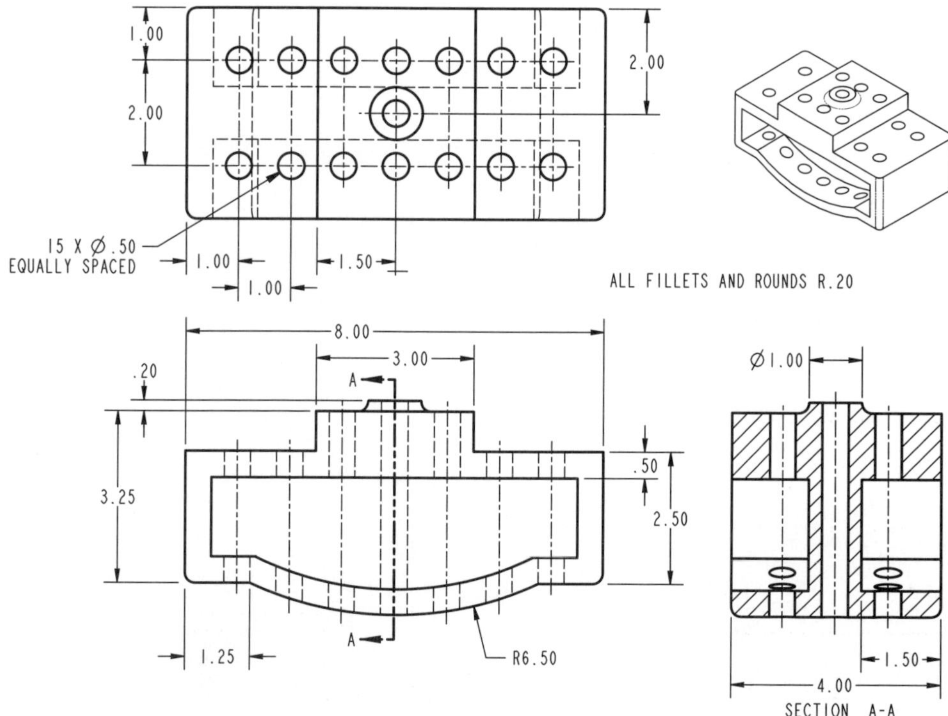

Figure 5-86 Problem four

Questions and Discussion

1. Describe the four different methods of placing a hole.
2. What is the difference between a Straight Hole, a Sketched Hole, and a Standard Hole? What are some uses of a Sketched Hole?
3. Describe methods for avoiding Round regeneration conflicts.
4. Describe the difference between a Simple Round and an Advanced Round.
5. How does a Neutral Plane Draft differ from a Neutral Curve Draft?
6. What is the purpose of a Neutral Plane?
7. Compare and contrast the Protrusion >> Extrude option with the Rib command.
8. Describe the assumptions associated with the following pattern categories:
   a. Identical
   b. Varying
   c. General
9. In regard to the Pattern command, what is a leader dimension?

CHAPTER

REVOLVED FEATURES

Introduction

Revolved feature construction techniques are common within Pro/ENGINEER. The most obvious revolved technique is the Revolve option found under the Protrusion and Cut commands. Other revolved feature construction techniques can be found with the sketched Hole option and the Shaft, Flange, and Neck commands. Upon finishing this chapter, you will be able to

- Construct a sketched hole.
- Model a shaft utilizing the Shaft command.
- Model a flange utilizing the Flange command.
- Cut a neck with the Neck command.
- Pattern a radial hole.
- Construct a revolved protrusion.
- Construct a revolved cut.

DEFINITIONS

Axis of Revolution The axis around which a section is revolved. Within Pro/ENGINEER, revolved features require a user-sketched centerline. This centerline serves as the Axis of Revolution

Through >> Axis Datum Plane A datum plane constructed through an axis.

REVOLVED FEATURE FUNDAMENTALS

A revolved feature is a section that is rotated around a centerline. For any type of revolved feature, within the sketcher environment the user sketches the profile of the section to be revolved and the centerline to revolve about (Figure 6–1). Revolved features may be positive space or negative space. A Revolved Protrusion is an example of a positive space feature. A Revolved Protrusion's negative space counterpart is the Revolved Cut. The Flange command is another example of a revolved positive space feature. Its counterpart negative space feature is the Neck command.

SKETCHING AND DIMENSIONING

As shown in Figure 6–1, the geometry of a revolved feature must be sketched on one side of the centerline, and the section must be closed. One centerline must be sketched. If multiple

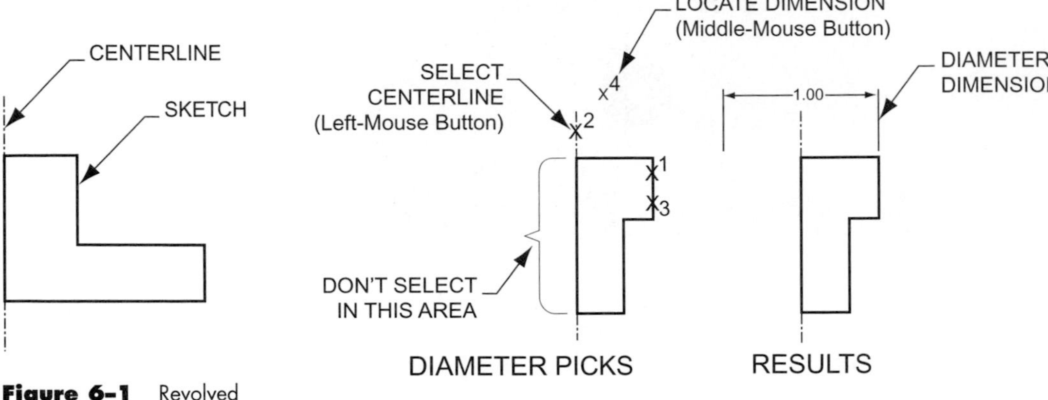

Figure 6–1 Revolved sketch

Figure 6–2 Diameter dimensions

centerlines exist in a section, the first one sketched will serve as the axis of revolution. Entities that lie on the axis of revolution will not serve as a replacement for the centerline.

Revolved features are often used to create cylindrical objects such as shafts and holes. Drafting standards require cylindrical objects to be dimensioned with a diameter value. This creates a unique situation within the sketcher environment. As shown in Figure 6–2, to dimension a revolved feature with a diameter value, perform the following steps:

1. Pick the geometry defining the outside edge of the feature.
2. Pick the centerline to serve as the axis of revolution.
3. Pick the geometry defining the outside edge of the feature.
4. Pick a location for the placement of the dimension text.

The resulting dimension should appear as shown in Figure 6–2.

MODELING POINT If a diameter dimension is not created, geometry was probably inadvertently selected instead of selecting the required centerline. Select the centerline on the work screen where it is clear that a centerline is the only entity residing (Figure 6–2).

Revolved Protrusions and Cuts

A Revolve option is available under the Protrusion and Cut commands. Revolved Protrusions are used to create positive space features while Revolved Cuts are used to create negative space features.

Revolved Feature Parameters

As with extruded Protrusions and Cuts, a variety of options are available for defining revolved feature parameters.

Revolve Direction

The Revolve direction attribute is similar to the Extrude direction attribute. Options are available for selecting a One Side or a Both Sides revolution. The One Side option will revolve the section from the sketching plane in one direction, while the Both Sides attribute will revolve the section both directions from the sketching plane.

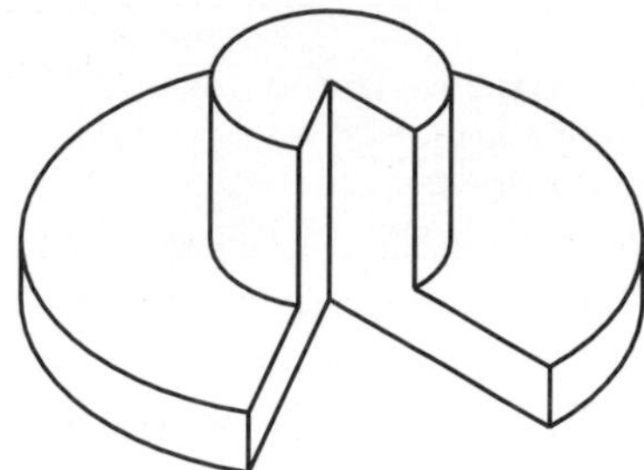

Figure 6–3 Variable angle of revolution

Angle of Revolution

The Angle of Revolution parameter is similar to the Extrude option's depth parameter. This option is used to specify the number of degrees that the section will be revolved about the axis of revolution. The following options are available:

- **Variable** The Variable option is used to specify an angle of revolution less than 360 degrees. The angular parameter specified is modifiable (Figure 6–3).
- **90** The 90 option is used to rotate a section at an angle of 90 degrees.
- **180** The 180 option is used to rotate a section at an angle of 180 degrees.
- **270** The 270 option is used to rotate a section at an angle of 270 degrees.
- **360** The 360 option is used to rotate a section a full 360 degrees.
- **UpToPnt/Vtx** The UpToPnt/Vtx option is used to revolve a section up to a selected point or vertex.
- **UpTo Plane** The UpTo Plane option is used to revolve a section up to a selected plane.

Creating a Revolved Protrusion

The Revolve option is used extensively for creating base geometric features. The following is a step-by-step guide for creating a revolved Protrusion:

Step 1: Select FEATURE >> CREATE >> PROTRUSION.

Step 2: Select REVOLVE on the Solid Options menu.

Step 3: Select between a SOLID and THIN feature, then select DONE.

Just like extruded features, revolved features can be created as a solid or a thin feature. Solid features have mass, while thin features have a user-defined wall thickness.

Step 4: Select either ONE SIDE or BOTH SIDES, then select DONE.

Step 5: Select a sketching plane and orient the sketcher environment.

Revolved features may be sketched on a part plane or datum plane. Additionally, an on-the-fly datum plane can be created with the Make Datum option.

Step 6: Use the CENTERLINE option to sketch the centerline (see Figure 6–4).

A sketched centerline is a requirement for any revolved feature. For Revolved Protrusions and Cuts, the centerline can be sketched at any orientation. For the first geometric feature of a part and when sketching on Pro/ENGINEER's default datum planes, it is recommend that you sketch the centerline of a revolved feature aligned with the edge of a datum plane. Consider design intent when sketching the centerline.

> **MODELING POINT** A centerline does not have to be the first entity sketched for a revolved feature. If a centerline is not present when selecting Done to exit the sketcher environment, Pro/ENGINEER will provide a warning message.

STEP 7: Sketch the section (Figure 6–4).

Use appropriate sketching tools to create the section. A revolved section has to be closed and must lie completely on one side of the centerline.

STEP 8: Select the Continue icon to exit the sketcher environment.

STEP 9: Specify an angle of revolution.

STEP 10: Preview the feature on the Feature Definition dialog box, then select OKAY to create the feature (Figure 6–5).

The Preview option is located on the Feature Definition dialog box. If the feature is not defined correctly, use the Define option to make changes to an element.

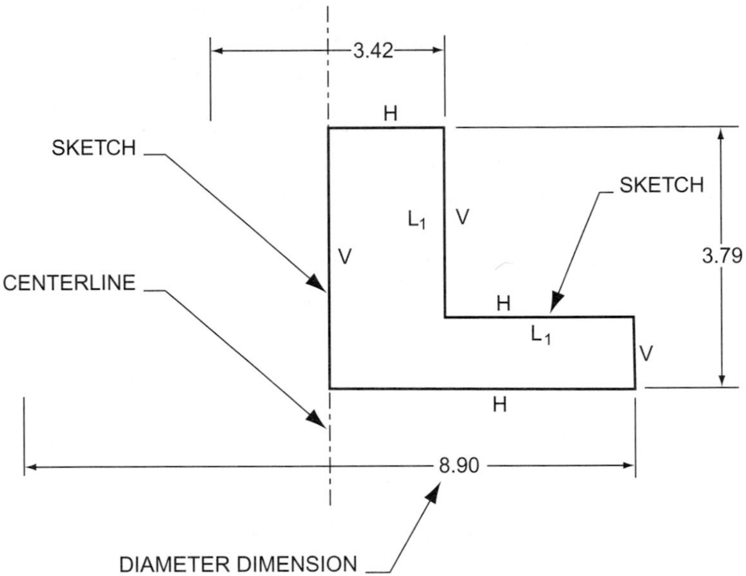

Figure 6–4 Section sketching

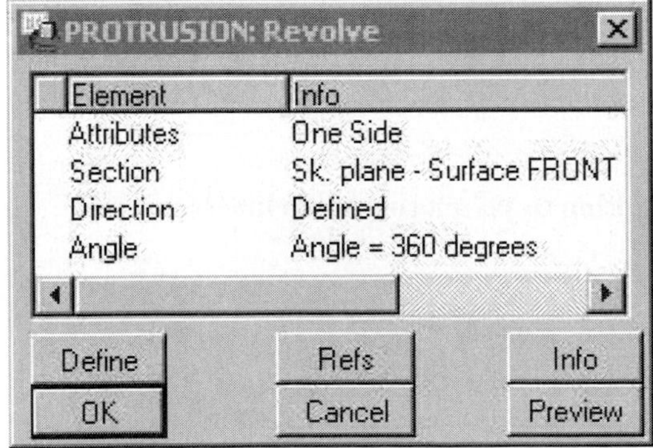

Figure 6–5 Revolve Feature Definition dialog box

Revolved Hole Options

Pro/ENGINEER provides three hole types: straight, sketched, and standard. Straight holes have a constant diameter throughout the length of the feature. Sketched holes are used to create unique profiles, such as those that exist with counterbored and countersunk holes. In addition to the three types of holes, Pro/ENGINEER provides five placement options: Linear, Coaxial, Radial, Diameter, and On Point. The linear option is used to locate a hole from two reference edges, while the Coaxial option is used to locate a hole's centerline coincident with an existing axis. The Radial option is used to locate a hole at a distance from an axis and at an angle to a reference plane (Figure 6–6). The hole's distance from the reference axis is defined by a radius value. Like the Radial option, the Diameter option is used to locate a hole at a distance from an axis and at an angle to a reference plane (Figure 6–7). With the Diameter option, the hole's distance from the reference axis is defined by a diameter value.

Sketched Holes

The Sketched Hole option requires the user to sketch the profile of a hole (Figure 6–8). Most normal sketching tools can be used. The hole sketcher environment does not provide an option for specifying references, though. A sketched hole is created originally independent of any specific part features and later placed according to the hole placement option being used (e.g., Linear, Coaxial, Diameter, Radial, or On Point).

When sketching a hole, a vertical, user-created centerline is required. Within the sketcher environment, use the Centerline icon to create this entity. All sketched entities must be created on one side of the centerline and must be closed. A geometric entity may be placed on top of the centerline, but the centerline cannot serve as an element of the hole profile. Additionally, one sketched entity must lie perpendicular to the centerline. This entity will be aligned with the placement plane when placing the hole. For sketched holes with multiple perpendicular lines, the uppermost line within the sketcher environment will serve as the placement reference.

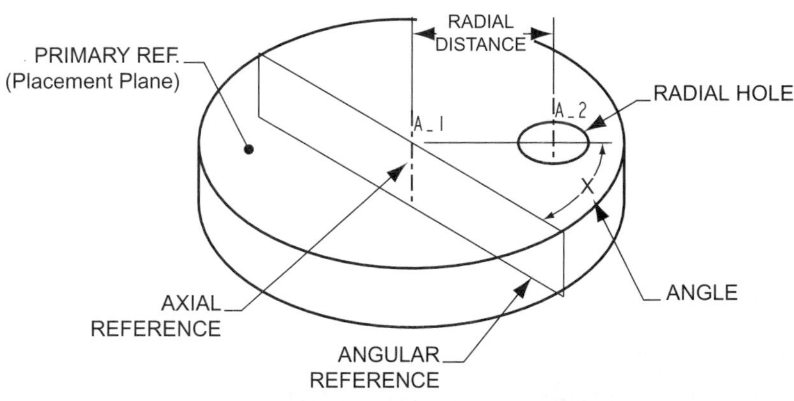

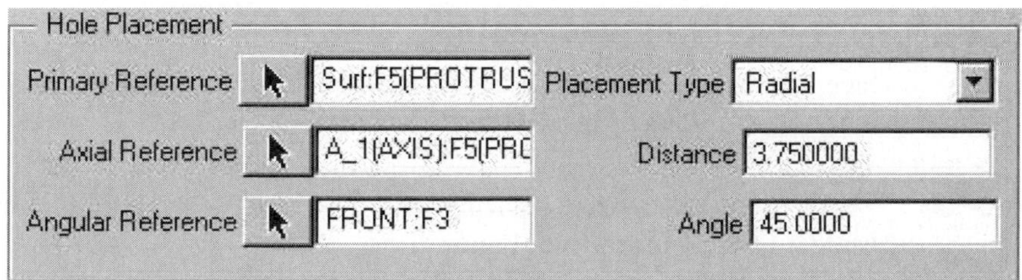

Figure 6–6 Radial hole placement

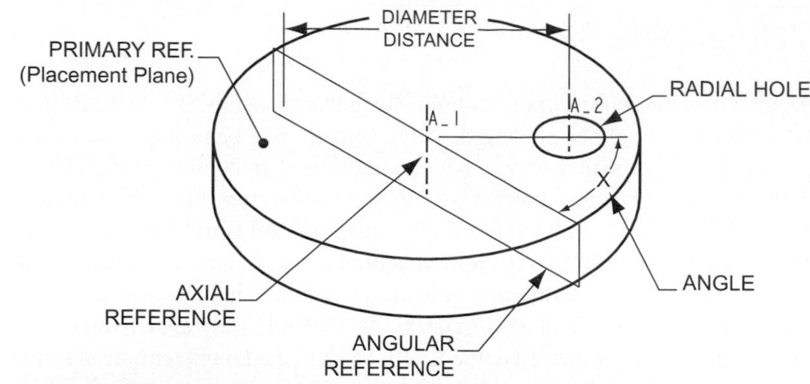

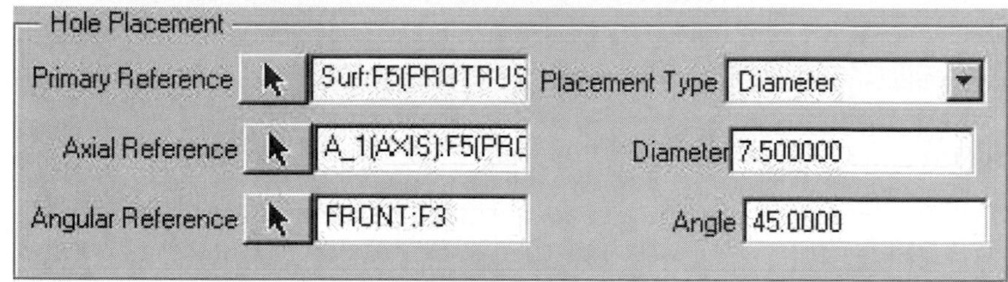

Figure 6–7 Diameter hole placement

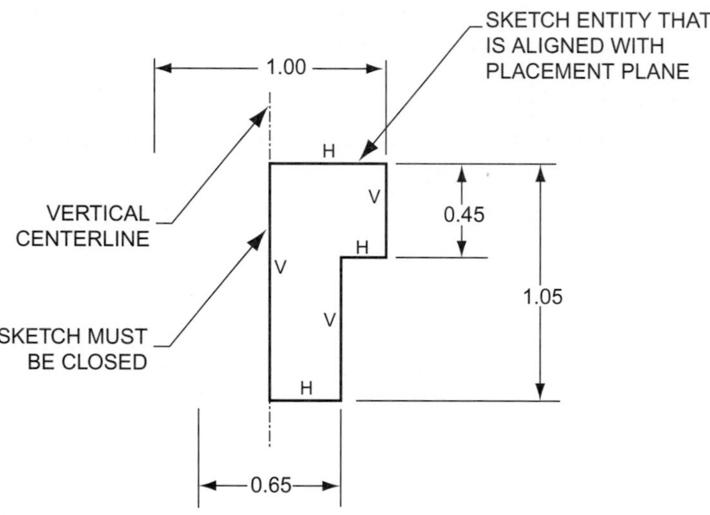

Figure 6–8 Sketching a counterbored hole

CREATING A SKETCHED HOLE

Perform the following steps to create a sketched hole.

STEP 1: Select FEATURE >> CREATE >> HOLE.

STEP 2: Select SKETCHED as the Hole Type.

Once the Sketched option is selected, Pro/ENGINEER will launch a sketcher environment for the creation of the hole's profile.

STEP 3: Within the sketcher environment, use the CENTERLINE icon to create a vertical centerline (Figure 6–8).

All revolved features require a centerline within the sketch. For a sketched hole, this centerline must be vertical.

STEP 4: Sketch the profile of the hole (Figure 6–8).

Sketched entities of the hole must be created on one side of the centerline and must be completely closed. The centerline will not serve as an entity of the sketch.

One sketched entity must be created perpendicular to the centerline. This entity will be used to align the hole with the placement plane. For multiple perpendicular entities, the uppermost one in the sketcher environment will serve this purpose.

STEP 5: Dimension the sketch to meet design intent.

Use the Dimension option to dimension the sketch to meet design intent. Since holes are defined with diameter values, use the diameter dimensioning technique described previously in this chapter.

STEP 6: Modify dimension values.

STEP 7: Select the Continue icon to exit the sketcher environment.

STEP 8: Dependent upon the hole's placement option, select the hole's Primary Reference.

Place the hole according to requirements for a Linear, Coaxial, Radial, Diameter, or On Point hole. The primary references for each placement type follow:

- **Linear** Placement plane.
- **Coaxial** Axial reference.
- **Radial** Placement plane.
- **Diameter** Placement plane.
- **On Point** Datum point reference.

STEP 9: Select the hole's placement type.

Hole placement types available include: Linear, Coaxial, Radial, Diameter, and On Point.

STEP 10: Select remaining placement references as required.

STEP 11: Select the Built Feature icon on the Hole dialog box.

RADIAL HOLE PLACEMENTS

The Radial and Diameter hole placement options are used frequently with the Pattern command to create a radial pattern of a hole (Figure 6–9). The Radial and Diameter options place a hole at a user-specified distance from an existing axis and at an angle to a reference plane (Figure 6–10). With patterned holes, this angular dimension is used as the leader dimension for creating the pattern.

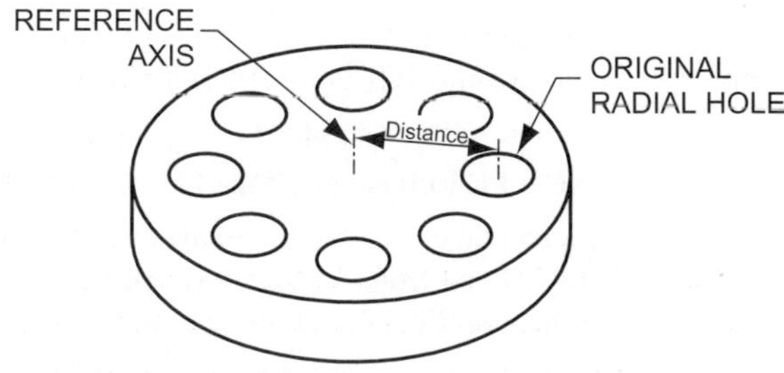

Figure 6–9 Patterned hole

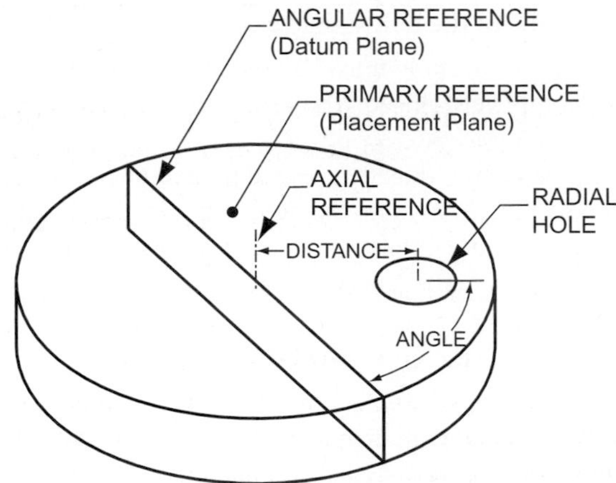

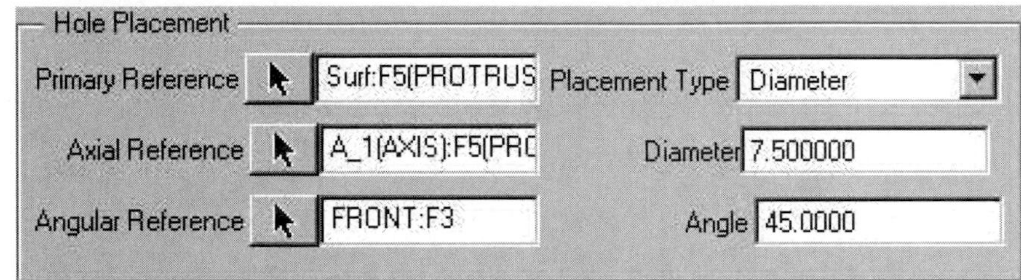

Figure 6–10 Radial hole creation

Creating a Straight-Diameter (Radial) Hole

Perform the following steps to place a Straight-Diameter Hole (see Figure 6–10).

Step 1: Select FEATURE >> CREATE >> HOLE.

Step 2: On the Hole dialog box, select STRAIGHT as the Hole Type.

Radial holes may be created with the Sketched and Standard hole types also.

Step 3: Enter the hole's Diameter value.

Step 4: Enter the hole's Depth One parameter (and value if required).

Step 5: Pick the hole's Placement Point (Primary Reference).

Step 6: Select DIAMETER (or Radial) as the hole's Placement Type.

Other options available include Linear and Coaxial. The Linear option will locate a hole from two reference edges, while the Coaxial option will place the hole's centerline coincident with an existing axis.

Step 7: Select the Axial Reference for the hole.

The hole will be located at a user-specified distance from the selected axis.

Step 8: Enter the Diameter distance value for the Radial Hole.

In the Diameter textbox, enter the distance value for the Diameter hole. The diameter value will be twice a corresponding radial value. As an example, if the hole is to be located three inches from the axial reference, then you should enter six as the diameter value. This option is often used to create a Bolt-Circle hole pattern.

Step 9: Select a Reference Plane, then enter an angular value from the plane.

Select an existing planar surface, then enter an angular value. The hole will be located from the reference plane at the specified angle. A datum plane or planar surface may be selected.

Step 10: Select the Build Feature checkmark.

The Shaft Option

The counterpart to the sketched hole option is the Shaft command. While the sketched Hole option creates a negative space feature, the Shaft command creates a positive space feature (Figure 6–11). The techniques and options for creating a shaft are the same as for creating a hole. As an example, shafts require a vertical centerline with entities sketched on one side only. Additionally, placement options for a shaft are the same as for a hole. However, when compared to the sketched hole option, one concern does exist. As with sketched holes, the uppermost line in the sketcher environment constructed perpendicular to the centerline is aligned with the placement plane. Thus, it is often necessary to sketch a shaft's section upside-down. To construct a shaft, perform the same steps for creating a sketched hole. The configuration file option *allow_anatomic_features* has to be set to Yes for the Shaft command to show on the Solid Options menu.

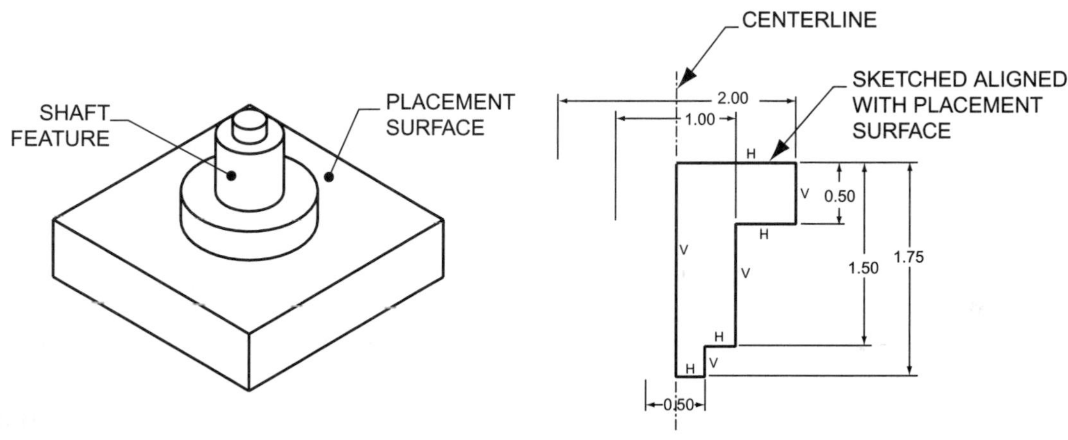

Figure 6–11 The shaft option

The Flange and Neck Options

A Flange is a revolved positive space feature created around an existing revolved part or feature (Figure 6–12). A Neck is a revolved negative space feature created around an existing revolved part or feature. The sketching plane for both a Flange feature and a Neck feature must be a Through >> Axis datum plane. As shown in Figure 6–13, when sketching a Flange or a Neck, the ends of the sketched section must be aligned with the surface of an existing revolved feature. The configuration file option *allow_anatomic_features* has to be set to Yes for the Flange and Neck commands to show on the Solid Options menu.

Perform the following steps to create either a Flange or a Neck.

Step 1: Select FEATURE >> CREATE >> FLANGE (or NECK).

Step 2: Specify the number of degrees of revolution.

Step 3: Select or create a datum plane that lies through the axis of revolution of a part or feature.

The plane for sketching a Flange or Neck section must be a Through >> Axis datum plane (Figure 6–14). If one does not exist, create one with the Make Datum option.

Step 4: Accept the view direction, then orient the sketcher environment.

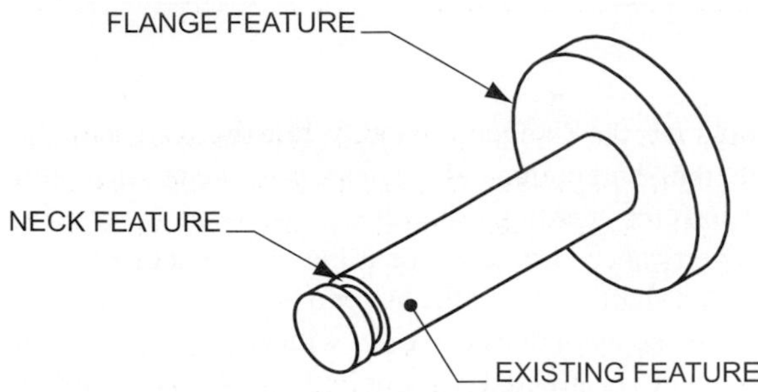

Figure 6–12 Flange and neck features

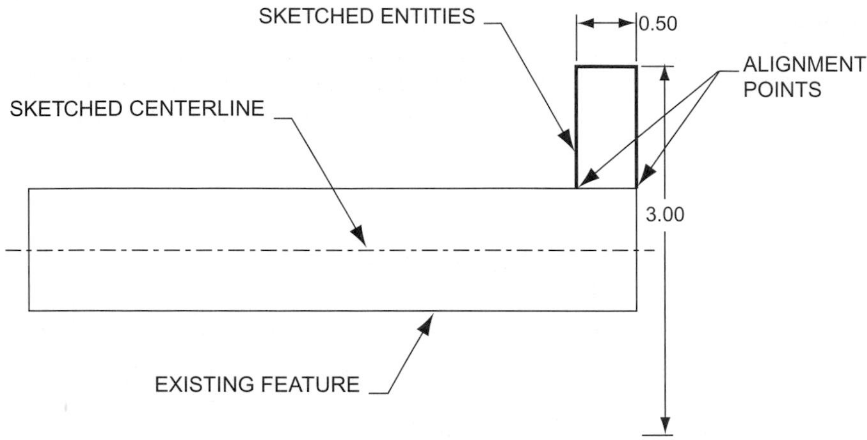

Figure 6–13 The sketched flange section

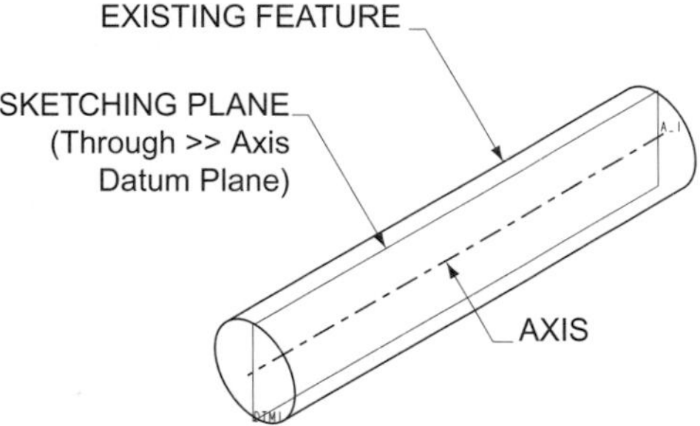

Figure 6–14 Selecting a datum plane

STEP 5: Use the CENTERLINE icon to create a centerline to revolve the Flange or Neck around (Figure 6–13).

Align the centerline with the axis of the existing revolved feature. The existing axis will not serve as the centerline.

STEP 6: Sketch the section of the Flange or Neck feature.

As shown in Figure 6–13, the section must be aligned with the surface of the existing revolved part or feature. Dimension the feature according to design intent.

STEP 7: Select DONE on the Sketcher menu to create the Flange or Neck.

When compared to most feature creation processes, the Neck and Flange options are unique. They do not utilize a Feature Definition dialog box.

Rotational Patterns

The Pattern command can be used to create a radial pattern of a Hole feature or a rotational copy of a sketched feature. To create a rotational pattern, an angular dimension must exist that defines the feature to be patterned. There are two common situations when a rotational pattern may be used:

Radial Pattern

The Pattern command can be used to create an angular copy of a radial hole (Figure 6–15). The Hole-Radial option places a feature by entering a distance from a selected axis and by entering an angle to a reference plane. This angle can be used as the leader dimension in the pattern creation process.

Rotational Pattern

The Pattern option can be used to create a rotational copy of a sketched feature (Figure 6–16). The feature's section has to reference a Through Axis datum plane

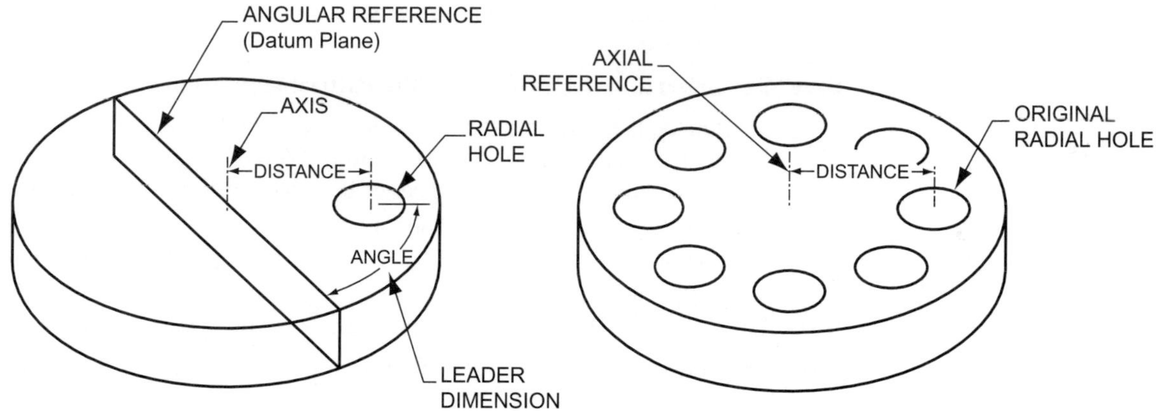

Figure 6–15 Radial pattern

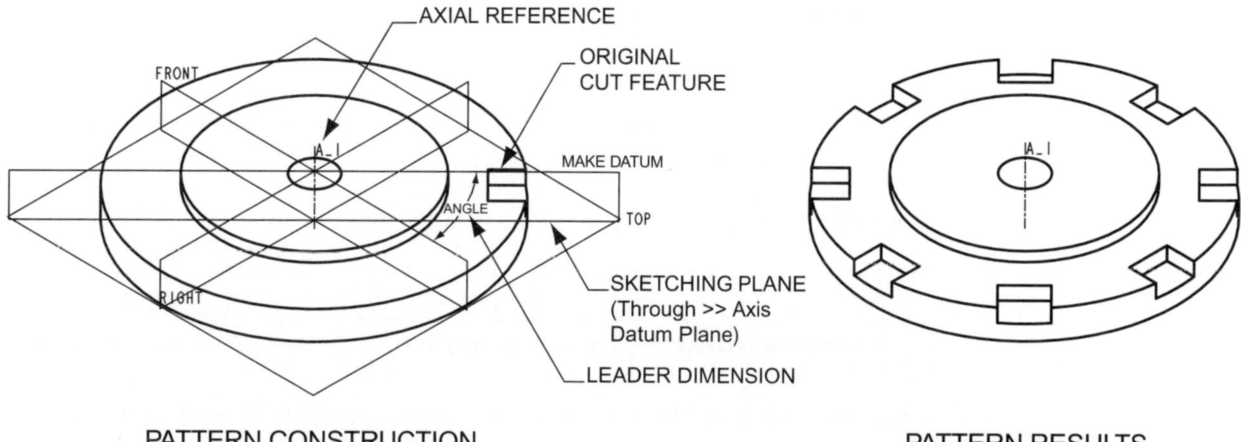

Figure 6–16 Rotational pattern

created with the Through and Angle constraint options or the feature has to be sketched on a Through Axis datum plane. The angular dimension defining the sketching plane is used to pattern the feature.

> **INSTRUCTIONAL NOTE** For more information on Patterns, see Chapter 6.

Creating a Radial Pattern

The following is a step-by-step approach for creating a Radial Pattern in one direction. Two directional patterns can be used with rotational patterns also. Additionally, dimensions other than the leader dimension can be varied.

Step 1: Select FEATURE >> PATTERN.

Step 2: Select the feature to pattern.

On the work screen or on the Model Tree, select a hole or sketched feature. The feature has to include an angular dimension that will serve as the leader dimension.

Step 3: Select a Pattern Option on the Pattern Options menu.

On the Pattern Options menu, select Identical, Varying, or General. Identical patterns require the most assumptions. With this option, instances of a pattern cannot intersect other instances or the edge of the placement plane. Additionally, an Identical pattern has to exist on one placement plane only. The General option does not require any assumptions, but takes longer to regenerate.

Step 4: Select VALUE on the Pattern Dimension Increment menu.

Step 5: Select an angular leader dimension for use in varying the feature (see Figures 6–15 and 6–16).

To create a rotational pattern of a feature, the feature must have an angular dimension. Holes and shafts placed with the Radial option incorporate a reference angle dimension to locate the hole. Features sketched on an on-the-fly datum plane can be patterned if the datum plane is defined with the Angle constraint.

Step 6: Enter a dimension increment value.

Enter the number of degrees that the feature will be incremented.

Step 7: Select DONE on the Exit menu.

Selecting Done will end the selection of varying dimensions for the first direction. Within a rotational pattern, dimensions other than the leader dimension can be varied also.

Step 8: Enter the number of instances in the first direction.

> **MODELING POINT** To create a bidirectional pattern, repeat steps 4–8. As with linear patterns, rotational patterns can be created in two directions.

Step 9: Select DONE to create the pattern.

DATUM AXES

Datum axes are used as references for the creation of features. As an example, they can be used to place a coaxial hole or a radial hole. Additionally, they are used often to create datum planes. When holes, cylinders, and revolve features are created, datum axes are created automatically. Datum axes created separately from a part feature are considered to be features. They are named in sequential order on the model tree starting with A_1.

CREATING DATUM AXES

Datum axes can be placed on parts or assemblies. Perform the following steps to create a datum axis in Part mode:

STEP 1: **On the Datum Toolbar, select the CREATE DATUM AXIS icon.**

STEP 2: **Select a CONSTRAINT OPTION on the Datum Axis menu, then pick geometry appropriate to the selected constraint option.**

Datum axes are created utilizing existing part features. Unlike datum planes, all datum axis constraint options are stand-alone. The following is a list of available constraint options:

THRU EDGE

The Through Edge option creates a datum axis through an existing edge of a part. The user has to select a geometric edge formed by two part surfaces.

NORMAL PLN

The Normal to Plane option creates a datum axis perpendicular to an existing plane. This plane may be a part or datum plane. Dimensional references are required to locate the axis. References are created by selecting two part edges and then providing dimensional values.

PNT NORM PLN

This constraint option places a datum axis perpendicular to a plane and through an existing datum point. The datum point does not have to lie on the selected plane.

THRU CYL

Through Cylinder places a datum axis at the center of an existing cylinder. The surface can be completely cylindrical or a partial cylinder, such as a round.

TWO PLANES

The Two Planes option places a datum axis at the intersection of two planar surfaces or two datum planes. A common use of this feature is to place an axis at the intersection of two of Pro/ENGINEER's default orthogonal datum planes.

TWO PNT/VTX

This constraint option places a datum axis between two datum points, between two vertexes, or between a datum point and a vertex.

PNT ON SURF

This constraint option places a datum axis perpendicular to a surface and through a point that lies on the surface.

TAN CURVE

Tangent Curve places a datum axis tangent to a curve or edge at a selected point. The point must exist prior to the use of this option.

> **MODELING POINT** Many constraint options require you to select multiple geometric features. Watch the message area carefully to know what object to select.

STEP 3: Select DONE.

The datum axis will be created when you select Done.

SUMMARY

Some of the most common features created within Pro/ENGINEER are revolved features. Options that utilize a form of a revolved feature include Revolved Protrusions, Revolved Cuts, Sketched Holes, Shafts, Flanges, and Necks. One of the requirements of any revolved feature is the sketching of a centerline. Within the sketcher environment for a feature option, entities of the sketch must lie completely on one side of the centerline. In addition to the typical options that are revolved, the Pattern command allows for the creation of rotational instances of a feature.

TUTORIAL

Revolved Features Tutorial

This tutorial exercise provides step-by-step instruction on how to model the part shown in Figure 6–17.

This tutorial will cover:

- Creating a revolved protrusion.
- Creating a radial sketched hole.
- Creating a hole radial pattern.
- Creating a revolved cut.
- Modifying the number of holes in a pattern.

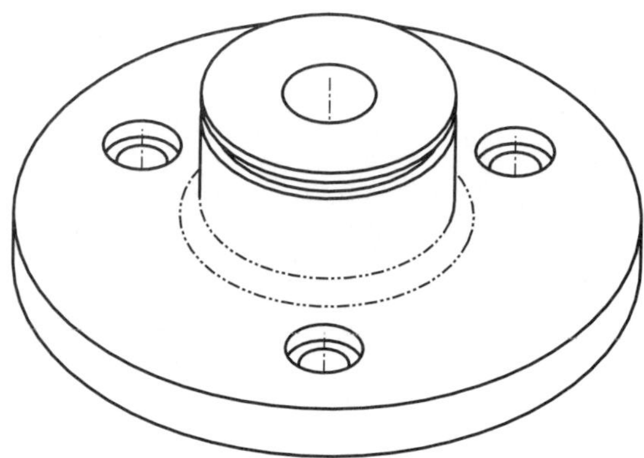

Figure 6–17 Finished part

Creating a Revolved Protrusion

The first section of the Revolved tutorial will create the base revolved Protrusion shown in Figure 6–18.

Step 1: Start Pro/ENGINEER.

Step 2: Establish an appropriate Working Directory.

Step 3: Start a New part file named *revolve1* (use the default template file).

Use either the File >> New option or the New icon to start a new part file named *revolve1*.

Step 4: Create a Revolved Protrusion with the following options:

- Protrusion command.
- Solid revolve.
- One sided.
- Datum plane FRONT as the sketching plane.
- Default sketch plane orientation.

Revolved positive space features are created under the Protrusion command. Revolved negative space features can be created under the Cut command.

Step 5: Close the References dialog box.

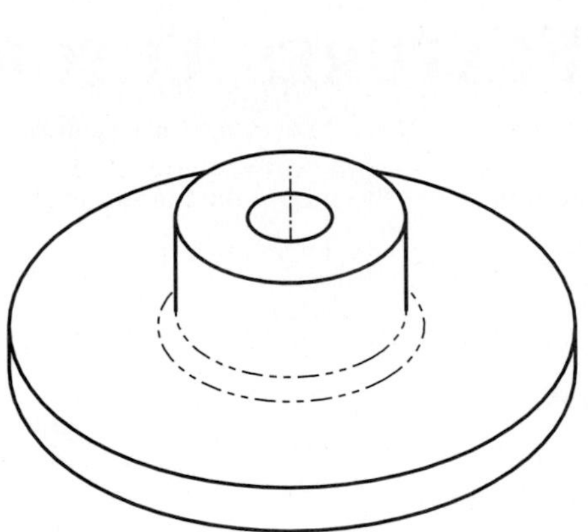

Figure 6–18 Revolved protrusion

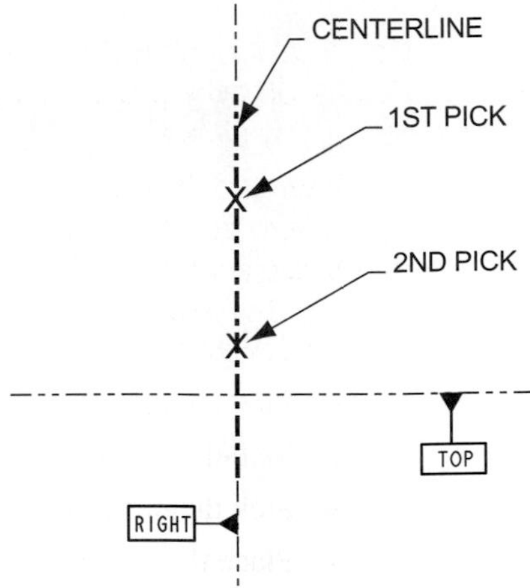

Figure 6–19 Sketching a centerline

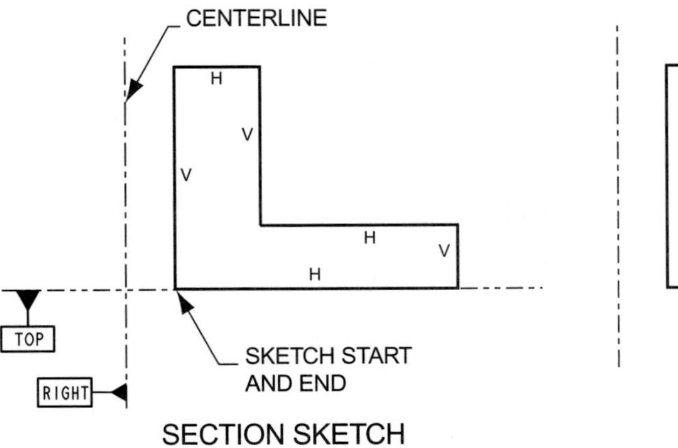

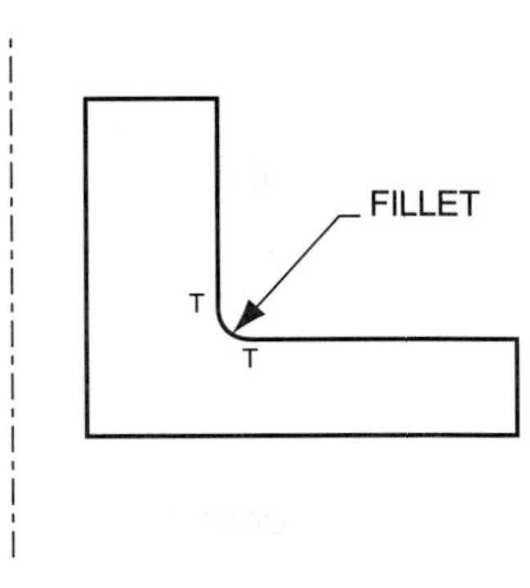

Figure 6–20 Sketching the section

STEP 6: Use the CENTERLINE option to sketch a vertical centerline aligned with datum plane RIGHT (Figure 6–19).

The Centerline icon can be found under the Line icon. Revolved features require a user-created centerline in the sketching environment. The sketched section will be revolved around this centerline. With Revolved Protrusions and Cuts, this centerline can be created at any angle.

STEP 7: Use the LINE option to sketch the section shown in Figure 6–20.

Sketch the section to the right of the centerline. Sketch all lines either horizontal or vertical as shown in the figure. Align the bottom of the sketch with datum plane TOP.

STEP 8: Use the CIRCULAR FILLET option to create the fillet shown in Figure 6–20.

When using the Circular Fillet option, select the two entities to fillet between.

STEP 9: Apply the dimensioning scheme shown in Figure 6–21. (Do not modify the Dimension Values within this step.)

Use the Dimension option to create the dimensioning scheme shown in Figure 6–21. Pro/ENGINEER does not know the design intent for a feature. Due to this, the dimensions created by Intent Manager may not match those necessary for the design.

To create the diameter dimensions, perform the following four selections (Figure 6–22):

1. Pick the outside edge of the entity (left mouse button).
2. Pick the centerline (left mouse button).
3. Pick the outside edge of the entity (left mouse button).
4. Place the dimension (middle mouse button).

STEP 10: Select the Modify option.

STEP 11: On the work screen, select the dimension defining the height of the flange (the 2.85 dimension in Figure 6–21).

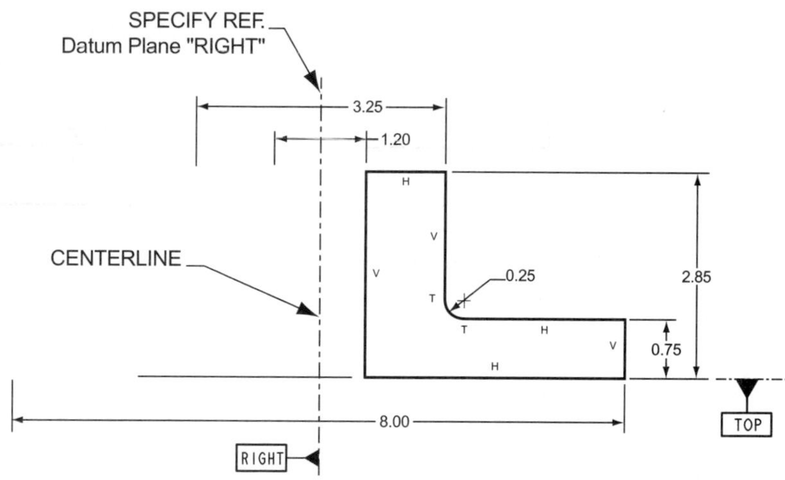

Figure 6–21 Dimensioning scheme

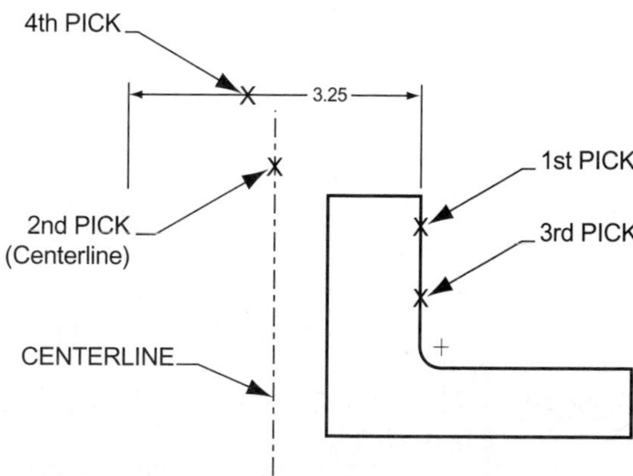

Figure 6–22 Diameter dimension

STEP 12: On the work screen, select the remaining five dimensions.

After selecting the six dimensions defining the section, your Modify Dimension dialog box should appear similar to Figure 6–23.

STEP 13: On the Modify Dimension dialog box, modify the flange's height dimension to have a value of 2.85 (select the tab key after entering value).

If you get unexpected results from your dimension modification, use Undo to correct any errors.

STEP 14: Uncheck the LOCK SCALE option.

STEP 15: On the dialog box, modify the remaining dimension values to match Figure 6–21.

STEP 16: When all dimension values match the illustration, select the Regenerate checkmark on the dialog box.

STEP 17: Select the Continue icon to exit the sketcher environment.

Do not select the Continue option until your section matches Figure 6–21.

STEP 18: Specify a 360-degree revolution, then select DONE.

STEP 19: Preview the feature, then select OKAY on the Feature Definition dialog box (Figure 6–24).

Your feature will look similar to Figure 6–25. The illustration shown displays the part with Tangent Edges set to Phantom and the

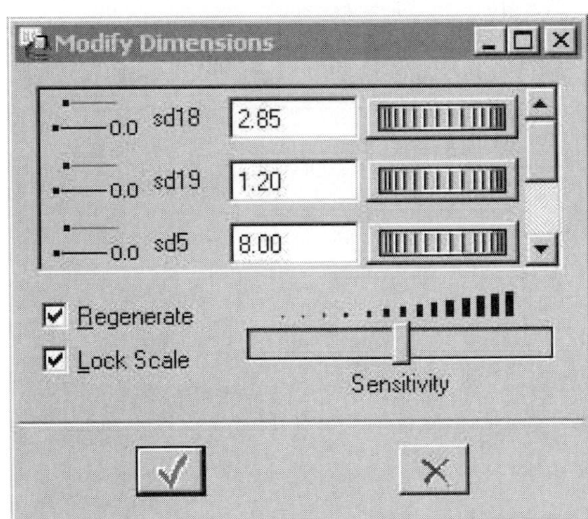

Figure 6–23 Modify Dimensions dialog box

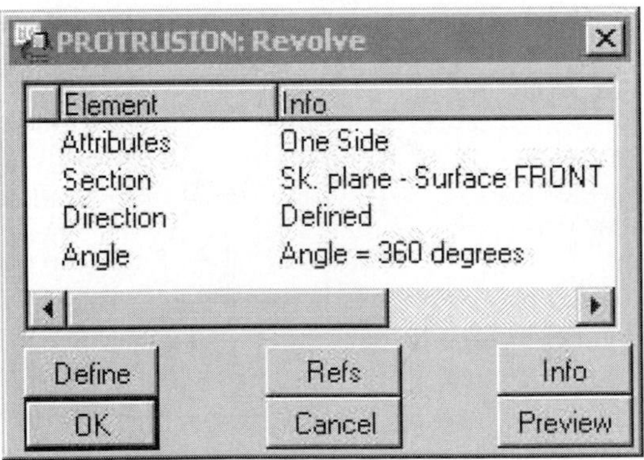

Figure 6–24 Revolve Feature Definition dialog box

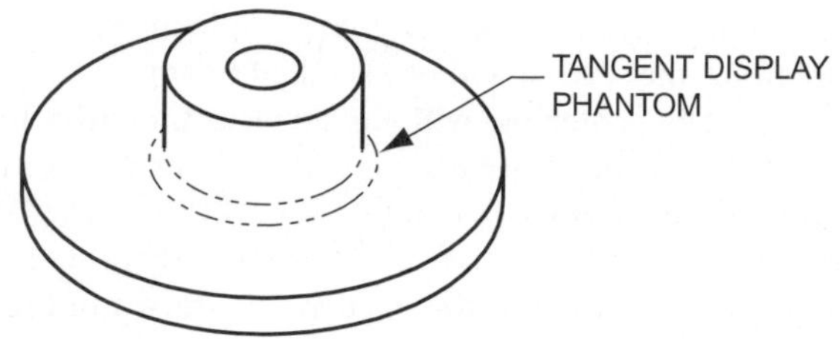

Figure 6–25 Finished feature

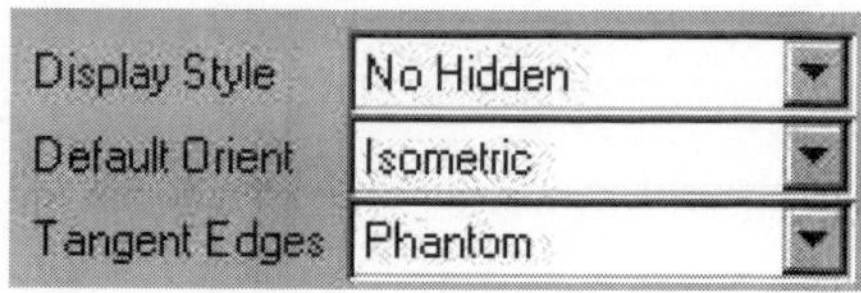

Figure 6–26 Environment options

default orientation set to Isometric. To change the Tangent Edge display and default orientation of a part, make the adjustments to the Environment dialog box (Utilities >> Environment), as shown in Figure 6–26.

CREATING A RADIAL SKETCHED HOLE

This segment of the tutorial create the sketched Radial Hole shown in Figure 6–27.

STEP 1: Select FEATURE >> CREATE >> HOLE.

STEP 2: Select SKETCHED as the hole type.

The Sketched option will require you to sketch the hole's cross-section. After you select this option, Pro/ENGINEER will open a sketcher environment.

STEP 3: Use the CENTERLINE option to create the vertical centerline shown in Figure 6–28.

A vertical centerline is a requirement for a sketched Hole feature. The section will be revolved around the centerline.

STEP 4: Sketch the section shown in Figure 6–29.

Use the Line option to sketch the section. For this specific section, sketch each entity either horizontally or vertically.

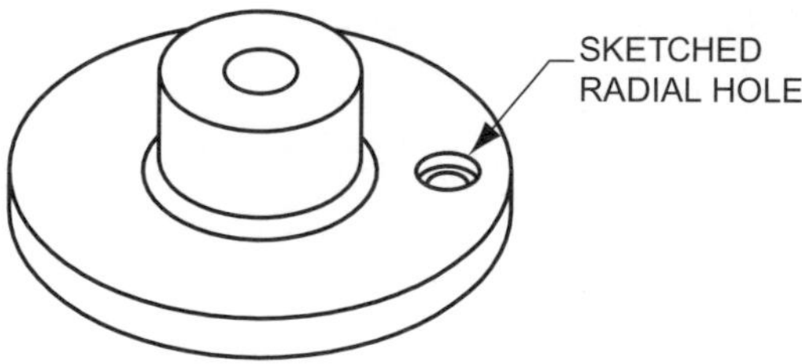

Figure 6–27 Finished hole feature

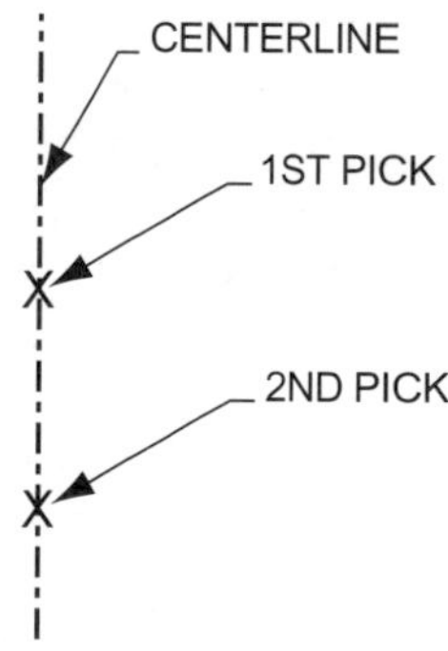

Figure 6–28 Sketching the centerline

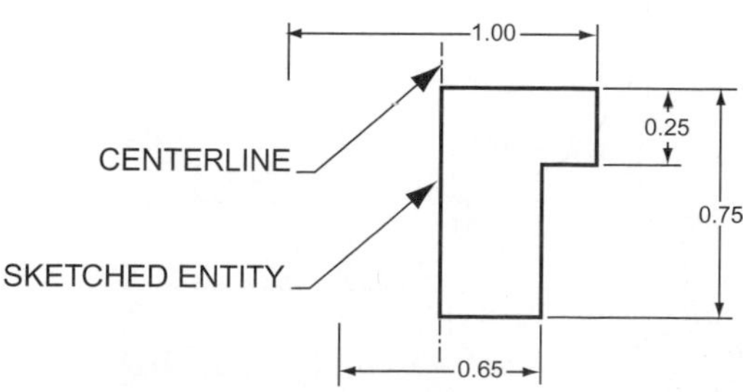

Figure 6–29 Sketched section

Sections sketched holes must be completely enclosed. As shown in Figure 6–29, the section will be sketched on top of the previously created centerline. **The centerline will not serve as an entity to close the section.** You will receive an error message when exiting the sketcher environment if the section is not closed.

STEP 5: Apply the dimensioning scheme as shown in Figure 6–29.

Use the Dimension option to match the dimensioning scheme shown in Figure 6–29. Holes are dimensioned with diameter values.

STEP 6: Modify dimension values to match Figure 6–29.

STEP 7: Select the Continue icon when the section is correct.

STEP 8: Pick the hole's placement plane (Primary Reference) as shown in Figure 6–30.

STEP 9: Select DIAMETER as the hole's Placement Type (Figure 6–30).

Using the Diameter option, the hole will be located at a distance (3 inches) from datum axis A_1 and at an angle to datum plane FRONT (45 degrees).

> **MODELING POINT** Creating a radial or diameter hole is the first step in modeling a rotationally patterned hole. The angular dimension created with a radial hole placement option is used as the leader dimension for patterning the hole around the center axis.

STEP 10: Select the hole's AXIAL REFERENCE.

As shown in Figure 6–30, select the reference axis. The hole will be located at a specified distance from the selected axis.

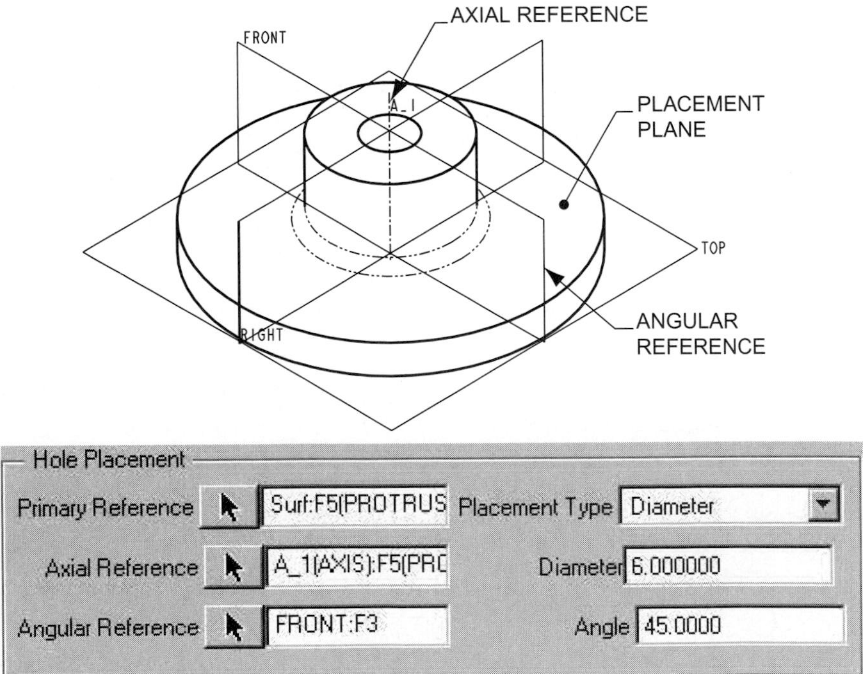

Figure 6–30 Hole placement references

STEP 11: Enter 6.00 as the Diameter value.

The Diameter value is double the distance from the Axial Reference to the center of the hole.

STEP 12: Select datum plane FRONT as the hole's ANGULAR REFERENCE plane.

The hole will be placed at an angle to the reference plane. As shown in Figure 6–30, select datum plane FRONT as the reference plane.

STEP 13: Enter 45 as the reference angle.

STEP 14: Preview the feature.

STEP 15: Select the Build Feature checkmark.

STEP 16: Save your part.

CREATING A RADIAL HOLE PATTERN

This segment of the tutorial will create a radial pattern of the hole created in the previous section of this tutorial. The 45-degree angular reference dimension used to locate the hole will be used as the leader dimension within the pattern. The finished pattern is shown in Figure 6–31.

STEP 1: Select FEATURE >> PATTERN.

> **MODELING POINT** The Pattern command will create a rotational or linear pattern of a single feature. To create a pattern of multiple features, each feature must be grouped using the Group command. Within the Group menu is a separate Pattern option.

STEP 2: On the Model Tree, select the previously created hole feature (Figure 6–32).

STEP 3: Select IDENTICAL >> DONE as the Pattern Option.

The Identical option allows several assumptions in the pattern creation process. Features patterned with this option must lie on the same placement plane, instances cannot intersect the placement plane's edge, and instances cannot intersect other instances. To allow instances to be placed on different planes and to allow instances to intersect an edge, use

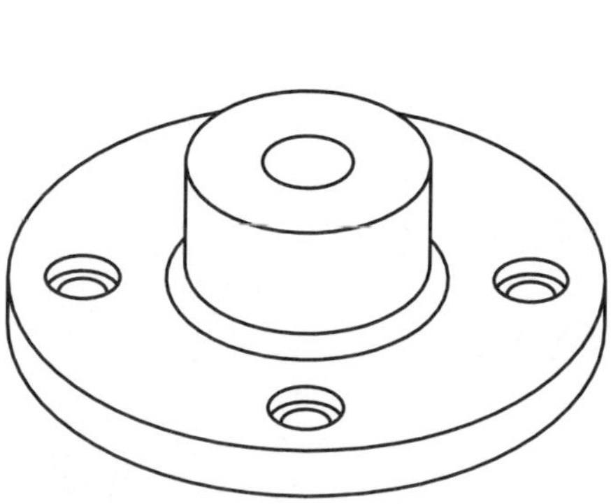

Figure 6–31 Finished pattern

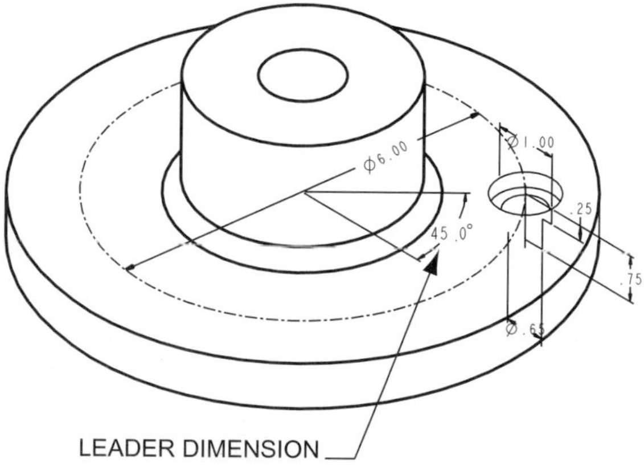

Figure 6–32 Feature parameters

either Varying or General option. To allow instances to intersect, use the General option.

STEP 4: As shown in Figure 6–32, pick the 45-degree angular dimension used to place the sketched radial hole.

The 45-degree angular dimension will be used as the leader dimension within the patterning process. This dimension will be varied to create the pattern.

STEP 5: Enter 90 as the increment value.

The leader dimension will be varied 90 degrees per instance of the hole feature. In other words, each hole within the pattern will be 90 degrees apart.

STEP 6: Select DONE on the Exit menu.

Selecting Done will end the varying process in the first direction of the pattern. Patterns may be unidirectional or bidirectional. For both cases, dimensions in addition to the leader dimension can be selected for varying. Selecting Done will exit the first direction of patterning.

STEP 7: Enter 4 as the number of instances in the first direction.

STEP 8: Select DONE on the Exit menu.

No instances of the hole will be created in the second direction. Selecting Done will create the pattern (Figure 6–31).

STEP 9: Save the part.

CREATING A REVOLVED CUT

This segment of the tutorial will create the Revolved Cut feature shown in Figure 6–33.

STEP 1: Select FEATURE >> CREATE >> CUT.

STEP 2: Select REVOLVE >> SOLID >> DONE.

STEP 3: Select ONE SIDE >> DONE.

The One Side option will revolve the section one direction from the sketching plane.

STEP 4: Select datum plane FRONT as the sketching plane (Figure 6–34), then select OKAY on the Direction menu.

STEP 5: Select DEFAULT to set the sketcher environment's orientation.

Optionally, you could select the Top option then pick the top of the flange. This would orient the flange toward the top of the sketcher environment.

STEP 6: Within the sketcher environment, specify the three references shown in Figure 6–35.

Use the References dialog box to specify the three references shown in the illustration.

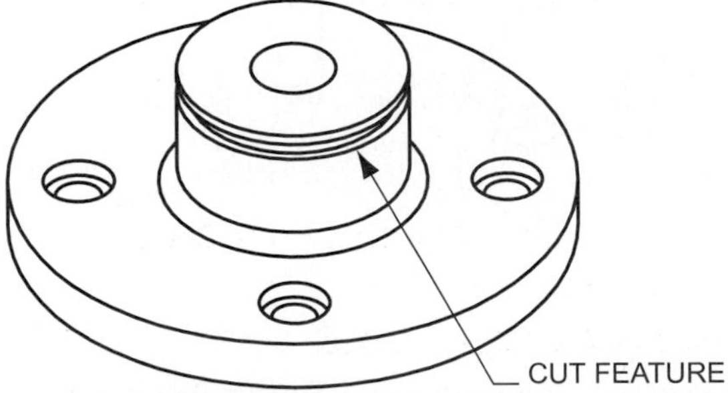

Figure 6-33 Revolved cut feature

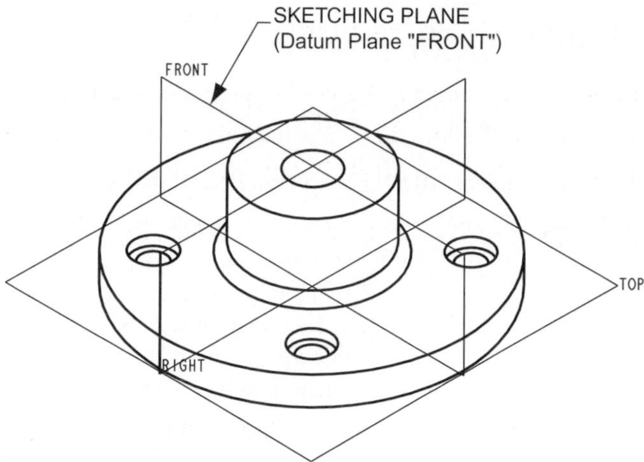

Figure 6–34 Sketching plane

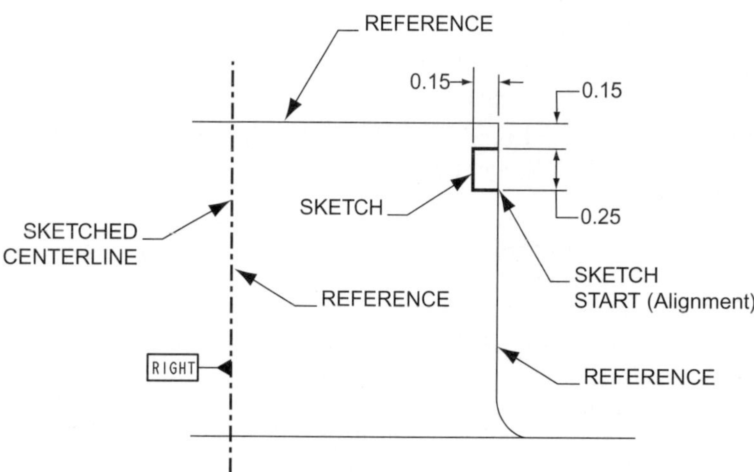

Figure 6–35 Section creation

STEP 7: Sketch the section shown in Figure 6–35.

Sketch the section as shown. To better view the entities being sketched, use a zoom option to match the zoom extents shown in Figure 6–35. By aligning with the outside edge of the existing flange, a closed section is not necessary. After sketching the section, apply the dimensioning scheme shown, then modify the dimension values accordingly.

STEP 8: Create a Centerline aligned with datum plane RIGHT (see Figure 6–35).

Sketch a vertical centerline aligned with datum plane RIGHT. This centerline will serve as the required axis of revolution.

STEP 9: Select the Continue icon to exit the sketcher environment.

STEP 10: Select the Default Material Removal Side.

STEP 11: Select 360 >> DONE as the Revolve Angular value.

STEP 12: Preview the feature, then select OKAY on the Feature Definition dialog box.

MODIFYING THE NUMBER OF HOLES

In this segment of the tutorial, you will use the Modify command to change the number of holes around the bolt-circle. The final pattern will appear as shown in Figure 6–36.

STEP 1: Select MODIFY on Pro/ENGINEER's Menu Manager.

STEP 2: As shown in Figure 6–37, select one of the three patterned instances of the original sketched hole.

Select one of the patterned instances of the hole, not the original. As shown, an instance of a rotated pattern will show the original leader dimension value (45 degrees for this part) and the increment value (90 degrees). Also shown is the number of instances of the patterned hole (4 holes).

STEP 3: Select the 90-degree value and modify it to equal 120.

STEP 4: Select the 4 HOLES parameter and modify it to equal 3.

When modifying the number of instances of a pattern, you must pick the actual text defining the value. In this case, select the number 4.

STEP 5: Select the 6.00 diameter dimension and modify it to equal 5.50.

STEP 6: REGENERATE the part.

Your part should appear similar to Figure 6–36.

STEP 7: Save your part.

STEP 8: Purge old versions of the part by using the FILE >> DELETE >> OLD VERSIONS option.

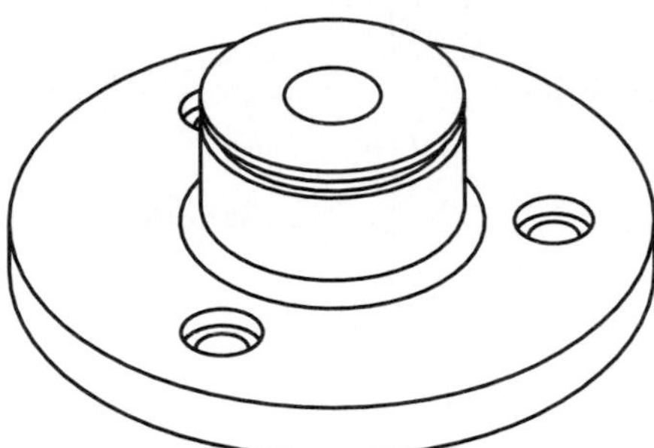

Figure 6–36 Finished part

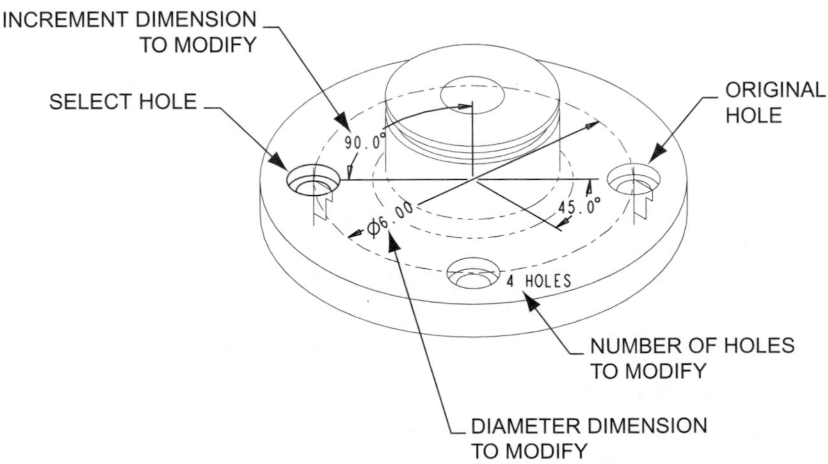

Figure 6–37 Pattern parameters

TUTORIAL

SHAFT TUTORIAL

This tutorial exercise provides step-by-step instruction on how to model the part shown in Figure 6–38. One of the features of this part will be a shaft. Since the Shaft command, by default, is not available on the Solid menu, the configuration file option *allow_anatomic_features* has to be set to Yes to display this command. Setting this option will be the first section in this tutorial. Chapter 2 provides more information on setting configuration file options.

This tutorial will cover:

- Setting configuration options.
- Creating a base extruded protrusion.
- Creating a shaft feature.
- Creating a cut feature utilizing an on-the-fly datum plane.
- Patterning a cut feature.

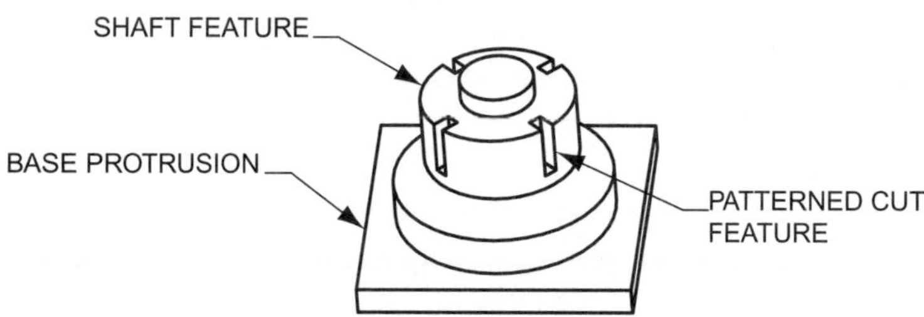

Figure 6–38 Final Part

SETTING CONFIGURATION OPTIONS

The Flange, Neck, Shaft, and Slot commands are not available by default on Pro/ENGINEER's solid menu. The configuration file option *allow_anatomic_features* set to Yes will reveal these commands. Perform the following steps to set this option.

STEP 1: On Pro/ENGINEER's menu bar, select UTILITIES >> OPTIONS.

Selecting Options will reveal the Options dialog box, which is used to set configurations settings. See Chapter 2 for more information on configuration file options.

STEP 2: In the Options textbox of the Preferences dialog box, enter *ALLOW_ANATOMIC_FEATURES.*

STEP 3: Set the value of the *allow_anatomic_features* configuration option to Yes.

STEP 4: Select the ADD/CHANGE option.

After selecting this option, notice how the allow_anatomic_features option is added to the configuration settings.

STEP 5: Select APPLY then CLOSE the dialog box.

Creating a Base Protrusion

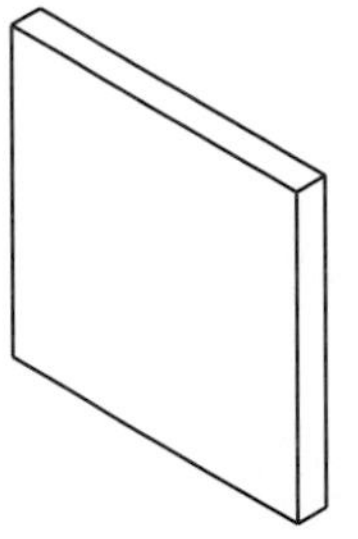

Figure 6–39 Base feature

The first segment of this tutorial will create the base feature upon which the Shaft feature will be placed. This base feature, shown in Figure 6–39, will be created as an Extruded Protrusion. This tutorial will not utilize Pro/ENGINEER's default datum planes.

STEP 1: Start Pro/ENGINEER.

STEP 2: Establish an appropriate Working Directory.

STEP 3: Select FILE >> NEW.

STEP 4: On the New dialog box, deselect the Use Default Template.

STEP 5: Enter *SHAFT1* as the name of the part file, then select OKAY.

STEP 6: On the New File Options dialog box, select the *EMPTY* template file, then select OKAY.

The Empty template has no default settings, parameters, features, or datums.

STEP 7: Select FEATURE >> CREATE >> PROTRUSION.

This tutorial will NOT utilize Pro/ENGINEER's default datum planes as the first feature of the part. When modeling this feature, notice the steps that are missing in the protrusion process.

STEP 8: Select EXTRUDE >> SOLID >> DONE on the Solid Options menu.

As the first feature of the part, there will be no option for selecting a sketching plane. Additionally, without a sketching plane, Pro/ENGINEER will allow a One Side Protrusion only.

STEP 9: Using appropriate sketching tools, sketch the section shown in Figure 6–40.

Use either the Line option or the Rectangle option to create the sketch. Modify the dimension values to match the figure.

STEP 10: After the section is complete, select the Continue icon to exit the sketcher environment.

As with any section, do not select continue until the section is complete.

STEP 11: Enter an Extrude Depth of .500.

STEP 12: Preview the feature, then select OKAY on the Feature Definition dialog box (Figure 6–41).

Notice on the Feature Definition dialog box how only two definitions are available. Without the use of Pro/ENGINEER's default datum planes, options such as extrude direction, placement references, and extrude side are not available.

STEP 13: Save the part.

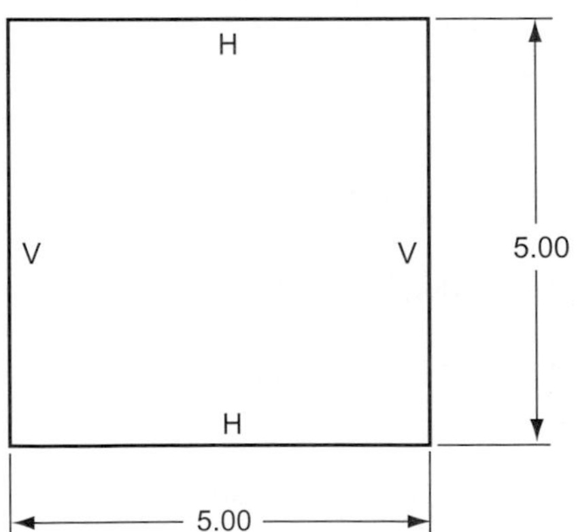

Figure 6–40 Regenerated section

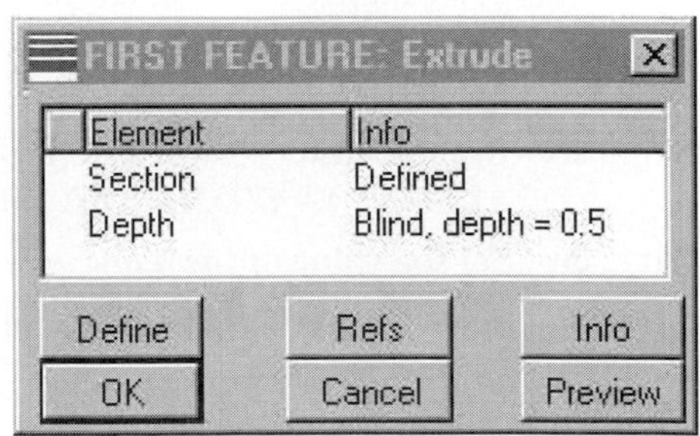

Figure 6–41 Feature Definition dialog box

CREATING A SHAFT

A Shaft will be the next feature created in this tutorial (Figure 6–42). The creation of a Shaft is similar to the creation of a sketched Hole feature. Unlike a sketched Hole, a Shaft feature has to be sketched upside-down. The configuration file option *allow_anatomic_features* has to be set to Yes for this command to show on the Solid menu.

STEP 1: Select FEATURE >> CREATE >> SHAFT.

The Shaft command creates a positive space feature. Its negative space counterpart is the sketched Hole.

STEP 2: Select LINEAR >> DONE as the Placement option.

The Linear option will place the shaft from two user-selected edges. Notice during the process of creating this feature how the options for a Shaft are the same as for a sketched Hole.

STEP 3: Within the independent sketcher environment, sketch a vertical centerline as shown in Figure 6–43.

All revolved features require a centerline. It is recommended for sketched holes and shafts that the centerline is constructed vertical. Pro/ENGINEER

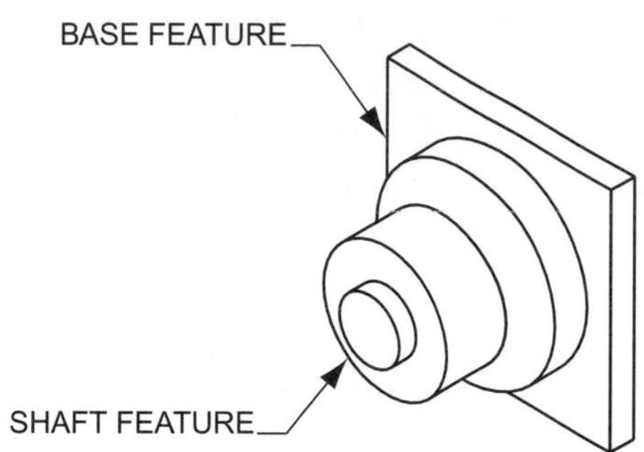

Figure 6–42 Shaft feature

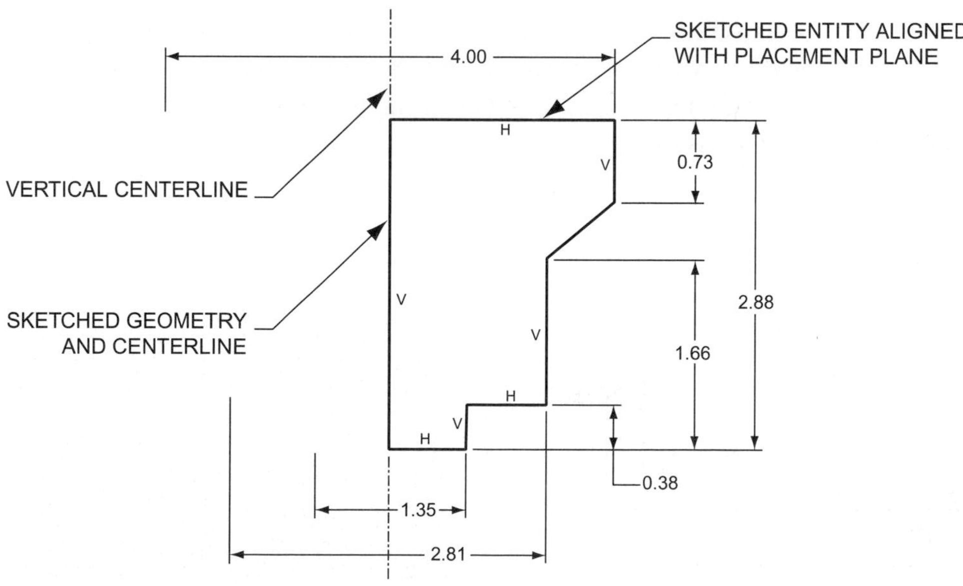

Figure 6–43 Sketched section

will allow you to exit the sketcher environment with a nonvertical centerline, but you will receive an error message when selecting OKAY on the Feature Definition dialog box.

STEP 4: **Sketch the section shown in Figure 6–43.**

Shafts are normally sketched upside-down. With a sketched Hole or Shaft, the uppermost line in the sketching environment perpendicular to the centerline will be the entity that aligns with the placement plane. Due to this, the portion of the section that aligns with the placement plane should be sketched at the top of the section.

STEP 5: **Dimension the sketch to match the design intent shown in Figure 6–43.**

STEP 6: **Modify the dimensioning scheme and values to match Figure 6–43.**

You can use either the Modify option or the Pick option to modify dimension values. To eliminate the possibility of a regeneration error, it is advisable to modify smaller dimension values first.

STEP 7: **Select the Continue option when the section is complete.**

STEP 8: **Select the top of the base feature as the Placement Plane (Figure 6–44).**

The placement plane for a shaft feature must be planar. Remember to observe Pro/ENGINEER's Message Area after each option's selection.

STEP 9: **Select one edge of the base feature as a location reference (Figure 6–44).**

The Linear placement option specified earlier in this tutorial will locate the shaft from two reference edges. This step requires the selection of one edge.

STEP 10: **Enter 2.50 as the reference distance.**

Enter the distance that the centerline of the shaft will lie from the first selected edge.

STEP 11: **Select a second edge of the base feature as a location reference (Figure 6–44).**

STEP 12: **Enter 2.50 as the second reference distance.**

STEP 13: **Select Preview on the Feature Definition dialog box.**

STEP 14: **If the Shaft is correct, select OKAY on the dialog box.**

STEP 15: **Save the part.**

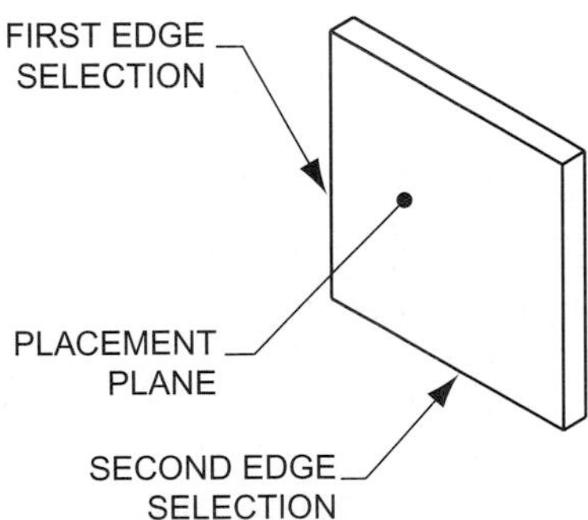

Figure 6–44 Shaft location

CREATING A CUT

This segment of the tutorial will create the Cut feature shown in Figure 6–45. This feature will be sketched on an on-the-fly datum plane. Additionally, the section will be extruded Both Sides from the sketching plane.

STEP 1: Select FEATURE >> CREATE >> CUT.

STEP 2: Select EXTRUDE >> DONE.

STEP 3: Select BOTH SIDES >> DONE on the Attributes menu.

The feature will be extruded both directions from the sketching plane.

STEP 4: Select MAKE DATUM on the Setup Plane menu.

The Make Datum option allows for the creation of an on-the-fly datum plane. On-the-fly datum planes are used to construct individual features and are not directly available for later features.

STEP 5: Select the THROUGH option, then on the work screen, pick Axis A_1 (Figure 6–46).

This sequence of options creates a Through >> Axis datum plane. Through >> Axis datum planes are necessary for the first constraint option when defining an angular datum plane.

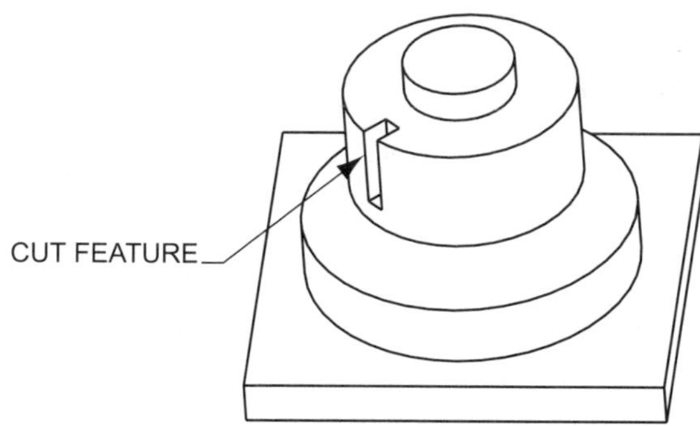

Figure 6–45 Cut feature

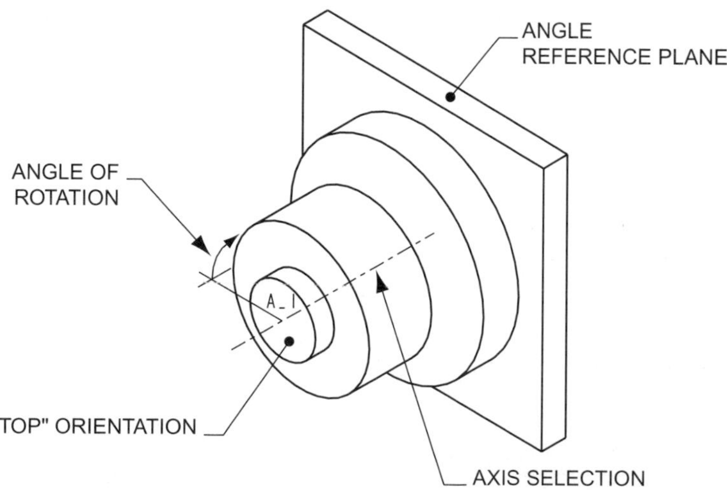

Figure 6–46 Datum creation

STEP 6: Select ANGLE on the Datum Plane menu, then pick the Reference plane shown in Figure 6–46.

The angle defined in this step will be used as the leader dimension in the rotation pattern created in the next section of this tutorial.

STEP 7: Select DONE on the Datum Plane menu.

STEP 8: Select ENTER VALUE on the Offset menu.

This option will create a datum plane with an angular reference value of 45 degrees. Notice on the work screen (and in Figure 6–46) the graphical display of the angle of rotation. The arrow points in the positive direction of rotation. A negative value can be entered if necessary.

STEP 9: Enter 45 as the angular value, then select OKAY to accept the default direction.

STEP 10: Select TOP, then pick the top of the shaft feature (Figure 6–46).

STEP 11: Specify the Two References shown in Figure 6–47, then close the References dialog box.

STEP 12: Create the section shown in Figure 6–47.

Only two lines are required to complete this section. After sketching the section, apply the dimensioning scheme shown in the figure, then modify the dimension values to match the figure. Remember that this is an extruded Cut. A centerline is not required.

STEP 13: Select the Continue option to exit the sketching environment.

STEP 14: Select OKAY to accept the default cut direction.

The direction of cut should be away from the model.

STEP 15: Select BLIND >> DONE as the cut depth, then enter .25 as the depth.

Since this is a Both Sides cut, the feature will extrude .125 inches in both directions from the sketching plane.

STEP 16: Preview the Cut, then select OKAY on the Feature Definition dialog box.

The Cut should appear similar to Figure 6–45.

STEP 17: Save the part.

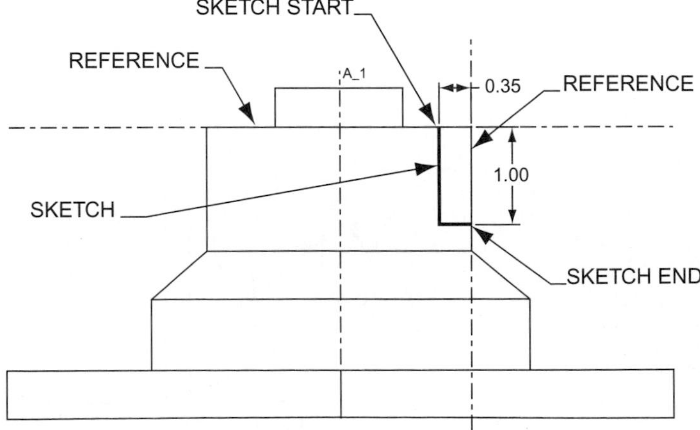

Figure 6–47 Sketched section

Creating a Pattern of the Cut

This segment of the tutorial will pattern the Cut feature previously created (Figure 6–48). When the Cut feature was constructed, it was sketch on an on-the-fly datum plane. This datum plane was created at an angle to a reference plane. This angular dimension will be used as the leader dimension for defining the pattern.

Step 1: Select FEATURE >> PATTERN.

Step 2: On the Model Tree, select the Cut feature.

Step 3: Select VARYING >> DONE as the Pattern Option.

The Varying option will allow patterned instances to intersect the edge of the placement plane and will allow instances to be placed on various planes. Why is the Identical option not available?

Step 4: Select the 45-degree dimension value that defines the reference angle of the Cut feature (Figure 6–49).

The 45-degree reference angle will serve as the pattern's leader dimension. This dimension will be used to define the first direction of the pattern.

Step 5: Enter 90 as the dimension increment value.

The leader dimension will be incremented at 90-degree intervals.

Step 6: Select DONE on the Exit menu.

In addition to the leader dimension, other dimension values can be varied. Within the Pattern command, you select Done to end the selection of dimensions to vary.

Step 7: Enter 4 as the number of instances of the cut in the first direction.

Step 8: Select DONE on the Exit menu.

Selecting Done will create the pattern.

Step 9: Save your part.

Step 10: Purge old versions of the part.

Use the File >> Delete >> Old Versions option to delete the older saved versions of the part.

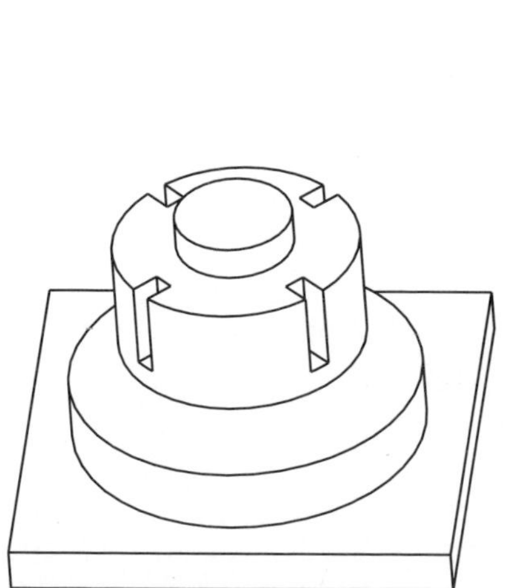

Figure 6–48 Patterned cut

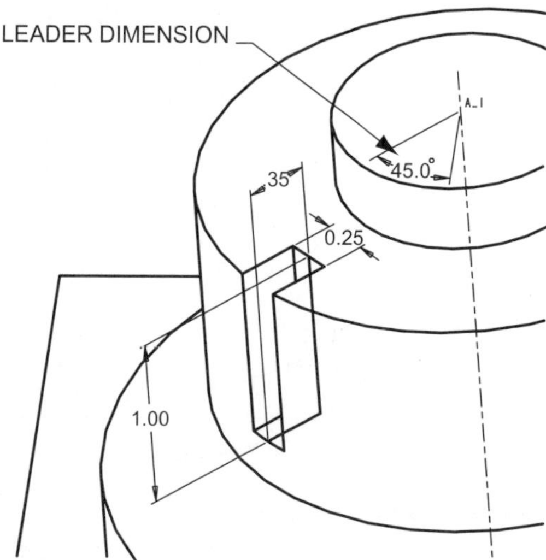

Figure 6–49 Leader selection

PROBLEMS

1. Using Pro/ENGINEER's Part mode, model the part shown in Figure 6–50. Create the base geometric feature as a revolved protrusion.
2. Using Pro/ENGINEER's Part mode, model the part shown in Figure 6–51. Create the base geometric feature as a revolved protrusion.

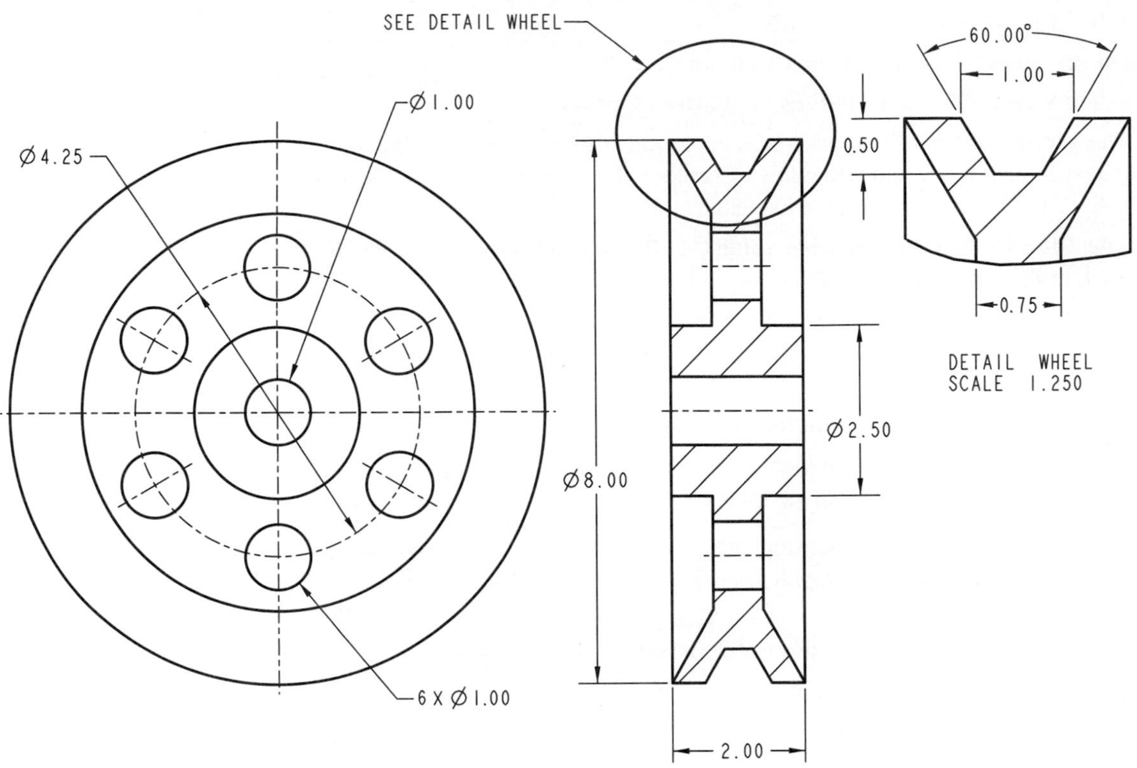

Figure 6–50 Problem one

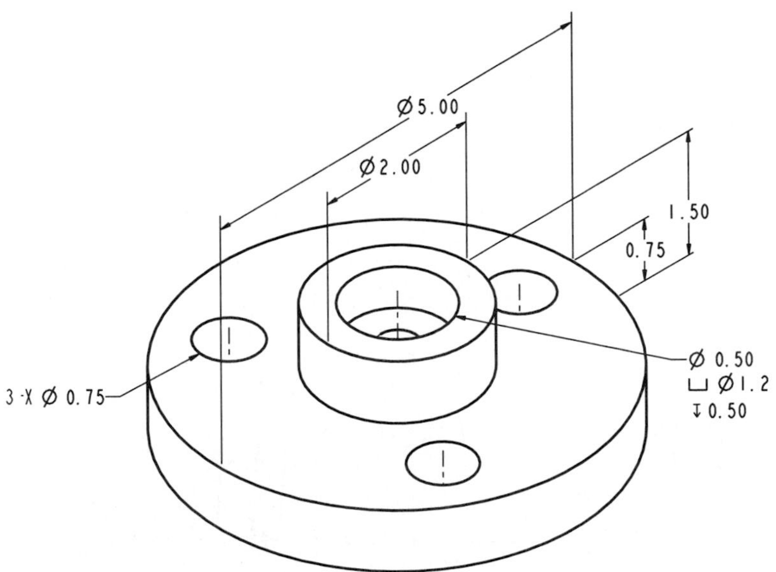

Figure 6–51 Problem two

3. Use Pro/ENGINEER to model the parts shown in Figures 6–52, 6–53, and 6–54. These parts will be used within an assembly model problem later in this textbook.

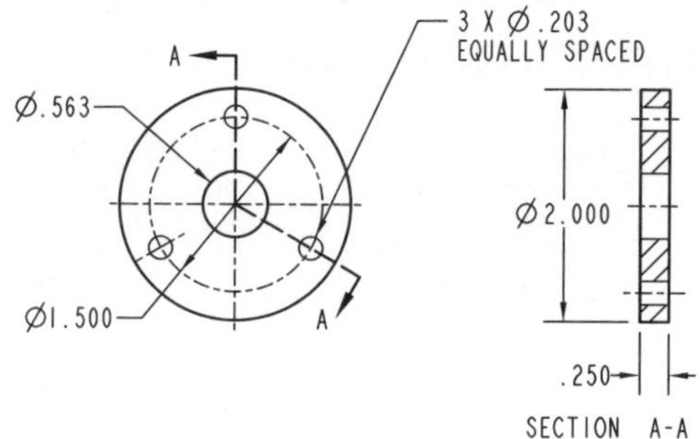

Figure 6-52 Retainer part

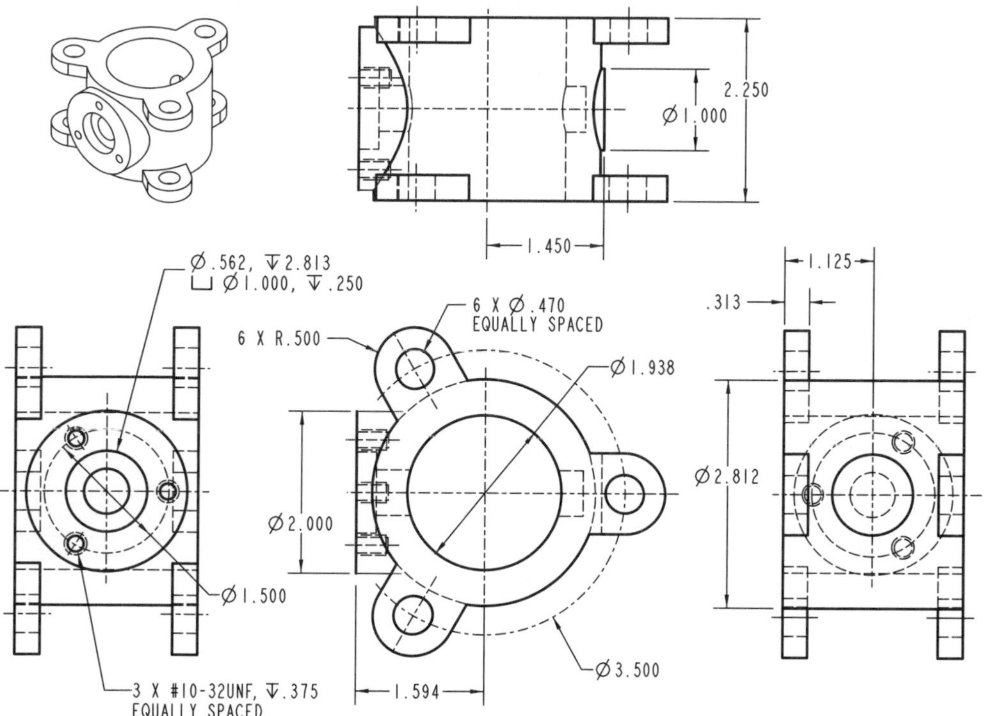

Figure 6-53 Body part

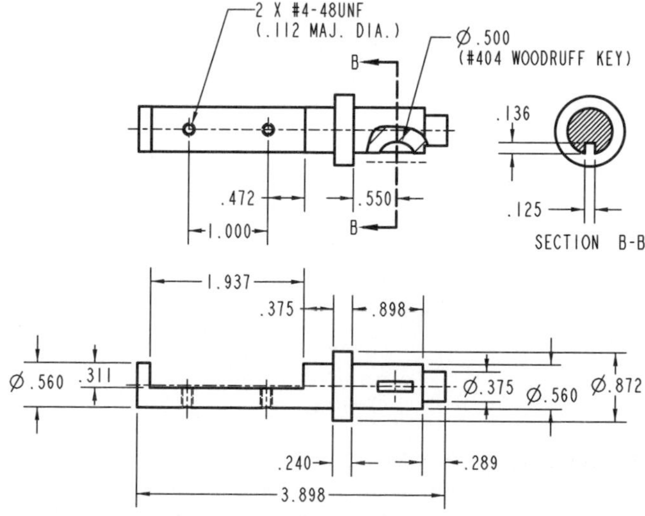

Figure 6-54 Shaft part

QUESTIONS AND DISCUSSION

1. List and describe various revolved features available within Pro/ENGINEER.
2. Describe the necessary requirements for sketching a revolved feature's section.
3. Describe the procedure for creating a diameter dimension within a section that will be revolved.
4. What revolved features require a vertical centerline?
5. Describe the difference between a hole placed with the Radial option and a hole placed with the Linear option.
6. What dimension type that defines a feature must be available to allow for the creation of a rotational pattern?

CHAPTER

7

FEATURE MANIPULATION TOOLS

Introduction

Commands such as Protrusion, Cut, Rib, Datum, and Hole are used to create part features. Pro/ENGINEER provides a variety of tools for manipulating existing features. As an example, multiple features can be combined with the Group option. When a group is created, it can be manipulated with options such Copy and Pattern. Other manipulation tools covered in this chapter include User-Defined Features, Relations, and Family Tables. Upon finishing this chapter, you will be able to

- Combine features as a local group.
- Pattern a local group.
- Copy features in a linear direction.
- Mirror features across a plane.
- Copy-rotate features.
- Copy features by specifying new references.
- Add a dimension relationship to a part.
- Create a family table.
- Create a User-Defined Feature.

DEFINITIONS

Family of parts A grouping of parts that are similar in shape, size, and geometry.

Family table A combination of parts that have similar features and geometry but vary slightly in selected items.

Group A collection of features combined to serve a common purpose.

Relation An explicit relationship that exists between dimensions and/or parameters.

User-defined feature A feature that is stored to disk as a group and can be used in other models.

User-defined feature library A computer directory location where User-Defined Features can be stored.

GROUPING FEATURES

Most Pro/ENGINEER modification options are utilized to manipulate individual features. However, it is often desirable to manipulate multiple features together. As an example, Figure 7–1 shows a part with a rotationally patterned boss, round, and hole. The normal

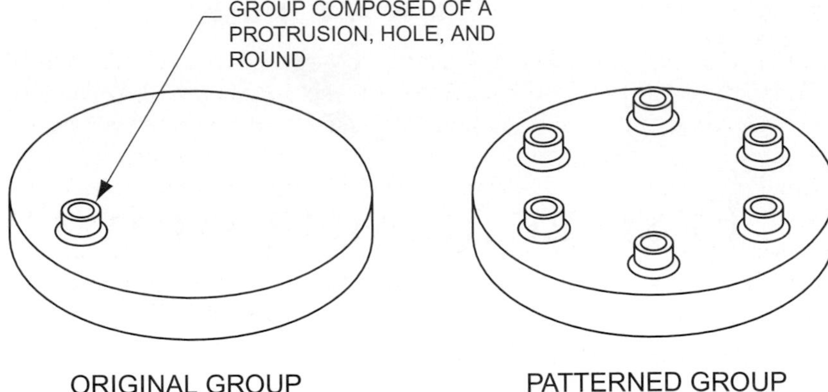

Figure 7-1 Patterned group

Pattern command is used on individual features, not multiple features. The grouped boss, round, and hole features shown in the illustration were patterned using the Group >> Pattern option.

THE GROUP MENU

The Group menu provides options for creating and manipulating groups. The following options exist:

CREATE

The Create option is used to create a group. Two options are available. The first option is to place a User-Defined Feature (UDF). When a UDF is placed in an object, it becomes a grouped feature. The second option is to create a Local Group. Local Groups are available only in the current model.

PATTERN

The Pattern option is used to create a rotational or linear pattern of a group. The Group >> Pattern option works similarly to the Feature >> Pattern command.

REPLACE

The Replace option is used to replace a UDF that has been placed in an object. The new UDF must have the same number and type of references. A Local Group cannot be replaced.

UNPATTERN

Grouped features are combined on the Model Tree as essentially one feature. Groups that are patterned are grouped together on the Model Tree as a patterned feature. The Unpattern option will break the pattern relationship, creating individual groups of each pattern instance.

UNGROUP

Features that are combined with the Group option can be ungrouped with the Ungroup option.

MODELING POINT The Group, Pattern, Unpattern, and Ungroup options can be used in sequence to copy multiple features in a way that will not require each feature to lose its individual identity. After a group has been patterned, it can be unpatterned with the Unpattern option. After performing an unpattern, each group that made up the pattern can be ungrouped with the Ungroup option.

Group Types

There are two types of **groups:** User-Defined Features and Local Groups. A **User-Defined Feature** (UDF) is a feature that has been grouped and saved to disk, often in a **UDF library.** A UDF can be retrieved and placed in the current working model or in another model. When a UDF is placed in an object, it becomes a grouped feature on the Model Tree.

A Local Group is a combination of features available within the current model only. The Group option allows multiple features to be grouped and patterned. Features that are combined to form a local group must be adjacent to each other in the order of regeneration. Thus, it is important during the modeling process to place intended group features next to each other on the Model Tree. To create a Local Group, perform the following steps:

Step 1: Select FEATURE >> GROUP >> CREATE.

After selecting Create, Pro/ENGINEER will launch the Open dialog box.

Step 2: Select CANCEL on the Open dialog box.

The Group >> Create option defaults to placing a User-Defined Feature. Select the Cancel option to exit from the Open dialog box.

Step 3: Select LOCAL GROUP on the Create Group menu, then enter a name for the group.

Step 4: On the Model Tree, select the features to include in the group.

Features may be selected directly on the work screen or from the model tree. Features in a group must be adjacent to each other on the model tree. When selecting features on the model tree, the beginning and ending features of the group may be selected.

Step 5: Select DONE on the Create Group menu.

Patterning a Group

A common reason to create groups is to pattern multiple features (Figure 7–2). Within the Group menu is a group Pattern option. Local Groups and User-Defined features can be patterned similarly to how individual features can be patterned with the Feature >> Pattern command. Linear and Rotational patterns may be created. While the Feature >> Pattern command has an option for selecting between an Identical, Varying, and General option, the Group >> Pattern option accepts the Identical option by default. Perform the following steps to pattern a group. The part (*group_pattern.prt*) shown in Figure 7–2 is available on the supplemental CD.

Step 1: Select FEATURE >> GROUP >> PATTERN.

When selecting the Group menu option from the Feature menu, Pro/ENGINEER will reveal the Open dialog box. Cancel this dialog box, and select Pattern on the Group menu.

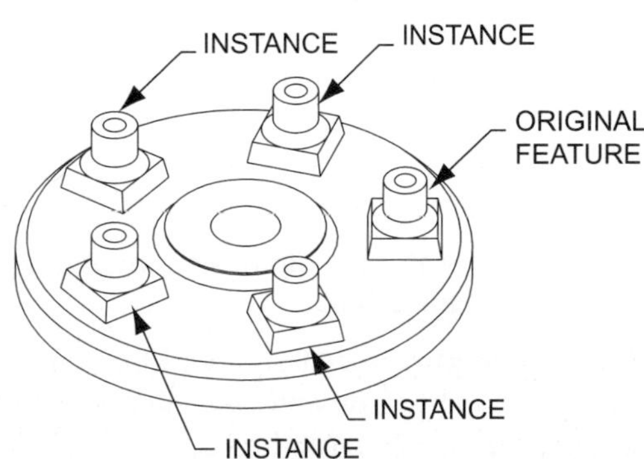

Figure 7–2 Pattern of features

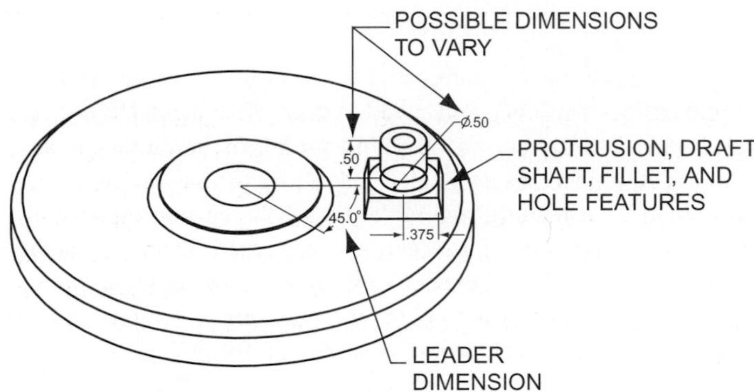

Figure 7-3 Dimension selection

STEP 2: Select the Group to be patterned.

The group to be patterned can be selected on the work screen or on the model tree. Only one group can be patterned at a time.

STEP 3: Select the leader dimension for the first direction of patterning (Figure 7–3).

The Group's Pattern option will allow groups to be patterned in two directions. Select the leader dimension that will drive the pattern in the first direction.

STEP 4: Enter the dimension increment for the leader dimension in the first direction.

Enter the amount that the leader dimension will vary in the first direction.

STEP 5: If required, select additional dimensions for patterning in the first direction and enter increment values (Figure 7–3).

This is an optional step. Multiple feature dimensions can be incremented in addition to the leader dimension. Select the dimensions to vary and enter increment values.

STEP 6: Select DONE to end the dimension selection in the first direction.

STEP 7: Enter the number of instances of the group in the first direction.

STEP 8: Select the leader dimension for the second direction of patterning.

INSTRUCTIONAL NOTE Groups can be patterned in one or two directions. The second direction is optional. If a pattern will be constructed in only one direction, select Done on the exit menu in lieu of a second leader dimension.

STEP 9: Enter the dimension increment for the leader dimension in the second direction.

STEP 10: If required, select additional dimensions for patterning in the second direction and enter increment values.

STEP 11: Select DONE to end the dimension selection in the second direction.

STEP 12: Enter the number of instances of the group in the second direction.

Copying Features

While the Pattern command is used to create multiple instances of a feature in a rotational or linear fashion, the Copy command is used to make a single copy of a feature or features. While the Pattern command creates new instances of a feature by varying dimensions that define the feature, the Copy command creates a copy by changing the placement of references and/or by changing dimension values. Figure 7–4 shows an example of a boss and coaxial hole that have been copied to another location. References required to define the location of the two features include a placement plane and two location edges. These references were changed to create the copy.

Copy Options

Unlike the Feature >> Pattern option, multiple features can be simultaneously copied with the Copy command. As shown in Figure 7–5, four basic types of copies can be created.

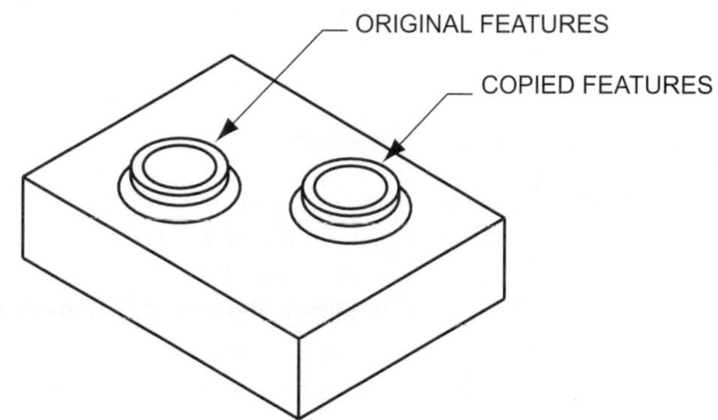

Figure 7–4 Copied features

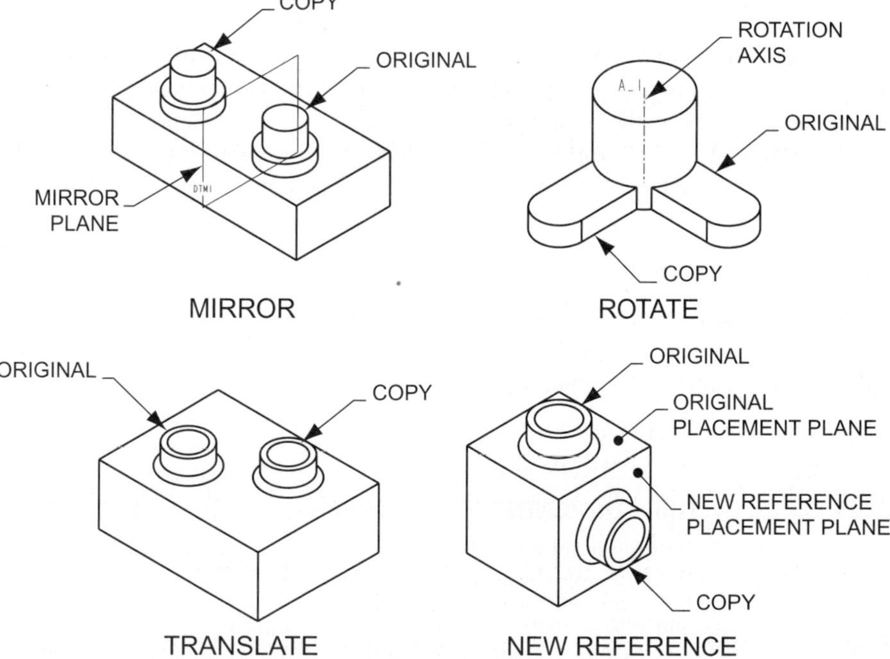

Figure 7–5 Copy options

MIRRORING FEATURES

The Mirror option is used to copy-mirror features about a plane. A mirrored image of the original feature is created.

ROTATIONAL COPIES

Rotate is a suboption under the Move option. Features can be copy-rotated around a datum curve, edge, axis, or coordinate system.

TRANSLATED FEATURES

Translate is a suboption under the Move option. Copied features are translated in a linear direction from the original features. The Same Refs option under the Copy menu creates a form of a translated copy.

NEW REFERENCES

Copies of features can be created by varying dimension values and by selecting new references. Examples of references include placement edges, reference axis, and placement plane. A feature can be copied with completely new references. The Same References (Same Refs) option creates a copy of selected features, but does not allow the selection of alternate references.

INDEPENDENT VERSUS DEPENDENT

Features can be copied independent or dependent of their parent features. When a feature is copied as dependent, its parent features' dimension values govern its dimension values. If a dimension is changed in the parent feature, the corresponding dimension is changed in the copy. The Independent option allows copied features to be independent of parent features. When a feature is copied as independent, the copy's dimensions will remain independent of its parents'. The dimensions of the copy can be changed, and any changes to the parents' dimensions will not affect the child's dimensions.

SELECTING A MODEL

It is a common procedure to copy a feature from one location on a model to another location. Options are available for copying features from a different version of a model or from a completely different model. The From-Different-Model (FromDifModel) option is used to copy features from a different model. Due to the change in references required to copy from one model to another, this option is available with the New Refs option only. The From-Different-Version (FromDifVers) option is used to copy features from a different version of the active model. The New Refs or the Same Refs option may be used.

MIRRORING FEATURES

The Mirror option creates a reflected image of selected features. To construct a mirror, select the features to be copied and then select a mirroring plane. Perform the following steps to create a mirrored feature.

STEP 1: Select FEATURE >> COPY.

STEP 2: Select the MIRROR option.

STEP 3: Select either DEPENDENT or INDEPENDENT, then select DONE.

The Independent option will make the new copy's dimension values independent of any parent features.

STEP 4: Select features to be mirrored.

Features can be selected by picking them on the work screen or on the Model Tree. Additionally, the All Feat option can be used to mirror every available

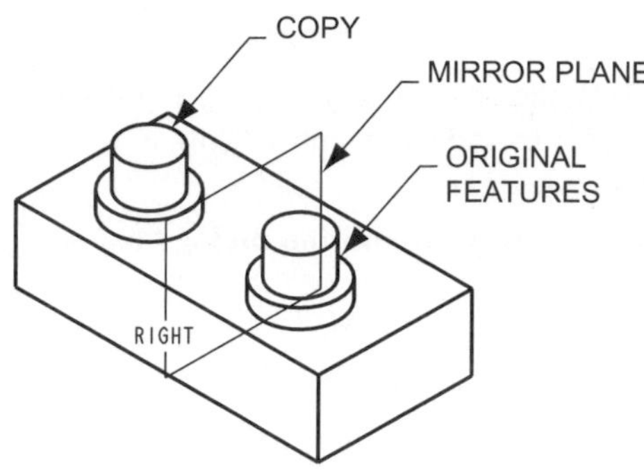

Figure 7–6 Mirroring features

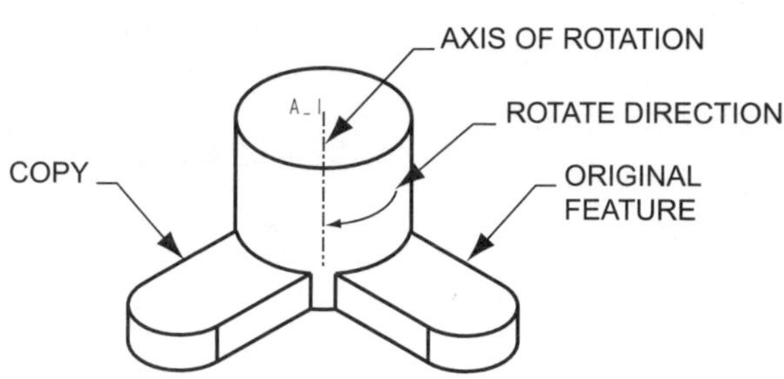

Figure 7–7 Rotating features

feature of a part. The Select option on the Copy menu can be used to select features during later steps.

STEP 5: Select DONE to finish selecting features.

STEP 6: Select a Mirroring Plane (Figure 7–6).

The mirroring plane can be any planar surface or datum plane. If a plane is not available, one can be created with the Make Datum option. The Mirror option does not utilize a Feature Definition dialog box. After the mirroring plane is selected, the copy is created.

ROTATING FEATURES

The Rotate option is located under the Move menu option. It is used to copy features by rotating them around an axis, edge, datum curve, or coordinate system. As with other Copy options, multiple features can be copied. Perform the following steps to copy-rotate selected features.

STEP 1: Select FEATURE >> COPY >> MOVE.

STEP 2: Select between INDEPENDENT and DEPENDENT.

When features are copied as Dependent, their dimension values are dependent upon the original feature. Any changes made to the parent feature will be reflected in the copy.

STEP 3: Select DONE on the Copy Feature menu.

STEP 4: Select features to copy, then select DONE (Figure 7–7).

STEP 5: Select ROTATE on the Move Feature menu (Figure 7–8).

Figure 7–8 Move feature menu

STEP 6: Select either CRV/EDG/AXIS or CSYS on the GEN SEL DIR menu, then select the appropriate entity on the work screen (Figure 7–9).

Features can be copy-rotated around a datum curve, edge, axis, or coordinate system. After choosing a rotation type, select the appropriate entity on the work screen.

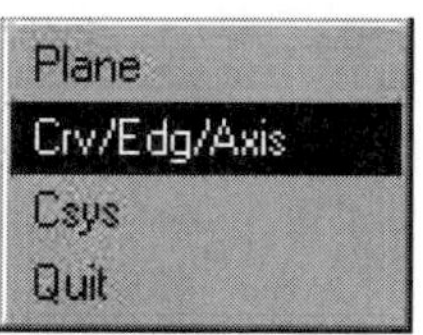

Figure 7–9 GEN SEL DIR menu

STEP 7: Select OKAY for direction of Rotation, or FLIP to change direction.

Pro/ENGINEER utilizes the right-hand rule to determine the direction of rotation. With the right-hand rule, point the thumb of your right hand in the direction of the arrow shown on the work screen. The remaining fingers of your right hand point in the direction of rotation.

STEP 8: Enter the degrees of rotation.

Enter the number of degrees that the selected features will be rotated.

STEP 9: Select DONE MOVE on the Move Feature menu (Figure 7–8).

STEP 10: Select DONE on the GP VAR DIMS menu.

Notice in Figure 7–10 the options for checking a dimension. The GP VAR DIMS menu allows for the selection and varying of dimension values during the copy process.

STEP 11: Select OKAY on the Feature Definition dialog box.

TRANSLATED FEATURES

Features can be copied in a linear direction using the Translate option (Figure 7–11). Features are copied perpendicular to a selected plane.

STEP 1: Select FEATURE >> COPY >> MOVE.

STEP 2: Select between INDEPENDENT and DEPENDENT.

When features are copied as Dependent, their dimension values are dependent upon the original feature. Any changes made to the parent feature will be reflected in the copy.

STEP 3: Select DONE on the Copy Feature menu.

STEP 4: Select features to copy, then select DONE.

STEP 5: Select TRANSLATE on the Move Feature menu.

STEP 6: Select PLANE on the GEN SEL DIR menu, then select a plane on the work screen (Figure 7–11).

Select a planar surface or a datum plane. Features will be copied perpendicular to the selected plane. The Make Datum option is available to create an on-the-fly datum plane.

STEP 7: Accept or FLIP the Translate Direction.

On the work screen, Pro/ENGINEER will graphically display the direction of translation.

STEP 8: Enter the translation value.

Enter the value that the copied features will be offset from the original features.

STEP 9: Select DONE MOVE on the Move Feature menu.

STEP 10: Select DONE on the GP VAR DIMS menu.

As with the Rotate option, the GP VAR DIMS menu allows for the selection and varying of dimension values during the copy process.

STEP 11: Select OKAY on the Feature Definition dialog box.

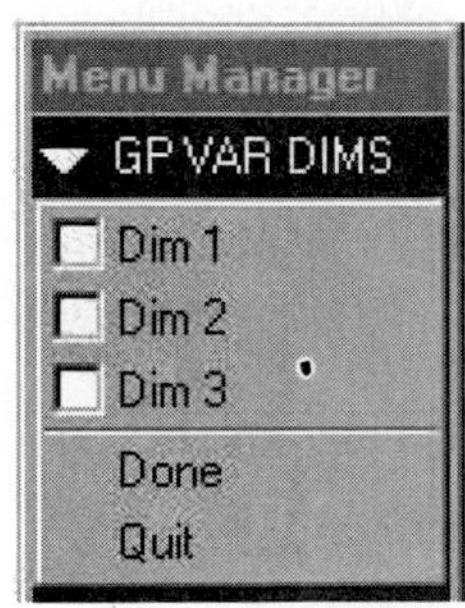

Figure 7–10 GP VAR DIMS menu

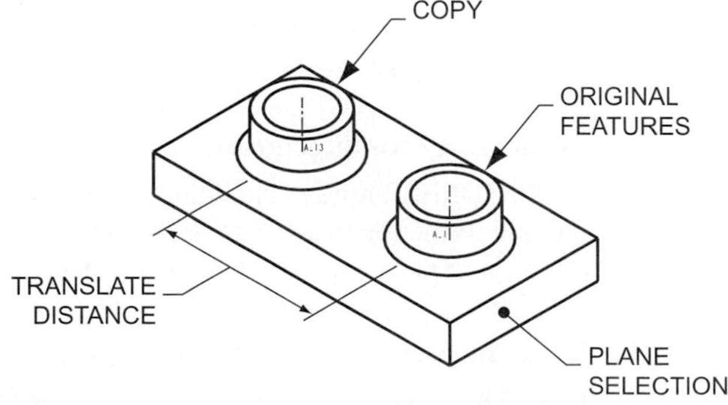

Figure 7–11 Translation

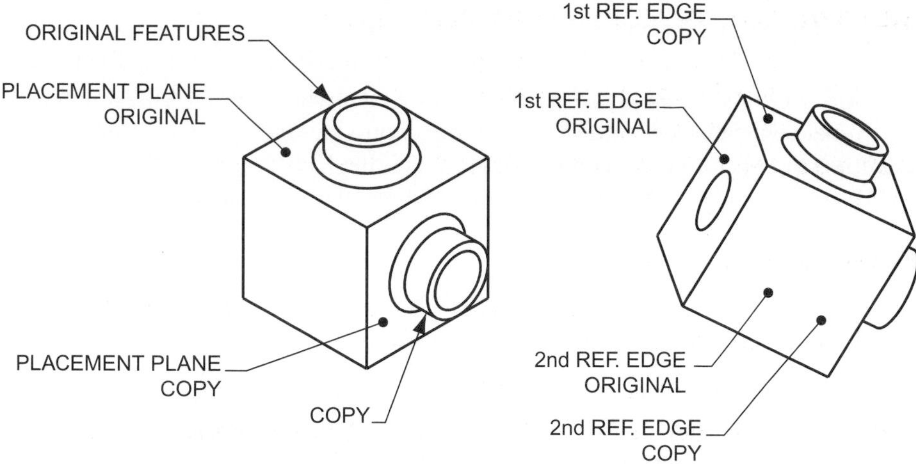

Figure 7-12 New reference option

Copying with New References

The New References (New Refs) option copies selected features by specifying new references and by varying feature dimensions (Figure 7–12). Examples of references that can be changed include sketching planes and reference edges. Perform the following steps to copy a feature with new references.

> **INSTRUCTIONAL NOTE** When Pro/ENGINEER requests the selection of a new reference for a copy (or other options such as Reroute), it highlights the old reference with the Section color setting. The Section color is set under the View >> Display Settings >> System Colors option. If the default Reference Color is hard to see, its setting should be changed.

Step 1: Select FEATURE >> COPY >> NEW REFS.

Step 2: Select between INDEPENDENT and DEPENDENT, then select DONE.

Step 3: Select features that will be copied, then select DONE.

Step 4: Select dimensions to vary (optional step).

Dimensions defining a feature can be varied during the copy process (Figure 7–13). Pro/ENGINEER will display the dimensions defining the selected features. On the work screen or on the GP VAR DIMS menu, select the dimensions to vary.

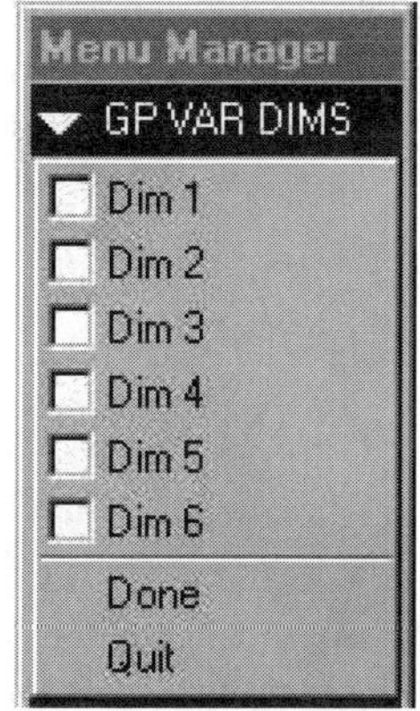

Figure 7-13 Variable dimension menu

Step 5: Select DONE on the GP VAR DIMS menu (Figure 7–13).

Step 6: Choose an option for each highlighted reference, then perform the appropriate reference selection.

In sequential order, Pro/ENGINEER will highlight each reference in the established Reference (section) Color. For each reference, you must perform one of the following options:

- **Alternate** This option requires the selection of a new reference. Select the appropriate new reference for the copy.
- **Same** This option keeps the highlighted reference for the copy.
- **Skip** This option allows you to skip the definition of a new reference for a copy. This reference has to be defined later with the Redefine option.
- **Ref Info** This option provides information about the reference.

Step 7: Select DONE on the Group Place menu.

THE MIRROR GEOMETRY COMMAND

The Mirror Geometry command is often confused with the Copy >> Mirror option. They are similar commands but while Copy >> Mirror provides the option of selecting features to mirror, the Mirror Geom command mirrors all the features of a part. After executing the Mirror Geometry command, the copied features will be created as a merged feature. To mirror an entire model, select the Mirror Geom option, then select a plane to mirror the part about.

USER-DEFINED FEATURES

A User-Defined Feature (UDF) can be saved to disk and used during the construction of future models. It is common practice to create a UDF library of features that are often used in an organization. User-Defined Features can be positive space (i.e., Protrusion) or negative space (i.e., Cut and Hole). Positive space User-Defined Features can be placed as the first feature of a part.

When creating a feature that will be saved as a UDF, careful consideration of references and the dimensioning scheme is critical. A UDF functions like a New Reference copy. When placing a UDF, Pro/ENGINEER prompts for the selection of new references for the copy.

A UDF can be Subordinate or Stand-Alone. If a UDF is created as subordinate, it depends upon the original model for its dimensional values. The original model stores the information defining the UDF. Any changes to the original model and to the original feature will be reflected in any placed UDF. Due to this dependence, the original model must exist. A Stand-Alone UDF stores the information associated with the feature in the copied-to-model. Once a stand-alone UDF is placed, it loses its associativity with its parent model.

A UDF can be created with one of three possible dimension types:

- **Variable dimensions** Dimension defined as variable can be modified during the copy process.
- **Invariable dimensions** Dimensions defined as invariable cannot be modified during the copy process.
- **Table-driven dimensions** Dimensions defined as table-driven receive their values from a family table.

THE UDF MENU

The UDF menu is accessible from the UDF Library menu. This menu is used to create, modify, and manipulate User-Defined Features. The following options are available:

CREATE

The Create option is used to create a new UDF. The UDF feature will be created in the current working directory. If a specific directory is used as a library to store User-Defined features, change the working directory before the creation process, or set the configuration file option *pro_group_dir* to specify the library directory.

MODIFY

The Modify option is used to modify an existing UDF.

LIST

The List option is used to list all available User-Defined Features in the current working directory.

DBMS

Users of Pro/ENGINEER before Release 20 will recognize Dbms as a Data-Based-Management option. With the UDF menu, Dbms allows for data-based management options for a UDF. Common file management functions such as Save, Save As, Rename, and Erase are available.

INTEGRATE

The Integrate option is used to define differences that might exist between the original UDF and the copied UDF.

CREATING A USER-DEFINED FEATURE

The following will show the step-by-step process for creating a UDF of the radial hole pattern shown in Figure 7–14. This part (*udf_part.prt*) is available on the supplemental CD. The original feature consists of a radial hole that has been patterned as a rotation. The references required to construct the UDF include the placement plane, reference axis, and reference planes. Perform the following steps to create a User-Defined Feature of this pattern.

STEP 1: Open or create the part with appropriate features.

STEP 2: Select the working directory where the UDF will be stored.

Use the File >> Working Directory option to select the directory in which to save the UDF.

STEP 3: Select FEATURE >> UDF LIBRARY.

STEP 4: Select CREATE on the UDF menu.

STEP 5: Enter *pattern1* as the UDF name.

STEP 6: Select STAND ALONE >> DONE.

A Stand-Alone UDF will be independent of the model from which the original feature was created. A Subordinate UDF will receive its parameters from the original model.

STEP 7: Select YES to include the reference part.

STEP 8: Select the hole pattern to include as a feature in the UDF.

On the work screen or on the Model Tree, select the hole pattern.

STEP 9: Select DONE on the Select Feature menu.

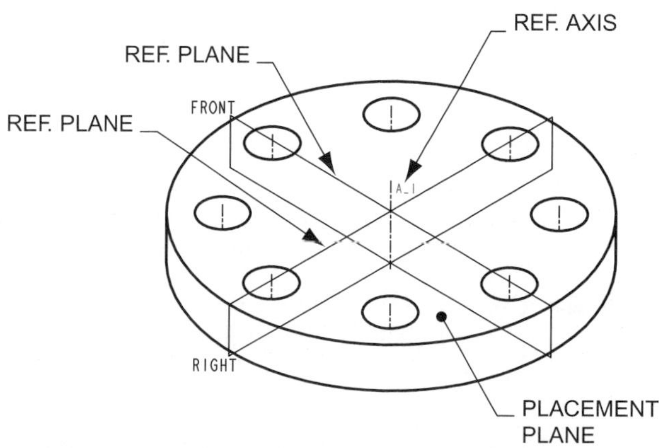

Figure 7–14 User-defined feature

STEP 10: Select DONE/RETURN on the UDF FEATS menu.

The Add option on the UDF Feature menu will allow you to add additional features to the new UDF, while the Remove option will allow you to remove features.

STEP 11: Enter a prompt for the highlighted axis.

Pro/ENGINEER will sequentially reveal each reference and request a prompt. The prompt entered by you will be displayed when placing the UDF into a new model. Enter the following prompt: *a Reference Axis to Place Patterned Holes About.*

(*Note:* Pro/ENGINEER will add the word "Select" to the front of the prompt.)

STEP 12: Enter a prompt for the highlighted reference plane.

Radial holes require the selection of a reference plane. This option will require you to establish a prompt that will request the selection of a new reference plane when the UDF is placed. Enter the following prompt for the highlighted reference plane: *a Reference Plane to Reference the Original Patterned Hole.*

STEP 13: Select SINGLE >> DONE/RETURN on the Prompts menu.

When Pro/ENGINEER encounters references that can share the same prompt, it will provide you with the option for selecting a Single prompt for all the references, or an option for selecting individual prompts (Multiple option) for each reference. In this example, the eight holes forming the pattern will each require a placement plane. Pro/ENGINEER recognizes this and gives you the chance to enter one prompt for each hole placement plane.

STEP 14: Enter a prompt for the highlighted placement plane.

This prompt will serve as the placement plane for all three holes. Enter the following prompt: *a Placement Plane for the Holes.*

STEP 15: Use the NEXT and PREVIOUS options on the Modify Prompts menu to toggle through prompts.

If a prompt needs to be changed, use the Enter Prompt option.

STEP 16: Select DONE/RETURN on the Modify Prompts menu.

STEP 17: Select OKAY on the UDF Creation dialog box.

PLACING A USER-DEFINED FEATURE

This guide will step you through the process for placing the UDF created in the previous section, "Creating a User-Defined Feature." This UDF (*pattern1.gph*) is also available on the supplemental CD. As shown in Figure 7–15, utilizing Pro/ENGINEER's default datum planes, model a Circular protrusion with a diameter of 8 inches and an extrusion depth of 1 inch. This part will serve as the base feature that the UDF feature will be attached to.

STEP 1: Select FEATURE >> CREATE >> USER DEFINED.

User-Defined features can be placed using the Group option also. User-Defined features and groups are closely related. All User-Defined Features are placed in a model as a group.

STEP 2: Using the OPEN dialog box, manipulate the directory to locate the *pattern1* UDF created in the previous guide.

User-Defined features are saved to disk with an **.gph* file extension. Locate the file *pattern1.gph* that was created in the "Creating a User-Defined Feature" section. This UDF can also be found on the supplemental CD.

STEP 3: Open the *PATTERN1* User-Defined Feature.

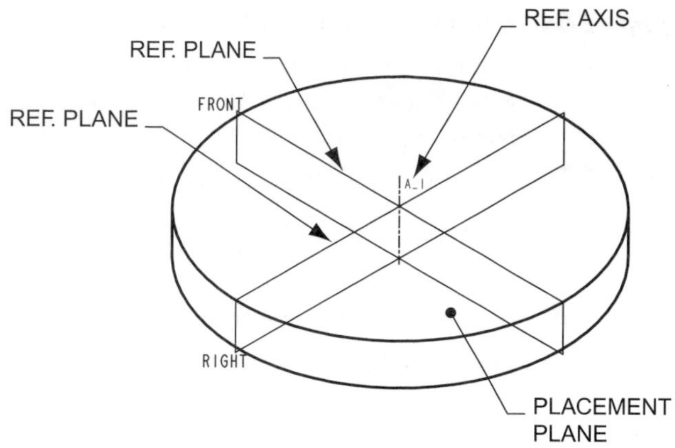

Figure 7-15 Placing a UDF

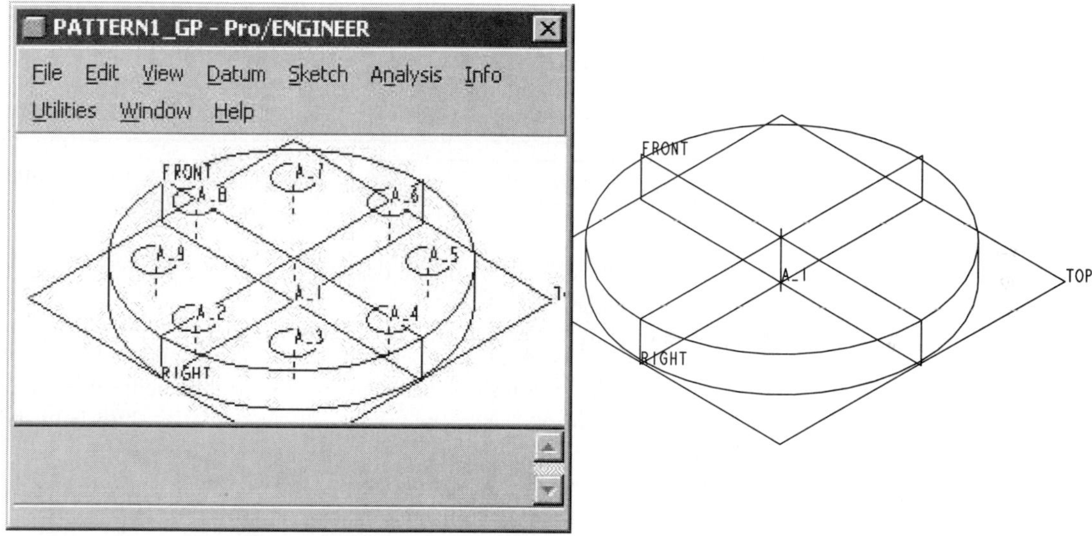

Figure 7-16 Reference part displayed

STEP 4: Select YES to retrieve the reference part.

When a UDF is created as Stand-Alone, the Yes option retrieves the original model from which the UDF was created. When Yes is selected, the original model will be displayed in a separate window (Figure 7–16). When selecting new references, the references on the original model will be highlighted.

STEP 5: Select INDEPENDENT >> DONE on the Placement Options menu.

This option will allow the placed feature to be independent of the part from which the UDF was created.

STEP 6: Select SAME DIMS >> DONE.

The Same Dims option will keep the dimension values that define the original UDF. The User Scale option will allow a copied UDF to be scaled a user-specified amount.

STEP 7: Select NORMAL >> DONE.

The Normal option will allow dimensions to be modifiable. The Read Only option will display dimensions, but they cannot be modified. The Blank option will hide all dimensions.

STEP 8: Select the Reference axis on the new model (Figure 7–15).

The prompt in the message area requests you to *Select a Reference Axis to Place Patterned Holes About*. This is the prompt entered in the previous section. Check the reference window to see if the Axis is highlighted. If is not, your reference colors might need changing.

The Alternate option on the Select Reference menu allows you to select a reference different from the one specified. As shown in Figure 7–15, select the axis.

STEP 9: Select a reference plane to reference the original patterned hole.

The prompt in the message area requests you to *Select a Reference Plane to Reference the Original Patterned Hole*. Select the datum as shown in Figure 7–15.

STEP 10: Select a placement plane for the holes.

The prompt is the message area requests you to *Select a Placement Plane for the Holes*. Select the top of the part as shown in Figure 7–15.

STEP 11: Reselect the placement plane to orient the holes.

STEP 12: Select DONE on the Group Placement menu.

RELATIONS

Mathematical and conditional **relationships** can be established between dimension values. Chapter 3 introduced the utilization of Relations within the sketcher environment. Relations within the sketcher environment are used to establish dimensional relationships between dimensions of a feature. The Relations command found under the Part menu is used to establish relationships between any two dimensions of a part. Within Assembly mode, the Relations option can be used to establish relationships between dimensions from different parts.

Dimensions can be shown with numeric values or as symbols. Figure 7–17 shows an example of a part with dimensions shown as symbols. Dimension values are displayed as a *d* followed by the dimension number (i.e., *d3*). Other parameters that can be displayed symbolically include Reference Dimensions (i.e., *rd3*), Plus-Minus Symmetrical Tolerance mode (i.e., *tpm4*), Positive Plus-Minus Tolerance mode (i.e., *tp4*), Negative Plus-Minus Tolerance mode (i.e., *tm4*), and number of instances of a feature (i.e., *p5*).

Most algebraic operators and functions can be used to define a relation, as can most comparison operators. Table 7–1 provides a list of mathematical operations, functions, and comparisons supported in relation statements. All trigonometric functions use degrees.

CONDITIONAL STATEMENTS

Pro/ENGINEER's Relations command has the ability to utilize conditional statements for the purpose of capturing design intent. As an example, Figure 7–18 shows a flange that incorporates holes on a bolt circle centerline. In a model such as this, the holes are created typically as a patterned radial hole. Imagine a situation where the number of holes and the diameter of the bolt circle are governed by the diameter of the flange. Suppose, in this example, that the bolt circle diameter is always 2 inches less than the diameter of the flange. Also, suppose that the design requires 4 holes on the bolt circle if the diameter of the flange is 10 inches or less and 6 holes if the flange diameter is greater than 10 inches. Using the following relational statements, parameters can be established that control this design intent for the bolt circle pattern.

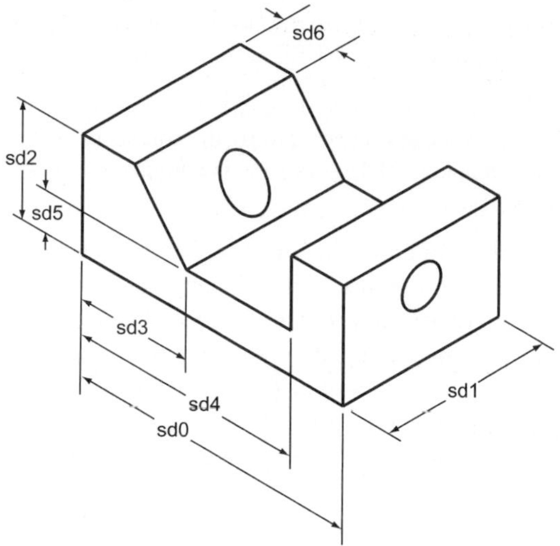

Figure 7-17 Dimension symbols

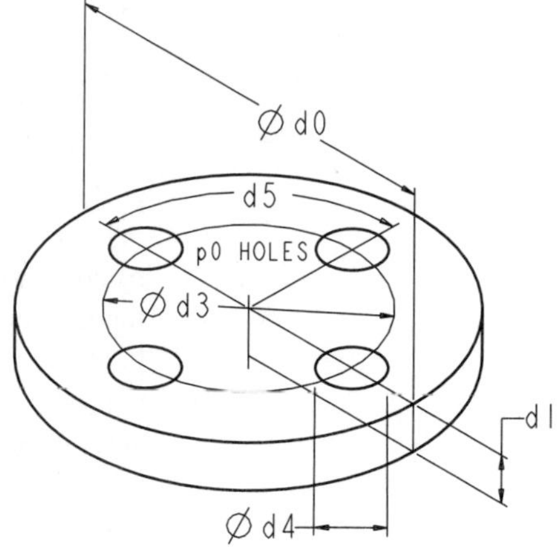

Figure 7-18 Dimension symbols

```
IF d0 <= 10      (Line 1)
p0 = 4           (Line 2)
d5 = 90          (Line 3)
ELSE             (Line 4)
p0 = 6           (Line 5)
d5 = 60          (Line 6)
ENDIF            (Line 7)
d3 = d0 - 2      (Line 8)
```

Table 7-1 Mathematical operations

Operator or Function	Meaning and Example
+	Addition d1 = d2 + d3
−	Subtraction d1 = d2 − d3
*	Multiplication d1 = d2 * d3
/	Division d1 = d2/d3
^	Exponentiation d1^2
( )	Group parentheses d1 = (d2 + d3)/d4
=	Equal to d1 = d2
cos ( )	cosine d1 = d2/cos(d3)
tan ( )	tangent d1 = d3 * tan (d4)
sin ( )	sine d2 = d3/sin (d2)
sqrt ( )	square root d1 = sqrt (d2) + sqrt (d3)
==	Equal to d1 == 5.0
>	Greater than d2 > d1
<	Less than d3 < d5
>=	Greater than or equal to d3 >= d4
<=	Less than or equal d5 <= d6
!=	Not equal to d1 != d2 * 5
|	Or (d1 * d2) | (d3 * d4)
&	And (d1 * d2) & (d3 * d4)
~	Not (d3 * d4) ~ (d5 * d6)

A conditional relation is defined between an "IF" statement and an "ENDIF" statement. The above example utilizes an IF-ELSE statement. With an IF-ELSE statement, if the condition is true the expressions following the IF statement will occur. If the condition is not true, the condition following the ELSE statement will occur. The above example reads as follows:

> If the flange diameter is less than or equal to 10, the number of holes is equal to 4 and the angular spacing between each hole is 90 degrees; else, the number of holes is equal to 6 and the angular spacing is equal to 60 degrees.

In the above example, the condition statement is the diameter of the flange (d0) being less than or equal to 10 (see line 1). If this is true, the number of holes on the bolt circle (p0) will be equal to 4, and the angle between each instance of the hole (d5) will be 90 degrees. If the flange diameter is not less than or equal to 10, the number of holes on the bolt circle will be 6, and the angle between instances will be 60 degrees. In a conditional statement, the ELSE expression shown in line 4 must occupy a line by itself. The expression ENDIF is used to end the conditional statement. Additionally, in the above example, line 8 is used to make the bolt circle diameter always 2 less than the flange diameter.

Adding and Editing Relations

Relations are added to an object using the Relations >> Add option. Dimension and parameter symbols are not automatically displayed upon selecting the Add option. Since these symbols are needed for the accurate creation of a relation, it is necessary to select each appropriate feature after selecting the Relations command. When a feature is selected in this manner, dimensions and parameters will be shown with their assigned dimension symbols.

Relations are added one relation at a time through a textbox. After typing a relation, use the enter key to input the relation. For conditional statements, input each line one at a time, also through the textbox. Relations are evaluated within a model in the order in which they are defined. In most cases of conflict, the later relation overrides the former.

The Show Relations (Show Rel) and the Edit Relations (Edit Rel) options can be used to display defined relations. While the Show Rel option allows for the viewing of relations, the Edit Rel option allows for the modification of existing relations and for the creation of new relations.

Perform the following steps to add relations to a model:

Step 1: Select RELATIONS on the Part menu.

Step 2: Select the Part or Feature applicable to the relation.

Selecting a model will display the dimensions defining the selected part or feature. While in the Relations menu, dimension will be displayed in the symbol format. You can use the Switch Dim command from the Info menu to change the display of a dimension.

Step 3: Select ADD on the Part menu.

Step 4: Type the appropriate relation in the textbox then select the ENTER key.

Step 5: To end the adding of relations, select the ENTER key with a blank textbox.

Step 6: Regenerate the model.

Defined relations will not take effect until the model is regenerated.

Family Tables

A **Family of Parts** comprises components that share common geometric features. An example of a family of parts is the Hex Head bolt. Hex Head bolts come in a variety of sizes, but share common characteristics. Bolts can have different lengths and diameters but share similar head features and thread parameters (Figure 7–19). **Family Tables** are groups

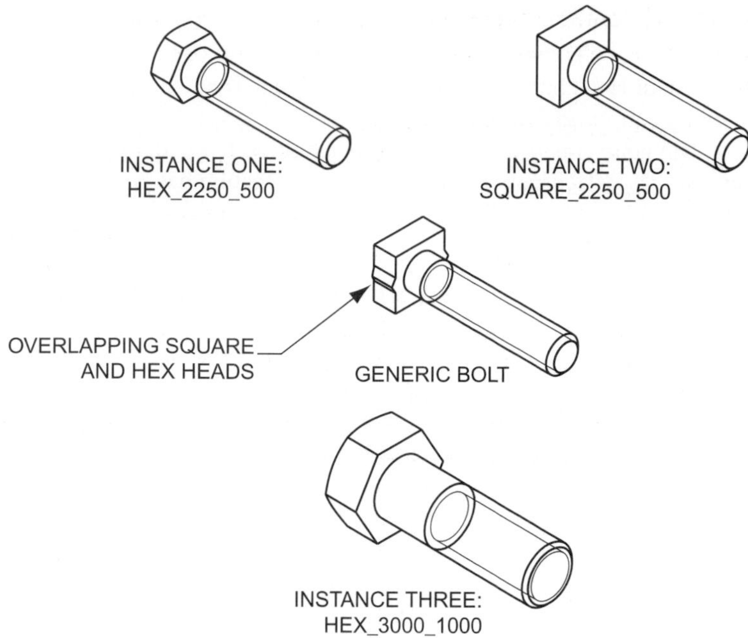

Figure 7–19 Family table example

of similar features, parts, or assemblies. Within companies, it is common to find parts and assemblies with common geometric characteristics. Examples can be found in the automotive industry. An automobile manufacturer probably has a large variety of cam shafts. A Family Table can be created that drives the construction of a company's line of cam shafts. Examples can be found with assemblies also. Imagine the number of alternators that a major automobile company produces. Alternators have similar characteristics but can vary in certain components and features. A Family Table can be created to control the design of a company's line of alternators.

Family Tables have many advantages. Storing and controlling large numbers of components can be difficult to manage and costly. For a family of parts with large numbers of variations, Family Tables allow components to be stored to disk while consuming significantly less storage space. Family Tables also save time in the modeling process. If a design is known to work effectively, a variation of the design can be created by changing a parameter within the Family Table. This technique also allows for the standardization of a design.

Adding Items to a Family Table

No specific option is used to create a Family Table. A Family Table is created automatically when an item is selected to add to the Family Table. Examples of items that can be added include dimensions, features, components, and user parameters. Other items that can be added include groups, pattern tables, and reference models. To add an item, select the Add Item option from the Family Table menu, then select the item type to add. Some items, such as groups, are better selected by utilizing the Model Tree.

Creating a Family Table

This section will introduce the creation of Family Tables by creating a Family Table of Hex Head and Square Head bolts. Three instances will be created: two Hex Head and one Square Head. Figure 7–19 shows the two instances of the Hex Head bolt, the instance of the Square Head bolt, and the generic part. The generic part is a standard Pro/ENGINEER part which is used to create the instances. This part (*generic_bolt.prt*) can be found on the supplemental CD. Notice in the figure the overlap of the heads of the Hex Head bolt and the Square Head bolt. As shown on the Model Tree in Figure 7–20, the Hex Head bolt is actually a grouped feature.

The generic part in this tutorial is a bolt with a major diameter of 0.500 inches and a bolt length of 2.25 inches. A cosmetic thread feature has been created with a minor diameter of 0.375 inches. Additionally, the following dimensional relations have been added to the generic part:

- The height of both bolt heads is set to 0.667 times the major diameter.
- The distance across the flats of both bolt heads is set to 1.5 times the major diameter.
- The cosmetic tread minor diameter is set to 0.75 times the major diameter.

This tutorial will add the bolt's major diameter and length dimensions to the Family Table. Additionally, the two grouped features that defined the bolt heads will be added. Three instances of the part will be created. The first instance will be a Hex Head bolt with the same dimensions as the generic part. The second instance will be a Square Head bolt, also with the same dimensions as the generic part. The third instance will be a Hex Head bolt with a 1 inch major diameter and a length of 3 inches.

Perform the following steps to create a Family Table of bolts:

STEP 1: Create the Bolt with the Hex Head (or Open the *generic_bolt* part from the supplemental CD).

Use appropriate modeling tools to create the bolt. This part, named *generic_bolt,* can be found on the supplemental CD also. If you elected to open the existing part, skip steps 2–4.

STEP 2: Group the features comprising the Hex Head, then suppress the group.

By grouping the features comprising the Hex Head, these features can be manipulated together. This group will be suppressed to allow for the creation of the Square Head.

STEP 3: Create the Square Head feature on the Bolt.

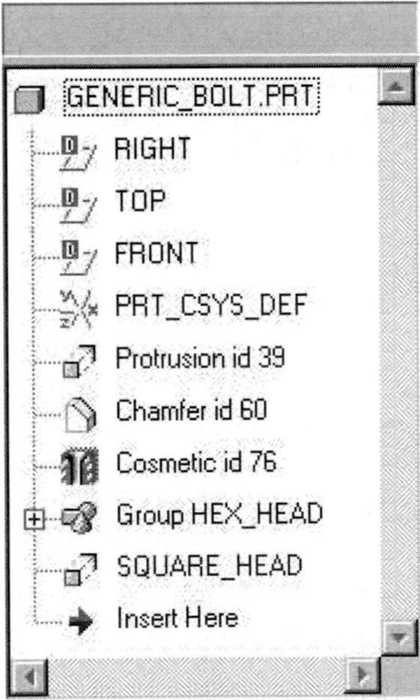

Figure 7–20 Model tree

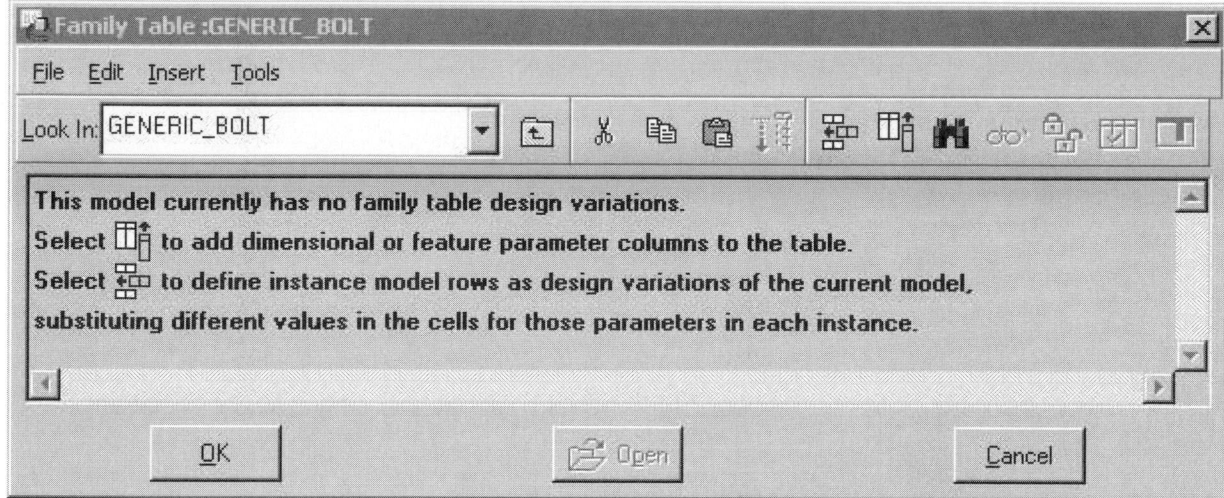

Figure 7-21 Family table dialog box

STEP 4: Use the RESUME option to unsuppress the Hex_Head group.

STEP 5: Select the FAMILY TAB option from the Part menu.

Each part is capable of one family table. The family table is created when the first item is added to it.

STEP 6: On the Family Table dialog box (see Figure 7–21), select the ADD COLUMN icon.

Examples of items that can be added to a family table include dimensions, features, and components.

STEP 7: Select the DIMENSION item on the Family Item dialog box (see Figure 7–22) then select the shaft of the bolt.

The first items that will be added to the family table include the dimensions that define the major diameter of the bolt and the length of the bolt.

STEP 8: As shown in Figure 7–23, select the dimensions defining the bolt's major diameter and its length.

Using the Dimension item type, selected dimensions will be added to the Family Table. As shown in Figure 7–22, symbols representing each dimension (e.g., d1 and d0) will be displayed on the dialog box within the Items area.

STEP 9: Select the FEATURE item on the Family Items dialog box.

The Feature option allows features, including grouped features, to be added to the Family Table. Two features will be added: The Hex_Head group feature and the Square_Head_Bolt feature.

STEP 10: On the Model Tree, select the HEX_HEAD group feature and the SQUARE_HEAD_BOLT feature.

STEP 11: Select OKAY on the Family Items dialog box.

STEP 12: On the Family Table dialog box, select the NEW INSTANCE icon.

Notice the addition of the *GENERIC_BOLT* instance in the table editor of the Family table dialog box (Figure 7–24). This name is derived from the part's name and represents the name of the generic instance of the bolt. This instance includes the part's default dimensional values and default features.

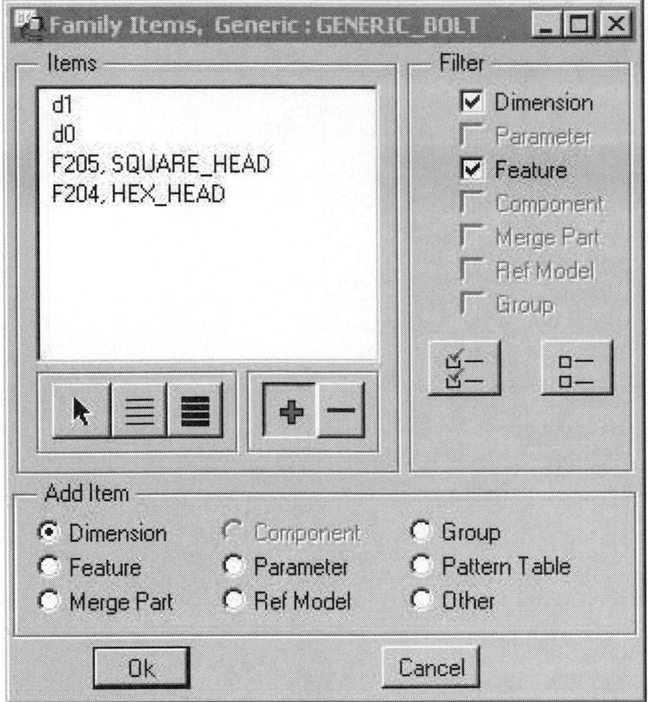

Figure 7-22 Family items dialog box

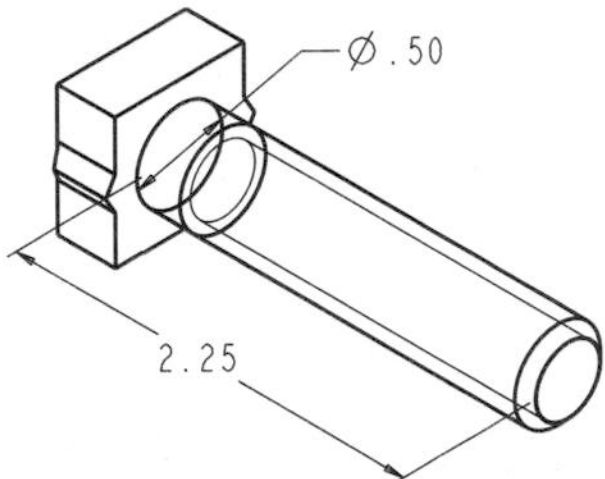

Figure 7-23 Bolt dimensions

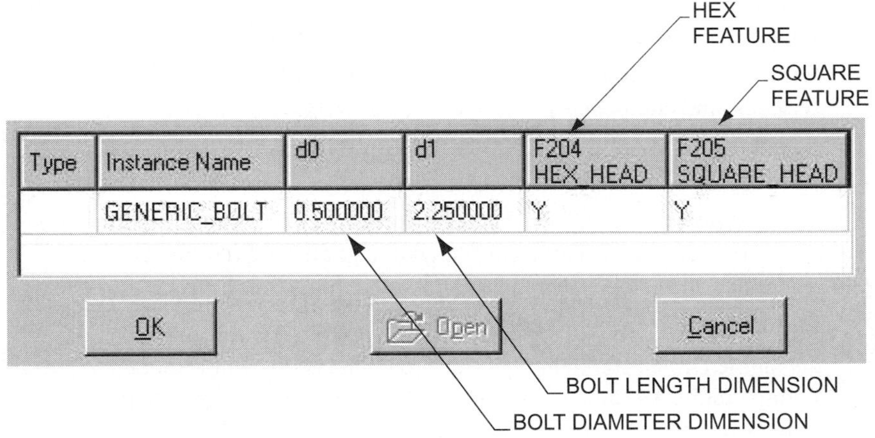

Type	Instance Name	d0	d1	F204 HEX_HEAD	F205 SQUARE_HEAD
	GENERIC_BOLT	0.500000	2.250000	Y	Y

Figure 7-24 Family table editor

STEP 13: As shown in Figure 7–25, enter the information in the table to create the instance *HEX_2250_500.*

The new instance's name will be *HEX_2250_500*. This name represents a Hex Head bolt that will be 2.250 inches in length and 0.500 inches in diameter. The table will not accept a period value or a space in the instance name, hence, the naming standard shown in this example. Enter 2.25 for the length of the instance and 0.5 as the bolt diameter. In the table, enter a Y value for the new instance in the column representing the Hex Head feature and a N value for the column representing the Square Head feature. A Y value will reveal the feature in the instance, and a N value will remove the feature.

Type	Instance Name	d0	d1	F204 HEX HEAD	F205 SQUARE HEAD
	GENERIC_BOLT	0.500000	2.250000	Y	Y
	HEX_2250_500	0.50000	2.250000	Y	N
	SQU_2250_500	0.50000	2.250000	N	Y

OK Open Cancel

Figure 7–25 Instance creation

STEP 14: As shown in Figure 7–25, enter parameters in the table to create the instance SQUARE_2250_500.

This instance will consist of a Square Head bolt with a major diameter of 0.500 inches and a bolt length of 2.250 inches.

STEP 15: On the dialog box, pick the row representing the HEX_2250_500 instance, then select the dialog box's OPEN icon.

Pro/ENGINEER will open a new window with the instance of the part shown. Selecting the Save option from the File menu will save the current instance as a part.

STEP 16: Close the instance from memory by selecting FILE >> ERASE >> CURRENT.

STEP 17: Activate the generic part (Window >> Activate), then select the FAMILY TAB menu option.

STEP 18: Create a HEX_3000_1000 instance.

Create an instance of the generic part that includes a hex head bolt that is 3 inches in length with a major diameter of 1 inch.

STEP 19: Close the Family Table dialog box.

STEP 20: Save the generic part.

CROSS SECTIONS

Traditional drafting standards utilize cross sections to display the internal details of models. Pro/ENGINEER utilizes cross sections within Drawing, Part, and Assembly modes for the same purpose. Additionally, within Pro/ENGINEER, mass properties of a cross section can be calculated.

Pro/ENGINEER provides the option of creating two types of cross sections: Planar and Offset. Planar Cross Sections are created within a model through a selected planar surface or datum plane. Figure 7–26 shows an example of a revolved part with a section through a datum plane. Offset Cross Sections are created by extruding a sketched section. Offset Cross Sections are constructed utilizing normal sketching techniques. Offset Sections must be created with an open section. Once created within Part or Assembly mode, Cross Sections are blanked from the screen. They can be redisplayed with the Show option.

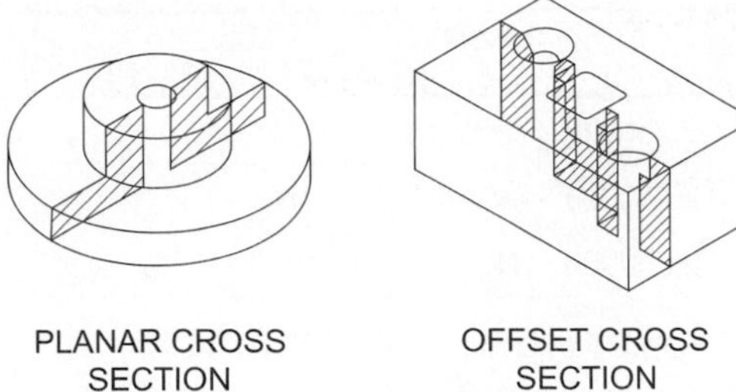

Figure 7-26 Cross section examples

MODIFYING CROSS SECTIONS

Cross Section parameters can be modified. The following is a list of parameters that can be modified or manipulated:

NAME

The creation process of a Cross Section requires the entering of a name. The Name command found under the Cross Section >> Modify menu can be used to rename a Cross Section.

HATCHING

A Cross Section creates a hatch pattern along the section or planar line. Parameters defining a hatch pattern such as Line Style, Color, Spacing and Offset can be changed using the Modify >> Hatching option. A commonly used Hatching pattern can be stored for later use by using the Hatching >> Save option. The configuration file option *pro_crosshatch_dir* is used to specify the directory where hatch patterns are saved.

DIMENSIONS

Dimension values defining a Cross Section can be modified with the Dim Values command. A model is automatically regenerated upon finishing the dimension modification process.

CREATING A PLANAR CROSS SECTION

Planar Cross Sections are created through Surface Planes or Datum planes. Perform the following steps to create a planar cross section.

STEP 1: Select the X-SECTION option on the Part menu.

> **INSTRUCTIONAL NOTE** Within Assembly mode, the X-Section option can be found under the Assembly >> Setup menu.

STEP 2: Select CREATE on the Cross Section menu.

STEP 3: Select MODEL >> PLANAR >> SINGLE >> DONE.

STEP 4: Enter a Name for the Cross Section.

Cross Sections created in Part and Assembly modes can be used for the creation of a section view in Drawing mode. Select a Cross Section name that would be appropriate for an engineering drawing.

STEP 5: Select a surface plane or datum plane.

The Cross Section will be created through the selected plane. The Make Datum option is available for on-the-fly datum planes.

STEP 6: Select DONE/RETURN on the CROSS SEC menu.

CREATING OFFSET CROSS SECTIONS

Offset Cross Sections are created by extruding a sketched cross section. Perform the following steps to create an Offset Cross Section:

STEP 1: Select the X-SECTION option on the Part menu.

STEP 2: Select CREATE on the Cross Section menu.

STEP 3: Select MODEL >> OFFSET.

STEP 4: Select either ONE SIDE or BOTH SIDES on the Cross Section Options menu, then select DONE.

The One Side option will create a cross section on one side of the sketched section, while the Both Sides option will create the section on both sides.

STEP 5: Enter a Name for the Cross Section.

STEP 6: Select a sketching plane and set up the sketcher environment according to normal sketching requirements.

STEP 7: Sketch the Cross Section using available sketcher tools (Figure 7–27).

The following rules and techniques apply to a Cross Section sketch.

- The section must be open, and the first and last entities must be straight line segments.
- Avoid using curved and spline entities for the section sketch. They create a nonmodifiable horizontal hatch pattern.
- When possible, sketch lines that are perpendicular or parallel to each other.
- When creating a cross section through a hole or shaft, use the axis of the feature as a reference.

STEP 8: When the section is complete, select the continue icon on the Sketcher toolbar.

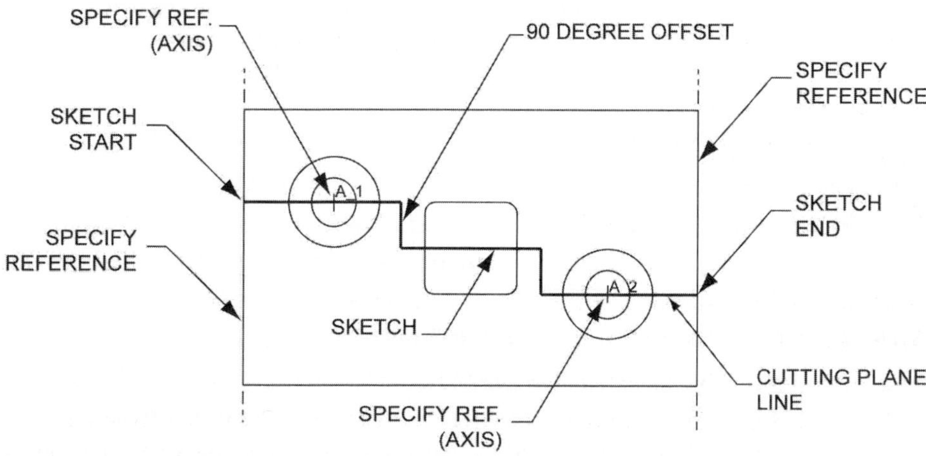

Figure 7–27 Sketching a cross section

MODEL TREE

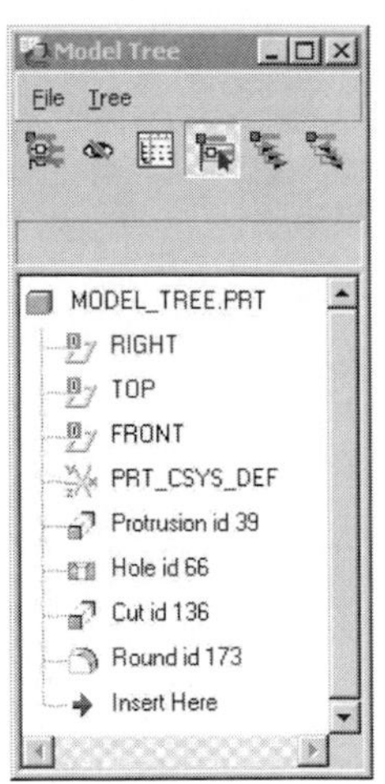

Figure 7-28 The model tree

Pro/ENGINEER's Model Tree (Figure 7–28) is a useful tool for the manipulation of features and parts. The following is a list of some of the uses of the Model Tree:

- Selecting features and parts that are difficult to select on the work screen.
- Deleting features and parts from the model.
- Redefining definitions assigned to a feature.
- Inserting features.
- Reordering features.
- Suppressing features.
- Rerouting features.

SUPPRESSING FEATURES

Complicated parts can slow regenerations. Suppressing a feature will remove it from the model tree and from regenerations. Additionally, features can complicate the modeling environment, cluttering the work screen. An uncluttered work screen can make selecting and sketching features easier. Perform the following steps to suppress features:

STEP 1: Select FEATURE >> SUPPRESS.

STEP 2: Select features to suppress.

On the Model Tree or on the work screen, select the features to be suppressed. When selecting a parent feature for suppression, all child features will be suppressed.

STEP 3: Select DONE on the Delete/Suppress menu.

MODELING POINT Many feature modification tools, such as Insert and Suppress, remove features from the Model Tree and from the regeneration sequence. The Resume command is used to return suppressed features to a part.

INSERTING FEATURES

Pro/ENGINEER adds new features to the end of the model tree. Insert Mode will allow a new feature to be added after an existing feature. This is useful for inserting features that will serve as references for future part features. Many times, it is necessary to create a feature that will serve as a reference for an existing feature. As an example, the sketch plane for a feature can be redefined. If it is the intent of the user to change the sketch plane to a datum plane that does not exist, this new datum plane will have to be inserted into the model tree before the feature under modification.

Perform the following steps to insert a new feature into the model tree:

STEP 1: Select FEATURE >> INSERT MODE >> ACTIVATE .

STEP 2: Select a Feature on the model tree to insert after.

After selecting a feature to insert after, all features after the selected feature will be suppressed on the model tree and on the work screen.

STEP 3: Create the feature or features to insert.

The insert mode will allow multiple features to be inserted into the model tree.

STEP 4: Select RESUME on the Feature menu to unsuppress features.

The Resume option will activate features that are suppressed. Selecting Cancel instead of Resume will activate suppressed features and delete any newly created features.

Reordering Features

Often, it is necessary to change the order in which features were created. As an example, during the modeling process, it might become apparent that features should be grouped and then patterned together. Features to be grouped should be adjacent to each other on the model tree. If the features to be grouped do not meet this requirement, then one or more of the features can be moved on the Model Tree using the Reorder command. Reordering a feature changes the order in which parts on the model tree are regenerated.

Due to parent-child relationships that might exist between features, care should be taken when reordering features. Adhere to the following rules when reordering features:

- Child features cannot be moved before parent features on the model tree.
- Parent features cannot be moved after child features on the model tree.

Perform the following steps to reorder a feature on the model tree.

Step 1: **Select FEATURE >> REORDER.**

Step 2: **Select features to reorder, then select DONE.**

There are three options available for selecting features:

- **Select** This option allows you to select each feature individually on the model tree or on the work screen. Choose Done Sel from the Select Feature menu when through selecting features.
- **Layer** This option allows you to select all features on a specific layer. Choose Done Sel from the Layer Select menu when through selecting layers.
- **Range** The Range option allows you to select a range of features by entering the range number of the first feature and the range number of the last feature to be selected.

Step 3: **Select either inserting BEFORE or AFTER a selected feature.**

The Reorder command will allow you to insert selected features Before or After an existing feature.

Step 4: **On the Model Tree, select feature to Insert Before or After.**

Rerouting Features

Rerouting a feature breaks the parent-child relationship that exists between it and another feature. There are two options available for rerouting a feature: Reroute Feat and Replace Ref.

Reroute Features Option The Reroute Features option will allow the selection of new or missing references for a feature. This option allows for the replacement of a reference for a child feature that has lost a parent feature. The focus of this option is on the child and its references. This option is available also in the event of a failed regeneration of a feature.

Perform the following steps to reroute a feature.

Step 1: **Select FEATURE >> REROUTE.**

Step 2: **Select REROUTE FEAT on the Reroute References menu.**

Step 3: **On the model tree, select the feature to reroute.**

Step 4: **Enter YES as the Roll Back option.**

This option is available for rolling back the part to just before the failed feature. This allows younger references that might cause clutter to be removed from the Model Tree and from the work screen.

STEP 5: Select a reference option for each feature reference.

Pro/ENGINEER will highlight each feature reference in the assigned section color. The following options are available for each reference in turn:

- **Alternate** This option allows for the selection of a new reference.
- **Same Ref** This option allows the same reference to remain.
- **Ref Info** This option shows information about a reference.
- **Done** Select this option when the rerouting process is finished.
- **Quit Reroute** This option allows for the termination of the reroute process.

STEP 6: Select DONE from the Reroute menu.

REPLACE REFERENCE OPTION The Replace References option allows for the replacement of a reference for child features. This option is used when a parent reference must be replaced. An option is available for replacing the reference with one new reference or selecting multiple replacement references. The focus of this option is on replacing the child feature of a parent feature. Perform the following steps to replace references.

STEP 1: Select FEATURE >> REROUTE.

STEP 2: Select REPLACE REF on the Reroute References menu.

STEP 3: Select either FEATURE or INDIV ENTITY on the Select Type menu.

The Feature option allows for the replacement of all references of a feature. The Individual Entity option allows for the replacement of selected individual entities.

STEP 4: Select either SEL FEAT or ALL CHILDREN.

The Select Feat option allows for the selection of new references for each child feature. The All Children option allows all child features to be referenced to one new feature.

REGENERATING FEATURES

Pro/ENGINEER is a feature based parametric modeling package. Features are displayed within Pro/ENGINEER on the Model Tree in the order in which they are created. During part regeneration, the geometry defining each element is recalculated sequentially from the first feature to the last.

Pro/ENGINEER requires a regeneration when dimensions are modified or when dimensional relationships are established. During dimension and feature modification, a part is regenerated from the last modified feature.

REGENERATION FAILURES

Within parametric design, features have explicit relationships with other features. These established feature references cannot conflict. When a conflict is detected, a failed regeneration occurs. There are several reasons why features may conflict. First, section entities cannot overlap. This is a common conflict with swept features. Second, features with bad edges will fail during regeneration. There are two categories of failed regenerations: failed regeneration during feature modeling and failed regeneration during feature modification.

Most feature creation processes utilize a Feature Definition dialog box. For these features, if a conflict exists during feature creation or redefinition, a Resolve button will appear in the dialog box (Figure 7–29). Two options are available for solving the regeneration conflict. The first and most practical solution is to redefine the definitions associated with the feature. As an example, Sweep sections often overlap, resulting in a conflict. The Sweep's section or trajectory can be redefined to solve the overlap problem.

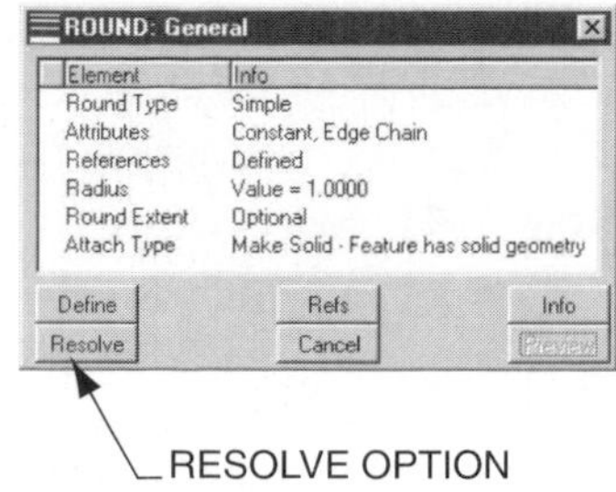

Figure 7-29 Resolve option

In another example, the Round command utilizes a dialog box. Often, an inappropriate round radius value will create a failed regeneration. A radius value can be redefined through the Feature Definition dialog box.

An alternative to redefinition is to access the resolve environment by selecting the Resolve button. The resolve environment provides diagnostic information and avenues for modifying other features. The following options are found under the Resolve Feature menu:

- **Undo changes** The undo changes option rolls the modeling process back to the last successful regeneration.
- **Investigate** This option will investigate possible causes of the regeneration failure.
- **Fix model** The fix model option rolls the modeling process back to before the failure and provides tools for fixing the problem.
- **Quick fix** The quick fix option provides avenues for redefining, rerouting, suppressing, or deleting the failed feature.

Features that fail and do not utilize a Feature Definition dialog box (e.g., Neck and Rib options) require the use of a Feature Failed menu. The following options are available with the Feature Failed menu:

- **Redefine** This option allows for redefinition of the feature.
- **Show ref** The show reference option shows references associated with a failed feature. Many failures occur due to conflicts with references. Evaluating feature references might highlight the regeneration conflict.
- **Geom check** The geometry check option checks geometry for overlapping and misalignment problems.
- **Feat info** This option displays information about the failed feature.

FEATURE SIMPLIFIED REPRESENTATIONS

Parts with multiple features and Assemblies with multiple parts can become complex. Within Part mode, Simplified Representations allow for the creation of a simpler Pro/ENGINEER part by excluding selected features from the display. Multiple representations can be created for a part and set current whenever it is necessary. Perform the following steps to create a Simplified Representation for a part.

STEP 1: Select the SIMPLFD REP >> CREATE option on the Part menu.

STEP 2: Enter a name for the Simplified Representation.

STEP 3: Select either INCLUDE FEAT or EXCLUDE FEAT on the Representation Attribute menu.

- **Include feature** The Include Feature option includes all features in the simplified representation except for those specifically selected for exclusion.

- **Exclude feature** The Exclude Feature option removes all features from the simplified representation except for those specifically selected for inclusion.

STEP 4: Select either REGENERATE or ACCELERATE.

- **Regenerate** The simplified representation is created when the master model is regenerated.
- **Accelerate** This option uses an accelerator file to retrieve the simplified representation. Only one accelerate file can be created per part.

STEP 5: Select either WHOLE MODEL or GEOMSNPSHOT.

- **Whole model** The Whole Model option includes all part features, parameters, and definitions.
- **GeomSnpshot** The Geometric Snapshot option creates a snapshot of the part without modifiable dimensions, parameters, and definitions. This option is available only with the Accelerate option.

STEP 6: Select DONE on the Representation Attribute menu.

STEP 7: Select the FEATURES option on the Edit Method menu.

The Features option allows for the selection of part features for inclusion or exclusion in a simplified representation. Other options under the Edit Method menu include:

- **Attributes** The Attributes option allows for the selection of attributes for creation of simplified representations.
- **Work region** The Work Region option removes a portion of the part from the display.
- **Surface** The Surface option creates a simplified representation from copied part surfaces.

STEP 8: Select EXCLUDE or INCLUDE from the FEAT INC/EXC menu.

- **Exclude** This option excludes features from the simplified representation. This option is available when Include was selected in step 3.
- **Include** This option includes features from the simplified representation. This option is available when Exclude was selected in step 3.

STEP 9: Select features to include or exclude from the simplified representation.

STEP 10: Select DONE on the FEAT INC/EXC menu to end the selection process.

SUMMARY

The traditional way within Pro/ENGINEER to create features is to use commands such as Protrusion and Cut. However, options exist that optimize the modeling process by manipulating these features. Like most mid-range CAD packages, Pro/ENGINEER has commands for manipulating existing entities and features. Options, such as Pattern, allow for the copying and arraying of most features. Other options, such as Copy, also exist for creating instances of a feature. The understanding of these manipulation tools is critical for creating advanced parts within Pro/ENGINEER.

TUTORIAL

MANIPULATING TUTORIAL 1

This tutorial will create the part shown in Figure 7–30. The following topics will be covered:

- Creating a revolved protrusion.
- Creating an extruded protrusion.
- Creating a coaxial hole.
- Mirroring a feature.
- Rotating a feature.
- Adding dimensional relationships.

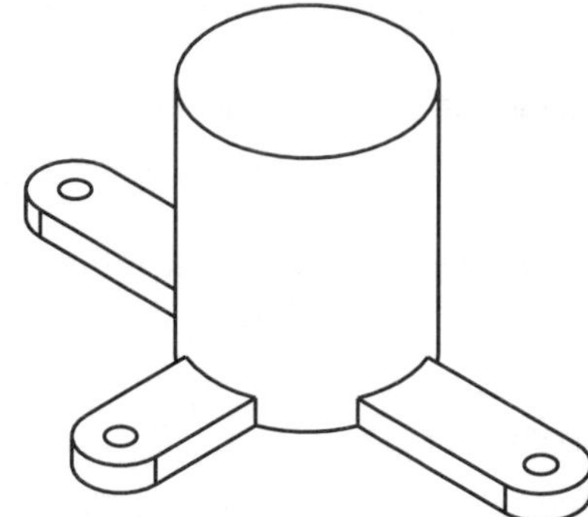

Figure 7-30 Finished part

CREATING THE BASE PROTRUSION

The first section of this tutorial will require you to create the Protrusion feature shown in Figure 7–31. To begin, start Pro/ENGINEER and create a part model named *Rotate*.

Create the base protrusion of the part as a Revolve with the following requirements:

- Revolve the section shown in Figure 7–31.
- Use Pro/ENGINEER's default datum planes.
- Sketch the Revolved Protrusion on datum plane FRONT.
- Use a 360-degree revolution parameter.

CREATING AN EXTRUDED PROTRUSION

This segment of the tutorial will create the Extruded Protrusion shown in Figure 7–32.

STEP 1: Setup a Solid Extruded PROTRUSION for the new feature shown in Figure 7–32 by using the following options.

- Use a One Side Extrude.
- Select datum plane TOP as the sketching plane.
- Use the Default sketching orientation.

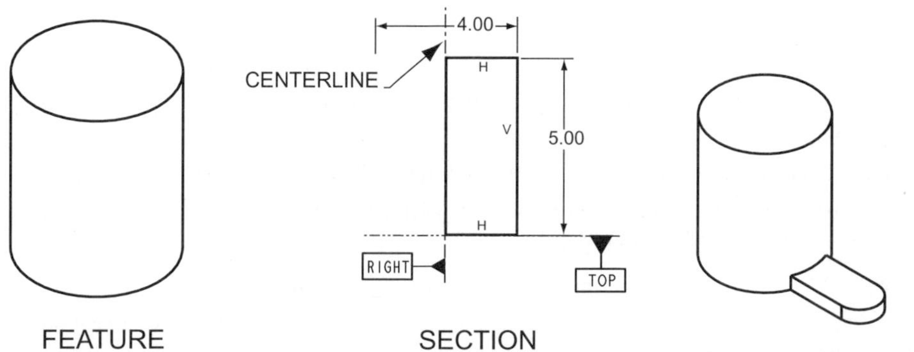

Figure 7-31 Base protrusion and section

Figure 7-32 Extruded protrusion

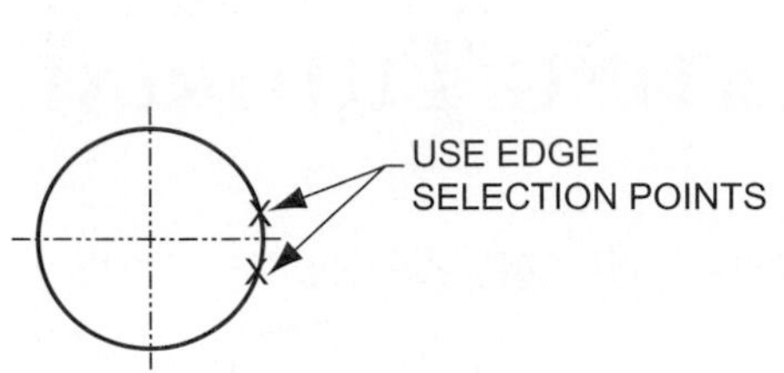

Figure 7–33 Selecting an edge

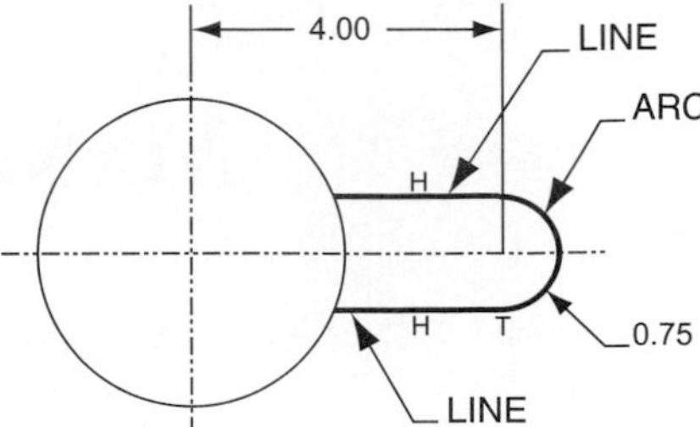

Figure 7–34 Sketched section

STEP 2: Once established in the sketcher environment, close the References dialog box.

STEP 3: Select the USE EDGE icon, then select the SINGLE option on the Type dialog box.

The Use Edge option is used to project existing feature edges as sketch entities within the current sketching environment.

STEP 4: As shown in Figure 7–33, select the outside edge of the base revolved feature in the two specified locations. (Only select each location once.)

The Use Edge option will project the outside edge of the revolved feature onto the sketching plane as sketcher entities. The selected edges become references within the sketching environment.

STEP 5: Sketch the two line entities and the arc entity shown in Figure 7–34.

Sketch the two new lines and the arc as shown. The ends of each line should be aligned with the edge of the existing revolved protrusion. Since the previous step of this tutorial turned the edges of the revolution into entities and into references, the lines will snap to the edge.

STEP 6: Modify the dimension values to match Figure 7–34.

STEP 7: Select the Trim Entities icon on the Sketcher toolbar.

Note: The Trim Entities icon is located behind the Dynamic Trim icon.

> **INSTRUCTIONAL NOTE** Ensure that you select the Trim Entities icon and not the Dynamic Trim icon. The Dynamic Trim option selects entities to trim by requiring the user to dynamically draw a construction spline.

With the Trim option, the two selected entities will be trimmed at their intersection point. You must pick the portion of each entity that should not be deleted.

STEP 8: Select the Four locations shown in Figure 7–35.

With the Trim option selected, the picked locations identify the entities to trim and the portion of each entity to keep.

STEP 9: When the section is complete, select the Continue icon.

If you get an error message upon selecting the Continue option, you probably have section entities that overlap. Within any section, you cannot have sections that branch or sketch entities that lie on top of other entities. If you

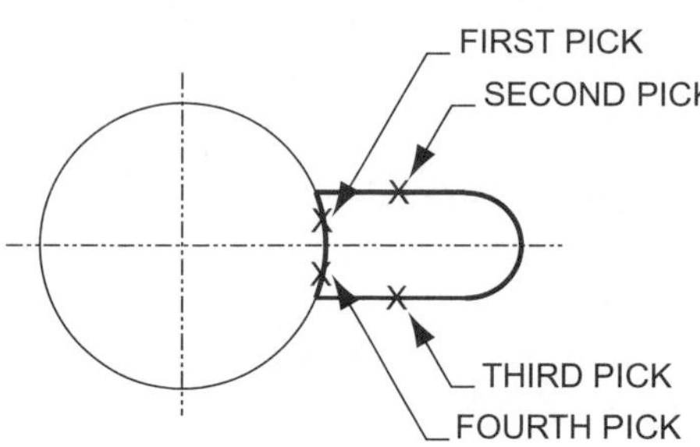

Figure 7–35 Trim selections

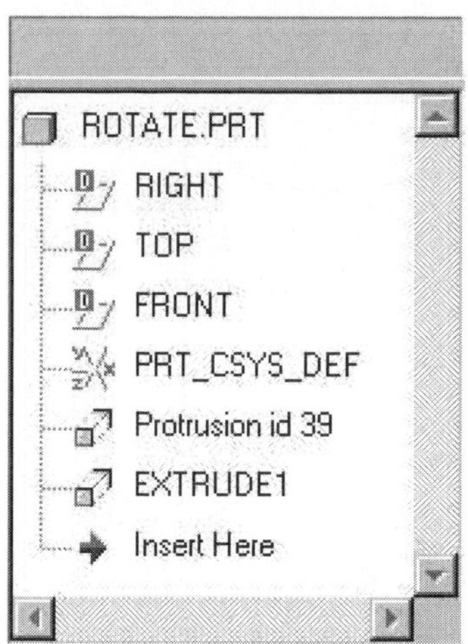

Figure 7–36 Naming a feature

do get an error, observe your sketch to see if there are any obvious problems. If necessary, try to delete entities created with the Use Edge option. You can use the Undo and Redo options to help with the manipulation. A final work-around solution would be to sketch the section's inside arc using the Arc option instead of the Use Edge option.

STEP 10: Enter a BLIND depth of .500.

STEP 11: Preview the protrusion, then select OKAY on the dialog box.

STEP 12: If necessary, select DONE on the Feature menu.

STEP 13: Select the SET UP >> NAME command from the Menu Manager.

STEP 14: On the Model Tree, select the Protrusion feature that defines the previously created extruded feature and rename it *EXTRUDE1.*

This step renames the last feature. Use the Name option to rename the feature *EXTRUDE1* (Figure 7–36).

Creating a Coaxial Hole

This segment of the tutorial will create the Coaxial Hole shown in Figure 7–37. To create the hole, the datum axis shown will have to be created first. This axis will be created with the Datum Axis command with the Through Cylinder constraint option.

STEP 1: Select the DATUM AXIS icon on the Datum toolbar.

STEP 2: Select THRU CYL as the constraint option.

The Thru Cylinder constraint option will place a datum axis through the center of an existing cylindrical feature. The cylindrical feature can be any solid feature with a rounded surface (e.g., Shaft, Round, extruded arc, etc.).

STEP 3: Select the *Extrude1* feature in the location shown in Figure 7–38.

STEP 4: Create the Coaxial Hole shown in Figure 7–39.

Use the following parameters when creating the hole:

- Create a Straight hole.
- Use a Hole diameter of 0.500.

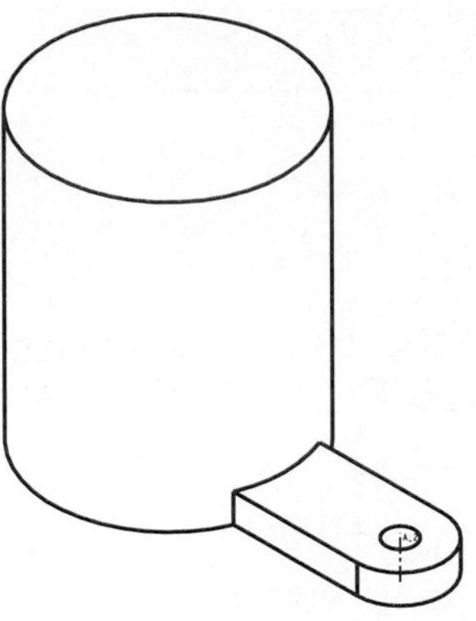

Figure 7–37 Coaxial hole feature

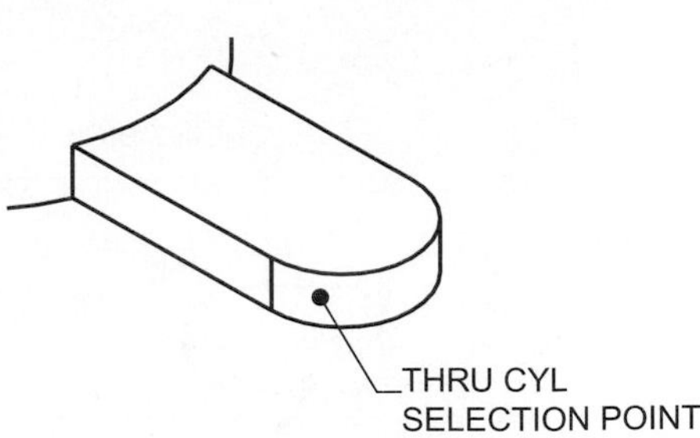

Figure 7–38 Cylinder selection

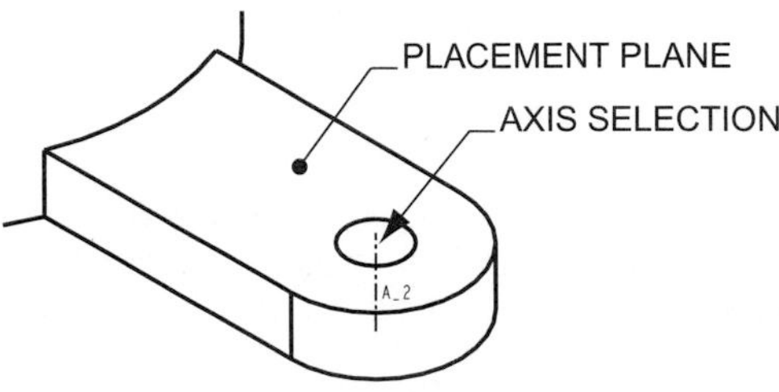

Figure 7–39 Coaxial hole creation

- Create the Hole with a Thru All depth one parameter.
- Select the previously created datum axis as the primary reference.
- Use the Coaxial placement option.

MIRRORING THE EXTRUDED FEATURE

This segment of the tutorial will mirror the extruded Protrusion, Coaxial Hole, and Datum Axis features (Figure 7–40). The Copy >> Mirror option requires the selection of a mirror plane. In this tutorial, one of Pro/ENGINEER's default datum planes will be used as this plane.

STEP 1: Select FEATURE >> COPY.

STEP 2: Select the MIRROR option from the Copy Feature menu.

STEP 3: Select DEPENDENT on the Copy Feature menu, then select DONE.

With the Dependent option, the copied features' dimensions are dependent upon their parent features' dimensions. If a dimension on a parent feature changes, the corresponding dimension on its child will change.

STEP 4: On the Model Tree, select the EXTRUDE1 feature, the Datum Axis feature, and the Coaxial Hole feature (Figure 7–41).

The last three features on the Model Tree will be copy-mirrored. Optionally, you can select the features on the work screen.

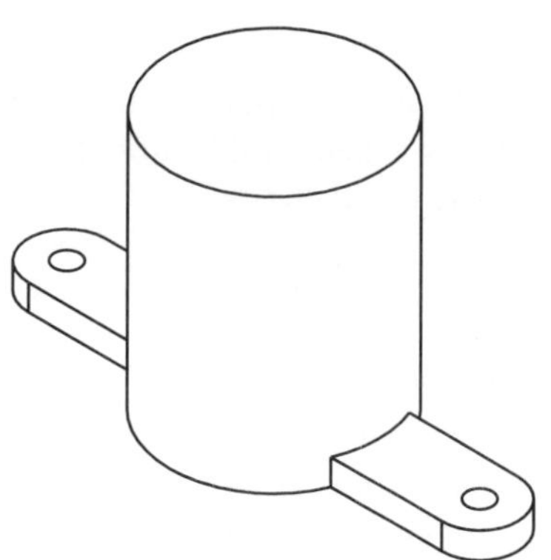

Figure 7–40 Mirrored features

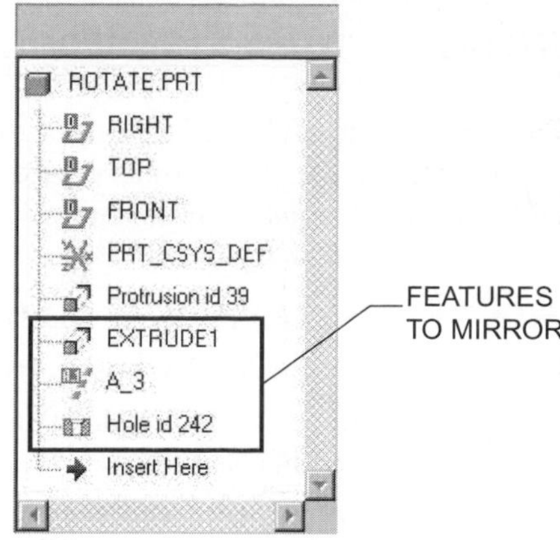

Figure 7–41 Feature selection

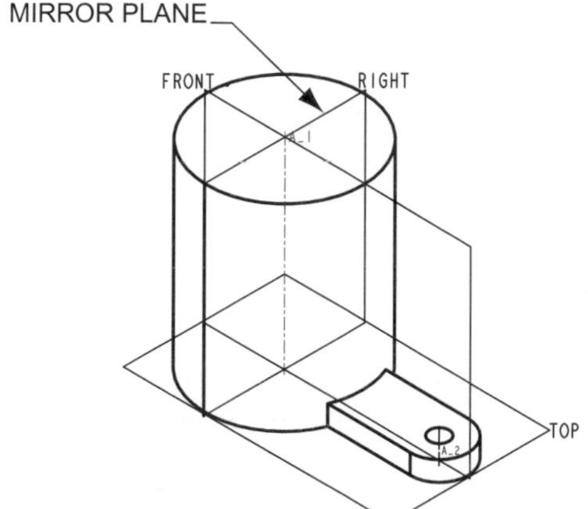

Figure 7–42 Mirror plane selection

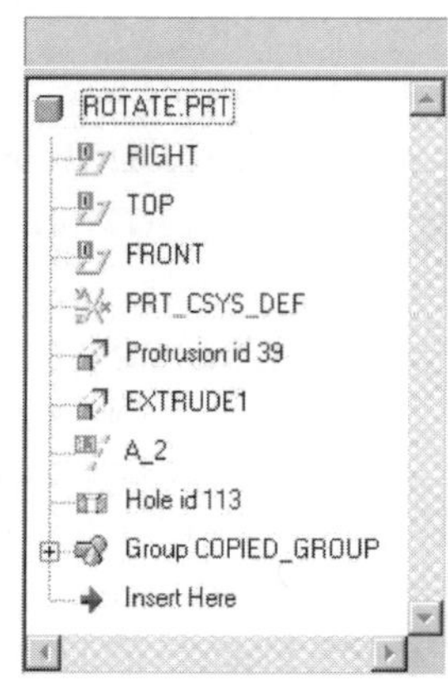

Figure 7–43 Model tree

STEP 5: Select DONE SEL on the Get Select menu.

Select Done Select to exit the feature selection process.

STEP 6: Select DONE on the Select Feature menu.

The Select Feature menu provides an option for selecting additional features to be copied. Select Done to end the selection process.

STEP 7: As shown in Figure 7–42, select datum plane RIGHT as the plane to mirror the selected features about.

After selecting the plane to mirror about, the mirrored grouped features will be created.

STEP 8: Observe your Model Tree.

Notice the group feature added to the Model Tree (Figure 7–43). By selecting the + icon next to the feature on the Model Tree, the group will be expanded to reveal the elements of the group.

ROTATING THE EXTRUDED FEATURE

The Copy command has an option for rotating features. Unlike the Pattern command, the Rotate option can array multiple features simultaneously. This segment of the tutorial you will rotate the Protrusion, Hole, and Axis features 90 degrees to create the grouped feature shown in Figure 7–44.

STEP 1: Select FEATURE >> COPY.

STEP 2: Select the MOVE option from the Copy Feature menu.

Other options available under the Copy Feature menu include New Refs, Same Refs, and Mirror.

STEP 3: Select INDEPENDENT >> DONE.

The Independent option will make the copied features' dimension values independent from their parent features' values. If a parent feature's dimension value changes, the copied feature's corresponding dimension value will not change.

STEP 4: On the Model Tree, select the EXTRUDE1 feature, the original Coaxial Hole, and the Axis locating the Coaxial Hole (Figure 7–45).

Select the features that were mirrored in the previous section of this tutorial.

STEP 5: Select DONE SEL on the Get Select menu.

STEP 6: Select DONE on the Select Feature menu.

STEP 7: Select ROTATE on the Move Feature menu.

The **Rotate** option will copy a feature about an edge, axis, or curve.

STEP 8: Select CRV/EDG/AXIS, then select the Axis defining the center of the base revolved protrusion.

As shown in Figure 7–46, the axis used to define the center of the revolved protrusion feature will be used as the center of the rotation. On the work screen, select this axis.

STEP 9: Select OKAY to accept the Direction for Translation (Figure 7–46).

On the work screen, an arrow points in the direction of translation. Pro/ENGINEER uses the Right-Hand rule to determine this direction. Using your right-hand, with the thumb pointed in the direction of the arrow, the copy will rotate the direction that the fingers of your hand are pointing.

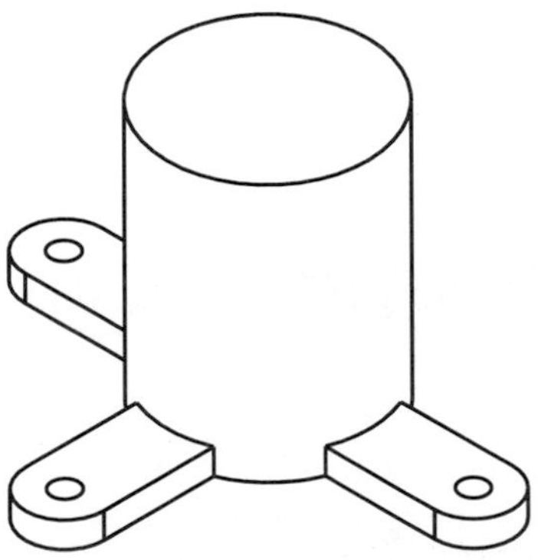

Figure 7–44 Rotated features

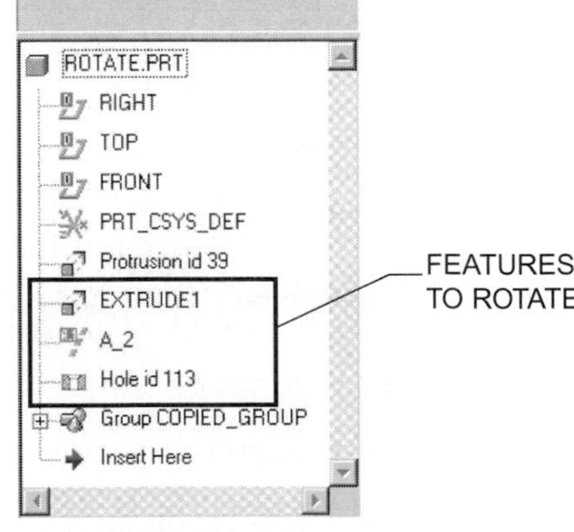

Figure 7–45 Feature selection

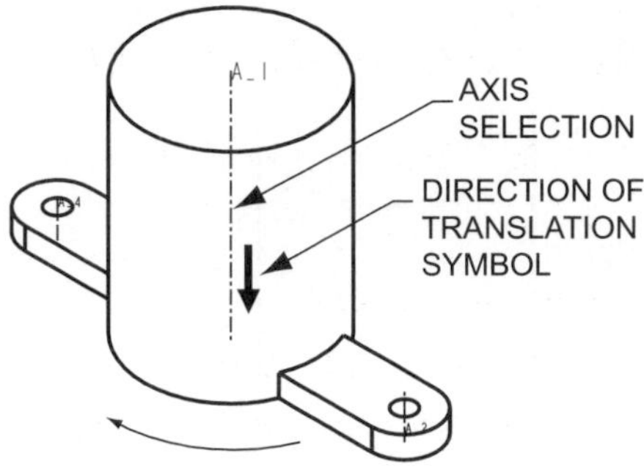

Figure 7–46 Axis selection

STEP 10: Enter 90 as the Rotation Angle.

STEP 11: Select DONE MOVE on the Move Feature menu.

STEP 12: Select DONE on the GP VAR DIMS menu.

The dimensions from the features being copied can be varied during the copy process. The GP VAR DIMS menu provides the option of selecting dimensions to vary during the copy.

STEP 13: Select OKAY on the dialog box.

STEP 14: If necessary, select DONE to exit the feature menu, then save the part.

ADDING RELATIONS

This segment of the tutorial will create a dimensional relationship that will make the distance from the center of the base protrusion to the center of the first axial hole equal to the height of the base protrusion. In Figure 7–47, dimension d12 will be set equal to dimension d1.

STEP 1: Select the RELATIONS option from the Menu Manager.

STEP 2: On the work screen, select the base revolved Protrusion and the EXTRUDE1 feature.

The dimensions shown in Figure 7–47 should appear similar to your model. Within the Relations menu, dimensions are shown symbolically. The symbols displayed in the illustration may or may not match your symbols. When following this tutorial, use the corresponding symbols from your model.

From Figure 7–47, you will add a relation that will make the dimension displayed with symbol d12 equal to the dimension displayed with the symbol d1.

STEP 3: Select the ADD option on the Relations menu.

STEP 4: With the symbols shown in Figure 7–47, add an equation that will make the d12 dimension equal to the d1 dimension.

For the symbols shown in Figure 7–47, add the equation:

$$d12 = d1$$

STEP 5: Select ENTER on the keyboard to quit the Add option.

STEP 6: Select DONE on the Model Relations menu.

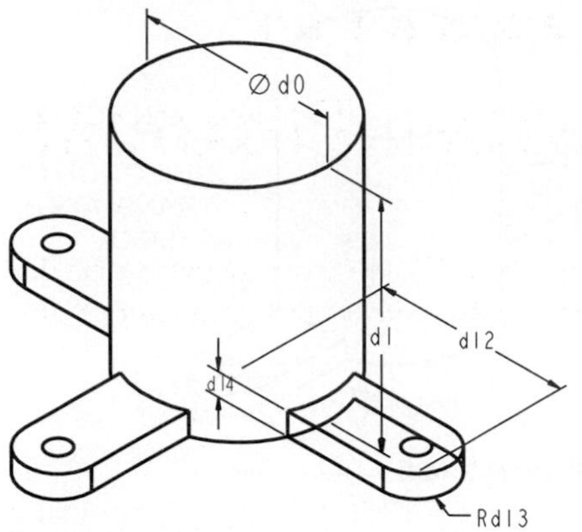

Figure 7–47 Dimension symbols

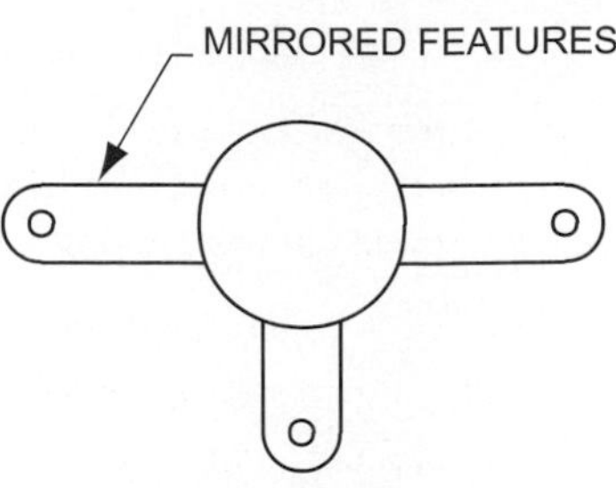

Figure 7–48 Finished features

STEP 7: Select REGENERATE on the Part menu.

After regenerating, notice how the EXTRUDE1 feature and the mirrored feature lengthen (Figure 7–48). Why did the second copy-rotated feature not lengthen with the other two features? The Independent option used in the creation of the rotated copy made its dimensions independent of its parent feature's dimensions.

TUTORIAL

MANIPULATING TUTORIAL 2

This tutorial will create the part shown in Figure 7–49. The primary objective of this tutorial is to demonstrate the Group command and the ability to pattern grouped features. Covered in this tutorial will be the following topics:

- Creating a Through >> Axis datum plane.
- Creating an extruded boss feature.
- Creating a coaxial hole and fillet.
- Grouping features.
- Patterning a group.
- Creating a conditional relation.

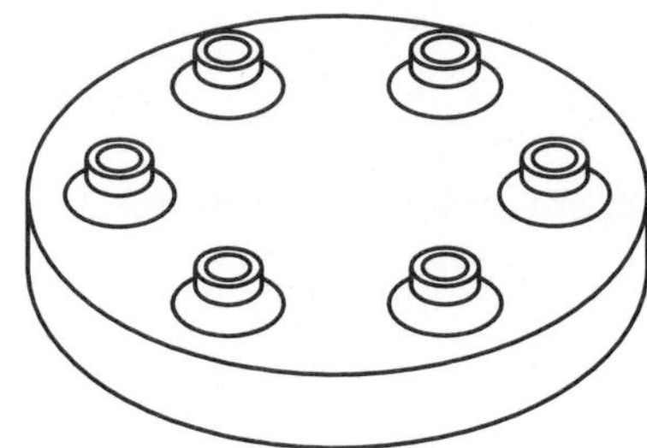

Figure 7–49 Finished part

CREATING THE BASE PROTRUSION

The base feature for this tutorial is shown in Figure 7–50. Using Pro/ENGINEER's default datum planes and the following requirements, create this feature.

- Revolve the section shown in Figure 7–50.
- Use Pro/ENGINEER's default datum planes.
- Sketch the Revolved Protrusion on datum plane FRONT.
- Enter a 360-degree angle of revolution.

CREATING A THROUGH >> AXIS DATUM PLANE

This tutorial will create a datum plane through the center axis of the base feature and at an angle to datum plane FRONT. The boss features shown in Figure 7–49 are patterned groups. To create a revolved patterned group, an angular dimension used within the definition of a feature in the group must be used as the leader dimension. This segment of the tutorial will create the angular datum plane shown in Figure 7–51. The angle defining this datum plane will be the leader dimension for the pattern.

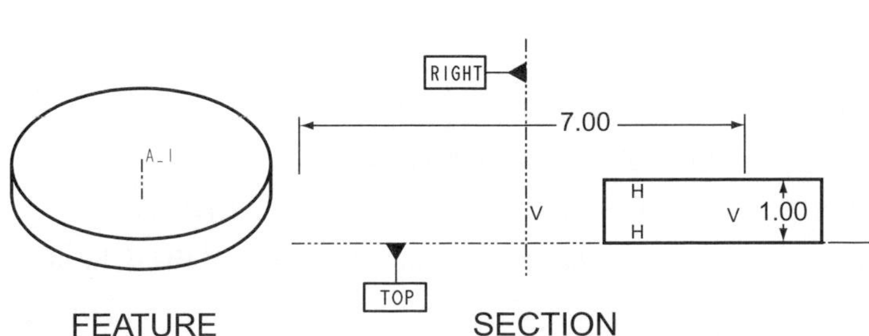

Figure 7–50 Base protrusion

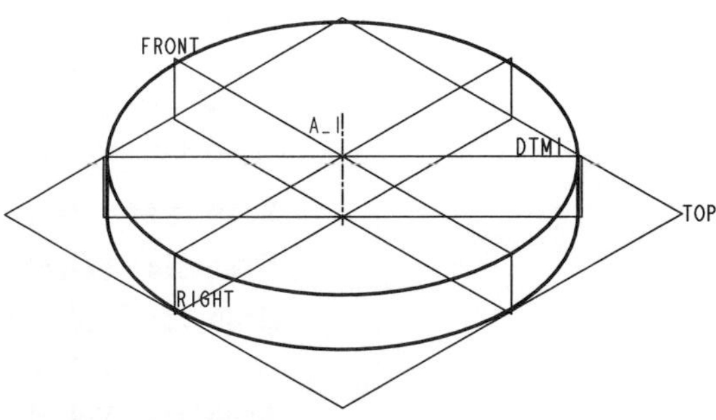

Figure 7–51 Through >> Axis datum plane

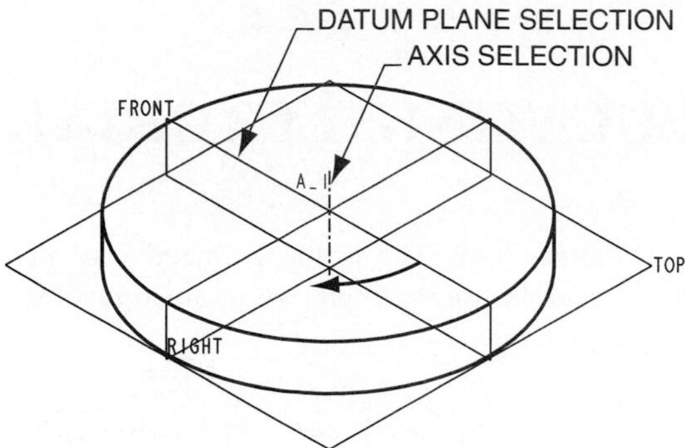

Figure 7–52 Datum creation

STEP 1: **Select VIEW >> DEFAULT from the Menu Bar.**

STEP 2: **Select the DATUM PLANE icon.**

STEP 3: **Select the THROUGH option from the Datum Plane menu, then select the axis at the center of the base protrusion (Figure 7–52).**

The Through option will construct a datum plane through an edge, axis, curve, point, vertex, plane, or cylinder.

STEP 4: **Select the ANGLE option from the Datum Plane menu, then select datum plane FRONT (Figure 7–52).**

When using the Through option, with the exception of the Through >> Plane constraint, a paired constraint option is required. The Angle constraint option will construct the datum plane at an angle to a selected plane. In combination, the Through >> Axis and Angle constraint options will construct this new datum plane through the center axis and at a specified angle to datum plane FRONT.

STEP 5: **Select DONE from the Datum Plane menu.**

STEP 6: **Select ENTER VALUE from the Offset menu.**

STEP 7: **In Pro/ENGINEER's textbox, enter −45 as the angular value.**

On the work screen, Pro/ENGINEER will graphically display the direction of rotation for the angular datum plane. In this example, entering a value of −45 will create the new datum plane at a 45-degree angle to FRONT.

CREATING THE BOSS FEATURE

This segment of the tutorial will create the Extruded Protrusion shown in Figure 7–53. This protrusion will serve as the first feature in the group to be patterned.

STEP 1: **Select FEATURE >> CREATE >> PROTRUSION.**

STEP 2: **Select EXTRUDE >> SOLID >> DONE >> ONE SIDE >> DONE.**

STEP 3: **Select the top of the part as the sketching plane, then accept the default extrusion direction.**

STEP 4: **Select TOP, then pick DTM1 to orient this plane toward the top of the sketcher environment (Figure 7–54).**

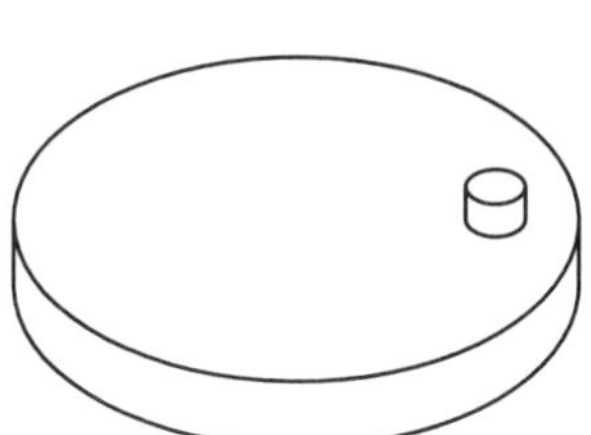

Figure 7–53 Boss feature

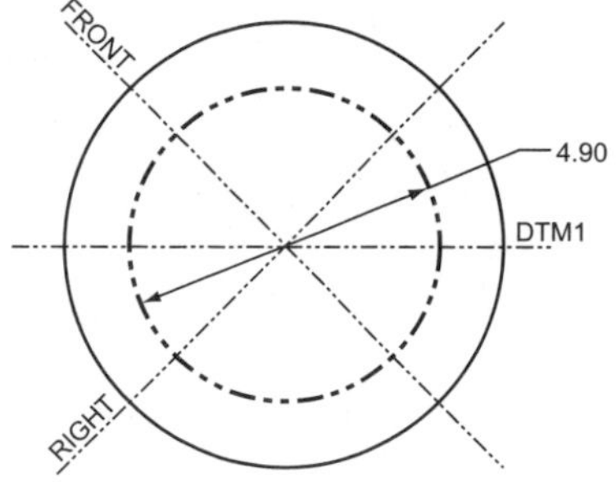

Figure 7–54 Construction circle

> **INSTRUCTION POINT** The orientation of datum plane DTM1 toward the top of the sketcher environment is an important step of this tutorial. Without this proper orientation, you will get an error message when patterning the group later in this tutorial. As shown in Figure 7–54, datum plane DTM1 should be horizontal on the work screen. If datum plane DTM1 is not horizontal, restart this segment of the tutorial.

STEP 5: Use the References dialog box to specify datum planes RIGHT, FRONT, and DTM1 as References.

STEP 6: Use the CIRCLE icon to create the circle shown in Figure 7–54.

STEP 7: With the previously created circle still picked, on Pro/ENGINEER's Menu Bar select EDIT >> TOGGLE CONSTRUCTION.

Elements must be picked before you can turn them into construction entities. If the circle is not selected, use the Pick icon to select it. Elements created as Construction entities will not extrude with geometry entities. This construction circle will be used to align and locate the extruded feature.

STEP 8: Modify the construction circle's diameter to equal a value of 4.90.

STEP 9: Use the CIRCLE option to create the circle entity shown in Figure 7–55.

Align the center of the new circle at the intersection of the construction circle and datum plane DTM1. Modify the Circle's diameter to equal a value of 0.750.

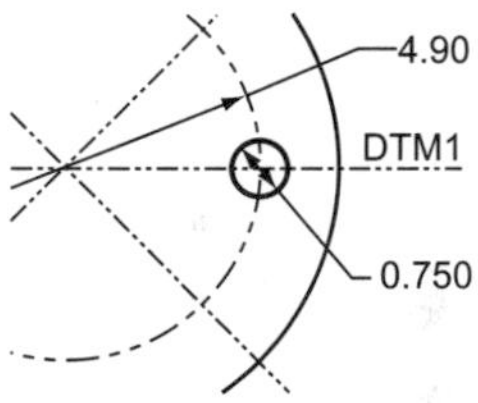

Figure 7–55 Circle creation

STEP 10: Exit the sketcher environment and extrude the Protrusion a BLIND distance of 0.500.

STEP 11: Preview the Protrusion, then select OKAY on the Feature Definition dialog box.

STEP 12: Select DONE to exit the Feature menu, then select the SET UP menu option.

The next three steps will give each Protrusion feature a more descriptive name.

STEP 13: Select the NAME option from the Part Setup menu.

STEP 14: On the Model Tree, select the first Protrusion feature and rename it *BASE* (Figure 7–56).

This will allow for the renaming of this feature to make it more descriptive on the Model Tree.

STEP 15: Use the NAME command to rename the last Protrusion feature *BOSS* (see Figure 7–56).

STEP 16: Save your part.

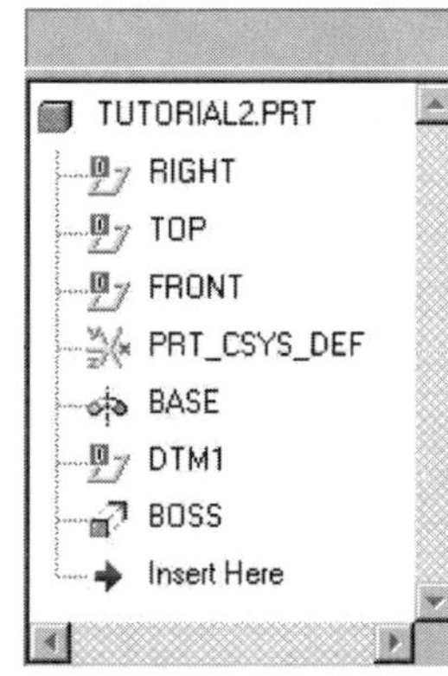

Figure 7–56 Model tree

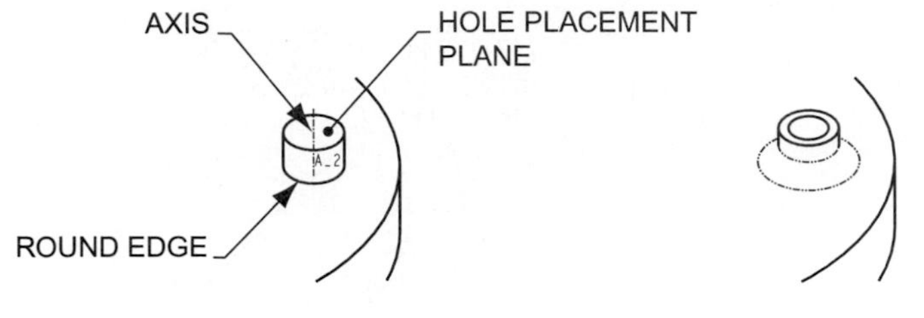

Figure 7–57 Hole and round features

Creating a Coaxial Hole and Round

This segment of the tutorial will create the Hole and Round features shown in Figure 7–57. These two features will be combined with the Boss feature and datum plane DTM1 to form a group.

Step 1: Create a COAXIAL HOLE as shown in Figure 7–57 by using the following options:

- Use the Hole command.
- Create a Straight hole.
- The hole will have a diameter value of 0.500.
- Create the hole using the Thru All depth option.
- Use the BOSS feature's axis as the primary reference.
- Place the hole with the Coaxial option.

Step 2: Create a ROUND as shown in Figure 7–57 using the following options:

- Use the Round Command.
- Create a Simple Round.
- Use a Constant radius value of 0.250.
- Use the Edge Chain and Tangent Chain options to select the base of the Boss feature.

Grouping Features

This section of the tutorial will group datum plane DTM1 with the Boss, Hole, and Round features. When features are grouped, they are turned essentially into one feature. The design of this part requires these features to be arrayed about a bolt-circle. Since the normal Pattern command arrays one feature at a time, these features have to be grouped first (Figure 7–58).

Step 1: Select FEATURE >> GROUP.

Step 2: Select LOCAL GROUP on the Create Group menu.

Local groups are available only within the model in which they were created.

Step 3: Enter *BOSS_GROUP* as the Name of the Local Group.

Step 4: On the Model Tree (Figure 7–59), select datum plane DTM1, the BOSS feature, the Hole feature, and the Round feature.

Each feature could be selected on the work screen. Since features have to be adjacent to each other in the order of regeneration, it is often easier to pick features on the Model Tree.

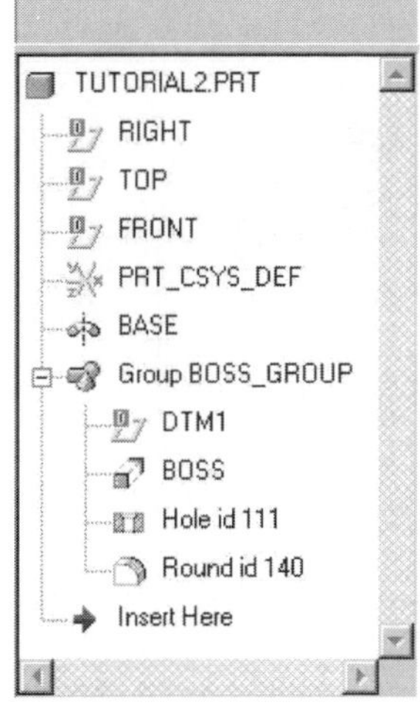

Figure 7–58 Model tree

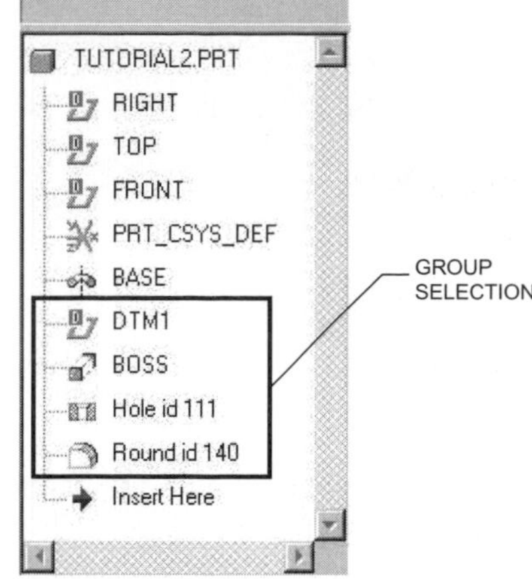

Figure 7–59 Feature selection

STEP 5: Select DONE on the Select Feature menu then observe the changes to the Model Tree.

Your Model Tree should appear as shown in Figure 7–58.

STEP 6: Select DONE/RETURN on the Group menu.

STEP 7: Save your model.

PATTERNING THE BOSS GROUP

The normal Pattern command can only pattern one feature at a time, and does not work with grouped features. The Group menu provides an option for patterning groups. This option will be used to create the rotational pattern shown in Figure 7–60.

STEP 1: Select FEATURE >> GROUP.

STEP 2: Select PATTERN on the Group menu.

The Group menu provides an option to pattern grouped features. The group Pattern option works similarly to the normal Pattern option. Unlike normal patterns, the Group >> Pattern option only allows Identical patterns. Due to this, patterned group instances must lie on the same placement plane, instances cannot intersect, and instances cannot intersect an edge.

STEP 3: On the Model Tree or on the work screen, select the *BOSS_GROUP* feature.

STEP 4: For the Leader dimension in the first direction, pick the 45-degree dimension used to define the angle of datum plane DTM1 (Figure 7–61).

Select the 45-degree dimension shown in Figure 7–61 as the first direction leader dimension. This dimension is the angular reference dimension obtained from the creation of datum plane DTM1. Since DTM1 was included in the group, this angular dimension is available for varying under the Group >> Pattern option. Since rotational patterns require an angular dimension, this is the necessary dimension required in the first direction of the pattern.

STEP 5: Enter 90 as the leader dimension's increment value in the first direction.

Entering 90 as the increment value will create each instance of the pattern 90 degrees apart.

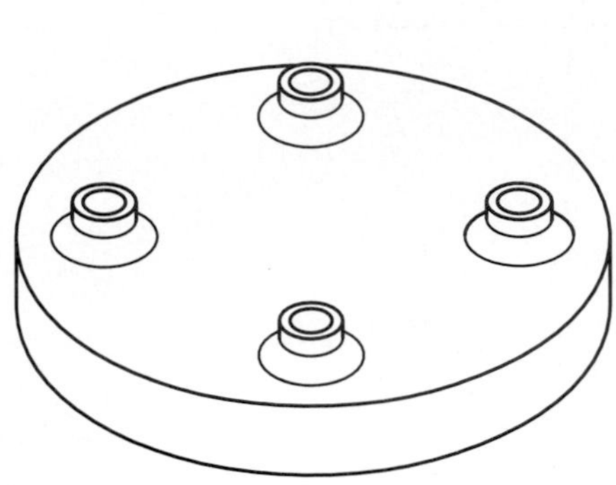

Figure 7–60 Patterned group

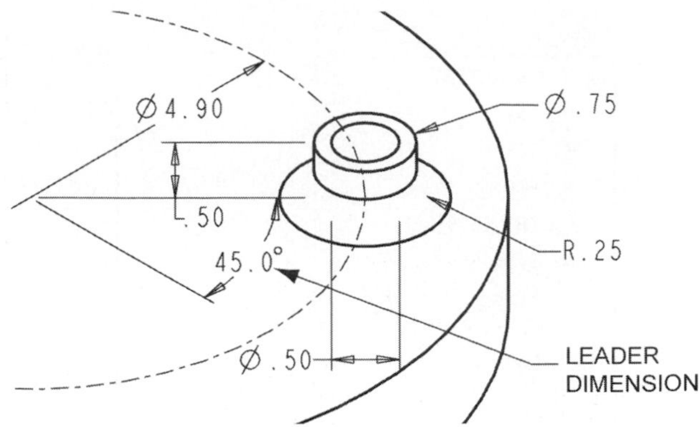

Figure 7–61 Leader dimension selection

STEP 6: Select DONE from the Exit menu.

Selecting Done will stop the selection of dimensions in the first direction.

STEP 7: Enter 4 as the number of instances in the first direction.

STEP 8: Select DONE to create the pattern.

This group will be patterned in one direction only. Due to this, the selection of a second leader dimension is not necessary. Selecting Done from the exit menu will create a pattern of the group.

STEP 9: Select DONE/RETURN from the Group menu.

STEP 10: Save your part.

ESTABLISHING A CONDITIONAL RELATIONSHIP

One of the powerful capabilities of a Feature-Based/Parametric modeling package is its ability to incorporate design intent into a model. One of the ways that intent can be built into a model is through the creation of dimensional relationships. A relation is an explicit mathematical relationship that exists between two dimensions. The flexibility of the Relations command allows conditional statements to be built into a relationship's equation. This tutorial will create a dimensional relationship that utilizes a conditional statement.

The current diameter of the base feature in this tutorial is 7 inches. This portion of the tutorial will add a conditional relationship that will drive the number of BOSS_GROUP features within the rotational pattern. The following design intent applies:

- A base feature diameter of 5 inches or less will have 4 equally spaced boss features
- A base feature diameter over 5 inches, but less than or equal to 10 inches, will have 6 equally spaced boss features
- A base feature diameter over 10 inches will have 8 equally spaced boss features
- The centerline diameter of the patterned groups will be 70 percent of the diameter of the base feature's diameter

As shown in Figure 7–62, the following dimensions with matching symbols will be used. Your symbols may be different.

- **Base feature diameter (d0)** This is the diameter value of the base feature. Initially, it is set to a value of 7 inches.
- **Number of grouped boss features (p0)** This is the number of instances that the grouped boss was patterned. Initially, there are 4 instances.

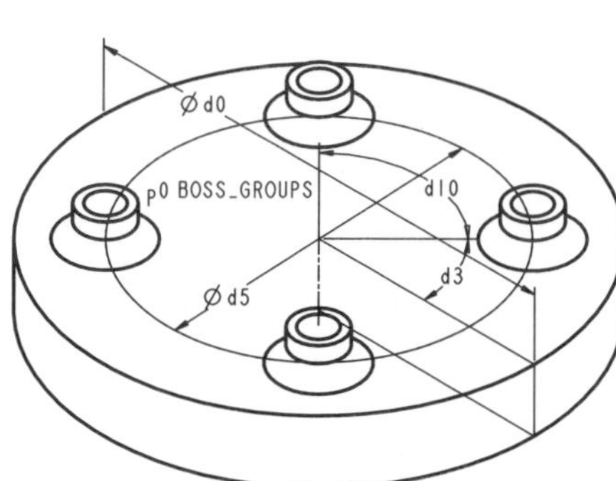

Figure 7–62 Dimension symbols

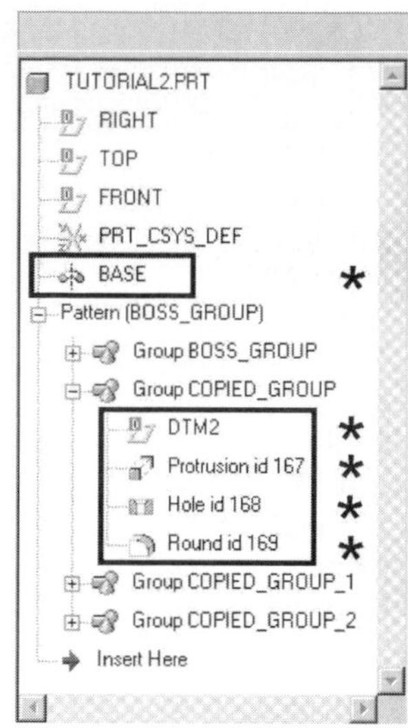

Figure 7–63 Model tree

- **Diameter of the pattern (d5)** This is the diameter value that defines the location of the first Boss feature. Initially, this value is set to 4.90 inches.
- **Increment value of the pattern (d10)** This is the number of degrees incremented between each instance of the pattern. Initially, this value is set to 90 degrees.

STEP 1: Select the RELATIONS option from the Part menu.

STEP 2: On the Model Tree, expand the Pattern feature and the second Group feature by selecting the + sign to the left of the required feature's name (Figure 7–63).

Selecting the + sign next to a feature's name will reveal components, features, and elements defining the feature.

STEP 3: On the Model Tree, select the BASE Protrusion feature and each feature defining the second BOSS_GROUP group (Figure 7–63).

Selecting features of the first grouped feature will not reveal the dimension symbol defining the pattern increment value. Before adding relations, make sure that the dimensions shown in Figure 7–62 are revealed on your work screen.

STEP 4: Select the ADD option on the Relations menu.

STEP 5: Enter the relation statements shown in Table 7.2.

Enter each statement line by line as shown. Selecting the Enter key at the end of each statement will allow for the entering of a new line. An alternative is to use the Edit Relations (Edit Rel) option (Figure 7–64). Make sure that you use the dimension symbols associated with your model. Your symbols may be different from those shown in Figure 7–62 and from those shown in the statements in Table 7–2.

> **INSTRUCTIONAL POINT** While entering relation equations in the textbox, you may get an error resulting in an apparent frozen textbox. In this case, enter the statement *ENDIF* into the textbox. This error is a result of an unsatisfactory ending to a conditional statement.

STEP 6: Select the REGENERATE option on the Part menu.

For relational changes to take effect, you have to regenerate the model.

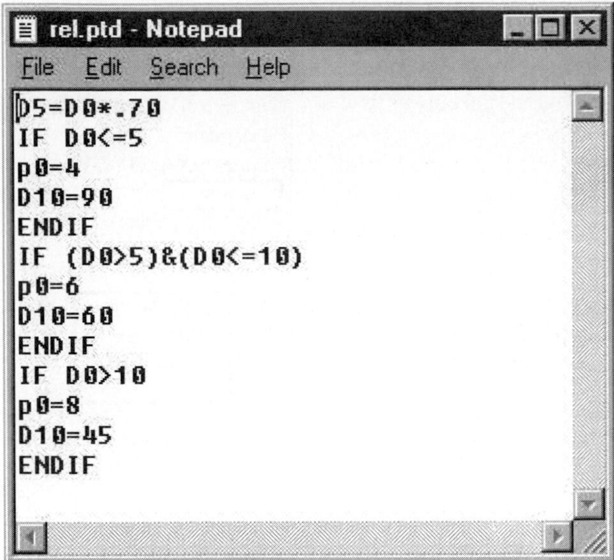

Figure 7-64 Edit relations option

Table 7-2 Statement line table

Statement Line	Meaning
d5 = d0 * .70	Diameter of the bolt-circle is 70% the diameter of the flange
IF d0 <= 5	If the diameter of the flange is less than or equal to 5.00
p0 = 4	The number of holes is equal to 4
d10 = 90	The angular increment value is equal to 90 degrees
ENDIF	End of the definition
IF (d0 > 5) & (d0 <= 10)	If the flange diameter is greater than 5.00 and less than or equal to 10.00
p0 = 6	The number of holes is equal to 6
d10 = 60	The angular increment value is equal to 60 degrees
ENDIF	End of the definition
IF d0 > 10	If the flange diameter is greater than 10.00
p0 = 8	The number of holes is equal to 8
d10 = 45	The angular increment value is equal to 45 degrees
ENDIF	End of the definition

STEP 7: Use the MODIFY >> VALUE option to change the base diameter dimension to a value of 12.00.

How many boss groups should the part have after regeneration?

> **INSTRUCTIONAL NOTE** For this particular part, changing the flange diameter to a value of 4.00 or less will produce a regeneration error. The Group >> Pattern option defaults to an Identical pattern option. Instances of an Identical pattern cannot intersect each other or the placement plane's edge. In this case, making the flange too small can violate both requirements.

STEP 8: Regenerate the model.

STEP 9: MODIFY the base diameter dimension to a value of 9.00, then REGENERATE the model.

You should have 6 boss groups.

STEP 10: MODIFY the base diameter dimension to a value of 4.75, then REGENERATE the model.

STEP 11: Save your part.

PROBLEMS

1. Using Pro/ENGINEER's Part mode, model the part shown in Figure 7–65. Construct the part using the following order of operations:
   a. Create the base geometric feature as a revolved Protrusion. Include the hole in the section.
   b. Model one leg feature (including the hole).
   c. Use the Copy >> Move >> Rotate option to create three instances of the leg feature.
   d. Create the remaining three leg instances.
2. Model the part shown Figure 7–66. Use the Copy-Rotate option to create the multiple instances of the Cut and Hole features.
3. Model the part shown in Figure 7–67.
4. Model the part shown in the drawing in Figure 7–68.
5. Model the part shown in the drawing in Figure 7–69.

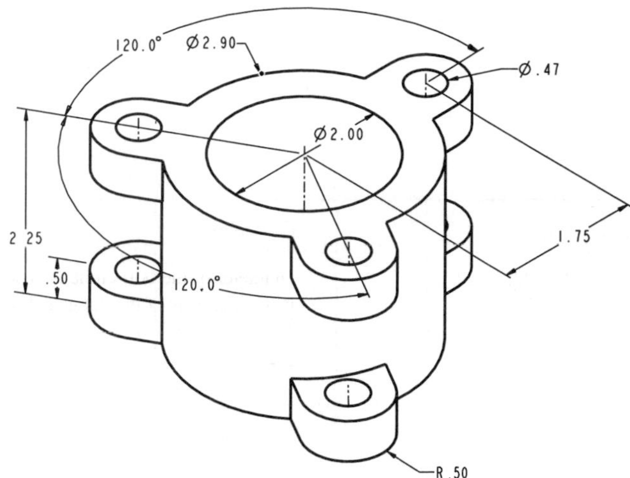

Figure 7–65 Problem one

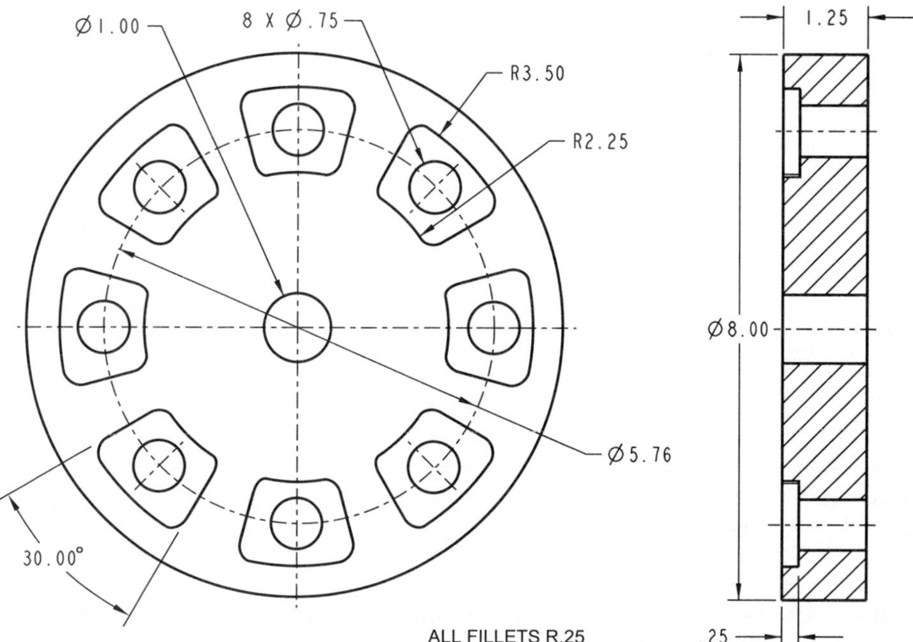

Figure 7–66 Problem two

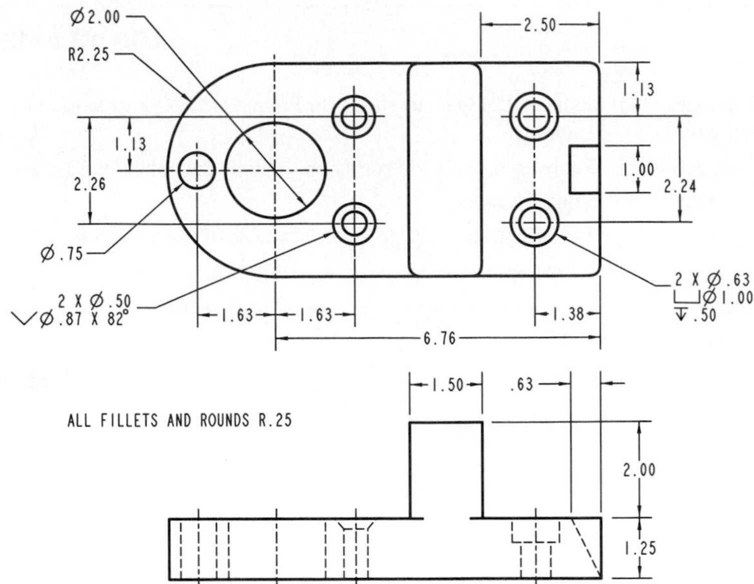

Figure 7-67 Problem three

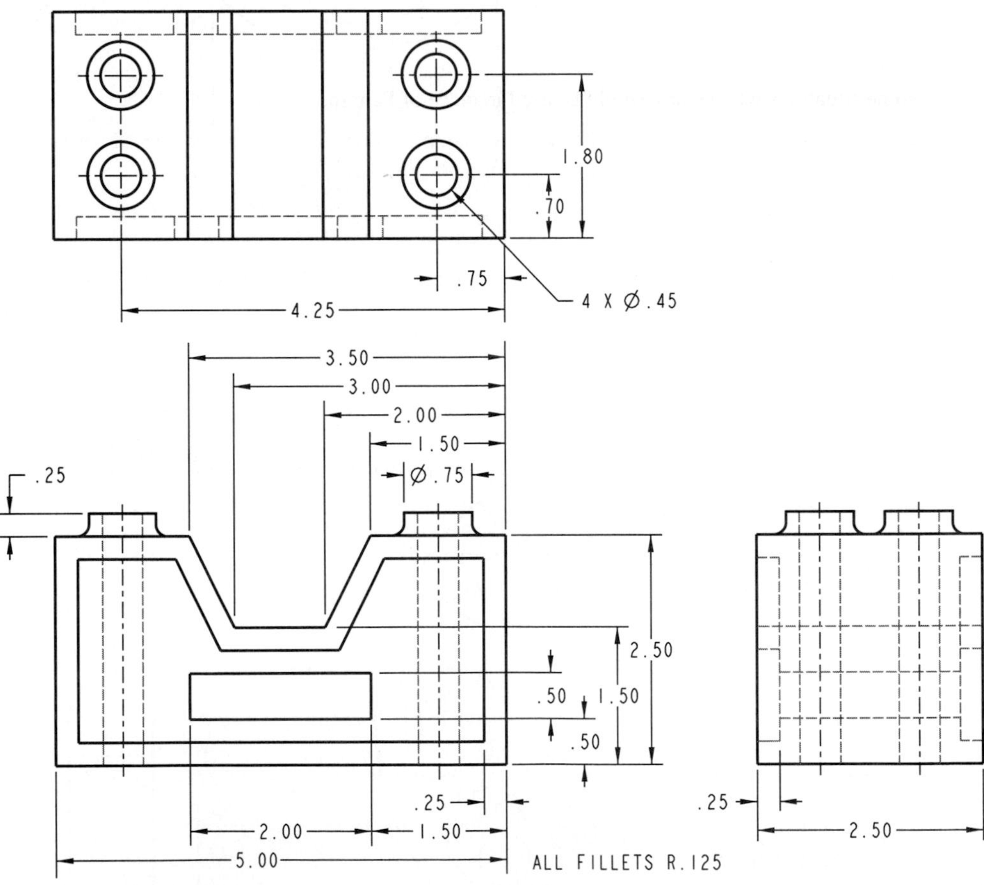

Figure 7-68 Problem four

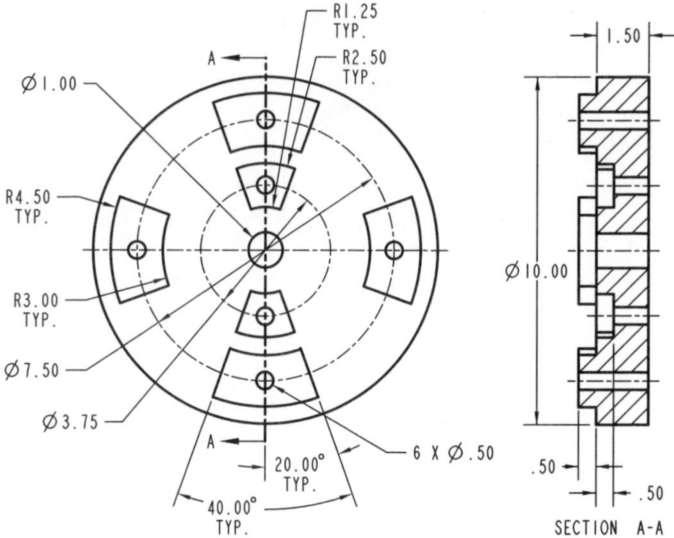

Figure 7-69 Problem five

QUESTIONS AND DISCUSSION

1. Describe the difference between the Pattern command and the Copy command. Describe some situations when Copy would be used in place of Pattern.
2. Describe the difference between the Copy >> Same Refs option and the Copy >> Move >> Translate option. Can you think of a situation when you might use one option over the other?
3. Describe the difference between copying features independently and copying features dependently.
4. When would the Mirror Geom option be considered in lieu of the Copy >> Mirror option?
5. Describe how a company might utilize User-Defined Features to streamline the modeling process.
6. Write relation statements for the following situations:
   - Dimension d0 is equal to two times dimension d1.
   - Dimension d0 is equal to the sum of dimension d1 and d2 divided by d3.
   - Dimension d0 is equal to the square root of d1.
7. List items types that can be included in a Family Table.
8. How does Pro/ENGINEER utilize Cross Sections?

CHAPTER

8

CREATING A PRO/ENGINEER DRAWING

Introduction

Drawing mode is used to produce detailed engineer drawings of parts and assemblies. Drawing mode has the capability to produce orthographic views to include section views and auxiliary views. This chapter introduces the fundamentals of Drawing mode and how to produce an annotated multiview drawing. Upon finishing this chapter, you will be able to

- Create orthographic views of an existing model within Drawing model.
- Setup, retrieve, and create sheet formats.
- Manipulate the settings of a drawing by changing Drawing Setup file options.
- Create a detailed view of a model.
- Apply parametric and nonparametric dimensions to a drawing.
- Show and erase entities such as geometric tolerances, centerlines, and datums.
- Create notes on a drawing, including notes with leaders and balloon notes.
- Create a table on a drawing.

DEFINITIONS

Associative dimension A parametric dimension available for viewing and/or modification in multiple modules of Pro/ENGINEER.

Drawing setup file A text file used to establish many of Drawing mode's default settings. As an example, the default text height for a drawing is set in the drawing setup file.

Format A Pro/ENGINEER module used to create drawing borders and title blocks. Formats can be added to a Pro/ENGINEER drawing.

Multiview drawing The use of multiple orthographic views to graphically display and communicate an engineering design.

Nonparametric dimension A dimension created in Drawing mode that is not used to construct a part or assembly. Nonparametric dimension values cannot be modified.

Parametric dimension A dimension used to define a part or assembly feature. Parametric dimensions can be modified or redefined.

DRAWING FUNDAMENTALS

Pro/ENGINEER is a fully integrated and associative engineer design package. Since it is an integrated package, an array of design, engineering, and manufacturing tools is available. Components of a design can be modeled within Part mode, grouped in Assembly

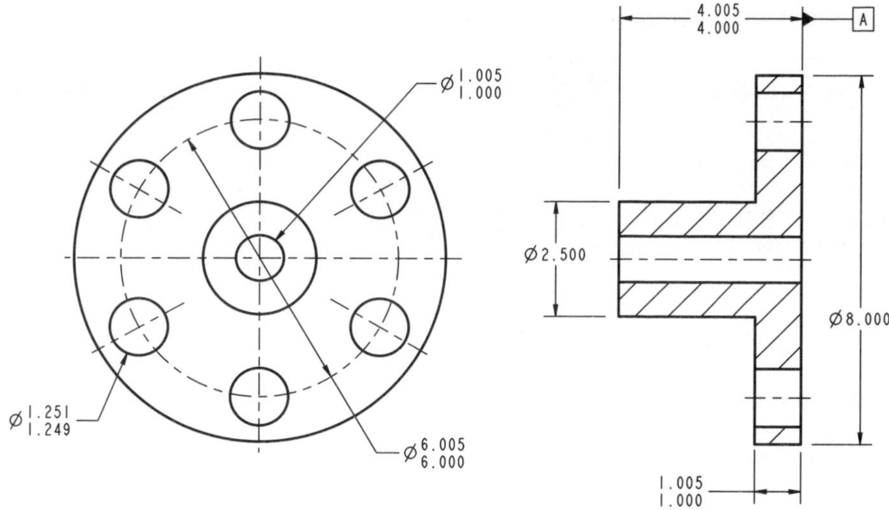

Figure 8–1 A Pro/ENGINEER drawing

mode, simulated and tested in Pro/MECHANICA, and machining code can be developed in Manufacturing mode (Pro/NC). The two-directional associativity that exists between modes allows changes made in one mode to be reflected in another. Combined within a strong computer network system, a true paperless manufacturing environment can be established.

Two-dimensional orthographic drawings were once considered one of the initial steps in the design process. A traditional approach to engineering design might require engineering drawings to be produced first, followed by engineering analysis and CNC code production. With Pro/ENGINEER, engineering drawings are considered "downstream" applications that occur after part modeling and analysis.

Pro/ENGINEER's basic package provides a module for creating orthographic drawings (Figure 8–1). Within Drawing mode, detailed multiview drawings can be created from existing models. Dimensions used to create a part (referred to as parametric dimensions) can be revealed in Drawing mode to document a design. Parametric dimensions can be modified in Drawing mode with the changes reflected in other modules of Pro/ENGINEER. Options are available for creating section views, detailed views, and auxiliary views. There are options for creating notes, leaders, nonparametric dimensions, and tables. Additionally, within Drawing mode, there exist a variety of two-dimension drafting tools.

Drawing mode can be used to create detailed drawings from existing parts and assemblies. When creating a new drawing, Pro/ENGINEER provides an option for selecting the model from which to create the drawing. Also, additional models can be added to a drawing from within Drawing mode. Multiple views of a model can be added to a drawing and annotations applied. Pro/ENGINEER provides existing sheet formats (e.g., A, B, C, and D) with detailed title blocks and borders. Format mode can be used to create additional sheet formats.

Drawing Setup File

Drawing mode uses **Drawing Setup Files** (DTL) to control the appearance of drawings. Pro/ENGINEER comes with default settings for a variety of drawing parameter options. Examples of parameters include text height, arrowhead size, arrowhead style, tolerance

Table 8–1 Common drawing setup file options

Option/Description	Default Value (Optional Value)
crossec_arrow_length Controls the length of cutting plane line arrowheads.	0.1875
crossec_arrow_width Controls the width of cutting-plane line arrowheads.	0.0625
dim_leader_length Controls the length of a dimension line when the dimension line arrowheads fall outside of the extension lines.	0.5000
draw_arrow_length Sets the length of dimension arrowheads.	0.1875
draw_arrow_style Sets the arrowhead style.	closed (Open or filled)
draw_arrow_width Sets the width of dimension arrowheads.	0.0625
drawing_text_height Sets the height of text in a drawing.	0.15625
drawing_units Sets units for a drawing.	inch (foot, mm, cm, or m)
gtol_datums Sets the display of geometric tolerance datum symbols.	Std_ansi (std_iso, std_jis, or std_ansi_mm)
leader_elbow_length Sets the length of a leader's elbow.	0.2500
radial_pattern_axis_circle Controls the display of rotational pattern features. Set to *yes* produces a bolt-circle centerline.	no (yes)
text_orientation Sets the orientation of dimension text.	horizontal (parallel or parallel_diam_horiz)
text_width_factor Sets the width factor for text.	0.8000 (0.25 through 8)
tol_display Sets the display of tolerance values.	no (yes)

display, and drawing units. Default values can be changed permanently or for individual drawings. Multiple drawing setup files can be created and stored for later use. Pro/ENGINEER'S default drawing setup file (*prodetail.dtl*) can be found in the <*pro_engineer load point*>*text*\ directory.

ADVANCED >> DRAW SETUP

New DTL Files are created or current drawing settings are changed with the Advanced >> Draw Setup option. Pro/ENGINEER utilizes an Options dialog box similar to the configuration file's Options dialog box for drawing setup changes (see Chapter 2). The configuration file option *drawing_setup_file* can be used to establish a specific DTL file. If this option is not set, Pro/ENGINEER uses the default DTL file. The configuration file option *pro_dtl_setup_dir* can be used to set the directory that Pro/ENGINEER searches for DTL files. Table 8–1 provides a list of common DTL file options with default values.

SHEET FORMATS

A **Format** is an overlay for a Pro/ENGINEER drawing. It can include a border, title block, notes, tables, and graphics. A sheet format has the file extension *.frm. Pro/ENGINEER provides a variety of predefined standard formats for use with ANSI and ISO sheet sizes

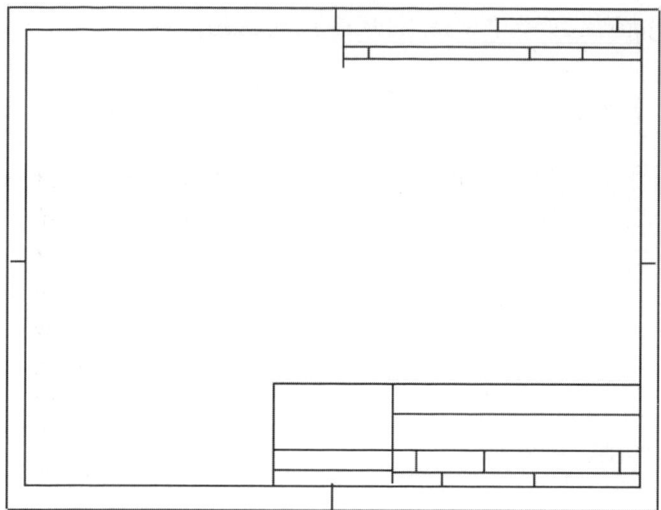

Figure 8–2 ANSI A size format and title block

(e.g., A, B, C, and D size sheets). These standard formats can be modified to produce a customized format. Additionally, Pro/ENGINEER's Format mode can be used to create a new sheet format. Figure 8–2 shows an example of an A size sheet format. A library of standard sheets can be created. The configuration file option *pro_format_dir* can be used to specify the directory path where standard sheets are stored.

Modifying Formats

An existing sheet format can be modified to create a customized format. Pro/ENGINEER's existing formats are located in the Format directory under the Pro/ENGINEER's program directory. Perform the following steps to modify an existing format:

Step 1: Start Pro/ENGINEER.

Step 2: Select FILE >> OPEN.

A drawing format can be opened from most modes of Pro/ENGINEER.

Step 3: Locate and Open the format to modify.

Manipulate the Look In directory structure under the File Open dialog box to locate the format to open. On the dialog box, the file Type option can be changed to only display Format files.

Step 4: Use available sketch creation and modification tools to modify the format.

> **INSTRUCTIONAL NOTE** See the sections on "Two-Dimensional Drafting" and "Modifying a Drawing" found later in this chapter for information on creating and modifying a sketch.

Drawing Mode provides a variety of tools for creating two-dimensional drawings. Options are available for creating lines, arcs, circles, splines, and ellipses. Construction options such as copy, mirror, intersect, trim, and offset are also available.

Step 5: Select the SAVE option from the File menu.

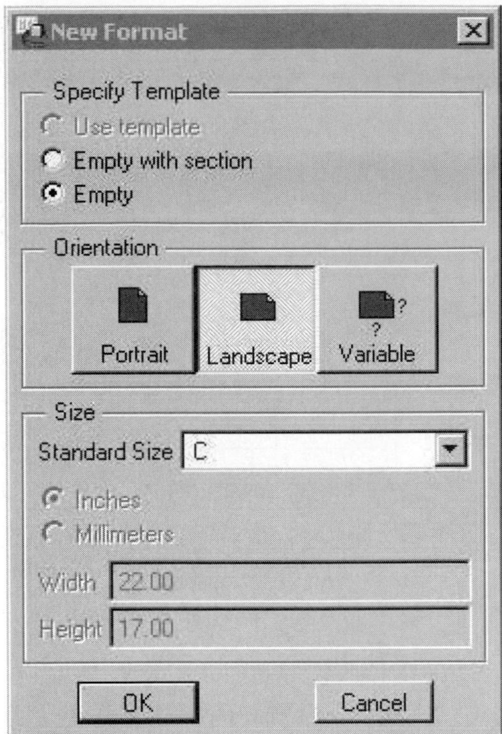

Figure 8–3 New Format dialog box

Creating Formats

Formats can be created from scratch. Format mode is the Pro/ENGINEER foundation module used for the creation of standard sheet formats. Sketching tools such as line, circle, arc, and note can be used to create the format section. A new format object can be created using the File >> New option. When a new format is initially created, Pro/ENGINEER reveals the New Format dialog box (Figure 8–3). The following options exist on this dialog box:

Specify Template

Through the Empty sub-option, the Specify Template option allows for the selection of a standard sheet size (e.g., A, B, C, etc.). The Empty with Section sub-option allows an existing section file (*.sec) to be incorporated within the new format.

Orientation

The Orientation option is available concurrently with the Empty option. Portrait, Landscape, and Variable sheet orientations are available.

Size

The Size option is available concurrently with the Empty option. This option allows for the selection of a standard sheet size or for the selection of a user-defined sheet size. User-defined sheet sizes can be established using inches or millimeters.

Creating a New Drawing

The Drawing mode option from the New dialog box is used to create new drawings. When Drawing mode is selected and a file name entered, Pro/ENGINEER introduces the New Drawing dialog box (Figure 8–4). If a part or assembly is currently active in session memory, Pro/ENGINEER defaults to this part or assembly as the model from which to

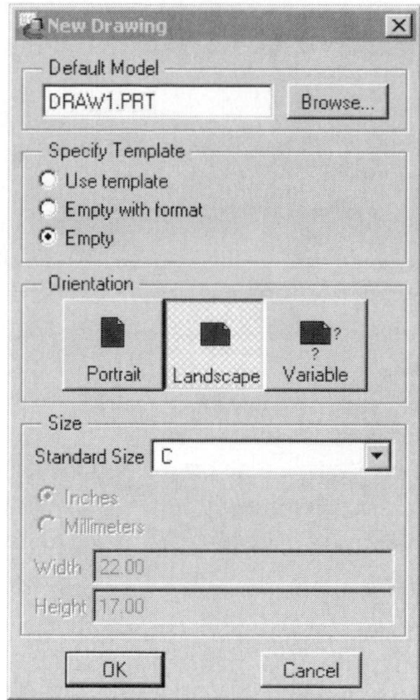

Figure 8-4 New drawing dialog box

create the drawing. An option is available for browsing to find other existing models. The New Drawing dialog box provides the option for specifying a standard sheet size or for retrieving an existing format.

Three types of items can be added to a drawing: Formats, 2D draft entities, and Model views. Formats are placed in a drawing using the Sheets >> Format >> Add/Replace option from the Drawing menu. Pro/ENGINEER provides an Open dialog box for browsing to find an appropriate format. Multiple sketching tools allow for the creation of 2D draft entities. Other options are available for adding dimensions and notes. Model views can be added to a drawing with the Views >> Add option. General, section, projected, auxiliary, revolved, and detail views can be added.

The following will show step-by-step how to create a new drawing. Several options are available that can vary the method for creating a new drawing. This process will allow for the creation of a new drawing without specifying a standard sheet size from the New Drawing dialog box.

STEP 1: Select FILE >> NEW.

STEP 2: Select Drawing mode from the New dialog box, enter a name for the new drawing, then select OKAY. (Use the default template file.)

STEP 3: On the New Drawing dialog box, select the DEFAULT MODEL from which to create the drawing.

Pro/ENGINEER will default to the current active model. You can use the Browse option to search for any existing object.

STEP 4: Select the EMPTY button from the Specify Sheet option.

The Empty option will require the defining of a sheet size and orientation. Alternately, the Use Template option is usable for the selection of an existing format, while the Empty-with-Format option is used to create a drawing without an established size or sheet format.

STEP 5: Select a sheet Orientation and a sheet Size, then select OKAY.

The Empty option requires the selection of a sheet orientation and size. A Portrait, Landscape, or Variable orientation is available. Standard sheet sizes and user-defined sheets sizes are also available.

Drawing Views

Pro/ENGINEER's Drawing mode provides options for creating a variety of orthographic views to include section views, auxiliary views, detailed views, revolved views, broken views, and partial views. As many views as necessary to fully describe a model can be added to a drawing sheet. Once inserted into a drawing, parameters associated with the model, to include dimension values, can be created. Views are also fully associated. This allows any model changes made in a drawing to be reflected in the part or assembly model. Additionally, changes in one view of a drawing will reflect accordingly in all views of the model.

The Views Menu

The Views menu option is used to create views of an existing Pro/ENGINEER model. The Views menu has options for manipulating and modifying existing views. The following menu options are available:

Add View

The Add View option is used to create new views. The first view created must be a General view. From a General view, other views of the model can be added.

Move View

The Move View option is used to move views on the work screen. When placing a view, Pro/ENGINEER requests a Center Point for the drawing view. The view is set initially at this location. The Move View option can be used to reposition this or any other view.

Modify View

The View Modify menu option provides a variety of tools for modifying a view. The following is a partial list of the available options:

- **View type** The View Type option is available to change the type of view. As an example, a Projection view can be changed to a General or Auxiliary view.
- **Change scale** This option is used to change the scale of a nonchild view. The view cannot be a child view of another view, and the view must have been inserted with a specific, user-defined scale value.
- **View name** Pro/ENGINEER provides each view with a unique name. This option is used to rename a view.
- **Reorient** When inserting a view, Pro/ENGINEER provides the option of orienting the view. As an example, a view can be oriented to allow for a proper front view vantage point. The Reorient option allows this orientation to be changed.
- **X-Section** This option allows a Cross Section to be replaced.
- **Z-Clipping** The Z-Clipping option allows for the exclusion of all graphics on a view behind a selected plane. This option is advantageous for views with background graphics that can clutter the drawing.

Erase View

By erasing a view, you temporarily remove it from the drawing screen.
Erasing a view removes it from the regeneration but does not affect any other view, including child views. To return a view to the drawing screen, use the Resume View option.

Delete View

The Delete view option permanently removes a view from the drawing. Views that are parents of other views cannot be deleted.

Relate View

The Relate View option is used to assign draft entities to a selected view. As an example, a note might be placed into a drawing using the Create >> Note option. This note can be assigned to a specific view with the Relate View option.

Display Mode

This menu option is used to modify the display of a selected view. The following options are available:

- **View display** This option is used to change the display of lines on a drawing view. A view's hidden lines can be specified as Wireframe, Hidden, or No Hidden. Additionally, tangent edge lines can be specified to display as a solid line, a centerline, a phantom line, a dimmed line, or set not to be displayed.
- **Edge display** The Edge Display option is similar to the View Display option. With the Edge Display option, individual hidden or tangent edges can be changed.
- **Member display** This option is used to control the display of assembly views.

Drawing Models

Multiple models can be accessed within one drawing. The Drawing Models (Dwg Models) option allows additional models to be added to the current drawing. The Set option allows a specific model to be set as the active model in the drawing.

View Types

Pro/ENGINEER provides a variety of view types to serve the documentation needs of a model. The following is a list of the available types:

General Views

The General view is the basic view type available. It is required as the first view placed into a drawing and is used by other types as a parent view. General views require a user-defined orientation. A General view is first placed into a drawing using Pro/ENGINEER's current default view orientation. The Orientation dialog box (Figure 8–5) is used to orient the view to match orthographic view projection requirements. In Figure 8–6, the Front view of the object was inserted as a General view.

Projection Views

The Projection view is an orthographic projection from a General view or from an existing view. Projection views follow normal lines of projection according to conventional drafting standards. As an example, a Projection view would be used to create a right-side view off an existing front view. Projection views become child views of the view from which they are projected. In Figure 8–6, the Top and Right-Side views were created as Projection views.

Auxiliary Views

Auxiliary views are used to project a view when normal lines of projection will not work. They are used to show the true size of a surface that cannot be shown from one of the six primary views. Auxiliary views are projected from a selected edge or axis. The auxiliary view shown in Figure 8–6 is projected off the Front view.

Detailed Views

Often, features are too small to describe fully with a standard projection view. In such a case, it is common practice to enlarge portions of a drawing to allow for more accurate detailing. Figure 8–6 shows an example of a detailed view.

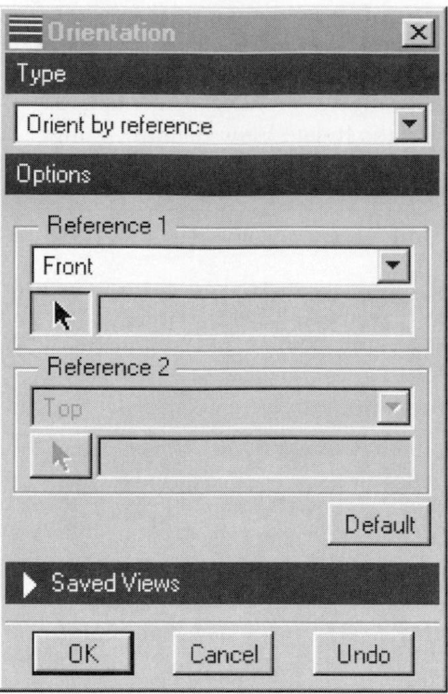

Figure 8-5 Orientation dialog box

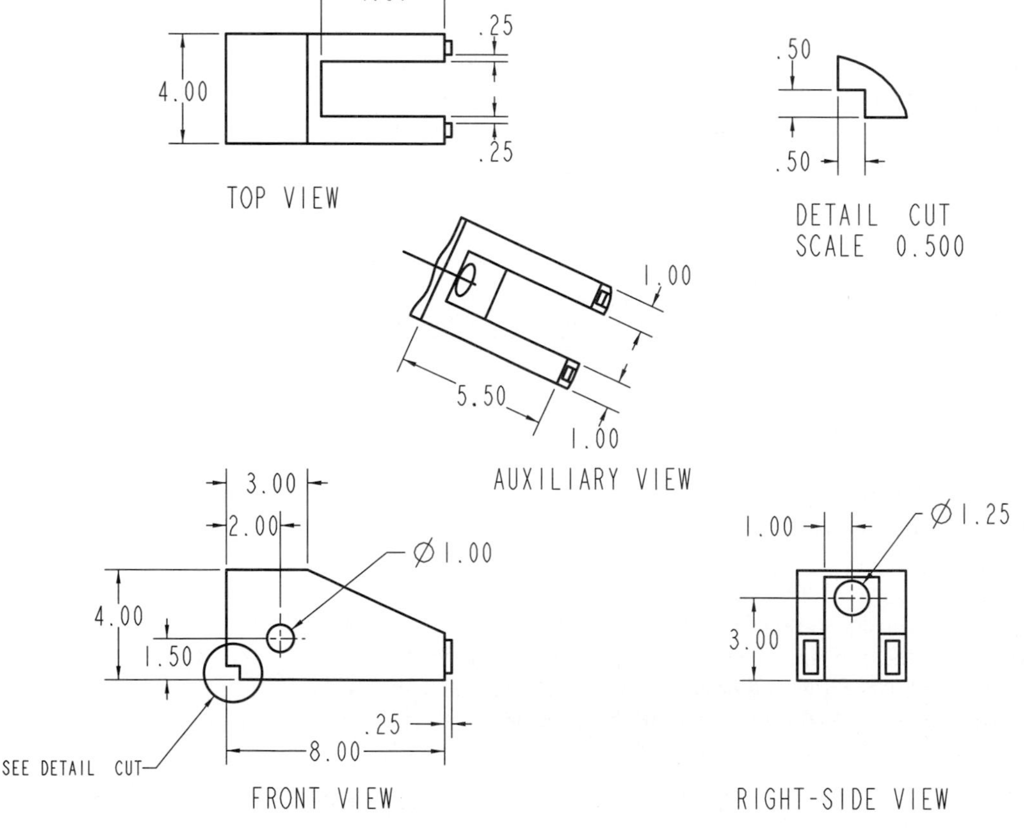

Figure 8-6 View examples

Revolved Views

Revolved views are used to show the Cross Section of a part or feature. A Cross Section is required and can be retrieved or constructed from within the Revolve option. Once selected, the Cross Section is revolved 90 degrees from the cutting plane. A Revolved view can be either a Full view or a Partial view.

View Visibilities

In addition to view types such as General, Projection, and Auxiliary, Drawing mode provides the option of controlling the visibility of selected portions of a view. The following View Visibility options are available:

Half Views

Often, it is not necessary to show an entire model in a view. A good example would be a symmetrical object. Half views remove the portion of a model on one side of a selected datum plane or planar surface. Figure 8–7 shows an example of a Half view. Half views can be used with General, Projection, and Auxiliary views only.

Partial Views

It is not necessary to document an entire view when only a small portion of a feature needs to be detailed. The Partial view option allows a selected small portion of a view to be created. The area to be revealed is enclosed in a sketched spline. Figure 8–7 shows an example of a Partial Section view. Unlike Detailed views, Partial views must follow normal lines of projection. Partial views can be used with General, Projection, and Auxiliary views only.

Broken Views

As with Partial and Half views, in many views it is not necessary to show the entire object. As shown in Figure 8–7, Broken views are used often with long, consistent cross sections. With a Broken view, multiple horizontal or vertical breaks may be used if necessary. The Drawing Setup file option *broken_view_offset* is used to set the offset distance between portions of a view. The Move View option can be used to move a portion of a Broken view.

Single-Surface Views

A single-surface view is a projection of an individual part surface. They may be used with any view type except Detailed and Revolved. When creating a single-surface view, only the selected surface is revealed. The Of Surface option is used to create a single-surface view.

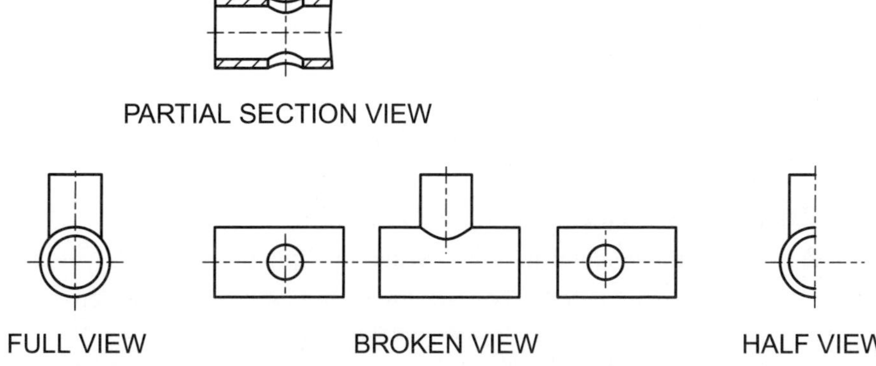

Figure 8–7 View types

Multiple Models

When starting a new drawing, Pro/ENGINEER provides the option of selecting a model (part or assembly) from which to create the drawing. Using the Views >> Add View option, views can be created of this model. Pro/ENGINEER also provides an option for adding additional models to a drawing. By having multiple models associated with a drawing, multiple models can be detailed on the same drawing sheet.

A model must be added to a drawing before a view of the model can be created. To add a new model to a drawing, use the Views >> Dwg Models >> Add Model option. When a model is added to a drawing, it becomes the current model. To set another model current, use the Set Model option. To delete a model from a drawing, use the Del Model option.

Creating a General View

Views are created with the Views >> Add View menu option. Options are available for selecting a View type, a view visibility, and a scale. Additionally, a section view of a model can be specified. By default, when no view exists in a drawing, General is the only view type available. After a General view has been added to the drawing, Projection becomes the default view type. Perform the following steps to add a full nonsectioned General view to a drawing.

Step 1: Select VIEWS >> ADD VIEW.

Step 2: Select GENERAL as the View Type to add.

Other options include Projection, Auxiliary, Detailed, and Revolved.

Step 3: Select FULL VIEW as the view visibility.

As an alternative to a standard Full view, a Half, Partial, or Broken view can be created for a part drawing. For assembly drawings, an additional option for creating an exploded view is available.

Step 4: Select NoXsec as the Cross Section type.

The following section options are available: Section, NoXsec, and Of Surface. The Section option is used to create a section view of a model, while the Of Surface option is used to create a single-surface view. For views without a section, use the NoXsec option.

Step 5: Select SCALE >> DONE.

This option is available for General views only. The Scale option allows for a user-specified scale for the view. The No Scale option fits the view based on the size of the model. The configuration file option *default_draw_scale* is used to set a default initial scale factor.

Step 6: On the work screen, select a location for the view.

After selecting the location, the view will be placed with the default view orientation.

Step 7: Enter a scale factor for the view.

Step 8: Orient the model using the Orientation dialog box.

The Orientation dialog box (see Figure 8–5) requires the selection of two perpendicular, planar surfaces or datum planes. The first reference is set by default to Front while the second reference is set to Top. Additional reference options available include Back, Bottom, Left, Right, Horizontal Axis, and Vertical Axis. A selected plane will face in the direction of the set reference option. As an example, if Top is selected as the reference, a selected plane will orient toward the Top of the work screen.

STEP 9: Add additional views if necessary.

SETTING A DISPLAY MODE

Views may be displayed as Wireframe, Hidden, or No Hidden. Views set with the Hidden option will print dashed hidden lines. By default, the display mode of a view is dependent upon the display mode set for the object. An object's display mode can be set on the toolbar or in the Environment dialog box. Individual display modes can be set for each view of a drawing using the Views >> Disp Mode option. The display of edges and assembly members can be manipulated with the Display Mode option also.

Tangent edges can be displayed in a variety of styles on a model. The Environment dialog box setting *Tangent Edges* is used to set a default style. The configuration file option *tangent_edge_display* can be used to set a default tangent display style. The following tangent edge styles are available within Pro/ENGINEER:

- **Solid** Tangent edges are displayed as a solid line.
- **No display** Tangent edges are not displayed.
- **Phantom** Tangent edges are displayed with a phantom line.
- **Centerline** Tangent edges are displayed with a centerline.
- **Dimmed** Tangent edges are displayed in the color set for dimmed entities, menus, and commands.
- **Tan default** Tangent edges are displayed as set in the Environment dialog box.

MODELING POINT Pro/ENGINEER has a default color for command lines or entities that are dimmed. As an example, if a command is inappropriate for a specific operation, it will appear dimmed on the menu. The configuration file option *system_dimmed_menu_color* is used to set the color of dimmed entities and commands.

To change the display mode of an individual view, perform the following steps:

STEP 1: Select VIEWS >> DISP MODE >> VIEW DISP.

STEP 2: Select individual views to modify, then select DONE SEL on the Get Select menu.

STEP 3: Select a Display Mode for the selected views.

Select either Wireframe, Hidden Line, No Hidden, or Default from the View Display menu.

STEP 4: Select a Tangent Edge display style.

From the View Display menu, select a Tangent edge display style.

STEP 5: Select DONE on the View Display menu.

DETAILED VIEWS

Detailed views are used on drawings to highlight portions of a component. A component might have a section with small and complicated features that would be hard to detail within the realm of the drawing's scale. Other sections of a drawing might require a

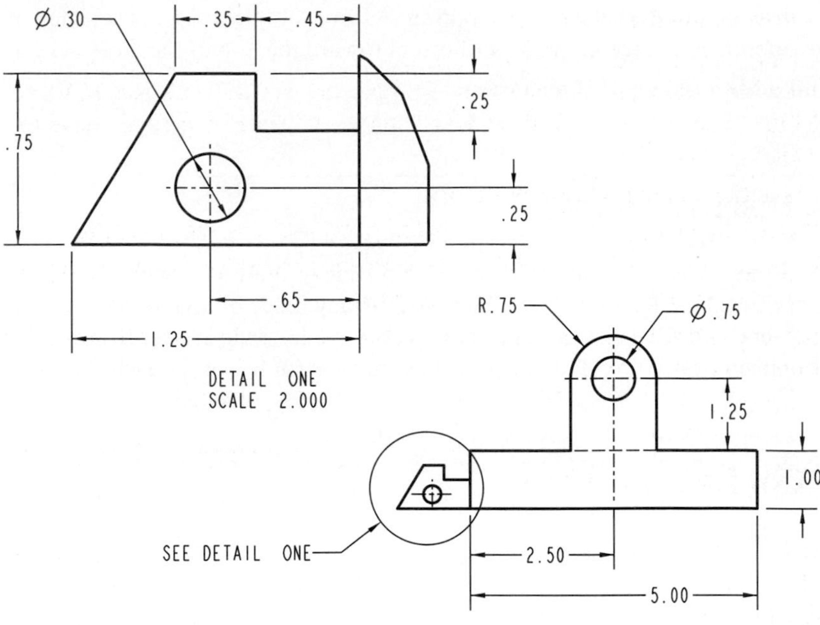

Figure 8–8 Detailed view

detailed view due to a specific importance factor. Figure 8–8 shows an example of a drawing with a detailed view.

Within Pro/ENGINEER, a Detailed view can be created any time after a general view has been placed. The scale of a detailed view is independent of its parent view. Additionally, Detailed views do not lie along normal lines of projection. By default, a detailed view reflects its parent view. The display mode of a detailed view (line and tangent display) is the same as its parent view. Any cross sections shown in a parent view will be reflected in its detailed views. This view relationship can be broken with the Views >> View Disp >> Det Indep option.

Perform the following steps to create a detailed view:

STEP 1: Select VIEWS >> ADD VIEW.

STEP 2: Select DETAILED >> DONE.

STEP 3: Select the location for the view.

On the work screen, select the location for the detailed view.

STEP 4: Enter a scale factor for the detailed view.

If the scale factor is not appropriate, it can be changed at a later time with the Edit >> Value option.

STEP 5: Select a reference point on the edge of an entity in the parent view.

Select the entity in the view that includes the portion of the drawing used to create the detailed view. You have to select the edge of an entity. Pro/ENGINEER uses this reference point to regenerate the detailed view.

STEP 6: Draw a Spline around the geometry to include in the detailed view.

Use the left mouse button to select points on the spline. Use the middle mouse button to close the spline. Objects included inside the spline boundary will be included in the detailed view.

STEP 7: Enter a Name for the detailed view.

As shown in Figure 8–8, Pro/ENGINEER will label the detailed view with the provided name.

STEP 8: Select a Boundary Type.

On the parent view, Pro/ENGINEER will enclose the detailed portion in a boundary. There are four boundary types available: Circle, Ellipse, H/V Ellipse, and Spline. For Ellipse, H/V Ellipse, and Spline types, you have to sketch the boundary.

STEP 9: Place the detailed view's name label.

For detailed views with a circle as the boundary type, the view will be complete at this point. For Ellipse, H/V Ellipse, and Spline boundary types, you have to select a leader attachment point.

STEP 10: Select the leader attachment point (Ellipse, H/V Ellipse, and Spline boundary types only).

SHOWING AND ERASING ITEMS

Drawing mode is used to create annotated presentations of models created in Part and Assembly modes. One of the uses of Drawing mode is to create orthographic views of a model. Orthographic drawings frequently consist of items such as dimensions, centerlines, notes, and geometric tolerances. Within Part mode, features are fully defined through the use of dimensions, constraints, and references. Dimensions used to define a part are considered parametric. When a feature is created, the dimensioning scheme defining the feature should match the intent for the design of the part. These parametric dimensions can be used in Drawing mode to annotate the drawing of the model.

Other items created in Part or Assembly mode can be used in Drawing mode. Drawing mode's Show/Erase option is used to show or erase an item created in Part or Assembly mode. The Erase selection within the Show/Erase option can be misleading. Since the Show/Erase option is used to manipulate the display of items created in Part or Assembly mode, these items cannot be deleted in Drawing mode. A more descriptive name for the Show/Erase option would be Show/No-Show. Items that are erased are actually hidden from display.

The manipulation of the display of items is accomplished through the Show/Erase dialog box (Figure 8–9). Listed are the items that can be shown or erased. To show or erase an item type, the button associated with the item type has to be selected. Multiple item types can be selected at one time. Several options are available for controlling the items that are displayed:

- **Show all** The Show All option will display all items of a particular type. As an example, if Show All is selected with the Dimension item type, all parametric dimensions defining a model will be displayed.
- **Feature** The Feature option will display items on a selected feature. The user must select the feature on the work screen.
- **Part** The Part option will display items on a selected part. The user must select the part on the work screen or on the model tree. This option is useful for assembly drawings where items on individual parts must be displayed.
- **View** The View option will display items on a selected view. The user must select the view on the work screen.
- **Feat & view** The Feature and View option will display items on a selected feature within the view from which the feature was selected.
- **Part & view** The Part and View option will display items on a selected part within the view from which the part was selected.
- **Erased** The Erased option will show only items that have previously been erased.
- **Never shown** The Never Shown option will show only items that have never been previously shown.

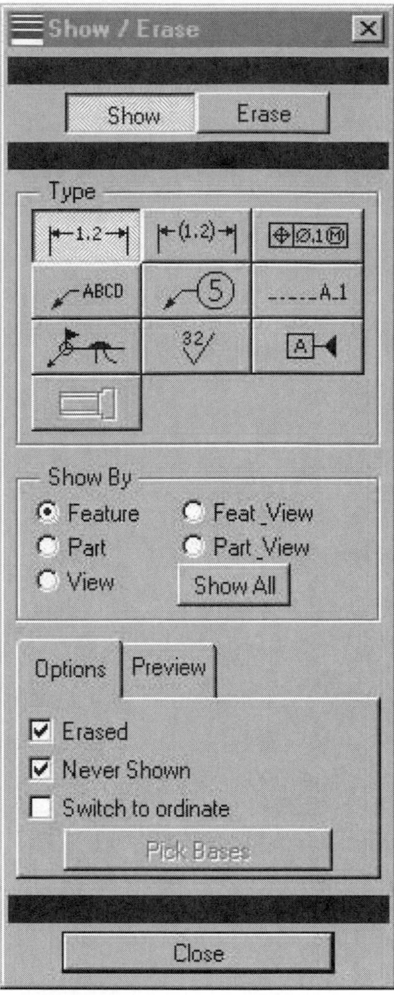

Figure 8-9 Show/erase dialog box

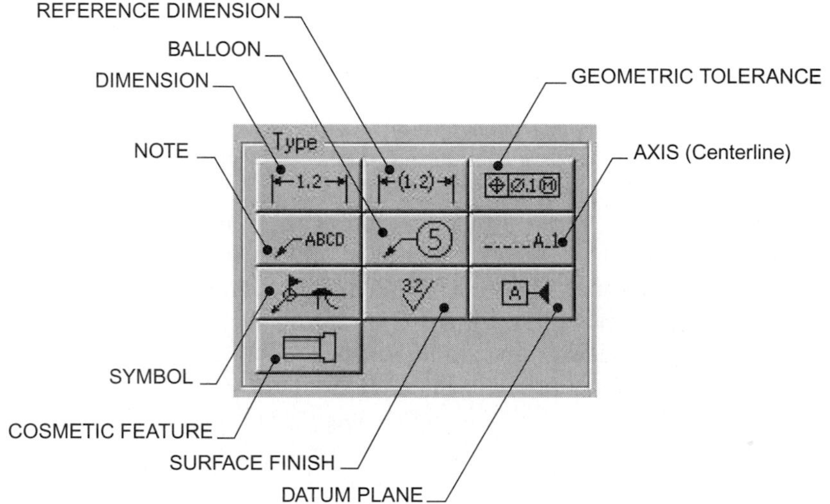

Figure 8-10 Item types

- **Preview** Located under the Preview tab is the Preview option. The Preview option is used to preview items before they are displayed on the work screen. Options are available for accepting or not accepting an item.

Showing All Item Types

The Show All option is used to show all items of a selected type. As an example, you can show all Geometric Tolerances, or you can show all Centerlines. Perform the following steps to Show All of a selected item.

Step 1: On the menu bar, select VIEW >> SHOW AND ERASE.

Step 2: Select the SHOW option on the Show/Erase dialog box.

The Show option will show items of a select type. The Show and Erase options cannot be selected at the same time.

Step 3: Select an Item Type to show (Figure 8–10).

Examples of item types to show include Dimensions, Centerlines (Axis option), Datums, and Geometric Tolerances.

STEP 4: Select the SHOW ALL option.

The Show All option will display all items of a selected type. This is a useful option for displaying item types that will not overwhelm a drawing, such as centerlines. When using this option for dimensions, a drawing can become cluttered.

STEP 5: Select OKAY to confirm the use of Show All.

After selecting OKAY, the items will be displayed on the work screen.

STEP 6: Select Preview option (if required), then Close the dialog box.

SHOWING/ERASING LIMITED ITEM TYPES

While the Show All option is used to show all items of a selected type, options are available for you to show or erase limited numbers of an item type. Perform the following steps to show or erase a limited number of items.

STEP 1: On the menu bar, select VIEW >> SHOW AND ERASE.

STEP 2: Select either the SHOW or the ERASE option on the Show/Erase dialog box.

The Show option will show items of a selected type while the Erase option will hide items of a selected type. The Show and Erase options cannot be selected at the same time.

STEP 3: Select an Item Type to show (Figure 8–10).

Examples of item types to show include Dimensions, Axes, Datum Planes, and Geometric Tolerances.

STEP 4: Select an option for showing items.

You can select either Feature, Part, View, Feat & View, or Part & View. Feature will show items on a selected feature, the Part option will show items on a selected part, and View will show items in a selected view.

STEP 5: Select DONE SEL on the Get Select menu or select the middle mouse button.

POP-UP MENU

Drawing mode provides a Pop-Up menu for the modification of drawing mode entities to include dimension text and dimension properties. Using the Pop-Up menu, an item can be selected for modification during the drawing process. To modify an item with the Pop-Up menu, select the item with the left mouse button then with the right mouse button reselect the item. The reselection of the item with the right mouse button will reveal the pop-up menu:

DIMENSIONS

Dimensions can be modified in a multiple of ways with the Pop-Up menu. The following dimension modification options are available:

- Dimensions may be moved.
- Dimensions may be switched to another view.
- Dimensions may be jogged.
- Arrows may be flipped.
- Witness lines may be shown or erased.
- Arrow styles may be changed.

- Dimension nominal values may be changed.
- Dimensions may be erased or unerased.
- Dimension values may be changed.

Geometric tolerances

Geometric tolerances created can be modified or manipulated using the Pop-Up menu. The following options are available:

- Geometric tolerances may be moved on the work screen.
- Geometric tolerance attachments can be modified.
- Geometric tolerances can be switched to another view.
- The leader type of a Geometric tolerance can be changed.
- Geometric tolerances can be redefined.
- Geometric tolerances can be erased or unerased.
- Geometric tolerances can be deleted.

Notes

Notes can be modified and/or manipulated in the following ways using the Pop-Up menu:

- Notes may be moved on the work screen.
- Notes may be switched to another view.
- Notes text styles can be modified.
- Text can be modified.
- Notes can be erased or unerased.
- Notes can be deleted.

Views

Views can be modified with the Pop-Up menu in the following ways:

- Views may be moved on the work screen.
- The scale of a view can be changed.
- A view's cross section can be replaced.
- The view text can be modified.

Dimensioning and Tolerancing

Pro/ENGINEER's Drawing mode has the capability to display dimensions created in Part or Assembly mode and to create dimensions within Drawing mode itself. Dimensions created in Part mode can be displayed using the Show/Erase option. These dimensions are associative and can be modified. Due to the associativity between modes, any parametric dimension modified in Drawing mode is also modified in other modes, such as Part and Assembly modes. While parametric dimensions can be hidden with the Show/Erase option, they cannot be deleted.

Driven Dimensions can be created within Drawing mode using the Insert >> Dimension option. Dimensions created in this fashion are not associative and are not modifiable. The model geometry drives the value of each dimension. This dimension option provides the following suboptions:

- **On entity** Attaches the dimension's witness line to the point where the entity was picked.

- **Midpoint** Attaches the dimension's witness line to the midpoint of a selected entity.
- **Intersect** Attaches the dimension's witness line to the closest intersection point of two entities.
- **Center** Attaches the dimension's witness line to the center of a circle, arc, or ellipse.
- **Tangent** Attaches the dimension's witness line tangent to a circle, arc, or ellipse.
- **Horizontal** Creates a dimension that is horizontal according to the drawing environment's orientation.
- **Vertical** Creates a dimension that is vertical according to the drawing environment's orientation.
- **Slanted** Creates a dimension that measures the distance between two points.
- **Parallel** Creates a dimension between two points that is parallel to a selected entity.
- **Normal** Creates a dimension between two points that is normal to a selected entity.

Manipulating Dimensions

Drawing mode provides a variety of options for manipulating dimensions. These tools are used to create a clear and readable engineering drawing.

 Move

Drawing mode provides two tools for moving drawing entities. For the manipulation of dimensions and views, the Move icon is available. An alternative to the Move icon is the dragging of drawing entities after their selection. When a dimension is selected, grip points are provided that allow an entire dimension to be repositioned including the dimension's text, dimension line, and witness lines. Additional grip points are provided that allow for the moving of only the dimension's number value.

 Switch View

The Switch View icon is used to switch the view in which a dimension is located (Figure 8–11). An alternative to the Switch View icon is the Switch View option available on the pop-up menu. The Switch View option is used often after the Show/Erase >> Show All command. When showing all parametric dimensions in a drawing, Pro/ENGINEER will randomly place dimensions in a view. The placement view may not be the appropriate view according to drafting dimensioning standards.

Flip Arrow

The Flip Arrow option is used to change the direction that an arrowhead points. This option can be used with linear and radial dimensions (Figure 8–11). The Flip Arrow option is accessible through the pop-up menu by preselecting the dimension text with the left mouse button, followed by reselecting the dimension text with the right mouse button.

Make Jog

Dimensions can become too confined when annotating small features. On a normal dimension, the distance between two witness lines is the same as the nominal size of the dimension. The Make Jog option, available by selecting Insert >> Job, can be used to create a jog in a witness line (Figure 8–12).

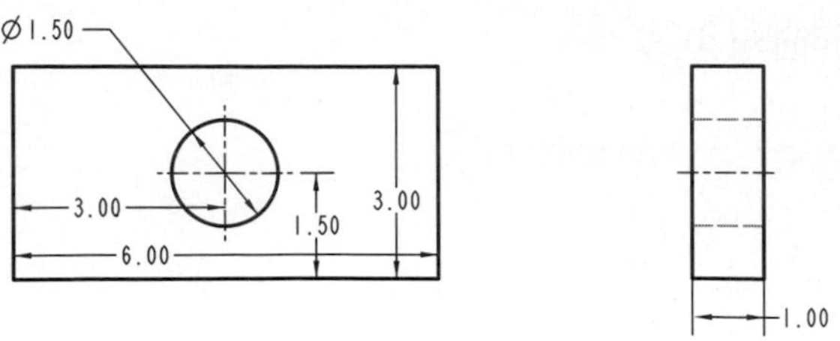

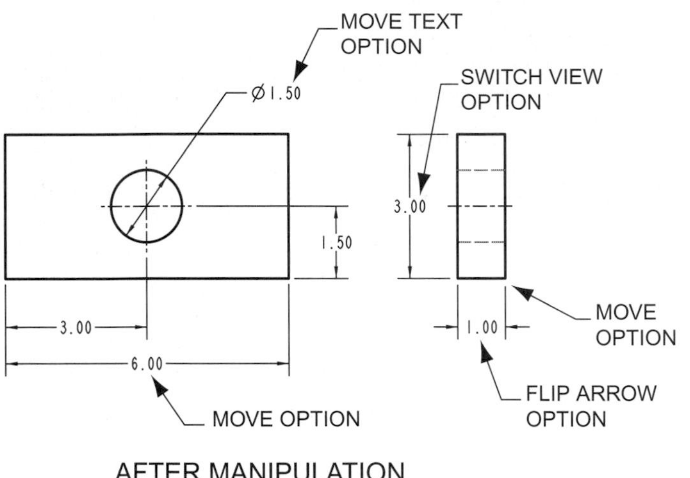

Figure 8–11 Manipulation tools

Figure 8–12 Make Jog option

Dimension Tolerances and Modification

Pro/ENGINEER allows for the creation of dimensions that adhere to ANSI or ISO tolerance standards. See Chapter 2 for more information on setting a tolerance standard. Within Drawing mode, tolerances can be displayed in a variety of formats. Available formats include Limits, PlusMinus, and PlusMinusSymmetric. In Drawing mode, before a dimension can be displayed as a tolerance, the Drawing Setup file option *Tol_Display* has to be set to Yes. Use the Advanced >> Draw Setup option and set the *Tol_Display* option to modify the tolerance display for an individual drawing.

When utilizing ANSI as the tolerance standard, tolerance values and formats can be set with the Properties option on Drawings mode's Pop-Up menu, which displays the Modify Dimension dialog box (Figure 8–13). The Modify Dimension dialog box is used to set the following dimension options:

- **Tolerance mode** Nominal, PlusMinus, and PlusMinusSymmetric can be selected.
- **Tolerance values** Tolerance values associated with a particular format can be entered.
- **Decimal places** The number of decimal places of a dimension can be selected.
- **Basic dimension** When utilizing geometric tolerances, a dimension can be set as basic.
- **Dimension Text** The Dimension Text tab (Figure 8–13) is used to add text and symbols around a dimension value. When adding text, the dimension value cannot be deleted. The symbol pallet is available for the selection of symbols (Fig. 8–14). Figure 8–15 shows an example of how a typical dimension text note can be changed into a counterbored hole note.

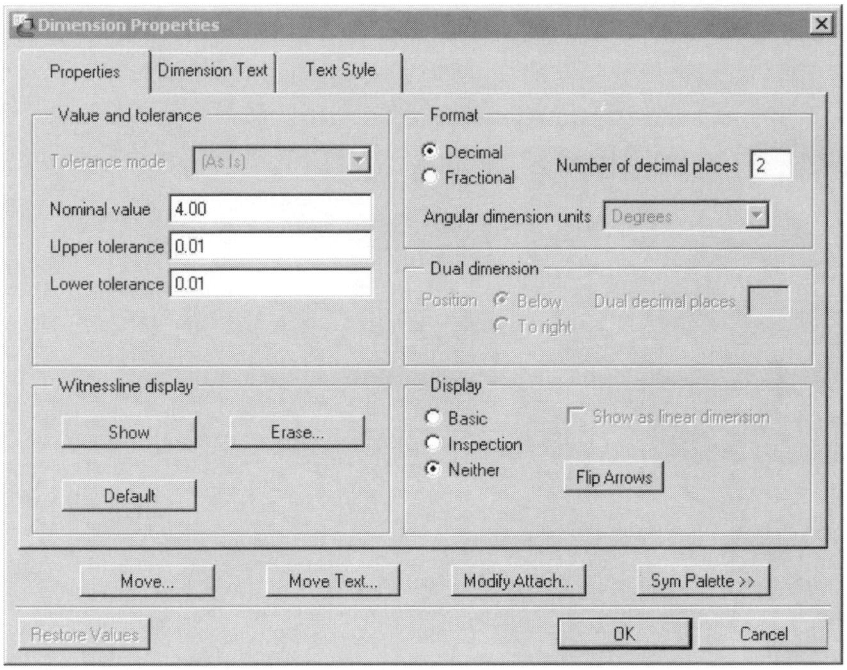

Dimension Properties
Properties
Dimension Text
Text Style
Value and tolerance
Tolerance mode
(As Is)
Nominal value
4.00
Upper tolerance
0.01
Lower tolerance
0.01
Format
Decimal
Fractional
Number of decimal places
2
Angular dimension units
Degrees
Dual dimension
Position
Below
To right
Dual decimal places
Witnessline display
Show
Erase...
Default
Display
Basic
Inspection
Neither
Show as linear dimension
Flip Arrows
Move...
Move Text...
Modify Attach...
Sym Palette >>
Restore Values
OK
Cancel

Figure 8-13 Modify Dimension dialog box

Symbol Palette
Close

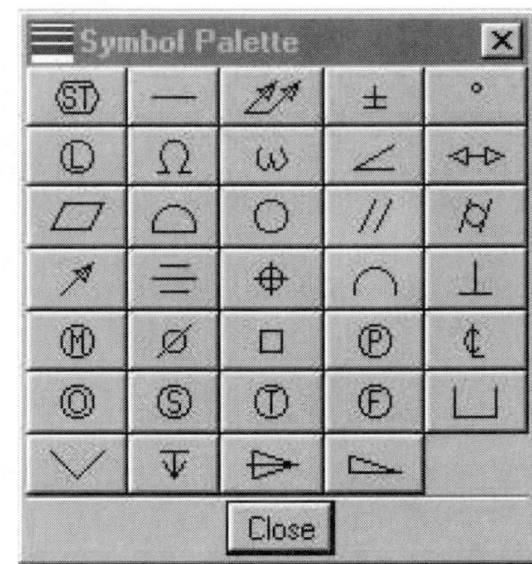

Figure 8-14 Symbol pallet

Ø2.25
Ø1.50
Ø1.500, ⌴Ø2.25, ↧.25
BEFORE MODIFICATION
AFTER MODIFICATION
Properties
Dimension Text
Text Style
Ø&d4, ⌴{0:Ø}{1:@D}, ↧&d6
Depth Added to Note
Original Parametric Hole Dimension
Hole Dimension Added to Note
Name
d3
Prefix
Postfix
MODIFY DIMENSION DIALOG BOX (Dim Text Tab)

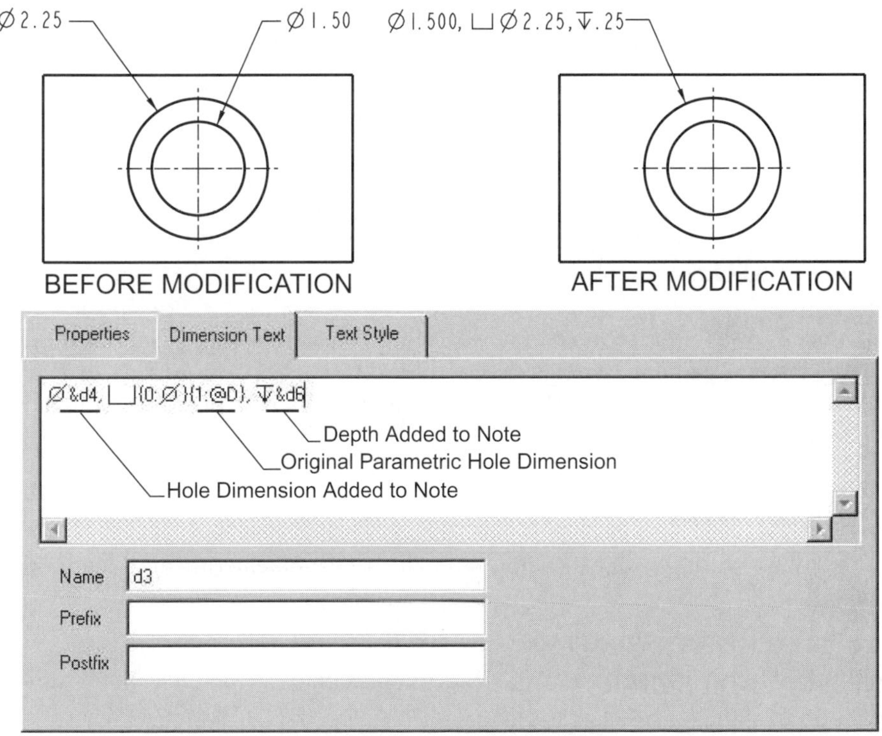

Figure 8-15 Dimension text modification

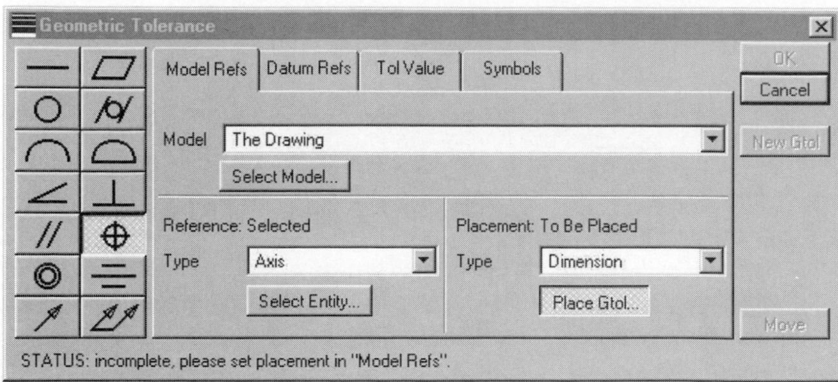

Figure 8-16 Geometric tolerance dialog box

MODELING POINT Notice in Figure 8–15 the 1.50 diameter hole displayed in the drawing before modification. This value is a parametric dimension used to define the size of the hole. This dimension can be modified, which after regeneration will modify the part's hole diameter. By entering the hole's assigned dimension symbol in the Dim Text tab (e.g., *&d11*) instead of its dimension value, the "after modification" hole note can be used to modify its corresponding parametric dimension.

Geometric Tolerances

Geometric tolerances are used to control geometric form, orientation, and location. An example of a geometric tolerance would be specifying that a planar surface is flat to within a tolerance value. The Geometric Tolerance dialog box is used to create a geometric tolerance characteristic (Figure 8–16). This dialog box can be accessed through the Insert >> Geometric Tolerance menu option. See Chapter 2 for more information on establishing a geometric tolerance. Before a Datum can be utilized in a drawing, it first must be set through the Edit >> Properties option on Pro/ENGINEER's menu bar.

Creating Notes

Drawing mode allows for the creation of notes. Notes may be stand-alone, or they may be attached to a leader. Additionally, notes can be entered from the keyboard, or they can be input from a text file. Pro/ENGINEER provides the Symbol Palette (Fig. 8–16) for adding symbols to a note.

Note without Leader

Perform the following steps to create a note without a leader by entering the note through the keyboard.

Step 1: Select INSERT >> NOTE.

Step 2: Select NO LEADER from the Note Types menu.

Step 3: Select a Format/Placement type.

Available Format/Placement types include Horizontal, Vertical, Angled, Left-Justify, Right-Justify, Center-Justify, and Related to Dimension Text.

Step 4: Select MAKE NOTE.

STEP 5: On the work screen, select the location for the note.

Using the mouse cursor, select the location for the text. After creating the text, the text can be repositioned with the Move option.

STEP 6: Enter note in Pro/ENGINEER's textbox.

When creating a note through the keyboard, Pro/ENGINEER will provide a textbox for the creation of the note. Select Enter on the keyboard to end the creation of the note. For the creation of symbols within the note, Pro/ENGINEER provides a Symbol Palette.

NOTE WITH A STANDARD LEADER

Perform the following steps to create a note with a standard leader by entering the note through the keyboard.

STEP 1: Select INSERT >> NOTE.

STEP 2: Select LEADER from the Note Types menu.

STEP 3: Select a Leader Type.

Pro/ENGINEER provides the option of creating either a Standard Leader, a Normal Leader, or a Tangent Leader. The Normal Leader (Normal Ldr) option will create the leader perpendicular to the selected entity, while Tangent Leader (Tangent Ldr) will create the leader tangent to the selected arc, circle, or ellipse. Standard is the default and allows for multiple leader attachment points.

STEP 4: Select a Format/Placement type.

Available Format/Placement types include Horizontal, Vertical, Angled, Left-Justify, Right-Justify, Center-Justify, and Related to Dimension Text.

STEP 5: Select the MAKE NOTE option.

STEP 6: Select an Attachment Type (for Standard Leaders only), then select an appropriate attachment point.

When placing a standard leader, Pro/ENGINEER provides the following leader attachment types:

- **On entity** Attaches a leader to a model or to a draft entity.
- **On surface** Attaches a leader to a surface. The surface may be a model surface, datum plane, or cosmetic thread.
- **Free point** Attaches a leader anywhere on the drawing. With the Free Point option, the leader does not have to be attached to an entity or surface.
- **Midpoint** Attaches a leader at the midpoint of a model edge or draft entity.
- **Intersect** Attaches a leader at the intersection of two entities.

STEP 7: Select DONE SEL to end attachment point locations.

STEP 8: Enter note in Pro/ENGINEER's textbox.

When creating a note through the keyboard, Pro/ENGINEER will provide a textbox for the creation of the note. Select Enter on the keyboard to end the creation of the note. For the creation input of symbols, Pro/ENGINEER provides the Symbol Palette.

CREATING DRAWING TABLES

A drawing table is similar to a table that might be created in a popular word processing application (Figure 8–17). A drawing table is composed of cells arranged in columns and

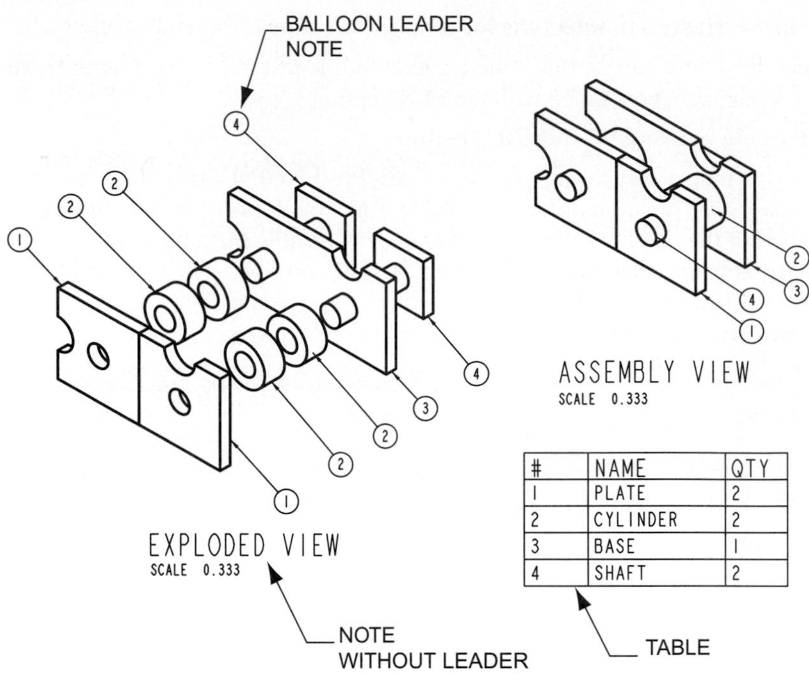

Figure 8–17 Notes, leaders, and tables

rows. Standard text can be entered into a text cell. Perform the following steps to create a table:

STEP 1: Select TABLE >> CREATE.

STEP 2: Select a Table Direction Creation method.

The following creation methods are available:

- **Descending** Creates the table from the top down.
- **Ascending** Creates the table from the bottom up.
- **Rightward** Creates the table from the left to right.
- **Leftward** Creates the table from the right to left.

Select either Descending or Ascending, then select either Rightward or Leftward.

STEP 3: Select BY NUM CHARS as the Cell Creation Method.

The following cell creation methods are available:

- **By Num Chars** Creates a table by graphically picking the number of characters to include in each cell (Figure 8–18).
- **By length** Creates a table by specifying the size of each in drawing units.

STEP 4: On the work screen, select the location for the cell.

This option will locate the table dependent upon the table direction creation method selected previously. As an example, when creating a table with the

Figure 8–18 Creating a table by the number of characters

Descending and Leftward options, you will select the upper-right corner of where the table is to be located. When creating a table with the Ascending and Rightward options, you will select the lower-left corner of where the table is to be located.

STEP 5: On the work screen, mark off the number of characters to include in each column of the table (Figure 8–18).

STEP 6: Select DONE to finish the number of columns.

STEP 7: On the work screen, mark off the number of characters to include in each row of the table.

> **MODELING POINT** Tables can be modified with the Table >> Modify Table option. Options are available for adding rows and columns, for merging cells, and for modifying text justification. Other useful manipulation tools include Rotating a table and resizing rows and columns.

STEP 8: Select DONE to finish the number of rows.

Selecting DONE will create the table.

TWO-DIMENSIONAL DRAFTING

While Pro/ENGINEER is considered primarily a three-dimensional design application with Drawing mode used to create annotated detailed drawings of 3D models, Drawing mode can be used strictly to create 2D drawings. Any two-dimensional geometry created in Drawing mode is nonparametric and cannot be associated with any other Pro/ENGINEER mode.

Pro/ENGINEER's drafting capabilities are similar to two-dimensional modeling tools found in many popular mid-range computer-aided drafting applications. The following options are available:

- **Draft geometry** Used to create lines, circles, arcs, splines, ellipses, points, and chamfers.
- **Construction geometry** Lines and circles used to create draft geometry. They are displayed in phantom font.
- **Draft dimensions** Used to dimension two-dimensional geometry.
- **Draft cross sections** Used to create section lines.

DRAFT GEOMETRY

Pro/ENGINEER provides a variety of tools for creating two-dimensional draft geometry. By default, Drawing mode allows for the creation of individual entities only. As an example, when sketching a line between two points, only one line will be created between the two points. To continue the line creation process, the start and ends of the next line will have to be selected. To connect entities during the sketching process, Drawing mode provides the Chain option. On the Drawing toolbar, selecting the Chain icon will allow drafted entities to be connected. Using this option, the end of one entity becomes the start of the next. To terminate a chain, select End Chain.

The Pop-up menu (right mouse button) provides several options for creating precise draft entities. As an example, for lines, polar or relative coordinates can be entered. For a circle entity, the Pop-up menu provides an option for entering the circle's radius. The

Pop-up menu provides the Select References option for constraining an entity to a selected existing entity. One example of this would be constraining a new line entity perpendicular to an existing line. Another example would be constraining a new circle tangent to an existing line. An existing entity has to be specified as a reference before a new entity can be constrained to it.

The following options are available for the creation of draft entities. Drafting in Drawing mode is similar to sketching in a sketcher environment. Refer to Chapter 3 for specific information on creating sketch entities. Each of the following options are available on either the Drawing toolbar or the Sketch menu.

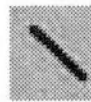

LINES

Lines can be created through the Line icon or through the Sketch line option. For a chain of lines, the Chain icon or the Sketch >> Chain option is available. The Pop-up menu provided by the right mouse button provides options for creating a line at a specific angle and with either polar coordinates or relative coordinates.

CIRCLES

Circles can be created through the Circle icon or the Sketch >> Circle option. The Pop-up menu provides an option for specifying a precise radius for the circle.

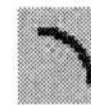

ARCS

Arcs can be created through the use of one of two Arc icons or through the Sketch >> Arc suboptions. Two types of arcs are available: a three point arc and a center/endpoints arc.

SPLINES

The Spline option is available on the Drawing toolbar or the Sketch menu. To create a spline, locate the spline's start point first, intermediate points along the spline second, then the spline's endpoint last. The Pop-up menu provides an option for specifying a spline's initial start angle.

ELLIPSES

Pro/ENGINEER provides two options for creating ellipses. The first option creates a spline through the selection of the endpoints of the spline's axes. The second option creates a spline through the definition of the spline's center vertex, followed by the spline's major and minor axes definitions.

POINTS

The Point option is used to create a point on the work screen.

CHAMFERS

The Chamfer option is used to create a line that intersects and trims two nonparallel lines. Chamfer option available in Drawing mode is similar to options available under the Chamfer command in Part mode.

CONSTRUCTION GEOMETRY

Construction geometry is used within Pro/ENGINEER for the construction of two-dimensional draft geometry. Options are provided for creating construction lines and construction circles. Figure 8–19 shows an example of a two-dimensional drawing after the creation of construction geometry and after the creation of draft geometry. Construction geometry appears as phantom line font on the work screen, but will not print.

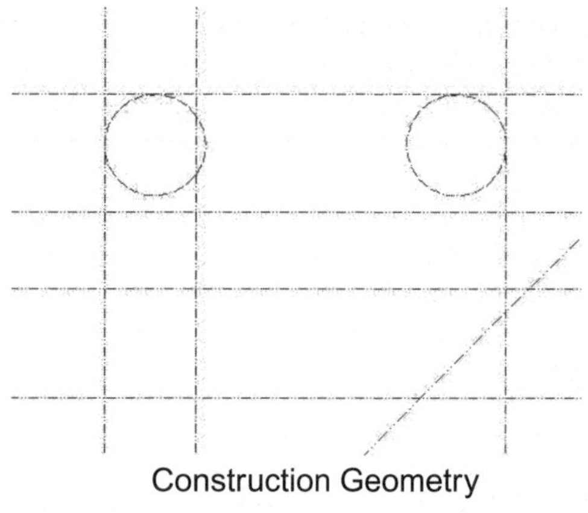

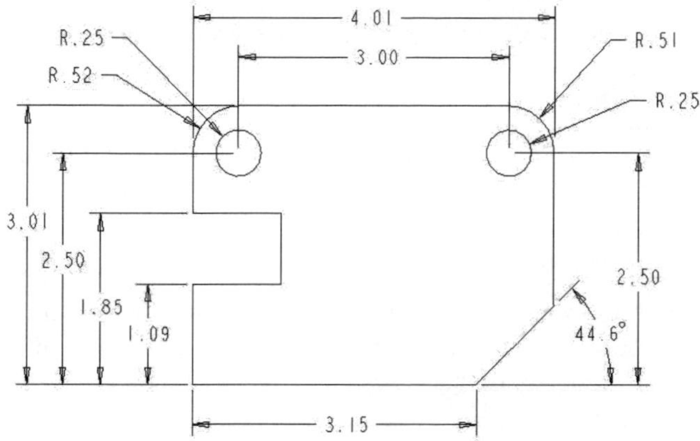

Figure 8–19 Construction geometry

LINE STYLES AND FONTS

When sketching draft geometry, the default line style is a continuous line. Within Pro/ENGINEER, a continuous object line does not have a line style but has a line font set as *Solidfont*. There is a slight difference between line styles and line fonts. A line style is a defined group consisting of a line font, color, and weight. As an example, a Phantom line style has a Phantomfont line font with a color of blue. The line styles available for draft geometry include Hidden, Geometry, Leader, Cut Plane, Phantom, Centerline, and None (Figure 8–20). Use the Format >> Default-Line-Style option to set a default line style for a specific line type (e.g., geometry, phantom, hidden, leader).

How a line style appears on the work screen may be different from how it appears when printed. As an example, a line defined as a hidden line style will appear on the work screen as a blue continuous line; it will print as a dashed hidden line, through. This is because the color of a line, in the absence of a set line font, will dictate the printed line font and the printed line weight. Figure 8–21 shows the color dialog box with available colors that can be assigned to a line. Each color within the dialog box corresponds to a particular line font, line weight, and pen number. As an example, the Geometry color (pen 1) will print a thick, continuous line. The Leader color (pen 2) will print a continuous, medium thick line, while the Hidden color (pen 3) will print a dashed, thin line.

A line font is the geometric definition of a line. The available fonts include Solidfont, Dotfont, Ctrlfont, Phantomfont, Dashfont, Ctrlfont_S_L, Ctrlfont_L_L, Ctrlfont_S_S, Dashfont_S_S, Phantomfont_S_S, and Ctrlfont_MID_L.

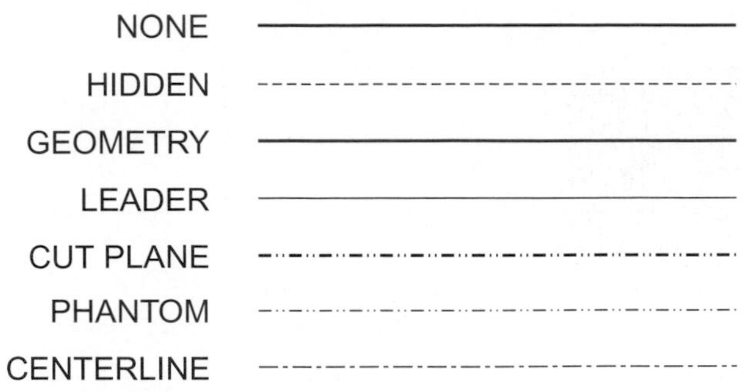

NONE
HIDDEN
GEOMETRY
LEADER
CUT PLANE
PHANTOM
CENTERLINE

Figure 8–20 Pro/ENGINEER linestyles

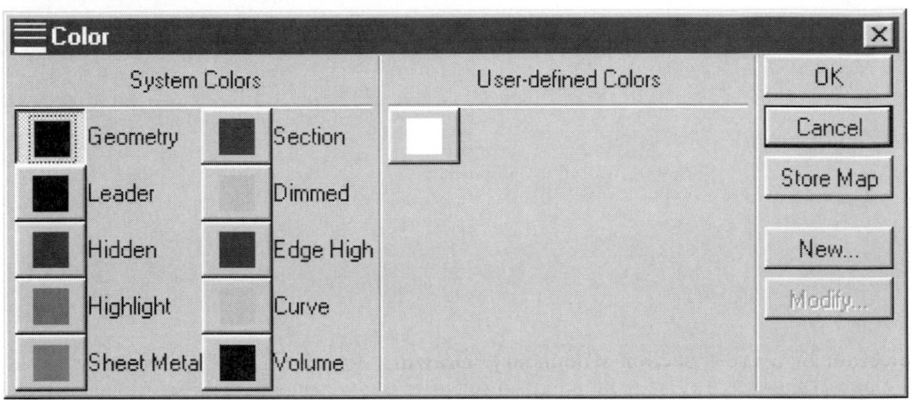

Figure 8–21 Linestyle color dialog box

MODELING POINT The weight of a printed line is governed by its pen number. Correspondingly, the pen number of a line is governed by its color. As an example, the Hidden color (blue) is assigned to pen one.

The line weight produced from a pen number can be modified with the configuration file option *pen#_line_weight* (where # is equal to the pen number). Each option can be set equal to a value ranging from 1–16. Each increment value equals a value of 0.005 inches. As an example, if *pen1_line_weight* is set to 2, the line weight will be equal to 0.010 inches (2 × 0.005 = 0.010).

Draft Dimensions

Draft dimensions (Insert >> Dimension) are used to annotate draft geometry created in drawing mode. The New References dimension option is similar to the standard dimensioning option found in a sketcher environment. See Chapter 3 for more information on creating draft dimensions.

Draft Cross Sections

Drawing mode's Edit >> Fill >> Hatched option is used to create section lining or hatch within an enclosed area. The Fill >> Solid can be used also to completely fill an area. Figure 8–22 shows an example of several types of cross sections that can be created. As shown in the first example, cross sections can include islands of unfilled areas.

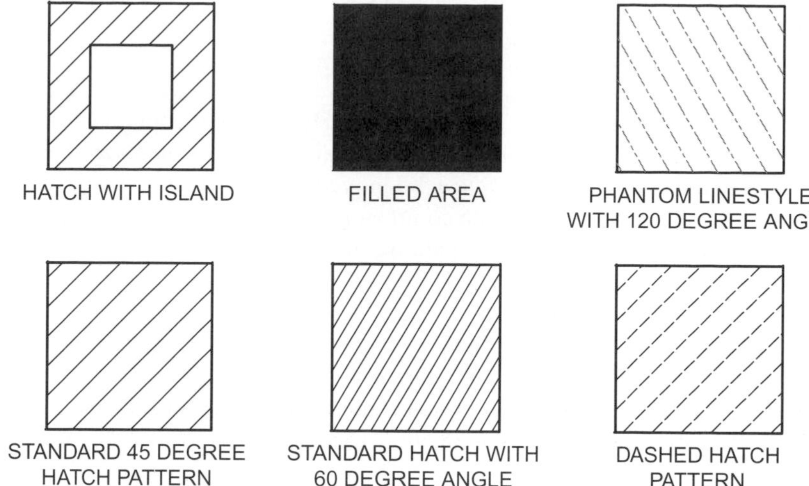

Figure 8–22 Hatch examples

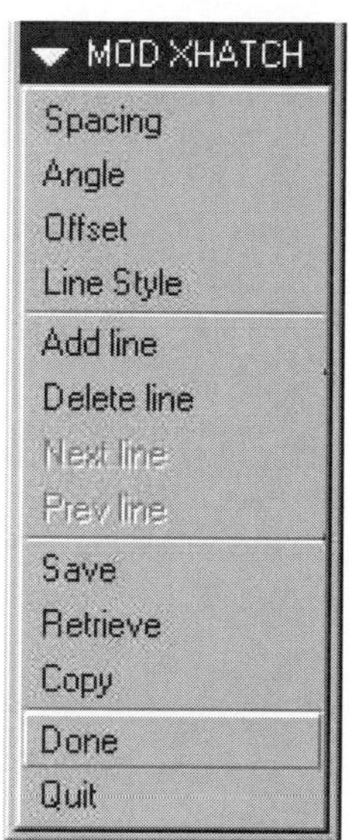

Figure 8–23 Modify Cross Hatch Menu

Draft cross sections are created with either the Hatched or Solid option by the preselecting of drafting entities. An area to incorporate section lining must be completely closed. The Solid option will fill the selected area with the color associated with the current line style. The Hatched option requires the entering of a cross section name. Upon completing the selection of a cross section's boundary, Drawing mode displays the Modify Cross Hatch menu (Figure 8–23). This menu allows for the modification of hatch spacing, angle, offset, and line style. This menu can also be accessed through the Edit >> Properties option.

Manipulating Draft Geometry

Most new users of two-dimensional computer-aided drafting applications use draft geometry techniques such as the line, arc, and circle options to create drawings. While it is obvious that these are central commands for the creation of draft geometry, other powerful options are available that allow for the manipulation of existing draft entities. These options can be used in combination with basic geometry creation techniques to construct a drawing. Within Drawing mode, many of these tools are identical or similar to options available within a typical Pro/ENGINEER sketching environment. Within drawing mode, the following manipulation options are available:

Copy

The Copy option can be found under the Tools menu. It is used to create identical instances of selected entities (Figure 8–24). Suboptions are available for either translating or rotating the copied entities. The Translate option has commands for

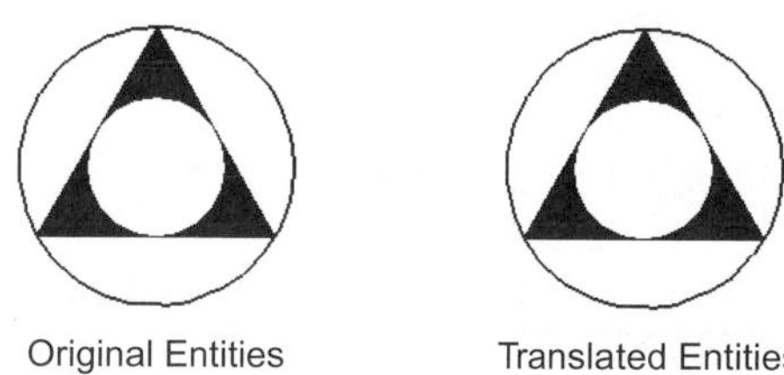

Figure 8–24 Horizontal translation

copying along the horizontal or vertical axis. An additional option is available for translating at an angle (measured counterclockwise from the three o'clock position) and distance from the original entities.

Group

The Group option can be found under the Tools menu. It is used to combine entities. One advantage of a group is that its entities can be manipulated together. As an example, all the entities in a group can be moved at once, or all can be deleted together. Use the Group >> Create command to build a group and use the Group >> Explode command to break a group into its individual entities.

Intersect

The Intersect option can be found under the Tools menu. It is used to break entities at their selected intersection point. As shown in Figure 8–25, selecting the intersection of two lines will create four entities. This option is identical to the Intersect option found within Part mode's sketching environment.

Mirror

The Mirror option can be found under the Tools menu. It is used to create an identical reflected image of selected entities (Figure 8–26). The user is required to select the entities to be mirrored and to select a draft line to mirror the draft geometry about.

Offset

The Offset option can be found under the Tools menu. It is used to create new draft geometry by offsetting from a selected entity at a user-defined distance (Figure 8–27). Single entities or chained entities may be offset. The Single Entity (Single Ent) option will offset only one entity at a time. Suboptions are available for tapering the offset and trimming the offset. The Entity Chain (Ent Chain) option allows a connected chain of entities to be offset at once.

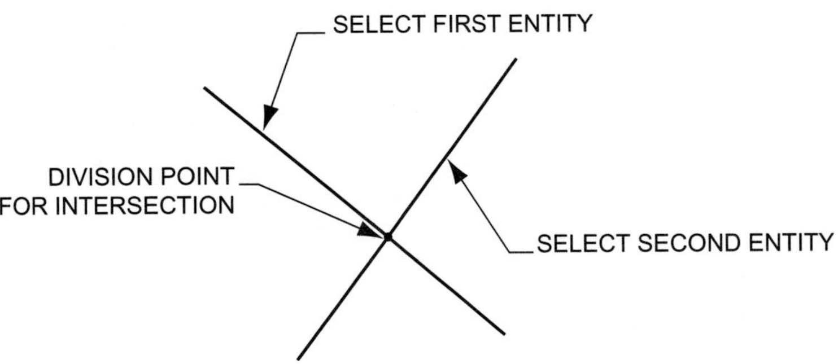

Figure 8–25 Intersection tool

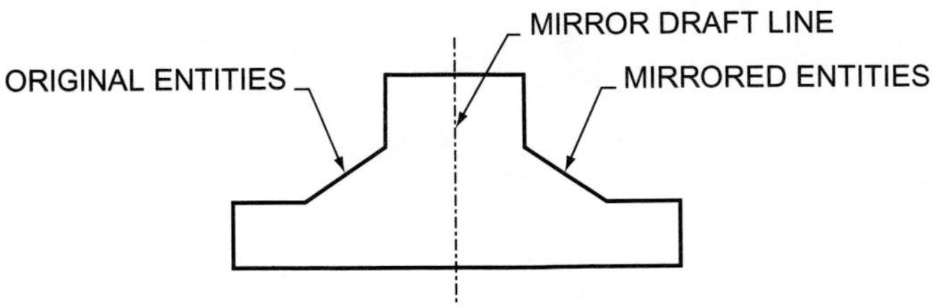

Figure 8–26 Mirrored entities

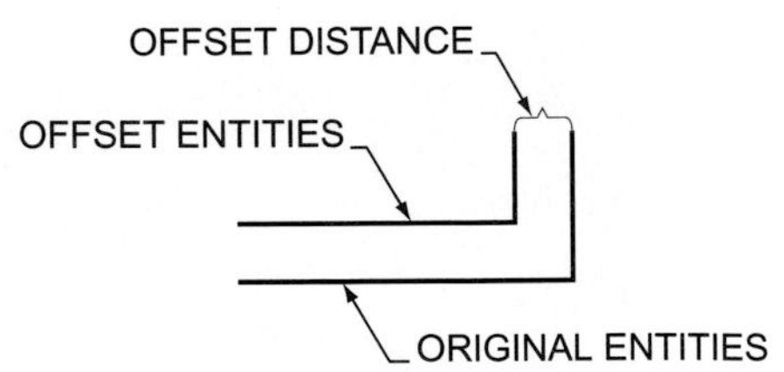

Figure 8–27 Offset tool

ROTATE

The Rotate option can be found under the Tools menu. It is used to rotate one or more draft entities.

TRANSLATE

The Translate option can be found under the Tools menu. The translate option functions similarly to the Copy >> Translate option, only no copies are created. Multiple entities can be translated at together.

TRIM

The Trim option can be found under the Tools menu. It is used to break and delete two entities at their intersection point. This option is identical to the Intersect option found in Part mode's sketcher environment.

SUMMARY

Drawing mode is used to create drawings of Pro/ENGINEER parts or assemblies. Multiple views of a model can be displayed to include projection views, section views, and partial views. Dimensions and other items such as cosmetic threads and geometric tolerance notes can be displayed in a drawing using the Show/Erase option. Drawing mode has the capability for the creation of nonassociative two-dimensional drawings. Notes, leaders, and nonparametric dimensions can be added to a drawing.

TUTORIAL

DRAWING TUTORIAL 1

This tutorial will demonstrate the creation of the drawing shown in Figure 8–28. The part used in this tutorial is shown in Figure 8–29. The start of this tutorial will require you to model this part. As shown in Figure 8–28, four different view types will be created: General, Projection, Auxiliary, and Detailed. This tutorial will cover

- Starting a drawing.
- Adding a drawing format.
- Creating a general view.
- Creating projection views.
- Creating a detail view.
- Creating notes.
- Modifying the drawing setup file.

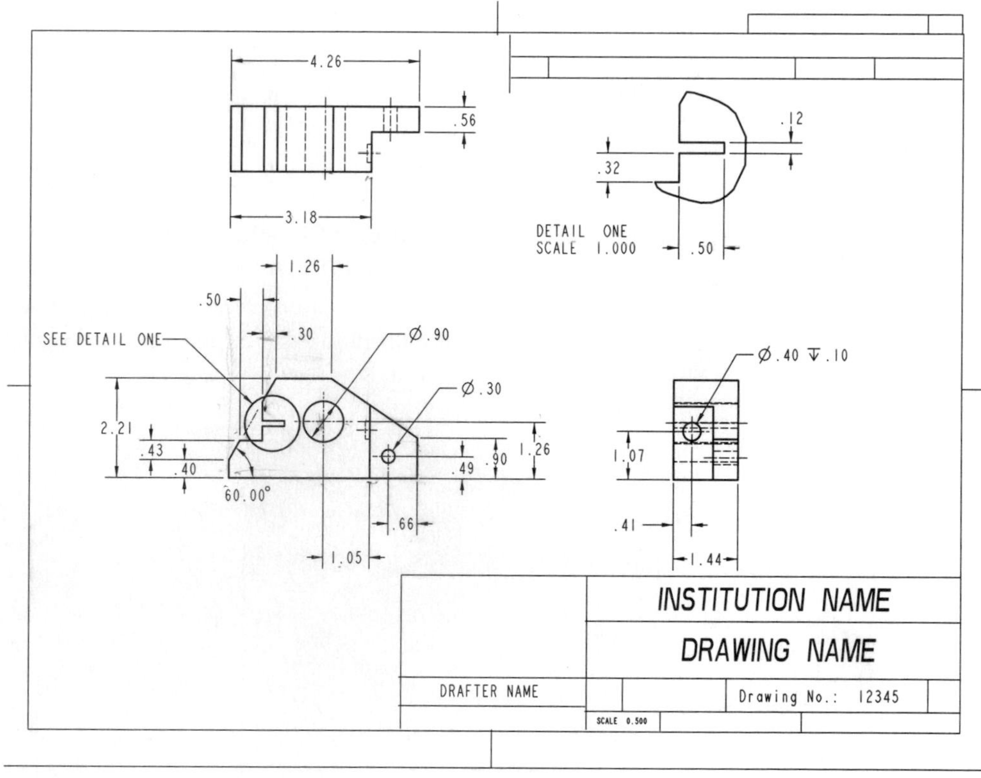

Figure 8–28 Multiview drawing

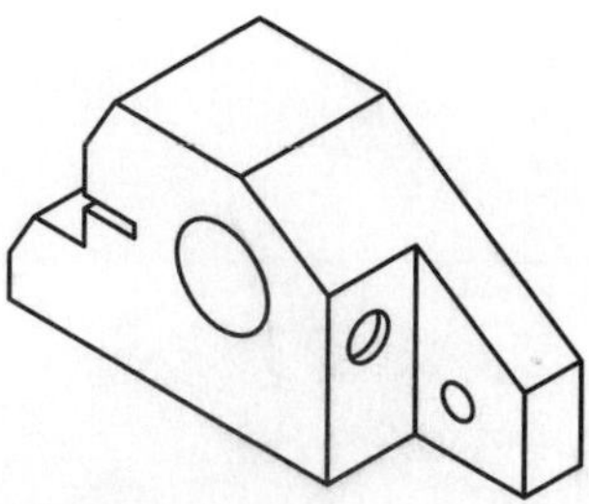

Figure 8–29 Model for drawing

Creating the Part

Model the part shown in Figures 8–28 and 8–29, naming it *view1*. The dimensioning scheme shown in Figure 8–28 matches the design intent for the part. When modeling this part, make sure that these dimensions are incorporated into your design.

Starting a Drawing

This section of the tutorial will create the object file for the drawing to be completed. Before starting this tutorial, ensure that you have completed the part model shown in Figure 8–29.

Step 1: Start Pro/ENGINEER.

If Pro/ENGINEER is not open, start the application.

Step 2: Set an appropriate Working Directory.

Step 3: Select FILE >> NEW.

Step 4: In the New dialog box, select DRAWING mode then enter *VIEW1* as the name of the drawing file (Figure 8–30).

Step 5: Select OKAY on the New dialog box.

After selecting OKAY on the dialog box, Pro/ENGINEER will reveal the New Drawing dialog box. This dialog box is used to select a model, a sheet size, and a format.

Step 6: Select the BROWSE option on the New Drawing dialog box and locate the *VIEW1* part (Figure 8–31).

Use the Browse option to locate the *view1* part created in the first section of this tutorial. The part will serve as the Default Model for the

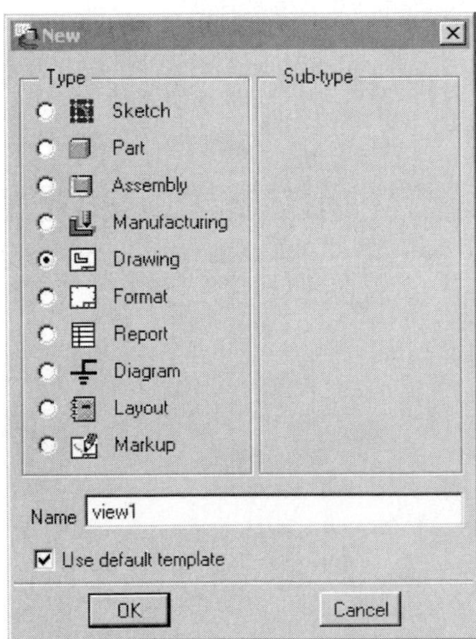

Figure 8–30 New dialog box

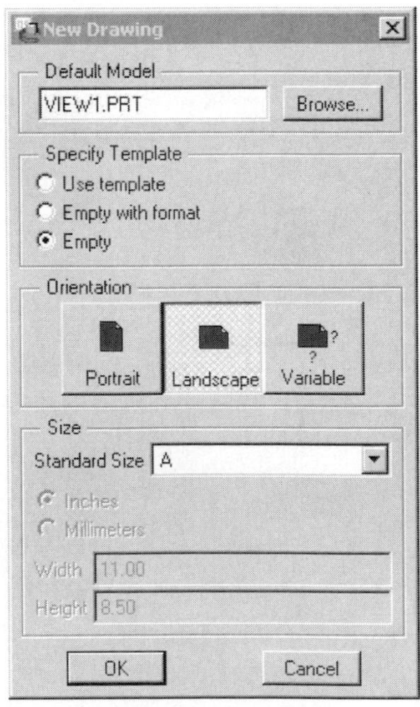

Figure 8–31 New Drawing dialog box

creation of drawing views. If *view1* is the active part during the creation of this drawing file, it will be displayed by default in the model selection box.

STEP 7: Select the EMPTY option (Figure 8–31) under the Specify Template option.

The Specify Template option allows for the selecting of a standard or user-defined sheet size. The Empty with Format option allows for the retrieval of a predefined sheet format. When retrieving a format, the size of the sheet defining the format will define the sheet size for the new drawing.

STEP 8: Select LANDSCAPE as the Orientation option.

STEP 9: Select A as the Standard Sheet Size (Figure 8–31).

Pro/ENGINEER provides a variety of standard sheet sizes (e.g., A, B, C, A1, A2, etc.). A unique sheet size can be entered in either inch or millimeter units.

STEP 10: Select OKAY on the New Drawing dialog box.

After selecting OKAY, Pro/ENGINEER will launch its Drawing mode.

ADDING A DRAWING FORMAT

This section of the tutorial will provide instruction on how to add a format to a drawing sheet. Formats can be retrieved from a variety of sources. Pro/ENGINEER provides sheet formats in the *format* subdirectory of the Pro/ENGINEER load point directory. Formats can also be imported through IGES and DXF. The configuration file option *pro_format_dir* is used to specify a default format directory.

STEP 1: Select SHEETS >> FORMAT >> ADD/REPLACE.

On the Menu Manager, select the Sheets menu option then select the Format option. The Add/Replace option is used to add formats to a drawing sheet or to replace an existing format.

STEP 2: Locate, select, and open an A size sheet format (Figure 8–32).

A format added to a drawing is associated with its original format file. Any changes made to the format through Format mode will be reflected in all drawings using that specific format.

STEP 3: Select DONE/RETURN to exit the Sheets menu.

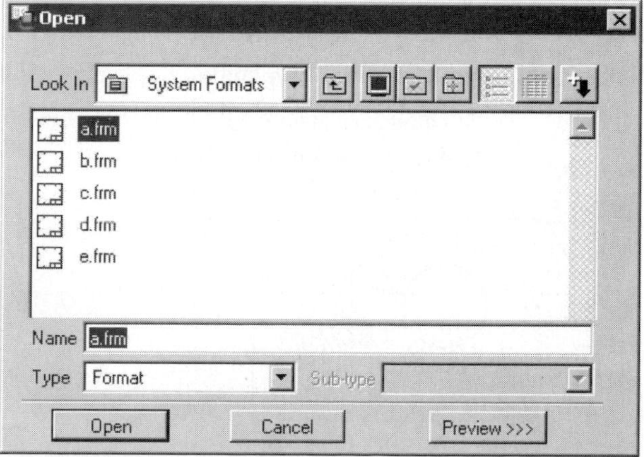

Figure 8–32 Opening an A size format

CREATING A GENERAL VIEW

General views serve as the parent view for all views projected off of it. This section of the tutorial will create a General view as the Front view of the drawing (Figure 8–33).

STEP 1: Select the VIEWS menu option.

STEP 2: Select ADD VIEW >> GENERAL >> FULL VIEW on the Views menu.

The Add View option is selected by default. Since no views currently exist within the drawing, General is the only view type available.

STEP 3: Select NoXsec >> SCALE on the View Type menu.

No-Cross-Section (No Xsec) is selected by default. The Section option will allow for the creation or retrieval of a section view. The Scale option allows for the entering of a scale value for the drawing view.

STEP 4: Select DONE on the View Type menu.

STEP 5: On the work screen, select the location for the drawing view.

The drawing under creation in this tutorial will consist of a Front, Right side, Top, and detailed view. The general view currently being defined will serve as the Front view. On the work screen, pick approximately where the front view will be located. The Views >> Move View option can be used to reposition this view's location.

STEP 6: In the textbox, enter .500 as the scale value for the view.

A scale value of 0.500 will create a view at half scale. The defining of a scale value is required due to the selection of the Scale option in step three.

After entering the scale value, Pro/ENGINEER will insert the model onto the work screen with the default orientation. The Orientation dialog box will be used to orient the view correctly.

STEP 7: On the Orientation dialog box, select FRONT as the Reference 1 option.

> **MODELING POINT** When selecting references to orient a model, it is often helpful to turn off all datum planes and to display the model as No Hidden (Figure 8–34). This technique provides clarity when selecting a reference surface.

A planar surface selected with the Front reference option orients the surface toward the front of the work screen. Other available reference options include Back, Top, Bottom, Left, Right, Vertical Axis, and Horizontal Axis.

STEP 8: Pick the front of the model (Figure 8–35).

STEP 9: On the Orientation Dialog box, select TOP as the Reference 2 option.

A planar surface selected with the Top reference option will orient the surface toward the top of the work screen.

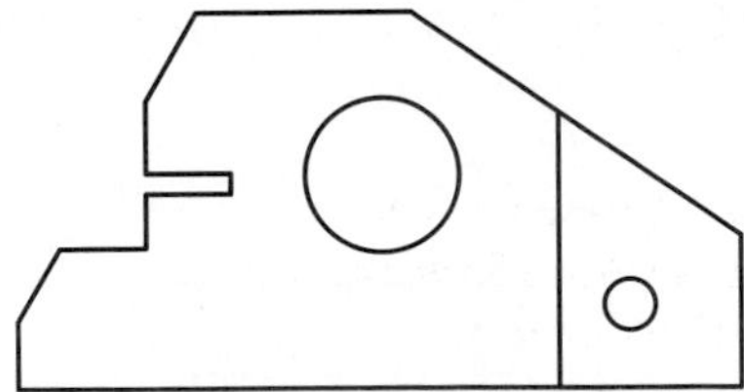

Figure 8–33 Front view of the part

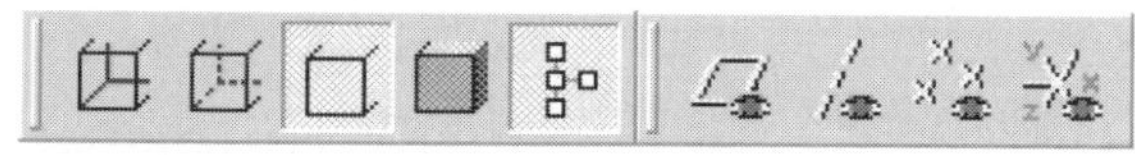
Figure 8–34 No hidden display and datum planes off

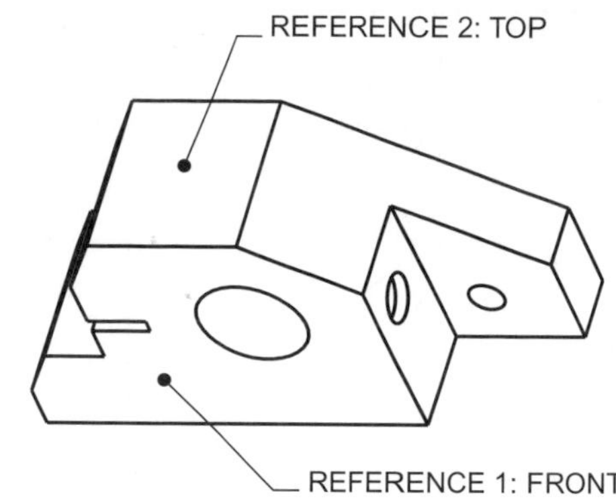

Figure 8–35 Orienting the model

STEP 10: Select the top of the model (Figure 8–35).

STEP 11: Select OKAY on the orientation dialog box.

Your front view should appear as shown in Figure 8–33.

CREATING PROJECTION VIEWS

Once a General view has been created, views can be projected off of it. Options exist for creating Projection, Auxiliary, and/or Section views. This segment of the tutorial will create a right-side view and a top view, both projected from the front view (Figure 8–36). The front view will be the parent view of these projected views.

STEP 1: Select VIEWS >> ADD VIEW.

STEP 2: Select PROJECTION >> FULL VIEW on the View Type menu.

The Projection option will project a view from an existing view. In this step of the tutorial, the right-side view will be projected from the front view.

STEP 3: Select NoXsec >> NO SCALE on the View Type menu.

Since this view will be projected from an existing view, it will take the scale value of its parent view.

STEP 4: Select DONE from the View Type menu.

Did you notice how steps 2 and 3 required the selection of default options? After a general view has been added to a drawing, Pro/ENGINEER defaults to the Projection >> Full View >> NoXsec >> No Scale options.

STEP 5: On the work screen, select the location for the Right-Side view.

Pro/ENGINEER will create a projected view based upon its parent view. Your view should appear as shown in Figure 8–37. The Views >> Move View option can be used to reposition a view.

STEP 6: Select VIEWS >> ADD VIEW.

Next, you will add the Top view.

STEP 7: Select DONE on the View Type menu.

You will take the default options found under the View Type menu (Projection >> Full View >> NoXsec >> No Scale).

STEP 8: On the work screen, select the location for the top view.

Your views should appear as shown in Figure 8–36.

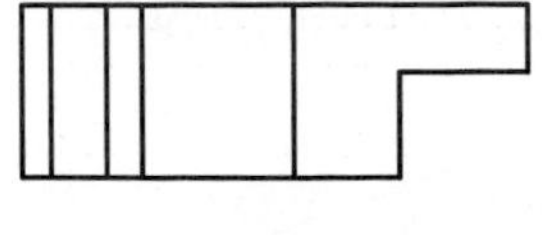

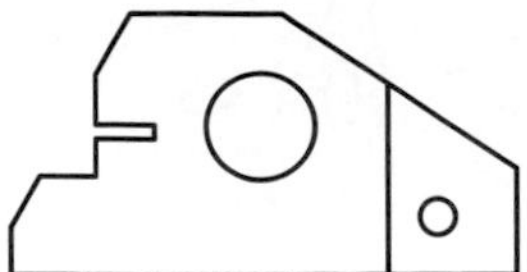

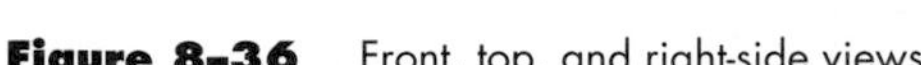

Figure 8–36 Front, top, and right-side views

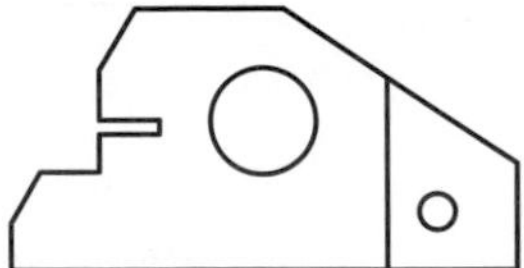

Figure 8–37 Front and right-side views

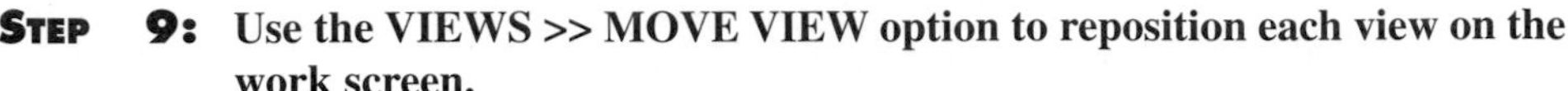

STEP 9: Use the VIEWS >> MOVE VIEW option to reposition each view on the work screen.

When moving a general view, all views projected from it will be repositioned to keep normal lines of projection. When moving a projected view, it will remain aligned with its parent view. Views can also be moved by dragging them on the screen with the left mouse button.

STEP 10: If necessary, select DONE/RETURN to exit the Views menu.

STEP 11: Save your drawing.

CREATING A DETAILED VIEW

Pro/ENGINEER's Drawing mode has an option for creating a detailed view. Detailed views are used to highlight and expand an area of an object that requires special attention. An example would be a small, complex feature that would be hard to dimension on a normal projection view. This segment of the tutorial will create the detailed view shown in Figure 8–38.

STEP 1: Select VIEWS >> ADD VIEW.

STEP 2: Select DETAILED as the view type to add.

STEP 3: Select DONE on the View Type menu.

STEP 4: On the work screen, select the location for the detailed view (Figure 8–38).

Using the mouse, select on the work screen where the detailed view will be located.

STEP 5: Enter 1.00 as the scale value for the detailed view.

This step will create a detailed view that is full scale.

STEP 6: On the front view, pick an entity centered within the detail area (Figure 8–39).

Pro/ENGINEER requires the selection of an entity in an existing view. This selection point is used to calculate the regeneration of the detailed view.

STEP 7: As shown in Figure 8–39, sketch a spline around the area to detail.

Using the mouse, sketch a spline that will include the area to be detailed. Use the left mouse button to select spline points and the middle mouse button to close the spline.

STEP 8: Enter *ONE* as the name for the detailed view.

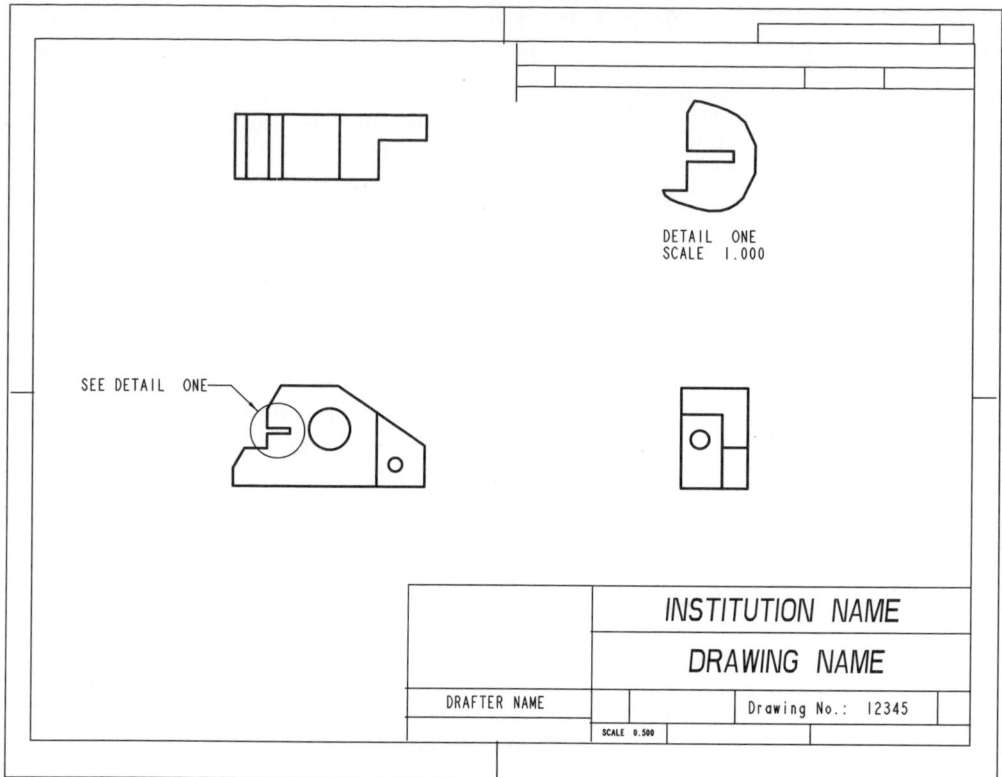

Figure 8–38 Detailed view of part

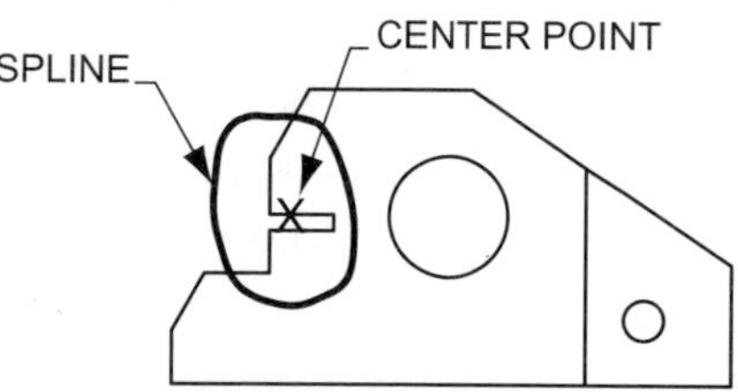

Figure 8–39 Sketching the spline

STEP 9: Select CIRCLE as the Detailed view boundary type.

The Circle option will include the area to be detailed in a circle (see Figure 8–38). Other options available include Ellipse, H/V Ellipse, and Spline.

STEP 10: Select a location on the work screen for the Detail Note.

As shown previously in Figure 8–38, this will locate the note *SEE DETAIL ONE*. This note can be moved later if necessary.

STEP 11: Use the VIEWS >> MOVE VIEW option to reposition the detailed view.

If necessary, reposition the detailed view or any other view.

STEP 12: Select DONE/RETURN on the Views menu.

Within Pro/ENGINEER, it is important for menu management to properly exit a menu. Always use an available Done or Done/Return option to properly exit any menu.

STEP 13: If necessary, select the detail view's note and drag it to an appropriate location.

Views, notes, dimensions, and other annotations can be moved by selecting the entity with the left mouse button. Your drawing should appear as shown in Figure 8–38.

STEP 14: Save your drawing.

ESTABLISHING DRAWING SETUP VALUES

Pro/ENGINEER's drawing setup file is used to set option values associated with a drawing. Examples of options that can be set include text height, arrowhead size, and geometric datum plane symbol. Multiple drawing setup files can be created. This segment of the tutorial will modify the values of the current drawing setup file.

STEP 1: Select ADVANCED >> DRAW SETUP.

After selecting Draw Setup, Pro/ENGINEER will open the active drawing settings within an Options dialog box. This dialog box is similar to the Preferences dialog box used to make changes to the configuration file option.

STEP 2: Change the Text and Arrowhead values for the current drawing.

Within the Options dialog box, make changes to the options shown in Table 8–2. To change an option's value, perform the following steps:

1. Type the option's name in the Option field.
2. Type or select the option's value in the Value field.
3. Select the Add/Change icon.

Table 8-2 Values for drawing setup

Drawing Setup Option	New Value
drawing_text_height	0.125
text_width_factor	0.750
dim_leader_length	0.175
dim_text_gap	0.125
draw_arrow_length	0.125
draw_arrow_style	filled
draw_arrow_width	0.0416

STEP 3: Save the modified values for the active drawing setup file, then Exit the text editor.

CREATING DIMENSIONS

Pro/ENGINEER provides options for creating two types of dimensions in Drawing mode: parametric and nonparametric. The Show/Erase dialog box can be used to show parametric dimensions that were created in Part and Assembly modes. During part modeling, a sketched feature has to be fully defined by utilizing dimensions, constraints, and references. The dimensions defining a feature can be revealed in Drawing mode through the Show and Erase option. The second option, Insert >> Dimension, can be used to create nonparametric dimensions.

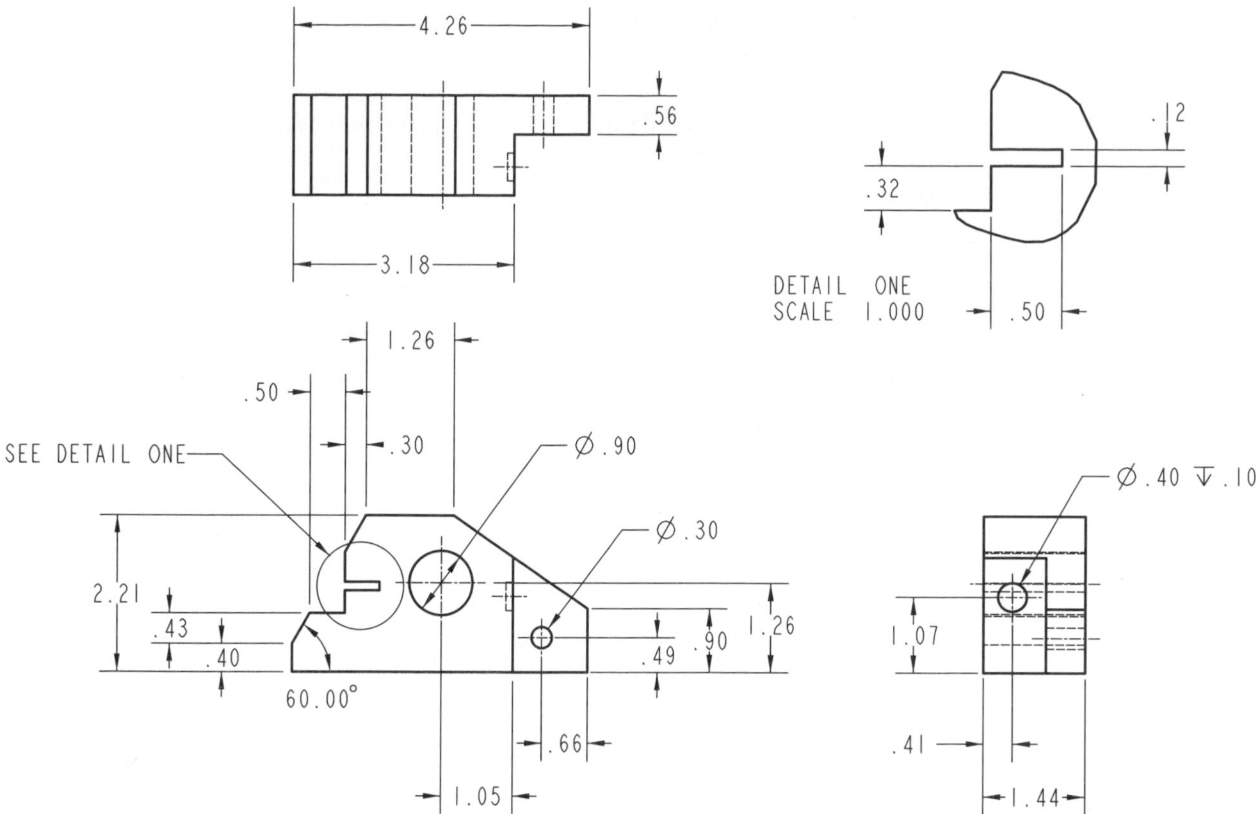

Figure 8–40 Dimensioned drawing

This segment of the tutorial will create the dimension annotations shown in Figure 8–40.

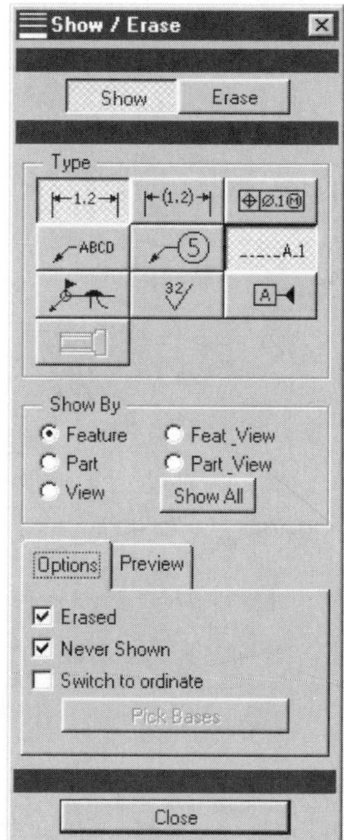

Figure 8–41 Show/Erase dialog box

STEP 1: On the menu bar, select VIEW >> SHOW AND ERASE.

The Show and Erase option is used to show and not show items that were defined in Part and/or Assembly modes. Figure 8–41 shows the Show/Erase dialog box. Items that can be shown include parametric dimensions, reference dimensions, geometric tolerances, notes, balloon notes, axes (centerlines), symbols, surface finish, datum planes, and cosmetic features.

STEP 2: On the Show/Erase dialog box, select the Dimension item type (Figure 8–42).

The dimension item type option will show parametric dimensions that were created in Part and Assembly modes.

STEP 3: On the Show/Erase dialog box, under the Options tab, check ERASED and NEVER SHOWN options (Figure 8–41).

The Erased option will show items that have previously been erased and the Never Shown option will show items that have never been shown.

STEP 4: On the Show/Erase dialog box, under the Preview tab, check the WITH PREVIEW option (Figure 8–43).

STEP 5: Select SHOW ALL (Figure 8–41) on the Show/Erase dialog box and Confirm the selection.

The Show All option will show all available item types. In this step of the tutorial, all dimensions available from the referenced model will be shown. As revealed in Figure 8–44, this can create a confusing and cluttered drawing.

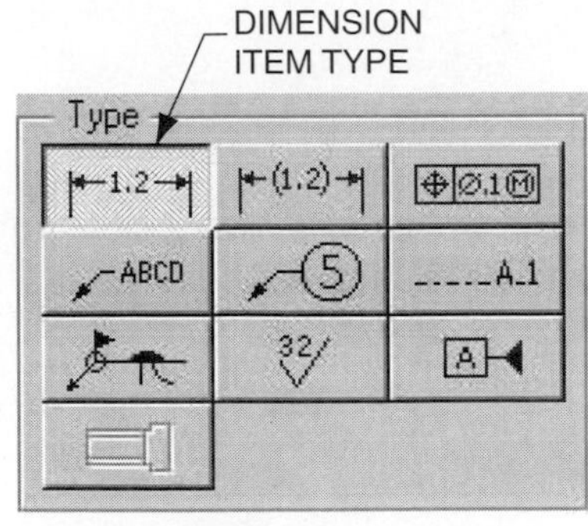

Figure 8–42 Dimension item type

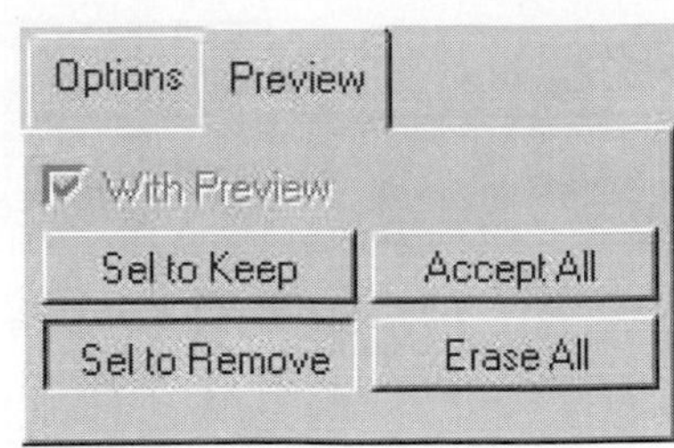

Figure 8–43 With Preview option

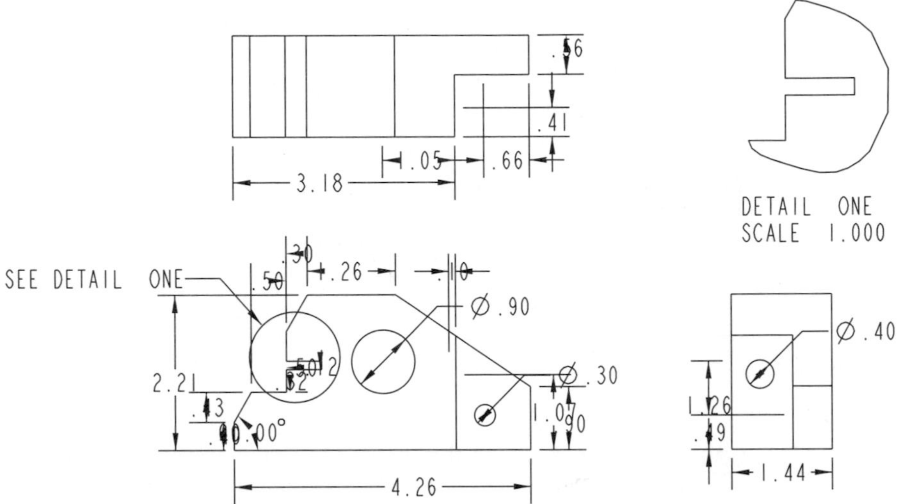

Figure 8–44 Dimensions shown on a drawing

You will use the Move, Switch View, Flip Arrows, and Erase options to clean the drawing.

STEP 6: Select ACCEPT ALL under the Preview option (Figure 8–43).

The previously selected With Preview option allows for the previewing of shown item types. Use the Accept All option to accept the shown dimensions.

STEP 7: CLOSE the Show/Erase dialog box.

STEP 8: Use Drawing mode's MOVE and MOVE TEXT capabilities and the Pop-Up menu's SWITCH VIEW and FLIP ARROWS options to reposition the dimensions to match Figure 8–45.

The Pop-Up menu is available by first selecting the text of a dimension with your left mouse button, followed by reselecting the same dimension text with your right mouse button (*Note:* A slight delay in releasing the right mouse button is necessary). A variety of modification and manipulation tools are available through the revealed pop-up menu.

Use the following options to reposition your dimensions on the work screen:

- **Move** The Move option is available through the selection of a dimension's text with the left mouse button. Once selected, the dimension, including text and witness lines, can be dragged with your curser.

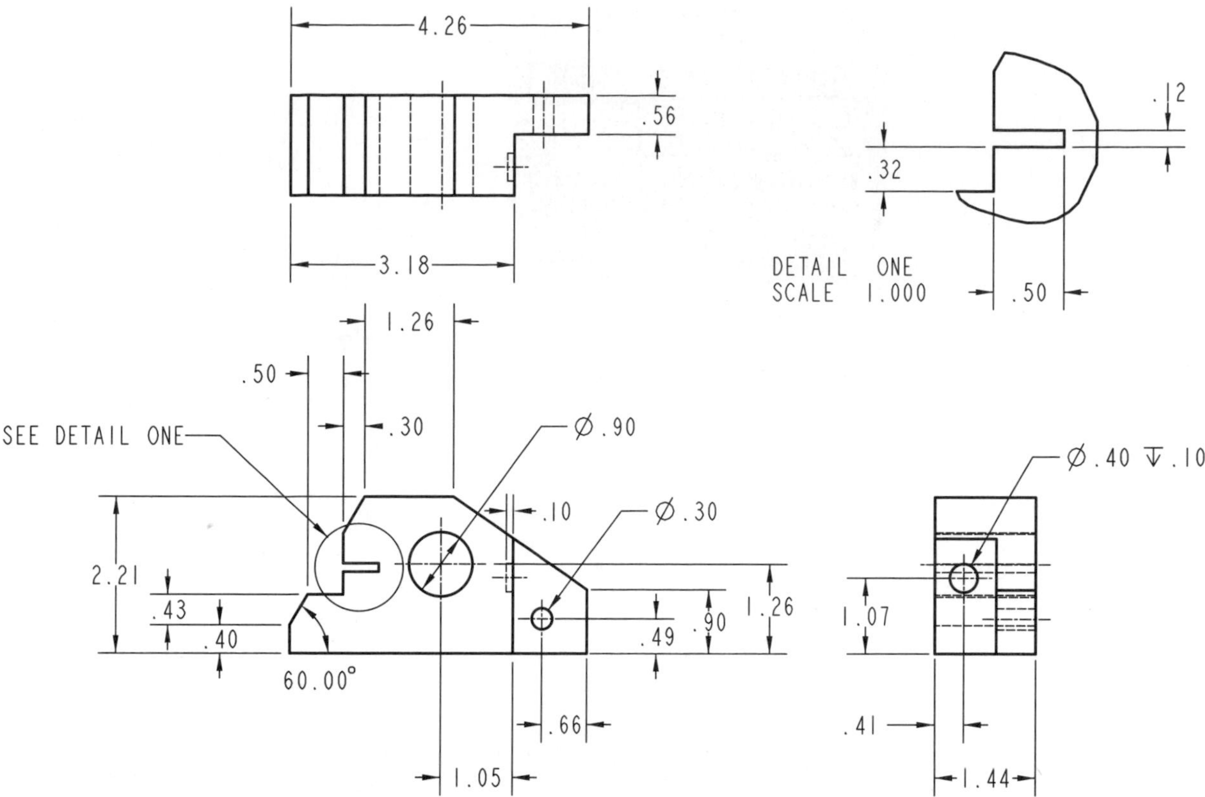

Figure 8-45 Repositioned dimensions

- **Move Text** The Move Text option is available through the selection of a dimension's text with the left mouse button. Once selected, grip points are available that will allow you to move only the text of a dimension.
- **Switch View** The Switch View option is used to switch a dimension from one view to another. It is available under the Pop-Up menu. Multiple dimensions can be selected by a combination of the Shift key and the left mouse button. Once dimensions are preselected, the Pop-Up menu is available through the right mouse button.
- **Flip Arrows** The Flip Arrows option is used to flip dimension arrowheads. It is available under the Pop-up menu.

MODELING POINT The Switch View icon is also available on the Drawing toolbar. It is a handy option for switching entities, to include dimension lines, from one view to a second view.

STEP 9: Use the SHOW/ERASE option to show all centerlines.

The Axis option (Figure 8–46) on the Show/Erase dialog box is used to show centerlines. On the Show/Erase dialog box, select the Axis option, then select Show All. Accept all available centerlines, then close the dialog box.

STEP 10: Use the left mouse button's move capability to create witness line gaps.

Where a dimension's witness line is linked to an entity (Figure 8–47), a visible gap (approx. 1/16-inch wide) is required. After selecting the

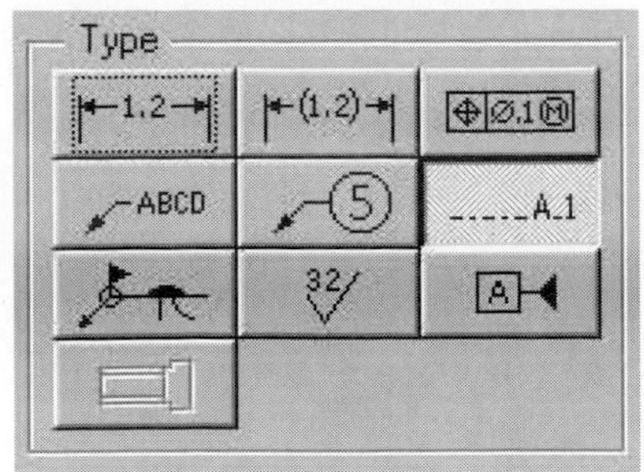

Figure 8–46 Axis item type

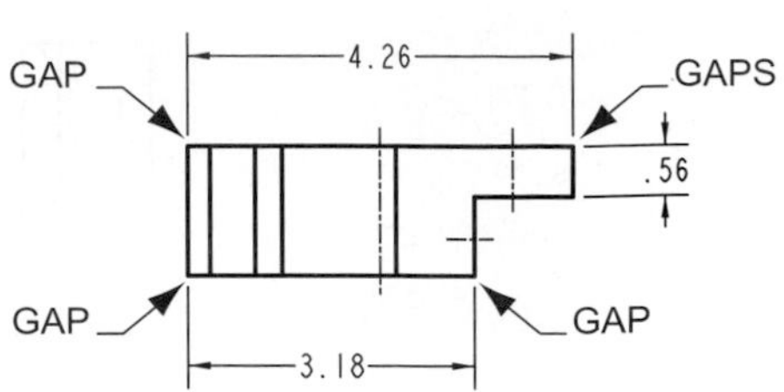

Figure 8–47 Witness line gaps

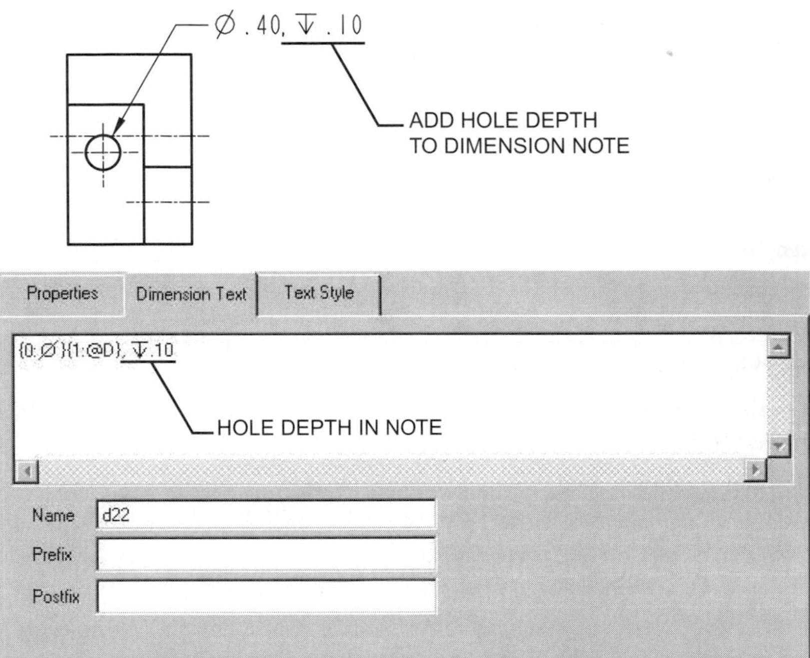

Figure 8–48 Add text and symbols to a dimension note

dimension's text with the left mouse button, drag the witness line's grip point to create the gap.

STEP 11: Pick the .50 inch diameter hole dimension, then select the Pop-Up menu's PROPERTIES option.

STEP 12: Use the Dimension Text tab on the Dimension Properties dialog box to modify the hole's note (Figure 8–48).

You will add the .10-inch depth value to the .40-inch diameter hole note. Under the Dimension Text tab, use the Sym Palette and the keyboard to add the hole depth attribute shown in the illustration.

STEP 13: Select the VIEW >> SHOW AND ERASE option.

The .10 dimension shown in Figure 8–49 is not needed in the drawing. The Show/Erase option can be used to remove it from the drawing.

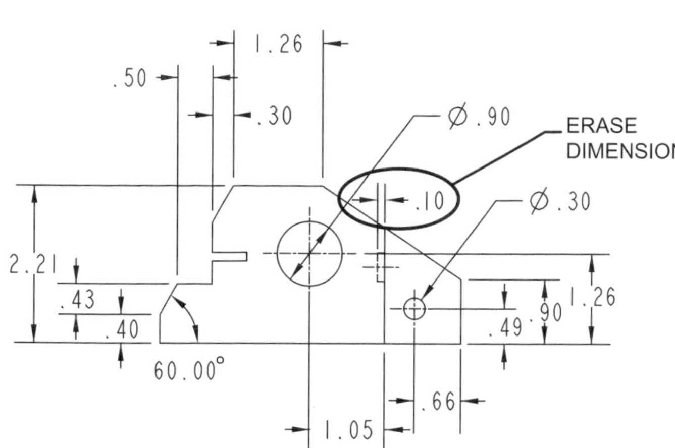

Figure 8-49 Erase dimension

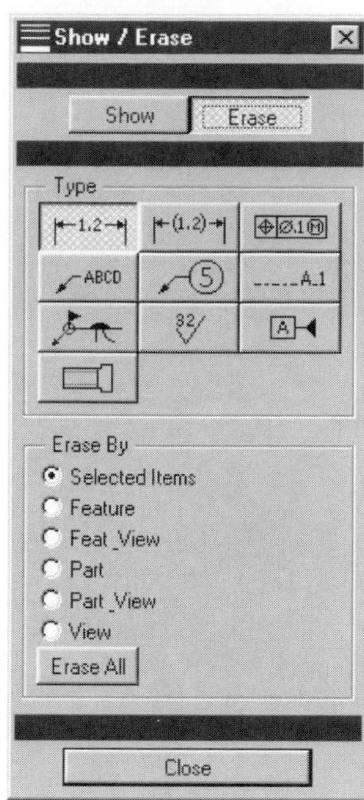

Figure 8-50 Dimension item type

STEP 14: On the Show/Erase dialog box, select the ERASE option, then select the DIMENSION item type (Figure 8–50).

STEP 15: On the work screen, select the .10 dimension text, then select the middle mouse button.

The Erase option does not delete a selected item. Instead, it removes it from the drawing screen. Any erased item can be redisplayed with the Show option.

STEP 16: Select the CLEAN DIMENSIONS icon on the Drawing toolbar.

The Clean Dimensions option will consistently space dimensions on a drawing view. The minimum distance between a dimension line and the object is 3/8 inch, while the minimum spacing between two dimension lines is 1/4 inch. These values can be increased when appropriate. After selecting the Clean Dims option, the Clean Dimensions dialog box will appear (Figure 8–51).

STEP 17: On the work screen, select the front, top, right-side, and detailed views, then select DONE SEL on the Get Select menu.

Pro/ENGINEER's default values for the Clean Dimensions dialog box are 0.500 for the object Offset setting and 0.375 for the dimension Increment setting. You will keep these default settings.

STEP 18: Select APPLY, then CLOSE the dialog box.

After applying the clean dimension settings, Pro/ENGINEER will create snap lines that will evenly space any dimension line. Depending upon the complexity of the drawing, the applied settings might not create an ideal drawing. Snap lines will not print when the drawing is plotted.

STEP 19: Use Drawing modification and manipulation tools to refine the placement of dimensions.

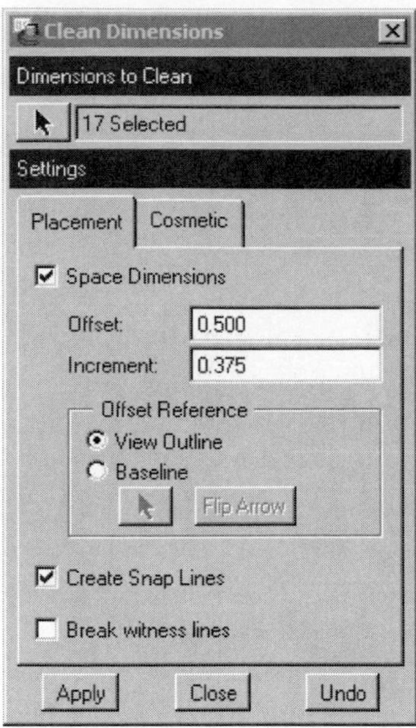

Figure 8–51 Clean Dimensions dialog box

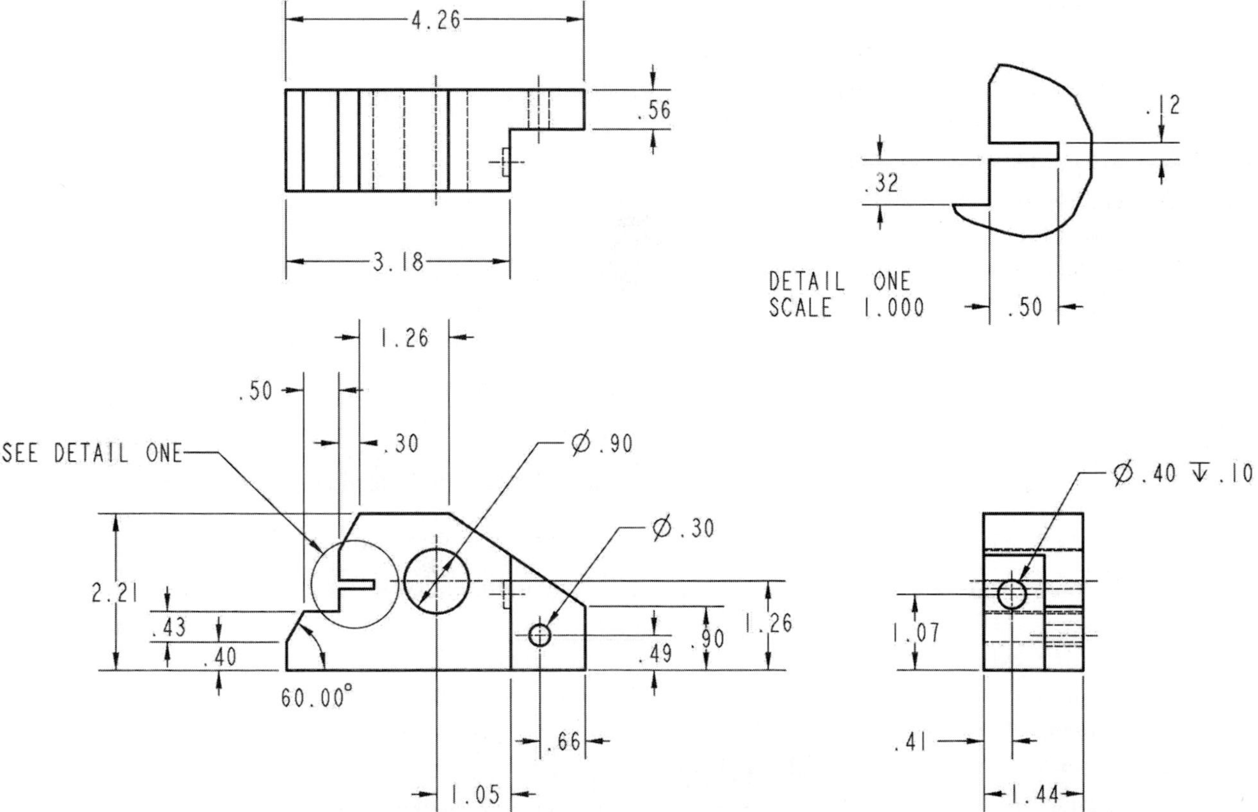

Figure 8–52 The final dimensioning scheme

Use available tools such as Move View, Move, Flip Arrows, and Move Text to tweak the placement of dimensions. Your final dimensioning scheme should appear as shown in Figure 8–52.

STEP 20: Save your drawing.

CREATING NOTES

Annotations can be added to a drawing using the Insert >> Note option. Within the Note Types menu, options are available for creating notes with or without leader lines. Also, options are available for justifying text and for entering text from a file. This segment of the tutorial will create text for the title block. The finished title block will appear as shown in Figure 8–53.

STEP 1: Using the Pop-Up menu's Modify Text Style option, modify the *SCALE 0.500* note to have a text height of 0.100 inches.

When a General view is added to a drawing, Pro/ENGINEER inserts a scale value. In this tutorial, the first general view has a scale value equal to 0.500. When selecting the *SCALE 0.500* note, you have to select both the word *SCALE* and the number *0.500*.

STEP 2: Move the *SCALE 0.500* note to the title block (Figure 8–53).

STEP 3: Select INSERT >> NOTE and create the text shown in the figure.

The Insert >> Note option is used to add text to a drawing. Use the following options to create each note:

- **No leader** Creates a note without a leader line.
- **Enter** Allows for the creation of a note through the keyboard.
- **Horizontal** Creates notes horizontal on the work screen.

STEP 4: Select CENTER on the Note Types menu.

The Center option will center-justify the note text. Other justification options available include Left, Right, and Default.

STEP 5: Select MAKE NOTE to enter the text for the note.

STEP 6: Within the work screen, select the location for the *INSTITUTION NAME* note.

STEP 7: In the textbox, enter your institution's name as the note text.

In Pro/ENGINEER's textbox, enter the text, then select the Enter key on the keyboard.

STEP 8: Select ENTER on the keyboard to end the Make Note option.

STEP 9: Repeat the note making steps to create the remaining notes.

STEP 10: Select DONE/RETURN from the Note Types menu to exit the note creation menu.

STEP 11: On the work screen, using a combination of the Shift key and the left mouse button, preselect the *INSTITUTION NAME* and the *DRAWING NAME* notes.

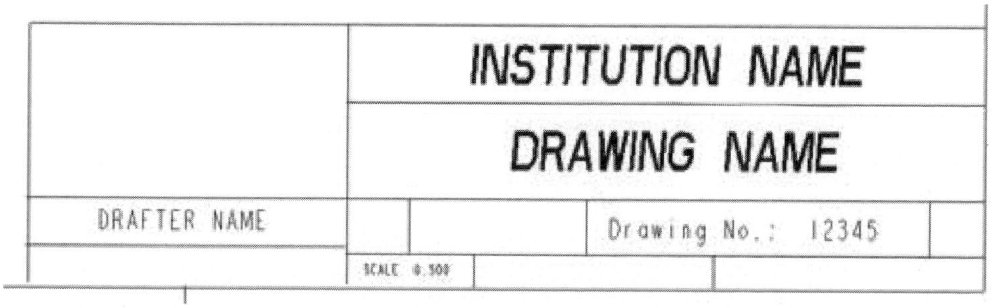

Figure 8–53 Title block information

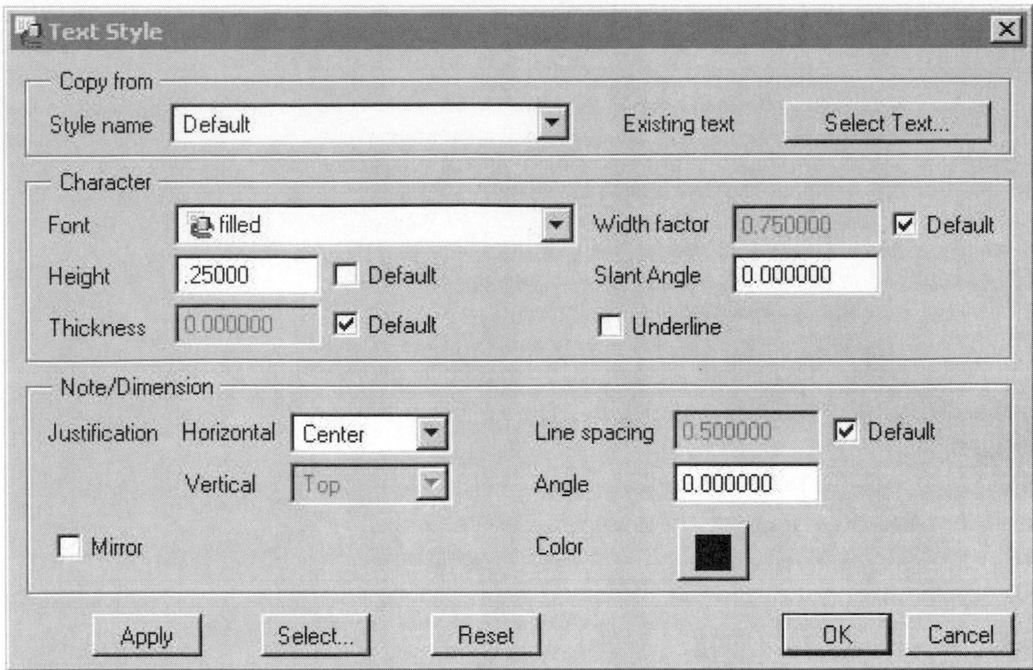

Figure 8–54 The Text Style dialog box

STEP 12: Access the Pop-Up menu's Modify Text Style option.

You will modify the text style and the text height of the *Institution Name* and *Drawing Name* notes.

STEP 13: On the Text Style dialog box, select FILLED as the Font (Figure 8–54).

STEP 14: On the Text Style dialog box, enter a text Height of 0.250.

STEP 15: Select OKAY on the Text Style dialog box.

STEP 16: Save your drawing.

SETTING DISPLAY MODES

The display of hidden lines on a drawing can be controlled with the display model options found on the toolbar (e.g., Wireframe, Hidden Line, and No Hidden). Additionally, the tangent edge display mode can be set with the Environment dialog box. The problem with using these two methods to set the display mode of a drawing is that a selected setting, such as hidden line display, will affect the entire drawing. Often, it is necessary to set a different display mode for a view or for an individual entity. The following steps will set the front, top, and right-side views with a Hidden Line display. Additionally, the detailed view will be set with a No Hidden display.

STEP 1: Select VIEWS >> DISP MODE >> VIEW DISP.

STEP 2: On the work screen, select the front, top, and right-side views.

STEP 3: Select DONE SEL on the Get Select menu (or select the middle mouse button).

STEP 4: Select HIDDEN LINE as the display mode.

Other options available include Wireframe, No Hidden, and Default. The Default setting assumes the display mode selected from the toolbar.

STEP 5: **Select DONE on the View Display menu.**

STEP 6: **Select VIEWS >> DISP MODE >> VIEW DISP.**

You will next set a No Hidden display mode for the Detailed view.

STEP 7: **On the work screen, select the Detailed view, then select DONE SEL.**

STEP 8: **Select DET INDEP on the View Display.**

The Detail Independent (Det Indep) selection makes the Detail's display mode independent from its parent's display mode.

STEP 9: **Select NO HIDDEN as the display mode.**

STEP 10: **Select DONE on the View Display menu.**

STEP 11: **Save your drawing and purge old versions of the drawing by using the FILE >> DELETE >> OLD VERSIONS option.**

TUTORIAL

DRAWING TUTORIAL 2

This tutorial will create the drawing shown in Figure 8–55. As with the first tutorial in this chapter, the start of this tutorial will require you to model the part. As shown in the figure, two different view types will be created: a General view and a Projection view.

As shown in Figure 8–55, this tutorial will demonstrate the creation of tolerances on a Pro/ENGINEER drawing. This tutorial will cover:

- Starting a drawing with a drawing format.
- Creating a general view.
- Creating a projection view.
- Adding geometric tolerances to a drawing.
- Setting dimensional tolerances.
- Modifying the drawing setup file.

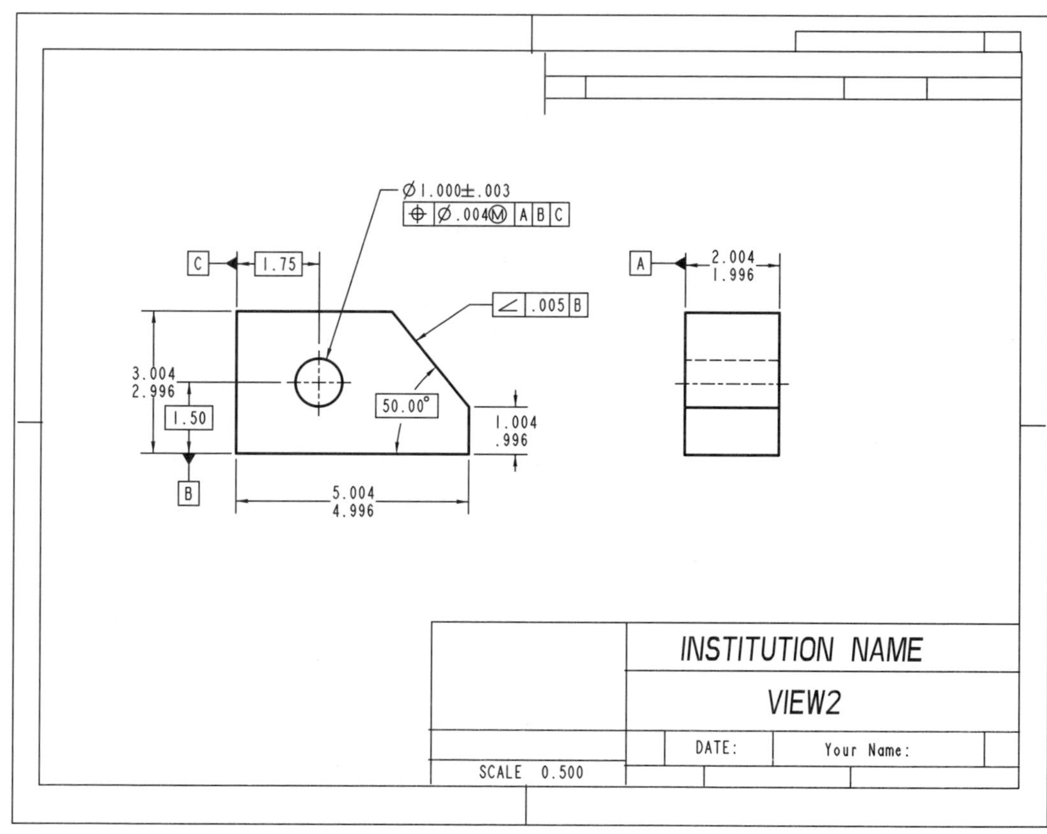

Figure 8–55 Pro/ENGINEER drawing with geometric tolerances

CREATING THE PART

Model the part shown in Figure 8–56, naming the part *VIEW2*. When modeling the part, make sure that the locations of your datum planes match the locations of the datum planes shown in the drawing. As a reference, later in this tutorial, the default datum planes

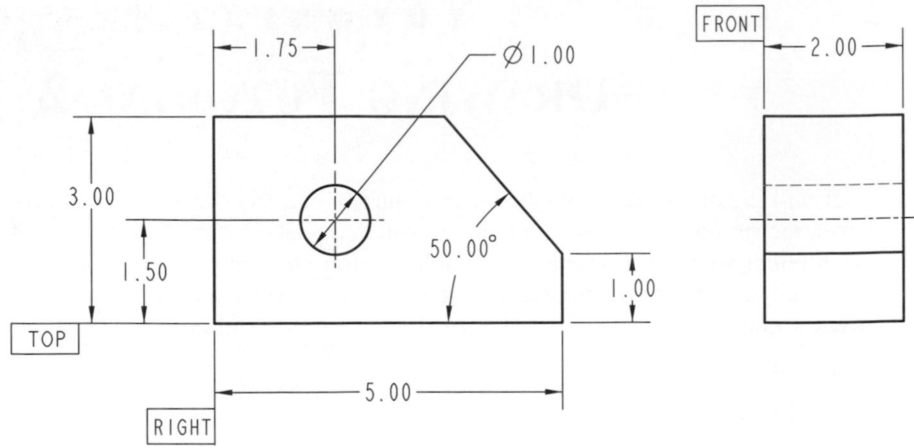

Figure 8–56 Part and dimensioning scheme

RIGHT, TOP, and FRONT will be renamed A, B, and C respectively. The required dimensioning design intent is shown. When modeling this part, make sure that these dimensions are incorporated into your design.

STARTING A DRAWING WITH A TEMPLATE

This section of the tutorial will create the object file for the drawing to be completed. You will utilize an existing drawing template file to set views, formats, and other drawing settings. Before starting this segment of the tutorial, ensure that the part shown in Figure 8–56 has been completed.

STEP 1: Start Pro/ENGINEER.

If Pro/ENGINEER is not open, start the application.

STEP 2: Set an appropriate Working Directory.

STEP 3: Select FILE >> NEW (Use the Default Template option).

STEP 4: In the New dialog box, select Drawing mode, then enter *VIEW2* as the name of the drawing file.

STEP 5: Select OKAY on the New dialog box.

After selecting OKAY on the dialog box, Pro/ENGINEER will reveal the New Drawing dialog box. This dialog box is used to select a model to create the drawing from and to set a sheet size.

STEP 6: On the New Drawing dialog box, select the BROWSE option and locate the *VIEW2* part (Figure 8–57).

Use the Browse option to locate the *VIEW2* part created in the first section of this tutorial. The part will serve as the Default Model for the creation of the drawing views.

STEP 7: On the New Drawing dialog box, under the Specify Template option, select the EMPTY WITH FORMAT option (Figure 8–57).

The Specify Sheet option allows for the selection of a standard or user-defined sheet size. The Empty-with-Format suboption allows for the retrieval of a predefined sheet format. When retrieving a format, the size of the sheet defining the format will define the sheet size for the new drawing.

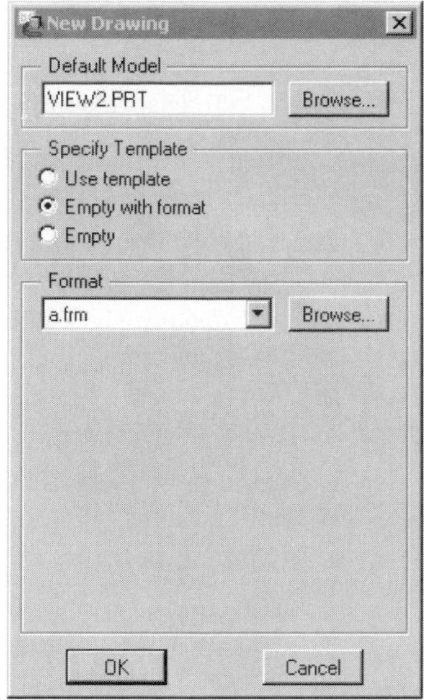

Figure 8-57 New Drawing dialog box

Template files under the Use Template option come with preestablished settings such as views and model display properties. As an example, the *a_drawing* template comes complete with preexisting front, top, and right-side views. Other information that can preexist in a template file includes drawing notes not derived from drawings model (e.g., notes and symbols) and parameter notes defined within the model.

STEP 8: Select the BROWSE option under the Format section of the New Drawing dialog box.

The second Browse option on the dialog box allows you to browse the directory structure to locate existing sheet formats.

STEP 9: Open "a.frm" as the standard Sheet format.

Pro/ENGINEER provides a variety of standard sheet formats (e.g., A, B, C, D, and E). A user-defined format can also be selected. When picking an existing format, the size of the format defines the size of the drawing sheet.

STEP 10: Select OKAY on the New Drawing dialog box.

ESTABLISHING DRAWING SETUP VALUES

This section of the tutorial will temporarily set Pro/ENGINEER's drawing settings. The Advanced >> Draw Setup option will be used to change the default settings for this specific drawing.

STEP 1: Select ADVANCED >> DRAW SETUP.

After selecting Draw Setup, Pro/ENGINEER will open the active drawing settings within an Options dialog box. This dialog box is similar to the Options dialog box used to make changes to configuration file options.

STEP 2: Change the Text, Arrowhead, and Datum values for the current drawing.

Within the Options dialog box, make changes to the options shown in Table 8–3. To change an option's value, perform the following steps:

1. Type the option's name in the Option field.
2. Type or select the option's value in the Value field.
3. Select the Add/Change icon.

The item *tol_display* will display dimensions in tolerance mode. The item *gtol_datums* when set to a value of *STD_ISO* will display datums in the ISO format.

Table 8–3 Values for drawing setup

Drawing Setup Option	New Value
drawing_text_height	0.125
text_width_factor	0.750
dim_leader_length	0.175
dim_text_gap	0.125
tol_display	YES
draw_arrow_length	0.125
draw_arrow_style	filled
draw_arrow_width	0.0416
gtol_datums	STD_ISO

STEP 3: Apply and Close the modified values for the active drawing setup file.

Save your drawing setup file values.

STEP 4: Select DONE/RETURN to close the Advanced Drawing Options menu.

CREATING THE GENERAL VIEW

General views serve as the parent view for all views projected off of it. This section of the tutorial will create a General view as the Front view of the drawing, as shown in Figure 8–58.

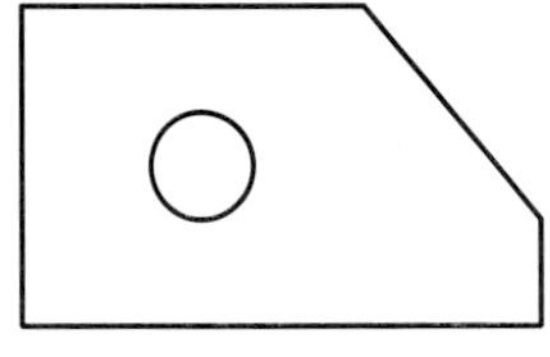

Figure 8–58 The front view

STEP 1: Select VIEWS >> ADD VIEW >> GENERAL >> FULL VIEW.

The Add View option is selected by default. Since no views currently exist within the drawing, General is the only view type available.

STEP 2: Select NoXsec >> SCALE >> DONE.

No-Cross-Section (No Xsec) is selected by default. The Section option will allow for the creation or retrieval of a section view. The Scale option allows for the entering of a scale value for the drawing view.

STEP 3: On the work screen, select the location for the Front view.

The drawing under creation in this tutorial will consist of a Front view and a Right-side view. The general view currently being defined will serve as the Front view. On the work screen, pick approximately where the Front view will be located. Once placed, the Views >> Move View option can be used to reposition the location of any view.

STEP 4: Enter .500 as the scale value for the view.

A scale value of 0.500 will create a view at half scale. The defining of a scale value is required due to the selection of the Scale option in step 2.

After entering the scale value, Pro/ENGINEER will insert the model into the work screen with the default orientation. The Orientation dialog box will be used to orient the view correctly.

INSTRUCTIONAL NOTE Depending upon your part construction method, your default orientation might be different from the orientation shown in Figure 8–59. Adjust your view orientation reference selections accordingly. Your final view should match the front view shown in Figure 8–58.

STEP 5: On the Orientation dialog box, select FRONT as the Reference 1 option.

MODELING POINT When selecting references to orient a model, it is often helpful to turn off all datum planes and to display the model as No Hidden. This technique will provide clarity when selecting a reference surface.

STEP 6: Select the front of the model (Figure 8–59).

A planar surface selected with the Front reference option will orient the surface toward the front of the work screen.

STEP 7: On the Orientation Dialog box, select TOP as the Reference 2 option.

STEP 8: Select the top of the model that will point toward the top of the work screen (Figure 8–59).

Select the surface shown in the figure. If your default model does not appear as shown in the illustration, adjust your reference selection accordingly. Your final orientation should appear as shown in Figure 8–58.

STEP 9: Select OKAY on the orientation dialog box.

Your front view should appear as shown in Figure 8–58.

STEP 10: Save your drawing.

Drawings are saved with an *.drw file extension.

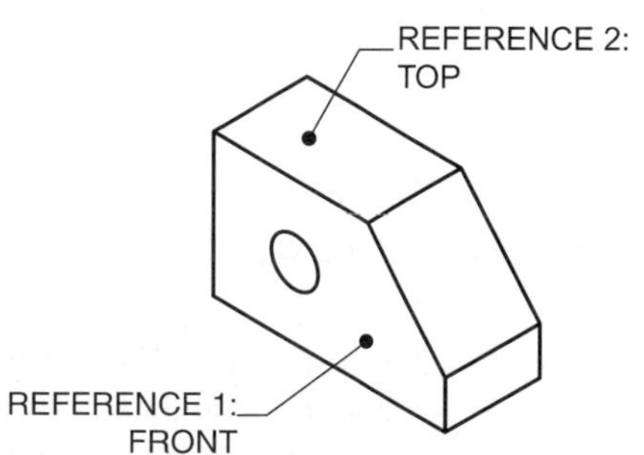

Figure 8–59 Orienting the model

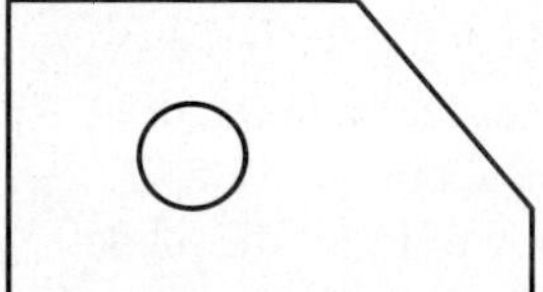

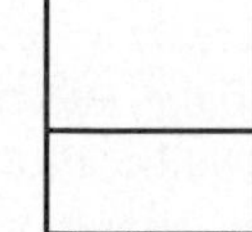

Figure 8–60 The front and right-side views

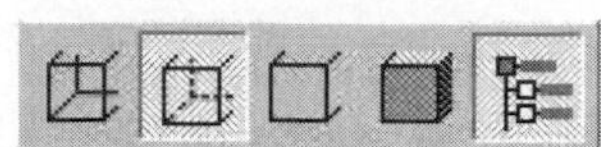

Figure 8–61 Model display toolbar

Creating the Right-Side View

This segment of the tutorial will create the right-side view of the part (Figure 8–60).

Step 1: Select VIEWS >> ADD VIEW.

Step 2: Select PROJECTION >> DONE.

Since Projection, Full View, and NoXsec are the default selections, you only have to select Done to perform this step.

Step 3: On the work screen, select the location for the Right-Side view.

Pro/ENGINEER will create a projected view based upon its parent view. Your view should appear as shown in Figure 8–60. The Views >> Move View option can be used to reposition a view.

Step 4: On the toolbar, select the Hidden Line display icon (Figure 8–61).

Selecting Hidden Line as the model display mode will produce hidden lines when the drawing is plotted.

Step 5: If necessary, select DONE/RETURN to exit the Views men.

Setting and Renaming Datum Planes

This section of the tutorial will set and rename your datum planes. To utilize datum planes within a drawing, and with geometric tolerances, they have to first be set. The Set Up >> Geom Tolerance option is used to set datums in Part mode. The Edit >> Properties option in Drawing mode will be used in this tutorial to set datums.

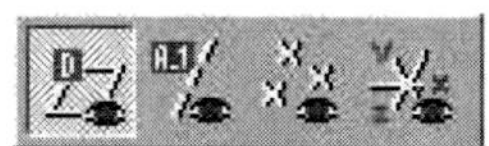

Figure 8–62 Datum display toolbar

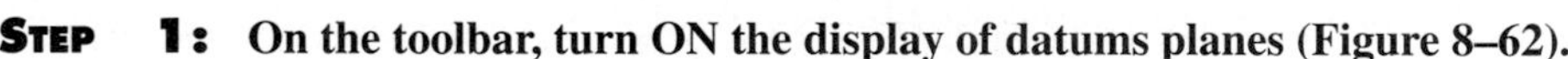

Step 1: On the toolbar, turn ON the display of datums planes (Figure 8–62).

Step 2: On the work screen, pick datum plane FRONT.

Select datum plane FRONT. If during the construction process the location of your datum plane FRONT does not match the location of FRONT as shown previously in Figure 8–56, adjust your datum selection to match this figure.

Step 3: On the menu bar, select EDIT >> PROPERTIES.

After selecting this datum plane, the Datum dialog box will appear (Figure 8–63).

Step 4: Rename this datum plane A.

As shown in Figure 8–63, in the Name textbox, rename FRONT to a value of A.

Step 5: On the Datum dialog box, set the datum by selecting the datum plane symbol button (Figure 8–63).

If your datum plane symbol does not appear as shown in the figure, you did not properly set the drawing setup file option *gtol_datums* to *STD_ISO* as required earlier in this tutorial.

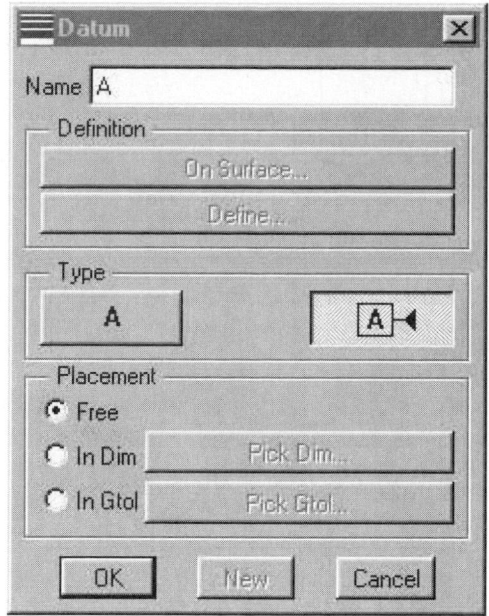

Figure 8–63 Datum dialog box

STEP 6: Select OKAY on the Datum dialog box to accept the values.

After selecting OKAY, notice on the work screen how the datum symbol is now displayed in the ISO format (Figure 8–64).

STEP 7: Drag datum Plane A to the location shown in Figure 8–64.

INSTRUCTIONAL NOTE If your location of datum plane A does not match the location shown in Figure 8–64, you probably extruded the base protrusion of the part the wrong direction. This can be fixed in Part mode by redefining the direction of extrusion.

STEP 8: Use the EDIT >> PROPERTIES option to set datum plane TOP and rename it B (Figure 8–65).

STEP 9: Use the EDIT >> PROPERTIES option to set datum plane RIGHT and rename it C (Figure 8–65).

STEP 10: Select the VIEW >> SHOW AND ERASE option.

As shown in Figure 8–66, two datum plane B symbols are available in the drawing. Except for the need of clarity, only one datum plane symbol is required. You will use the Show/Erase dialog box to erase the second datum plane B symbol.

STEP 11: On the Show/Erase dialog box, select the ERASE option and then select the DATUM type (Figure 8–67).

STEP 12: On the Show/Erase dialog box, make sure that the SELECTED ITEMS Erase By item type is selected.

The Selected Items option will erase only items that are selected on the work screen.

STEP 13: On the work screen, select the extra datum plane symbol B (Figure 8–66), then select Done Select (Done Sel) from the Get Select menu.

STEP 14: Close the Show/Erase dialog box.

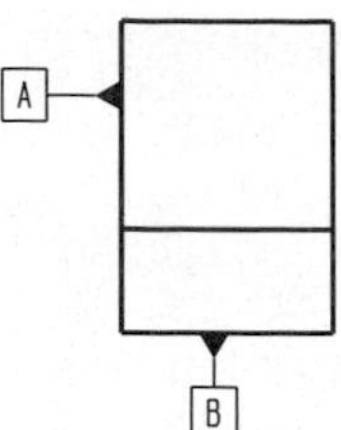

Figure 8–64 ISO datum plane symbol

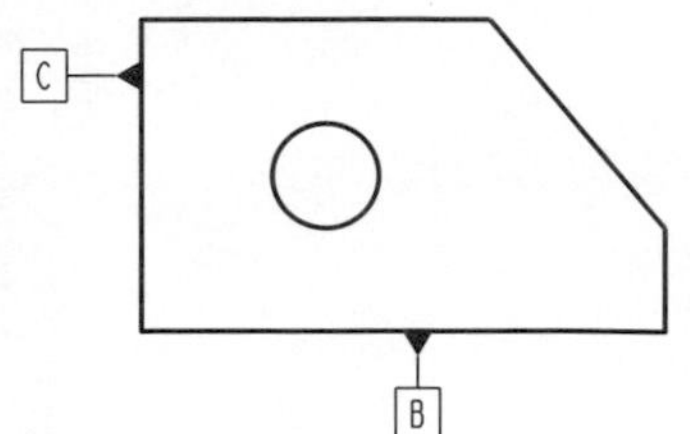

Figure 8–65 Setting datum planes

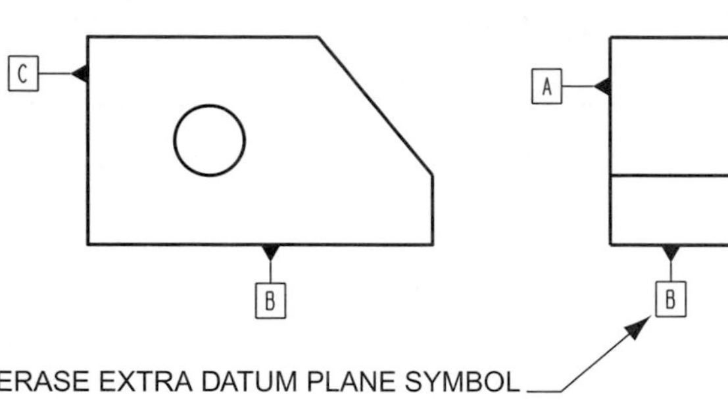

Figure 8–66 Erase the extra datum plane symbol

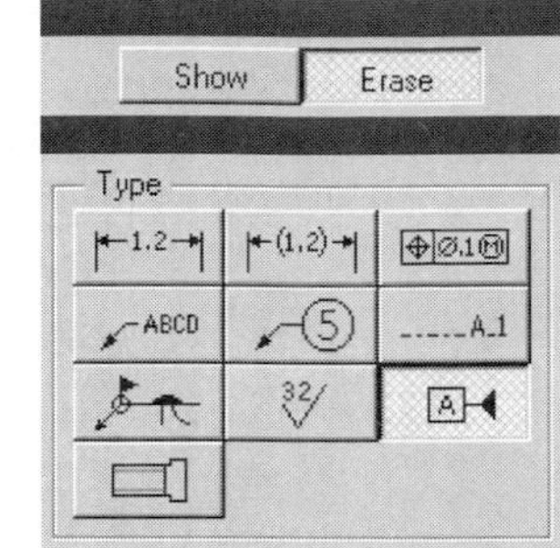

Figure 8–67 Datum plane item

CREATING DIMENSIONS

This segment of the tutorial will show the parametric dimensions that were created when the model was constructed in Part mode. You will use the Show/Erase option to show all dimensions and centerlines.

STEP 1: Select the SHOW/ERASE option.

The Show/Erase option is used to show or not show items that were defined in Part and/or Assembly modes.

STEP 2: On the Show/Erase dialog box, select the SHOW option and deselect the DATUM option (Figure 8–68).

From the previous segment of this tutorial, the Erase option and the Datum item type should still be selected. If they are, select the Show option and deselect the Datum option.

STEP 3: On the Show/Erase dialog box, select the DIMENSION and AXIS item types (Figure 8–68).

The Dimension item type option will show parametric dimensions that were created in Part mode. The Axis item type will create centerlines.

STEP 4: On the Show/Erase dialog box, under the Options tab, check the ERASED and NEVER SHOWN options.

The Erased option will show items that have previously been erased, and the Never Shown option will show items that have never been shown.

STEP 5: On the Show/Erase dialog box, under the Preview tab, uncheck the WITH PREVIEW option (Figure 8–69).

Due to the limited number of dimensions in this tutorial, you will not preview your dimensions.

Figure 8–68 The Show/Erase dialog box

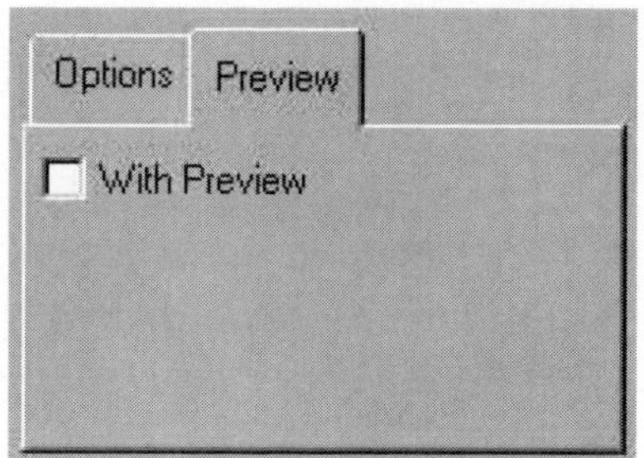

Figure 8-69 Showing without a preview

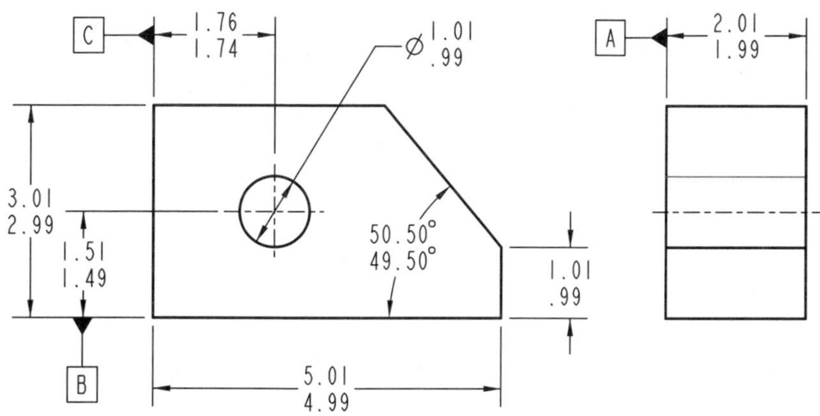

Figure 8-70 Drawing with dimensions and centerlines

STEP 6: Select the SHOW ALL option, then select YES to confirm the show all.

The Show All option will show all available item types. In this step of the tutorial, all dimensions available from the referenced model will be shown.

STEP 7: Close the Show/Erase dialog box.

Your drawing should appear similar to Figure 8–70. Previously, you set the Drawing Setup File option *tol_display* equal to a value of Yes. This created the tolerance display shown on your drawing.

STEP 8: Use Drawing mode's MOVE and MOVE TEXT capabilities and the Pop-Up menu's SWITCH VIEW and FLIP ARROWS options to reposition the dimensions to match Figure 8–70.

Use the following options from the Detail menu to reposition the dimensions on the work screen:

- **Move** The Move option is available through the selection of a dimension's text with the left mouse button. Once selected, the dimension, including text and witness lines, can be dragged with your curser.
- **Move text** The Move Text option is available through the selection of a dimension's text with the left mouse button. Once selected, grip points are available that will allow you to move only the text of a dimension.
- **Switch view** The Switch View option is used to switch a dimension from one view to another. It is available under the Pop-Up menu. Multiple dimensions can be selected by a combination of the Shift key and the left mouse button. Once dimensions are preselected, the Pop-Up menu is available through the right mouse button.

- **Flip arrows** The Flip Arrows option is used to flip dimension arrowheads. It is available under the Pop-Up menu.

STEP 9: Use Drawing mode's Move and Drag capabilities to create dimension witness line gaps.

Where a dimension's witness line is linked to an entity, a visible gap (approximately 1/16-inch wide) is required. Drag the end of each witness line to create these gaps in each view.

STEP 10: Select the CLEAN DIMENSIONS icon.

The Clean Dimension option will consistently space dimensions on a drawing view.

STEP 11: On the work screen, select the front and right-side views then select Done Select (Done Sel) from the Get Select menu.

Pro/ENGINEER's default values for the Clean Dimensions dialog box are 0.500 for the object Offset setting and 0.375 for the dimension Increment setting. You will keep these default settings.

STEP 12: Select APPLY, then CLOSE the dialog box.

After applying the clean dimension settings, Pro/ENGINEER will create snap lines that will evenly space linear dimension lines. Depending upon the complexity of the drawing, the applied settings might not create an ideal drawing.

STEP 13: Use Dimension Modification tools to refine the placement of dimensions and datums.

Use available tools such as Move View, Move, Flip Arrows, and Move Text to tweak the placement of dimensions. Your final dimensioning scheme should appear as shown in Figure 8–70.

STEP 14: Save your drawing.

SETTING GEOMETRIC TOLERANCES

Geometric tolerances are used to control the form and/or location of geometric features. Within Pro/ENGINEER, Geometric Tolerances can be incorporated into a model through Part, Assembly, or Drawing mode. This tutorial will first establish a Position tolerance (±.004 at MMC) for the 1-inch nominal diameter hole. Second, an Angular tolerance (±.005) will be provided for the 50-degree angled surface. Your final Geometric Tolerances should appear as shown in Figure 8–71.

STEP 1: On the menu bar, select INSERT >> GEOMETRIC TOLERANCE.

After selecting the Geometric Tolerance option, the Geometric Tolerance dialog box will be displayed. This is the same Geometric Tolerance dialog box that that is utilized in Part mode. From this step forward, the steps for creating Geometric Tolerances are the same in Drawing mode as they are in Part mode.

STEP 2: On the Geometric Tolerance dialog box, select the POSITION tolerance symbol (Figure 8–72).

STEP 3: On the Toolbar, turn on the display of Axes (Figure 8–73).

This portion of the tutorial will create a Position tolerance for the 1-inch nominal size hole. A Position tolerance affects the axis of a hole. Due to this, in the next step of this tutorial, you will be required to select the hole's axis.

STEP 4: On the Geometric Tolerance dialog box, change the Reference Type to AXIS (Figure 8–72).

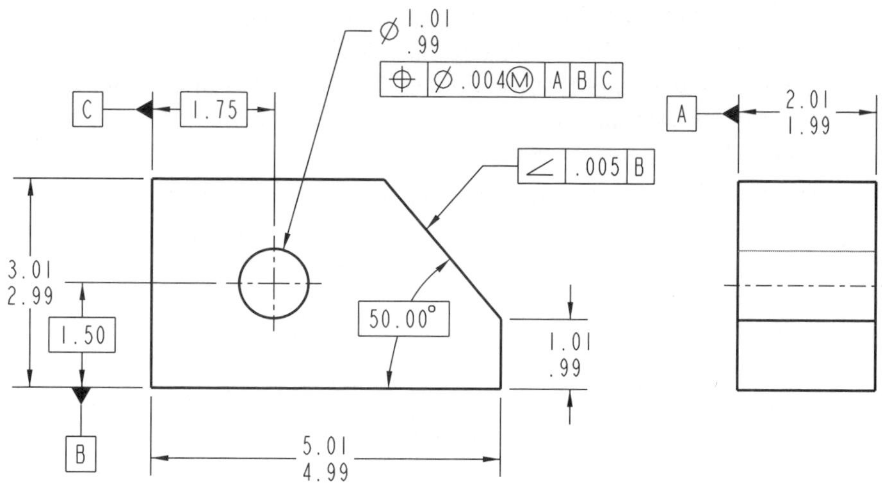

Figure 8–71 Drawing with geometric tolerances

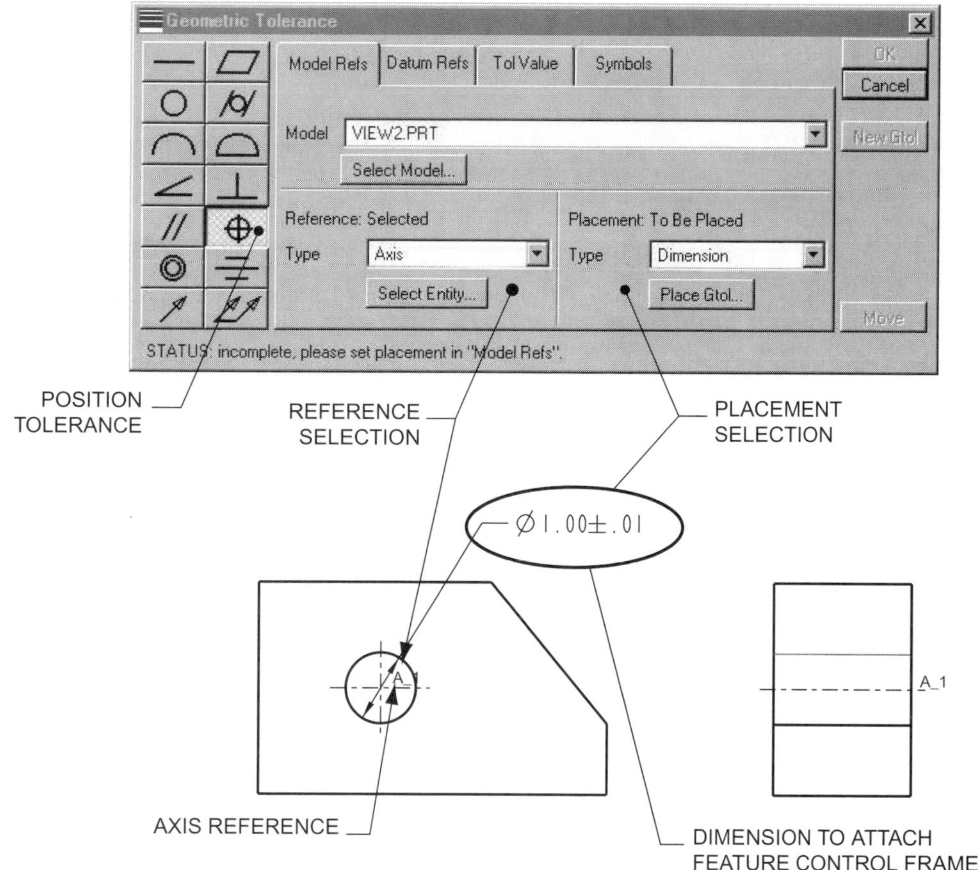

Figure 8–72 Geometric Tolerance dialog box

Figure 8–73 Datum axis display

STEP 5: **On the work screen, select the axis at the center of the hole (Figure 8–72).**

STEP 6: **On the Geometric Tolerance dialog box, set DIMENSION as the Placement Type (Figure 8–72).**

STEP 7: **If necessary, select the PLACE GTOL option on the Geometric Tolerance dialog box.**

STEP 8: **On the work screen select the 1.00 inch diameter dimension value (Figure 8–72).**

When selecting the dimension value, you will have to select the dimension text. The nominal size of the hole is 1 inch. The dimension should be currently displayed with a Limit Tolerance format. This will be modified in a later step.

After performing this step, notice on the work screen how the Feature Control Frame is now visible.

STEP 9: On the Geometric Tolerance dialog box, select the DATUM REFS tab.

STEP 10: On the Geometric Tolerance dialog box, under the PRIMARY Datum Reference tab, set the Basic datum to A (Figure 8–74).

Notice how the Feature Control Frame updates dynamically.

STEP 11: On the Geometric Tolerance dialog box, under the SECONDARY Datum Reference tab, set the Basic datum to B (Figure 8–75).

STEP 12: On the Geometric Tolerance dialog box, under the TERTIARY Datum Reference tab, set the Basic datum to C.

STEP 13: On the Geometric Tolerance dialog box, under the TOL VALUE tab, set the Overall Tolerance to a value of 0.004 and set the MATERIAL CONDITION to MMC (Figure 8–76).

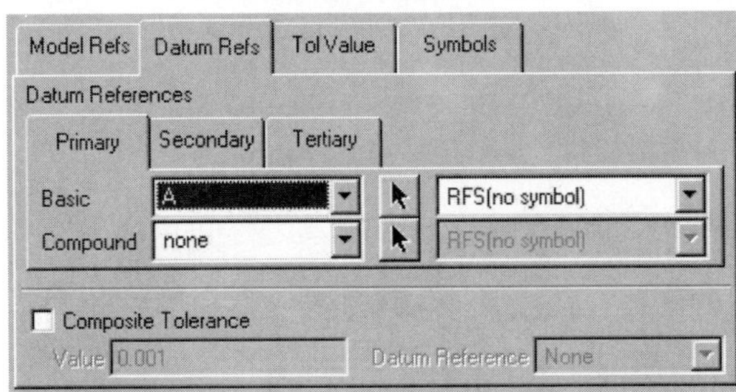

Figure 8–74 Primary datum selection

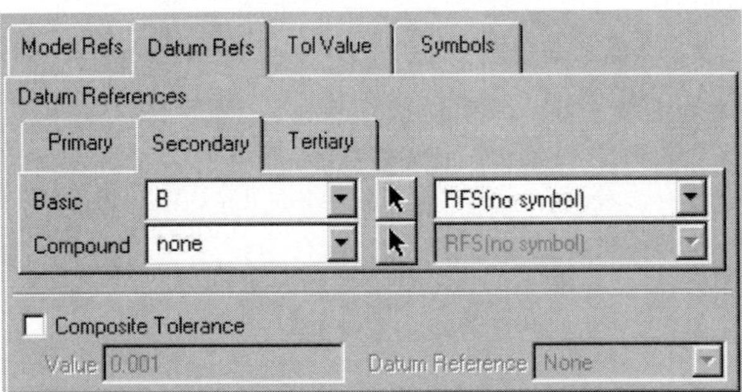

Figure 8–75 Secondary datum selection

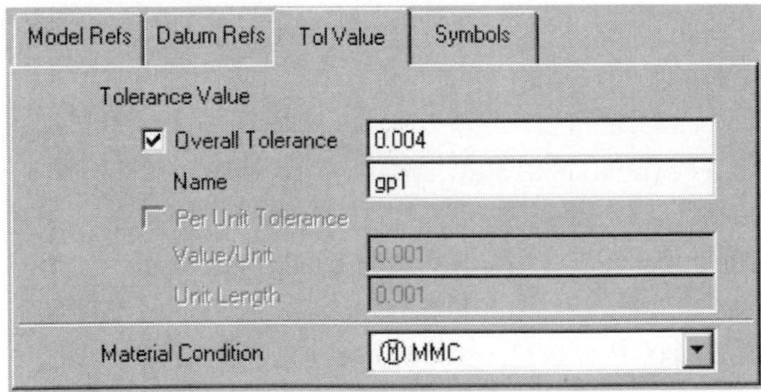

Figure 8–76 Tolerance value

STEP 14: On the Geometric Tolerance dialog box, under the SYMBOLS tab, check the Diameter Symbol option (Figure 8–77).

Since a Position tolerance for a hole creates a cylindrical tolerance zone, it requires a diameter symbol with the tolerance value. Notice how the Feature Control Frame updates dynamically.

STEP 15: Select OKAY on the Geometric Tolerance dialog box.

STEP 16: If prompted, select YES to confirm the setting of Basic dimensions (If Available).

Your drawing should appear as shown in Figure 8–78. Notice how the value of each location dimension is enclosed in a box. This box represents a basic dimension. The confirmation required in this step sets the hole's location dimensions to Basic. If you do not get this message, use the Dimension Properties dialog box (Pop-Up menu >> Properties) to set each location dimension as basic.

STEP 17: Select INSERT >> GEOMETRIC TOLERANCE.

The remaining steps of this segment of the tutorial will create the Angular Geometric Tolerance.

STEP 18: On the Geometric Tolerance dialog box, select the ANGULAR tolerance characteristic (Figure 8–79).

STEP 19: On the Geometric Tolerance dialog box, change the Reference type to SURFACE, then on the work screen select the edge of the angled surface.

STEP 20: Change the Placement Type to LEADERS (Figure 8–79).

The Feature Control Frame will be attached to the surface with a leader.

STEP 21: Select ON ENTITY >> ARROW HEAD on the Attachment Type menu.

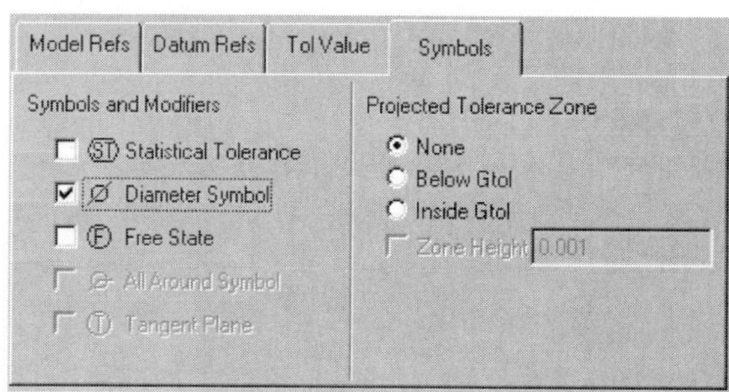

Figure 8-77 Diameter symbols selection

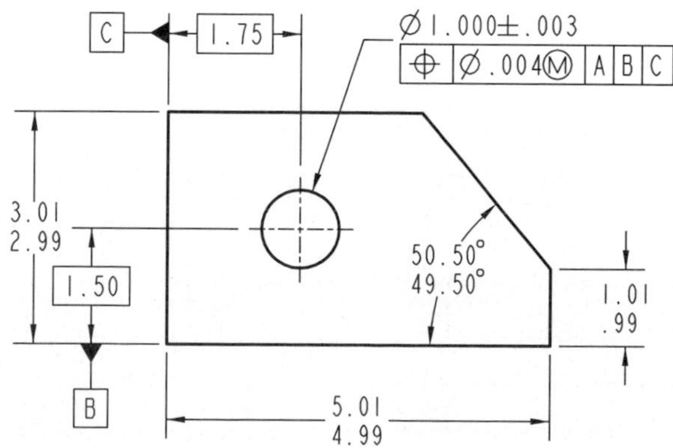

Figure 8-78 Position tolerance

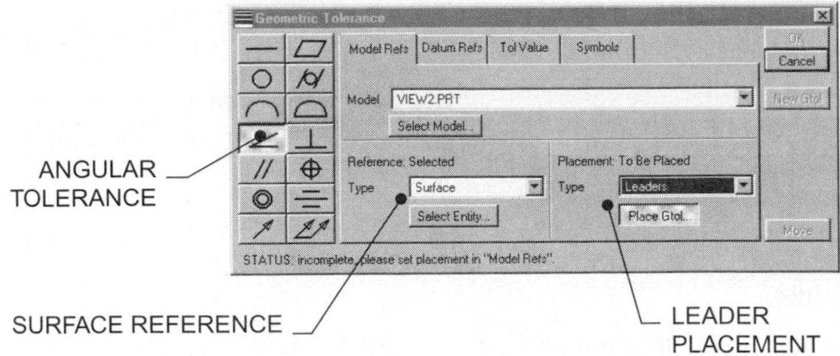

Figure 8–79 Geometric Tolerance dialog box

STEP 22: On the work screen, pick the edge of the angled surface (only select once).

The selected location on the edge will be the attachment point for the leader. Only pick the edge once.

STEP 23: Select the DONE option on the Attachment Type menu.

STEP 24: On the work screen, select a location for the Feature Control Frame.

STEP 25: On the Geometric Tolerance dialog box, under the Datum Refs tab, set the PRIMARY datum reference to B (Figure 8–80).

An Angular Geometric tolerance requires one datum reference. In this example, datum plane B will be used as the datum reference.

STEP 26: On the Geometric Tolerance dialog box, under the Datum Refs tab, set the SECONDARY and TERTIARY datum references to NONE (Figure 8–80).

STEP 27: On the Geometric Tolerance dialog box, under the Tol Value tab, set the Overall Tolerance to a value of 0.005 and set the Material Condition to RFS [no symbol] (Figure 8–81).

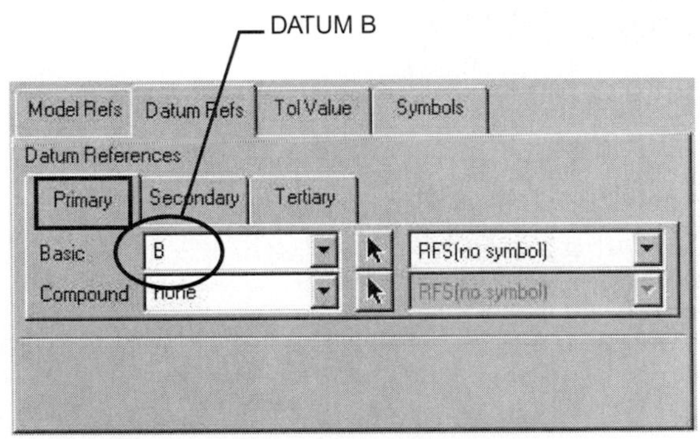

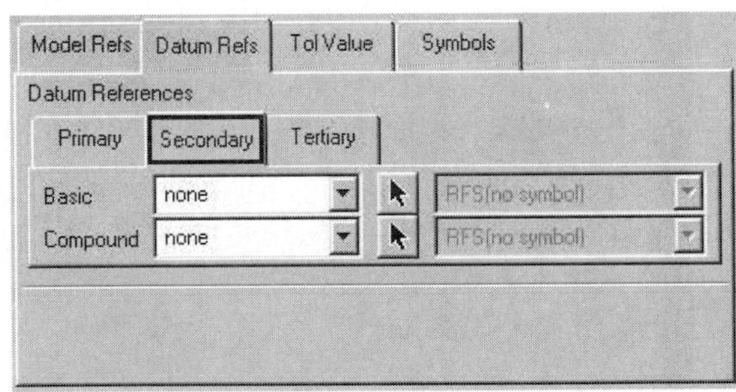

PRIMARY DATUM REFERENCE

SECONDARY DATUM REFERENCE

Figure 8–80 Datum reference selection

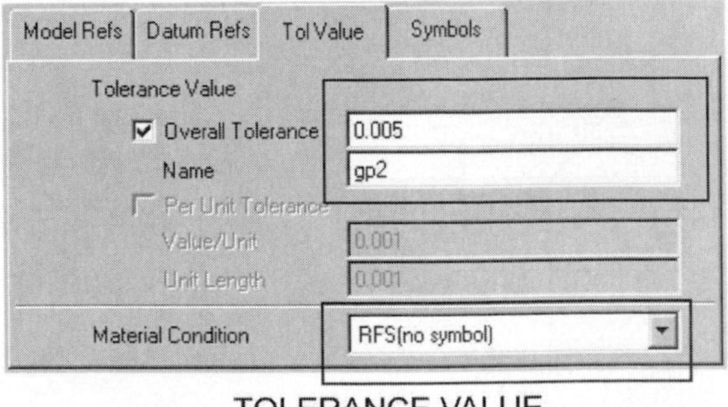

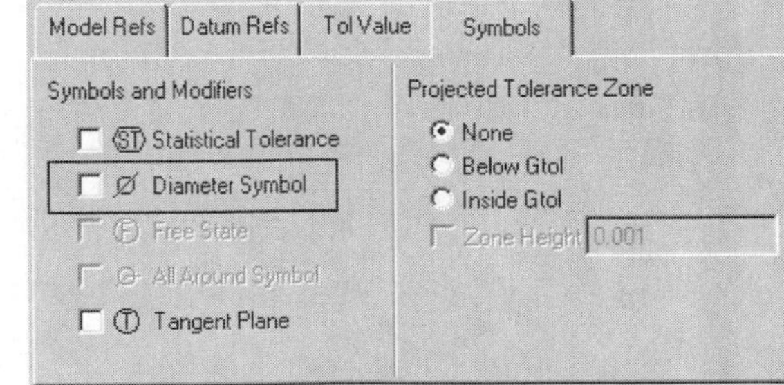

TOLERANCE VALUE

NO DIAMETER SYMBOL

Figure 8–81 Tolerance value and diameter symbol

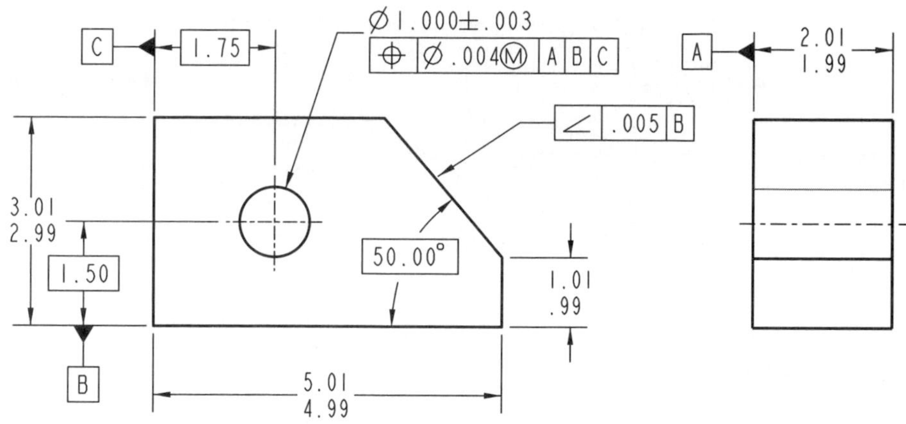

Figure 8–82 Geometric tolerances

STEP 28: On the Geometric Tolerance dialog box, under the Symbols tab, uncheck the DIAMETER SYMBOL button (Figure 8–81).

Angular Geometric tolerances do not create cylindrical tolerance zones. Hence, they do not require a diameter symbols.

STEP 29: Select OKAY on the Geometric Tolerance dialog box.

Due to the Angular Geometric tolerance, the dimension defining the size of the angle (50-degree nominal size) should be set as basic.

> **INSTRUCTIONAL NOTE** If you make a mistake with a geometric tolerance's feature control frame, use the Pop-Up menu's REDEFINE GTOL option to make necessary change.

STEP 30: Select the 50 angular dimension value with your left mouse button, then with your right mouse button, access the Pop-Up menu.

STEP 31: Select the PROPERTIES option on the Pop-Up menu to access the Dimension Properties dialog box.

STEP 32: On the Properties tab, select BASIC as the dimension display type.

STEP 33: Select OKAY to exit the dialog box.

Your drawing should appear as shown in Figure 8–82.

SETTING DIMENSIONAL TOLERANCES

This segment of the tutorial will create the dimensional tolerances shown in Figure 8–83. Previously in this exercise, you set the Drawing Setup File option *display_tol* to Yes. This option is used to display dimensions in a tolerance format.

STEP 1: On the work screen, select the hole's diameter dimension value then access the Pop-Up menu with the right mouse button.

This portion of the tutorial will create a tolerance dimension defining the size of the hole.

STEP 2: Select the PROPERTIES option on the Pop-Up menu.

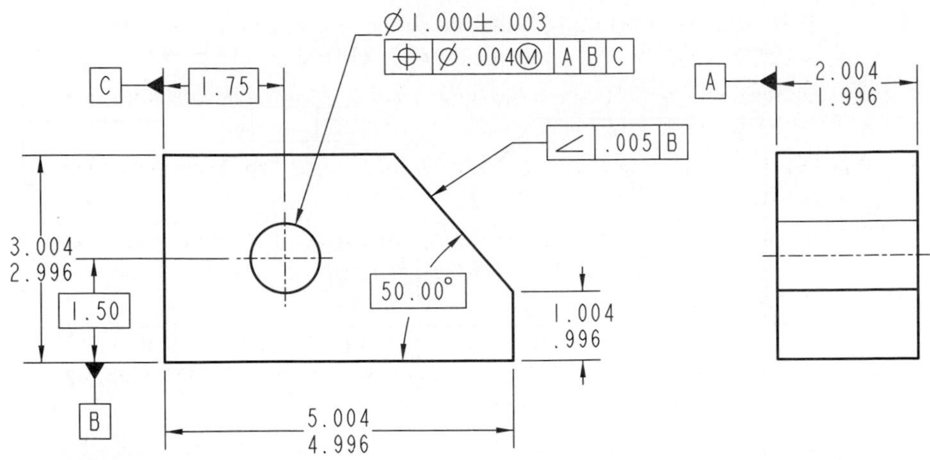

Figure 8-83 Dimensional tolerances

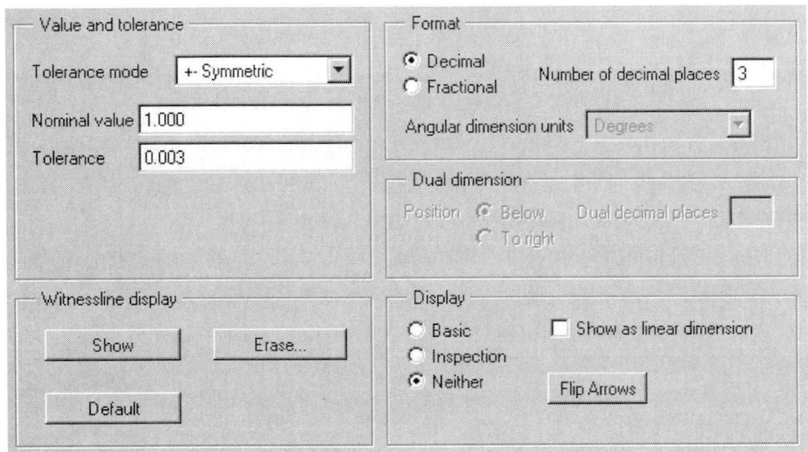

Figure 8-84 Modify Dimension dialog box

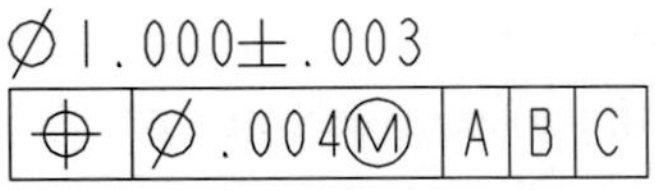

Ø1.000±.003

⌖	Ø.004Ⓜ	A	B	C

Figure 8-85 Hole dimension note

STEP 3: On the Dimension Properties dialog box, change the Tolerance Mode to +−SYMMETRIC (Figure 8–84).

STEP 4: On the Dimension Properties dialog box, change the Number of Digits to a value of 3 (Figure 8–84).

STEP 5: On the dialog box, change the Tolerance value to 0.003 (Figure 8–84).

This will change the tolerance of the hole dimension to plus-minus 0.003 inches.

STEP 6: Select OKAY to exit the Dimension Properties dialog box.

Your dimension note should appear as shown in Figure 8–85.

STEP 7: Using a combination of your SHIFT key and left mouse button, pick the remaining four dimensions that are currently displayed with a Limit format (see Figure 8–86).

This portion of the tutorial will modify the remaining dimension to have a tolerance value of plus-minus 0.004 with a Limit format.

STEP 8: Select EDIT >> PROPERTIES or select the Properties option on the Pop-Up menu.

STEP 9: On the Dimension Properties dialog box, change the Number of Digits to a value of 3 (Figure 8–87).

STEP 10: On the Dimension Properties dialog box, change the Tolerance Mode to +−SYMMETRIC, then change the Tolerance value to 0.004 (Figure 8–87).

Changing the Tolerance Mode to +−**Symmetric** will allow you to change the Tolerance value for all four dimensions at once. In the next step, you will change the Tolerance Mode back to Limits.

STEP 11: Change the Tolerance Mode to LIMITS (Figure 8–88), then select OKAY on the Modify Dimension dialog box.

Your drawing should appear as shown in Figure 8–89.

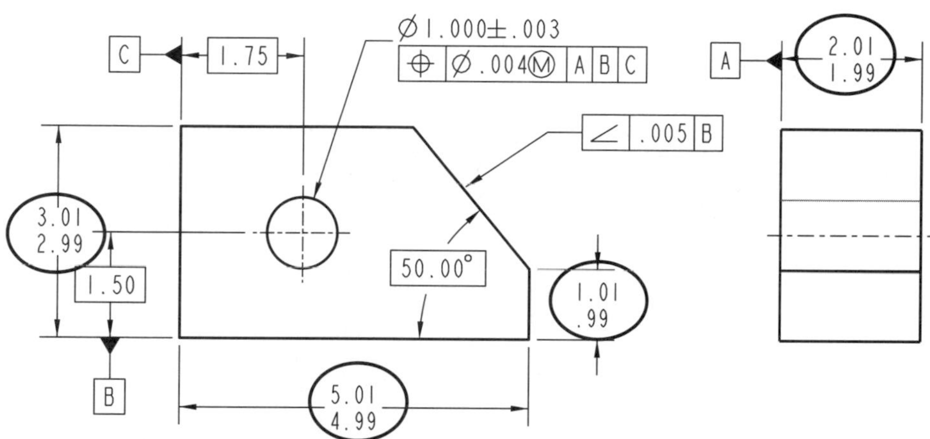

Figure 8–86 Dimensions to change format

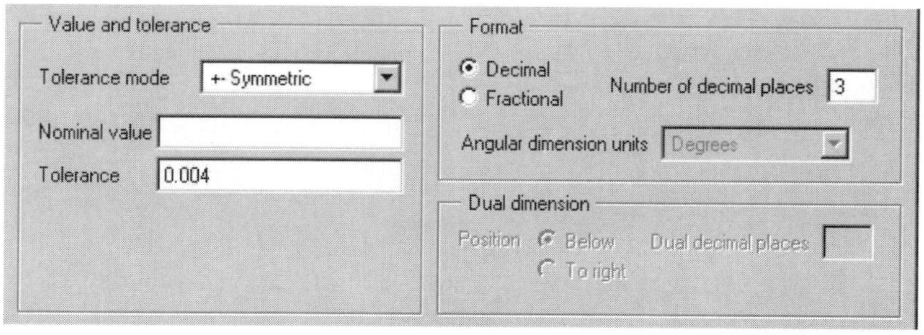

Value and tolerance
Tolerance mode: +- Symmetric
Nominal value:
Tolerance: 0.004
Format
Decimal
Fractional
Number of decimal places: 3
Angular dimension units: Degrees
Dual dimension
Position: Below / To right
Dual decimal places

Figure 8–87 Modify Dimension dialog box

Value and tolerance
Tolerance mode: Limits
Upper limit
Lower limit

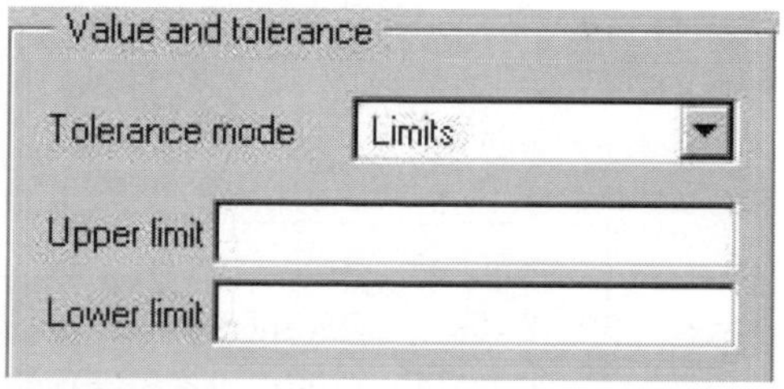

Figure 8–88 Limits tolerance format

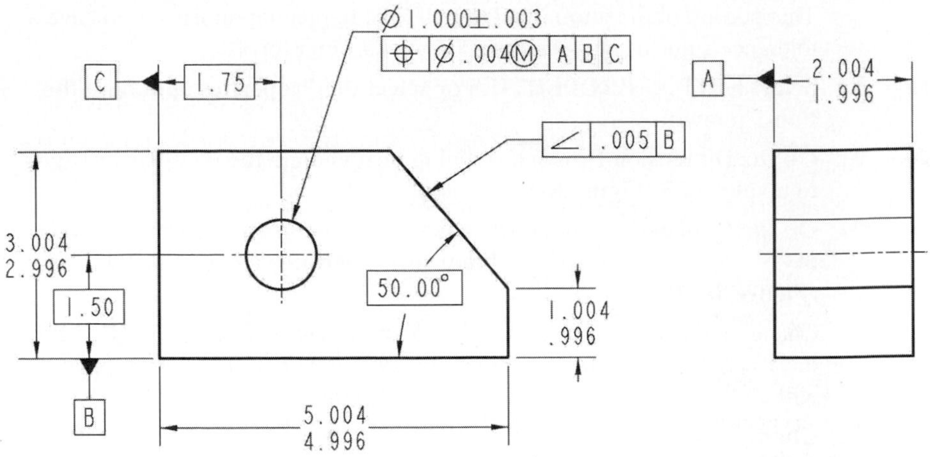

Figure 8–89 Final dimensioning and tolerance scheme

Creating the Title Block

Use the Insert >> Note option to create the title block information shown in Figure 8–90. Move the scale note (SCALE 0.5000) to the title block. The Pop-Up menu's Modify Text Style (Mod Text Style) option can be used to modify the height and style of any text.

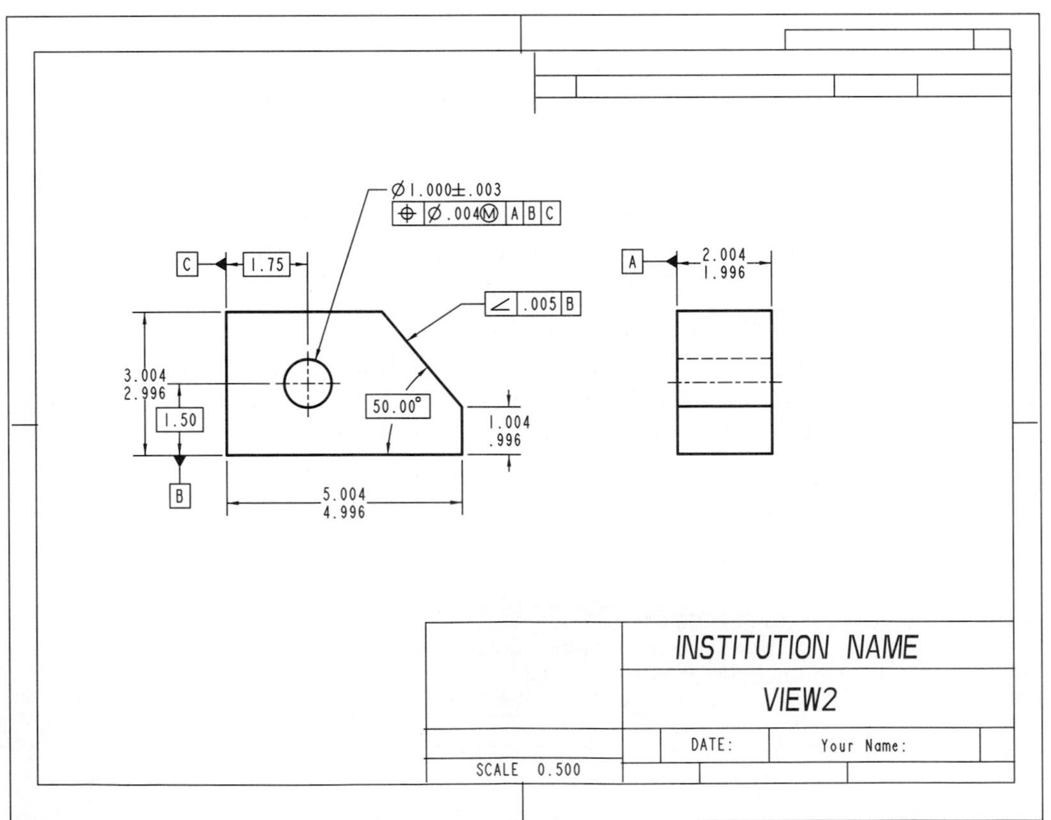

Figure 8–90 Finished drawing

PROBLEMS

1. Model the part shown in Figure 8–91. For this problem, meet the following requirements:
   - The dimensions shown in the figure meet design intent. During part modeling, incorporate these dimensions.
   - Create an engineering drawing with Front, Top, and Right-Side Views.
   - Use an A size sheet and format.
   - Fully dimension the engineering drawing using the part's parametric dimensions.
2. Model the part shown in Figure 8–92. For this problem, meet the following requirements:
   - The dimensions shown in the figure meet design intent. During part modeling, incorporate these dimensions.
   - Create an engineering drawing with Front, Top, and Right-Side Views.
   - Use an A size sheet and format.
   - Fully dimension the engineering drawing using the part's parametric dimensions.
3. Model the part shown in Figure 8–93. For this problem, meet the following requirements:
   - The dimensions shown in the figure meet design intent. During part modeling, incorporate these dimensions.
   - Create an engineering drawing with Front, Top, and Right-Side views.
   - Use an A size sheet and format.
   - Fully dimension the engineering drawing using the part's parametric dimensions.
   - Apply geometric tolerance annotations and ISO datum plane symbols as shown in the figure.

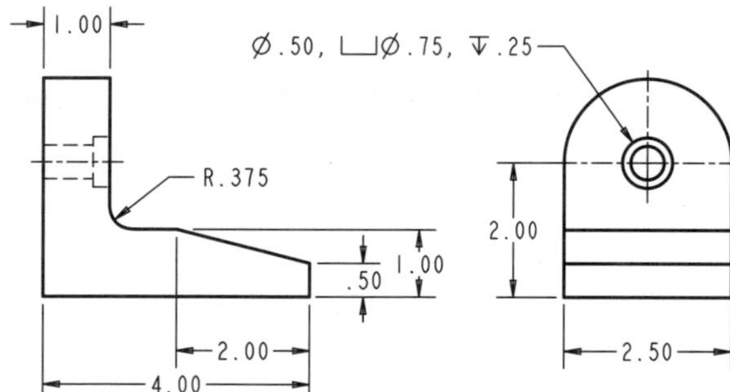

Figure 8-91 Problem one

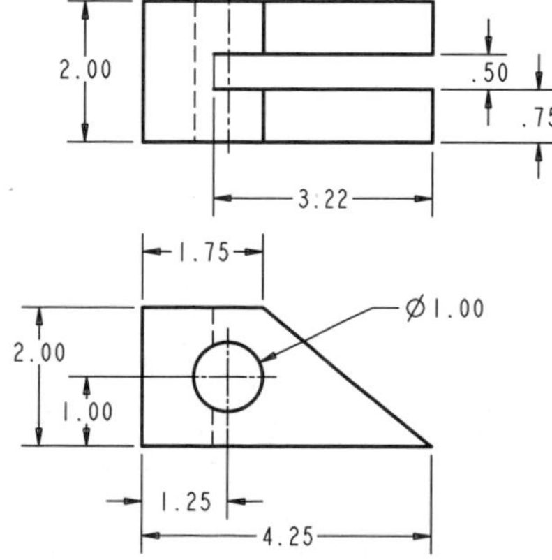

Figure 8-92 Problem two

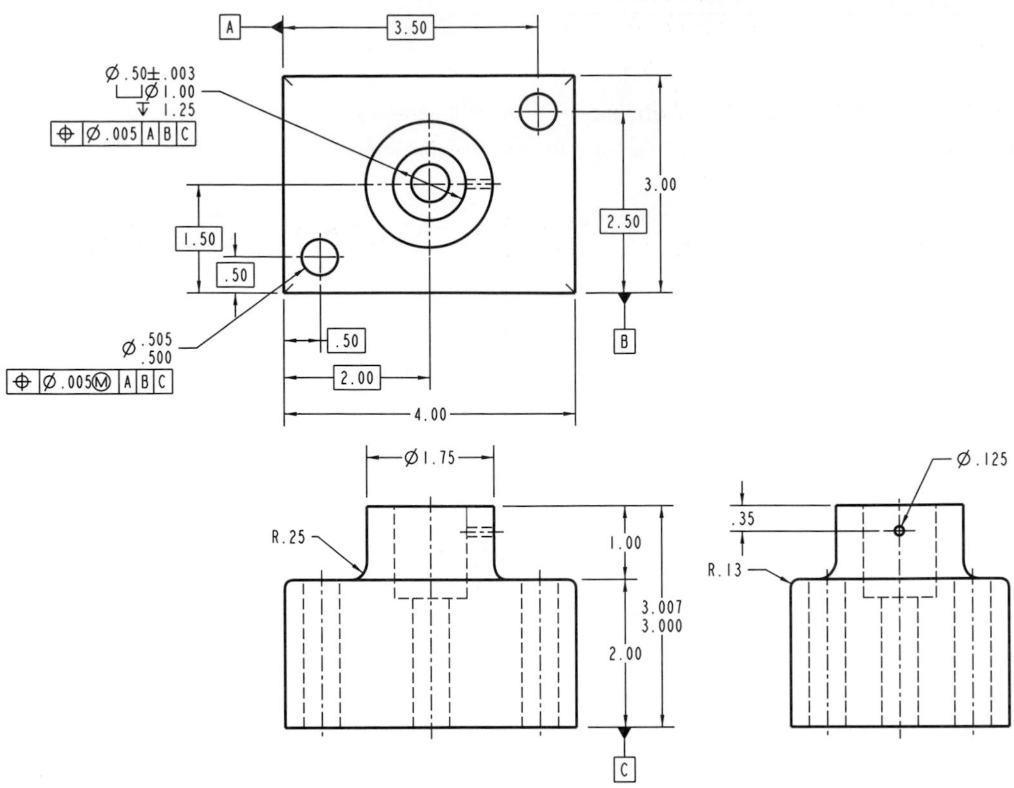

Figure 8-93 Problem three

Questions and Discussion

1. What is the default dimension text height of a Pro/ENGINEER drawing, and how can this text height be changed for an individual drawing? How can the default text height be changed permanently?
2. Describe the process used in Pro/ENGINEER's Drawing mode to add a border/title block overlay to a drawing. How are border/title block overlays created in Pro/ENGINEER?
3. Describe the difference between erasing a drawing view and deleting a drawing view.
4. Describe the difference between a General View and a Projection View.
5. Describe the difference between the Drawing mode views Half, Partial, and Broken.
6. How can a hidden line display be set permanently in a specific drawing view?
7. How can tangent edge display be turned off permanently in a specific drawing view?
8. Describe the process used in Drawing mode to display parametric dimensions and centerlines.
9. Describe the process for adding symbols to a dimension note.

CHAPTER

9

SECTIONS AND ADVANCED DRAWING VIEWS

Introduction

Pro/ENGINEER's Drawing mode provides a variety of options for creating orthographic and detailed drawings. This chapter will explore some of the advanced views available in this mode. Highlighted will be section views and auxiliary views. Upon finishing this chapter, you will be able to

- Create a full section view in drawing mode.
- Create a half section.
- Create an offset section.
- Create a broken out section.
- Create an aligned view.
- Create an auxiliary view.

DEFINITIONS

Auxilary view Any orthographic view that is not one of the six principle views.

Cutting plane line A thick line used to show the cutting pattern of a section view.

Drawing setup file A file used to establish environmental settings for a drawing. Examples of possible settings include text height and arrowhead size.

Offset section A section view whose cutting plane is offset to include more features within the section.

Section lining A pattern used on a section view to show where a model is cut.

SECTION VIEW FUNDAMENTALS

Section views are utilized within a working drawing to show details of a design that would be difficult to view using traditional orthographic projection. Figure 9–1 shows an example of a Full Section view. A section view simulates what a model would look like if it were actually cut apart. On the Section View, **section lining** is used to show where the part is cut. Industry standards provide section lining patterns for a variety of materials. The default section lining used in Pro/ENGINEER represents iron (see Figure 9–1). A section lining's line style, weight, spacing, and angle can be changed with the Edit >> Properties option.

Another important line type associated with a section view is the **cutting plane line** (Figure 9–1). Cutting plane lines typically lie in a view adjacent to the section view and represent the cutting path of the cross section (a Removed Section is an example of a view

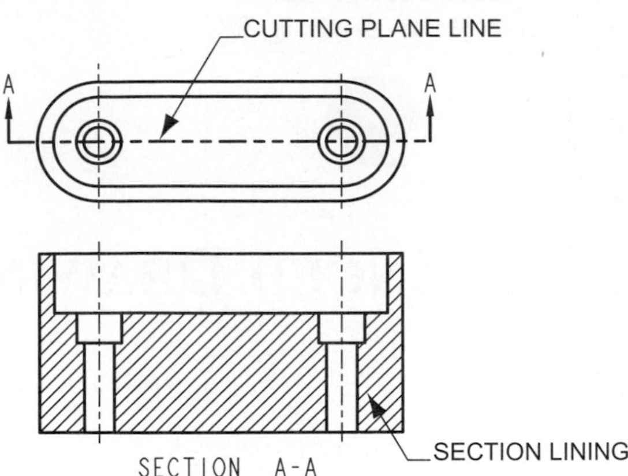

Figure 9–1 Full section view

where the cutting plane line does not lie in a view adjacent to its section view). Arrowheads terminate the ends of a cutting plane line and point in the viewing direction of the section. The **Drawing Setup File** options *crossec_arrow_length* and *crossec_arrow_width* control the size of a cutting plane line's arrowheads.

Within Pro/ENGINEER, section views are created from defined Cross Sections. Pro/ENGINEER provides options for creating Cross Sections in Part, Assembly, and Drawing modes. Chapter 7 provides details on how to create a Cross Section in Part mode. Cross Sections created in Part and Assembly modes can be retrieved and used to create section views in Drawing mode. While creating a section view, Drawing mode also provides an option for creating a Cross Section.

Section View Types

The most common section view used is the Full section. Other types of section views are available to serve of variety of documentation needs. The following is a description of the types of section views found under Pro/ENGINEER's Xsec Type menu:

Full Section

The Full section view is the traditional type of section view used on most engineering drawings. As shown in Figure 9–2, A Full Section passes completely through a model. A Full section is available for General, Projection, and Auxiliary views.

Half Section

The Half section view is similar to the Full section, except only half of the view is sectioned. As shown in Figure 9–2, for a symmetrical model Half sections provide the advantage of a section on half the view, while also presenting the other half with traditional projection. A Half section is available for General, Projection, and Auxiliary views. It is not available with the Half, Broken, and Partial view types.

Local

The Local Section option is used to create a Broken-Out section view. As shown in Figure 9–2, a Local section creates a section in a specific, user-defined area. Local sections are available for General, Projection, and Auxiliary views. It is not available with the Half and Broken view types.

Full & Local

The Full & Local Section option is a section view with both a Full section and a Local section (Figure 9–2). The Full Section is placed first.

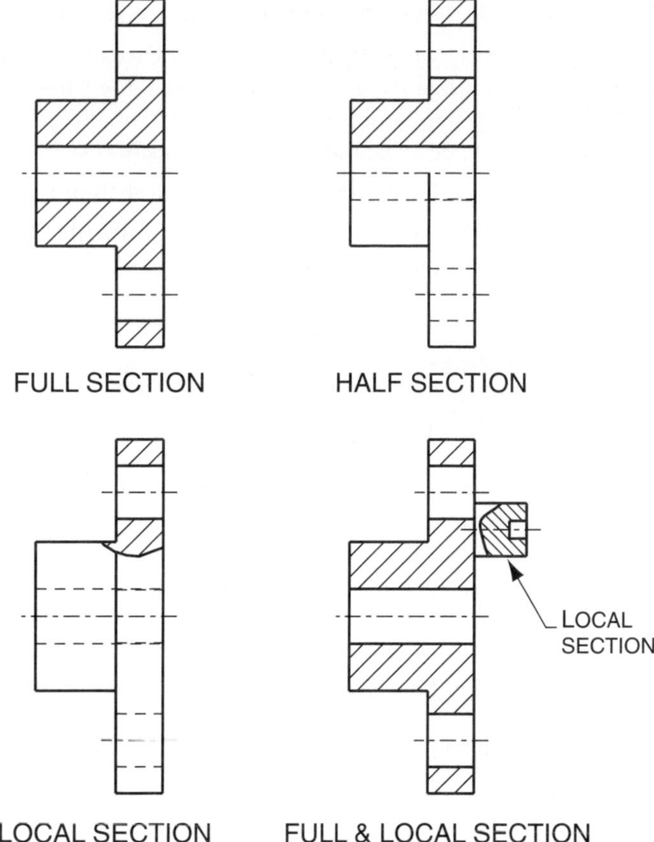

Figure 9–2 Section view types

FULL SECTIONS

Full Section views cut completely through an object and show the entire model. Figures 9–1 and 9–2 show examples of Full Section views. Cross Sections used to construct a Full Section are created along a planar surface. Often, a datum plane is used as this surface. Perform the following steps to construct a Full Section view.

STEP 1: Select VIEWS >> ADD VIEW.

STEP 2: Select a View Type.

Full Section views can be created as a Projection, Auxiliary, General, Detailed, or Revolved view.

STEP 3: Select FULL VIEW as a view type.

Full View will create a view of an entire model.

STEP 4: Select SECTION from the View Type menu.

STEP 5: Select either SCALE or NO SCALE from the View Type menu.

For General views, an option is available for either specifying a view scale or for taking the default. When creating a section view that is projected from an existing view (e.g., Projection or Auxiliary), the scale of the parent view is used by default.

STEP 6: Select DONE to accept the view types.

STEP 7: Select FULL as the section type.

A Full section is a section that runs completely through a model (Fig. 9–2).

STEP 8: Select TOTAL XSEC from the Cross Section Type menu.

As shown in Figure 9–3, A Total Cross Section (Total Xsec) is a section view that includes geometry that borders the cross section and other geometry of the model. An Area Cross Section (Area Xsec) only shows the geometry that borders the cross section.

STEP 9: Select DONE to accept the cross section type values.

STEP 10: On the work screen, select a location for the section view.

STEP 11: Select CREATE on the Cross Section Enter menu.

> **INSTRUCTIONAL NOTE** If an appropriate Cross Section already exists, use the Retrieve option to select the Cross Section.

Create will construct a Cross Section within the section view option, while Retrieve will select a Cross Section that has been previously created in Part or Assembly mode. The procedure for creating a Cross Section in Drawing mode is similar to the procedure for constructing a Cross Section in Part and Assembly modes. To construct a full section, a planar surface is needed. Datum planes are used often as this surface.

STEP 12: Select PLANAR >> DONE as the Cross Section creation method.

Planar will create a Cross Section through a part at the location of a planar surface. While Planar is used to create a straight Cross Section, the Offset option will create a Cross Section that does not lie along a straight line.

STEP 13: Enter a name for the section view.

In Pro/ENGINEER's textbox, enter a name for the section view. Within the area of mechanical drafting, sections are often named with an alphabetic character.

STEP 14: Select either a planar surface or a datum plane.

This plane will create the cross section that will define the section view. The plane has to lie parallel to the section view location.

STEP 15: Select a view to locate the Cutting Plane Line (Figure 9–4).

Pro/ENGINEER's message area is requesting a "view for arrows where the section is perpendicular." The Cutting Plane Line defining the Cross Section

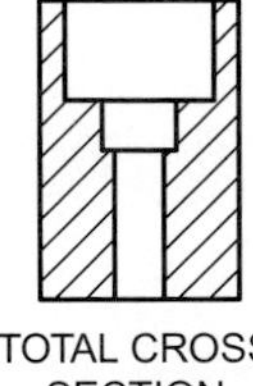

Figure 9–3 Total and area section views

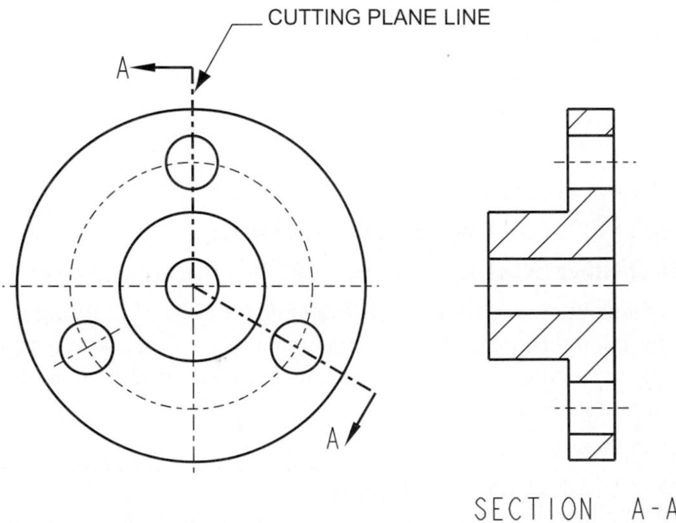

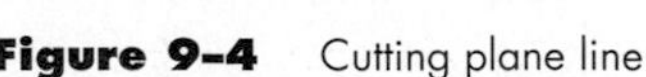

Figure 9–4 Cutting plane line

cut will be located in this view. If a Cutting Plane Line is not necessary, select the middle mouse button.

HALF SECTIONS

Section views are used to improve the clarity of an engineering design by providing an avenue where the interior details of a model can be viewed. Often, it is not necessary to create a Full section. Full sections show the entire view of a model as a section. Due to this, some details outside the section might lose their clarity. Half sections are used to create a section view through only half of a model. Figure 9–5 shows an example of a Half section. Perform the following steps to construct a Half section view.

STEP 1: Select VIEWS >> ADD VIEW.

STEP 2: Select a View Type.

Full Section views can be created as a Projection, Auxiliary, or General.

STEP 3: Select FULL VIEW as a view type.

Do not confuse a Half view of a model with a Half section view. Half section views are actually full views of a model with only half of the view sectioned.

STEP 4: Select SECTION from the View Type menu.

STEP 5: Select either SCALE or NO SCALE from the View Type menu.

For General views, the option is available for either specifying a view scale or for taking the default. When creating a section view that is projected from an existing view (e.g., Projection or Auxiliary), the scale of the parent view is used by default.

STEP 6: Select DONE to accept the view types.

STEP 7: Select HALF >> TOTAL XSEC >> DONE as the section type.

Pro/ENGINEER does not provide the option of creating an area or aligned cross section with a Half section.

STEP 8: On the work screen, select a location for the section view.

STEP 9: Select a reference plane that will form the division edge for the Half section (Figure 9–6).

On the section view, select a planar surface (usually a datum plane) that will serve as the point dividing the sectioned half of the view from the nonsectioned half.

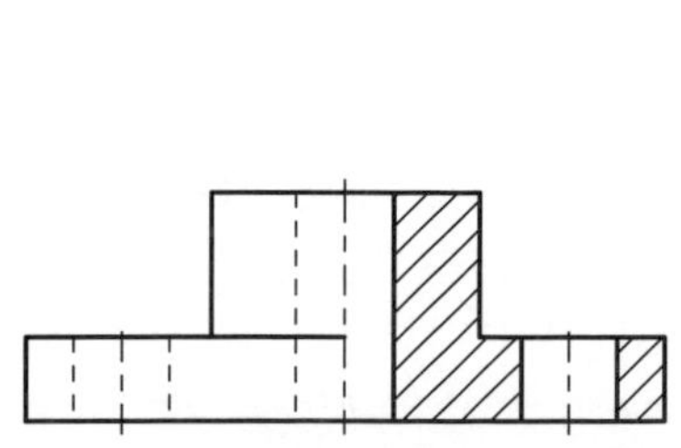

Figure 9-5 Half section view

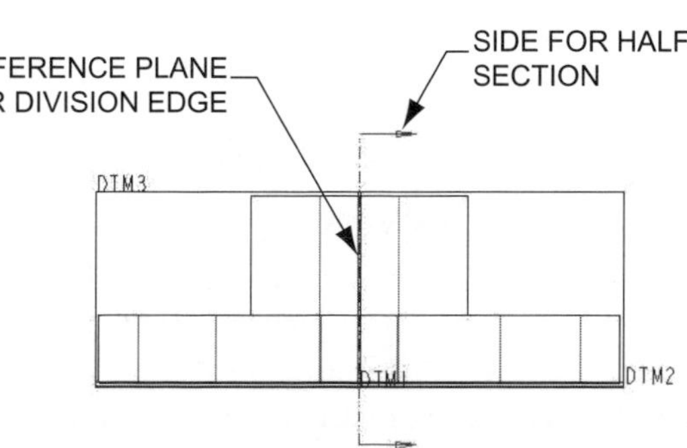

Figure 9-6 Reference plane selection

STEP 10: Choose either FLIP or OKAY from the Direction menu to select a side of the view that will be sectioned.

As shown in Figure 9–6, Pro/ENGINEER provides arrows that point toward the side of the model that will be sectioned. Choose either Okay to accept the default side or choose Flip to change the sectioned side.

STEP 11: Select CREATE from the Xsec Enter menu.

INSTRUCTIONAL NOTE If an appropriate Cross Section view already exists, use the Retrieve option to select the Cross Section.

Create will construct a Cross Section within the section view option, while Retrieve will select a Cross Section that has been previously created in Part or Assembly mode.

STEP 12: Select PLANAR >> DONE as the Cross Section creation method.

Planar will create a Cross Section through a part at the location of a planar surface. While Planar is used to create a straight Cross Section, the Offset option will create a Cross Section that does not lie along a straight line.

STEP 13: Enter a name for the section view.

In Pro/ENGINEER's textbox, enter a name for the section view. Within the area of mechanical drafting, sections are often named with an alphabetic character.

STEP 14: Select either a planar surface or a datum plane.

This plane will create the cross section that will define the section view. The plane has to lie parallel to the section view location.

STEP 15: Select a view to locate the Cutting Plane Line.

Pro/ENGINEER's message area is requesting a "view for arrows where the section is perpendicular." The Cutting Plane Line defining the Cross Section cut will be located in this view. If a Cutting Plane Line is not necessary, select the middle mouse button.

OFFSET SECTIONS

Most section views are situated along straight cutting planes. Often, features cannot be fully described through a straight cut. Drafting standards allow for an Offset Cutting Plane. Figure 9–7 shows an example of an **Offset section** view with its corresponding offset cutting plane. With an Offset section, the cutting plane is offset by the use of 90-degree bends that allow the cutting plane to pass through features that require sectioning. While normal section views are created within Pro/ENGINEER using a planar surface or datum plane, the cutting plane line for an offset section is sketched. The part file (*offset.prt*) and drawing file (*offset.drw*) for practicing this guide are available on the supplemental CD. Perform the following steps to construct a projected offset section view.

STEP 1: Select VIEWS >> ADD VIEW >> PROJECTION.

Offset Section views can be created as a Projection, Auxiliary, or General view. This guide will demonstrate the creation of a section view as a Projection.

STEP 2: Select FULL VIEW as a view type.

STEP 3: Select SECTION from the View Type menu.

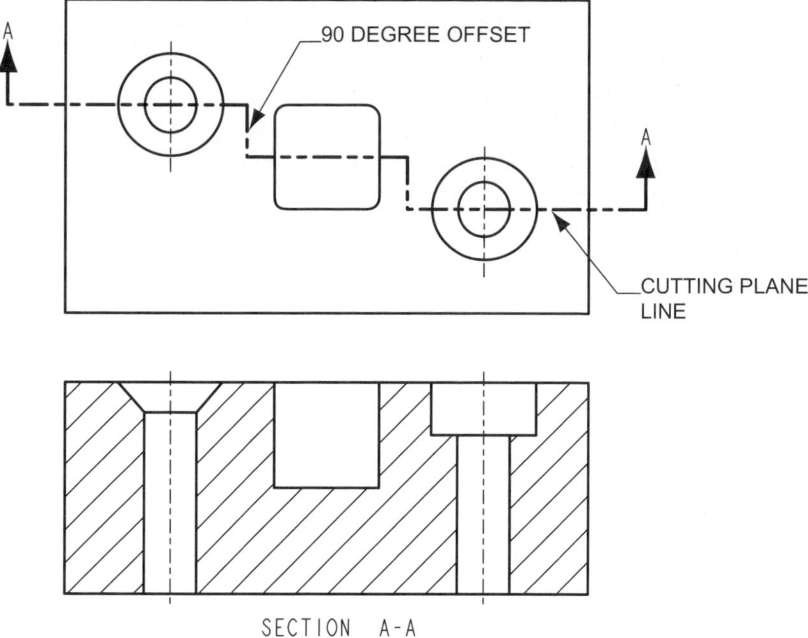

Figure 9-7 Offset section

STEP 4: Select NO SCALE from the View Type menu (default selection).

For General views, the option is available for either specifying a view scale or for taking the default. When creating a section view that is projected from an existing view (e.g., Projection or Auxiliary), the scale of the parent view is used by default.

STEP 5: Select DONE from the View Type menu to accept the view type values.

STEP 6: Select a Cross Section Type, then select DONE.

Offset Section views can be created as a Full, Half, or Local cross section type. Additionally, a Total Cross Section (TotalXsec), Area Cross Section (Area Xsec), Aligned Cross Section (Align Xsec), or Total Align Cross Section type can be used.

STEP 7: On the work screen, select the location of the section view.

Use the mouse cursor to locate the section view. When creating an offset section as part of a General view, you are required to orient the view.

STEP 8: Select CREATE from the Cross-Section Enter menu.

Offset cross sections can be created in Drawing mode directly within the adding of a view, or they can be created in Part mode and retrieved during the placing of a section view. This guide will demonstrate the creation of an offset cross section view in Drawing mode.

STEP 9: Select OFFSET from the Cross-Section Create menu.

STEP 10: Select BOTH SIDES >> SINGLE >> DONE.

STEP 11: In Pro/ENGINEER's textbox, enter a name for the section view.

STEP 12: Switch to the model's window.

If necessary, use the Windows Taskbar or Pro/ENGINEER's Application Manager to switch to the part or assembly model from which the drawing is being created. When constructing an Offset section, for projection views, you must select a sketching plane on the actual model. When selecting a

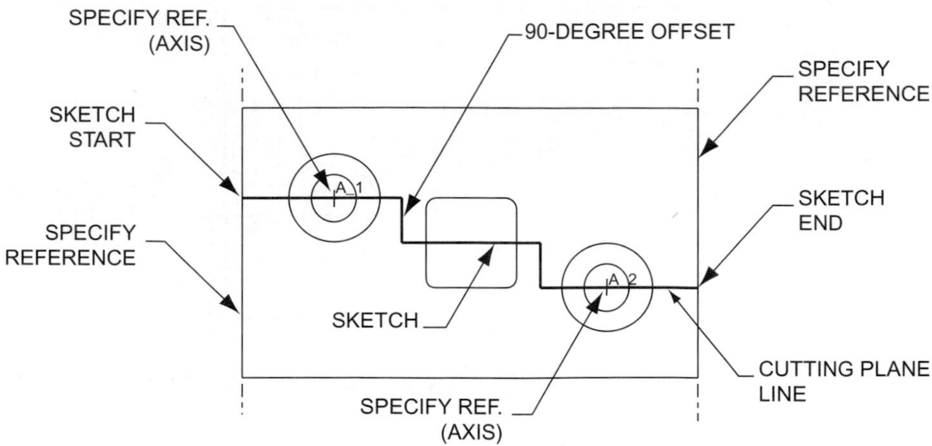

Figure 9–8 Sketching an offset section

sketching plane from a projection view, ensure that the model from which the drawing is being produced is in the active window. You must switch to this window at this step in the view creation process.

STEP 13: Select or create a sketching plane then orient the sketching environment.

Pro/ENGINEER creates offset sections by sketching a cutting plane line. This step requires you to select or create a planar surface that will be suitable for sketching the cutting plane line.

STEP 14: Sketch the cutting plane line (Figure 9–8).

Utilizing appropriate sketching tools, sketch the cutting plane line. As shown in the figure, when sketching the cutting plane line, ensure that you differentiate between the lines that form the cut and the lines that form the offset.

> **MODELING POINT** When sketching a cutting plane line with Intent Manager active, select the axis of each hole to include in the section as a reference. When sketching without Intent Manager, be sure to align the cutting plane line with the axis of each hole.

STEP 15: Select the Continue icon to exit the sketching environment.

STEP 16: On the drawing, select a view to display the cutting plane line.

STEP 17: Select OKAY to accept the direction of view, or use the FLIP option to change the direction.

BROKEN OUT SECTIONS

Broken Out sections allow for the display of internal details of a model without the creation of a Full section. Figure 9–9 shows an example of a typical Broken Out section. Broken Out sections do not utilize a cutting plane line. Within Pro/ENGINEER, a Broken Out section can be created with a General, Projection, or Detail view. The part file (*broken_out.prt*) and drawing file (*broken_out.drw*) for practicing this guide are available on the supplemental CD. Perform the following steps to create a Broken Out section.

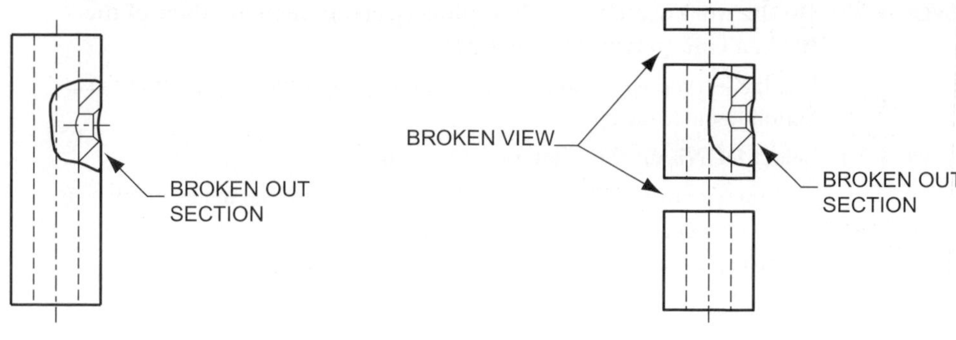

Figure 9-9 Broken out section

Figure 9-10 A broken out view versus a broken view

STEP 1: Select VIEWS >> ADD VIEW.

STEP 2: Select either GENERAL, PROJECTION, or DETAIL.

STEP 3: Select FULL VIEW >> SECTION.

Do not confuse the Broken View option with the creation of a Broken Out view. The Broken View option creates a view with break lines (Figure 9–10). Since this guide is demonstrating the creation of a Broken Out section on a full view, you should pick the Full View option.

STEP 4: If required, select either SCALE or NO SCALE.

STEP 5: Select DONE to accept view type values.

STEP 6: Select LOCAL >> TotalXsec >> DONE.

The Local option is the key selection for creating a Broken Out section. This option creates a section view within a sketched spline boundary. This boundary will be sketched in a later step.

STEP 7: On the work screen, select a location for the view, then, if required, enter a scale value.

In step 4, if you selected Scale, you must enter a scale value during this step.

STEP 8: Orient the Model (for a General View only).

For a General view, orient the model as required to properly display the Broken Out section.

STEP 9: Select ADD BREAKOUT >> SHOW OUTER on the View Boundary menu.

The Add Breakout option allows for the sketching of a spline boundary. The Show Outer option will show the remainder of the view outside of the sketched spline boundary.

STEP 10: Select CREATE on the Xsec Enter menu.

STEP 11: Select PLANAR >> SINGLE >> DONE.

STEP 12: Enter a name for the Broken Out section.

STEP 13: Select a plane for creating the Cross Section.

STEP 14: Select a view to place the cutting plane line or select the middle mouse button to not place.

STEP 15: On the work screen, select an entity approximately at the center of where the Broken Out view will be located (Figure 9–11).

This selection is required for the normal regeneration of the broken out portion of the sectioned view.

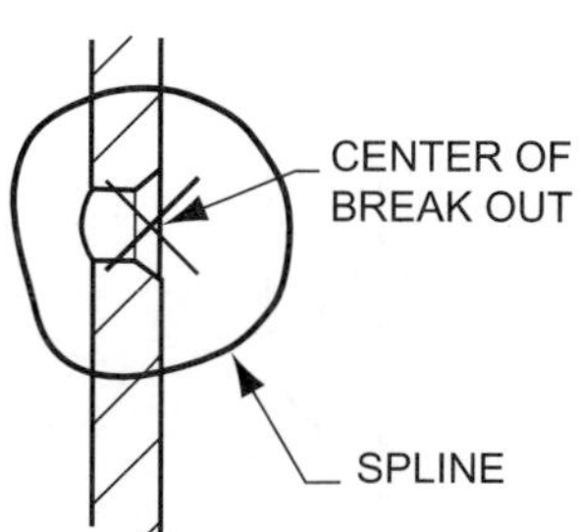

Figure 9-11 Creating the break out

STEP 16: On the work screen, sketch a spline to create the boundary of the Broken Out section (Figure 9–11).

Use the left mouse button to locate spline points and the middle mouse button to close the spline.

STEP 17: Select DONE on the View Boundary menu.

Multiple nonoverlapping splines may be sketched. Sketch as many as is necessary, then select the Done option. Your Broken Out section view will be created after selecting Done.

ALIGNED SECTION VIEWS

Engineering graphics is a language used to communicate design intent. Standards and conventions exist that govern the way designs are displayed on an engineering drawing. Within the realm of engineering graphics, designs are often displayed using multiview projection. Multiview projection does not always present the best display of a design. Figure 9–12 shows a design displayed using normal lines of projection. Normal lines of projection for a multiview drawing project at a 90-degree angle. With the drawing shown in Figure 9–12, the part does not form a 90-degree angle. This presents a projection problem. Clarity for this design can be improved with the use of an Aligned view. As shown in Figure 9–13, the angled feature of the model can be aligned with normal lines of projection to create a drawing with more clarity.

Pro/ENGINEER's Drawing mode has the capability of producing an Aligned Section view. The following guide shows how to produce the Aligned Section view shown in Figure 9–13. The part (*align.prt*) and drawing (*align.drw*) files for this guide are available on the supplemental CD.

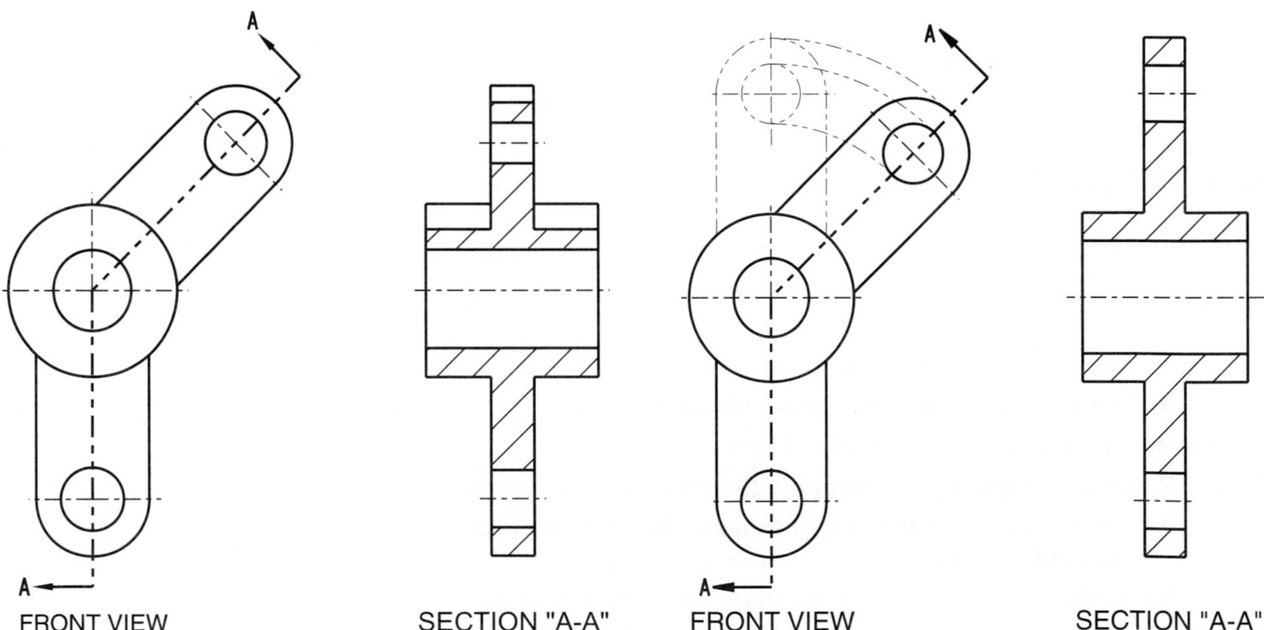

Figure 9–12 Normal lines of projection

Figure 9–13 Aligned view

INSTRUCTIONAL NOTE Cross Sections can be created in Part, Assembly, or Drawing mode. During the creation of a Section view, the Cross Section can be created during the View Add option. In this guide, the Cross Section was created in Part mode as an Offset Section. This Cross Section will be Retrieved during the view creation process.

STEP 1: Select VIEWS >> ADD VIEW.

STEP 2: Select PROJECTION >> FULL VIEW >> SECTION >> DONE.

This example shows the creation of an Aligned Section view projected from an existing view. Aligned Section views can be created as a General view.

STEP 3: Select FULL >> TOTAL ALIGN >> DONE on the Cross-Section Type menu.

The Full option will create the section view completely through the model. The Total Aligned option will create the aligned view. Another option could be Align Xsec. This option would produce the view shown in Figure 9–14.

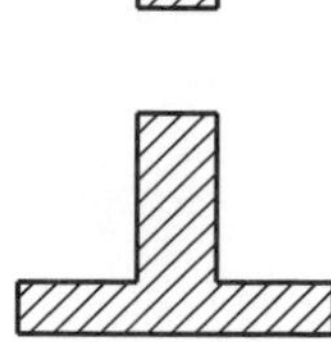

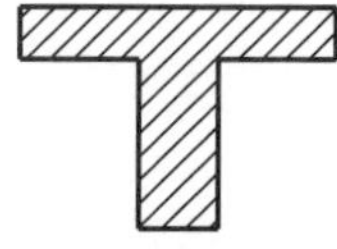

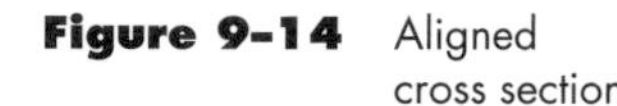

Figure 9–14 Aligned cross section

STEP 4: On the work screen, select the location for the view.

STEP 5: Retrieve the section view that was created in Part mode.

As previously mentioned, the Cross Section used in this guide was created in Part mode as an Offset Section. Pro/ENGINEER provides the capability for creating the identical Cross Section in Drawing mode through the use of the Create option.

STEP 6: Select an Axis to revolve the feature about.

This is the key step for the creation of an aligned view. In this example, select the axis shown in Figure 9–15. This axis serves as the rotation point for aligning the view.

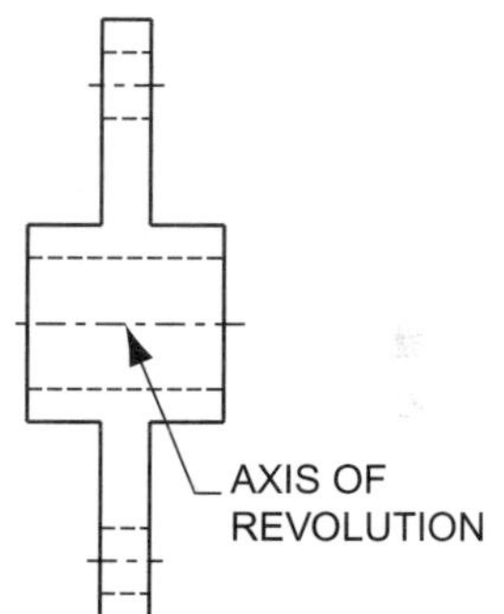

Figure 9–15 Axis of revolution

STEP 7: Select the view for the Cutting Plane Line.

In this example, the Cutting Plane line was placed in the Front view. If the cutting plane line is not desired, select the middle mouse button.

STEP 8: Select OKAY to accept the default viewing direction.

REVOLVED SECTIONS

Revolved Sections are used to show the cross section of a spoke, rail, or rib type feature. Additionally, they are used with features that are extruded, such as wide flange beams. Revolved sections are useful for representing the cross section of a feature without having to create a separate orthographic view. Revolved Sections are displayed by revolving the cross section 90 degrees. Figure 9–16 shows three different ways to locate the cross section in relation to its parent view. The following guide will demonstrate how to superimpose a Revolved section onto a view.

STEP 1: Create or Identify the view from which to obtain the Revolved Section.

A Revolved section can be created from a Projection, Auxiliary, or General view. Additionally, one can even be created from an existing Revolved view.

STEP 2: Select VIEWS >> ADD VIEW >> REVOLVED.

STEP 3: Select FULL VIEW >> DONE.

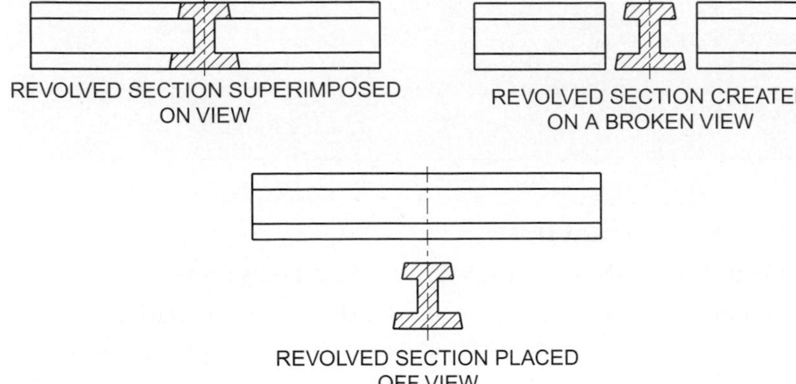

Figure 9-16 Revolved sections

The Revolved option allows only for either a Full View or a Partial View. A Revolved view by default has to be a section view.

STEP 4: On the work screen, select the location for the Revolved view.

For a superimposed Revolved section view, select the view used to create the Revolved section. The Move View option can be used to reposition the view.

STEP 5: Select the view from which to create the Revolved Section.

STEP 6: RETRIEVE an existing Cross Section or CREATE a new one.

The Cross Section created or retrieved in this step will be used as the Revolved Section.

STEP 7: Select a Symmetry Axis for the Revolved Section or select the middle mouse button to accept the default.

The Symmetry Axis is the location about which the Revolved Section view will be centered.

STEP 8: Use the VIEWS >> MOVE VIEW option to refine the placement of the Revolved Section.

AUXILIARY VIEWS

Within the language of engineering graphics, any object has six principle views. An **Auxiliary view** is any orthographic projected view that is not one of the six principle views. Auxiliary views are used frequently to show the true size of an inclined surface. Figure 9–17 shows an example of an auxiliary view and how it helps to better represent the inclined surface. This guide will demonstrate the creation of an Auxiliary view. The part file (*auxiliary.prt*) and drawing file (*auxiliary.drw*) for practicing this guide are available on the supplemental CD.

STEP 1: Select VIEWS >> ADD VIEW.

STEP 2: Select AUXILIARY as a view type.

Auxiliary is one of the five principle view types available in Drawing mode. An existing view is required before an Auxiliary view can be created.

STEP 3: Select FULL VIEW >> NoXsec.

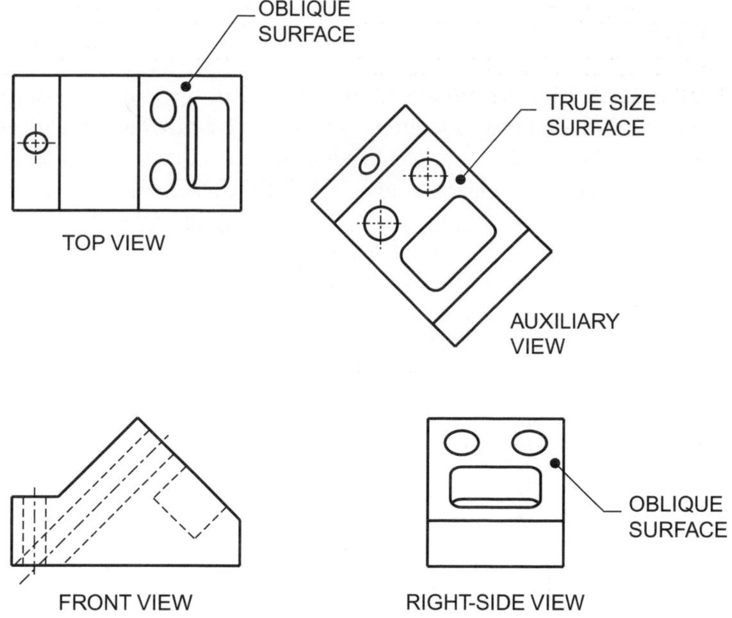

Figure 9-17 An auxiliary view

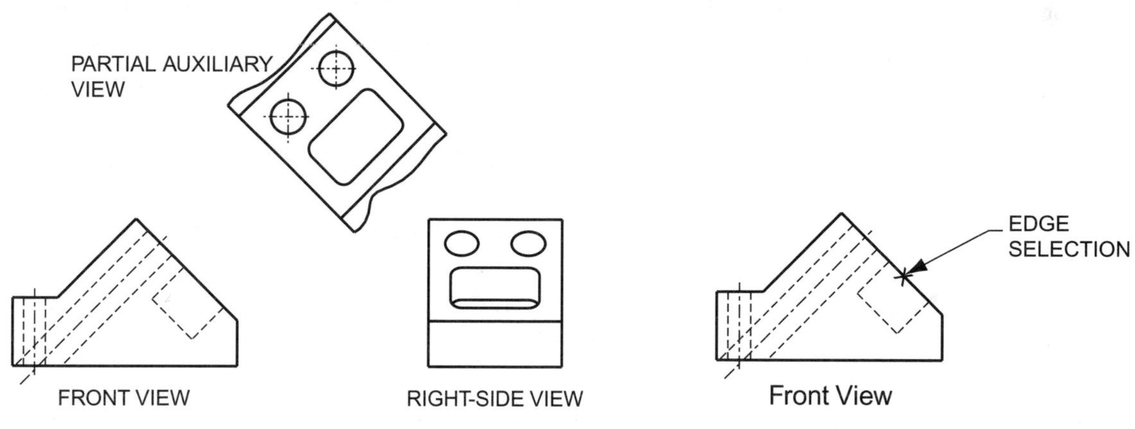

Figure 9-18 Partial auxiliary view

Figure 9-19 Edge selection

An Auxiliary view can be created as a Half view or as a Partial view. Figure 9–18 shows an example of a Partial Auxiliary view. Additionally, an Auxiliary view can be created as a section.

STEP 4: Select DONE from the View Type menu.

STEP 5: On the work screen, select a location for the Auxiliary view.

STEP 6: On the work screen, select an edge or an axis from which to project the Auxiliary view (Figure 9–19).

STEP 7: Use the VIEWS >> MOVE VIEW option to reposition the Auxiliary view.

SUMMARY

Pro/ENGINEER is an integrated engineering design tool that allows for a full range of applications including modeling, assembly, manufacturing, and analysis. Due to these tools, Pro/ENGINEER is a design package, not a drafting application. Despite this, there is still a need to document a design. Pro/ENGINEER's Drawing mode provides multiple tools for creating a high-quality engineering drawing. Included in these tools are the capability to create a variety of view types, such as section, detail, and auxiliary. Since Pro/ENGINEER is a fully associative computer-aided design application, models created in Part and Assembly modes can be used to create views in Drawing mode. Additionally, parametric dimensions that define a design can be used within Drawing mode to document the design.

TUTORIAL

Advanced Drawing Tutorial 1

This tutorial exercise will provide instruction on how to create the drawing shown in Figure 9–20. The first step in this tutorial will require you to model the part from which the drawing will be created.

Within this tutorial, the following topics will be covered:

- Starting a drawing.
- Establishing drawing setup values.
- Creating a general view.
- Creating an aligned section view.
- Creating a partial broken out section view.
- Annotating a drawing.

Creating the Part

Using Part mode, model the part shown in the drawing in Figure 9–20. Name the drawing file *section1*. Start the modeling process by creating Pro/ENGINEER's default datum planes. When defining features, the dimensioning scheme shown in figure matches the design intent for the part. Incorporate this intent into your model. Create the base feature as a revolved Protrusion. Create the bolt circle hole pattern using the Radial Hole option and the Pattern command. The cross section for the drawing will be created in Drawing mode.

Starting a Drawing

This section of the tutorial will create the object file for the drawing to be completed. Do not start this section until you have modeled the part portrayed in Figure 9–20.

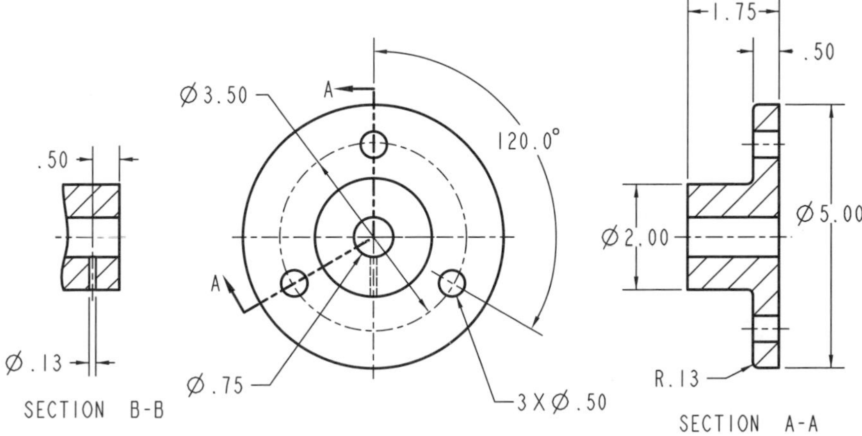

Figure 9–20 Completed views

STEP 1: Start Pro/ENGINEER.

If Pro/ENGINEER is not open, start the application.

STEP 2: Set an appropriate Working Directory.

STEP 3: Select FILE >> NEW.

STEP 4: In the New dialog box, select DRAWING mode, then enter *SECTION1* as the name of the drawing file.

STEP 5: Select OKAY on the New dialog box.

After selecting OKAY on the dialog box, Pro/ENGINEER will reveal the New Drawing dialog box. The New Drawing dialog box is used to select a model from which to create the drawing and to set a sheet size and format.

STEP 6: Under the Default Model textbox, select the BROWSE option and locate the *SECTION1* part (Figure 9–21).

Use the Browse option to locate the *section1* part created in the first section of this tutorial. The part will serve as the Default Model for the creation of drawing views.

STEP 7: Select the EMPTY WITH FORMAT option under the Specify Template section.

You will use the Specify Sheet option to select an A size format.

STEP 8: Select the BROWSE option under the Format option on the New Drawing dialog box, then open "a.frm" as the Standard Sheet format (see Figure 9–21).

The Browse option will allow you to browse the directory structure to locate an existing sheet format.

STEP 9: Select OKAY on the New Drawing dialog box.

After selecting OKAY, Pro/ENGINEER will launch its Drawing mode.

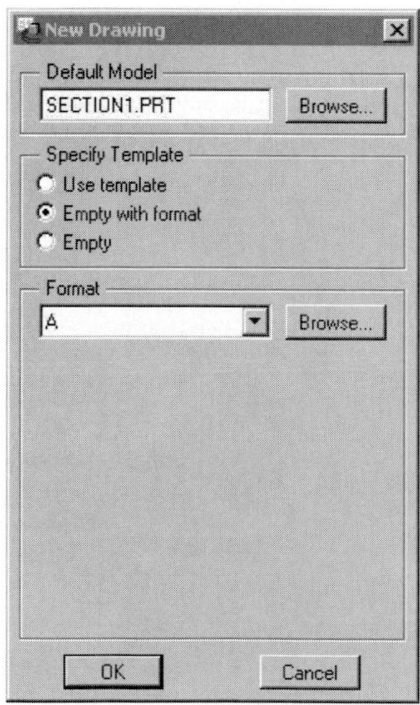

Figure 9–21 New Drawing dialog box

Establishing Drawing Setup Values

This section of the tutorial will temporarily set Pro/ENGINEER's drawing settings. The Advanced >> Draw Setup option will be used to change the default settings for this specific drawing.

Step 1: Select ADVANCED >> DRAW SETUP.

After selecting the Draw Setup option, Pro/ENGINEER will open the active drawing setup file with the Options dialog box.

Step 2: Change the Text and Arrowhead values for the current drawing.

Within the Options dialog box, make the changes shown in the following table. The *radial_pattern_axis_circle* item is used to create a bolt circle centerline around the hole pattern.

Table 9–1 Values for drawing setup

Drawing Setup Item	New Value
drawing_text_height	0.125
text_width_factor	0.750
dim_leader_length	0.175
dim_text_gap	0.125
draw_arrow_length	0.125
draw_arrow_style	filled
draw_arrow_width	0.0416
radial_pattern_axis_circle	YES

Step 3: Apply the new options and close the dialog box.

Creating the General View

General views serve as the parent view for all views projected off of it. This section of the tutorial will create a General view as the Front view of the drawing (see Figures 9–20 and 9–22).

Step 1: Select the VIEWS menu option from the Drawing menu.

Step 2: Select ADD VIEW >> GENERAL >> FULL VIEW >> NoXsec >> SCALE.

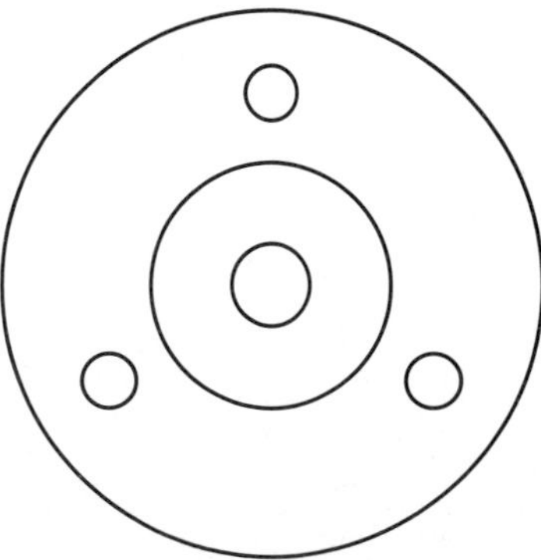

Figure 9–22 The front view

STEP 3: Select DONE on the View Type menu.

STEP 4: On the work screen, select the location for the Front view.

The drawing under creation in this tutorial will consist of a Front view and a Full Section Right-side view. The general view currently being defined will serve as the Front view. On the work screen, pick approximately where the Front view will be located. Once placed, the Views >> Move View option can be used to reposition the location of the view.

STEP 5: In the textbox, enter .500 as the scale value for the view.

STEP 6: Use the Orientation dialog box to orient the view to match Figure 9–22.

On the Orientation dialog box, use the appropriate references to set the view shown in Figure 9–22. Due to the nature of the part, you will have to select a datum plane as a reference. Pay careful attention that your hole locations match those shown in the illustration. If necessary, you can use the Views >> Modify View >> Reorient option to reorient the part.

STEP 7: Save your drawing.

Drawings are saved with an **.drw* file extension.

CREATING AN ALIGNED SECTION VIEW

This segment of the tutorial will create the Aligned Full Section view shown in Figure 9–23. The cutting plane for the section will match the path of the Cutting Plane Line shown. The Cross Section defining this section view will be created within Drawing mode as a part of this tutorial. Perform the following steps to create this view.

STEP 1: Select VIEWS >> ADD VIEW.

STEP 2: Select PROJECTION >> FULL VIEW >> SECTION >> DONE.

The Full View option will create a full projection view, while Section will provide options for creating a section view from the projected view.

STEP 3: Select the FULL option on the Cross-Section Type menu.

Do not confuse a Full section view with a Full View view type. The Full option from the Cross Section Type menu will create a section completely through the available view. This view could be a full, half, broken, or partial view.

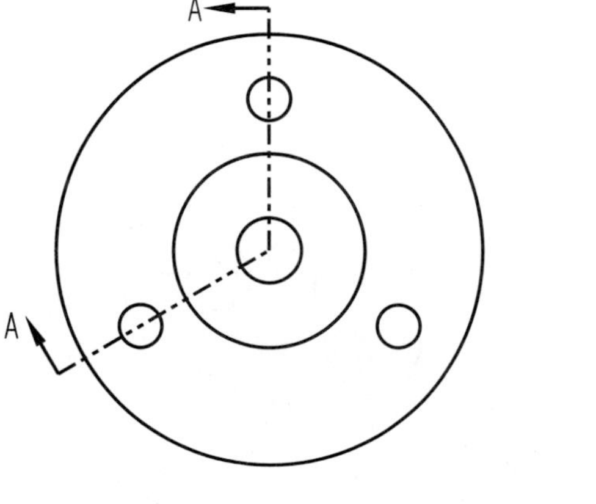

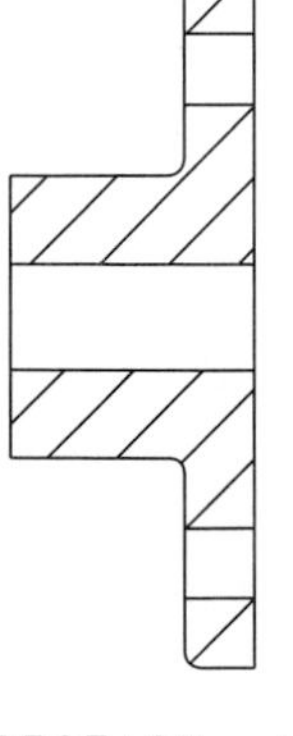

Figure 9–23 Aligned section view

STEP 4: Select TOTAL ALIGN >> DONE on the Xsec Type menu.

The Total Align option allows for the creation of the Aligned Section view. The TotalXsec option would create a Full Section view, but the view would not be aligned. The Aligned Xsec option creates an aligned view also, but only the sectioning is displayed.

STEP 5: On the work screen, select a location for the Right-Side view.

STEP 6: Select the CREATE option on the Cross-Section Enter menu.

Pro/ENGINEER allows you to either create a new Cross Section or retrieve one that was previously created. In this tutorial you will create the Cross Section as an Offset.

STEP 7: Select OFFSET >> BOTH SIDES >> DONE.

Since the cutting plane for this section view does not follow a straight path, you will have to create an Offset section. An offset section requires the sketching of the cutting path.

STEP 8: In Pro/ENGINEER's textbox, enter A as the name for the Cross Section.

STEP 9: Change to the window including the *section1* part (when using a Windows operating system, use the Taskbar to change windows).

Pro/ENGINEER requires you to sketch the cutting plane line within Part mode. If the part is currently open in a window, use the Windows Taskbar (or Pro/ENGINEER's Application Manager) to switch to the part's window. If the part is not open in a window prior to creating an offset section, Pro/ENGINEER will automatically open it for you.

> **MODELING POINT** When a drawing is created from an existing model, the model (part or assembly) is placed in Pro/ENGINEER's session memory. Since the drawing is referencing the model, the model cannot be erased from session memory. Additionally, a model can be in memory and not be included in a window.

STEP 10: As shown in Figure 9–24, select the top of the part to use as the sketching plane, then select OKAY to accept the direction of viewing.

STEP 11: Select DEFAULT from the Sketch View menu to accept the default sketching environment orientation.

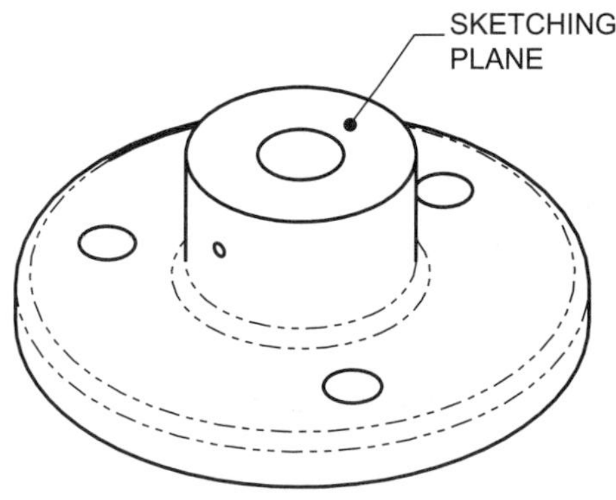

Figure 9–24 Sketching plane

STEP 12: Specify the four references shown in Figure 9–25.

Notice in the figure how the three axes are specified as references. These axes will be used to define the path of the cutting plane. Dynamically rotating the model will make the selection of each axis easier (control key and middle mouse button).

> **INSTRUCTIONAL POINT** If your part model was not opened before starting this segment of the tutorial, Pro/ENGINEER will open it automatically within its own subwindow. You will have to use the menu bar's Sketch >> References option to define the necessary references.

STEP 13: Use the SKETCH >> LINE option to sketch the cutting path shown in Figure 9–25.

When sketching the path, make sure that each line is aligned with the references that were specified.

STEP 14: Select the Continue icon (or Sketch >> Done) to end the cutting plane's definition.

STEP 15: Switch to the Window including the drawing.

STEP 16: As shown in Figure 9–26, select the center axis to define the alignment rotation axis.

This selection will determine the pivot axis for the alignment.

STEP 17: On the work screen, select the Front view as the view to display the Cutting Plane Line.

In the message area, Pro/ENGINEER is asking you to "pick a view for arrows where the section is perp. MIDDLE button for none." Select the view to locate the Cutting Plane Line. If a Cutting Plane Line is not required, select the middle mouse button.

STEP 18: Select OKAY to accept the viewing direction.

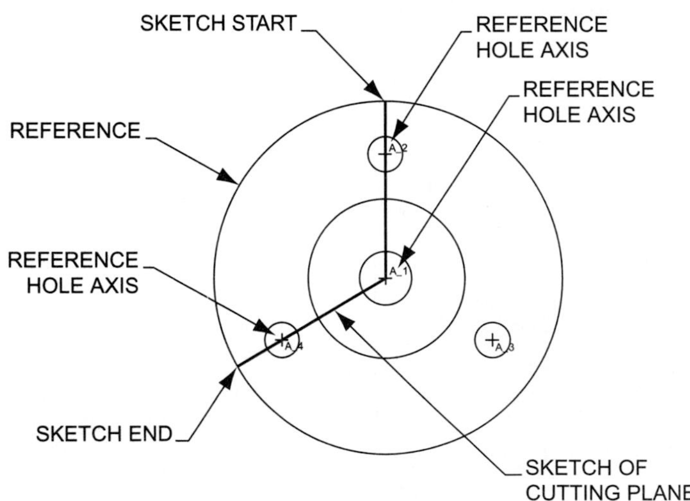

Figure 9–25 Sketched cutting plane

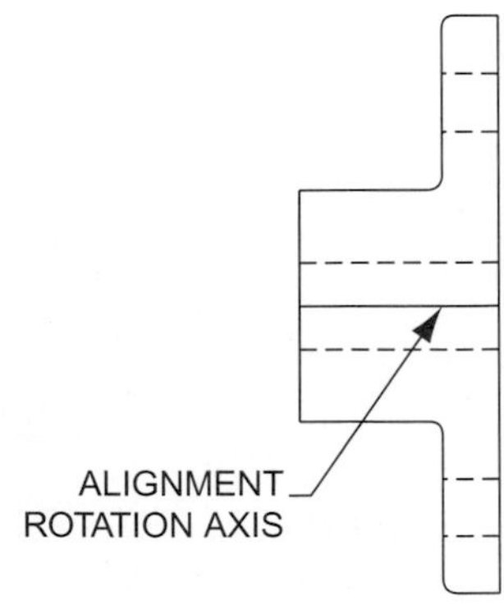

Figure 9–26 Axis selection

Creating a Partial Broken Out Section View

This segment of the tutorial will create the Partial Broken Out view shown in Figure 9–27. Partial views are available with sectioned and nonsectioned views. A Broken Out section view is created with the Local section type option and can be created as a Full, Broken, or Partial view.

Step 1: Select VIEWS >> ADD VIEW.

Step 2: Select PROJECTION >> PARTIAL VIEW from the View Type menu.

The Partial View option requires you to sketch a spline to define the area that will be included in the view. The Local option that creates the Broken Out section also requires you to sketch a spline. This spline defines the area to include in the Broken Out section. When combining the Partial View and Local options, one sketched spline will serve for both functions.

Step 3: Select SECTION >> DONE .

Step 4: Select LOCAL >> DONE on the Cross-Section Type menu.

The Local option is used to create a Broken Out section view.

Step 5: On the work screen, select the location for the view.

As shown in Figure 9–27, select a location that will project a Left-Side view.

Step 6: Select ADD BREAKOUT >> SHOW OUTER on the View Boundary menu.

These options are selected by default. The Add Breakout option is used to allow for the sketching of a spline boundary while the Show Outer option will show the sketched boundary in the view. At this time, do not select Done.

Step 7: Select the CREATE option on the Cross-Section Enter menu.

You will create the Cross Section at this point in the tutorial.

Step 8: Select PLANAR >> SINGLE >> DONE on the Cross-Section Create menu.

Step 9: In Pro/ENGINEER's textbox, enter B as the name for the Cross Section.

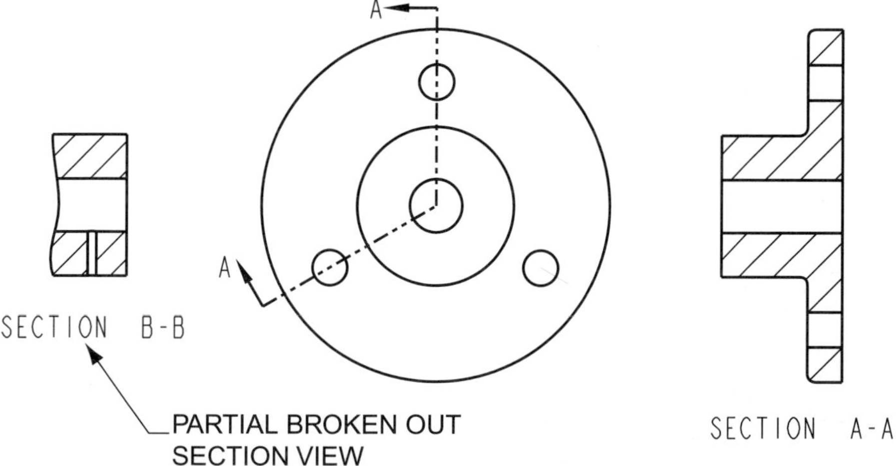

Figure 9–27 Partial broken out section

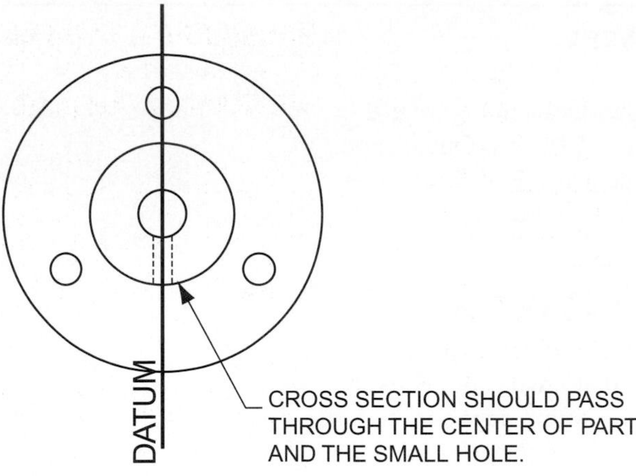

Figure 9–28 Cross section definition

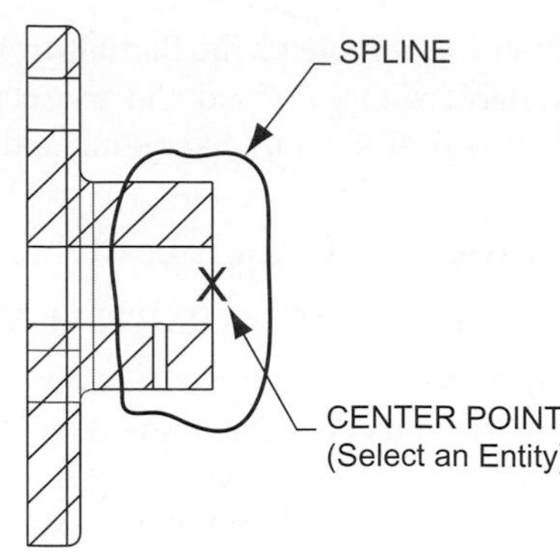

Figure 9–29 Sketching the boundary

STEP 10: In the Front view of the drawing, select the datum plane that runs vertical through the drawing (see Figure 9–28).

The datum as shown in the figure will serve as the planar surface that defines the cross section. Select a corresponding datum plane. If one does not currently exist on your model, use the MAKE DATUM option to create one on-the-fly.

STEP 11: Select the middle mouse button.

Pro/ENGINEER is requesting in the message area that you either select a view for the cutting plane line or select the "MIDDLE button for none." You will not use a cutting plane line for this section view.

STEP 12: On the new Left-Side view, within the area to be sectioned, select a Center point for the outer boundary (Figure 9–29).

On the work screen, in the Left-Side view, select a point on an entity that will reside in the partial section view. Pro/ENGINEER uses this selection to regenerate the section.

Note: If you have trouble selecting a point, dynamically zoom in on the view.

STEP 13: As shown in Figure 9–29, sketch a spline that will define the Partial View and the Local section.

The left mouse button is used to select points on the spline, while the middle mouse button is used to terminate and close the spline.

STEP 14: On the View Boundary menu, select DONE.

CENTERLINES AND DIMENSIONS

Within this segment of the tutorial, you will first show the centerlines shown in Figure 9–30. Second, you will display the parametric dimensions that define the part. Finally, you will set specific display modes for each view.

STEP 1: On the menu bar, select VIEW >> SHOW AND ERASE.

STEP 2: On the Show/Erase dialog box, select the SHOW option and the AXIS Item Type (Figure 9–31).

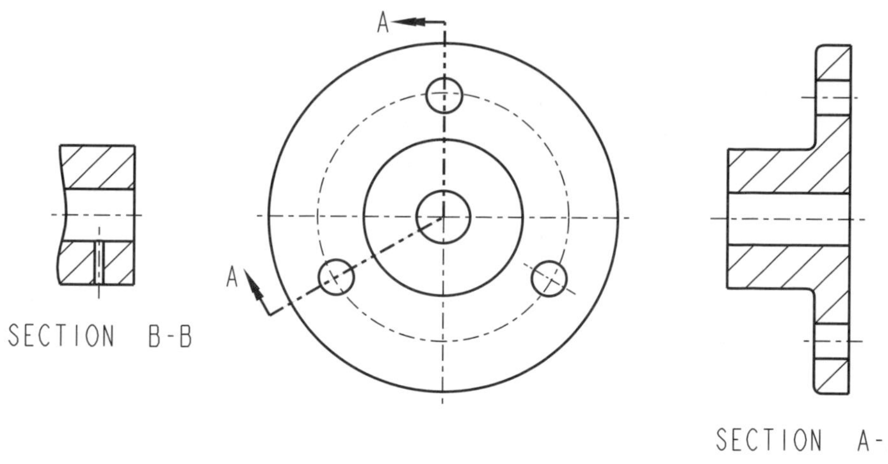

Figure 9-30 Centerlines

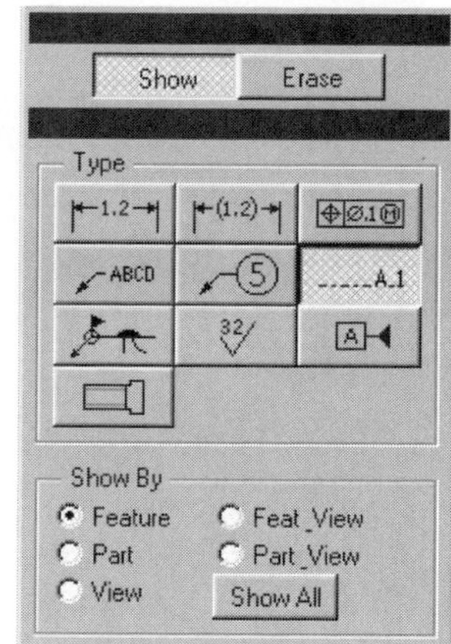

Figure 9-31 Show/Erase dialog box

STEP 3: On the Show/Erase dialog box, select the SHOW ALL option, then select YES to confirm the selection

The Show All option will show all of a selected item type. In this example, all axes from the part will be projected as centerlines. Make sure that the Never Shown option is selected.

Note: If the With Preview option is selected, accept all of the centerlines shown on the work screen.

STEP 4: Use the Show/Erase dialog box to show all available dimensions.

STEP 5: Use Drawing mode's MOVE and MOVE TEXT capabilities and the Pop-Up menu's SWITCH VIEW and FLIP ARROWS options to reposition the dimensions to match Figure 9–32.

The Pop-Up menu is available by first selecting the text of a dimension with your left mouse button, followed by reselecting the same dimension

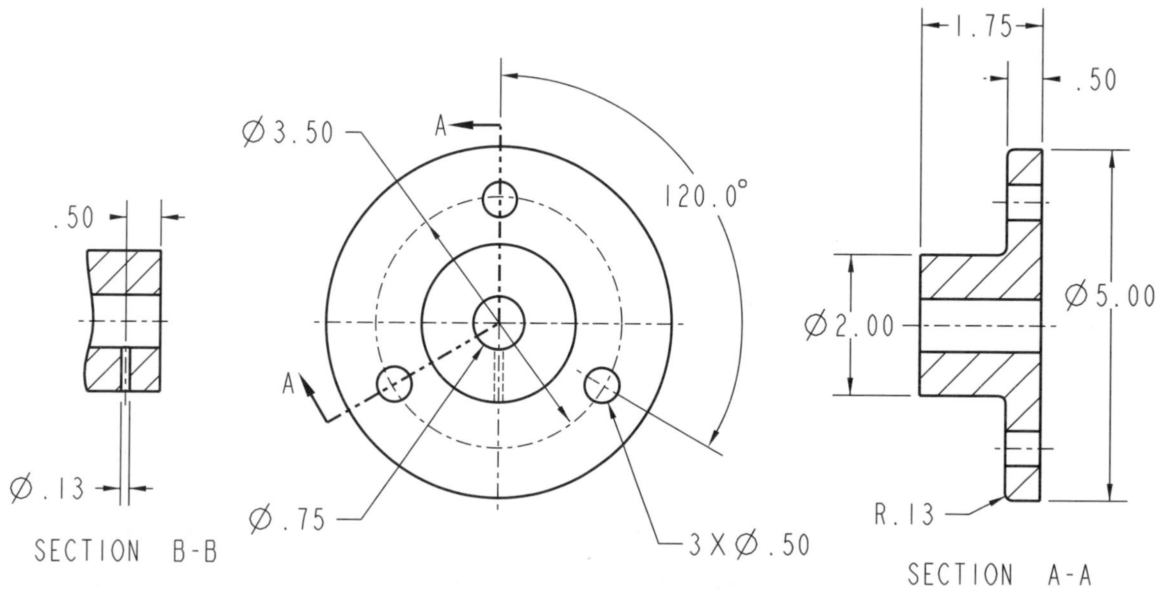

Figure 9-32 Drawing dimensions

text with your right mouse button (*Note:* A slight delay in releasing the right mouse button is necessary). A variety of modification and manipulation tools are available through the revealed pop-up menu.

Use the following options to reposition your dimensions on the work screen:

- **Move** The Move option is available through the selection of a dimension's text with the left mouse button. Once selected, the dimension, including text and witness lines, can be dragged with your curser.
- **Move text** The Move Text option is available through the selection of a dimension's text with the left mouse button. Once selected, grip points are available that will allow you to move only the text of a dimension.
- **Switch view** The Switch View option is used to switch a dimension from one view to another. It is available under the Pop-Up menu. Multiple dimensions can be selected by a combination of the Shift key and the left mouse button. Once dimensions are preselected, the Pop-Up menu is available through the right mouse button.
- **Flip arrows** The Flip Arrows option is used to flip dimension arrowheads. It is available under the Pop-Up menu.

STEP 6: Use the Show/Erase dialog box's ERASE option to hide dimensions not shown in Figure 9–32.

STEP 7: Select the Clean Dimensions icon to set dimension spacing.

The Clean Dimensions dialog box is used to consistently space dimensions.

STEP 8: On the work screen, select the two section views, then select Done Select (Done Sel) on the Get Select menu.

STEP 9: On the Clean Dimensions dialog box, sets the options shown in Figure 9–33, then APPLY the settings.

STEP 10: CLOSE the Clean Dimensions dialog box.

STEP 11: Select VIEWS >> DISP MODE >> VIEW DISP.

In the next exercise in this tutorial, you will set the display mode for each view. In this example, you will set the two section views with a No Hidden display. You will also set the Front view with a Hidden display and with a No Display Tangent display.

STEP 12: On the work screen, select the two section views, then select Done Select (Done Sel) on the Get Select menu.

You will set the two section views to not display hidden lines.

STEP 13: Select NO HIDDEN >> DONE on the View Display menu.

STEP 14: Select VIEWS >> DISP MODE >> VIEW DISP.

STEP 15: On the work screen, select the front view, then select Done Select.

STEP 16: Select HIDDEN LINE >> NO DISP TAN >> DONE.

You will set the front view to display hidden lines and to not display tangent edges.

STEP 17: With your left mouse button, pick the 0.50 diameter hole dimension, then access the Pop-Up menu with your right mouse button.

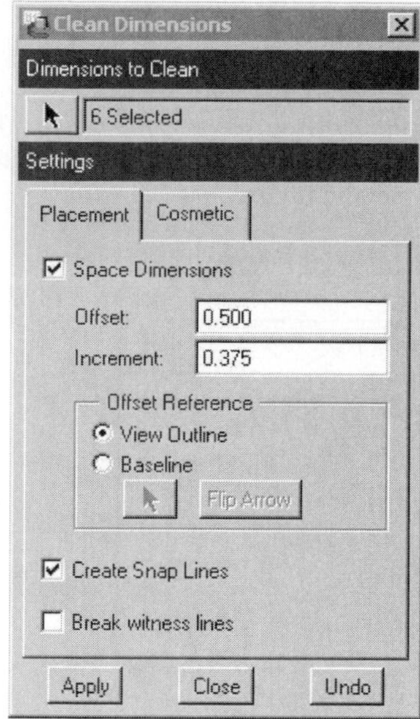

Figure 9–33 Clean Dimensions dialog box

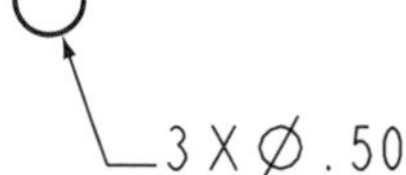

Figure 9–34 Dimension text modification

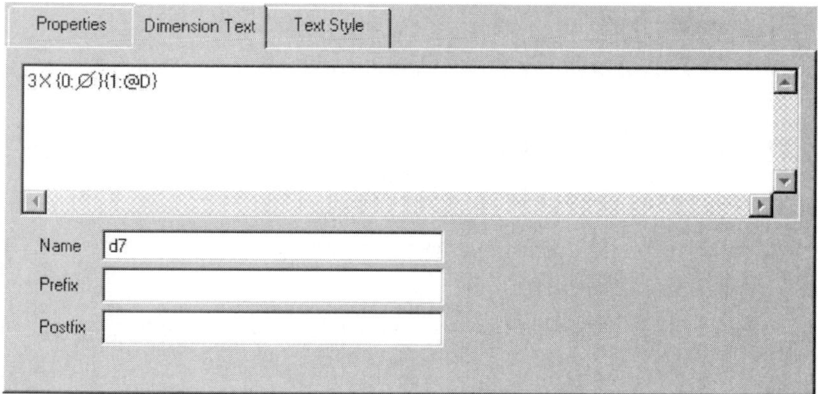

Figure 9–35 Dimension Text tab

The next exercise in this tutorial will require you to modify the dimension text of the 0.500 diameter hole dimension to match Figure 9–34.

STEP 18: **Select the PROPERTIES option on the Pop-Up menu.**

STEP 19: **On the Dimension Properties dialog box, select the Dimension Text tab (Figure 9–35).**

STEP 20: **Modify the dimension parameters in the Dimension Text box to add a 3X in front of the existing text.**

STEP 21: **Select OKAY to exit the Dimension Properties dialog box.**

STEP 22: **Save your drawing file.**

Title Block Notes

In this segment of the tutorial, create the notes for the title block as shown in Figure 9–36. Use the Insert >> Note menu option to create the required text and the Pop-Up menu's Modify Text Style option to make text style and text height adjustments.

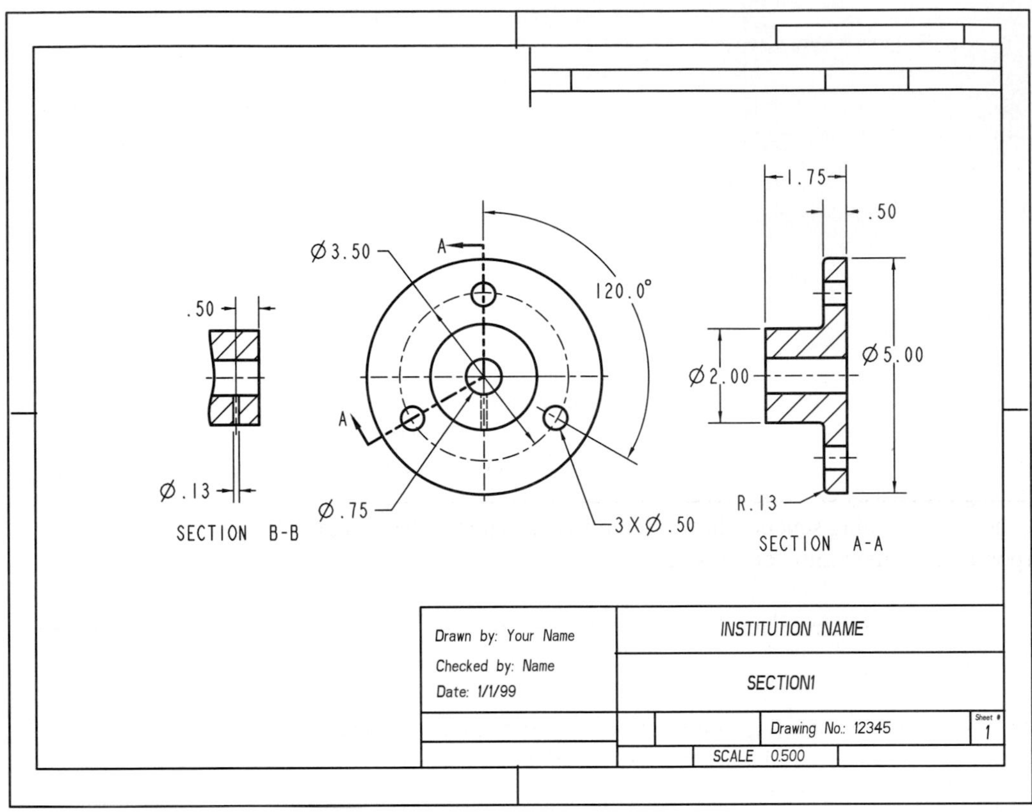

Figure 9–36 Finished drawing with title block information

TUTORIAL

ADVANCED DRAWING TUTORIAL 2

This tutorial exercise will provide instruction on how to create the drawing shown in Figure 9–37. The first step in this tutorial will require you to model the part from which the drawing will be created.

Within this tutorial, the following topics will be covered:

- Starting a drawing.
- Establishing drawing setup values.
- Creating a broken view.
- Creating a partial auxiliary view.
- Annotating a drawing.
- Modifying dimension values.

CREATING THE PART

Using Part mode, model the part shown in the drawing in Figure 9–37. Name your part file *auxiliary1*. Start the modeling process by creating Pro/ENGINEER's default datum planes. When defining features, the dimensioning scheme shown in the figure matches the design intent for the part. Incorporate this intent into your model.

STARTING A DRAWING

This section of the tutorial will create the new object file for the drawing to be completed. Do not start this section until the part shown in Figure 9–37 is complete.

STEP 1: Start Pro/ENGINEER.

STEP 2: Set an appropriate Working Directory.

STEP 3: Select FILE >> NEW.

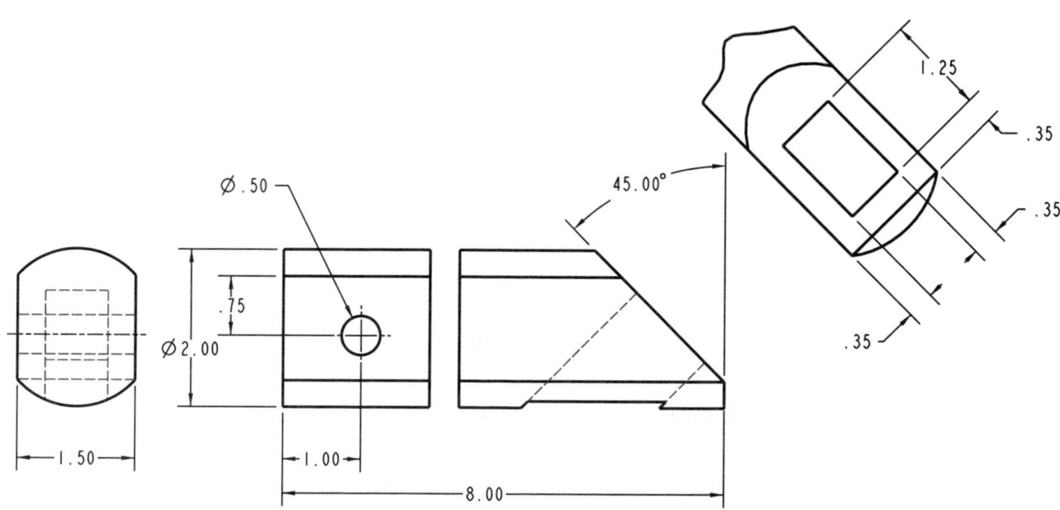

Figure 9-37 Completed views

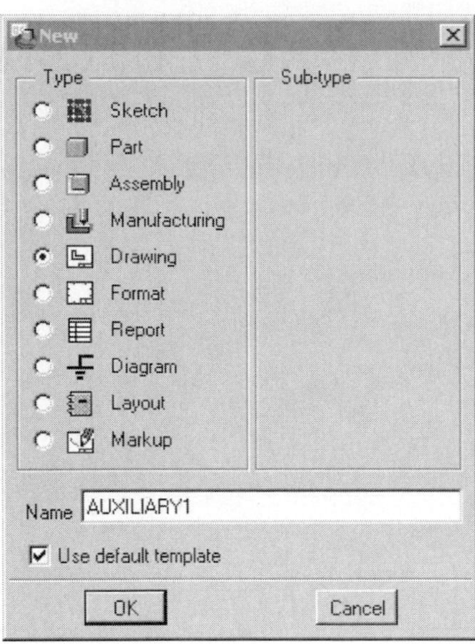

Figure 9–38 New dialog box

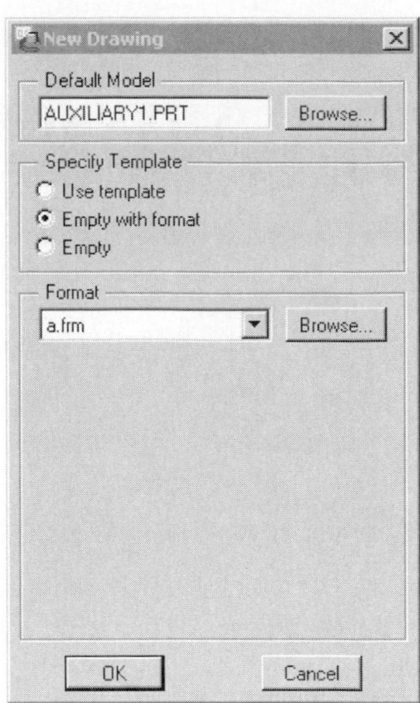

Figure 9–39 New Drawing dialog box

STEP 4: In the New dialog box, select DRAWING mode, then enter *AUXILIARY1* as the name of the drawing file (Figure 9–38).

STEP 5: Using the Default Template file, select OKAY on the New dialog box.

After selecting OKAY on the dialog box, Pro/ENGINEER will reveal the New Drawing dialog box. This dialog box is used to select a model to create the drawing from and to set a sheet size and format.

STEP 6: Under the Default Model option, select the BROWSE option and locate the *AUXILIARY1* part (Figure 9–39).

Use the Browse option to locate the *auxiliary1* part created in the first section of this tutorial. The part will serve as the default model for the creation of drawing views.

STEP 7: Under the Specify Template option, select the EMPTY WITH FORMAT option.

You will use the Empty-with-Format option to select an A size format.

STEP 8: Under the Format option on the New Drawing dialog box, select the BROWSE option (Figure 9–39).

The Browse option will allow you to browse the directory structure to locate an existing sheet format.

STEP 9: Open "a.frm" as the Standard Sheet format (or select a format as assigned).

STEP 10: Select OKAY on the New Drawing dialog box.

After selecting OKAY, Pro/ENGINEER will launch its Drawing mode.

ESTABLISHING DRAWING SETUP VALUES

This section of the tutorial will temporarily set Pro/ENGINEER's drawing settings. The Advanced >> Draw Setup option will be used to change the default settings for this specific drawing.

STEP 1: Select ADVANCED >> DRAW SETUP.

After selecting the Draw Setup option, Pro/ENGINEER will open the active drawing setup file with an Options dialog box.

STEP 2: Change the Text and Arrowhead values for the current drawing.

Within Options dialog box, make the changes shown below.

Table 9-2 Values for drawing setup

Drawing Setup Item	New Value
drawing_text_height	0.125
text_width_factor	0.750
dim_leader_length	0.175
dim_text_gap	0.125
draw_arrow_length	0.125
draw_arrow_style	filled
draw_arrow_width	0.0416
radial_pattern_axis_circle	YES

STEP 3: Apply the settings, then close the Options dialog box.

STEP 4: Select DONE/RETURN to exit the Advanced Drawing Options menu.

CREATING THE BROKEN FRONT VIEW

This section of the tutorial will create a Broken General view as the Front view of the drawing (Figure 9–40).

STEP 1: Select VIEWS >> ADD VIEW >> GENERAL.

STEP 2: Select BROKEN VIEW >> NoXsec >> SCALE >> DONE.

STEP 3: On the work screen, select the location for the Front view.

The drawing under creation in this tutorial will consist of a Broken Front view, a Left-Side view, and an Auxiliary view.

STEP 4: In the textbox, enter .750 as the scale value for the view.

STEP 5: Use the Orientation dialog box to orient the view to match Figure 9–41.

On the Orientation dialog box, use appropriate references to set the view shown in Figure 9–41. If necessary, you can use the Views >> Modify View >> Reorient option to reorient the part.

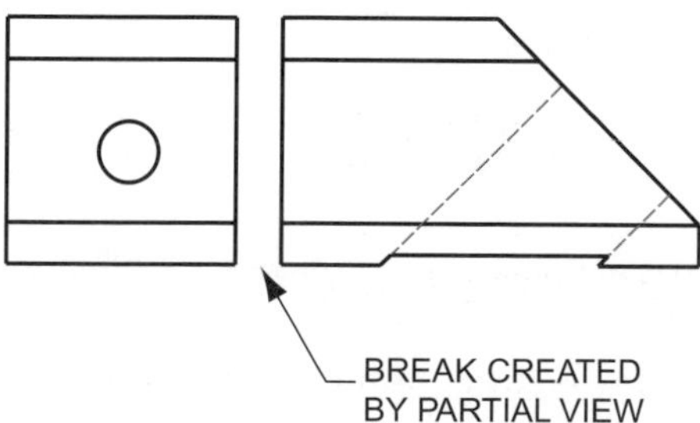

Figure 9–40 The front view

Figure 9–41 Oriented view

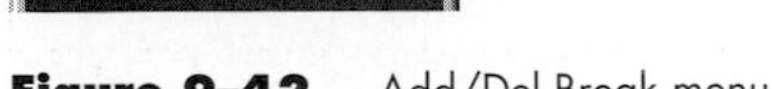

Figure 9–42 Add/Del Break menu

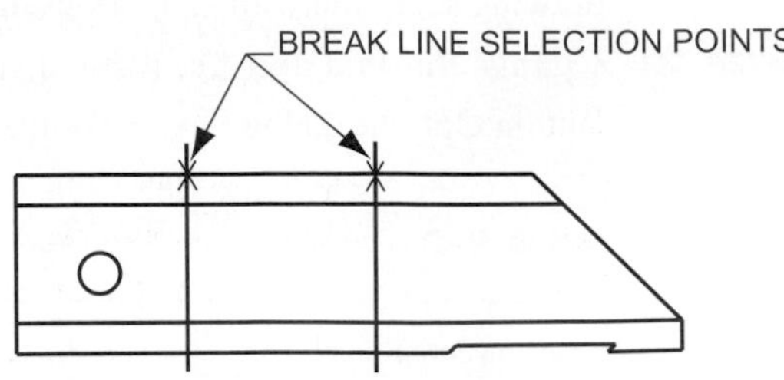

Figure 9–43 Break line selection

STEP 6: Select the ADD >> VERTICAL option on the Add/Del Break menu (Figure 9–42).

The Add option will add a break line to the view, while the Delete option will remove a break line. The Vertical option creates vertical break lines, while the Horizontal option creates horizontal break lines. In this tutorial, you will create two vertical break lines.

STEP 7: As shown on Figure 9–43, select the locations for the two break lines.

STEP 8: Select DONE on the Add/Del Break menu.

STEP 9: Select DONE on the Exit menu.

STEP 10: Use the VIEWS >> MOVE VIEW option to reposition each section of the view.

Each broken section of the view can be moved with the Move View option or by dragging with the left mouse button.

STEP 11: Select DONE/RETURN to exit the Views menu.

STEP 12: Save your drawing.

CREATING A PARTIAL AUXILIARY VIEW AND LEFT-SIDE VIEW

This section of the tutorial will create the Partial Auxiliary view and the Left-Side View shown in Figure 9–44.

STEP 1: Select VIEWS >> ADD VIEW.

STEP 2: Select AUXILIARY >> PARTIAL VIEW >> NoXsec >> DONE.

STEP 3: On the work screen, pick the location for the Auxiliary view (Figure 9–44).

The drawing under creation in this tutorial will consist of a Broken Front view, a Left-Side view, and an Auxiliary view.

STEP 4: Select the edge shown in Figure 9–45.

When creating an Auxiliary view, Pro/ENGINEER requires the selection of an existing Edge or Axis. The auxiliary view will be projected from this edge.

STEP 5: On the Auxiliary view, select a point on an entity that will lie within the boundary of the Partial view (Figure 9–45).

STEP 6: On the Auxiliary view, sketch a spline that will serve as the boundary for the Partial view (Figure 9–45).

Use the left mouse button to select spline points and the middle mouse button to close the spline.

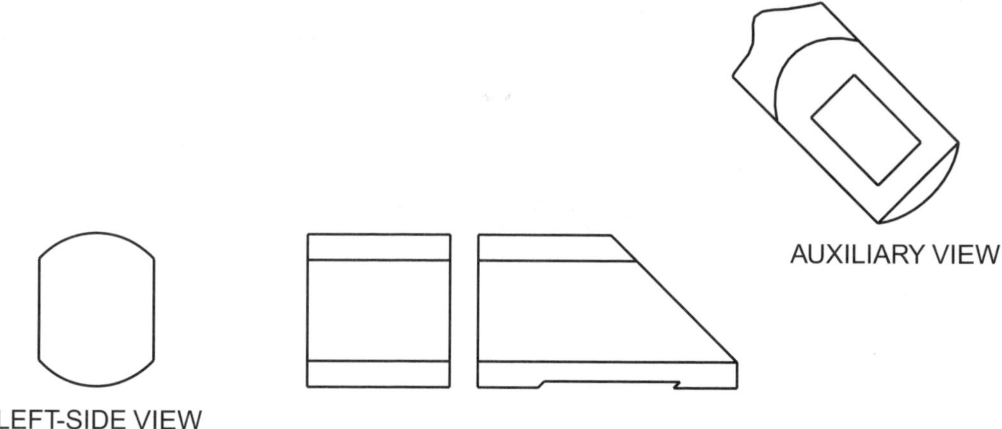

Figure 9–44 Partial auxiliary view and left-side view

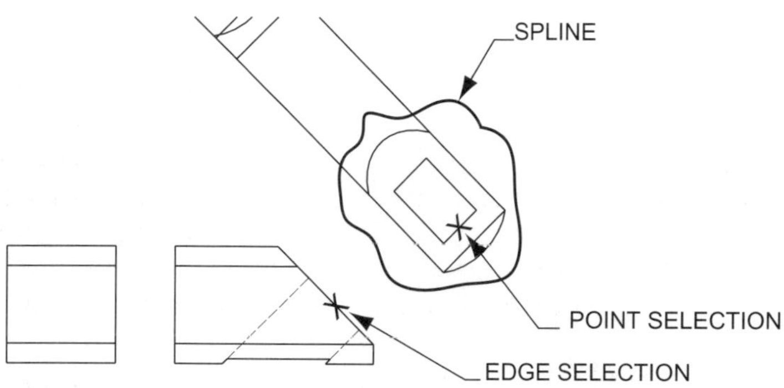

Figure 9–45 Auxiliary view creation

STEP 7: **Use the ADD VIEW >> PROJECTION >> FULL VIEW options to create the Left-Side view (Figure 9–44).**

STEP 8: **Reposition the views as shown in Figure 9–44.**

STEP 9: **Save your drawing.**

ADDING DIMENSIONS AND CENTERLINES

This segment of the tutorial will add dimensions and centerlines to the drawing (Figure 9–46).

STEP 1: **On the menu bar, select VIEW >> SHOW AND ERASE.**

STEP 2: **Select the DIMENSION and AXIS item types (Figure 9–47).**

The Dimension item type will show parametric dimensions, while the axis option will create centerlines from existing axes.

STEP 3: **Under the Preview Tab, deselect the WITH PREVIEW option (Figure 9–48).**

STEP 4: **On the Show/Erase dialog box, select the SHOW ALL option; then select YES to confirm the selection.**

STEP 5: **Close the Show/Erase dialog box.**

STEP 6: **Use Drawing mode's MOVE and MOVE TEXT capabilities and the Pop-Up menu's SWITCH VIEW and FLIP ARROWS options to reposition the dimensions to match Figure 9–46.**

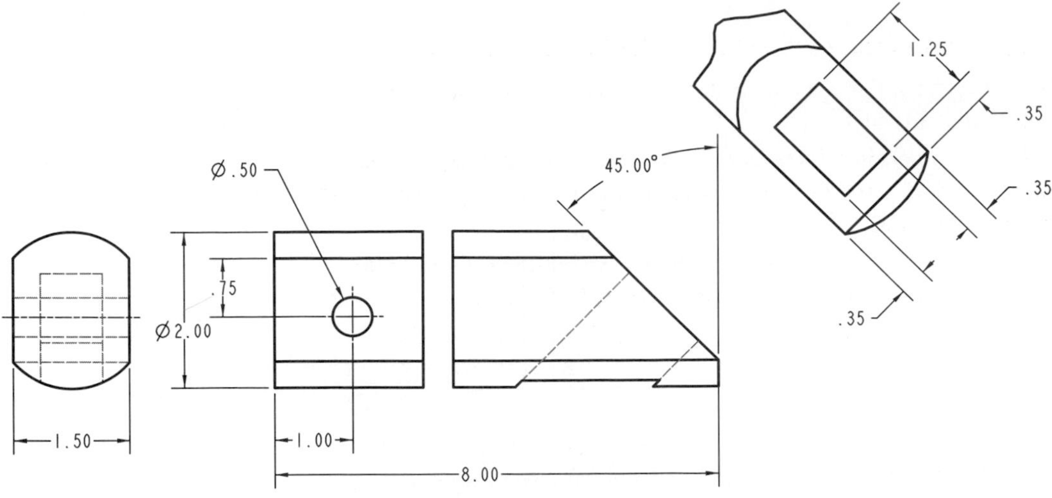

Figure 9-46 Dimensions and centerlines

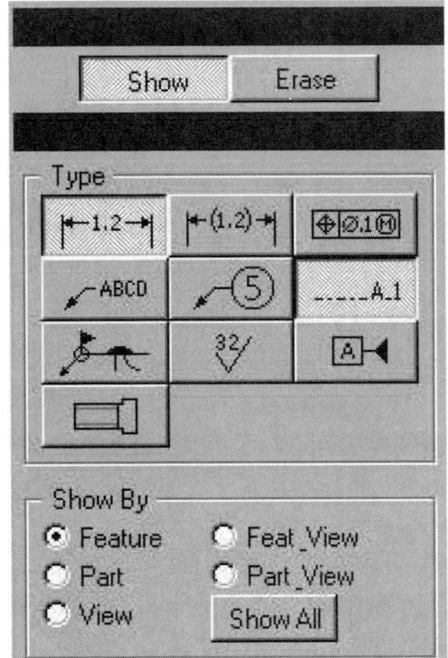

Figure 9-47 Show/Erase item types

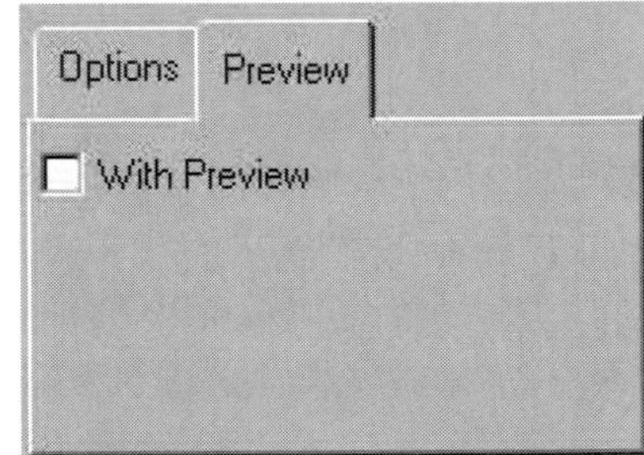

Figure 9-48 With Preview option

Use the following options to reposition your dimensions on the work screen:

- **Move** The Move option is available through the selection of a dimension's text with the left mouse button. Once selected, the dimension, including text and witness lines, can be dragged with your curser.
- **Move text** The Move Text option is available through the selection of a dimension's text with the left mouse button. Once selected, grip points are available that will allow you to move only the text of a dimension.
- **Switch view** The Switch View option is used to switch a dimension from one view to another. It is available under the Pop-Up menu. Multiple dimensions can be selected by a combination of the Shift key and the left mouse button. Once dimensions are preselected, the Pop-Up menu is available through the right mouse button.
- **Flip arrows** The Flip Arrows option is used to flip dimension arrowheads. It is available under the Pop-Up menu.

STEP 7: Select the Clean Dimensions icon on the Drawing toolbar.

You will use the Clean Dimensions dialog box to consistently space the dimensions on the drawing.

STEP 8: **On the work screen, select the Front, Left-Side, and Auxiliary views, then select the middle mouse button (the middle mouse button serves as the Done Select option).**

STEP 9: **On the Clean Dimensions dialog box, take the default values of 0.500 (Offset) and 0.375 (Increment), then select the APPLY option.**

The Offset value will set the distance of the first dimension from the model. The Increment option will set the distance between each dimension.

STEP 10: **Close the Clean Dimensions dialog box.**

STEP 11: **Use Drawing mode's MOVE and MOVE TEXT capabilities and the Pop-Up menu's SWITCH VIEW and FLIP ARROWS options to reposition the dimensions to match Figure 9–46.**

The Clean Dimensions dialog box will not perfectly locate every dimension. Depending upon the complexity of a drawing, you will probably have to reposition dimensions after using the Clean Dimensions option.

STEP 12: **On the Toolbar, select the HIDDEN LINE display option to display hidden lines on each view.**

STEP 13: **Save your drawing.**

MODIFYING DIMENSION VALUES

On a drawing, dimensions displayed using the Show/Erase dialog box are parametric dimensions that are associated with other modes of Pro/ENGINEER. In this example, the dimensions displayed on the drawing are associated with the dimensions from the model in Part mode. Pro/ENGINEER is a bi-associative parametric modeling application. Dimension value changes made in one mode of Pro/ENGINEER will reflect in all modes. As an example, if you change the 8-inch length of the part used in this drawing, this change will be reflected in the part and in the drawing. In this segment of the tutorial, you will Modify the 1.25-inch dimension found in the Auxiliary view.

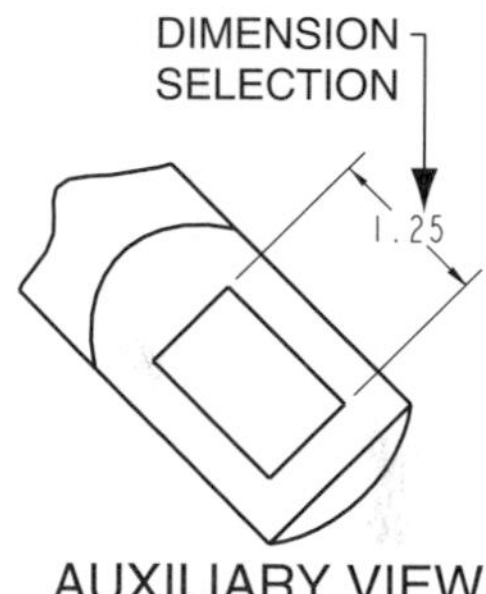

Figure 9–49 Dimension selection

STEP 1: **In the Auxiliary view, with your left mouse button, select the 1.25-inch dimension (Figure 9–49).**

STEP 2: **Access the Pop-Up menu by selecting the dimension with your right mouse button.**

STEP 3: **On the Pop-Up menu, select the NOMINAL VAL option.**

The Nominal Value option is used to modify dimension values. The Properties option with the Dimension Properties dialog box can be used to modify dimension values also.

STEP 4: **In Pro/ENGINEER's textbox, change the dimension value to equal 1.50.**

STEP 5: **Select REGENERATE >> MODEL on the Drawing menu.**

The changes in the part and in the drawing will be reflected after regenerating the model (Figure 9–50).

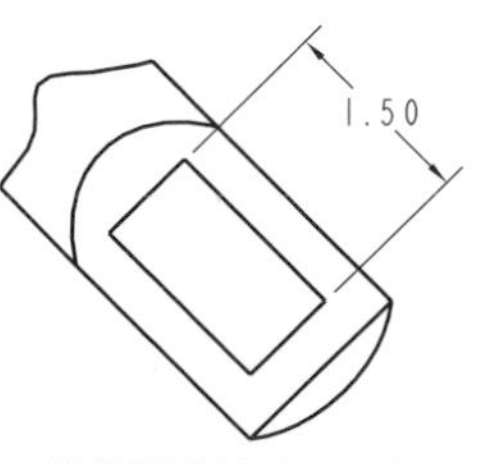

Figure 9–50 Dimension changed

STEP 6: **Open the object in Part mode to observe the changes to the model.**

STEP 7: **In Drawing mode, use the MODIFY >> VALUE option to change the value of the previously modified dimension back to a value of 1.25.**

STEP 8: **In Drawing mode, regenerate the model.**

STEP 9: **Save your drawing.**

Title Block Information

Use the Insert >> Note option to create the title block information shown in Figure 9–51. Use the Pop-Up menu's Modify Text Style option to make changes to text style and height. Your final drawing should appear as shown in Figure 9–52.

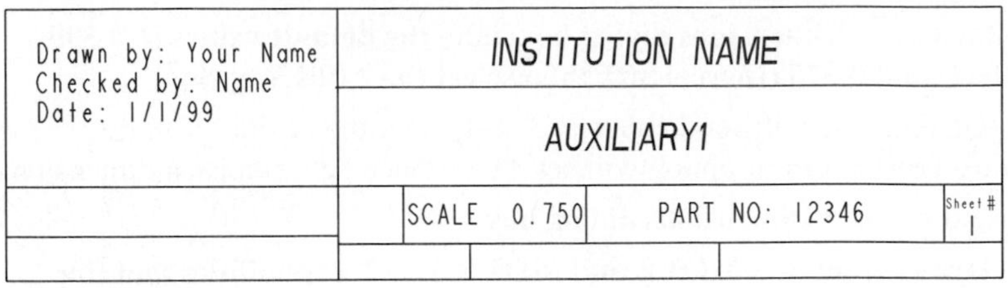

Drawn by: Your Name
Checked by: Name
Date: 1/1/99
INSTITUTION NAME
AUXILIARY1
SCALE 0.750
PART NO: 12346
Sheet #
1

Figure 9–51 Title block information

1.25
.35
.35
.35
45.00°
Ø.50
.75
Ø2.00
1.50
1.00
8.00
Drawn by: Your Name
Checked by: Name
Date: 1/1/99
INSTITUTION NAME
AUXILIARY1
SCALE 0.750
PART NO: 12346
Sheet #
1

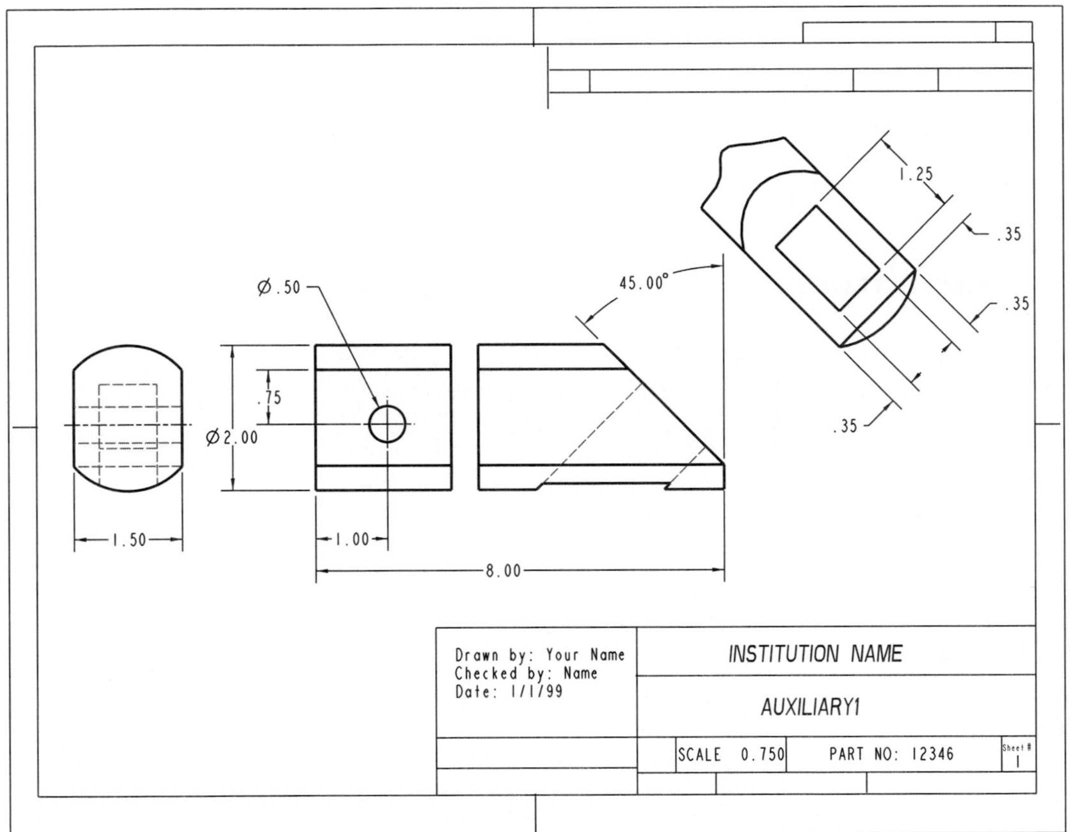

Figure 9–52 Final drawing

Problems

1. Model the part shown in Figure 9–53, then create a detailed drawing of the part. When completing this problem, meet the following requirements:
   - The dimensions shown in the figure meet design intent. During part modeling, incorporate these dimensions.
   - Create an engineering drawing with Front and Top views. The Front view should be a Full Section view.

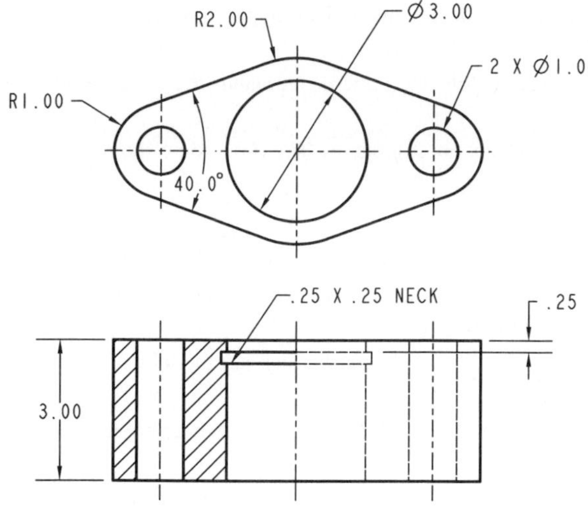

Figure 9-53 Problem one

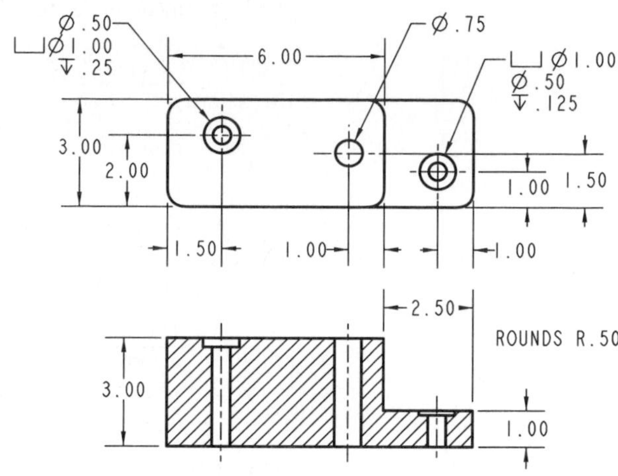

Figure 9-54 Problem two

- Use an A size sheet and format.
- Fully dimension the engineering drawing using the part's parametric dimensions.

2. Model the part shown in Figure 9–54, then create a detailed drawing of the part. When completing the problem, meet the following requirements:

- The dimensions shown in the figure meet design intent. During part modeling, incorporate these dimensions.
- Create an engineering drawing with Front and Top views. The Front view should be an Offset Full Section view.
- Use an "A" size sheet and format.
- Fully dimension the engineering drawing using the part's parametric dimensions.

3. Model the part shown in Figure 9–55, then create a detailed drawing of the part. When completing this problem, meet the following requirements:

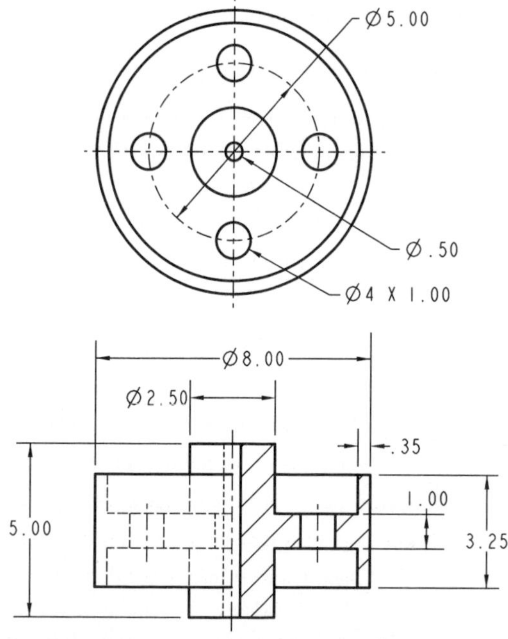

Figure 9-55 Problem three

- The dimensions shown in the figure meet design intent. During part modeling, incorporate these dimensions.
- When modeling the part, use the Radial Hole and Pattern commands to create the bolt-circle pattern.
- Create an engineering drawing with Front and Top views. The Front view should be a Half Section view.
- Use an "A" size sheet and format.
- Fully dimension the engineering drawing using the part's parametric dimensions.

4. Model the part shown in Figure 9–56, then create a detailed drawing of the part. When completing this problem, meet the following requirements:
   - The dimensions shown in the figure meet design intent. During part modeling, incorporate these dimensions.
   - When modeling the part, use the Radial Hole and Pattern commands to create the bolt-circle pattern.
   - Create an engineering drawing with Front, Top, Right-Side, and Auxiliary views. Create the Auxiliary view with the Of-Surface view type option.
   - Use an "A" size sheet and format.
   - Fully dimension the engineering drawing using the part's parametric dimensions.

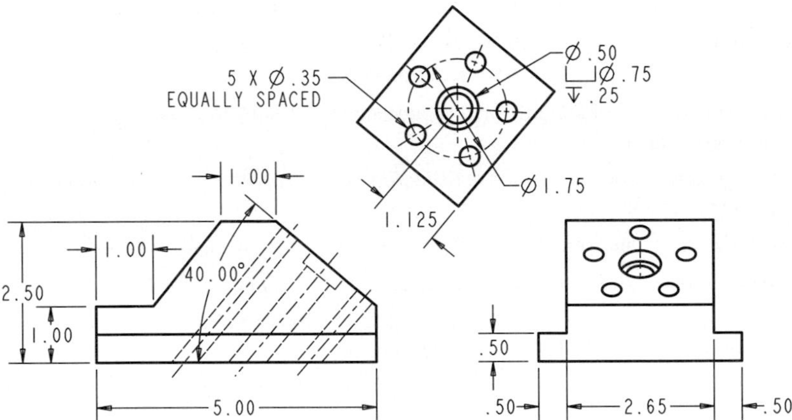

Figure 9–56 Problem four

QUESTIONS AND DISCUSSION

1. Describe possible uses of a section view.
2. Describe different section view types available within Pro/ENGINEER.
3. What is the difference between a Total Cross Section and an Area Cross Section.
4. What is the purpose of a Cutting Plane Line?
5. In Pro/ENGINEER, how is the cutting plane created for an Offset Section?
6. In Pro/ENGINEER, what is the difference between a Broken view and a Broken Out view?
7. What is a Revolved section when used on an engineering drawing?
8. Describe uses of an Auxiliary view.

CHAPTER

10

SWEPT AND BLENDED FEATURES

Introduction

Protrusion and Cut are the two basic feature solid model creation tools available within Pro/ENGINEER. Covered previously in this textbook, the Extrude and Revolve options are powerful tools used within each command for creating either straight or revolved features. This chapter expands upon the Protrusion and Cut commands by introducing the Sweep and Blend options. In addition, the datum curve and coordinate system options will be covered. Upon finishing this chapter, you will be able to create

- A swept positive space feature.
- A blended positive space feature.
- A swept negative space feature.
- A swept positive space feature.
- A datum curve.
- A coordinate system.

DEFINITIONS

Blended feature A feature with two or more user-sketched sections.

Swept feature A feature with a user-sketched section and a user-defined trajectory.

Trajectory The extrude path of a swept feature. Sweep trajectories can be either sketched by the user or selected on the work screen.

SWEEP AND BLEND FUNDAMENTALS

Previously covered in this textbook, the Extrude and Revolve options can be used to create either positive space or negative space features. This chapter expands upon the Protrusion and Cut commands by introducing the Sweep and Blend options. The Sweep option can be compared to the Extrude option. While the Extrude option creates a feature by protruding a section along a straight trajectory, the Sweep option creates a section along a user-defined trajectory (Figure 10–1). This trajectory can be either user-sketched or selected on the work screen. The Blend option can also be compared to the Extrude option. Primarily, the Blend option creates a feature by protruding along a straight trajectory between two or more user-defined sections. A partially revolved Blend can be created also.

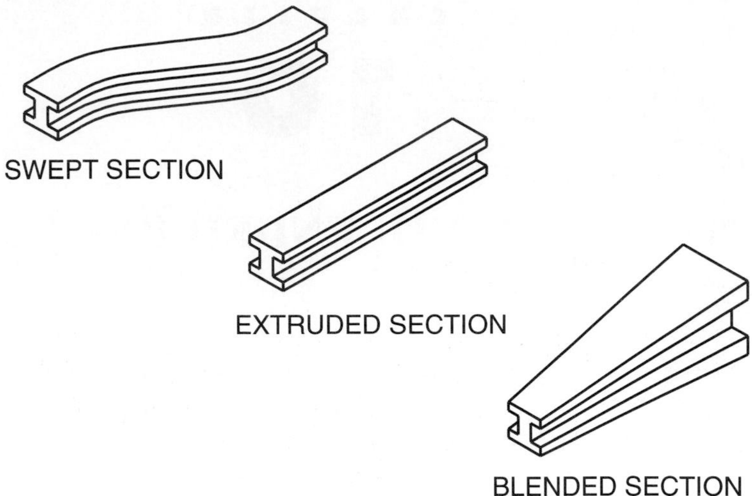

Figure 10–1 Swept, extruded, and blended features

Swept Features

Protruding a section along a user-defined trajectory defines the geometric definition of a **swept feature.** Figure 10–1 shows an example of a Sweep. The section in this feature is protruded along a curved trajectory. Also in Figure 10–1, the same section is shown protruded along a straight trajectory with the Extrude option. Like the Extrude and Revolve options, the Sweep option is available with the Protrusion and Cut commands.

The Sweep option requires the definition of one section and one trajectory. The order of operation requires the creation of the trajectory first followed by the creation of the section. A Sweep's trajectory can be either sketched or selected. For most situations, sketching the trajectory is the preferred method. When sketching the trajectory, normal sketching tools are used. The trajectory can be opened or closed. Any planar surface or datum plane can be used to sketch the trajectory. Due to the nature of the sketching environment, a sketched trajectory is two-dimensional only. A selected trajectory can be used to define a three-dimensional path. A part edge or datum curve can be used as the selected trajectory.

Within the Sweep option, when a section is swept, cross sections of the section are created normal to the trajectory's path. Along a curved path, when the feature is regenerated, these normal sections cannot overlap. An overlap can occur by having a section that is too large for its corresponding trajectory, or it can occur by having a trajectory that is too small for its corresponding section. Both situations often occur when a radius in a trajectory is too small. Figure 10–2a shows a swept feature with a small section. When the section is increased to a larger size (Figure 10–2b), an overlap in the feature's sections can occur. In Figure 10–2b, the section is on the verge of being too large. In this situation, in order to increase the section size, you will have to increase the radii of the trajectory's arcs. If the section should overlap in the bend of the trajectory, you will get a failed regeneration. There are three ways to resolve this problem:

- Create a smaller section.
- Create a larger trajectory.
- Make the arcs on a trajectory larger.

When a swept feature's open trajectory meets with an existing feature or features, the user is provided with the option of either merging the ends of the trajectory or leaving the ends free (Figure 10–3). The Merge Ends option will smoothly merge the new swept feature with adjacent solid features. The trajectory has to be aligned with the existing features.

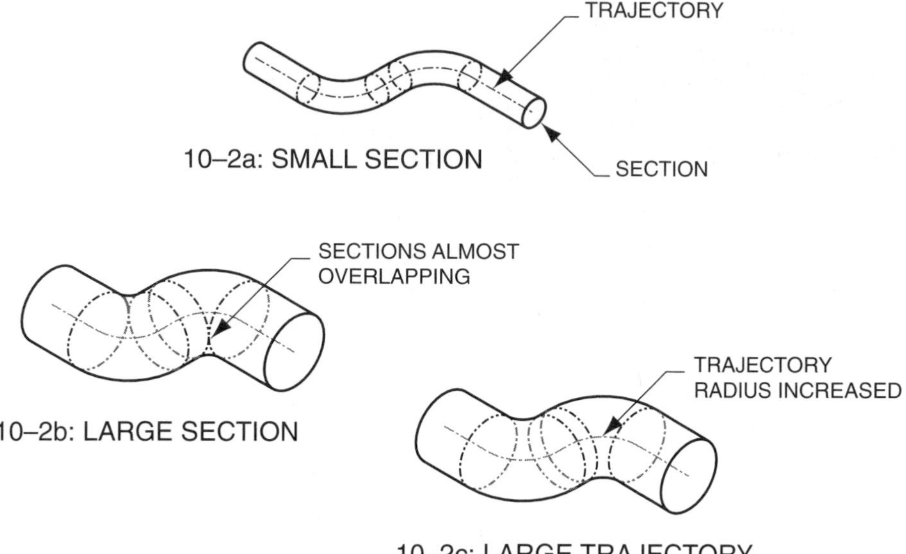

Figure 10-2 Overcoming regeneration failures

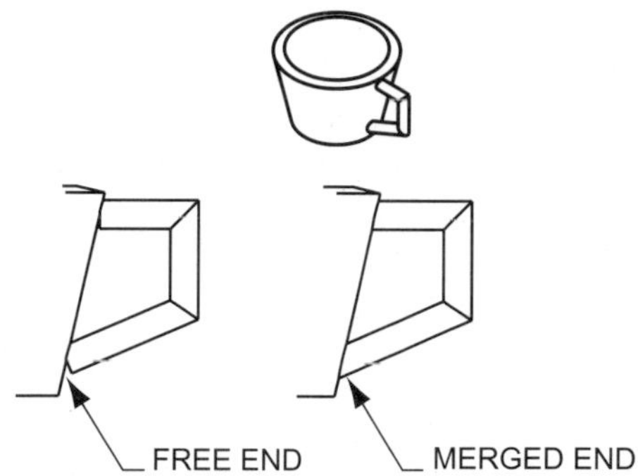

Figure 10-3 Trajectory attributes

Creating a Sweep with a Sketched Section

The Sweep option is available with the Protrusion and Cut commands. An example of a swept cut is shown in Figure 10–4. Notice how a Thin feature is also available with this option. The following guide will demonstrate how to create the swept Protrusion feature shown in Figure 10–5. The same steps are valid for Cut features. When creating a swept feature, the first step is to define the trajectory. In this example, the sketching plane for the trajectory will be one of Pro/ENGINEER's default datum planes.

Step 1: Setup the PROTRUSION command with the following options.

- Sweep option
- Solid feature
- Sketched trajectory

NOTE: A trajectory can be either user-sketched or selected on the work screen. Sketched trajectories are two-dimensional only.

Step 2: Select a Sketching Plane then Orient the Sketcher environment.

The process for selecting a sketching plane is the same within the Sweep option as it is within most Pro/ENGINEER feature options. Any planar surface or datum plane can be selected.

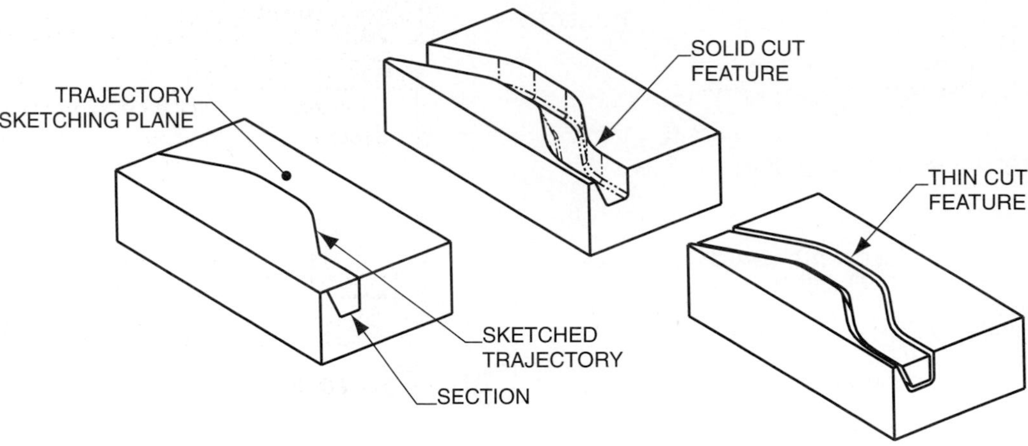

Figure 10-4 Sweep options

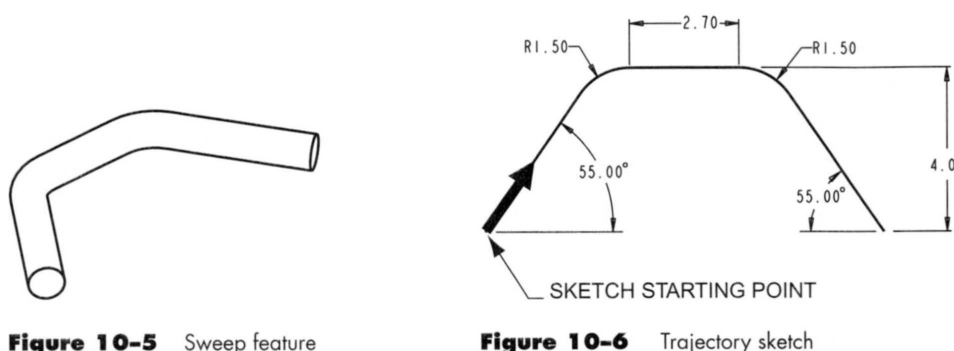

Figure 10-5 Sweep feature

Figure 10-6 Trajectory sketch

STEP 3: Sketch the Trajectory (Figure 10–6).

Use normal Pro/ENGINEER sketching tools to create the trajectory. A trajectory's path can be open or closed. When sketching the trajectory, avoid small radii arcs. Keep in mind that the section for the feature will be sketched at the starting point of the trajectory and normal to the trajectory at this point. By preselecting a vertex, the Pop-Up menu provides an option for defining the selected point as the starting point for the section.

STEP 4: Select the Continue icon to exit the trajectory sketching environment.

STEP 5: Select either FREE ENDS or MERGE ENDS (not available for first geometric feature).

The Merge Ends option will smoothly merge the swept feature with existing geometry. The trajectory has to be aligned with an existing feature. This option is not available for the first geometric feature of a part.

> **INSTRUCTIONAL NOTE** This guide demonstrates the creation of a sweep as the first geometric feature of a part. The Free Ends/Merge Ends option will not be available for this example.

STEP 6: Sketch the section for the swept feature (Figure 10–7).

Upon entering the sketching environment for the swept feature's cross section, the environment will be oriented normal to the trajectory's start point.

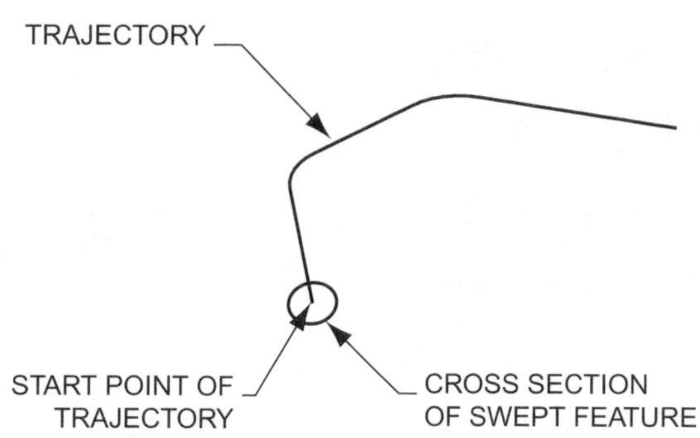

Figure 10–7 Sketching cross section

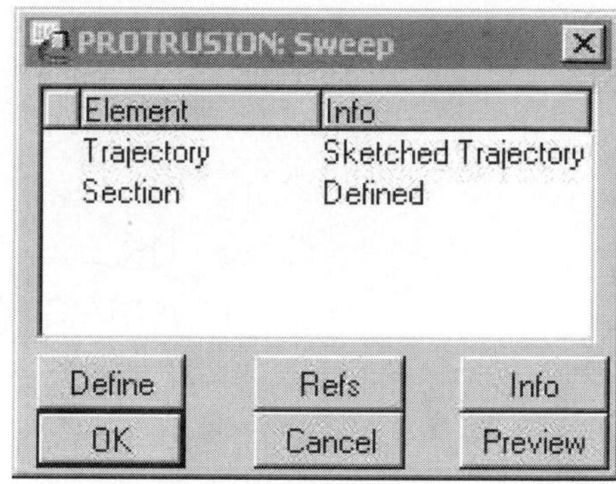

Figure 10–8 Sweep Definition dialog box

Often, it is helpful to dynamically rotate the sketching environment to better visualize the position of this starting point. Use normal sketching tools to create the section. The section for a Sweep has to be closed.

STEP 7: Select the Continue icon to exit the sketching environment.

STEP 8: Preview the sweep on the Feature Definition dialog box (Figure 10–8).

Previewing a swept feature is important. If a failed regeneration occurs, either enlarge your trajectory (especially radii) or decrease the size of the section.

STEP 9: Select OKAY on the dialog box.

BLENDED FEATURES

A **Blend** is a feature created from two or more planar sections. Blend sections are joined together at their edges to form one feature. Three types of Blends are available: Parallel, Rotational, and General. An example of a Parallel Blend is shown in Figure 10–9. With this type of Blend, each section of the feature is sketched in the same sketching environment. The sections defining the feature are shown in the figure. Additionally, Pro/ENGINEER provides the attribute option for creating either a Smooth Blend or a Straight Blend. Both examples are shown in Figure 10–9.

The second blend type is a Rotational Blend. Rotational Blends are created by sketching two or more sections with a rotational angle defined between each section. Within each section, the user must create a coordinate system. This coordinate system defines the pivot point for each section with sections revolved around the Y-Axis of the coordinate system. Figure 10–10 shows an example of a Rotational Blend with four sections. Each section has an angular spacing of 45 degrees. The maximum possible angle between two sections is 120 degrees. Smooth and Straight blended features for these sections are shown.

The Parallel Blend option will create a feature by joining two or more parallel sections, and the Rotation Blend will create a feature by joining two or more sections by revolving the sections about a coordinate system. The third Blend option, General, will create a feature by rotating sections about all three axes of each section's coordinate system. The construction process for a General Blend is similar to the Rotational Blend. Unlike a Rotational Blend, where sections are rotated about the Y-axis, the General option will allow simultaneous rotations about all three axes. Figure 10–11 shows an example of a General Blend with three sections. The second section is rotated about the Y-axis with an angle of 90 degrees. The third section is rotated about the X-axis and Y-axis an angle of 90 degrees for both axes.

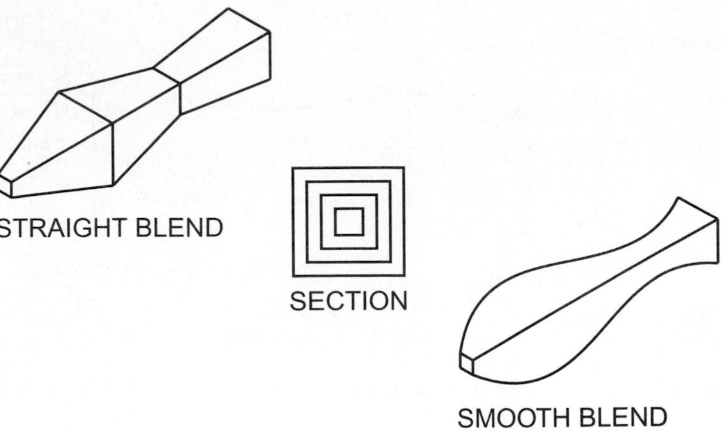

Figure 10-9 Parallel blend examples

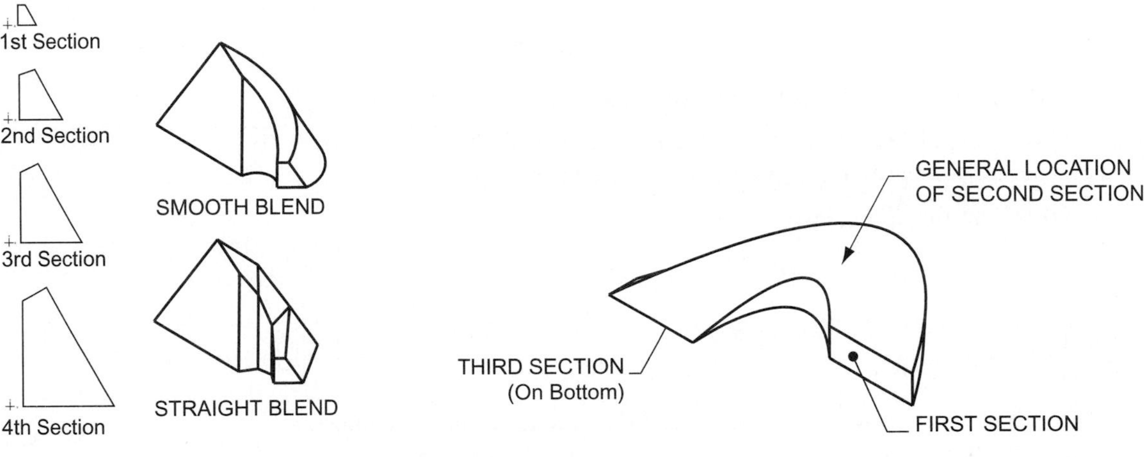

Figure 10-10 A rotational blend

Figure 10-11 A general blend

Creating a Parallel Blend

This tutorial guide will demonstrate the modeling of the blended feature shown in Figure 10–12. This feature will be sketched on one of Pro/ENGINEER's default datum planes. There are several points to remember when creating a Parallel Blend:

- A blended feature has to have two or more sections. A user-defined distance will separate each section.
- In most situations, each section of a blend has to have the same number of entities. The only exception to this rule is if one section has a single point entity. Notice in Figure 10–12 that the first two sections consist of line and arc entities. The outside section has eight line entities, while the middle section consists of four lines and four arcs. The third section is composed of a single point entity.
- The starting point for each section has to be in the same general location and normally should be pointing in the same direction. Figure 10–13 shows the starting point for each section (the point section does not have a visible starting point). Notice the arrows that graphically display the starting point and direction for each section. Figure 10–13 shows what would happen to a feature if the starting points were not placed correctly.

Perform the following steps to create the blended feature shown in Figure 10–12.

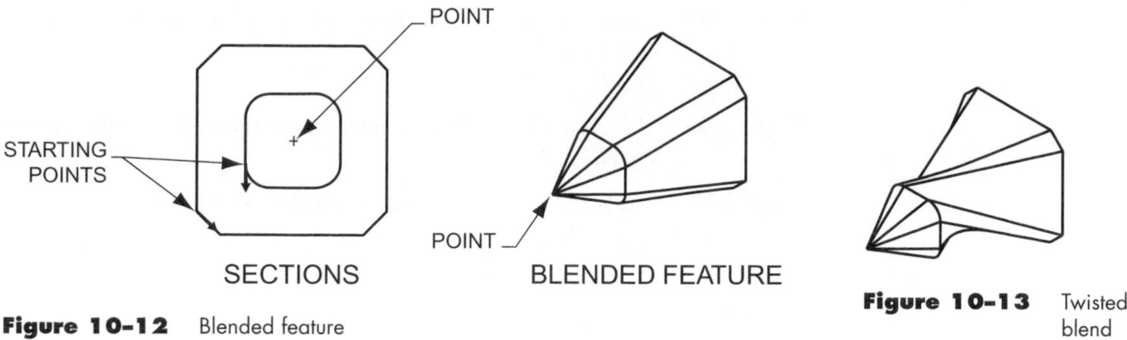

Figure 10-12 Blended feature

Figure 10-13 Twisted blend

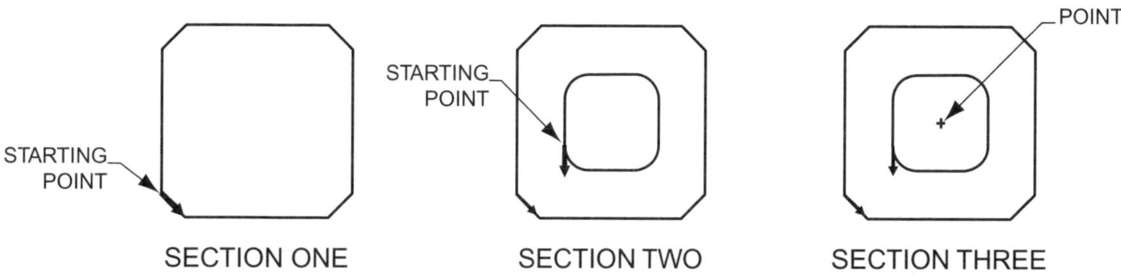

Figure 10-14 Blended feature sections

STEP 1: Setup the PROTRUSION command with the following options.

- Parallel Blend Option (with a Regular Sketched Section).
- Solid Feature.
- One Sided Extrusion.
- Straight Blend Attribute.
- Sketched Section.

STEP 2: Select and Orient the Sketching Plane.

STEP 3: Use appropriate sketching tools to sketch Section One (Figure 10–14).

STEP 4: Use the Pop-Up menu's START POINT option to place the section's starting point as shown. (You must preselect the new start point's vertex location before accessing the Pop-Up menu.)

On the menu bar, the SKETCH >> FEATURE TOOLS >> START POINT option is also available for changing a section's start point. As with the Pop-Up menu's start point option, the vertex for the new starting point has to be preselected.

STEP 5: On the menu bar, select SKETCH >> FEATURE TOOLS >> TOGGLE SECTION (or select Toggle Section on the Pop-Up menu).

Within the Parallel option, sections are sketched within the same sketching environment. The Toggle Section option is used to move from one section to another. To end the creation of sections for a parallel blend, select Toggle Section without sketching any entities.

STEP 6: Use appropriate sketching tools to sketch Section Two (Figure 10–14).

STEP 7: Use the Pop-Up menu's START POINT option to place the section's starting point as shown in the figure.

STEP 8: **On the menu bar, select SKETCH >> FEATURE TOOLS >> TOGGLE SECTION.**

STEP 9: **Use the POINT icon to create the Point entity shown in Section Three.**

STEP 10: **Select the Continue icon to exit the sketching environment.**

STEP 11: **Enter a depth value to define the distance between the First section and the Second section.**

STEP 12: **Enter a depth between the Second Section and the Third Section.**

STEP 13: **Preview the feature on the Blend Feature Definition dialog box.**

STEP 14: **Select OKAY on the dialog box.**

DATUM CURVES

Datum curves are used extensively for the creation of swept features. Additionally, they are used for the creation of surface features. Datum curves are considered features within Pro/ENGINEER and are label with a *Curve_id* on the model tree.

CREATING DATUM CURVES

Datum curves can be sketched using normal sketching tools. For surfacing operations, it is often necessary to construct single curves made of many segments. Datum curves may be opened or closed. Perform the following steps to create a datum curve:

STEP 1: **Select the Datum Curve icon on the Datum Toolbar.**

STEP 2: **Select a Constraint Option on the Curve Options menu.**

The following are examples of constraint options that are available:

SKETCH

The Sketch option allows a user to sketch geometry defining a datum curve. Normal Pro/ENGINEER sketch tools are used. The section can be open and multiple loops may be used.

INTERSECTION OF SURFACES (Intr. Surfs)

The Intersection of Surfaces option places a datum curve at the intersection of two surfaces or quilts. Options are available for picking individual surfaces or whole quilts.

THROUGH POINTS (Thru Points)

The Through Points option places a datum curve through existing datum points. The curve created can be splined or can have user defined radii. Individual datum points can be picked or a datum point array can be selected using the Whole option.

FROM FILE

This option imports a datum curve from an IGES, VDA, SET, or Pro/ENGINEER **.ibl* file.

STEP 3: **Select DONE, then pick geometry appropriate to the selected constraint option.**

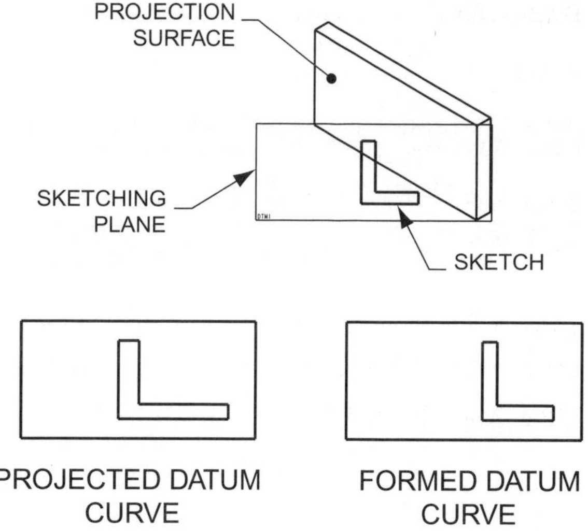

Figure 10–15 Projected and formed datum curves

Projected and Formed Datum Curves

Projected, Formed, and Split curves are datum curves that are constructed by sketching a section or selecting an existing datum curve and projecting it onto one or more surfaces.

Projected Datum Curve

A projected datum curve forms a true projection of the curve being projected. Figure 10–15 shows how a projected datum curve appears on the receiving surface.

Formed Datum Curve

A formed datum curve does not form a true projection when projected onto the receiving surface. As shown in Figure 10–15, the shape of the original curve is pasted onto the receiving surface as opposed to an actual projection.

The following are terms associated with projected, formed, and split curves:

Normal to Sketch (Norm to Sket)

The Normal to Sketch option projects a section or curve perpendicular to the original sketch.

Normal to Surface (Norm to Surf)

The Normal to Surface option projects a section or curve perpendicular to the plane or surface receiving the projection. This option is used with projected datum planes.

Along Direction (Along Dir)

This option projects a curve along a specified direction. The following suboptions are available:

- **Plane** Similar to the Norm to Surf option, the Plane option projects a curve perpendicular to a selected plane or surface.
- **Crv/Edg/Axis** This option projects a curve along the direction of a selected curve, part edge, or axis.
- **Csys** The Coordinate System option projects a curve along the direction of a selected coordinate axis.

Creating Datum Points

Perform the following steps to create a datum point:

Step 1: **Select the Datum Point icon on the Datum Toolbar.**

Datum points, like datum planes and axes, are considered part features.

Step 2: **Select a Constraint option on the Datum Point menu, then pick geometry appropriate to the selected constraint option.**

The following constraint options are available for datum points:

On Surface

The On Surface option places a datum point on an existing surface or plane. The user must select the surface, then the location point on the selected surface. The point is located by selecting two reference edges and providing a distance to each edge.

Offset Surface (Offset Surf)

The Offset Surface option works in the same sequence as the On Surface option. The user must add an additional offset value from the selected placement surface.

Curve Intersection Surface (Curve X Srf)

The Curve-at-the-Intersection-of-a-Surface option places a datum point at the intersection of a curve and surface. The following elements are available for selection as a curve or surface:

- **Curve** Part edge, part curve, part axis, or datum curve.
- **Surface** Part surface or datum plane.

On Vertex

This option places a datum point at the vertex of a datum curve or part edge.

Offset Coordinate System (Offset Csys)

The Offset Coordinate System options places a datum point at an absolute Cartesian, spherical, or cylindrical coordinate point. A coordinate system may exist before using this option or one can be created on-the-fly. The following options are available:

- **With dims or without dims** The With Dimensions option labels the datum points created with a default datum point symbol (PNTx). The Without Dimensions option does not label the datum points created.
- **Coordinate system** Three coordinate systems are available: Cartesian, cylindrical, and spherical.
- **Enter, edit, or read points** The user may place datum points directly on the work screen, or the user may enter point coordinate values by editing a table. Additionally, points may be read from an ASCII file.

MODELING POINT The Offset Coordinate system option is used to create a Datum Point Array. Datum point arrays are multiple datum points created within the same step.

THREE SURFACES (Three Srf)

The Three Surfaces option places a datum point at the intersection of three surfaces and/or planes.

AT CENTER

This option places a datum point at the center of a circle or arc.

ON CURVE

The On Curve option places a datum point on a curve. The curve may be a part edge, part curve, or datum curve. Three options are available:

- **Offset** When utilizing the offset option, after select the curve upon which to place the datum point, the user must select a planar surface to reference the point location.
- **Length ratio** When utilizing the length ratio option, a ratio value is entered to represent the location on the curve of the datum point from the curve's vertex. The ratio value must be between 0.0 and 1.0.
- **Actual len** The Actual Length option locates the datum point on the selected curve at a user provided distance from the curve's vertex. The distance cannot be greater than the curve length.

CURVE INTERSECTING CURVE (Crv X Crv)

The Curve Intersecting Curve option places a datum point on a selected curve at location that represents the minimum distance from a second selected curve. The created datum point represents the location where the two selected curves are at their closest point to each other. Curves under this option do not have to intersect.

OFFSET POINT

This option places a datum point on a straight object (axis, edge, or straight datum curve) at a user-provided distance from an existing datum point.

STEP 3: Select DONE.

The datum point will be created when you select Done on the menu.

COORDINATE SYSTEMS

Coordinate systems are useful for a variety of applications. Due to their feature-based modeling and sketching capabilities, most parametric design applications, such as Pro/ENGINEER, only passively use coordinate systems. Even though Pro/ENGINEER does not utilize a coordinate system as a primary means of creating geometry, many modeling tasks do require the presence of one. A computer-numeric-control machine tool is another example of a manufacturing application that utilizes this type of coordinate system. The following are some of the Pro/ENGINEER tasks that require a coordinate system.

- Many modeling tools, such as the Move/Rotate option under the Copy command, can utilize a coordinate system. Many such applications also provide options that do not require a coordinate system.
- Many Finite Element Analysis applications require a coordinate system.
- Pro/NC (Pro/ENGINEER's manufacturing module) requires a coordinate system to help in the definition of tool paths.

- Datum point arrays require the use of a coordinate system.
- Model mass properties calculations can utilize a coordinate system.

Types of Coordinate Systems

Three types of coordinate systems are utilized within Pro/ENGINEER. Of the three, the Cartesian coordinate system is the one used primarily. The following is a description of each coordinate system.

Cartesian Coordinate System

A Cartesian coordinate system is defined by three orthogonal axes. A typical Cartesian coordinate system's axes are labeled X, Y, and Z. Each axis of the coordinate system is divided into system units. Each axis intersects at a common point, and units along each axis may be positive or negative based on the direction of the axis. Elements, such as datum points, are defined by a distance along each axis.

Cylindrical Coordinate System

The cylindrical coordinate system is built around a type of Cartesian coordinate system. Like the Cartesian coordinate system, this coordinate system also uses three values to define a location. Instead of three axes meeting at one common absolute point, this system is defined by a radius value, a theta value, and a Z value. The radius value is the distance from the absolute point on the X-Y plane, and the theta value is the angle from a selected reference. The Z value is identical to the Cartesian coordinate system's Z value.

Spherical Coordinate System

Like the cylindrical coordinate system, the spherical coordinate system is built around a type of Cartesian coordinate system. Also like the cylindrical coordinate system, a radius value and a theta value are used to define the first two values of the element point. The third value is defined by phi. Phi equals an angle from the Z axis.

Creating a Cartesian Coordinate System

Perform the following steps to create a coordinate system

Step 1: Select the Coordinate System icon on the Datum Toolbar.

Step 2: Select a constraint option on the Options menu and then pick geometry appropriate to the selected constraint option.

3 Planes

With this option, the user selects three planes to create the coordinate system. The planes selected may be datum planes or part surfaces. Planes selected do not have to be perpendicular.

> **MODELING POINT** The **3 Planes** option is commonly used to create a coordinate system at the intersection of Pro/ENGINEER's default datum planes. It is also used to place a coordinate system at the corner of a part.

Point and Two Axes (Pnt + 2Axes)

The Point-Plus-Two-Axes option will place the absolute point of a coordinate system at an existing datum point. Two of the three coordinate system axes are defined by selecting existing datum axes. These datum axes do not have to pass through the datum point.

2 Axes

With the 2 Axes option, the user must select two axes. The origin will lie at the intersection of the two axes. The user must define the orientation of a plane through the origin and first axis.

Offset

The Offset option creates a coordinate system referenced from a second coordinate system. Offset values are entered along each axis of the referenced coordinate system.

Offset by View (Offs by View)

With the Offset-by-View option, the coordinate system will be orthogonal to the work screen. The user must select an existing coordinate system to reference from. The user must enter two rotation angles and a translation value.

Plane and Two Axes (Pln + 2Axes)

The Plane-Plus-Two-Axes option places the origin of the coordinate system at the intersection of a selected plane and axis. The user must also select a third axis to orient the coordinate system.

Origin and Z Axis (Orig + Zaxis)

With the Origin-Plus-Z-Axis option, a selected point and a selected axis define the coordinate system. The origin will lie on the selected point, and the Z-Axis will be oriented by the selected axis.

From File

With the From File option, a data file is used to define the coordinate system relative to an existing coordinate system.

Default

The Default option defines a coordinate system based on the section of a solid feature. The X-axis of the coordinate system is defined relative to horizontal within the sketcher environment. The Y-axis is defined relative to vertical within the sketcher environment. The origin of the coordinate system lies at the section's anchor point.

STEP 3: Define the axes of the coordinate system.

Pro/ENGINEER will project three axes from the origin of the coordinate system. One axis will be highlighted in red. The user is required to select from the following options to define the highlighted axes

- **X-axis** The highlighted axis will be defined as the X-Axis.
- **Y-axis** The highlighted axis will be defined as the Y-Axis.
- **Z-axis** The highlighted axis will be defined as the Z-Axis.
- **Next** This option will highlight the next arrow.
- **Previous** This option will highlight the previous arrow.
- **Reverse** This option reverses the direction of the highlighted arrow.

SUMMARY

The Extrude and Revolve options are useful for creating basic geometric features. Many parts, though, are comprised of shapes that do not follow normal extrude or revolve directions. The Sweep and Blend options are available for more advanced shapes. Both options can be compared to the Extrude option. While Extrude protrudes a section in a straight direction, the Sweep option protrudes a section along a user-defined trajectory. The Blend option protrudes two or more sections to form a feature. Later in this text, the Swept Blend option will be introduced. This option combines the functions of the Sweep option and the Blend option.

TUTORIAL

Blend Tutorial

This tutorial exercise will provide instruction on how to create the part shown in Figure 10–16. This part consists of an extruded Protrusion and two Blends. The first step in this tutorial will require you to construct the base extruded Protrusion.

Within this tutorial, the following topics will be covered:

- Creating a base protrusion.
- Creating a blend.
- Creating a section with the Use Edge option.

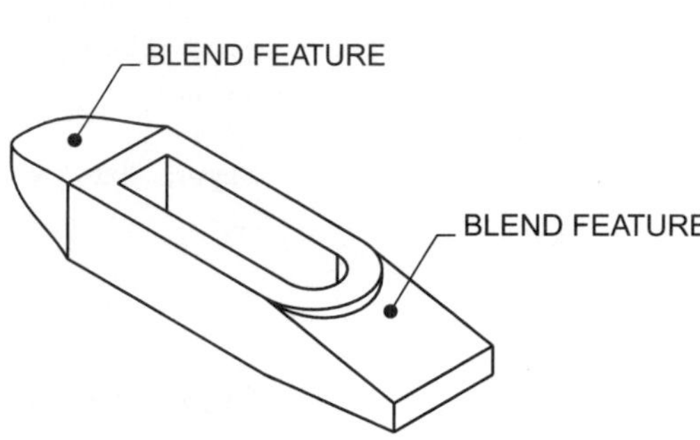

Figure 10–16 Completed part

Creating the Base Feature

Use Pro/ENGINEER's default datum planes to create this part. Name the part *Blend1*. Create the feature shown in Figure 10–17 as the base geometric feature of the part. Sketch the feature's section on datum plane FRONT. Extrude the feature one direction a distance of 2 inches.

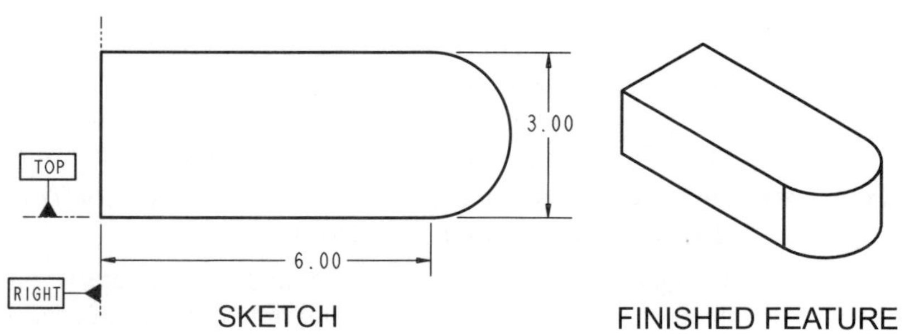

Figure 10–17 Base geometric feature

Creating a Blend

This segment of the tutorial will create the blended feature shown in Figure 10–18. The sketching plane for this feature will be an on-the-fly datum plane constructed through the two tangent edges formed on the base feature.

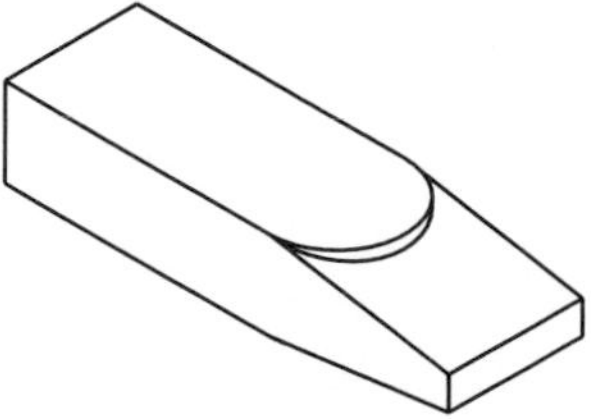

Figure 10–18 Blended feature

Step 1: Select FEATURE >> CREATE >> PROTRUSION.

Step 2: Select BLEND >> SOLID >> DONE.

A blend can be created as either a Solid feature or as a Thin feature.

Step 3: Select PARALLEL >> REGULAR SEC >> DONE on the Blend Options menu.

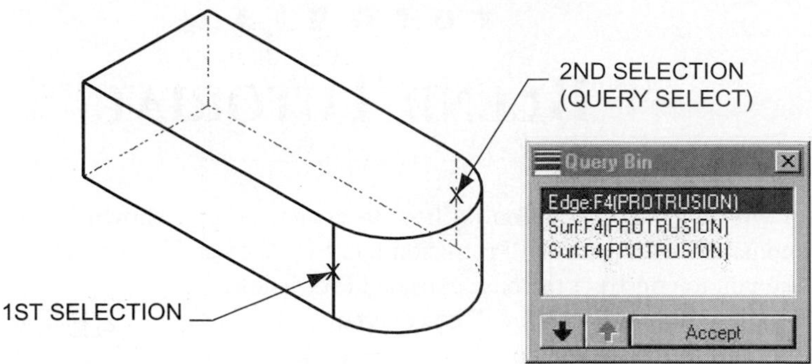

Figure 10–19 Edge selection

A Parallel Blend consists of two or more sections extruded along a straight trajectory.

STEP 4: Select STRAIGHT >> DONE on the Attributes menu.

STEP 5: Select the MAKE DATUM option.

You will create an on-the-fly datum plane through the tangent edges shown in Figure 10–19.

STEP 6: Select the THROUGH constraint option, then select the first location shown in Figure 10–19.

The Through constraint option will place a datum plane through an axis, edge, curve, point, plane, or cylinder. In this tutorial, you will place the datum plane through an edge. The edge selection requires a paired constraint option.

STEP 7: Select the THROUGH constraint option (do not select the edge).

STEP 8: On the Get Select menu, select the Query Select (Query Sel) option, then select the second location shown in Figure 10–19.

Elements and entities are often hard to select for complex or hidden features. In this example, you could dynamically rotate the part or use the Query Select option to pick the tangent edge. Query Select will reveal available entities, features, and elements at a selection location.

STEP 9: On the Query Bin dialog box, select the Edge:F4(PROTRUSION) element, then ACCEPT the selection.

The Query Select option will reveal valid elements at a selected location. In this example, three possible elements are available: an edge and two surfaces.

> **MODELING POINT** The Query Bin dialog box is available through the selection of the Query Bin option in the Environment dialog box.

STEP 10: Select DONE on the Datum Plane menu.

STEP 11: Select a direction of feature creation that will extrude the feature away from the first protrusion feature.

If necessary, use the Flip option to change the direction of protrusion.

STEP 12: Orient the sketching environment to match Figure 10–20.

STEP 13: In the sketcher environment, close the References dialog box.

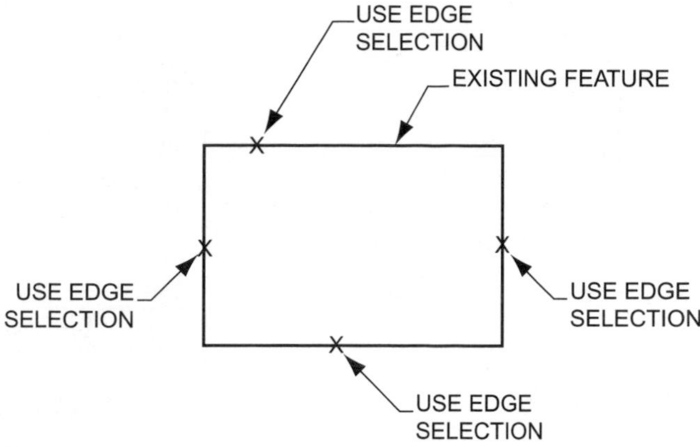

Figure 10-20 First section creation

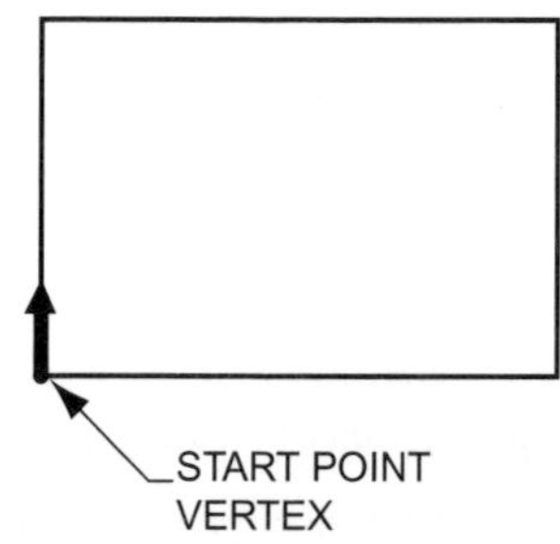

Figure 10-21 Start point selection

STEP 14: **On the Sketcher toolbar, select the USE EDGE icon.**

The Use Edge option is used to project feature elements onto the sketching plane for use in the current sketching environment. In this tutorial, you will use the edges of the existing part feature and turn these edges into sketcher entities.

STEP 15: **Select the four edge locations shown in Figure 10–20.**

On the work screen, select each edge once and notice the change to the sketching environment.

STEP 16: **If necessary, preselect the start point vertex shown in Figure 10–21, then utilize the Pop-Up menu to set this vertex as the section's start point.**

Blended features consist of two or more sections. The start points for each section should be in approximately the some locate for each section. Within this segment of the tutorial, the start point will be in the lower right-hand corner of each section. By preselecting a vertex, the Pop-up menu has an option for changing a section's start point.

STEP 17: **On the menu bar, select SKETCH >> FEATURE TOOLS >> TOGGLE SECTION (only select once).**

Within the construction of a blended feature, the Toggle Section option will be used to toggle between feature sections. The Pop-Up menu also provides an option for toggling between sections.

Parallel blend sections are created in one sketching environment. The Toggle Section option will allow you to switch to the sketching environment for the next section of the blend. When you select Toggle Section, notice the change of the existing sketcher entities

NOTE: **Only select Toggle Section once.**

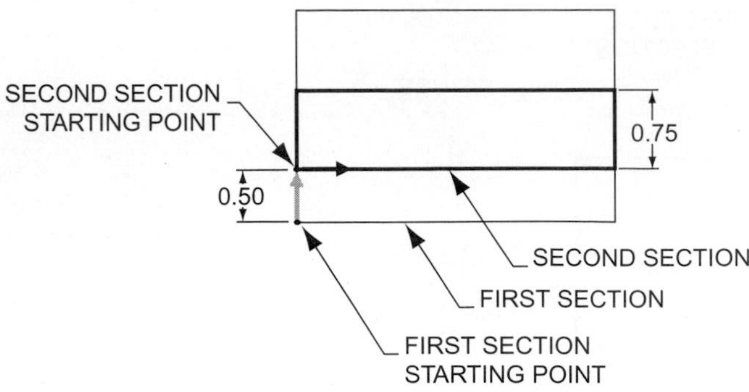

Figure 10-22 Second section

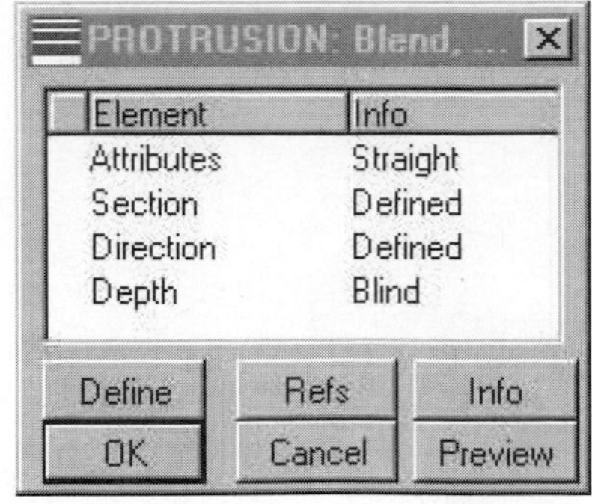

Figure 10-23 Feature Definition dialog box

STEP 18: **Use the RECTANGLE icon to sketch the entities shown in Figure 10–22.**

The design intent for this feature requires the left and right edges of this section to be aligned with the left and right edges of the base feature. Use the Start Point option on the Pop-Up menu to set this section's start point to the vertex shown in the figure. Remember to preselect the vertex before attempting to access the menu.

STEP 19: **Modify the section's dimensioning scheme and values to match Figure 10–22.**

STEP 20: **Select the Continue icon to exit the sketcher environment.**

STEP 21: **Select BLIND >> DONE.**

STEP 22: **Enter 4.00 as the depth of the second section.**

At this point in the Parallel Blend option, Pro/ENGINEER will require you to enter the distance between each section. Since there are only two sections, you will enter one depth distance.

STEP 23: **Preview the feature on the Feature Definition dialog box, then select OKAY (Figure 10–23).**

CREATING A SECOND BLEND

This segment of the tutorial will create the second blended feature shown in Figure 10–24. This blend will consist of four sections. The first section will be created from existing feature edges while the second and third sections will consist of sketched rectangular entities. The fourth section will consist of a single point entity.

STEP 1: **Select PROTRUSION >> BLEND >> DONE.**

STEP 2: **Select PARALLEL >> REGULAR SEC >> DONE.**

STEP 3: **Select SMOOTH >> DONE.**

While the Straight option will blend sections with straight surfaces, the Smooth option will create a smooth blend between each section.

STEP 4: **Select the sketching plane shown in Figure 10–25 and orient the sketching environment.**

STEP 5: **Close the References dialog box.**

STEP 6: **Create the Four Sections shown in Figure 10–26.**

Sketch the four sections that compose the blended feature. Utilize the Use Edge option to create the first section and the Rectangle option to create the

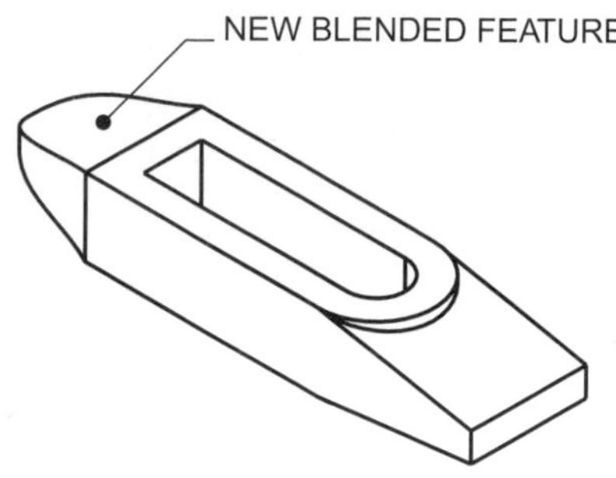

Figure 10–24 Blend feature

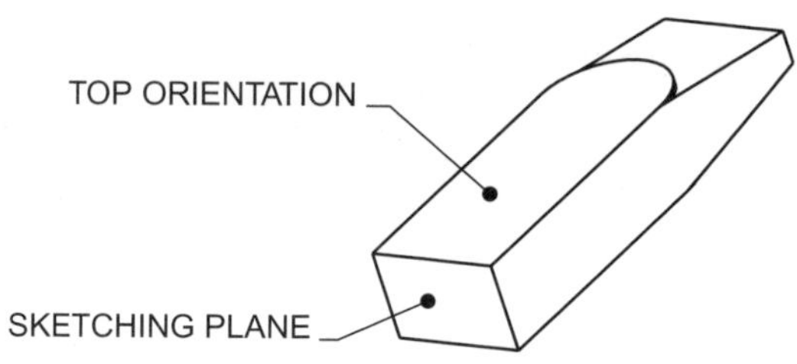

Figure 10–25 Sketching plane and orientation

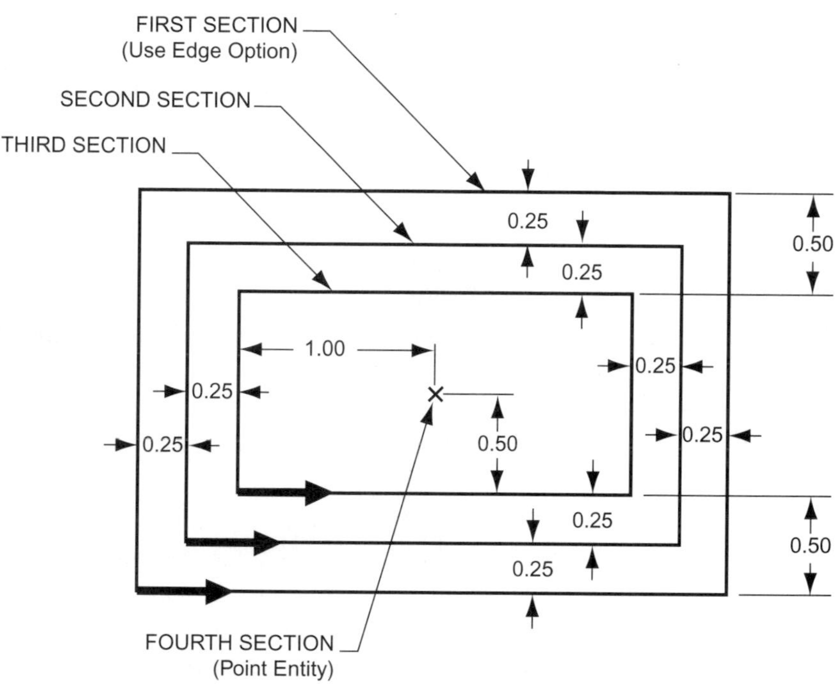

Figure 10–26 Blend sections

second and third. The Section Toggle option is required to switch between sections. Ensure that each section's start point is in the location shown in the figure. Create the fourth section with the Point option.

STEP 7: After the creation of the four required sections, select the continue option to exit the sketching environment.

STEP 8: Enter a Blind depth distance of 1 inch between each section.

STEP 9: Preview the Feature on the Feature Definition dialog box, then select OKAY.

CREATING A CUT FEATURE

This section of the tutorial will create the Cut feature shown in Figure 10–27. You will utilize the Offset Edge option to create the sketch.

STEP 1: Select CUT >> EXTRUDE >> DONE.

STEP 2: Setup a sketching environment for a One Side Cut with the sketching plane and orientation shown in Figure 10–27.

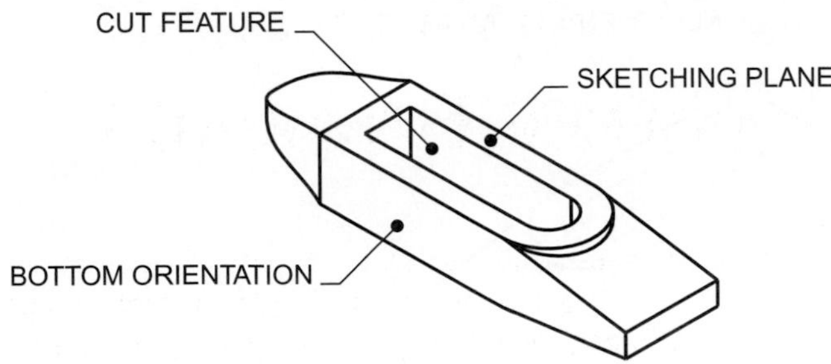

Figure 10–27 Cut feature

STEP 3: Close the References dialog box.

STEP 4: Select the OFFSET EDGE icon on the Sketcher toolbar.

NOTE: The Offset Edge icon is located behind the Use Edge icon.

STEP 5: On the Type dialog box, select the CHAIN option, then pick the two locations shown in Figure 10–28.

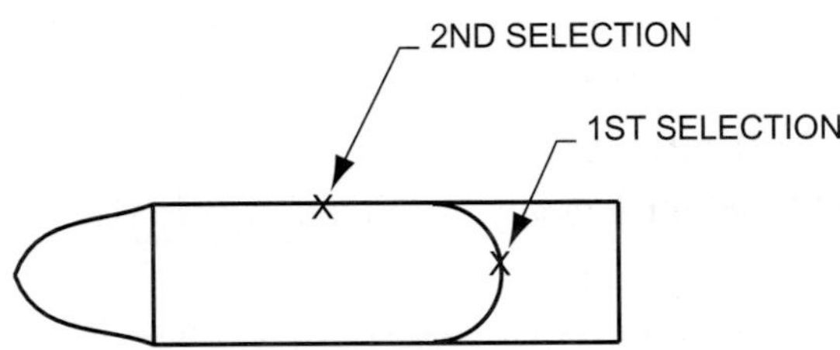

Figure 10–28 Offset Edge option selection

STEP 6: On the Choose menu, use the NEXT and PREVIOUS options to switch between possible chain options, then ACCEPT the chain that will create the correct Cut.

STEP 7: Enter an Offset Distance of 0.75 inches (enter a −0.75 if necessary).

When entering the offset distance, observe the offset side on the work screen. You can enter a negative value to change the side of offset.

STEP 8: Select the Continue option to exit the sketching environment.

STEP 9: Accept the direction of Cut, then enter a THRU ALL depth.

STEP 10: Preview the feature, then select OKAY on the Feature Definition dialog box.

STEP 11: Save your part.

TUTORIAL

SWEEP TUTORIAL 1

This tutorial exercise will provide instruction on how to create the part shown in Figure 10–29. The first segment of this tutorial will consist of creating the pipe feature through the use of a Swept Protrusion. The flange features will be created as extruded Protrusions with a patterned radial hole.

Within this tutorial, the following topics will be covered:

- Creating a swept protrusion.
- Creating an extruded protrusion.
- Creating a patterned radial hole.

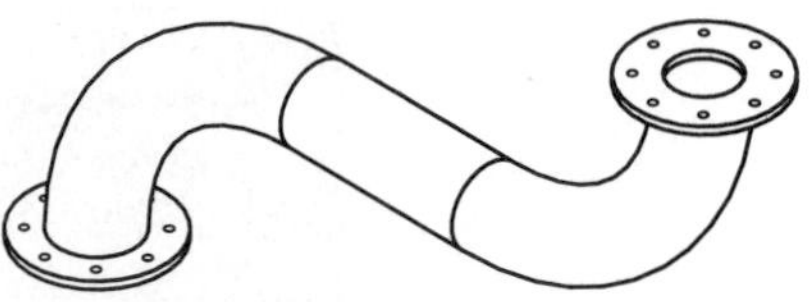

Figure 10-29 Completed part

CREATING THE BASE FEATURE

This segment of the tutorial will create the swept Protrusion feature shown in Figure 10–30. Start a part model utilizing the default template and named *Sweep1*. You will create this feature through the use of a sketched trajectory and a sketched section.

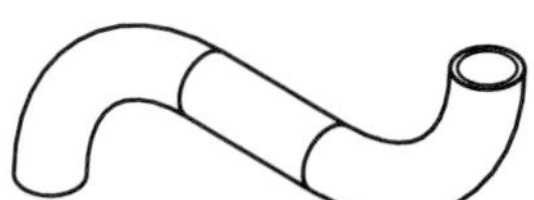

Figure 10-30 Swept feature

STEP 1: Select PROTRUSION >> SWEEP >> DONE.

The Sweep option is available under the Protrusion and Cut commands.

STEP 2: Select the SKETCH TRAJECTORY (Sketch Traj) option.

Within the Sweep option, the trajectory is created first, followed by the Sweep's section. A trajectory can be either sketched in a sketcher environment, or it can be selected on the work screen. Datum curves make valuable features that can be selected as a sweep's trajectory.

STEP 3: Select datum plane FRONT as the sketching plane, and orient datum plane TOP toward the top of the sketcher environment.

STEP 4: In the sketching environment, use the ARC and LINE options to create the section shown in Figure 10–31.

When starting the sketch for a trajectory, notice the arrow that is created. This arrow denotes the trajectory's start point and the direction of sweep.

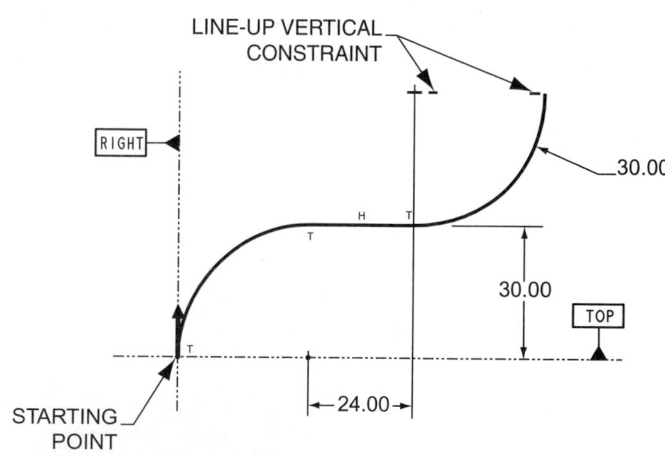

Figure 10-31 Trajectory sketch

A starting point can be changed with the Sketch >> Feature Tools >> Start Point option or with the Pop-Up menu's Start Point option.

The dimensioning scheme shown in Figure 10–31 matches the design intent for this feature. Use the Dimension option to adjust your dimensions to match this scheme. Use the Modify option to change your dimension values.

STEP 5: When the section is complete, select the Continue option to exit the sketcher environment.

> **MODELING POINT** In this tutorial, this swept feature is the first geometric feature of the part. If other features were to exist at this point, you would have the option of creating the Sweep with merged or free ends. The Merge End option will merge the sweep with any attached feature.

STEP 6: Sketch the swept feature's cross section as shown in Figure 10–32.

Upon finishing the trajectory's section, Pro/ENGINEER will open a sketching environment for the creation of the feature's cross section. Within this sketching environment, Pro/ENGINEER will orient the work screen normal to the trajectory at the trajectory's start point. Often, it is useful to dynamically rotate the work screen (control key and middle mouse button) to better visualize the start point's location (Figure 10–33).

Use the Circle option to create the two circles shown in Figure 10–32.

STEP 7: Select the Continue option to exit the sketching environment.

STEP 8: Preview the feature on the Feature Definition dialog box (Figure 10–34).

Notice on the dialog box the Trajectory and Section elements. The Sweep's trajectory can be redefined with the trajectory option, and the feature's cross section can be redefined with the section option.

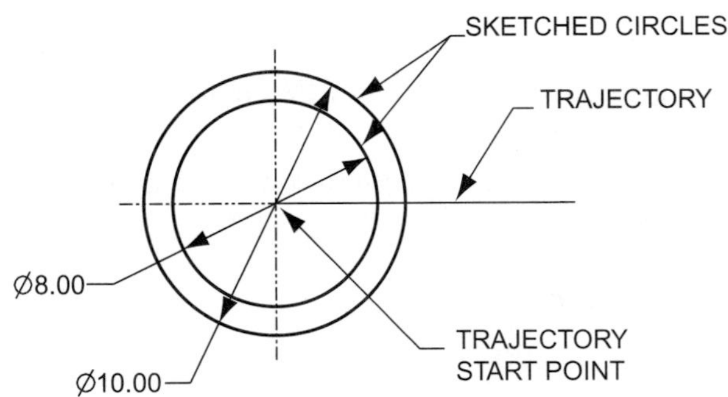

Figure 10–32 Cross section of sweep

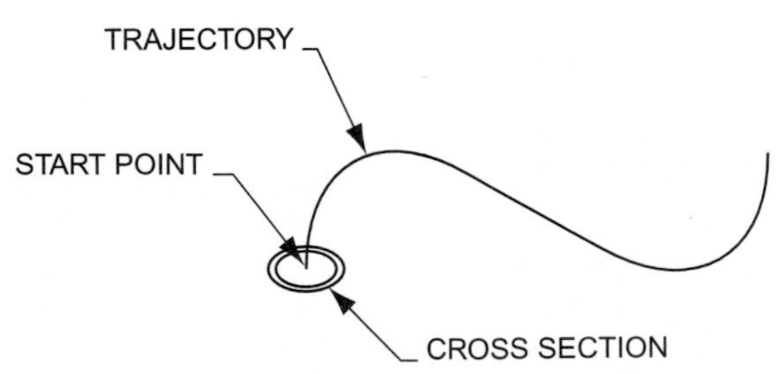

Figure 10–33 Cross section start point

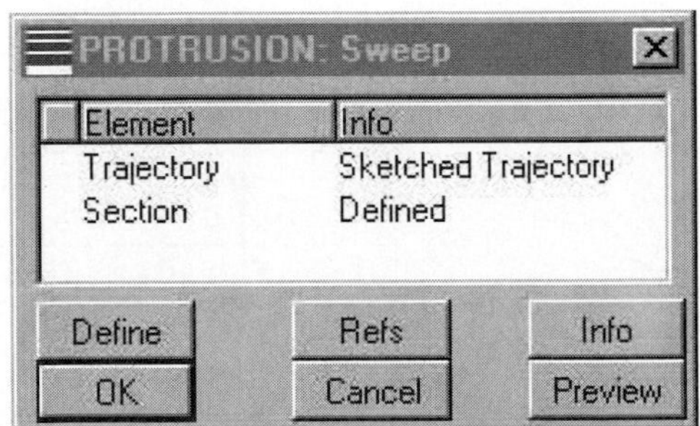

Figure 10–34 Feature Definition dialog box

STEP 9: **On the dialog box, select OKAY to create the feature.**

STEP 10: **Save your part.**

CREATING FLANGE FEATURES

This part of the tutorial will create the two flanges shown in Figure 10–35. The base feature of each flange will be constructed as an extruded Protrusion. The bolt-circle hole pattern will be created from a patterned radial hole.

STEP 1: **Setup an Extruded Protrusion.**

Use the Protrusion >> Extrude options to create the base feature of the first flange. Ensure that you create this feature at the end of the pipe feature that incorporates the default datum planes. Use the following settings:

- Solid Feature.
- One Side extrusion.
- Sketching Plane (as shown in Figure 10–36).

STEP 2: **Sketch the section shown in Figure 10–37.**

Specify the inside diameter of the swept pipe feature and datum planes RIGHT and FRONT as references. Sketch the two circles shown in Figure 10–37. Notice that the inside hole of the flange is the same diameter as the inside diameter of the swept pipe feature. The design intent calls for these two diameters to remain the same. Incorporate this into your sketch.

STEP 3: **Select the Continue option to exit the sketching environment, then extrude the feature a Blind distance of 1 inch.**

STEP 4: **Preview the feature, then select OKAY on the Feature Definition dialog box.**

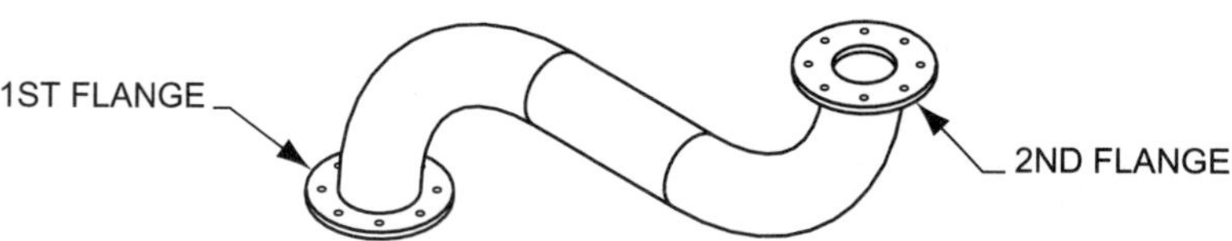

Figure 10–35 Flange features

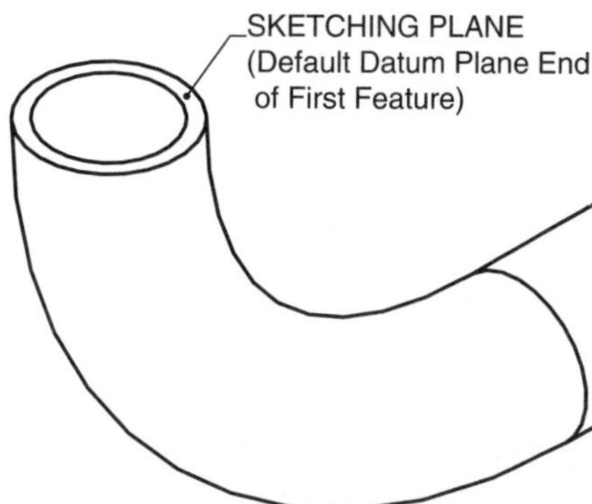

Figure 10–36 Sketching plane

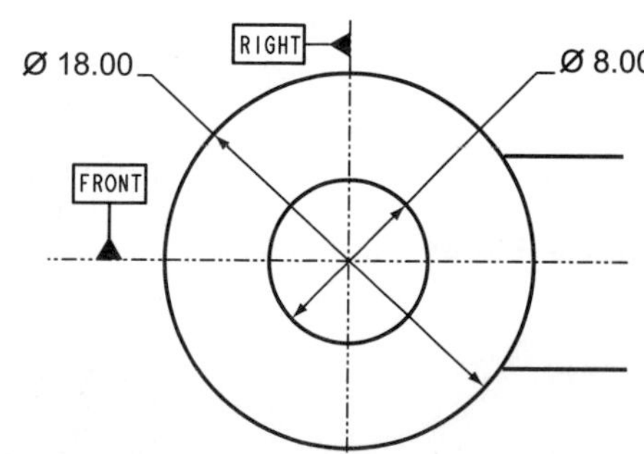

Figure 10–37 Sketched section

STEP 5: Create the Radial Hole feature shown in Figure 10–38.

Create the hole shown in the figure. This hole will be patterned to create the remaining holes on the flange. Use the following options for the hole:

- Straight hole.
- Hole diameter = 1 inch.
- Through All Depth option (one sided).
- Primary reference = top of flange.
- Diameter placement type with a bolt-circle dia. = 14 inch.
- Datum plane RIGHT as the angular reference (zero degree angle).

STEP 6: Use the PATTERN command to create a pattern of the Radial Hole (Figure 10–39).

Use the angular dimension that defines the radial hole as the leader dimension for the pattern. Increment eight holes an angular distance of 45 degrees.

STEP 7: Use the methods from the previous steps of this tutorial segment to create the second flange (Figure 10–40).

As with the first flange, use the Protrusion >> Extrude option to create the base feature of the flange. Your references will be slightly different from the first flange feature. Pattern a Radial Hole to create the flange holes.

STEP 8: Save your part.

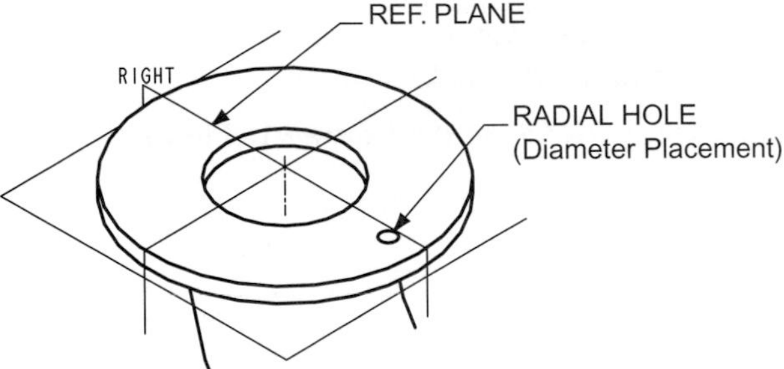

Figure 10–38 Radial hole

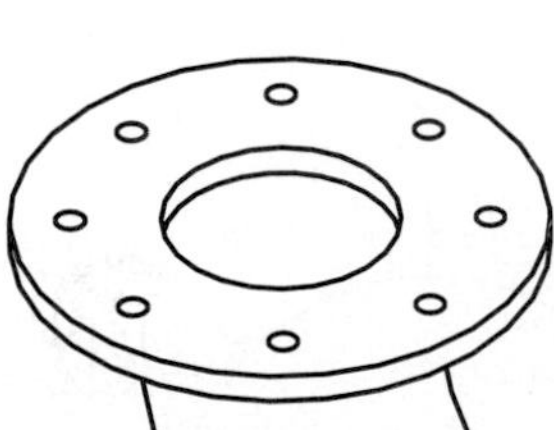

Figure 10–39 Patterned hole

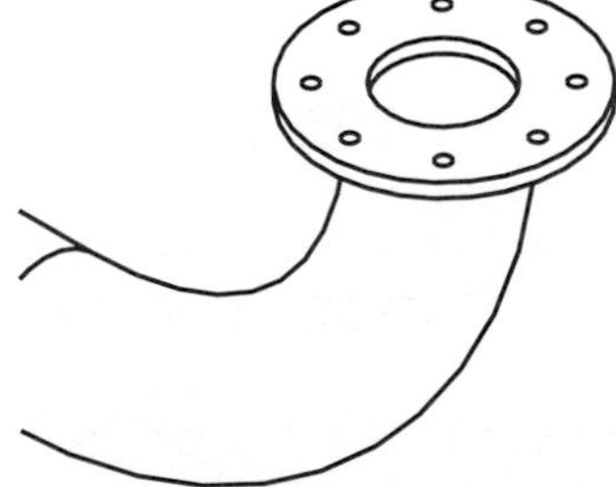

Figure 10–40 Second flange feature

TUTORIAL

SWEEP TUTORIAL 2

This tutorial exercise will provide instruction on how to create the wheel part shown in Figure 10–41. The first segment will consist of creating the hub and grip for the wheel. Both features will be created as Revolved Protrusions. The second segment will consist of creating a Datum Curve. This curve will be used as the trajectory for the Swept spoke feature created in the final segment.

Within this tutorial, the following topics will be covered:

- Creating a revolved protrusion.
- Creating a datum curve.
- Creating a swept feature.
- Creating a round.

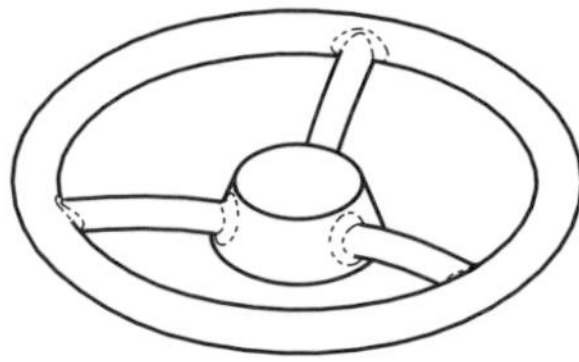

Figure 10–41 Completed part

CREATING THE BASE FEATURE

The base geometric feature of the part, the hub of the wheel, consists of a Revolved Protrusion. Figure 10–42 shows the sketch for the feature. Perform the following options when creating this feature:

- One side revolved protrusion.
- Sketching plane is datum plane RIGHT.
- Sketch the section as shown in Figure 10–42.
- 360-degree revolution.

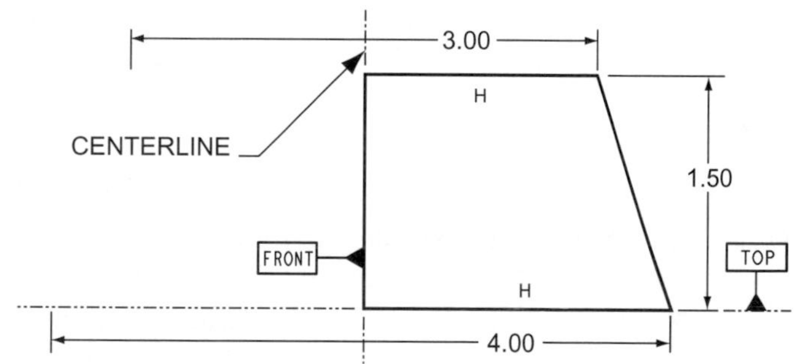

Figure 10–42 Hub sketch

CREATING THE WHEEL HANDLE

This segment of the tutorial will create the handle of the wheel part. As with the hub feature, this feature will be created as a revolved protrusion. Figure 10–43 shows the sketch for the feature. Perform the following options when creating the feature:

- One side revolved protrusion.
- Sketching plane is datum plane RIGHT.
- Sketch section as shown in Figure 10–43.
- 360-degree revolution.

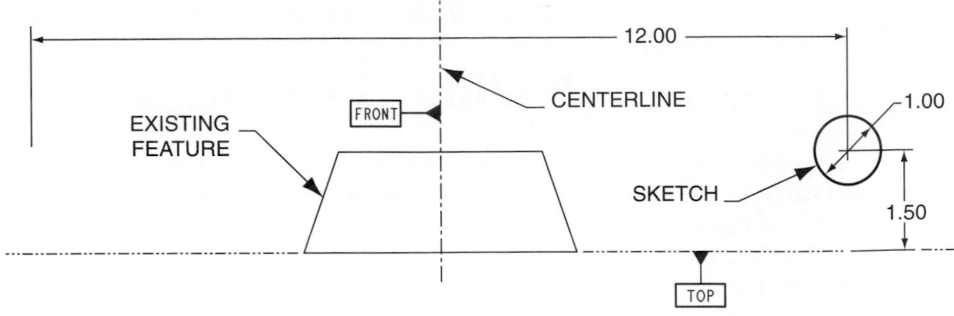

Figure 10–43 Wheel handle feature

Creating a Datum Curve

If a datum can be defined as a theoretically exact object, a datum curve is not actually a true datum. Datum curves are used in Pro/ENGINEER for the construction of features. As an example, datum curves are often used for trajectories in swept features. While Pro/ENGINEER provides multiple ways of creating datum curves (e.g., Thru Points, Projected, Formed, From Curve, etc.), the Sketch option is probably the most used. This segment of the tutorial will create a datum curve that will be used as the trajectory for the swept spoke feature of the wheel part. This trajectory could be sketched within the Sweep option. An advantage of sketching it as a separate feature is the flexibility it provides when having to adjust the trajectory's path. Perform the following steps to create the datum curve feature:

Step 1: **Select the DATUM CURVE icon on the datum toolbar.**

Step 2: **Select SKETCH >> DONE on the Curve Options menu.**

Step 3: **Set up the sketching environment.**

Your sketching environment should appear as shown in Figure 10–44. Establish the following options for the datum curve's sketching environment:

- Sketch on datum plane FRONT.
- Orient datum plane TOP to the top of the work screen.

Step 4: **While in the sketcher environment, turn off the display of datum planes and set Hidden as the model's display mode.**

Step 5: **In the sketcher environment, using the References dialog box, specify the four references shown in Figure 10–44.**

Specify datum planes RIGHT and TOP as references. Also, specify the two feature edges shown in Figure 10–44 as references.

> **MODELING POINT** If you inadvertently close the References dialog box or realize that it is needed at a later point in the sketching process, it can be accessed with the Sketch >> References option on the menu bar.

Step 6: **Use the POINT icon to create the three points shown in Figure 10–45.**

The datum curve will be sketched with the Spline option. While a spline can be sketched without points, the establishment of points allows for better control of the definition of the spline. Create the three points with the

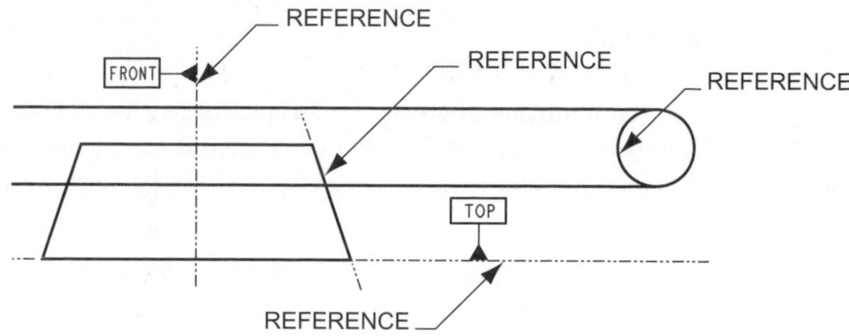

Figure 10-44 Specifying references

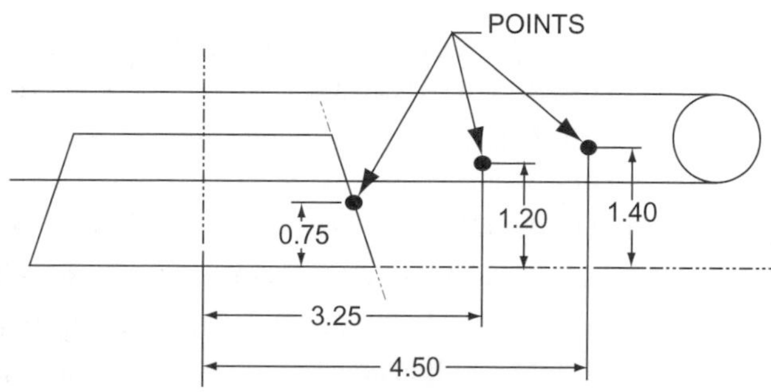

Figure 10-45 Point creation

dimensioning scheme shown in Figure 10–45. Notice that one point is aligned with existing geometry.

STEP 7: Use the SPLINE icon to create the spline entity shown in Figure 10–46.

When creating the spline, sketch from the left to the right by sketching through each control point. Align the right end of the spline with the wheel entity at the quadrant shown in the figure. After sketching the spline, the dimensions defining the control points can be used to adjust the spline geometry.

STEP 8: Select the Continue icon to exit the sketching environment.

STEP 9: Preview the datum curve, then select OKAY on the Feature Definition dialog box.

STEP 10: Save your part.

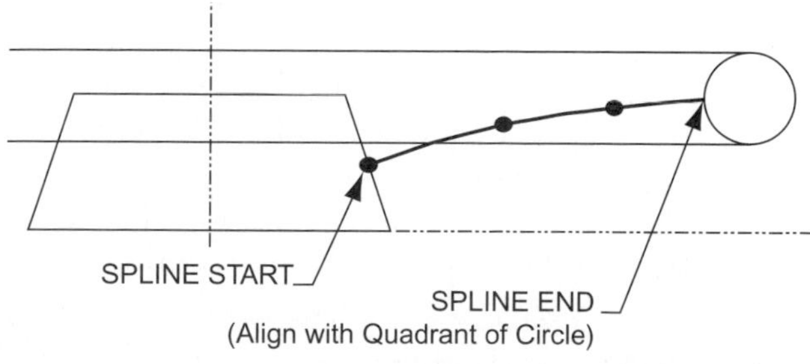

Figure 10-46 Spline creation

SWEEP CREATION

This segment of the tutorial will create the swept feature that defines one spoke of the wheel. The datum curve created previously in this tutorial will be used as the trajectory for this feature. Later in this tutorial, this swept feature will be Copy-Rotated to create the remaining spoke features.

STEP 1: Select FEATURE >> CREATE >> PROTRUSION.

STEP 2: Select SWEEP >> SOLID >> DONE.

STEP 3: Pick the Select Trajectory (Select Traj) option on the Sweep Trajectory menu.

This tutorial will require you to select the datum curve created previously as the trajectory for this swept feature.

STEP 4: With the ONE-BY-ONE option selected, on the work screen select the datum curve feature.

The One-By-One selection option requires you to select individual entities of a chain. In this example, the datum curve is only one entity.

STEP 5: On the Chain menu, select the DONE option.

STEP 6: From the Attributes menu, select the MERGE ENDS option; then select DONE.

The Merge Ends option will merge the ends of the swept feature with any existing features that the trajectory is aligned with. In this tutorial, the datum curve trajectory is aligned with the hub and the wheel features.

STEP 7: In the sketching environment, create the circle entity shown in Figure 10–47.

No additional references are needed for this section. If the orientation of the sketching environment is unclear, dynamically rotate the work screen to see the starting point of the section. The Orient Sketch icon on the sketcher toolbar will return you to the normal two-dimensional sketching environment. Align the center of the sketched circle with the center of the trajectory.

STEP 8: Select the Continue icon to exit the sketching environment.

Do not attempt to exit the sketcher environment until your sketch matches Figure 10–47.

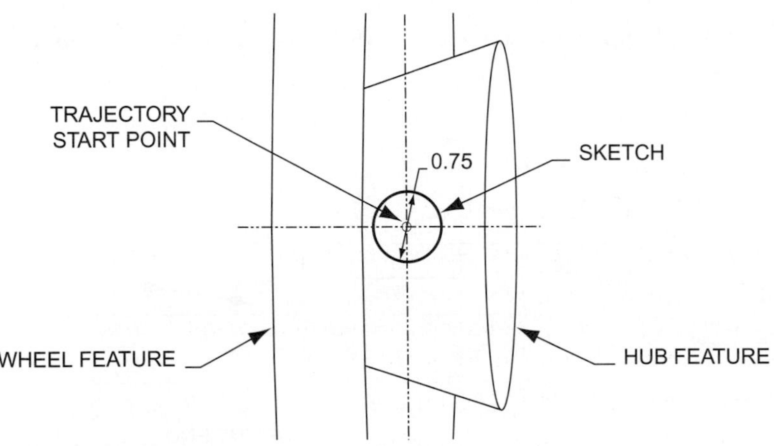

Figure 10–47 Sweep section

STEP 9: Preview the Sweep, then select OKAY on the Feature Definition dialog box.

STEP 10: Save your part.

CREATING A ROUND

In this segment of the tutorial, use the Round command (Feature >> Create >> Round) to create the two fillets shown in Figure 10–48. Use the following options when creating the fillets:

- Create a simple round.
- Create each fillet as a single feature.
- Select the edges shown in Figure 10–48.
- Use Tangent Edge as the edge selection technique.
- Use a constant round radius of 0.25 inches.

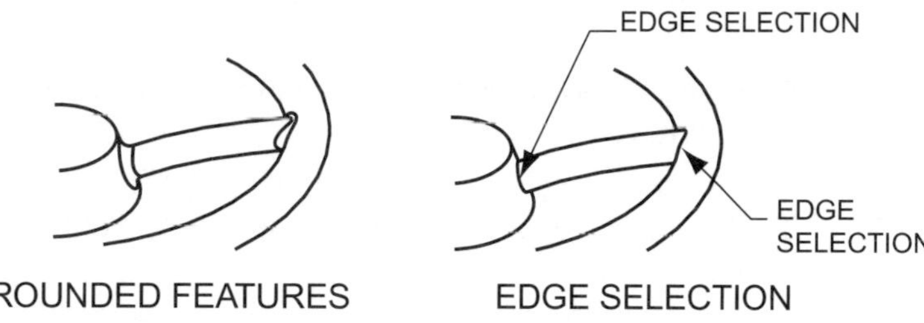

Figure 10–48 Fillet creation

GROUPING FEATURES

The Datum Curve trajectory feature, swept feature, and round features will be Copy-Rotated to create the three instances of the spoke. In this tutorial, you will group these features and use the Copy/Rotate option to make the two copies. Within Pro/ENGINEER, grouping the features is not a requirement of the Copy command. You will group the features in this example for two reasons. First, a copy of multiple features creates a group of the copies. In other words, the copied instances of the features will be groups. Second, grouping the features helps to keep them organized on the model tree. Use the Feature >> Group option to create a local group of the three features comprising the spoke.

STEP 1: Select FEATURE >> GROUP.

STEP 2: Select LOCAL GROUP on the Group menu.

STEP 3: In Pro/ENGINEER's input window, enter *SPOKE* as the name for the group.

STEP 4: On the Model Tree, select the datum curve feature, swept protrusion feature, and round features to define the group.

STEP 5: Select DONE on the Feature Select menu.

STEP 6: Select DONE/RETURN on the Group menu.

STEP 7: Observe the changes to your features on the model tree.

Copying the Spoke Group

This segment of the tutorial will create two copies of the spoke group (Figure 10–49). Perform the following steps to create each copy:

Step 1: Select FEATURE >> COPY.

Step 2: Select MOVE >> DEPENDENT >> DONE on the Copy Feature menu.

The Rotate option is located under the Move menu option. The Dependent option will make the copied feature's dimensions dependent upon its parent feature. In other words, any copies of the spoke group will remain the same size as its parent group's dimensions.

Step 3: On the model tree or on the work screen, select the group to copy then select DONE.

Step 4: Select ROTATE on the Move Feature menu.

Step 5: Select the CRV/EDG/AXIS option, then select the axis shown in Figure 10–50.

As the name implies, the Curve/Edge/Axis option will rotate a feature around a curve, edge, or axis. In this example, you will rotate the features around the center axis of the model.

Step 6: If necessary, FLIP the direction of rotation indicator arrow to match Figure 10–50.

The Rotate option uses the Right-Hand rule to rotate features. Using your right hand, your thumb points in the direction of the arrow on the work screen and your fingers point in the direction of rotation.

Step 7: Select OKAY to accept the rotate direction shown in Figure 10–50.

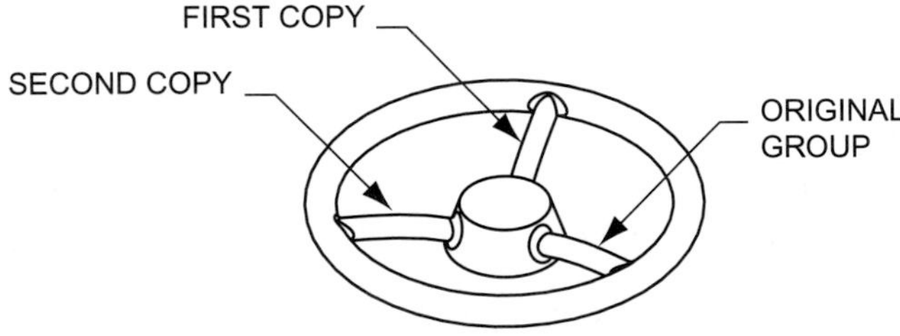

Figure 10–49 Copied features

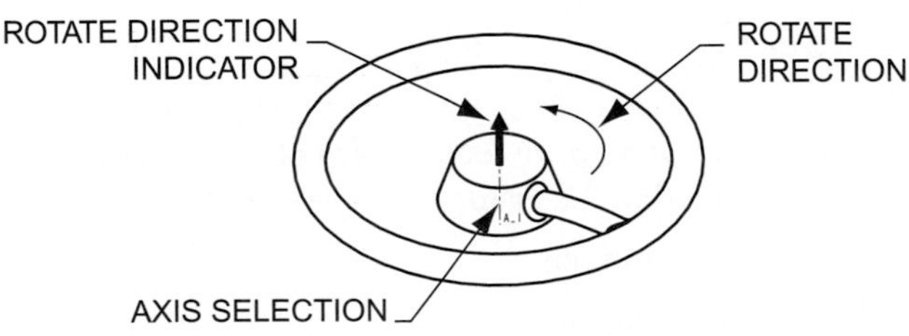

Figure 10–50 Copy creation

STEP 8: In Pro/ENGINEER's textbox, enter 120 as the number of degrees to rotate the group.

STEP 9: Select the DONE MOVE option on the Move Feature menu to finish the move process.

STEP 10: Select the DONE option on the Group Vary Dimension menu.

This step of the Copy command allows individual dimensions to be varied during the copy process. Within this tutorial, you will not vary any dimensions.

STEP 11: Select OKAY on the Feature Definition dialog box.

The Copy command does not have a Preview option on its Feature Definition dialog box.

STEP 12: Create the Second Spoke copy.

Use the previous steps of this tutorial to create the second copy of the spoke. You can use either the original group or the newly copied group as the group to copy.

STEP 13: Save your part.

STEP 14: Use the FILE >> DELETE >> OLD VERSIONS option to delete old versions of the part.

PROBLEMS

1. Use Pro/ENGINEER's Part mode to model the part shown in Figure 10–51.

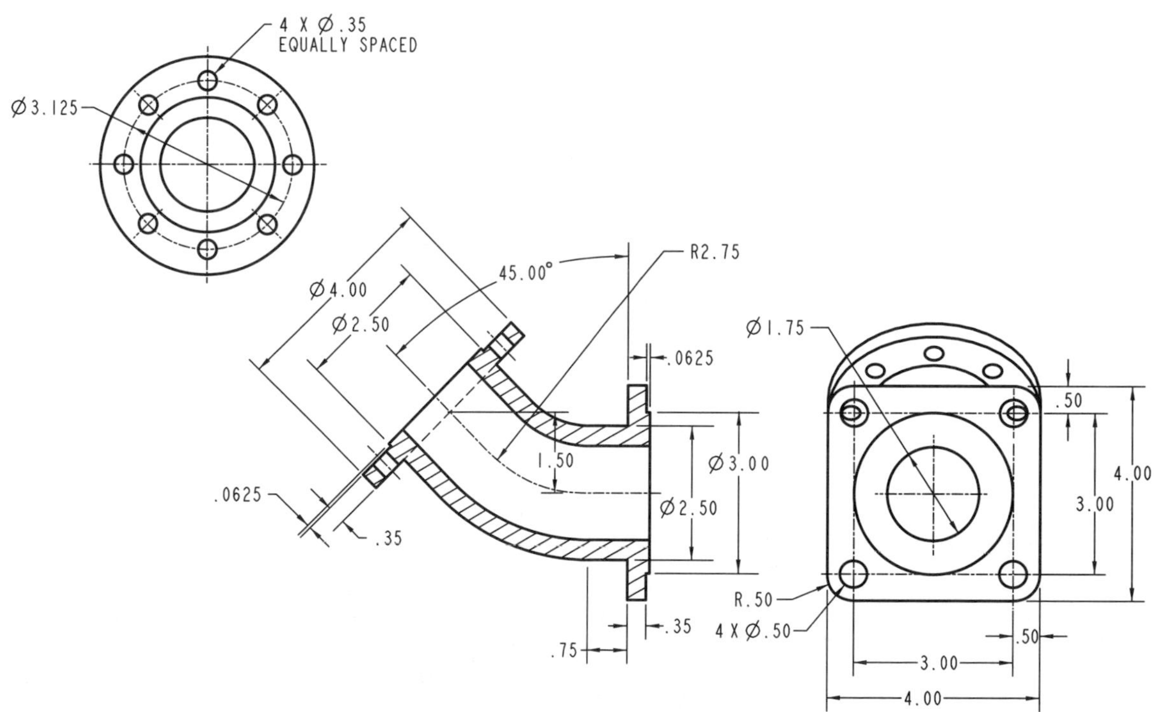

Figure 10–51 Problem one

2. Use Pro/ENGINEER's Part mode to model the part shown in Figure 10–52. Using the Protrusion >> Sweep command, create this part with only one geometric feature. Sketch the trajectory as an oval. With a closed trajectory, Pro/ENGINEER will give you the attribute option of adding or not adding inside faces (Add Inn Fcs and No Inn Fcs). Select the Add Inn Fcs option.

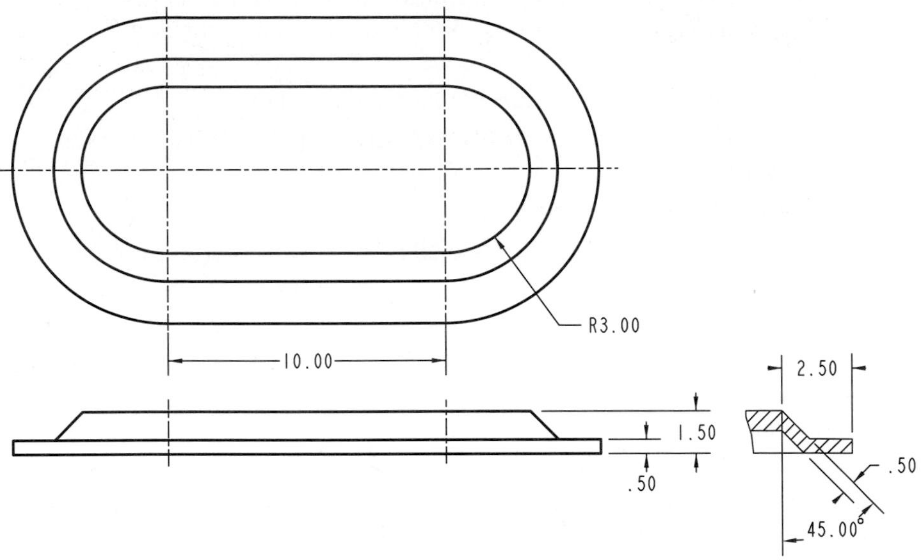

Figure 10–52 Problem two

3. Use Pro/ENGINEER's Part mode to model the part shown in Figure 10–53.

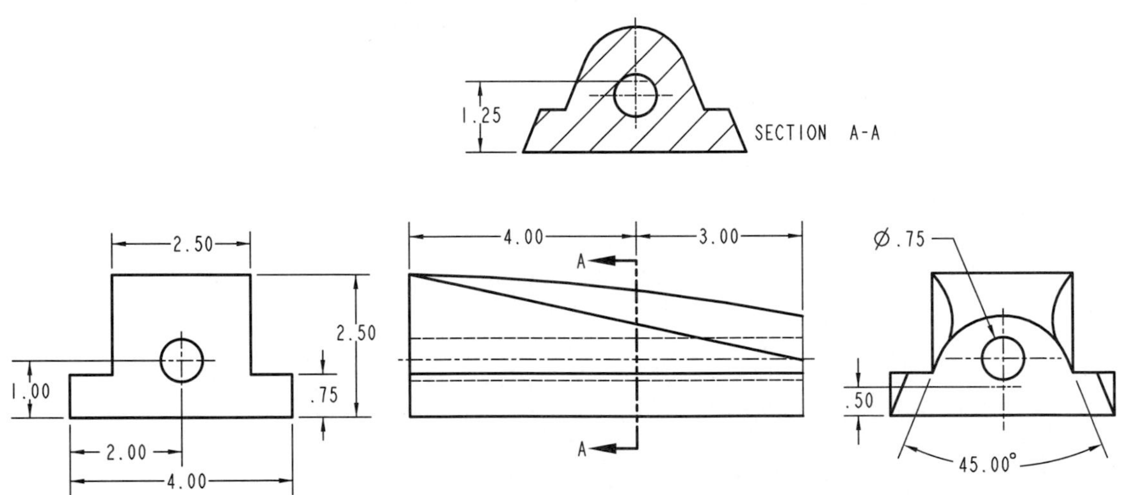

Figure 10–53 Problem three

4. Use Pro/ENGINEER's Part mode to model the propeller part shown in Figure 10–54. The part consists of a revolved feature (the hub) and three blended features (the propellers). Sketch the hub feature and the first propeller feature on Pro/ENGINEER's default datum plane FRONT using appropriate dimensions of your choosing. For the blended propeller feature, use the two sections shown in the figure with 24 inches separating each section. Use the Copy/Rotate option to create the remaining propeller features.

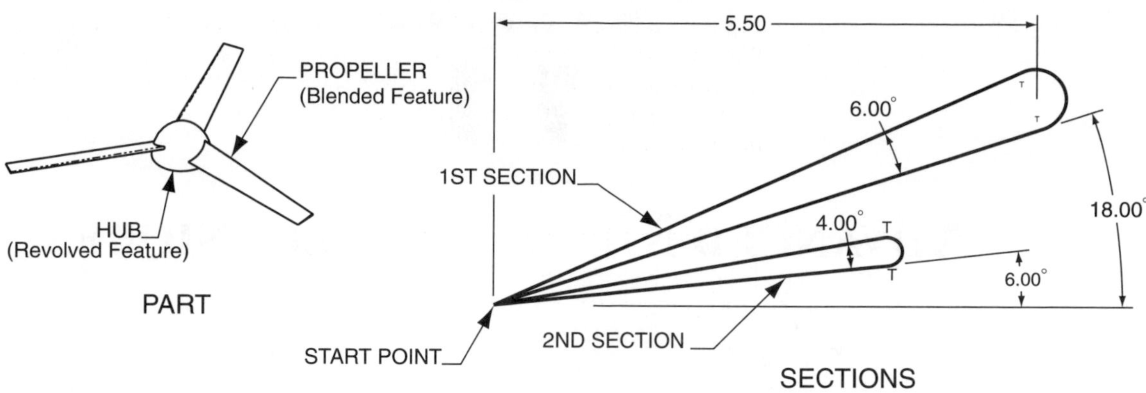

Figure 10-54 Problem four

Questions and Discussion

1. What two methods are available for defining a sweep's trajectory? Describe appropriate situations for using each method.
2. During the creation of a sweep, if a regeneration error is encountered during the preview of the feature, what are some possible causes of this failure?
3. What is the difference between a merged ends sweep and a free ends sweep?
4. During the creation of a swept feature, which is defined first, the trajectory or the cross section?
5. Describe the three types of blends available within the Blend option.
6. Describe several points that should be adhered to when creating a Parallel Blend.
7. When sketching the sections for a Parallel Blend, what menu option is used to switch from one section to the next? What process is used to move between sections for a Rotational Blend?
8. What is a Datum Curve, and what are some of its uses?
9. Describe the difference between a Projected Datum Curve and a Formed Datum Curve.

CHAPTER

11

ADVANCED MODELING TECHNIQUES

Introduction

Within the Protrusion and Cut commands, solid creation options are available to perform basic modeling operations. Included are the Extrude, Revolve, Sweep, Blend, and Use Quilt options. Also included under both the Protrusion and Cut commands are options for creating advanced solid features. Available are advanced feature options such as Swept Blends and Helical Sweeps. Upon finishing this chapter, you will be able to

- Create a swept blend.
- Create a variable section sweep.
- Create a spring feature with the Helical Sweep option.
- Create a bolt using the Helical Sweep option.

DEFINITIONS

Normal-to-Trajectory The Normal to Trajectory (Norm to Traj) option keeps the feature's cross sections normal to a selected trajectory. This option is available under Swept Blend and Variable Section Sweep.

Normal-To-Origin-Trajectory The Normal to Origin Trajectory (NrmToOriginTraj) option keeps the feature's cross sections normal to the defined Origin Trajectory. This option is available under the Swept Blend and Variable Section Sweep options.

Pitch On a thread, the pitch is the inverse of the number of threads per inch and is defined as the distance between a point on one thread to a corresponding point on the next thread. The Helical Sweep option requires the definition of either a constant pitch or a variable pitch.

Pivot Direction The Pivot Direction (Pivot Dir) option keeps the feature's cross sections normal to a selected planar pivot plane, edge, curve, or axis. This option is available under Swept Blend and Variable Section Sweep.

SWEPT BLEND OPTION

A swept blend is a combination of a sweep and a blend. A swept feature is a section protruded along a defined trajectory. This trajectory can be either sketched or selected. A parallel blended feature is a feature protruded along a straight trajectory between two or more user-defined sections. A *Swept Blend feature* is two or more sections protruded along a user-defined trajectory (Figure 11–1). Like a swept feature, a Swept Blend's trajectory can be either sketched or selected. In addition, sections can be either sketched or selected.

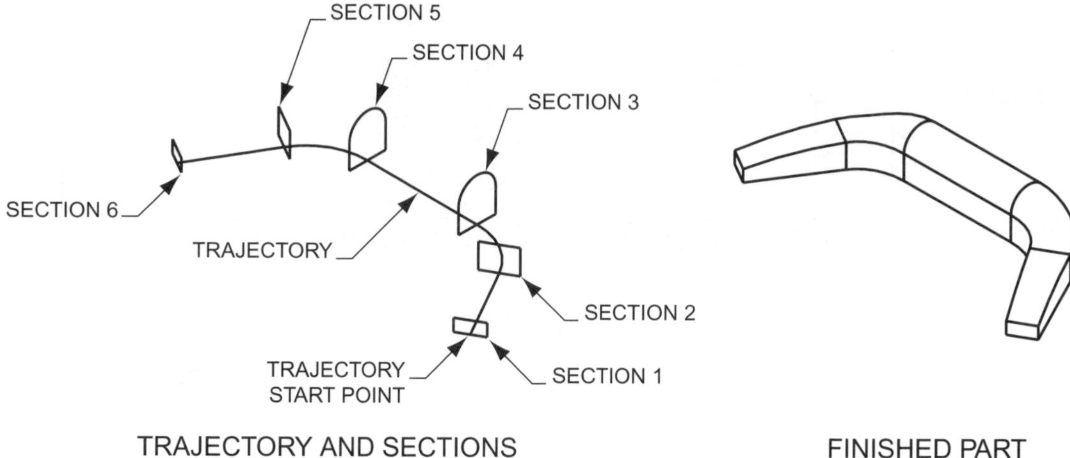

Figure 11–1 Swept blend creation

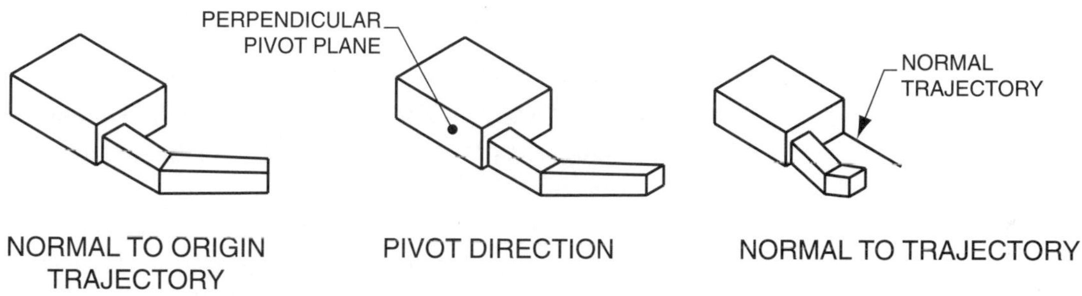

Figure 11–2 Swept blend types

Pro/ENGINEER provides three Swept Blend options. The following is a description of each.

Normal to Origin Trajectory

The **Normal-to-Origin-Trajectory** (NrmToOriginTraj) option keeps each of the feature's cross sections normal to the trajectory of the feature (Figure 11–2). This type of Swept Blend requires the definition of a trajectory (sketched or selected) and the definition of one or more sections. Each section is created normal to a vertex of the trajectory or normal to a datum point on the trajectory.

Pivot Direction

The Pivot Direction (Pivot Dir) option keeps the feature's cross sections normal to a selected planar pivot plane, edge, curve, or axis (Figure 11–2). Like the Normal-to-Origin-Trajectory option, this type of Swept Blend requires the definition of a trajectory and the definition of one or more sections. As shown in the figure, each section of the feature is created normal to the selected pivot plane.

Normal to Trajectory

The **Normal-to-Trajectory** (Norm To Traj) option keeps the feature's cross sections normal to a second trajectory (Figure 11–2). The section's trajectory can be sketched or selected. This option requires the definition of a sweep trajectory, a normal trajectory, and two or more sections. As shown in Figure 11–3, each section of the feature is created perpendicular to the normal trajectory.

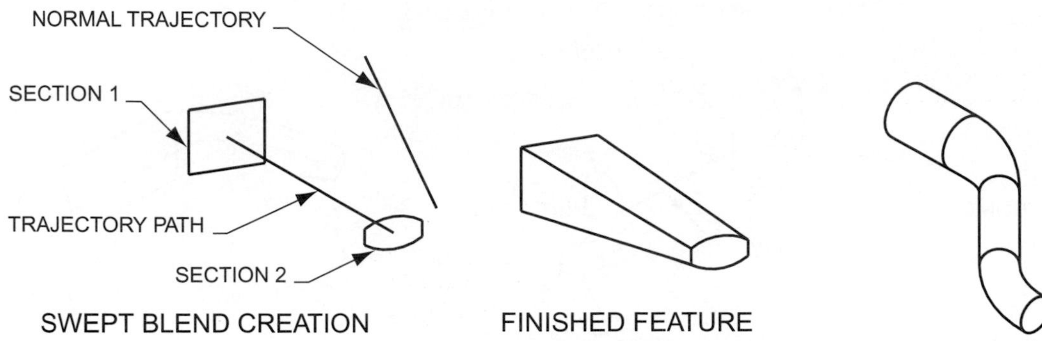

Figure 11–3 Norm to Traj creation

Figure 11–4 Swept blend feature

Creating a Swept Blend

The Swept Blend option is available within the Protrusion and Cut commands. This guide will demonstrate the creation of the Swept Blend feature shown in Figure 11–4. Pro/ENGINEER's default datum planes will be used as the base for the feature. Perform the following steps to create this feature.

Step 1: Select FEATURE >> CREATE >> PROTRUSION.

Step 2: Select ADVANCED >> DONE on the Solid Options menu.

Step 3: Select SWEPT BLEND >> DONE on the Advanced Feature Options menu.

Step 4: Select SKETCH SEC >> NrmToOriginTraj >> DONE.

The sections for a Swept Blend can be either sketched or selected. The Sketch Section (Sketch Sec) option will allow you to sketch the section within the sketching environment. The Normal-To-Origin-Trajectory (NrmToOriginTraj) option will create each cross section (sketched or selected) normal to the feature's trajectory.

Step 5: Select the SKETCH TRAJ option.

Like a Swept Blend's sections, the trajectory can be either sketched or selected.

> **INSTRUCTIONAL NOTE** If you choose to select a trajectory on the work screen, Pro/ENGINEER will provide the Section Orientation menu with the Pick XVector, Automatic, and Normal-to-Surface options. The Automatic option will automatically orient each section's sketching environment, while the Pick XVector option will allow the user to select the sketching environment's X-Axis. The Normal-to-Surface option will place each section perpendicular to any adjacent surface.
>
> Additionally, when selecting a trajectory, Pro/ENGINEER will provide the Confirm menu to accept or reject vertex points. On this menu, use the Accept option to accept vertex points and the Next option to skip a vertex point.

Step 6: Select and Orient the sketching plane.

Step 7: Sketch the trajectory (see Figure 11–5).

Using appropriate sketching tools, sketch the trajectory's section as shown in Figure 11–5. The dimensions shown in the figure match the design intent for the feature.

Step 8: Select the Continue icon to exit the sketching environment.

When the trajectory's section matches the sketch shown in Figure 11–5, exit the sketching environment with the Continue or Done option.

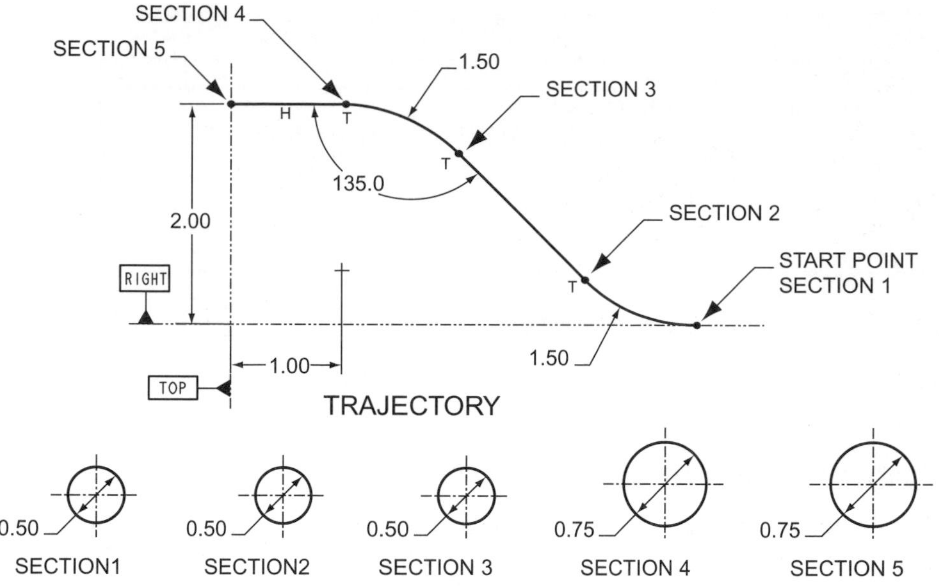

Figure 11–5 Trajectory and sections

STEP 9: Use the Confirm Selection menu to ACCEPT section vertices (see Figure 11–5).

The acceptance of a vertex will require the defining of a section at that particular vertex. In this example, you should accept each available vertex.

STEP 10: Enter a Z-Axis rotation angle (enter 0.00 in this example).

When using the Normal-To-Origin-Trajectory option, each section is sketched on an X-Y plane. Within the creation of a Swept Blend section, the trajectory's path is the Z-Axis and is normal to the X-Y plane. Each section of a trajectory is typically sketched with a zero degree rotation angle. This step gives you the option of rotating a section about the Z-Axis up to 120 degrees in either the positive or negative direction. Each section is still kept perpendicular to the trajectory.

STEP 11: Sketch the first section of the Swept Blend.

When a trajectory is sketched, Pro/ENGINEER will place sections at any accepted vertex or point. Notice in Figure 11–5 the start point of the trajectory. This will be the location for the first cross section. It is often helpful to dynamically rotate the sketching environment to get a better understanding of the sketch plane in relation to the trajectory. Also notice on the trajectory in the figure the location of each cross section. In this example, there will be five sections. As shown in the figure, the first section is a 0.500 diameter circle.

STEP 12: Select the Continue icon to exit the sketching environment for the first section.

STEP 13: Enter the Z-Axis rotation value for the section (usually 0.00).

STEP 14: Sketch the second section.

Sketch the second section according to the intent of the design. If necessary, dynamically rotate the sketching environment.

STEP 15: Select the Continue icon to exit the sketching environment for the second section.

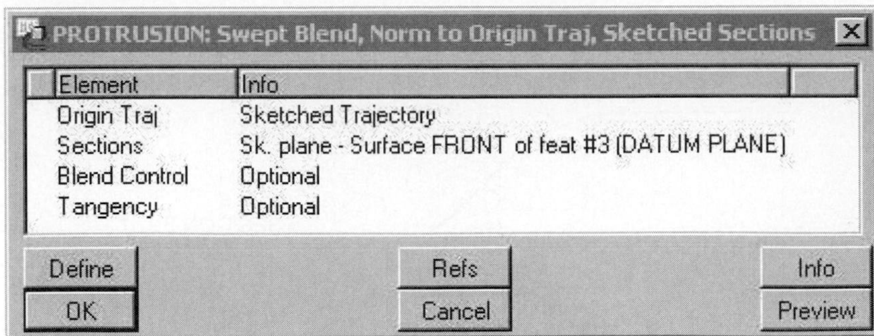
PROTRUSION: Swept Blend, Norm to Origin Traj, Sketched Sections

Element	Info
Origin Traj	Sketched Trajectory
Sections	Sk. plane - Surface FRONT of feat #3 (DATUM PLANE)
Blend Control	Optional
Tangency	Optional

Figure 11–6 Swept Blend dialog box

STEP 16: Sketch remaining sections for feature.

When the last vertex or datum point is reached on the trajectory, Pro/ENGINEER will automatically take you to the Swept Blend Feature Definition dialog box.

STEP 17: Preview the Swept Blend on the Feature Definition dialog box (Figure 11–6), then select OKAY.

VARIABLE SECTION SWEEP

The Variable Section Sweep option is an advanced feature modeling tool used for the creation of complex geometric shapes. This option sweeps a section along one or more trajectories. It is available within the Protrusion and Cut commands.

Pro/ENGINEER provides several degrees of control with the Variable Section Sweep option. First, the orientation of a section can be controlled with the Normal-To-Origin-Trajectory (NrmToOriginTraj), Pivot Direction (Pivot Dir), and Normal-To-Trajectory (Norm To Traj) options. These options are virtually identical to the options by the same name in the Swept Blend command. Second, sections can be aligned to one or more trajectories to vary the shape of the section (see Figure 11–7). Finally, the varying of a section can be controlled in the Feature Definition dialog box with the Constant and Variable options. The Constant option is used to maintain a section's size.

Figure 11–7 shows the construction of a Variable-Section-Sweep feature. This feature was constructed with the Normal-To-Origin-Trajectory option. Notice in the figure the Origin Trajectory. All Variable Section Sweeps require the selection of an Origin Trajectory that defines the direction of sweep. This trajectory is similar to the trajectory of a Swept Feature. As with a Swept Feature, this trajectory requires the definition of a starting point. In this feature, an X-Trajectory is also constructed. An X-Trajectory defines the horizontal vector of a section. If the feature's section is aligned with an X-Trajectory, as it is in this example, the section will vary along the path of the trajectory. Also shown in Figure 11–7 is the section for the feature. Within the construction of a Variable Section Sweep feature, the Origin Trajectory, X-Trajectory, additional trajectories, and the section can be either sketched or selected.

The normality of a Variable Section Sweep's section can be controlled with the Normal-To-Origin-Trajectory, Pivot-Direction, and Norm-To-Trajectory options. In the Variable Section Sweep command, these options work the same as the same Swept Blend command. The following is a description of each option.

NORMAL TO ORIGIN TRAJECTORY

The Normal-To-Origin-Trajectory (NrmToOriginTraj) option keeps each of the feature's cross sections normal to the defined Origin Trajectory (Figure 11–8). As seen in the figure, this option requires the sketching of an Origin Trajectory and an

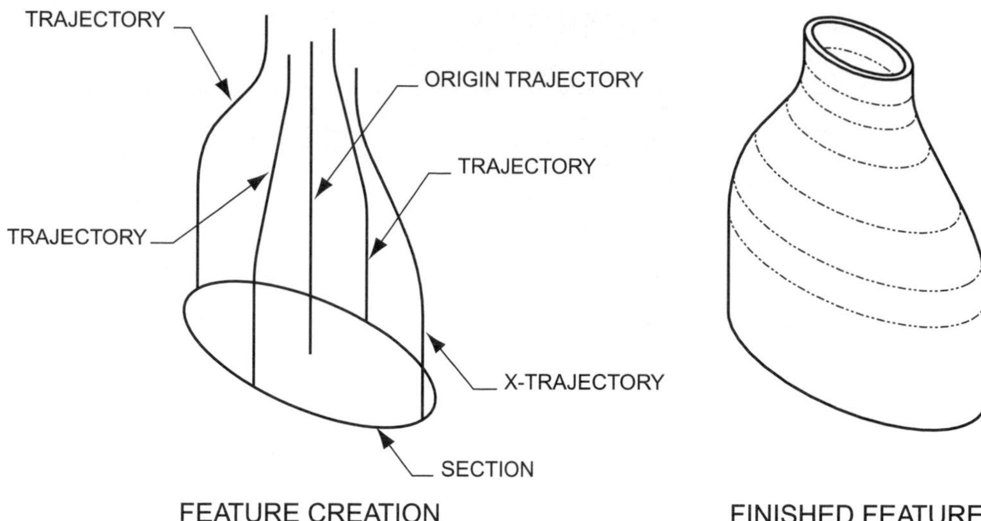

Figure 11–7 Variable section sweep construction

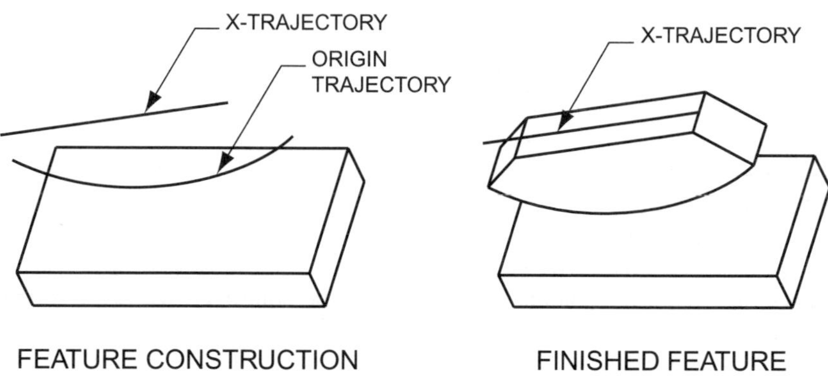

Figure 11–8 Normal to origin trajectory

X-Trajectory. Notice how the cross sections of the finished part are normal to the Origin Trajectory.

Pivot Direction

The **Pivot-Direction** (Pivot Dir) option keeps the feature's cross sections normal to a selected planar pivot plane, edge, curve, or axis (Figure 11–9). Like the Normal-To-Origin-Trajectory option, this type of Variable Section Sweep requires the definition of an Origin Trajectory and the definition of one section. Unlike the Normal-To-Origin-Trajectory option, the defining of an X-Trajectory is optional. Without an X-Trajectory, this option performs similarly to the Swept-Blend/Pivot-Direction option. Figure 11–9 shows the construct of a Variable Section Sweep with the Pivot Direction option. Notice, in the finished feature, how the feature's cross section remains normal to the selected pivot plane.

Normal to Trajectory

The Normal-To-Trajectory (Norm To Traj) option keeps the feature's cross sections normal to a second trajectory throughout the sweep (Figure 11–10). This option requires the definition of an Origin Trajectory, a Normal To Trajectory, and one section. Notice in Figure 11–10 how the feature's cross sections remain normal to the selected trajectory.

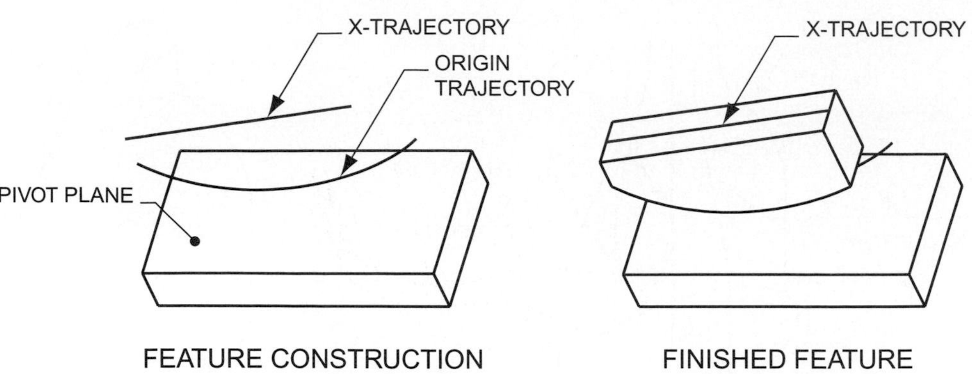

Figure 11–9 Pivot direction

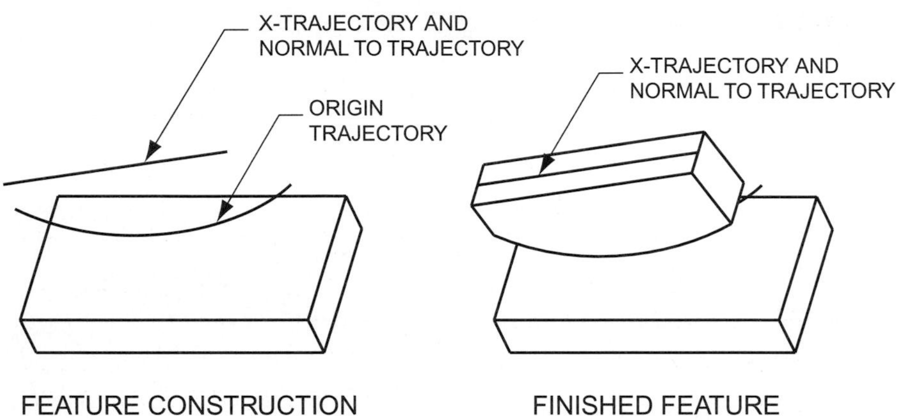

Figure 11–10 Normal to trajectory

Creating a Variable Section Sweep

The Variable Section Sweep option can be found under the Advanced menu option of the Protrusion and Cut commands. This option is used to sweep a section along multiple trajectories. Like all Protrusion and Cut features, the Variable Section Sweep feature can be created solid or thin. The following guide demonstrates the creation of the feature shown in Figure 11–11. This feature is constructed by sweeping a rectangular section along the four datum curves shown in the figure. A model file (*var_sec_swp.prt*) containing the part's predefined datum curves can be found on the supplemental CD and will serve as the base part for this guide. Perform the following steps to create the Variable Section Sweep feature shown.

Step 1: Select FEATURE >> CREATE >> PROTRUSION.

The Variable Section Sweep option can be found under the Protrusion and Cut commands.

Step 2: Select ADVANCED >> THIN >> DONE on the Solid Options menu.

Notice in Figure 11–11 that the feature has been created as a thin feature.

Step 3: Select VAR SEC SWEEP >> DONE on the Advanced Feature Options menu.

Step 4: Select NrmToOriginTraj >> DONE.

The Normal-To-Origin-Trajectory (NrmToOriginTraj) option keeps the feature's cross section perpendicular to the selected Origin Trajectory. The Pivot-Direction (Pivot Dir) option is used to keep the feature's cross section

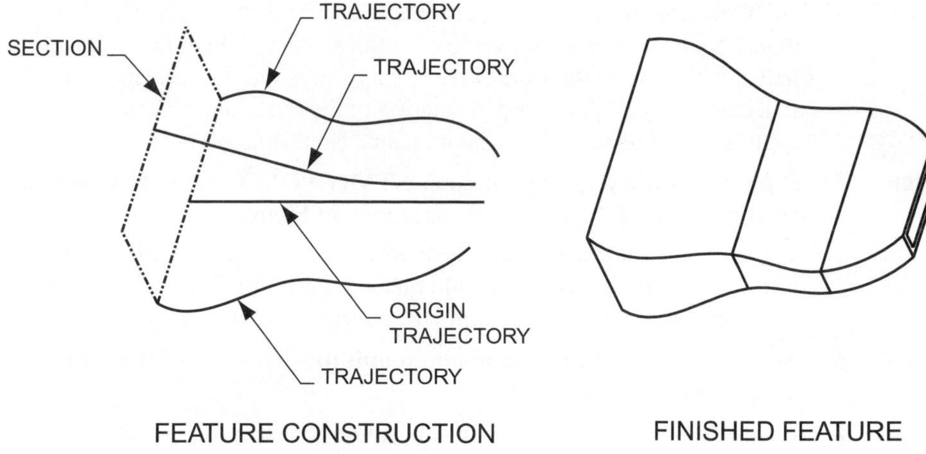

Figure 11-11 Variable section sweep feature

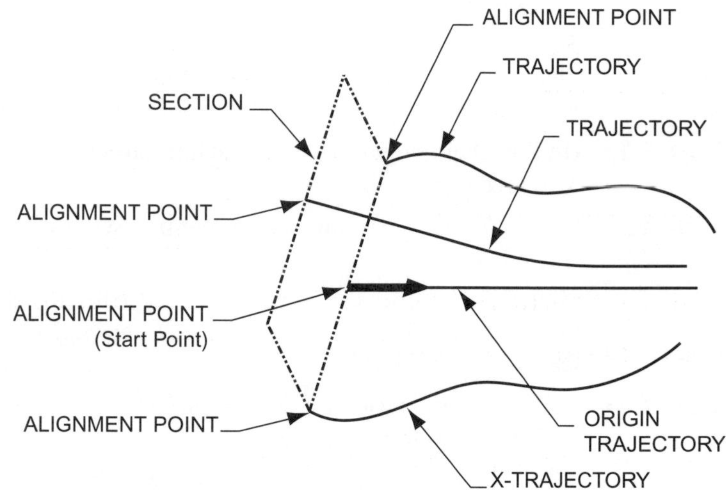

Figure 11-12 Feature construction

perpendicular to a select plane, axis, edge, or curve. The Normal-To-Trajectory (Norm To Traj) option will keep the cross section perpendicular to a selected trajectory other than the Origin Trajectory. In this guide, the Origin-Trajectory shown in Figure 11–11 will be used.

STEP 5: Choose the SELECT TRAJ option on the Variable Section Sweep menu.

The Variable Section Sweep menu is used to define trajectories. Within the Variable Section Sweep option, trajectories can be selected on the work screen or sketched. When sketching a trajectory, normal sketching tools are used.

STEP 6: Using the ONE BY ONE option, select the Origin Trajectory (Figure 11–12), then select the DONE SEL option on the Get Select menu.

Notice in the message area that Pro/ENGINEER is requesting you to "Specify trajectory that defines section origin." The first trajectory selected is the Origin Trajectory. This trajectory is used to define the start point for the feature's section. When using the Normal-To-Origin-Trajectory option, the feature's cross section will be perpendicular to this selected entity.

Only one entity is required for the Origin Trajectory. Use the Done Select option to quit the selection of entities. After picking Done Select, notice the arrow on the work screen. This arrow defines the start point of the trajectory. This graphical representation of the start point will not be available after selecting the Chain menu's Done option.

STEP 7: If necessary use the Chain menu's START POINT option to select the starting point of the trajectory as shown in Figure 11–12.

The start point of the trajectory is denoted on the selected entity by an arrow. The arrow's location should match Figure 11–12. If the location is different, use the Start Point option to select a different vertex.

STEP 8: Select DONE on the Chain menu to quit the definition of the Origin Trajectory.

STEP 9: Choose the SELECT TRAJ option on the Variable Section Sweep menu.

STEP 10: Using the CURVE CHAIN option, select the X-Trajectory shown in Figure 11–12; then from the Chain Options menu, pick the SELECT ALL option.

The X-Trajectory defines the horizontal vector of the section. The Curve Chain option allows you to select a chain of possible entities along with a *From Vertex* and a *To Vertex*. In this problem, the entire entity is selected.

STEP 11: On the Chain menu, select the DONE option to quit the definition of the X-Trajectory.

STEP 12: Use the Variable Section Sweep menu to select or sketch remaining trajectories.

Multiple trajectories can be selected or sketched to define a Variable Section Sweep feature. Select or sketch as many trajectories as is necessary for the feature.

STEP 13: When all trajectories are defined, select the DONE option on the Variable Section Sweep menu.

After selecting Done, Pro/ENGINEER will enter the sketching environment used for the creation of the feature's section. This environment will be normal to the Origin Trajectory and located at the Start Point.

STEP 14: Sketch the Section for the feature (Figures 11–12 and 11–13).

Pro/ENGINEER will provide a normal sketching environment for creating the section. When sketching a section for a Variable Section Sweep, if design intent requires it, it is important to align the sketch with the trajectories of the feature. Refer to the alignment points on Figure 11–12. Notice how they correspond to the alignment points on the sketch in Figure 11–13. By aligning with one or more trajectories, the section is allowed to vary.

STEP 15: Select the Continue icon to exit the sketching environment and finish the feature creation process.

STEP 16: Preview the feature on the Feature Definition dialog box (Figure 11–14).

On the dialog box, notice the many different elements that have been defined for the feature. Also notice the three different types of trajectories: Origin Trajectory, X-Trajectory, and Trajectories. The Origin Trajectory was the first trajectory selected and defines the section start point and direction of sweep. The X-Trajectory was the second trajectory selected, and it defines the horizontal variation of the section. Any trajectory selected after the X-Trajectory is defined under the Trajectory element.

STEP 17: Select OKAY to create the feature.

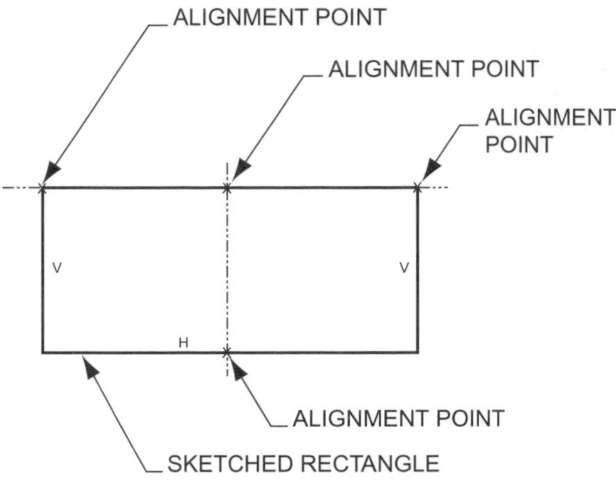

Figure 11-13 Sketched section

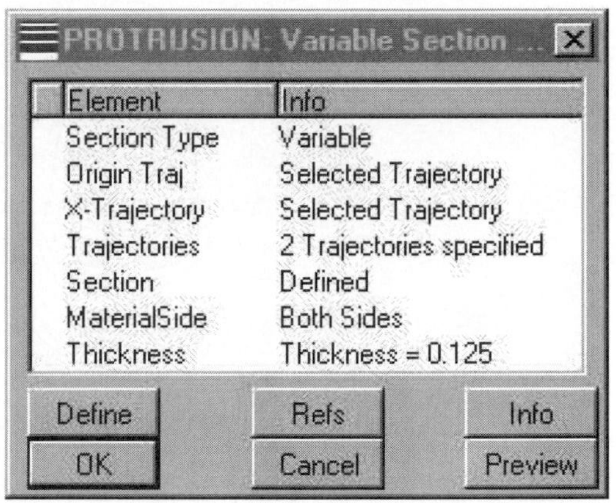

Figure 11-14 Variable Section Sweep dialog box

Helical Sweeps

As its name implies, the Helical Sweep option is useful for creating parts that consist of helical features. As shown in Figures 11–15 and 11–16, two features often created with the Helical Sweep options are the spring and thread. A spring feature is created with the Protrusion command, and a thread feature is created with the Cut command. Gears can also be defined with the Helical Sweep option.

Various definitions and attributes are available within the Helical Sweep option. Threads and springs are defined with a **pitch** definition. With a thread, the pitch is the inverse of the number of threads per inch and is the distance between a point on one thread to a corresponding point on the next thread. For a spring, the pitch is the distance between a point on one wire to a corresponding point on an adjacent wire. Pro/ENGINEER provides an option for either creating a helical feature with a constant pitch or a feature with a varying pitch. Figure 11–15 shows two examples of constant pitch spring features and two examples of variable pitch spring features. In the figure, notice in the third spring feature (C) that the spring's pitch varies in the middle of the spring. When creating the sweep

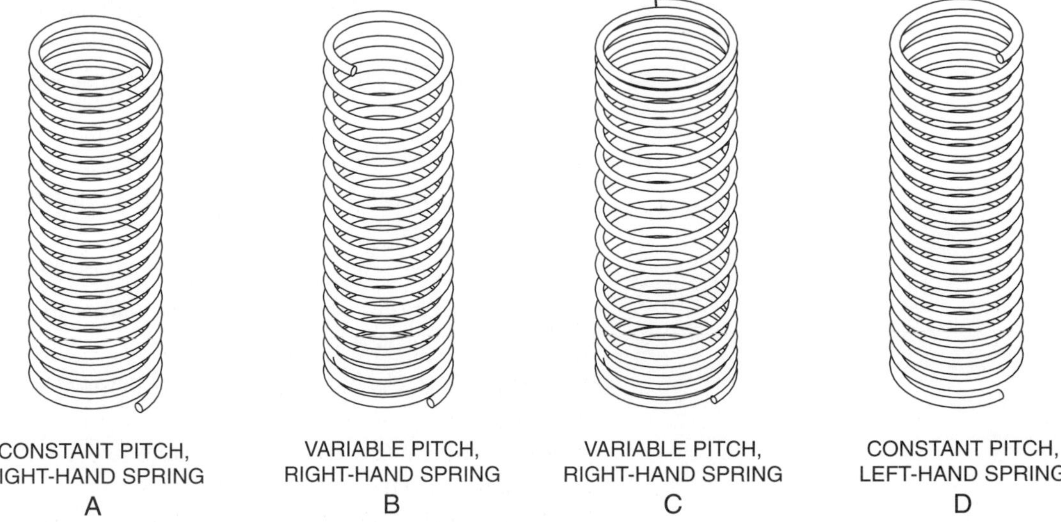

Figure 11-15 Helical sweep options

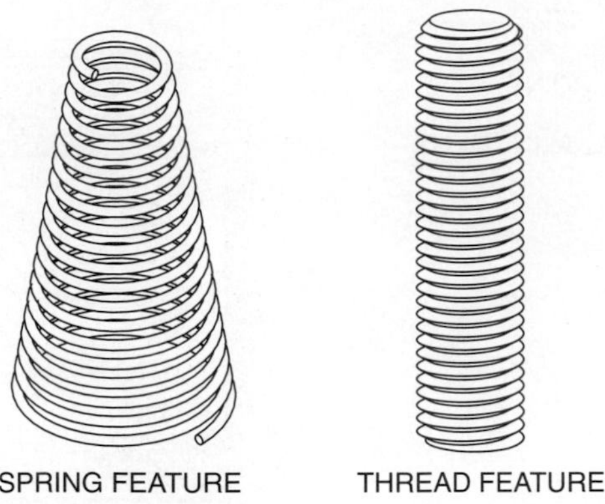

Figure 11–16 Helical sweeps

trajectory for this feature, points can be applied along the trajectory and the pitch at each point adjusted through the use of a pitch graph window.

Another definition available within the Helical Sweep option is the ability to create either a Right-Hand helical feature or a Left-Hand helical feature. Standard screw threads are right-handed. Some exceptions to this rule exist, such as threads assigned to an oxygen system. Figure 11–15a shows an example of a Right-Hand Constant Pitch spring, and Figure 11–15d shows an example of a Left-Hand Constant Pitch spring. The thread shown in Figure 11–16 is a Right-Hand thread.

Creating a Constant Pitch Helical Sweep Feature

The Helical Sweep option is often used to create spring and tread features. This guide will provide general information on the steps for creating a constant pitch helical feature. This option is available under the Protrusion and Cut commands.

Step 1: Select FEATURE >> CREATE >> PROTRUSION (or CUT).

Since a spring is a positive space feature, it is created with the Protrusion command. Detail threads are created with the Cut command.

Step 2: Select ADVANCED >> SOLID >> DONE on the Solid Options menu.

The Thin option is also available with a helical feature.

Step 3: Select HELICAL SWP >> DONE on the Advanced Feature Option menu.

Step 4: Select appropriate attributes on the Attributes menu, then select DONE

The following attributes are available:

- **Constant** Creates a helical feature with a constant pitch (Figure 11–15a).
- **Variable** Creates a helical feature with a variable pitch (Figure 11–15b).
- **Thru axis** Creates a helical feature around an axis. The axis is sketched as the first trajectory of the feature.
- **Norm to traj** Creates a helical feature perpendicular to a sketched trajectory.
- **Right-handed** Creates a helical feature swept to the right (Figure 11–15a).
- **Left-handed** Creates a helical feature swept to the left (Figure 11–15d).

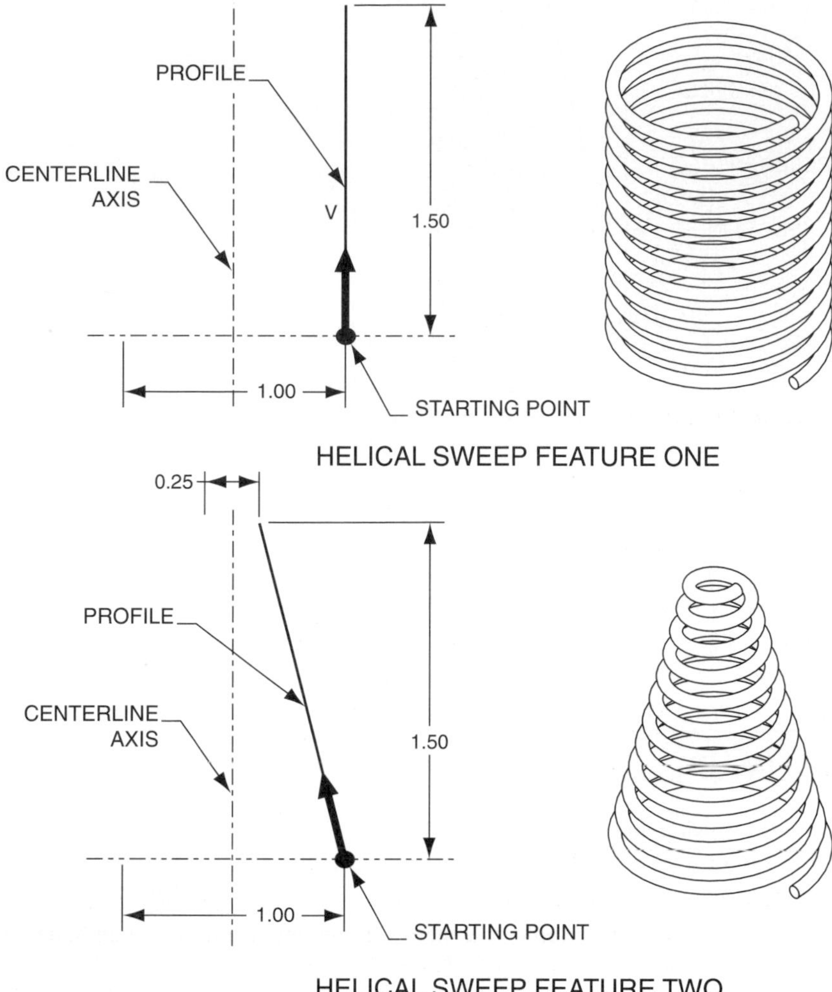

Figure 11–17 Helical sweep feature profile construction

STEP 5: Select then orient a sketching plane for the creation of the feature's profile.

STEP 6: Sketch the Profile of the Helical Sweep feature to include a required centerline (Figure 11–17).

The first geometric entity sketched defines the profile of the sweep feature. Figure 11–17 shows two examples of profiles and their respective helical features. A Helical Sweep feature requires the sketching of a centerline that defines the axis of revolution. The first centerline sketched serves this purpose.

STEP 7: Select the Continue icon to exit the sketching environment.

STEP 8: Enter the Pitch value for the feature.

STEP 9: Sketch the cross section of the helical feature (Figure 11–18).

Use appropriate sketching tools to create the section. Cross sections of Helical Sweep features are typically sketched around or near the start point of the profile of the feature. When sketching the cross section, the sketch plane is behind the sketching environment used to create the profile. It is often helpful to dynamically rotate the object to better visualize the sketching environment.

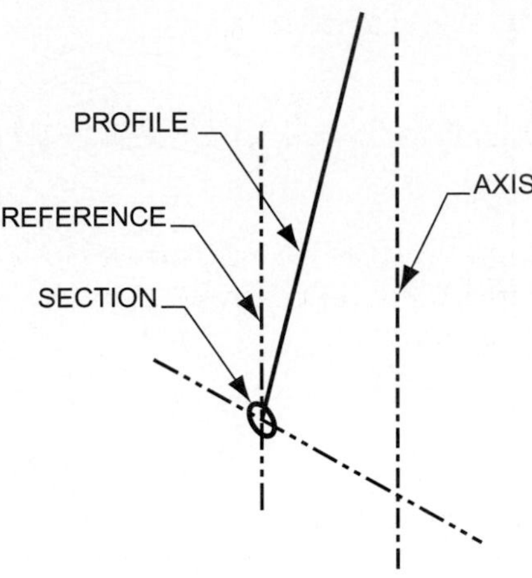

Figure 11-18 Isometric view of cross section construction

STEP 10: Select the Continue icon to exit the sketcher environment.

STEP 11: Preview the feature on the dialog box, then select OKAY.

SUMMARY

Most Protrusion and Cut features in Pro/ENGINEER can be created with relatively intuitive options such as Extrude, Revolve, Sweep, and Blend. Other common feature construction commands include Hole, Round, Chamfer, and Rib. While these commands can create most standard geometric shapes, complex shapes are more difficult to model. Under the Protrusion and Cut commands, Pro/ENGINEER provides the Advanced Feature Options menu for the construction of unique and/or more complex features. Available under this menu are options for creating Variable Section Sweeps, Swept Blends, and Helical Sweeps.

TUTORIAL

SWEPT BLEND TUTORIAL

This tutorial exercise will provide instruction on how to create the part shown in Figure 11–19. This part consists of three features: two extruded protrusions and one swept blend.

Within this tutorial, the following topics will be covered:

- Creating an extruded protrusion.
- Creating a swept blend.

THE FIRST FEATURE

The first feature created in this part will be the base extruded Protrusion. Create this feature on Pro/ENGINEER's default datum planes extruded in one direction with the dimensions shown in Figure 11–20. Sketch the section on datum plane TOP. Notice in the figure how the feature consists of four small holes and one large hole. Create all five holes and the base rectangular feature within the same section.

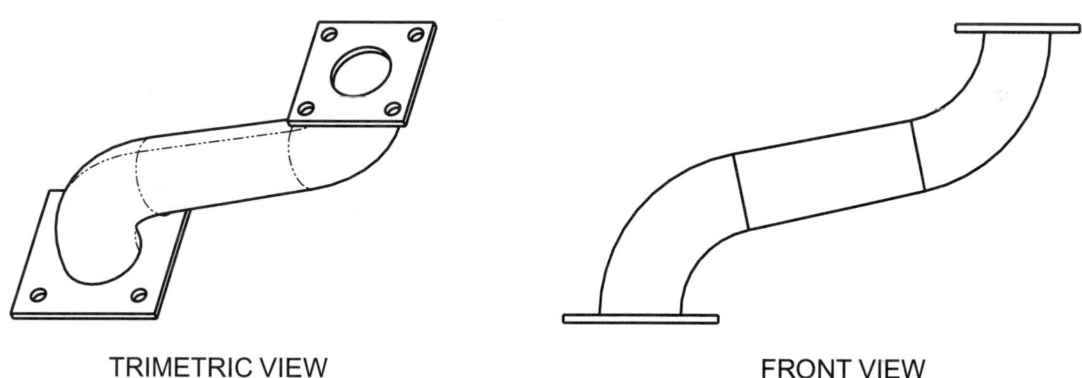

Figure 11–19 Completed part

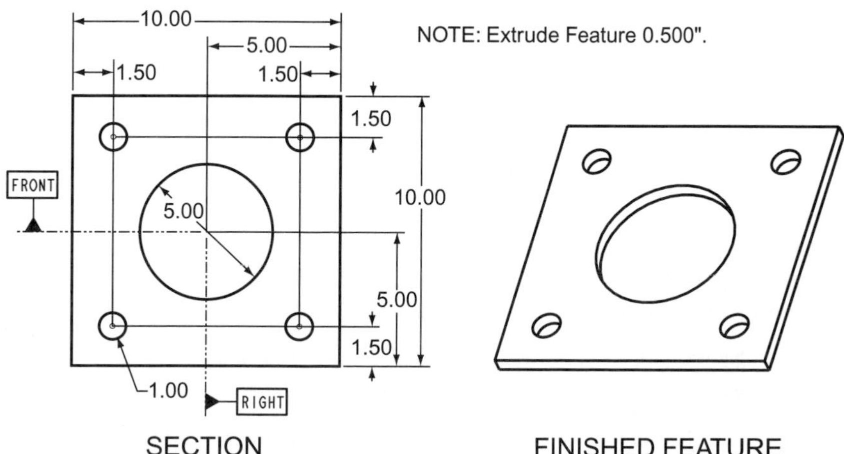

Figure 11–20 First feature

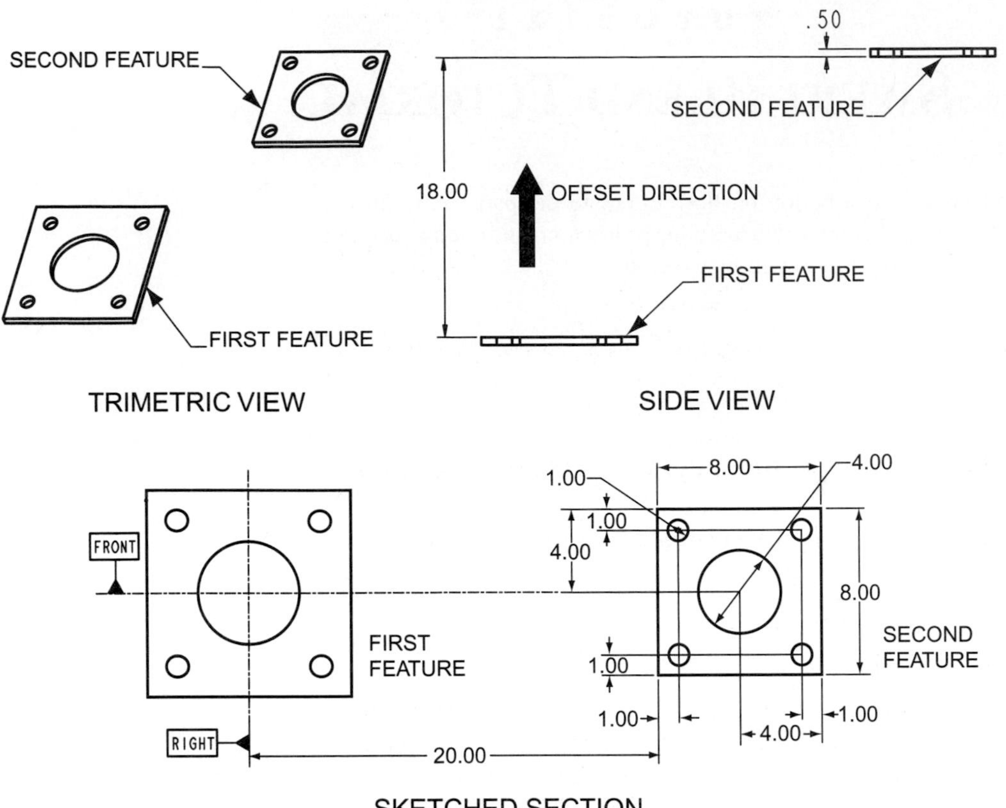

Figure 11-21 Offset second feature

The Second Feature

The second feature is similar to the first feature and is constructed in a similar manner (Figure 11–21). This feature's section is sketched on an on-the-fly datum plane offset 18 inches from the top of the first feature. As shown in the figure, the second feature's section is smaller than the first. Use the information shown in Figure 11–21 to create the second Protrusion feature. The dimensions shown match design intent.

Swept Blend Feature

This segment of the tutorial will use the Swept Blend option to create the object connecting the first feature with the second. The Swept Blend option is used to sweep multiple sections along a single trajectory. This option requires the definition of one trajectory and two or more sections. A trajectory can be either sketched or selected. Along the trajectory's path, vertices and datum points are used to define the location of each section. With the exception of the first and last vertex, the user has the option of accepting or rejecting a datum point or vertex for use in the swept blend. In this exercise, only the first and last vertices will be used.

Perform the following steps to create the swept blend feature shown in Figure 11–22.

Step 1: Select FEATURE >> CREATE >> PROTRUSION.

Step 2: Select ADVANCED >> THIN >> DONE.

The object under construction in this segment is a pipe feature (not to be confused with the Pipe command). The pipe effect will be created with the Thin option.

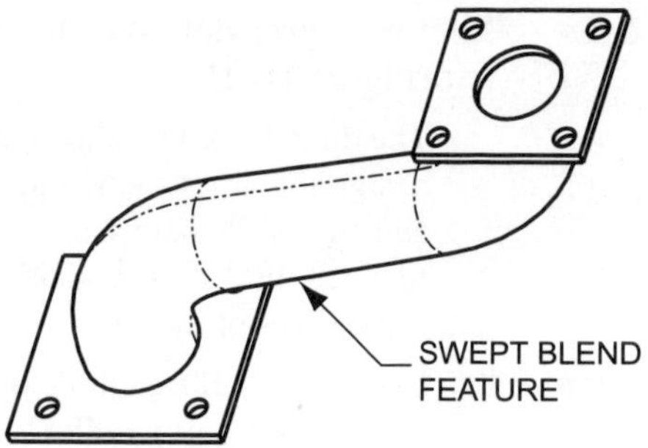

Figure 11-22 Swept blend feature

STEP 3: Select SWEPT BLEND >> DONE on the Advanced Feature Options menu.

STEP 4: Select NrmToOriginTraj >> DONE on the Blend Options menu.

The Normal-To-Origin-Trajectory (NrmToOriginTraj) option will create the feature with each cross section normal to the feature's Origin Trajectory.

STEP 5: Select the SKETCH TRAJ option on the Sweep Trajectory menu.

The Origin Trajectory of a Swept Blend feature can be either sketched or selected. Existing feature edges and datum curves are useful features for selecting. This tutorial will require the sketching of the trajectory.

STEP 6: Select datum plane FRONT as the sketching plane, then orient the sketching environment to match Figure 11–23.

STEP 7: Use the References dialog box to specify the references shown in Figure 11–23.

Four specific references are important for this feature: the top of the base protrusion, the bottom of the second protrusion feature, the center axis of the base protrusion's hole entity, and the center axis of the second protrusion's hole entity. If necessary, selecting Sketch >> References on the menu bar will allow you to access the References dialog box.

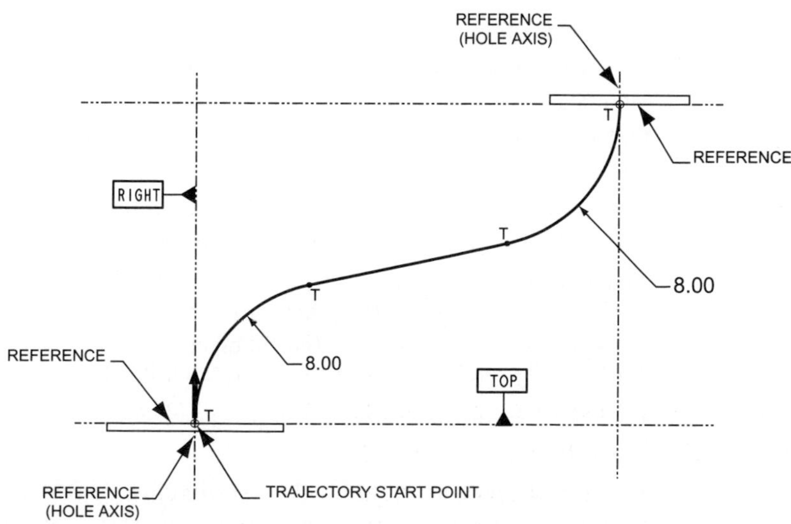

Figure 11-23 Trajectory sketch

STEP 8: Use appropriate sketching tools to create the sketch of the trajectory (Figure 11–23).

The dimensioning scheme and constraints shown in the figure match the design intent for the feature. When sketching the trajectory, align the start and end of the path with the axes from the large center holes of the first two features. In addition, your arc entities should be constrained tangent to the center axes of their respective holes.

STEP 9: Use the START POINT option on the Pop-Up menu to set the trajectory's start point to the location shown in Figure 11–23.

Your start point should match the figure. Preselect the vertex of the required start point, then access the Pop-Up menu with the right mouse button.

STEP 10: Select the Continue icon to complete the trajectory's section.

Notice on the work screen and in Figure 11–23 the two vertices located along the path of the trajectory. Pro/ENGINEER provides the option of creating a cross section at each vertex. In addition, a cross section at the start and at the end of the trajectory's path is required. The next step will give you the opportunity to either accept or not accept a vertex point. Notice on the work screen how the first vertex is highlighted in the reference color. Pro/ENGINEER is asking if you want to use this point as a cross section.

STEP 11: Select NEXT on the Confirm Select menu to skip the first vertex.

In this tutorial, you will use only the start and end of the trajectory as cross section locations. Selecting Next will skip to the next vertex or datum point.

STEP 12: Select NEXT on the Confirm Select menu to skip the second vertex.

The Previous option will return you to the previous vertex or datum point.

STEP 13: Enter 0.00 as the Z-Axis rotation angle for the first section.

The path of a trajectory is the Z-Axis for each cross section. Each section's sketching environment can be rotated about the Z-Axis. You will enter 0.00 as the rotation angle.

Upon entering the Z-Axis rotation angle, Pro/ENGINEER will take you to the sketching environment for the first cross section. Remember the defined starting point from your trajectory? This will be the location of the first sketched section.

STEP 14: Select the USE EDGE icon.

The first section will be created from the edge of the first feature's largest hole. The design intent for this feature requires the inside diameter of the pipe to equal the selected hole's diameter. The Use Edge option will help you to meet this intent.

STEP 15: Select the edges of the large diameter hole in the two locations shown in Figure 11–24 (only pick each location once).

You will have to select the edge of the hole in two locations. After selecting the locations, notice the start point arrow. This arrow serves the same purpose as the start point for a typical blended feature. As in the Blend option, the Swept Blend option requires the following rules:

- Start Point arrows have to be in the same general location for each section.
- In most situations, Start Point arrows have to point in the same direction.
- Each section of the Swept blend has to have the same number of entities.

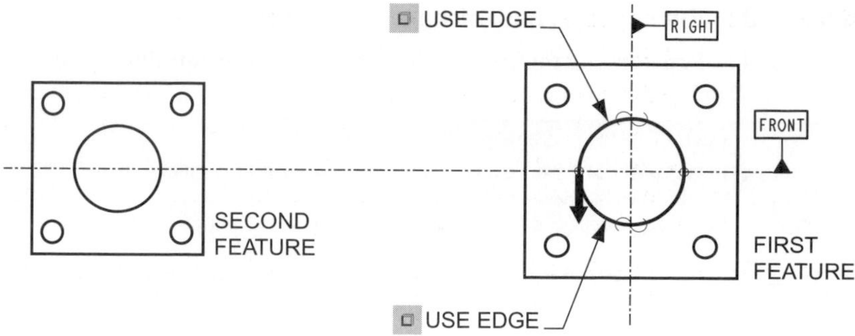

Figure 11–24 First section creation

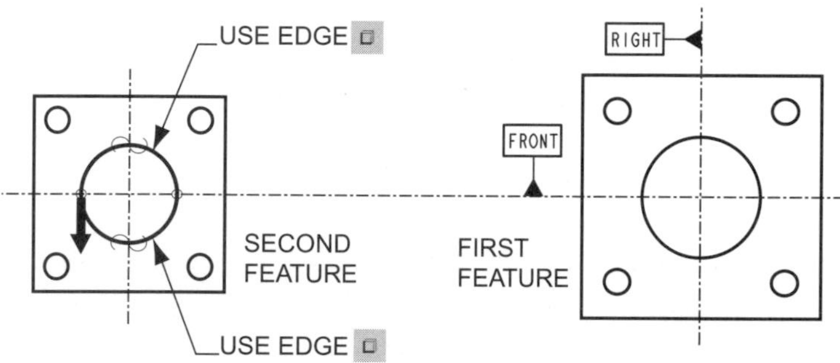

Figure 11–25 Second section creation

STEP 16: **Select the Continue icon when the section is complete.**

STEP 17: **Select the FLIP option then select OKAY.**

Recall from an earlier step of this tutorial that this feature was created with the Thin attribute. The Thin attribute allows you to define the side of the sketch on which the thin layer of material will be created. Using the Flip option will flip the direction of the arrow. This will allow the section to serve as the pipe feature's inside diameter. Your arrow should point away from the section. The Both option creates material on both sides of the sketch. If you accidentally skipped the step to create a Thin feature (step 2 in this tutorial), Pro/ENGINEER does not provide an option for redefining this attribute.

STEP 18: **Enter 0.00 as the Z-Axis rotation angle for the second cross section.**

STEP 19: **In the sketching environment, select the USE EDGE option to create the section shown in Figure 11–25.**

As with the first section, you will utilize the Use Edge option to create the section. On the second feature (Figure 11–25), select the edge of the largest diameter hole. Make sure that your start point matches the start point from the first cross section.

STEP 20: **Select the Continue icon to exit the sketching environment.**

STEP 21: **On the Thin Options menu, select the FLIP option; then select OKAY.**

Your arrow should point away from the section.

STEP 22: Enter .125 as the Thickness of the thin feature.

STEP 23: Preview the feature on the Feature Definition dialog box.

> **INSTRUCTIONAL NOTE** If your preview reveals a twisting effect to your Swept-Blend feature, then the Start Points for each section were not in the same general location. On your Feature Definition dialog box, Redefine one of the sections (the Sections element) to set a Start Point corresponding to the other section's Start Point.

STEP 24: When the feature is correct, select the OKAY option on the Feature Definition dialog box.

TUTORIAL

SPRING TUTORIAL

The spring feature shown in Figure 11–26 was created with the Helical Sweep option. Helical sweeps are created by first constructing (sketching or selecting) a trajectory followed by sketching the helical feature's cross section. Helical sweeps can be created with a constant pitch or a variable pitch. Within a spring feature, the pitch is the distance between corresponding points on the wire. Notice in the figure how this feature will have a variable pitch. Perform the following steps to create this feature.

STEP 1: Start a new part file utilizing Pro/ENGINEER's default datum planes.

STEP 2: Select FEATURE >> CREATE >> PROTRUSION.

The Helical Sweep option can be found under the Protrusion and Cut commands.

STEP 3: Select ADVANCED >> SOLID >> DONE on the Solid Options menu.

STEP 4: Select HELICAL SWP >> DONE on the Advanced Feature Options menu.

STEP 5: Select VARIABLE >> THRU AXIS >> RIGHT HANDED >> DONE on the Attributes menu.

The Variable option creates a helical feature with a variable pitch, while the Thru Axis option creates a helical sweep feature around a sketched axis. Helical Sweep features can be created either Right-Handed or Left-Handed. A standard thread is an example of a Right Hand helical feature.

STEP 6: Select datum plane FRONT as the sketching plane, then orient the sketching environment to match Figure 11–27.

STEP 7: Sketch the Sweep Profile shown in Figure 11–27.

The Helical Sweep option is a type of revolved feature. Due to this, it requires the sketching of a centerline that will serve as the feature's axis of revolution. Sketch the centerline entity aligned with the edge of datum plane RIGHT. In addition, notice the location of the sketched trajectory and the dimensioning scheme.

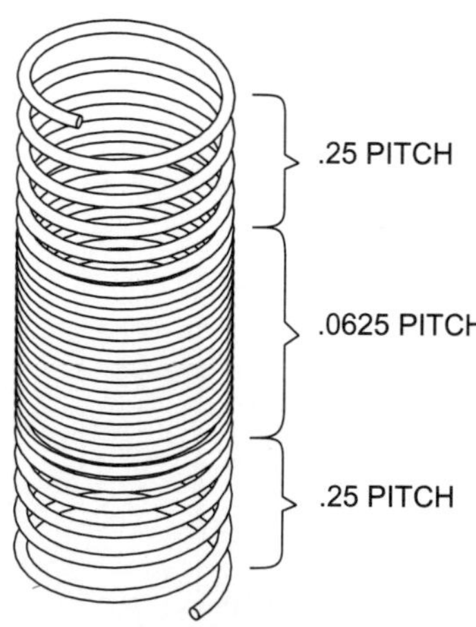

Figure 11-26 Spring feature

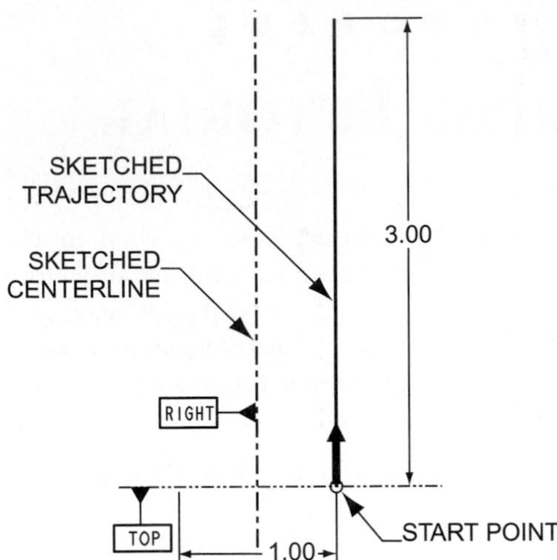

Figure 11-27 Sweep profile construction

STEP 8: Use the POINT icon to create the two points shown in Figure 11–28.

Points can be used to control variable pitches on a helical feature. As shown previously in Figure 11–26 and as shown in Figure 11–28, these points will serve as the division point for the changes in pitch along the feature's trajectory.

STEP 9: If necessary, use the Pop-Up menu's Start Point option to change the profile's Start Point to the location shown in Figure 11–27.

STEP 10: Select the Continue icon to exit the sketching environment.

STEP 11: In Pro/ENGINEER's Textbox, enter .25 as the Pitch at the start of the trajectory.

This value will be the pitch of the spring at the Start Point of the feature.

STEP 12: Enter .25 as the Pitch at the end of the trajectory.

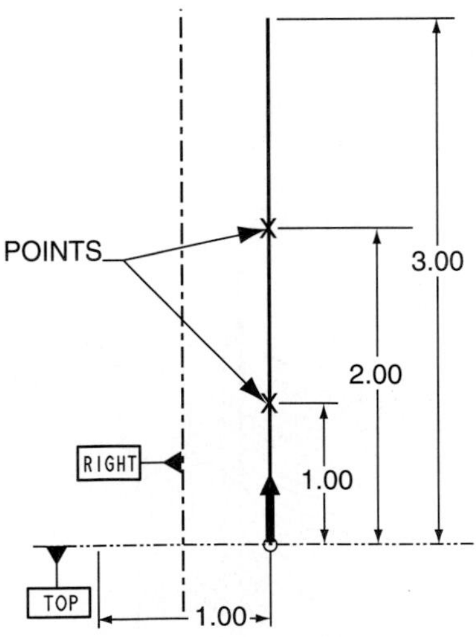

Figure 11-28 Point construction

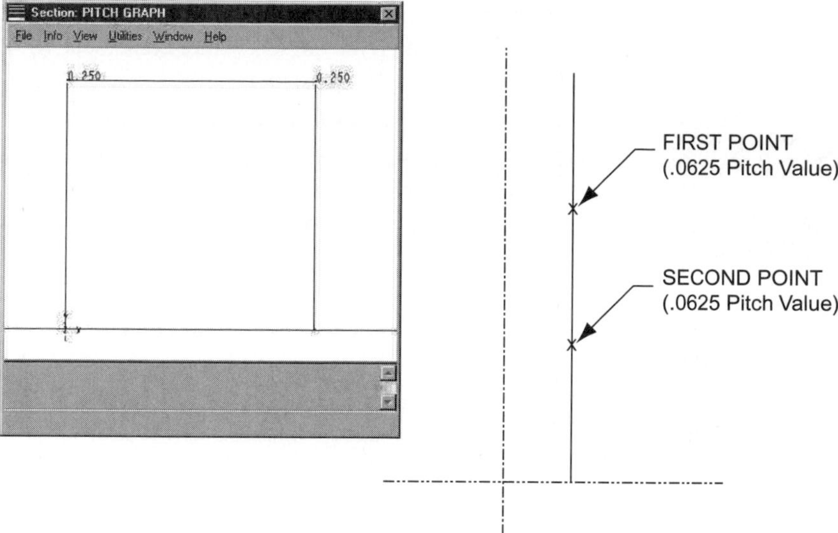

Figure 11–29 Graph window

STEP 13: On the Graph menu, select the DEFINE option; then on the work screen select the first point (Figure 11–29).

Pro/ENGINEER provides a graph window for the definition of pitch values along the Helical feature's path. The graph window serves as a tool to view the changes in the pitch values along the feature. Within the graph window, notice that the current values (0.250) of each point along the trajectory are displayed. On the work screen (not the graph window), select the first point for the redefinition of its pitch value.

STEP 14: In Pro/ENGINEER's textbox, enter .0625 as the new pitch value for the first point.

After entering the new pitch value, notice the change in the graph window.

STEP 15: Select the second point and enter .0625 as its pitch value.

STEP 16: On the Graph menu, select the DONE option.

STEP 17: In the sketcher environment, use the CIRCLE icon to create the circle entity shown in Figure 11–30.

Sections for swept features are sketched from the opposite side of the trajectory's sketching plane. Notice on your work screen and in Figure 11–30 how the trajectory's path is now to the left of the axis of

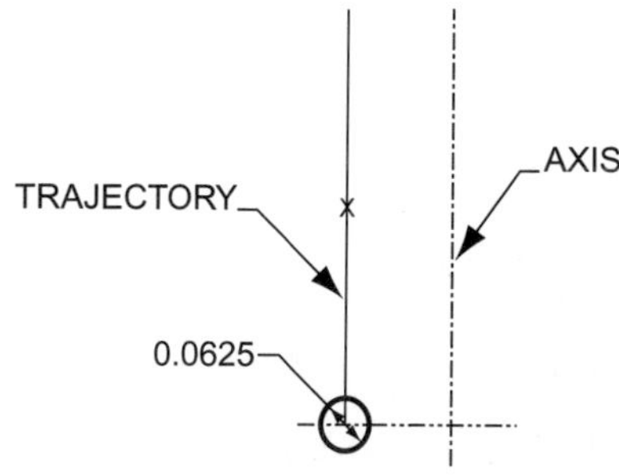

Figure 11–30 Sketched section

revolution. Dynamically rotate the work screen to better visualize the sketching environment. Sketch the circle's center coincident with the start point of the trajectory.

STEP 18: Select the Continue icon to exit the sketcher environment.

STEP 19: Preview the feature on the Feature Definition dialog box, then select OKAY to create the feature (Figure 11–31).

Notice on the dialog box the various elements that have been defined. Each can be modified at this point with the Define option.

STEP 20: Save your part.

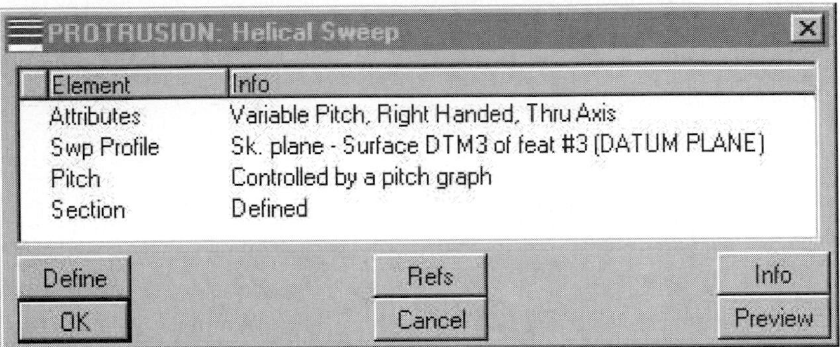

Figure 11–31 Feature Definition dialog box

TUTORIAL

Bolt Tutorial

This tutorial will create the bolt shown in Figure 11–32. Several steps are involved in the creation of this part. The first three steps involve the construction of the bolt shaft and the threads. The threads will be cut with the Helical Sweep option. The final two steps involve the modeling of the bolt's head. The following will be covered in this tutorial:

- Creating a helical sweep cut.
- Creating an extruded protrusion.
- Creating a revolved cut.

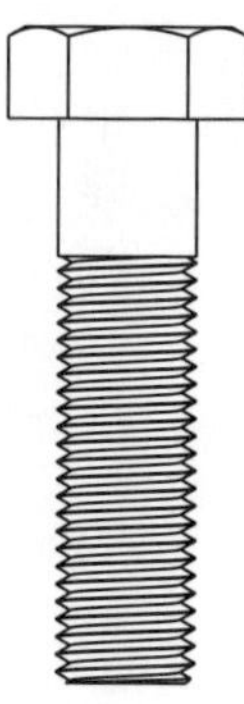

Figure 11–32 Bolt part

Creating the Bolt's Shaft

The first feature included in the bolt consists of a Revolved Protrusion. Use Pro/ENGINEER's default datum planes and sketch the revolved feature on datum plane FRONT. The finished feature and the sketch are shown in Figure 11–33. The following are some pointers for creating this feature:

- Create the feature as a revolved protrusion.
- Use Pro/ENGINEER's default datum planes and align the sketch with the datum planes as shown in Figure 11–33.
- Include the chamfer within the sketch.
- Revolve the protrusion 360 degrees.

Bolt Threads

This segment of the tutorial will use the Cut >> Helical Sweep option to create the threads for the bolt (Figure 11–34). The bolt will have eight threads per inch. Perform the following steps to create the Helical Cut:

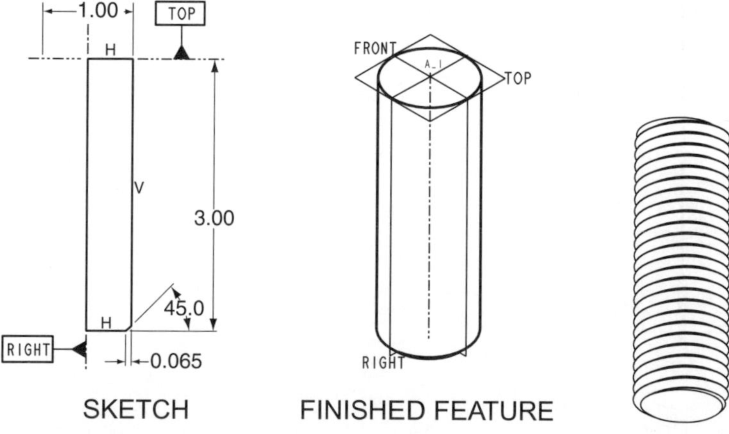

Figure 11–33 Revolved protrusion feature

Figure 11–34 Thread feature

STEP 1: **Select FEATURE >> CREATE >> CUT.**

STEP 2: **Select ADVANCED >> SOLID >> DONE on the Solid Options menu.**

STEP 3: **Select HELICAL SWP >> DONE on the Advanced Feature Options menu.**

STEP 4: **Select CONSTANT >> THRU AXIS >> RIGHT HANDED >> DONE as Attributes for the thread.**

STEP 5: **Select datum plane FRONT as the sketching plane then orient the sketch then orient the sketching environment to match Figure 11–35.**

STEP 6: **Use the References dialog box to specify the right edge of the shaft as a reference.**

STEP 7: **Define the Sweep Path by creating the section shown in Figure 11–35 (including the centerline).**

Sketch the single line entity shown in the figure. This line will serve as the trajectory path for the helical feature. When sketching the line, extend each end past the existing part. In addition to the line entity, sketch the centerline entity to serve as the axis of revolution.

STEP 8: **Define the dimensioning scheme shown in Figure 11–35.**

STEP 9: **Select the Continue icon to exit the sketcher environment.**

STEP 10: **Enter .125 as the Pitch Value of the thread.**

The bolt in this tutorial will have eight threads per inch. By definition, the pitch is the inverse of the number of threads per inch. The inverse of eight is 0.125.

STEP 11: **Sketch the Section defining the cut.**

Notice in Figure 11–36 how the sketch commences at the start point of the trajectory of the feature. The profile of the sketch is also shown in the figure. The profile is a typical 60-degree United National Standard thread form. In the figure, notice the small 0.005 inch gap. The Pitch of the helical sweep is 0.125 inches. This gap forms a flat thread ridge as would be typically found on a bolt.

STEP 12: **Select the Continue option to exit the sketcher environment.**

STEP 13: **Select an appropriate material removal direction.**

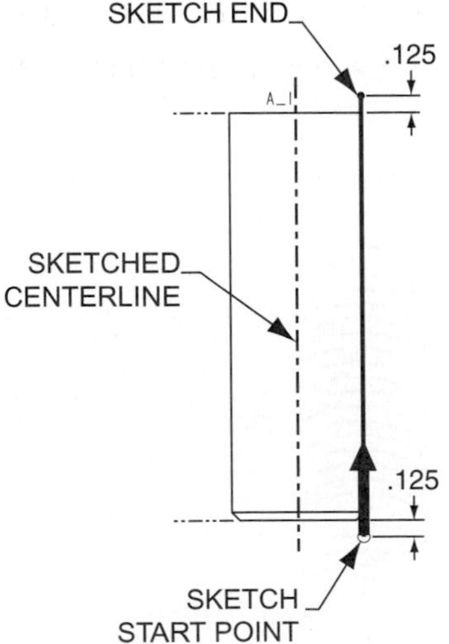

Figure 11–35 Sweep path section

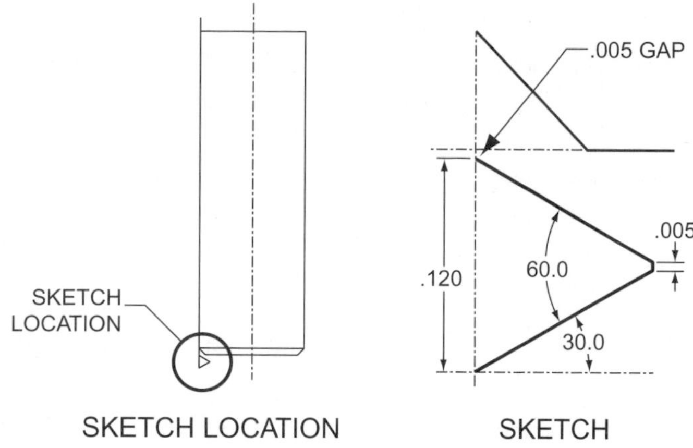

Figure 11–36 Helical sweep section

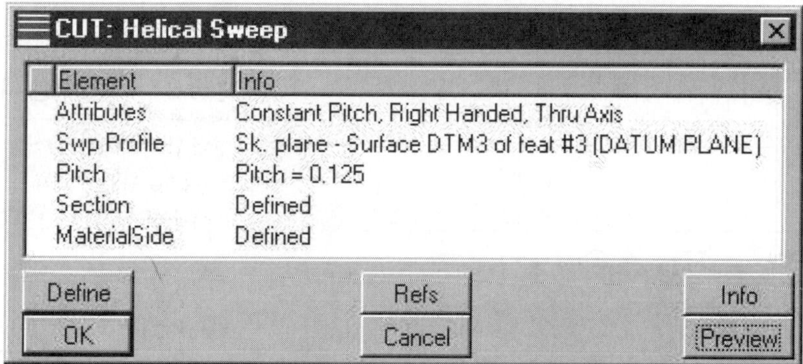

CUT: Helical Sweep

Element	Info
Attributes	Constant Pitch, Right Handed, Thru Axis
Swp Profile	Sk. plane - Surface DTM3 of feat #3 (DATUM PLANE)
Pitch	Pitch = 0.125
Section	Defined
MaterialSide	Defined

Define | Refs | Info
OK | Cancel | Preview

Figure 11–37 Feature Definition dialog box

STEP 14: Preview the feature, then select OKAY on the Feature Definition dialog box (Figure 11–37).

STEP 15: Save your part.

EXTRUSION EXTENSION

The first extruded Protrusion created within this part serves as the surface for forming the thread feature. The bolt in this tutorial is 4.00 inches long. The original Protrusion could have been created 4.00 inches in depth, but the termination of the Cut Helical Sweep at the 3.00 point (the thread length) would result in a poor thread representation. Adding the additional protrusion partially solves this problem.

Figure 11–38 shows the finished extruded Protrusion, the sketching plane, and the feature section. Create this feature using the following options:

- Create the feature using the Protrusion >> Extrude option.
- Extrude the feature toward one side.
- Use datum plane FRONT as the sketching plane.
- Within the sketching environment, utilize the Use-Edge option to create the section.

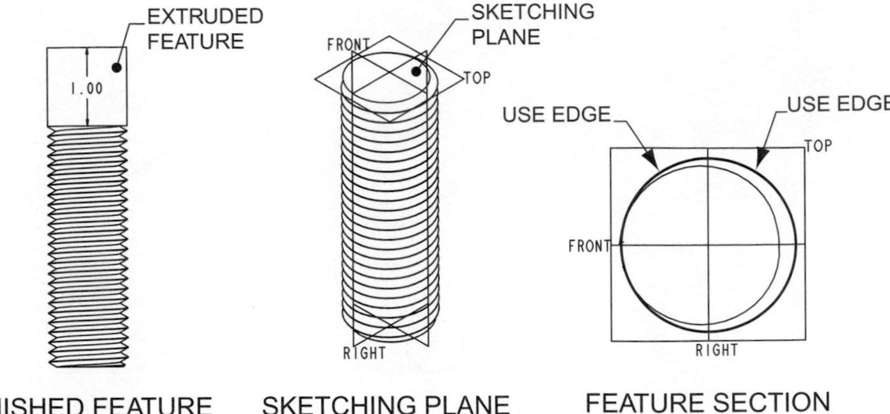

Figure 11-38 Feature construction

> **DESIGN INTENT** Utilizing the Use Edge option to create the section for this feature helps to capture the design intent of the part. Within this bolt part, the diameter of the shaft should be the same throughout the length of the bolt. The Use Edge option will tie the diameter of this feature with the first Protrusion feature.

BOLT HEAD CREATION

The head of the bolt requires the creation of two features. The first feature is an extruded Protrusion and is shown in Figure 11–39. Use the nonthreaded end of the existing part as the sketching plane. Extrude the feature one direction away from the sketching plane a distance of 0.667 inches.

Sketch the feature's section as shown in Figures 11–39 and 11–40. The sketching of the section can be challenging. The technique shown in the figures requires sketching one-quarter of the section then mirroring the remainder of the section using the Mirror icon. The Mirror option requires the existence of a centerline to mirror about. Entities to be mirrored must be preselected before the Mirror icon is accessible.

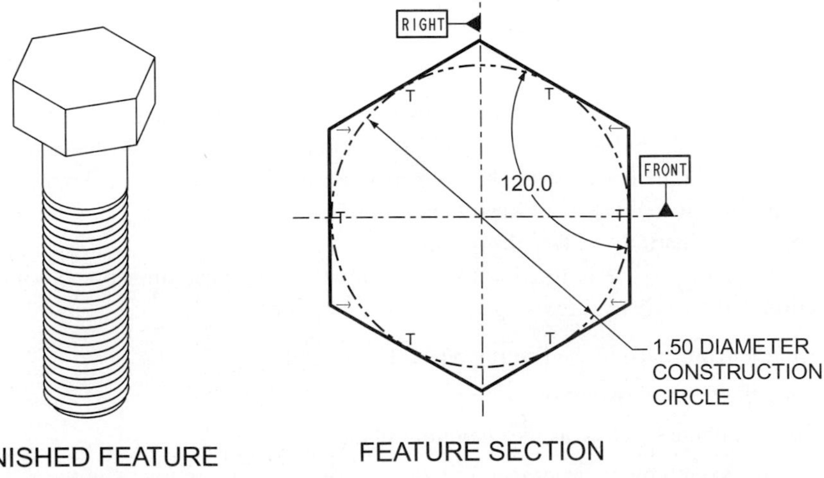

Figure 11-39 Bolt head protrusion

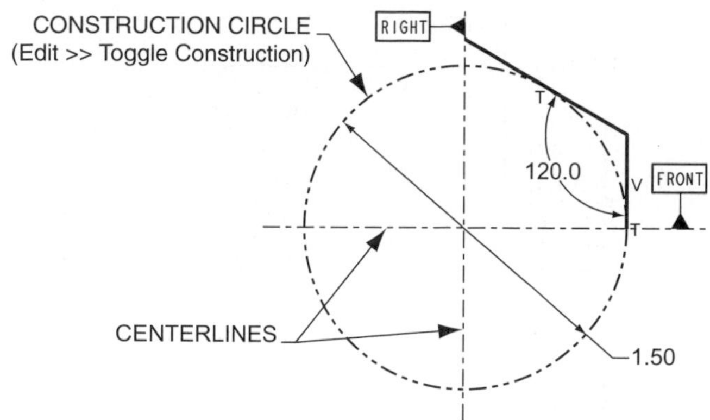

Figure 11–40 Section construction

Notice the construction circle in both figures. A sketched entity, such as a circle, can be converted to a construct entity with the Edit >> Toggle Construction option. A construction circle will not extrude with sketched geometric entities. The construction circle in this example defines the distance across the flats of the hex-head bolt.

Bolt Head Cut

The second feature required within the construction of the bolt's head is a revolved cut (Figure 11–41). Perform the following steps to create this feature:

Step 1: Select FEATURE >> CREATE >> CUT.

Step 2: Select REVOLVE >> SOLID >> DONE.

Step 3: Select ONE SIDE >> DONE as an Attribute for the Cut feature.

Step 4: For the sketching plane, select the datum plane that extends between the corners of the bolt head's extruded protrusion (datum plane RIGHT in Figure 11–41).

Step 5: Orient the sketcher environment to match Figure 11–42.

Step 6: Sketch the section shown in Figure 11–42.

Specify the top and right edges of the bolt head as references. The section requires one geometric line entity and one centerline entity. The centerline will serve as the axis of revolution.

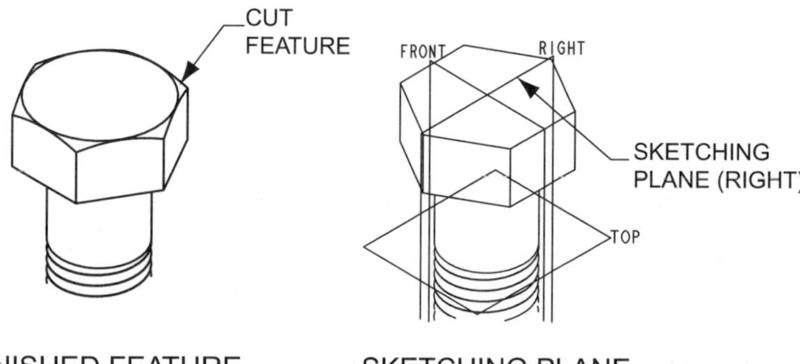

Figure 11–41 Bolt head cut

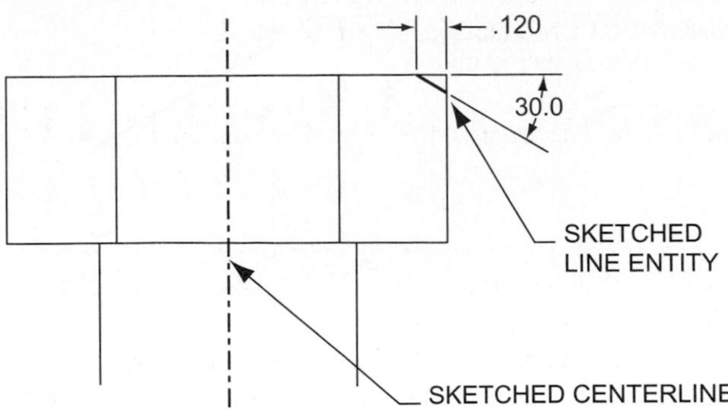

Figure 11-42 Cut section

STEP 7: Select the Continue icon to exit the sketching environment.

STEP 8: Select an appropriate Direction of Cut.

STEP 9: Select 360 as the degrees of revolution, then select DONE.

STEP 10: Preview the feature, then select OKAY on the Feature Definition dialog box.

STEP 11: Save your part.

TUTORIAL

VARIABLE SECTION SWEEP TUTORIAL

The Variable Section Sweep option is used to sweep a section along multiple trajectories. It is a useful tool for creating advanced geometric shapes. This tutorial will cover the process for creating a simple Variable Section Sweep.

The trajectories for a Variable Section Sweep can be either selected on the work screen or sketched. Datum curves are often created for use as trajectories within the construction process. This tutorial will use datum curves for this purpose (Figure 11–43). The following will be covered in this tutorial:

- Creating datum curves.
- Creating a variable section sweep feature.
- Creating a swept protrusion.
- Creating an extruded protrusion.

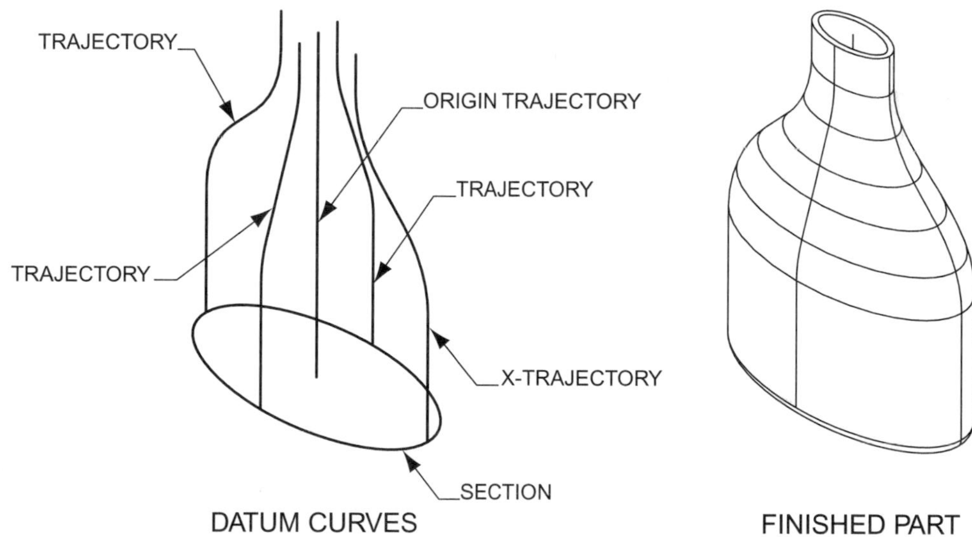

Figure 11–43 Variable section sweep

CREATING THE ORIGIN TRAJECTORY

In this segment of the tutorial, you will create the Origin Trajectory for the feature. The origin trajectory is used to define the start point of the trajectory, the direction of sweep, and the depth of the sweep. It can be either sketched or selected. In this tutorial, it will be created as a datum curve then selected within the creation of the Variable Section Sweep.

Sketch the Origin Trajectory on datum plane FRONT at the intersection of datum plane RIGHT. Create this feature using the Datum Curve icon on the Datum toolbar. Sketch as shown in the Sketch View of Figure 11–44.

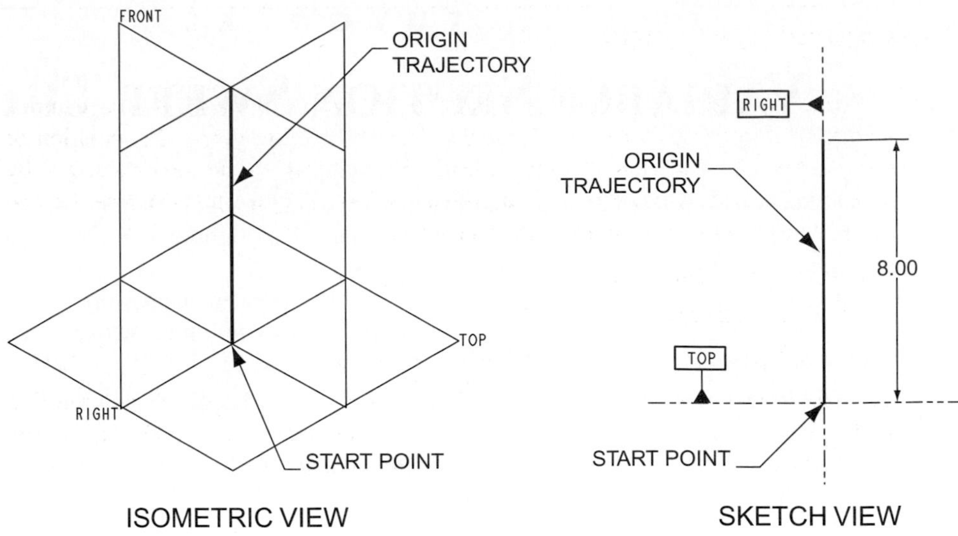

Figure 11–44 Origin trajectory construction

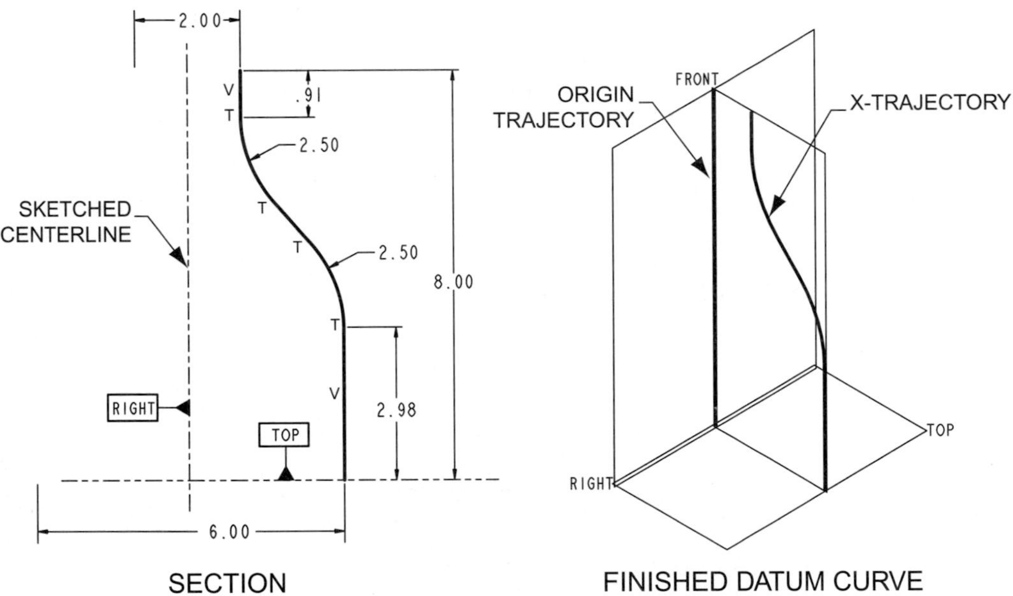

Figure 11–45 X-trajectory creation

X-Trajectory Creation

The X-Trajectory is used to define the variation in the horizontal vector of the Variable Section Sweep's section. The X-Trajectory for this tutorial will be defined by a sketched datum curve that will be selected during the construction of the Variable Section Sweep feature. Sketch the datum curve on datum plane FRONT as shown in Figure 11–45. Align the bottom of the sketch with datum plane TOP. Sketch a centerline aligned with datum plane RIGHT, and create the diameter dimensions as portrayed in the figure.

TRAJECTORY CREATION

Variable Section Sweep trajectories serve basically the same purpose as the feature's X-Trajectory. Multiple trajectories can be selected or sketched to define the variation of the section along the sweep path. In this tutorial, you will create the next trajectory by copy-mirroring the X-Trajectory over datum plane RIGHT (Figure 11–46). Use the following information when creating this new trajectory:

- Use the Copy >> Mirror option.
- Use the Dependent suboption.
- Copy the X-Trajectory over datum plane RIGHT.

Two additional trajectories are required and are shown in Figure 11–47. Sketch the first curve as shown in the figure then copy-mirror this curve to create the second. The bottom of the sketched datum curve should be aligned with datum plane TOP.

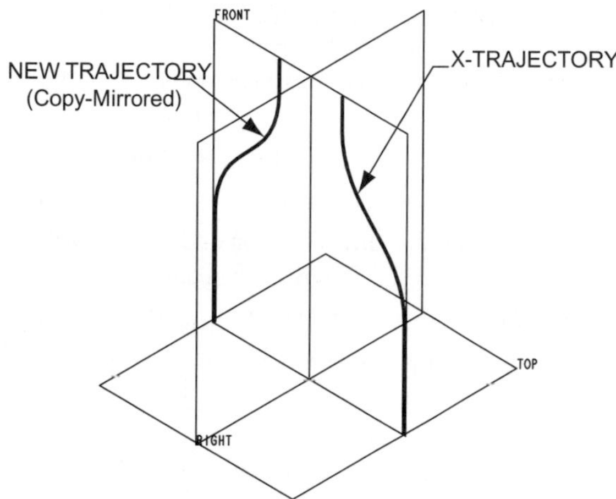

Figure 11–46 Trajectory creation

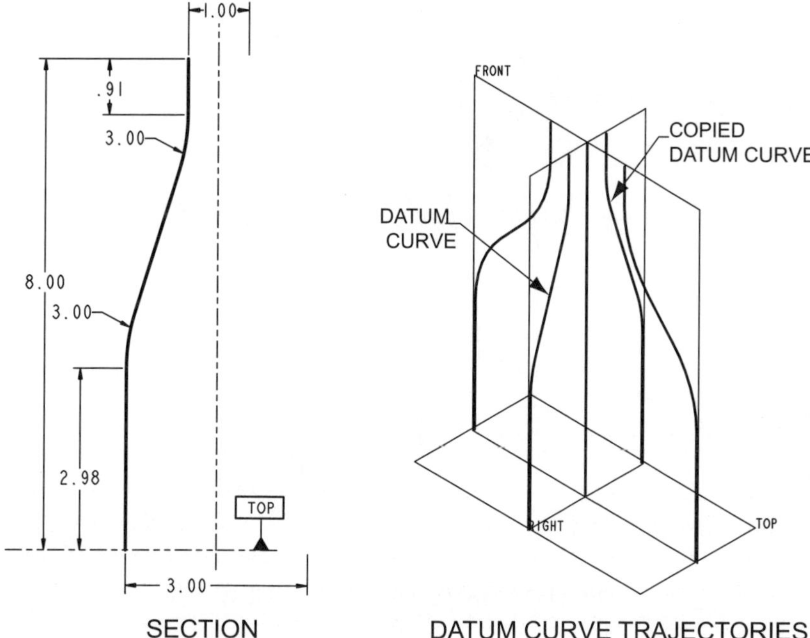

Figure 11–47 Trajectory creation

Variable Section Sweep Feature

This step of the tutorial will create the Thin Variable Section Sweep feature shown previously in Figure 11–43. A Variable Section Sweep feature requires the definition of one Origin Trajectory, one X-Trajectory, and one Section. Multiple trajectories in addition to the Origin Trajectory and the X-Trajectory can be defined. In this tutorial, a total of five trajectories will be utilized. Perform the following steps to create this feature.

Step 1: Select FEATURE >> CREATE >> PROTRUSION.

Step 2: Select ADVANCED >> THIN >> DONE.

Step 3: Select the VAR SEC SWP (Variable Section Sweep) option, then select DONE.

Step 4: NrmToOriginTraj >> DONE.

The Normal-To-Origin-Trajectory (NrmToOriginTraj) option will create the feature's cross section normal to the defined Origin Trajectory. The Pivot Direction option will create the feature's cross section normal to a selected plane or edge, while the Normal to Trajectory option will create the feature's cross section normal to a selected trajectory.

Step 5: Pick the SELECT TRAJ option on the Variable Section Sweep menu.

The first trajectory that is defined is the Origin Trajectory. This trajectory defines the direction and sweep of the feature. It can be either sketched or selected. It will be selected in this exercise.

Step 6: Select ONE BY ONE as the chain option, then select the Origin Trajectory (see Figure 11–48).

The first trajectory selected defines the Origin Trajectory. As shown in Figure, on the work screen select the Origin Trajectory on the end of the entity denoted by the start point arrowhead.

Step 7: Select DONE SEL on the Get Select menu.

By selecting the Done Select option, Pro/ENGINEER on the work screen will display the start point for the trajectory. If the start point is not aligned

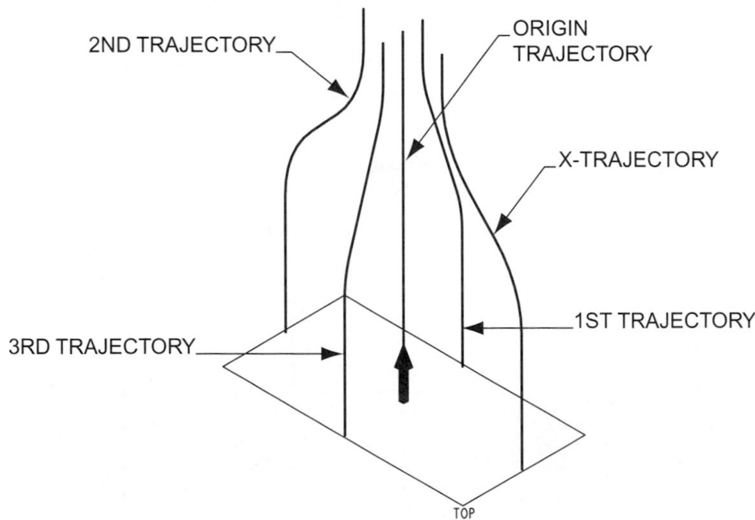

Figure 11–48 Trajectory selection

with datum plane TOP, use the Start Point option to select the opposite end of the entity.

STEP 8: On the Chain menu, select the DONE option.

The Done option will end the selection of the Origin Trajectory.

STEP 9: On the Variable Section Sweep menu, pick the SELECT TRAJ option.

STEP 10: On the Chain menu, select the CURVE CHAIN option, then select the X-Trajectory entity shown in Figure 11–48.

The second trajectory defined is the X-Trajectory. This trajectory is used to create the variation in the sweep section.

STEP 11: Pick the SELECT ALL option on the Chain Options menu.

STEP 12: On the Chain menu, select the DONE option to end the selection of the X-Trajectory.

STEP 13: On the Variable Section Sweep menu, pick the SELECT TRAJ option.

STEP 14: On the Chain menu, select the CURVE CHAIN option, then select the 1st trajectory entity shown in Figure 11–48.

STEP 15: Pick the SELECT ALL option on the Chain Options menu.

STEP 16: On the Chain menu, select the DONE option.

STEP 17: Select the remaining trajectories shown in Figure 11–48.

Use the Select Trajectory (Select Traj) option to select the remaining trajectories. Only one trajectory can be selected per selection option.

STEP 18: When all trajectories have been picked, select the DONE option on the Variable Section Sweep menu.

Upon selecting Done, Pro/ENGINEER will enter the sketching environment for the creation of the section.

STEP 19: Sketch the section using the Ellipse option (Figure 11–49).

Pro/ENGINEER will orient the sketching environment to lie perpendicular to the Origin Trajectory. You can dynamically rotate the work screen to better visualize the sketching plane. Notice in the figure the location of the alignment points. An ellipse is created by first locating its center-point followed by dragging its major and minor diameters. When locating your

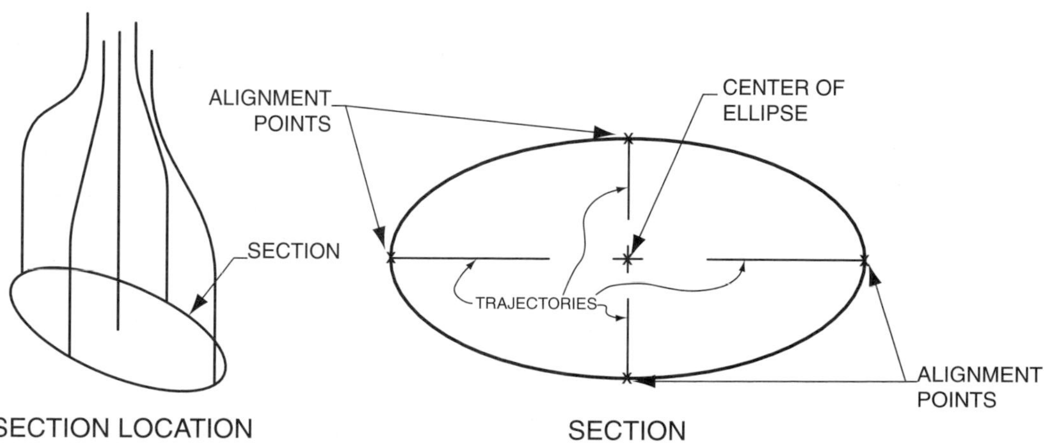

Figure 11–49 Section sketch

major and minor diameters, make sure your ellipse connects to each required alignment point.

STEP 20: Select the Continue option when the section is complete.

STEP 21: Select OKAY to allow the thin option to create material toward the inside of the section.

Recall that this segment of the tutorial requires you to construct a thin Variable-Section-Sweep. If you did not set thin as an attribute of this feature, you must start the feature creation process from the beginning. Thin is not a redefinable element of a solid feature.

STEP 22: Enter 0.125 as the width of the thin feature.

STEP 23: Preview the feature on the Feature Definition dialog box, then select OKAY to create the feature.

STEP 24: Save your part.

SWEPT FEATURE

Within this segment of the tutorial, you will create the base of the part as a swept protrusion (see Figure 11–50). As shown in Figure 11–51, the outside of the base of the existing part will be selected as the trajectory, and the section will be sketched on datum plane FRONT.

STEP 1: Select FEATURE >> CREATE >> PROTRUSION.

STEP 2: Select SWEEP >> SOLID >> DONE.

STEP 3: Pick the SELECT TRAJ option as the trajectory definition option.

Within this tutorial, you will select the edge shown in Figure 11–51 as the trajectory for this swept feature.

STEP 4: Select TANGENT CHAIN on the Chain menu, then pick the trajectory edge shown in Figure 11–51.

The Tangent Chain option will select a picked entity and any other entity tangent to the selected entity. In this step of the process, you will select the outside edge of the variable-section-sweep feature as shown in the figure.

NOTE: After selecting the edge, your model should display the start point location shown in Figure 11–52.

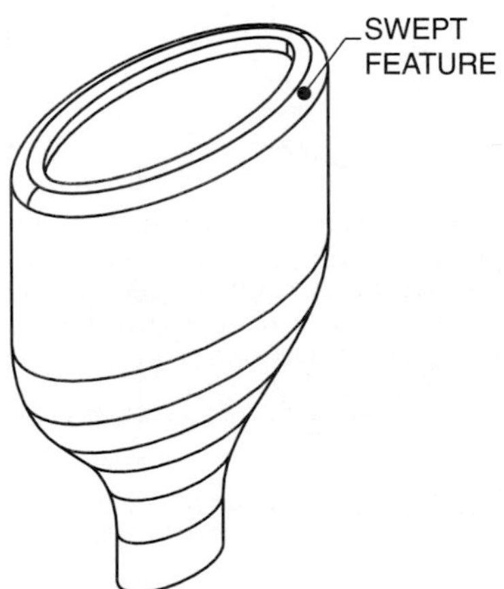

Figure 11–50 Swept feature

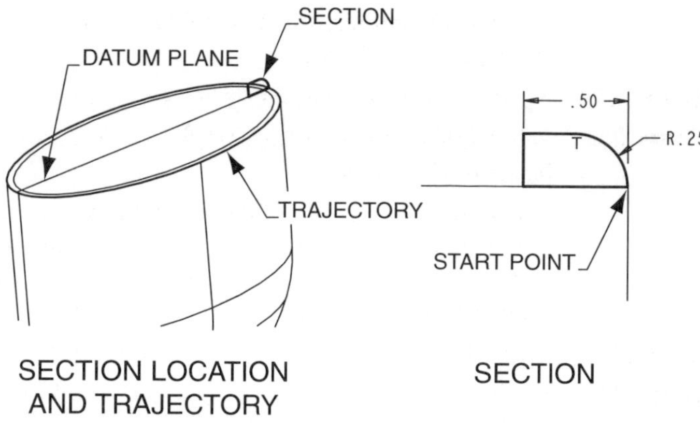

Figure 11–51 Swept feature creation

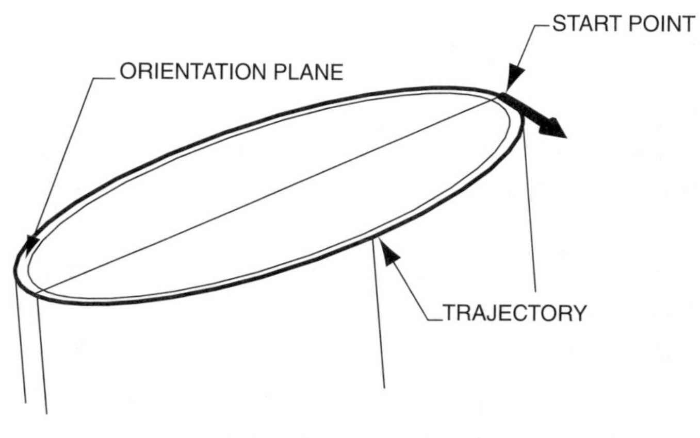

Figure 11–52 Trajectory and start point locations

> **INSTRUCTIONAL NOTE** The Tangent Chain option should select the entire perimeter edge of the part. If it doesn't, you can use the One-By-One option to continue picking necessary edges. A successful selection will reveal a start-point arrow at the location of the trajectory's start point. Observe your screen to confirm the presents of this start point.

STEP 5: Select DONE to exit the Chain menu.

> **INSTRUCTIONAL NOTE** The next step in this tutorial will require you to identify a planar surface to orient toward the top of the sketching environment. This is a similar process to orienting the sketching environment with the TOP option during the creation of an extruded protrusion.

STEP 6: Using the Choose menu, identify and ACCEPT the Orientation Plane shown in Figure 11–52.

The accepted surface will be oriented toward the top of your sketching environment.

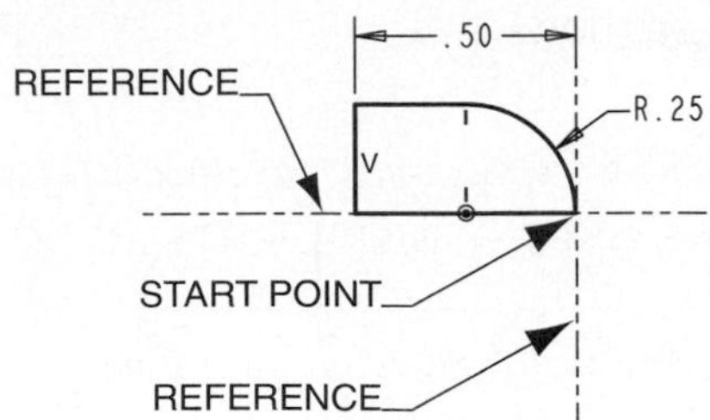

Figure 11-53 Sweep section

STEP 7: Select OKAY to accept the default orientation of the cross section.

STEP 8: Turn off the display of your datum planes.

STEP 9: Ensure the presence of the two references shown in Figure 11–53.

STEP 10: Sketch the section shown in Figure 11–53.

Refer to Figures 11–51, 11–52, and 11–53 when sketching the section. Incorporate any necessary dimensions and constraints.

STEP 11: When your section matches Figure 11–53, exit the sketcher environment.

STEP 12: Preview your feature on the Feature Definition dialog box.

STEP 13: Select OKAY to create feature.

EXTRUDED PROTRUSION

As shown in Figure 11–54, your part has a very obvious hole that needs to be plugged. This segment of the tutorial will solve this problem by filling this hole with an extruded protrusion. Within the sketcher environment of this feature, you will utilize the Use Edge option to create all necessary sketch entities.

STEP 1: Setup an extruded protrusion to create the feature shown in Figure 11–54.

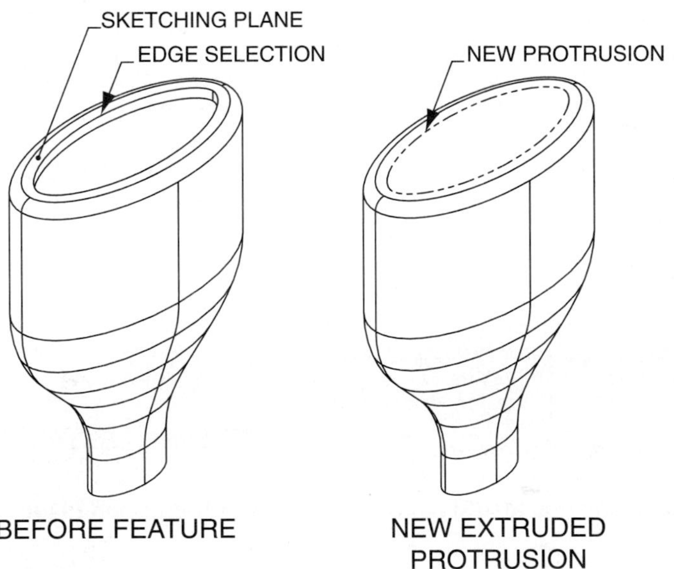

Figure 11-54 Extruded protrusion

Use the following options when establishing your protrusion:

- One-sided extrusion.
- Sketching plane as shown in Figure 11–54 (the bottom of the part).
- Orient datum plane FRONT toward the bottom of the sketcher environment as shown in Figure 11–55.
- The feature creation direction should extrude toward the inside of the part. You might have to "Flip" the default direction.

STEP 2: On the Environment dialog box (Utilities >> Environment), set NO DISPLAY as the Tangent Edges display style (see Figure 11–56).

As previously mentioned, you will utilize the Use Edge option to create sketcher entities. Within your current model, tangent edges can clutter the work screen and make edge selection difficult. By turning off the display of tangent edges, the selection of the edges in this segment of the tutorial should be easier.

STEP 3: On the Display toolbar, set HIDDEN as the model's display style.

STEP 4: (Optional Step) Specify datum planes FRONT and RIGHT as references.

This step is not a requirement for this section. Since the Use Edge option will be used to create all necessary sketcher entities, you actually do not need to specify references.

STEP 5: Select the USE EDGE icon on the Sketcher toolbar.

STEP 6: Pick the two edges shown in Figure 11–55 (only select each edge once).

STEP 7: Exit the sketcher environment.

If you receive an "Incomplete Section" error message, you probably picked an edge more than once in the previous step. You will need to delete any extra edges. Another possible cause of the error message is an incomplete closure of your section. If this is the case, you should observe your section and use the Use Edge option to pick edges that were excluded from the previous pick process.

STEP 8: Extrude your feature a Blind distance of 0.25.

STEP 9: Turn the display of Tangent Edges on.

STEP 10: Preview your feature, then select OKAY on the Feature Definition dialog box (Figure 11–56).

STEP 11: Save your part file.

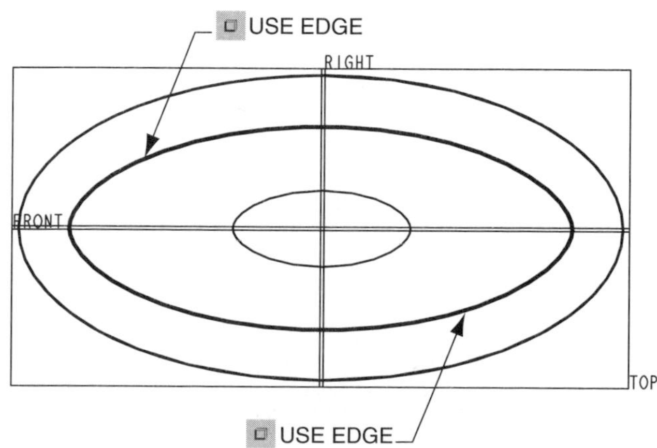

Figure 11–55 Sketcher environment and Use Edge option

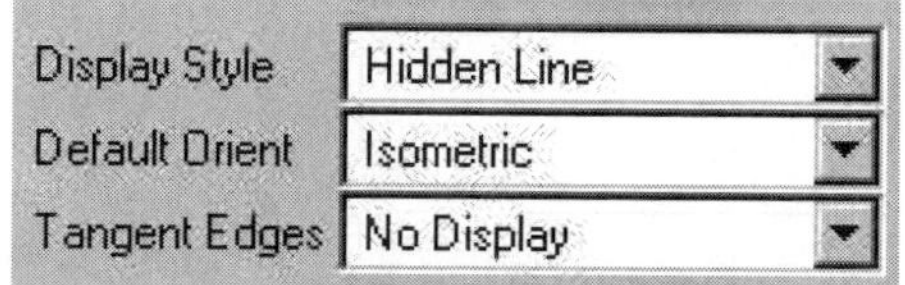

Figure 11–56 Tangent edges display

PROBLEMS

1. Model the part shown in Figure 11–57.
2. Use Pro/ENGINEER to model a 1.25-inch nominal diameter Hex Head bolt. The hex head has a 1.875-inch distance across its flats and a height of 0.844 inches. The bolt length is 3 inches with a thread length of 2.25 inches. Create the thread as a Cut Helical Sweep.
3. Use Pro/ENGINEER to model a 1.50-inch nominal diameter Hex Head bolt. The hex head has a 2.25 inch distance across its flats and a height of 1 inch. The bolt length is 4 inches with a thread length of 3.25 inches.
4. Model the spring feature shown in Figure 11–58.
5. Model the bottle part shown in Figure 11–59. Create the bottle as a Thin Variable Section Sweep feature. The trajectories for the feature are shown in the figure. The dimensions for the X-Trajectory are also shown. Create the X-Trajectory as a datum curve, then use the Copy >> Move >> Rotate option to create three more curves rotated 90-degrees apart. Create the remaining four trajectories by first sketching a datum curve on an on-the-fly datum plane

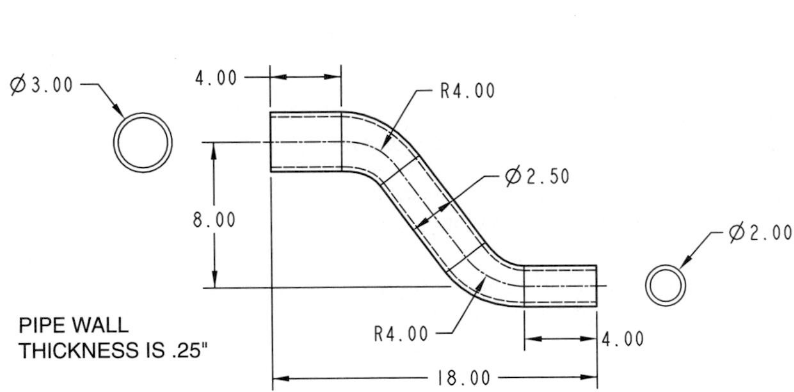

Figure 11–57 Problem one

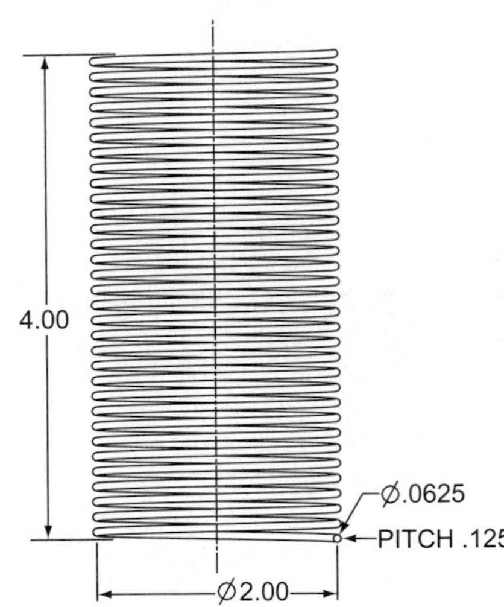

Figure 11–58 Problem four

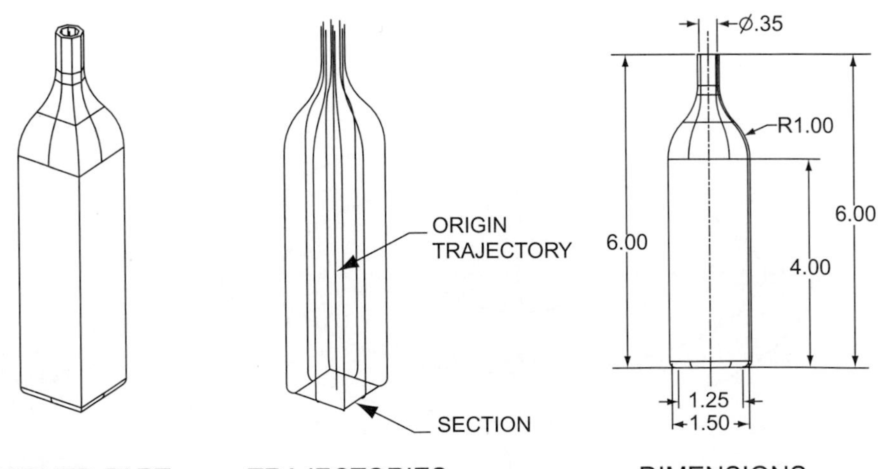

Figure 11–59 Problem five

at a 45-degree angle to one of Pro/ENGINEER's default datum planes. Create the section by projecting one of the existing datum curves with the Use Edge option. Use the Copy >> Move >> Rotate option to create the final three trajectories.

6. Use the Hex-Head bolt created in Problem 2 as the generic part of a family table of Hex-Head bolts. The following rules apply to the creation of Hex-Head bolts:
   - The distance across the flats of the bolt head is 1.5 times the bolt diameter.
   - The height of the bolt head is 0.667 times the bolt diameter.
   - The tread length is 2 times the diameter plus 0.25 inch.

   Refer to the section on family tables in Chapter 7. Create the following instances:
   - 1-inch nominal size bolt with a 3-inch bolt length.
   - 2-inch nominal size bolt with a 5-inch bolt length.
   - 0.5-inch nominal size bolt with a 2-inch bolt length.

Questions and Discussion

1. Compare and contrast the Normal-To-Origin-Trajectory (NrmToOriginTraj), Pivot Direction (Pivot Dir), and Normal-To-Trajectory (Norm To Traj) options.
2. Describe the basic principles behind a Swept Blend feature.
3. Name two methods for creating the trajectory for a Swept Blend feature.
4. When creating a Swept Blend, what is Pro/ENGINEER requesting when it asks for the Z-Axis rotation angle?
5. When creating a Variable Section Sweep feature, what is the first trajectory that must be defined? What is the purpose of this trajectory? What is the second trajectory that Pro/ENGINEER requires?
6. How many trajectories does Pro/ENGINEER require within the Variable Section Sweep option? How many can be defined?
7. Define the meaning behind the Pitch of a thread.
8. What two geometric features are often created with the Helical Sweep feature?

CHAPTER

12

ASSEMBLY MODELING

Introduction

Pro/ENGINEER is considered a design and engineering tool. The various modules of Pro/ENGINEER provide designers, engineers, and manufacturers with the tools necessary to take a design from the conceptual stage through the final manufacturing process. One of the most powerful tools of Pro/ENGINEER is its Assembly module. Within this application, existing components can be grouped as part of an assembly or as part of a subassembly. In addition, Pro/ENGINEER is capable of completing a design from the top down by allowing parts to be modeled with Assembly mode. Upon finishing this chapter, you will be able to

- Create an assembly through the placement of existing parts.
- Create parts within Assembly mode.
- Modify parts within an assembly.
- Apply dimensional relationships between parts in Assembly mode.
- Create assembly features.
- Create an assembly simplified representation.
- Create and animate an assembly mechanism.

DEFINITIONS

Assembly A collection of components that forms a complete design or a major end item.

Bottom-up design The placing of exiting components within an assembly.

Constraint The explicit relationship defined between components of an assembly.

Component A part or subassembly.

Package A component that has not been fully constrained within an assembly.

Parametric assembly An assembly with parts constrained to other parts.

Subassembly A collection of parts and/or smaller subassemblies that forms a subcomponent of a complete design or major end item.

Top-down design The designing of components within an assembly.

INTRODUCTION TO ASSEMBLY MODE

Pro/ENGINEER's **Assembly** mode is used to group components to meet the requirements of a design. **Components** can consist of existing parts and subassemblies, or components can be created directly within Assembly mode. Placing existing components to form an assembly is referred to as **bottom-up assembly design,** while creating parts within Assembly mode is a tool used within **top-down assembly design.**

Parts in Assembly mode maintain their associativity with their separate part files. Within Part mode, if a dimension value is modified, the part instance in Assembly mode is modified. Correspondingly, if an instance of a part is modified in Assembly mode, the component in Part mode is modified. In addition, when a part is created within Assembly mode using Top-Down assembly design, a new part file is created that can be modified separately within Part mode. When a component is placed into an assembly, the component's separate part or assembly file is placed into memory and remains there until the parent assembly is erased from memory.

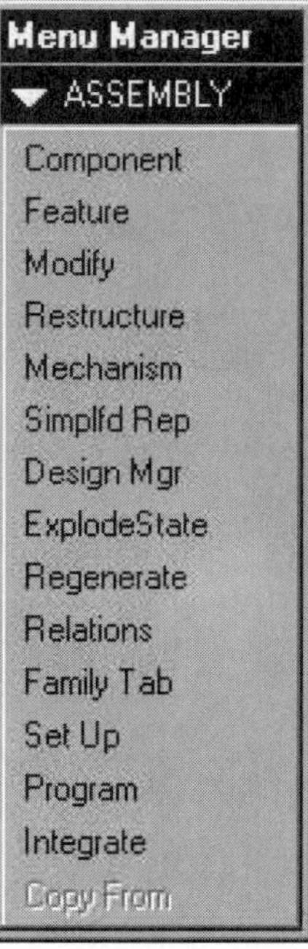

Figure 12–1 Assembly menu

Assembly Mode's Menu

The Assembly menu is considerably different from Part mode's menu (Figure 12–1). The Component menu option is used to place and create components. It has two submenus: assemble and create. The Assemble option is used to place existing parts and assemblies. Selecting this option will launch the Open dialog box. The Create option is available for the creation of new parts, subassemblies, skeleton models, and bulk items. Any component created in this menu will be saved as a separate object file when the primary assembly object is saved.

The Assembly menu's Feature option is used to create features specific to the assembly model. Suboptions common to Part mode such as Datum, Hole, Round, Chamfer, Cut, Protrusion, and Rib are available.

The Modify menu option is used to modify models that currently exist within the assembly. Suboptions are available for modifying existing parts, skeleton models, and assemblies. Like Part mode's Modify menu, options are available for modifying dimension values, dimension cosmetics, and geometric tolerances.

Placing Components

Existing parts and subassemblies can be placed into an assembly model. Placing an existing component is often referred to as Bottom-up assembly design. The Component >> Assembly and Component >> Package options are used to locate and open components. When a component is opened, its associated object file is opened into Pro/ENGINEER's memory (but not into a separate window). When an assembly is saved, objects within the assembly are saved to their separate object files. Individual components cannot be erased from memory as long as an associated assembly object is open.

A component can be placed into an assembly at any point during the assembly creation process, including as the first element of an assembly. When placed as the first component and before the creation of any assembly features, the object is placed without any defined constraints. When placed after a component or after an assembly feature, Pro/ENGINEER will launch the Component Placement dialog box (Figure 12–2). This dialog box has two tabs. The Place tab is used to establish constraints and mechanism connections. Constraints and connections define the relationship between components of an assembly. The Move tab is used to adjust the placement of a component during the placement process.

Assembly Constraints

When a component is placed into an assembly using the Assemble option, it can be fully constrained to existing components and features. This type of assembly is referred to as a

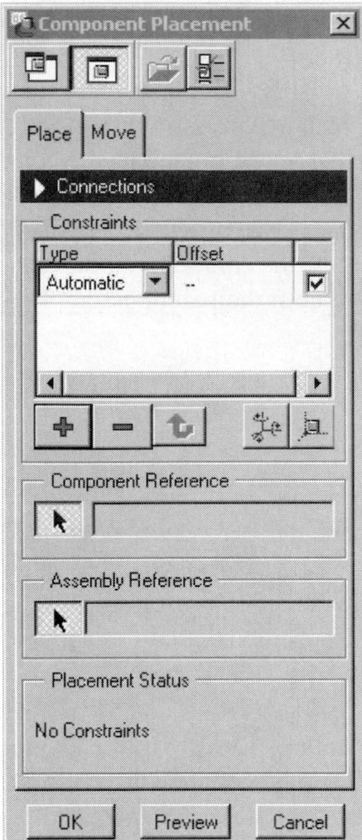

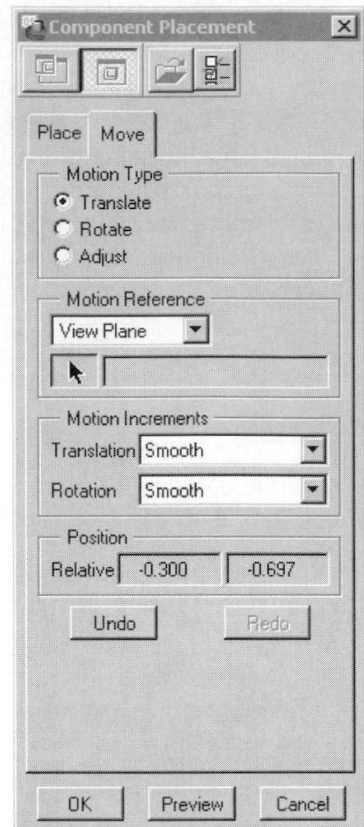

Figure 12–2 Component Placement dialog box

parametric assembly. Pro/ENGINEER provides a variety of **constraint** types for the placement of components (Figure 12–3). The following is a description of each:

AUTOMATIC

The Automatic constraint type is not actually a constraint type but a quick tool incorporated within the available constraint options to expedite the constraint defining process. It is the default constraint type when accessing the Component Placement dialog box. With the Automatic option, references are selected for both the component and the assembly. Pro/ENGINEER will determine the constraint to apply, but will provide you the option of selecting an alternative constraint. As an example, when mating two surfaces, with the automatic option, you must pick each surface. Pro/ENGINEER might give you an Align constraint with the option of changing your constraint to a Mate.

MATE

The Mate constraint type is used to place two surfaces coplanar. Any datum plane, part plane, or planar surface may be used. As shown in Figures 12–3 and 12–4, selected surface faces are placed along a common plane, but do not actually have to touch.

MATE OFFSET

Offset is a suboption under the Mate constraint type. Mating surfaces are placed coincident by default with the Mate constraint. The offset option places a user-specified offset distance between the selected surfaces (Figure 12–5). The distance can be modified at a later time with the Modify >> Value option.

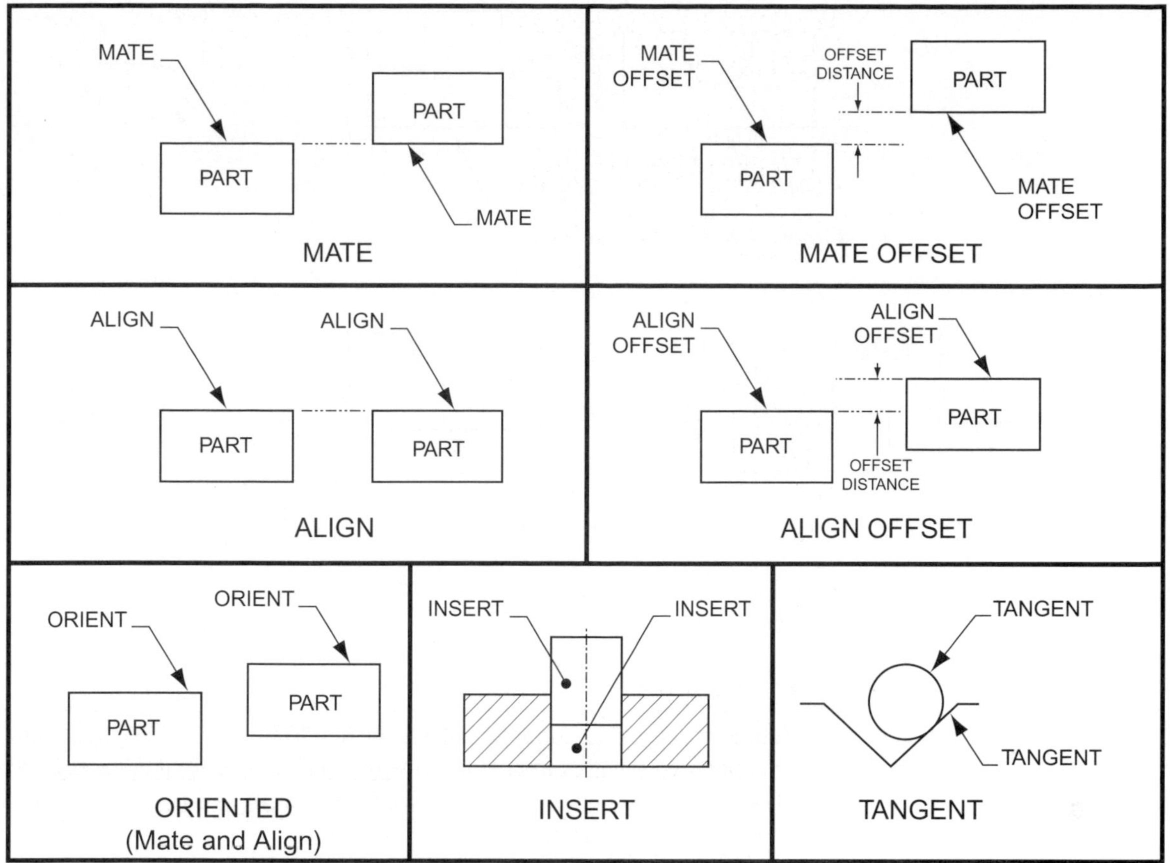

Figure 12–3 Constraint types

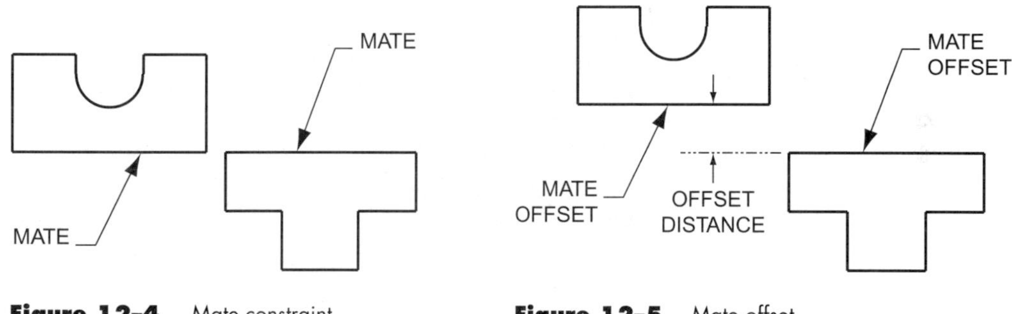

Figure 12–4 Mate constraint

Figure 12–5 Mate offset

Align

The Align constraint type is used to place surfaces coplanar and facing in the same direction (Figure 12–6). Like the Mate constraint, the surfaces do not have to touch. In addition, the Align constraint type is used to align axes, edges, and curves.

Align Offset

Like the Mate Offset option, Align has a suboption for offsetting two aligned surfaces. Similar to the Mate constraint, aligned surfaces are placed coincident by default. These surfaces can be offset a user-specified distance (Figure 12–7).

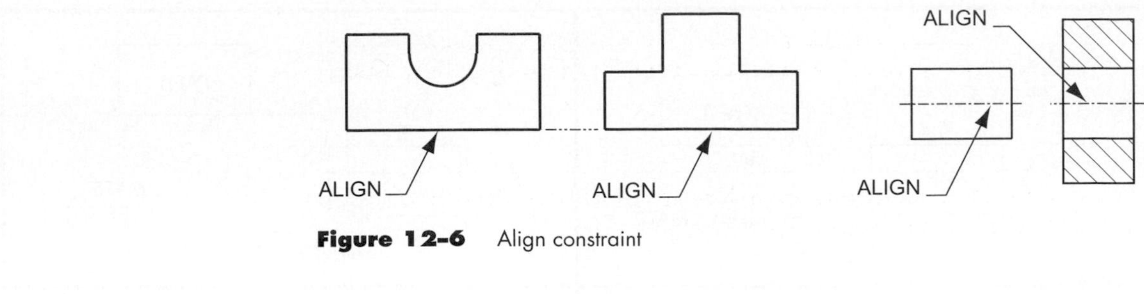

Figure 12-6 Align constraint

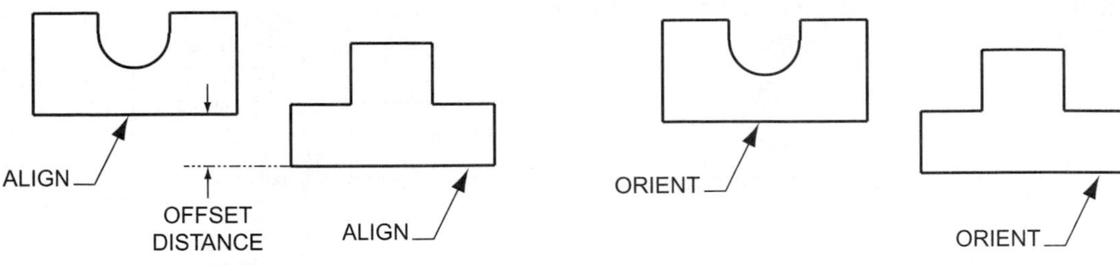

Figure 12-7 Align offset

Figure 12-8 Orient constraint

Oriented

Similar to the Offset suboption, the Align and Mate constraints have a suboption for orienting two surfaces in the same direction (Figure 12–8). Unlike coincident and offset aligns and mates, two constrained surfaces are not coplanar and have no specified offset distance.

Insert

The Insert constraint type makes the axes of two revolved features coincident (Figure 12–9). The user is required to select the surface of each feature. It is often used with shafts and holes for the alignment of each feature's centerline.

Tangent

The Tangent constraint type makes a cylindrical surface tangent to another surface (Figure 12–10). The user is required to select the surfaces of each feature.

Coordinate System

The Coordinate System constraint type aligns the coordinate systems of two parts. Within this constraint type, the axis of one coordinate system is aligned with the corresponding axis of a second coordinate system (e.g., X-axis with X-axis).

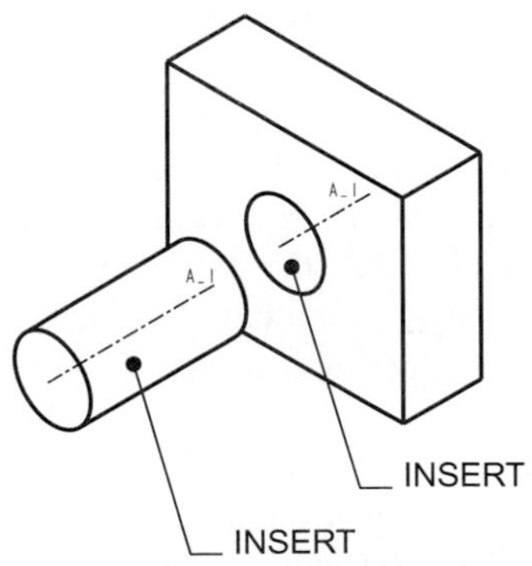

Figure 12-9 Insert constraint

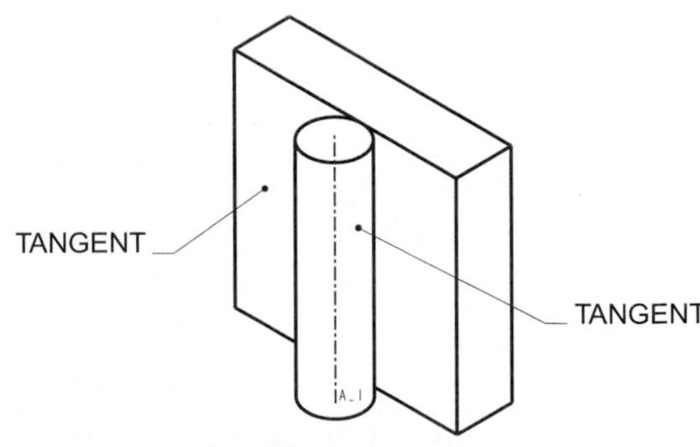

Figure 12-10 Tangent constraint

Point on Line

The Point On Line constraint type aligns a datum point with an existing edge, datum curve, or axis.

Point on Surface

The Point On Surface constraint type aligns a datum point with a surface. A surface can be any part surface or datum plane.

Edge on Surface

The Edge On Surface constraint type aligns the edge of a component with a surface. A surface can be any part surface or datum plane.

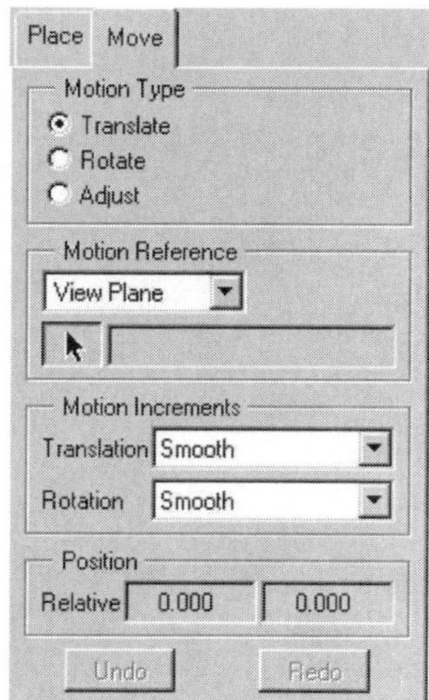

Figure 12–11 Move tab

Moving Components

When a component is placed into a scene, its default location might overlap existing components (making it hard to visualize), or it might be out of position for the type of constraints to apply. The Move tab (Figure 12–11) allows a partially constrained component to be moved on the work screen. The component can only be moved within the degrees of freedom allowed by existing constraints.

There are three Motion Types available: Translate, Rotate, and Adjust. The Translate option will move the component within the motion reference and the Rotate option will rotate the component around the selected motion reference. The Adjust option works similarly to available constraint options. This option allows the moving component to be mated and aligned with existing components.

When a Motion Type is selected, the relative motion is based on the Motion Reference selected. The following Motion References are available:

- **View plane** The motion will be relative to the current screen orientation.
- **Sel plane** The motion will be relative to a selected plane.
- **Entity/edge** The motion will be relative to a selected axis, edge, or curve.
- **Plane normal** The motion will be perpendicular to a selected plane.
- **2 points** Two selected vertices on the work screen are used to created the relative motion.
- **Csys** The motion will be relative to the X-axis of a selected coordinate system.

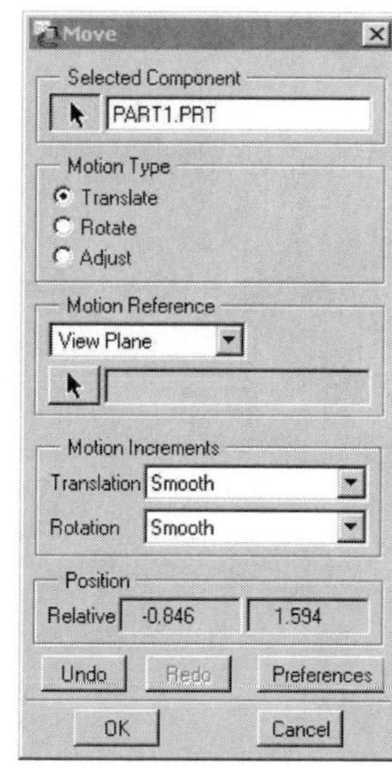

Figure 12–12 Move dialog box

Packaged Components

When a part or subassembly is placed with constraints using the Assemble option, it is considered a **parametric assembly.** Components of a parametric assembly must be fully constrained. If a component is only partially constrained, it is considered a **packaged component.** Pro/ENGINEER provides the option of placing a component directly into a model as a package with the Package option. A packaged component is considered nonparametric.

When using the Package >> Add >> Open option, a component is placed and repositioned with the Move dialog box. Notice in Figure 12–12 how this dialog looks very similar to the Move tab under the Component Placement dialog box. The options available are basically the same.

Placing a Parametric Component

Perform the following steps to place a parametric component

Step 1: From the Component menu, select the ASSEMBLE option.

Step 2: Use the Open dialog box to open a part or assembly.

STEP 3: Use Constraint Types available under the Component Placement dialog box to fully constrain the component.

A parametric assembly must be fully constrained. The Placement Status box on the dialog box tells the current constraint status of the component. The Move Tab is available for the translation and rotation of the component along existing degrees of freedom. A partially constrained component can be placed as a packaged assembly.

STEP 4: Select OKAY on the dialog box to finish the placement process.

MECHANISM DESIGN

Placing components into an assembly using traditional assembly constraints is a powerful and intuitive process. One of the disadvantages of this form of assembly modeling is the necessity to fully constrain (or fix) parent components. Fixed components have zero degrees of freedom to move. When a component is not fully constrained it is considered packaged.

Traditional component placement controls assembly by limiting the degrees of freedom of a component. When placing a component as a connection within mechanism, the component's placement is defined by the definition of a joint type, with the joint type limiting specific degrees of freedom. In many cases, the component is packaged by design. The following joint types are available:

PIN

A Pin joint is the most basic connection (Figure 12–13). It is primarily defined through the alignment of two axes. An additional requirement is the mating of two planes and/or two points to define the joint's translation limitation. This joint type provides one rotational degree of freedom.

CYLINDER

The Cylinder joint is similar to the Pin joint (Figure 12–14). Like the Pin joint, this joint is defined through the alignment of two axes. Unlike the Pin joint, no translation limitation is provided. Two degrees of freedom are provided: one rotational and one linear.

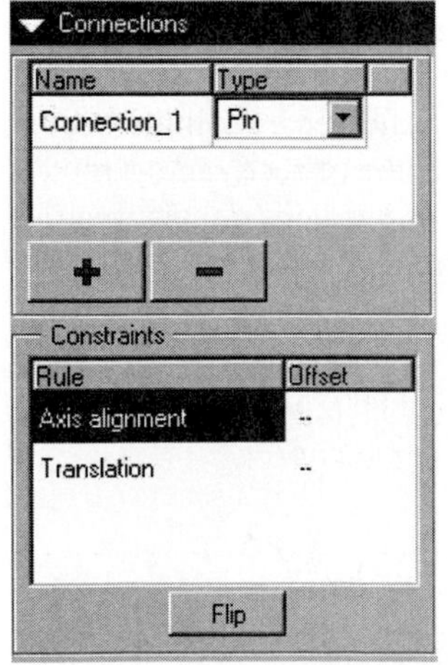

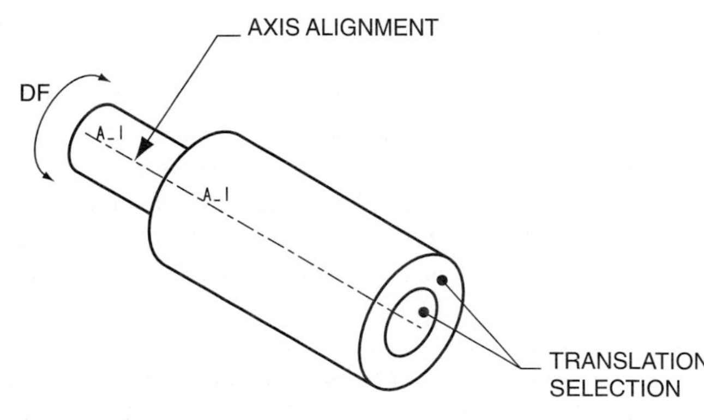

Figure 12–13 Pin joint

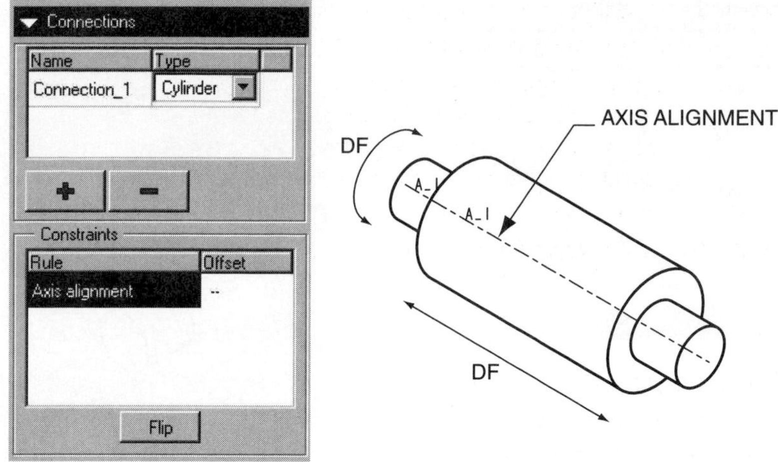

Figure 12–14 Cylinder joint

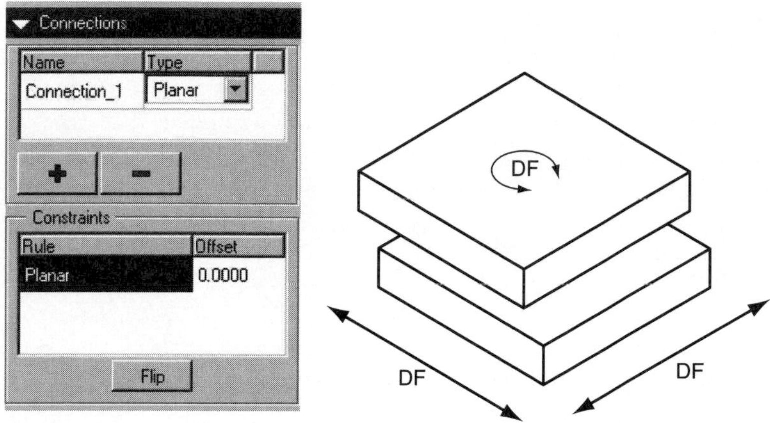

Figure 12–15 Planar joint

PLANAR

The Planar joint connects two planar surfaces (Figure 12–15). It is defined through the selection of a planar surface on each component. It provides three degrees of freedom: two linear and one rotational.

SLIDER

The Slider joint is similar to the Cylinder joint, except the Slider joint does not provide any rotational degrees of freedom (Figure 12–16). Like the Cylinder joint, the Slider joint requires the aligning of two axes. To control the rotational degrees of freedom, two planar surfaces must be mated and/or aligned. Any defining planes must lie parallel to the selected axes. One linear degree of freedom is provided.

BALL

The Ball joint provides three rotational degrees of freedom, but no linear (Figure 12–17). This joint requires the selection of two datum points.

BEARING

The Bearing joint is a combination of the Ball joint and the Slider joint (Figure 12–18). Within this joint, the ball definition is allowed to slide along a selected axis. This joint requires the alignment of two axes and two points. A Bearing joint has four degrees of freedom: one linear and three rotational.

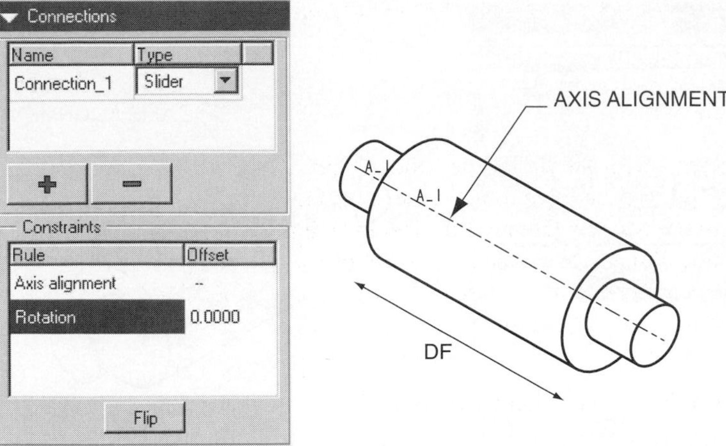

Figure 12-16 Slider joint

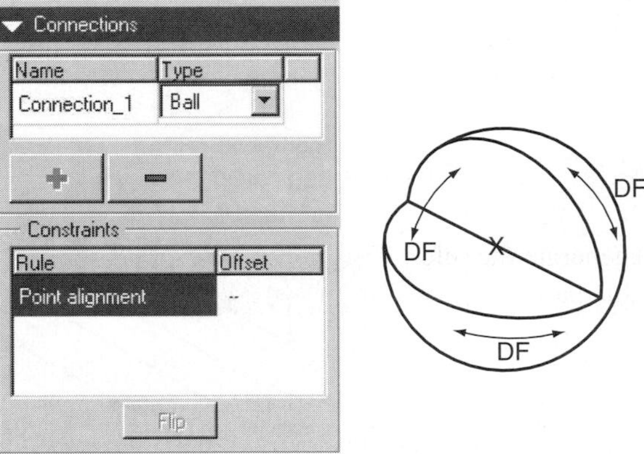

Figure 12-17 Ball joint

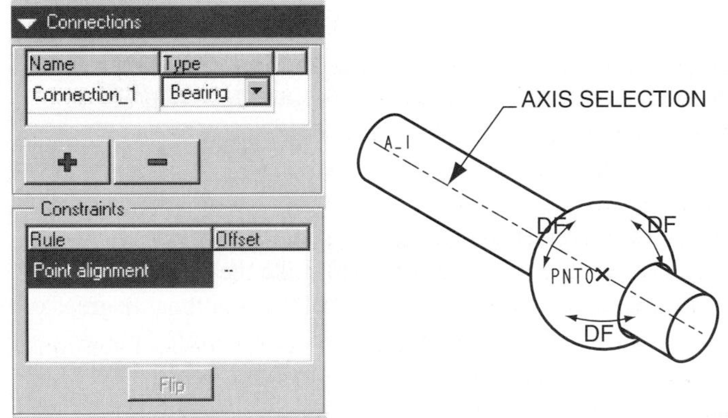

Figure 12-18 Bearing joint

Modifying Assemblies and Parts

Pro/ENGINEER's Assembly mode provides a variety of options for the manipulation and modification of assembly components. Within an assembly, modifiable components can include parts, skeleton models, subassemblies, assemblies, dimensions, and exploded models. Additionally, options are available for modifying dimensions, features, and components.

Modifying Dimensions

Pro/ENGINEER provides an option within the Modify Part, Skeleton, Subassembly, and Assembly menu options for the modification of dimension values. Figure 12–19 shows the Modify Part menu with the Modify Dimension (Modify Dim) option highlighted. When this option (or any other dimension modification option) is selected, Pro/ENGINEER reveals the Modify menu. The Modify menu is similar to the dimension modify menu found in part mode and functions in a similar manner.

Perform the following steps to modify a dimension value

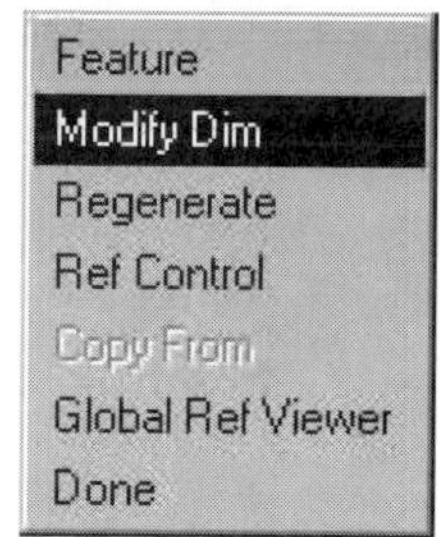

Figure 12–19 Modify Dimension menu

STEP 1: On the Assembly menu, select the MODIFY menu option.

STEP 2: Select the MOD DIM >> VALUE options.

STEP 3: On the work screen, select a dimensions to modify, then enter an appropriate new value.

STEP 4: On the Assembly Modify menu, select the DONE/RETURN option.

STEP 5: On the Assembly menu, select the REGENERATE option.

As in Part mode, Pro/ENGINEER will not update a dimension until the model has been regenerated.

STEP 6: On the Part to Regenerate menu, pick the SELECT >> PICK PART options, then pick the part to regenerate

The Pick Part option will regenerate the selected part. The Automatic option will regenerate every available part.

> **MODELING POINT** A component can be modified by right-mouse picking the component on the model tree. Options available include Modify, Redefine, Reroute, Replace, Suppress, and Delete.

Creating New Part Features

Within Assembly mode, features can be added to parts and to skeleton models. The Modify >> Mod Part >> Feature option is used for creating features within a selected part, and the Modify >> Mod Skel >> Feature option is used for creating features within a selected skeleton model. When a feature is created within a part or skeleton model, it is considered a component feature and will be created in the individual part or skeleton model object file. When creating a component feature in this manner, other components within the assembly can be used as references. This is referred to as an *External Reference*. External References should be used only when design intent dictates.

The Part Feature menu is shown in Figure 12–20 and has various options for creating features. Notice in the figure the available options. Are there any options recognizable from Part mode? The options within the Part Feature menu are virtually identical to the same options available under the Feature menu in Part mode. Solid creation options such as Hole, Round, Chamfer, Cut and Protrusion are available.

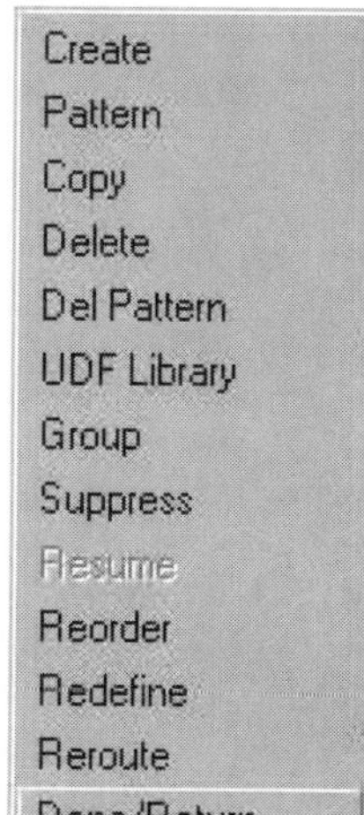

Figure 12–20 Part Feature menu

Redefining a Component Feature

The Redefine option is used in Part mode to modify attributes and elements associated with a part feature. This option is also available for the modification of part and skeleton models within Assembly mode. As with the modification of component dimensions within Assembly mode, features redefined in Assembly mode will also be redefined within their respective component object files. Perform the following steps to redefine a component feature.

STEP 1: **On the Assembly menu, select the MODIFY option.**

STEP 2: **On the Assembly Modify menu, select either the Modify Part (Mod Part) option or the Modify Skeleton (Mod Skel) option.**

STEP 3: **Select a Part or Skeleton model to redefine.**

STEP 4: **Select FEATURE >> REDEFINE.**

STEP 5: **On the Part or Skeleton model, select the feature to redefine.**

After the feature is selected, Pro/ENGINEER will launch the Feature Definition dialog box associated with the selected feature. On the dialog box, select the element to modify, then select the Define option.

CREATING ASSEMBLY FEATURES

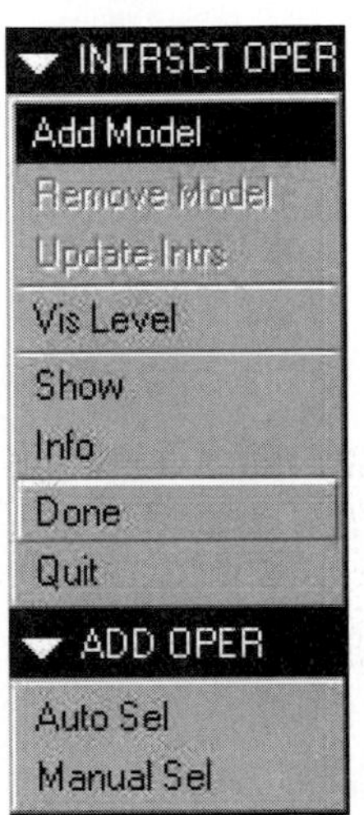

Figure 12-21 Intersection Operations menu

Pro/ENGINEER provides the Feature menu option to create assembly features. Within this option, only negative space solid features such as holes and cuts can be utilized. Datum and surface options are available also. When an assembly feature is created, its visibility level can be set at the assembly level or at the component level. When it is set at the assembly level, the feature is visual only in the assembly model. In other words, each part when opened in Part mode will not be affected. When the visibility level is set at the part level, the feature is created within each part.

When creating an assembly feature such as a hole or a cut, the process for constructing the feature is similar to the process for constructing the same feature in Part mode. The only additional requirement is to specify the feature's intersection components. Pro/ENGINEER provides the Intersection Operation menu to serve this purpose (Fig. 12–21). As shown in the figure, Pro/ENGINEER provides the Automatic Select (Auto Sel) and Manual Select (Manual Sel) options for selecting components to include in the selection set. The Visibility Level (Vis Level) option is used to set the visibility of the feature at the assembly or part level.

TOP-DOWN ASSEMBLY DESIGN

Grouping existing components to form an assembly is referred to as *bottom-up assembly design.* This is a common assembly construction technique when a design consists of existing parts. Figure 12–22 shows a graphical representation of this method of creating an assembly. Within such a technique, existing parts govern the final assembly design.

However, rarely in true conceptual design do individual components govern the makeup of an assembly. As an example, suppose an aircraft company is designing a new style of light single engine aircraft. Input would be provided from a variety of sources to include potential customers, industrial designers, aerospace engineers, manufacturing engineers, and subcontractors. The conceptual design would start with the hull design of the aircraft. The aircraft would have major subassemblies such as the power plant, the constant speed propeller assembly, control linkages, fuel system, and seating. Within top-down assembly design, each major subassembly would be built around and within the conceptual aircraft hull. Each major subassembly would have smaller subassemblies. These subassemblies would be designed around and within their larger parent assemblies. At the end of the top-down design process would be parts. Parts would be designed around the lowest level subassembly in which they reside. This form of design is shown graphically in Figure 12–22.

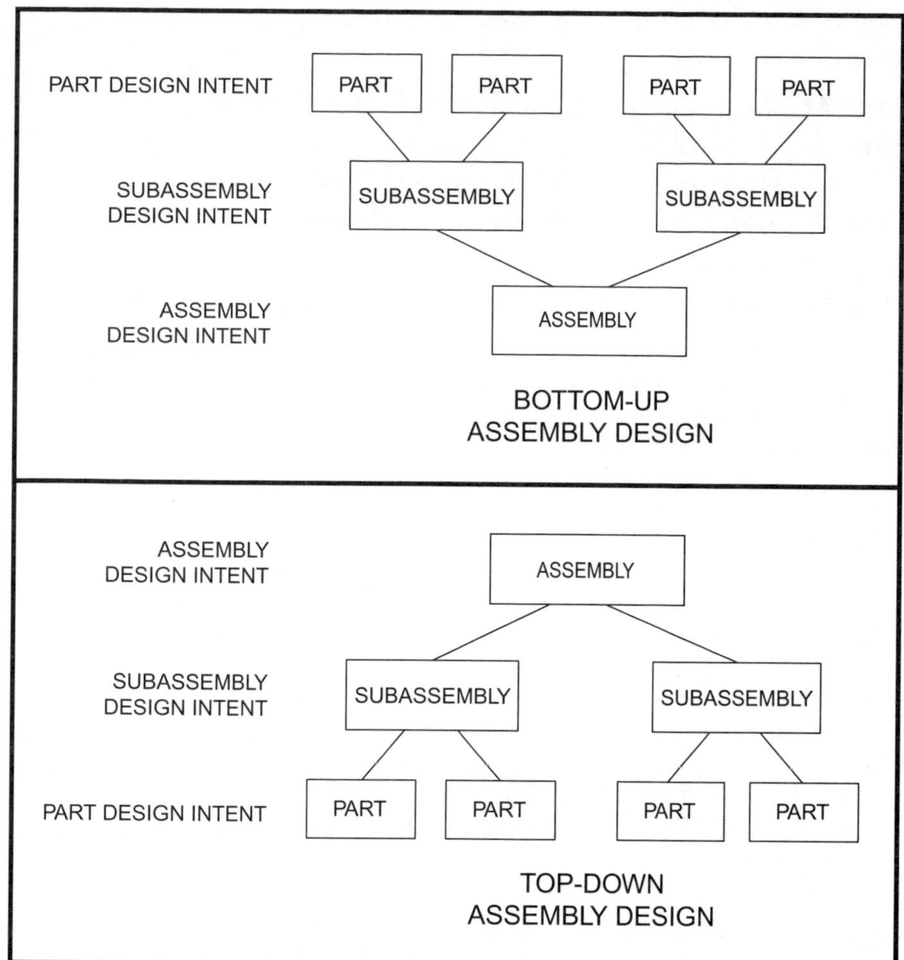

Figure 12-22 Top-down assembly design versus bottom-up

In real world assembly design, a combination of top-down and bottom-up assembly design is utilized. Manufacturing companies often use components common to other designs. Examples of this can be found in the automotive industry. Automotive companies often design cars that share identical major subassemblies, such as the power train. Within Pro/ENGINEER, assemblies can be designed using both top-down and bottom-up design.

Pro/ENGINEER provides multiple capabilities to enhance top-down assembly design. One of these capabilities is the ability to create components within Assembly mode. Another is the utilization of skeleton models.

Creating Parts in Assembly Mode

One of the strengths of Pro/ENGINEER is its ability to create parts directly within assembly mode. Creating parts for an assembly can be tedious at best when modeling within Part mode. Pro/ENGINEER provides the ability to create components directly in an assembly where they will actually function. Figure 12–23 shows an example of a part created in Assembly mode. The first illustration shows an existing part with the sketched section of the part under construction. Within the sketching environment, Pro/ENGINEER allows existing parts, features, datum planes, and axes to be referenced. In this example, the axis of the existing shaft will be referenced and used as the axis of rotation for the revolved pulley feature.

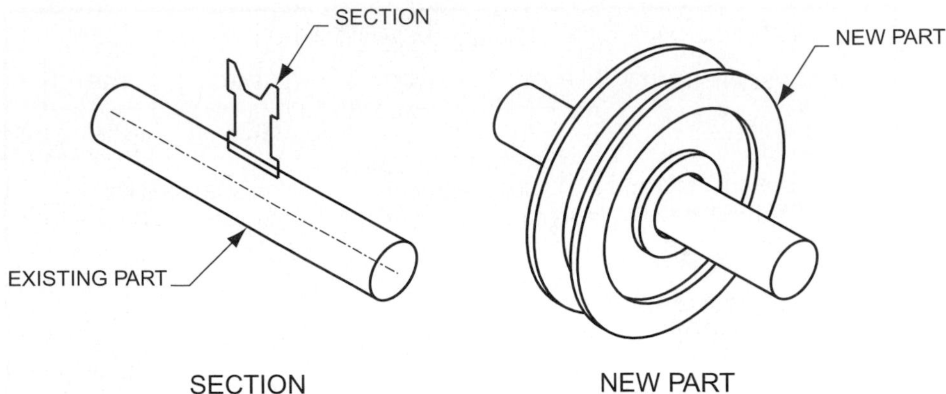

Figure 12–23 Part creation

When creating features within Part mode, new features can reference existing features. When a feature is referenced, it becomes a parent feature for the feature under construction. This same parent/child relationship can exist between parts and features in Assembly mode. When a part within an assembly references another part, an external reference is formed. When creating a part within Assembly mode, care should be taken when creating external references. The Info >> Global ReferenceViewer option can be used to check for the existence of references between parts.

Components can be created in Assembly mode using the Component >> Create option. Once a component (part, subassembly, or skeleton model) has been created, it becomes a part within the assembly. When the assembly is saved, the component is saved as its own object file. Perform the following steps to create a part within Assembly mode.

STEP 1: On the Menu bar, select UTILITIES >> REFERENCE CONTROL.

It is advisable to avoid the creation of external references within the creation of a new component. An external reference is created when one component references another. If a component does have an external reference, the component being referenced must reside in memory to pass necessary reference information.

STEP 2: On the Reference Control dialog box, select NONE and deselect BACKUP FORBIDDEN REFERENCES (as shown in Figure 12–24).

The None option will not allow a component to reference another component.

STEP 3: Select OKAY to exit the dialog box.

STEP 4: Select COMPONENT >> CREATE menu option.

Components consist of parts, subassemblies, and skeleton models. The Component menu is used to assemble existing components or to create new components. Additionally, many of the same options found in Part mode's Feature menu can be found in the Component menu. Examples include Delete, Suppress, Redefine, Reroute, Reorder, Insert Mode, and Pattern. Upon selecting the Create option, Pro/ENGINEER will launch the Component Create dialog box (Fig. 12–25).

MODELING POINT The Component >> Create option can also be access by using the right mouse to select the assembly model's name on the Model Tree.

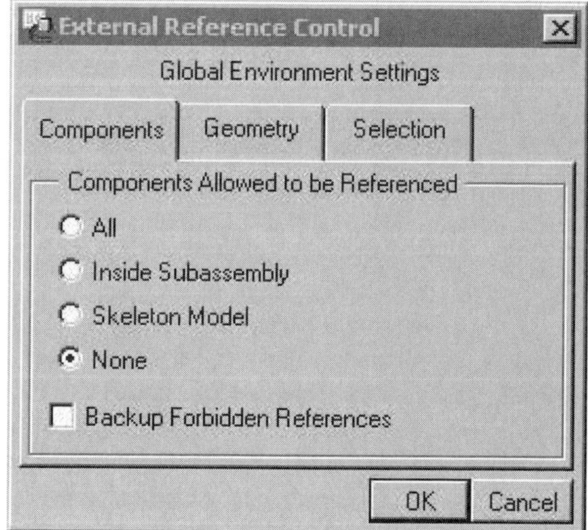

Figure 12-24 External reference control

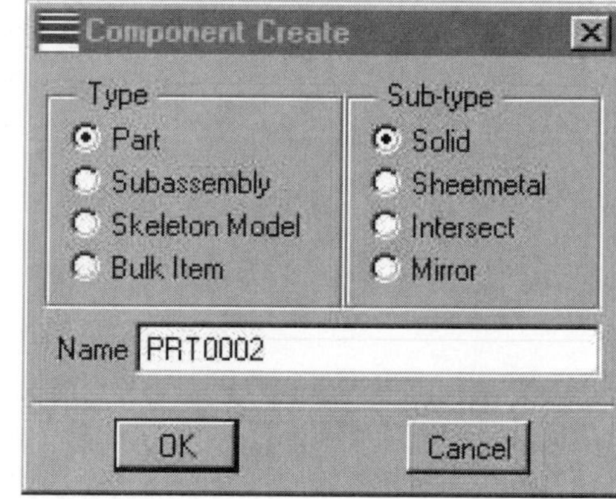

Figure 12-25 Component Create dialog box

STEP 5: On the Component Create dialog box, select PART (Figure 12–25).

The Component Create dialog box allows for the creation of parts, subassembly, and skeleton models. In addition, Bulk Items can be created. Bulk Items are features that are not suitable for modeling but are necessary for the assembly.

STEP 6: Enter a name for the component, then select OKAY.

STEP 7: On the Creation Options dialog box select a component creation method (Figure 12–26).

The following Creation Method options are available:

- **Copy from existing** This option will create a new part from an existing part. If this option is selected, use the Copy-From text box to enter the name of the part to copy. New parts lose their associativity with the parts from which they are copied.
- **Create first feature** This option allows for the creation of the first feature of the part. Typical feature creation options such as Solid, Surface, Datum, and Protrusion are available.
- **Locate default datums** This option creates a new part with its own set of default datum planes.
- **Empty** The Empty option creates a part definition with no geometric definition. The part will be listed on the model tree. Features can be added at a later time.

STEP 8: Use Pro/ENGINEER's component creation tools to create the part.

The feature creation tools available under the Create menu function identically to the same tools found in Part mode. Options available include Solid, Surface, Datum, Protrusion, and User Defined. Existing planar surfaces can be used as sketching planes. Since this is the first geometry feature that defines the part, no negative space feature creation options are available. Existing components and assembly features can be used as references.

INSTRUCTIONAL NOTE For parts with multiple features, continue with the following steps.

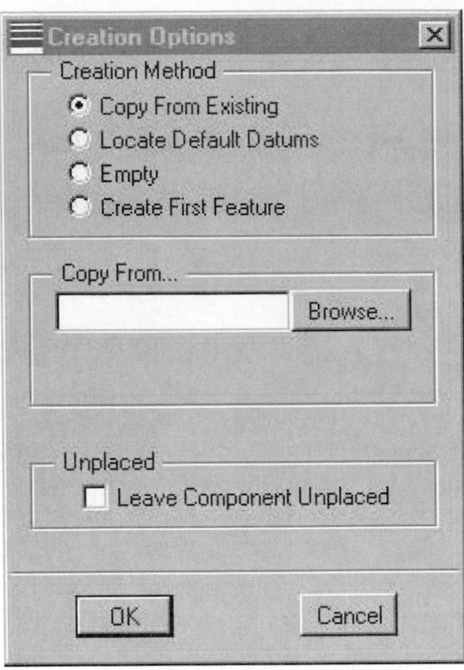

Figure 12-26 Creation Options menu

STEP 9: On the Model Tree, select the part with the right mouse button.

STEP 10: Select FEATURE CREATE on the Pop-Up menu.

STEP 11: Use the Feature Class menu to create additional part features.

SKELETON MODELS

Skeleton models are important components for the proper utilization of the top-down design process. Skeleton models serve as a form of three-dimensional layout and are utilized in a variety of ways. The following is a discussion of the possible uses of a skeleton model.

TOP-DOWN DESIGN CONTROL

Within the realm of top-down assembly modeling, the intent of a design is passed from the upper levels of the design to the individual components. Skeleton models are used to portray and transfer the upper level design intent during the modeling of subassemblies and components.

ASSEMBLY SPACE CLAIM

Top-down assembly design usually requires larger and more exterior components to be designed before the design of small and more specific components. As an example, an automobile's exterior would probably be designed before the front passenger seats. During the designing of the seats, a designer must work within the space allocated by the exterior design. Skeleton models can be utilized to represent claimed space for major design subassemblies.

SHARING OF INFORMATION

In large manufacturing companies, different teams design major subassemblies of a design. Skeleton models can be used to pass design information from one major subassembly to another. As an example, within the design of a new automobile, a company might elect to use an existing design for the engine, but use a new design for the transmission. The existing engine can be incorporated into the transmission

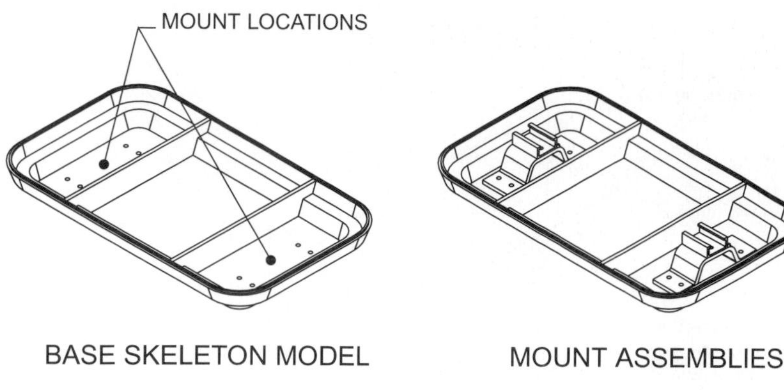

Figure 12-27 Skeleton model

design as a skeleton model. Feature information such as hole and shaft locations can be passed to the new design without the worry of creating external references.

Motion Control

The motion of an assembly can be designed and controlled through the use of skeleton models. True skeleton components constructed from datum axes, curves, and components can be created that serve as the "skeleton" of a subassembly. The relative motion of each component can be designed and modified using the skeleton components. When the design is optimized, the actual components can be created around the skeleton.

An example of the use of a skeleton model is shown in Figure 12–27. In this example, the base skeleton model represents the external shell of a product. It was designed separately from internal components. The problem presented here requires the designing and modeling of the mount assembly. Locations for the two required mount assemblies are shown in the first illustration. Within Pro/ENGINEER's Assembly mode, the base model was created in the assembly as a skeleton model using the Copy-From-Existing creation option. This option places a copy of the base model as a skeleton within the assembly. This allows the mount bracket to be modeled within its allocated space and around the existing mount holes. By using the skeleton model, no external references are created between the original base model and the mount assembly.

As shown in the above example, a skeleton model can be created as a copy from an existing part. The Component >> Create option has the capability to create skeleton models from the first feature also. Most of the modeling tools available for the creation of a part in assembly mode are available for the creation of a skeleton model. Only one skeleton model can be included in an assembly.

Assembly Relations

In Part mode, the Relations option is used to create a relationship between two dimension values. In Assembly mode, the Relations option can be used to create dimensional relationships between dimensions within a part or between two parts. The rules used to define relations in Assembly mode are similar to the rules for applying relations in Part mode.

One difference between adding relations in an assembly when compared to a part is in the dimension symbols. In Part mode, a dimension symbol consists of the letter *d* followed by the dimension number. In Assembly mode, the dimension symbol is followed by the Session_ID for the component (Figure 12–28).

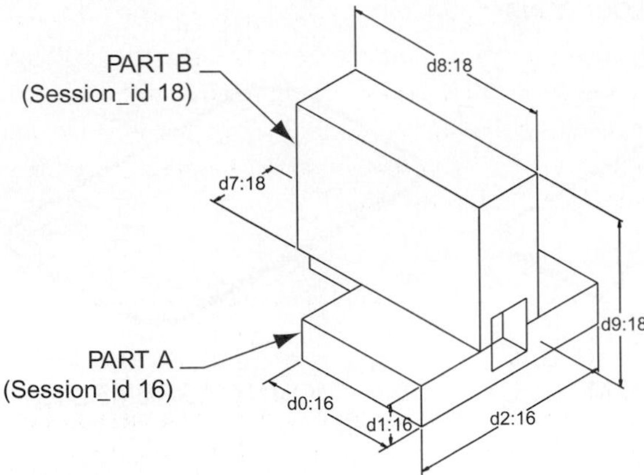

Figure 12-28 Dimension symbols

Notice in Figure 12–28 the format for the dimension symbols in Assembly mode. The assembly consists of two parts: Part A and Part B. The Session_ID for Part A is 16, and the Session_ID for Part B is 18. When a dimensional relationship is created for a part, only the dimension symbol needs to be entered. In this example, the Part Relations (Part Rel) suboption would be selected. When a dimensional relationship is created for an assembly, the full dimension symbol, including the Session_ID, needs to be entered. In this example, the Assembly Relations (Assem Rel) suboption would be used.

LAYOUT MODE

Layout mode is used to create two-dimensional layouts of an engineering design. It is helpful for the capture of design intent and for developing the general layout of a design. The sketch creation options available in Layout mode work identically to the two-dimensional entity creation options found in Drawing mode. See Chapter 8 for detailed information on creating a two-dimensional drawing.

A layout can be created and used to drive the dimensional values of a part or assembly. Figure 12–29 shows an example of a layout and an associated assembly model. When constructing a layout, dimension values are created and assigned a dimension symbol and value. Unlike Part and Assembly modes, Layout mode requires the definition of a user-defined symbol name. By declaring a layout in an assembly, selected dimensions of an

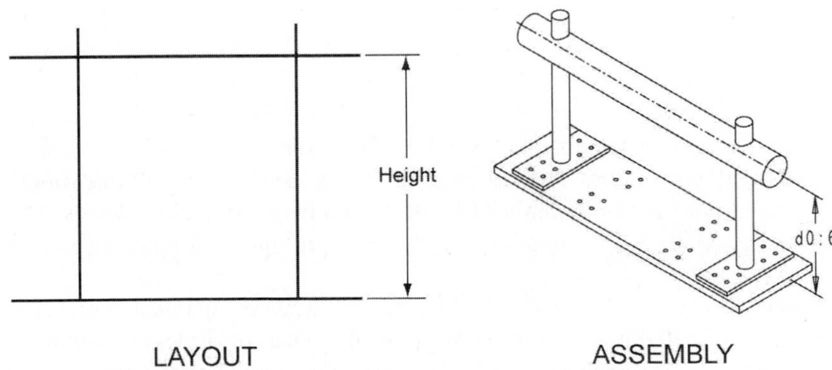

Figure 12-29 Layout with assembly

assembly can be related to the dimensions within a layout. As an example, the "Height" symbol shown in the layout it used to drive the "d0:6" dimension in the assembly. When the Height dimension's value is changed in the layout, the driven dimension in the assembly is changed. In Assembly mode, the Setup >> Declare >> Declare Lay option is used to link a layout with an assembly. Once a layout as been declared, the layout's dimension symbols can be used with Assembly mode's Relations option.

Simplified Representation

Assembly modeling necessitates the combining of many different parts and subassemblies. The more components added to an assembly and the more complex an assembly, the longer Pro/ENGINEER takes to regenerate and retrieve models. Assemblies are often composed of major subassemblies, minor subassemblies, and parts. During the design process, certain subassemblies and parts may not be needed. As an example, when designing an aircraft, a design team modeling a seat assembly probably would not need the power plant assembly. Pro/ENGINEER's Simplified Representation option allows selected components to be removed from the display. This allows for faster retrieval and regenerations.

Simplified Representations are created in Assembly mode with the Simplified Representation (Simplfd Rep) option. There are three types of representations available: master representation, graphics representation, and geometry representation.

Master Representation

The Master Representation option excludes selected components from the display. Components that are accepted for exclusion are removed from the work screen but remain on the model tree. Figure 12–30 shows the model tree during the exclusion of selected components.

Graphics Representation

The Graphics Representation option includes selected components on the work screen, with the remaining components displayed in a wireframe display format. Figure 12–31 shows an example of an assembly with a graphics representation. The parts in the shaded display format were selected for exclusion from the simplified

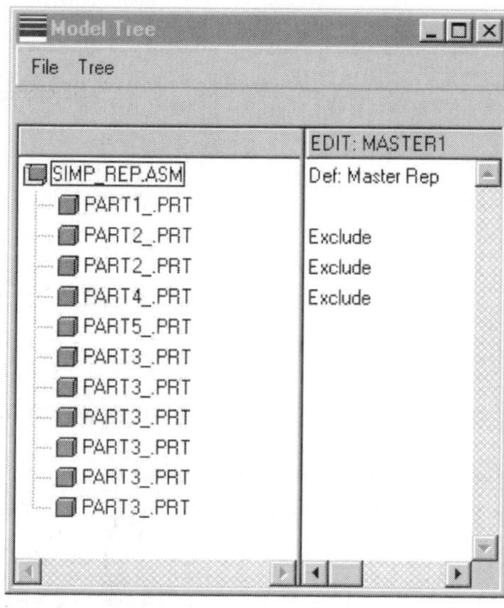

Figure 12–30 Model tree

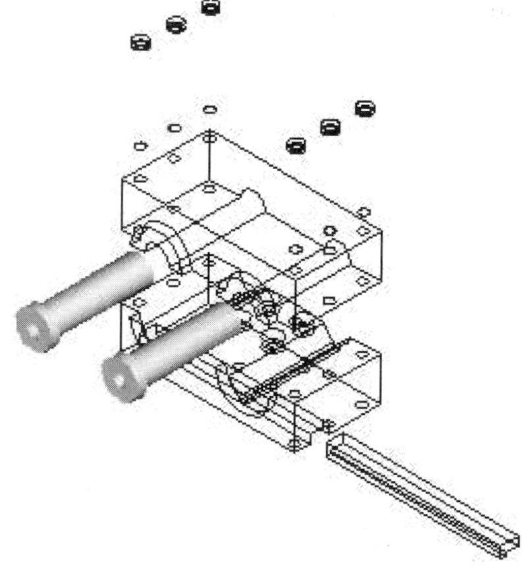

Figure 12–31 Graphics representation

representation and can be modified. The wireframe models cannot be modified. The wireframe display is a Pro/ENGINEER default and can be changed with the configuration file option *save_model_display.*

Geometry Representation

The Geometry Representation option includes selected components on the work screen, with the remaining components displayed in Pro/ENGINEER's current display style. Components excluded from the master representation can be referenced within an assembly. This type of representation takes longer to regenerate when compared to the graphics representation.

Creating a Simplified Representation

Perform the following steps to create a simplified representation.

Step 1: Select SIMPLFD REP >> CREATE.

Other options available under the Simplified Representation menu include Set Current, Copy, Redefine, Delete, and List. The Set Current option is used to set an established simplified representation current within the work screen.

Step 2: In Pro/ENGINEER's textbox, enter a name for the simplified representation.

Step 3: Select MASTER REP as the default rule.

The Master Rep option will exclude selected components from the display.

Step 4: With the EXCLUDE option selected, on the work screen or on the model tree pick components to exclude from the display.

Step 5: Select the DONE option.

Step 6: Use the Simplified Representation menu's SET CURRENT option to set a specific representation.

Exploded Assemblies

When components are added to an assembly, they are placed in their functional orientation and located. Often, this state of viewing an assembly can be confusing and less descriptive. Within the technical language of engineering graphics, assembly drawings are used to display the location of assembled components. To make the assembly drawing legible, the assembly can be exploded to separate components. Figure 12–32 shows an illustration of an exploded and an unexploded assembly. The View >> Explode option is used to explode a view. The ExplodeState option of the Assembly is used to create and set explode states. Multiple explode states can be created. The Set Current option is used to set a specific explode state.

Creating an Explode State

Perform the following steps to create an explode state for an assembly.

Step 1: On the Assembly menu, select the EXPLODESTATE option.

Step 2: Select CREATE on the Explode State menu.

Step 3: In Pro/ENGINEER's textbox, enter a name for the explode state.

Step 4: On the Explode Position dialog box, select TRANSLATE as the Motion type (Figure 12–32).

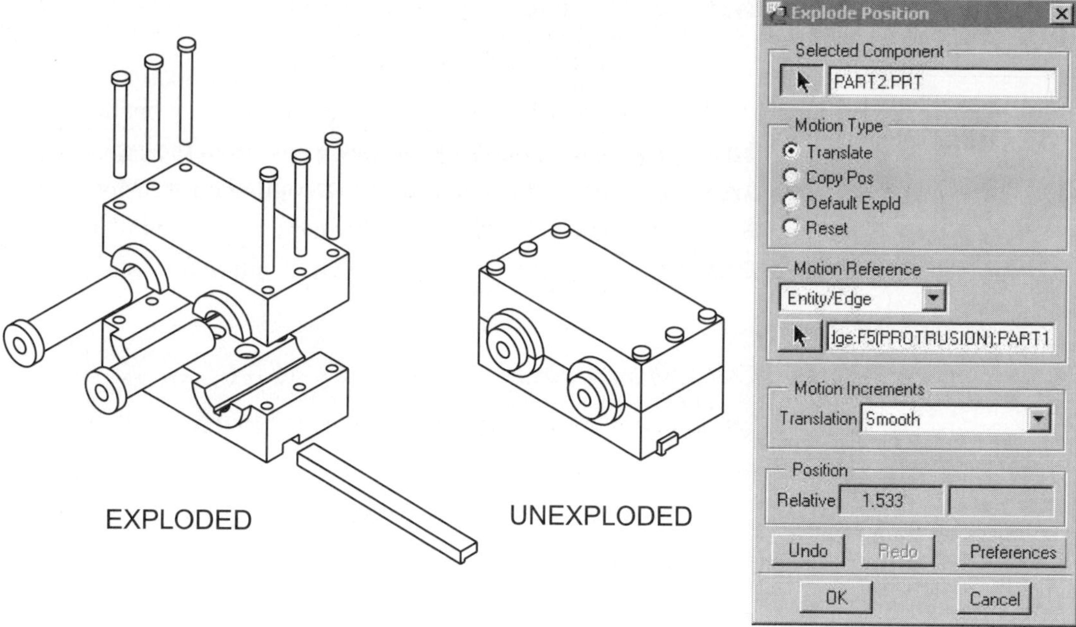

Figure 12-32 Exploded view

STEP 5: On the Explode Position dialog box, select a Motion Reference.

The setting of an explode state's components is similar to the movement of packaged components. Pro/ENGINEER provides the following translation options:

- **View plane** The motion will be relative to the current screen orientation. When this type is selected, any selected components movement will be parallel to the work screen.
- **Sel plane** When this type is selected, the motion will be parallel to a selected plane. Planar surfaces and datum planes can be selected.
- **Entity/edge** When this type is selected, the motion will be along the path of a selected axis, edge, or curve. The type is useful for confining the movement of a component along one axis.
- **Plane normal** When this type is selected, the motion will be perpendicular to a selected plane.
- **2 points** When this type is selected, two selected points on the work screen are used to created the relative motion. This type is useful for confining the movement of a component along one axis.
- **Csys** The motion will be relative to the X-axis of a selected coordinate system.

STEP 6: On the work screen, select an entity or plane relevant to the motion reference.

Your selection on the assembly model is based upon the motion reference selected in step 5. With the exception of the View Plane reference, you will select an entity, point, or plane on the work screen relevant to the current reference. As an example, if you select the Entity/Edge motion reference, you will select either an axis, edge, or curve. The selected motion reference will remain current until it is changed.

STEP 7: On the work screen, select and move a component.

STEP 8: Continue to move components on the work screen or change motion types.

Continue moving components on the work screen until the explode state is complete. You can change motion types to optimize the state.

STEP 9: Select OKAY on the dialog box when the explode state is complete.

STEP 10: Select the DONE/RETURN option on the Modify Explode menu.

STEP 11: Select DONE/RETURN on the Explode State menu.

STEP 12: Use the VIEW >> EXPLODE option to explode the view.

> **MODELING POINT** An exploded view can be added as a view in Drawing mode using the Add View >> Exploded option.

SUMMARY

Pro/ENGINEER is more than just a modeling application. It is a true engineering design package. One of the capabilities that separates Pro/ENGINEER from mid-ranged computer-aided design and drafting applications is its ability to model an entire engineering design. With appropriate modules of Pro/ENGINEER, a product can be designed, modeled, detailed, simulated, and manufactured. While Pro/ENGINEER's Part mode is the main component for low end design work, Assembly mode is the module for complete product design.

TUTORIAL

ASSEMBLY TUTORIAL

This tutorial will explore Pro/ENGINEER's basic bottom-up assembly modeling capabilities. In addition, the creation of exploded views and the creation of an exploded assembly drawing will be covered. The final product of this tutorial is shown in Figure 12–33. Within this tutorial, the following topics will be covered:

- Creating components for an assembly.
- Placing components into an assembly.
- Creating an exploded assembly.
- Creating an assembly drawing.

CREATING COMPONENTS FOR AN ASSEMBLY

During bottom-up assembly design, parts for an assembly are created using normal part modeling tools and techniques. As with any parametric model, the intent of the design needs to be considered. In addition, how components of an assembly will be parametrically linked is another important consideration. Often, extra datums will have to be added to a component to simulate the correct function of the design.

The first segment of this tutorial will require the modeling of the six parts that comprise the assembly. Notice in Figure 12–33 that the actual assembly consists of 11 parts. Two of the parts are used multiple times. Within Pro/ENGINEER, a component can be

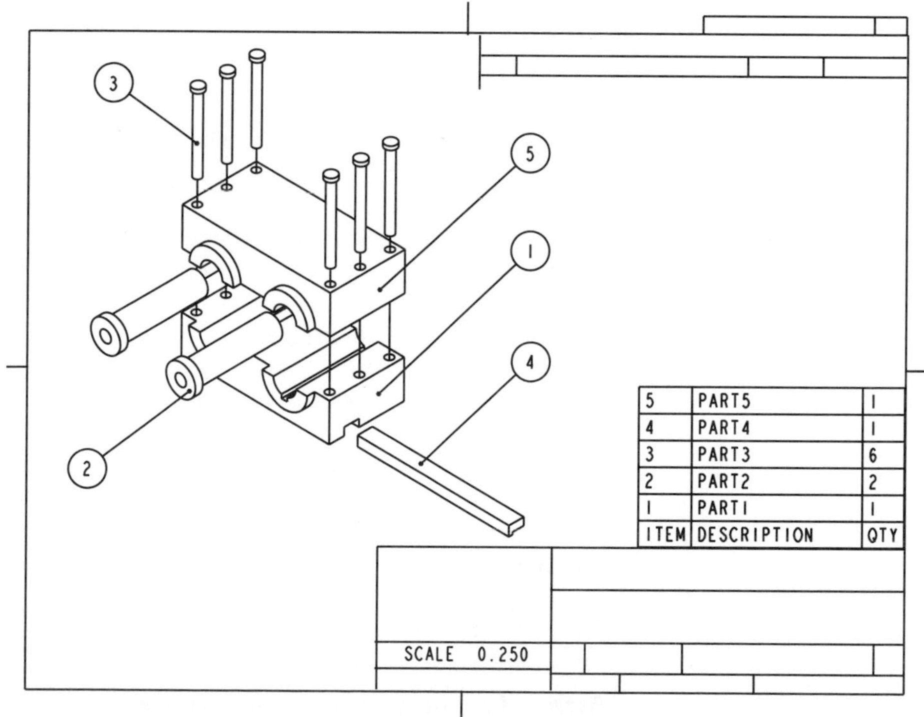

5	PART5	1
4	PART4	1
3	PART3	6
2	PART2	2
1	PART1	1
ITEM	DESCRIPTION	QTY

Figure 12–33 Assembly drawing

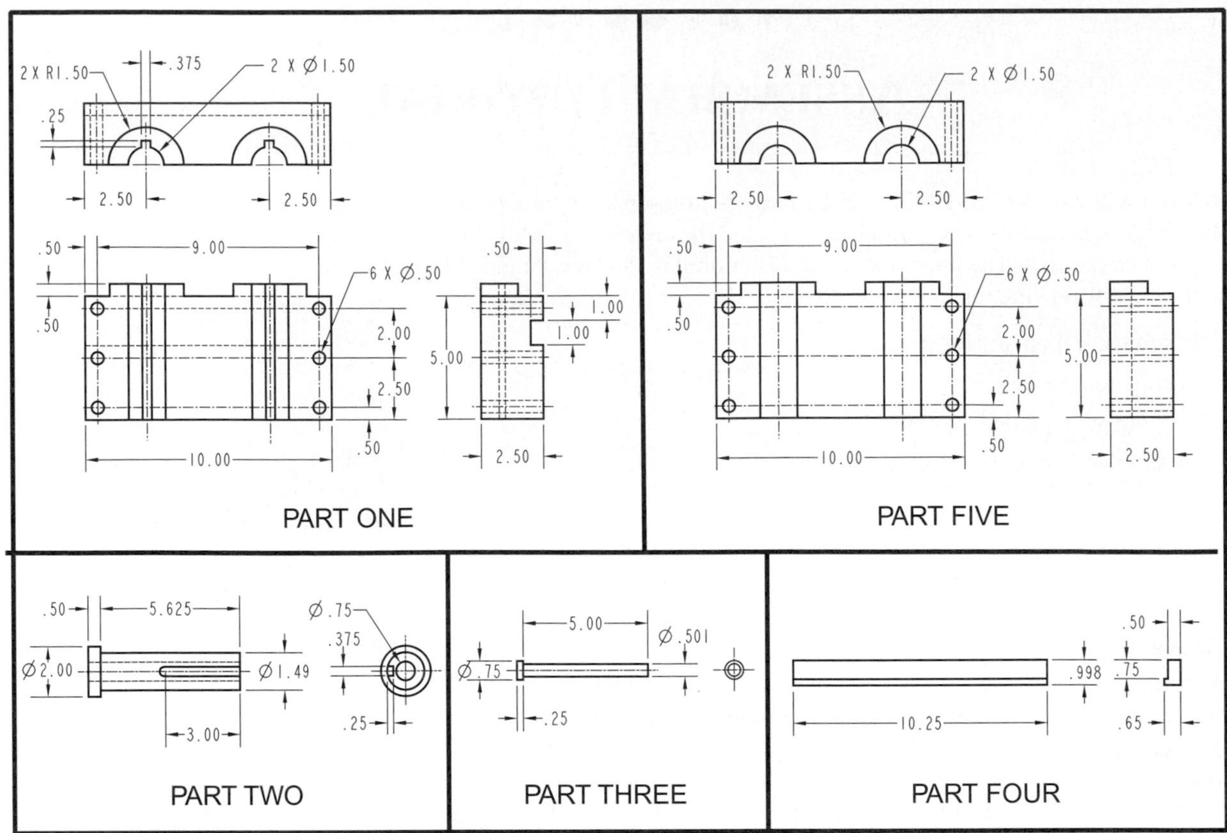

Figure 12–34 Assembly parts

placed multiple times into an assembly. Figure 12–34 shows the parts used in this tutorial. Use Part mode to model each part. The dimensions shown in each illustration represent the design intent. Incorporate this intent into each part.

Modeling Pointers

Notice the similarities between Part One and Part Five (Figure 12–34). The only difference between the two parts is the absence of three cut features in Part Five. If Part One is created first, you can use the New File Options dialog box as shown in Figure 12–35 to create Part Five as a copy of Part One. After creating the copy, delete the cut features. The New File Options dialog box is accessible by deselecting the Use-Default-Template on the New dialog box.

Create the six 0.50-inch diameter holes in Part One and Part Five as a patterned hole. Within the assembly, Part Three will be inserted into each hole. The Component >> Pattern >> Reference Pattern option can be used to pattern the first instance of Part Three to place the remaining five. This saves time in the component placement process and also meets the design intent for this assembly.

Placing Components into an Assembly

This segment of the tutorial will place the parts into the assembly. Part One will be placed first followed by Part Five, Part Two, Part Three, then Part Four.

Step 1: Use the New option to create a new Assembly object named *main_assembly* (use the Default Template file).

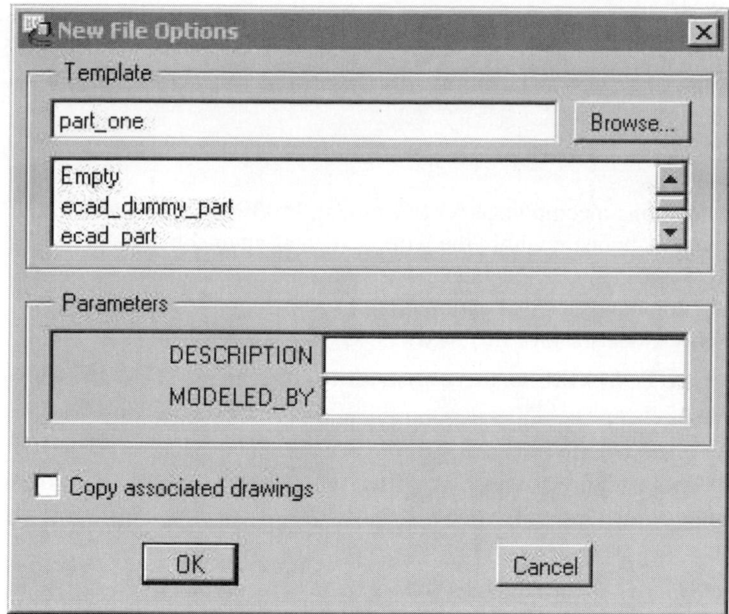

Figure 12–35 New dialog box

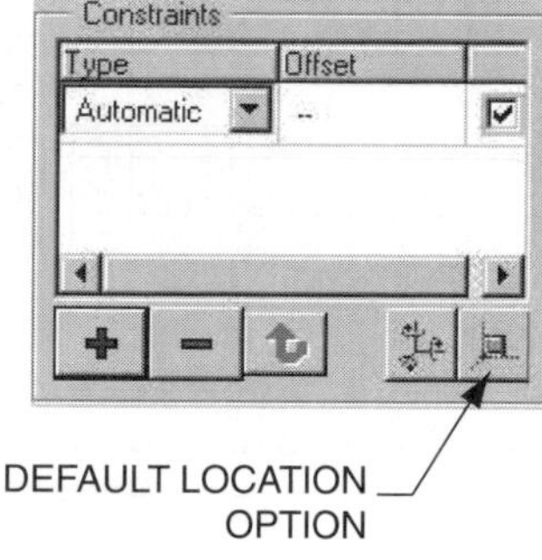

Figure 12–36 Default Location option

STEP 2: Select the COMPONENT option on the Assembly menu.

The Component menu is used to assembly, create, or manipulate parts and subassemblies. Since no components currently exist within the assembly model, only two options are available: assemble and create.

STEP 3: Select the ASSEMBLE option on the Component menu.

The Assemble option is used to place existing components (parts and subassemblies). The Create option is used to create components (parts, subassemblies, and skeleton models) within the context of the assembly object.

STEP 4: Use the Open dialog box to open the *Part_One* component.

After opening the part, Pro/ENGINEER will launch the Component placement dialog box. This dialog box is used to constrain components within the context of the assembly. The Part_One component will be constrained to the set of default datum planes.

STEP 5: Select the Default Location icon on the Component Placement dialog box (Figure 12–36).

The Default Location option will align your component's default datum planes with the assembly model's default datum planes.

STEP 6: Select OKAY on the Component Placement dialog box.

STEP 7: Select the COMPONENT >> ASSEMBLY option.

On the Component menu, notice the options that are available that were not available during the first component placement. The Delete, Suppress, Insert Mode, and Pattern options function similarly to the same options in Part mode. The Package option is used to place a component without parametric constraints.

STEP 8: Use the Open dialog box to open the *Part_Five* component.

INSTRUCTIONAL POINT When a component is opened, it is placed into the work screen referenced to how it was created in Part mode. The initial component placement illustrations within this tutorial may not match how your components are initially oriented.

After selecting a component for placement, Pro/ENGINEER will initially place the component within the work screen as an unconstrained model. Pro/ENGINEER will also launch the Component Placement dialog box. This dialog box is used to parametrically constrain a model to existing assembly models and/or features.

STEP 9: Select AUTOMATIC as the Constraint type (Figure 12–37).

Automatic is the default constraint type on the Component Placement dialog box. With this constraint, references are selected for both the component and the assembly. Pro/ENGINEER will determine the constraint to apply, but will provide you with the option of selecting an alternative constraint, such as Mate in this example.

STEP 10: Select the two Mate surfaces shown in Figure 12–38.

STEP 11: Under the Constraints section of the dialog box, change the previously created constraint from an aligned constraint to a MATE constraint (see Figures 12–38 and 12–39).

The Automatic option will normally assign an Align constraint between two selected surfaces. You should change this constraint type to a Mate.

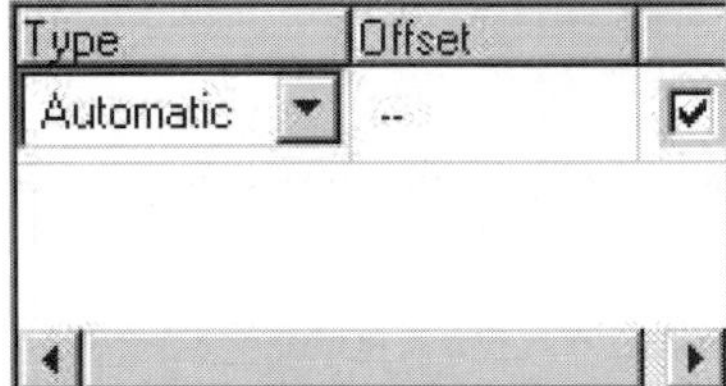

Figure 12–37 Automatic constraint selection

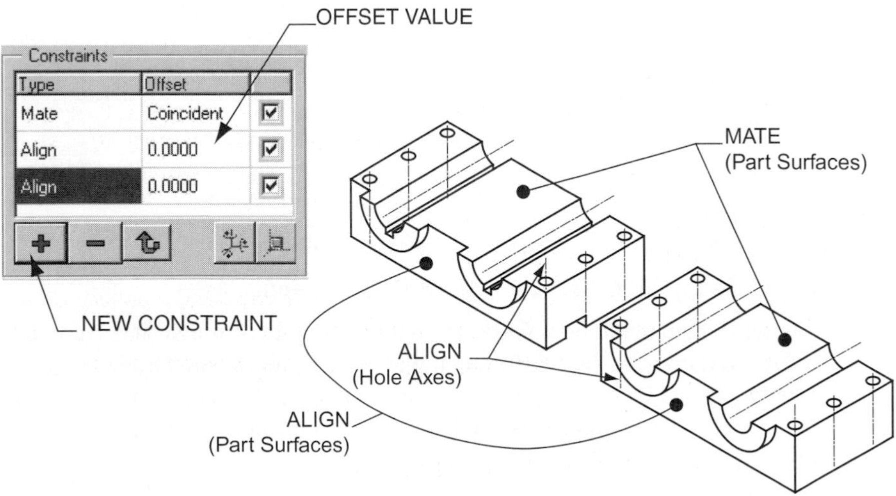

Figure 12–38 Constraint type selection and placement

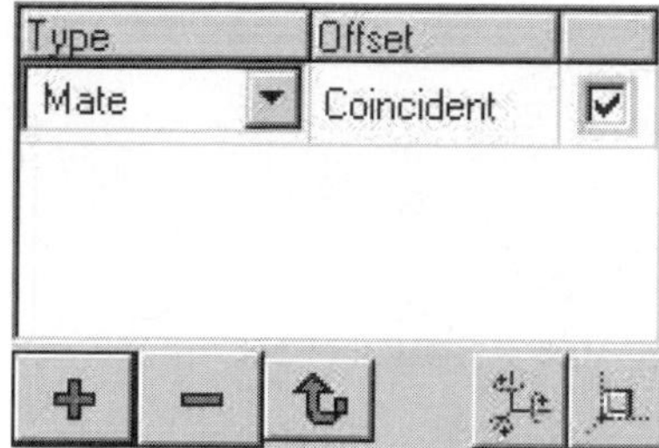

Figure 12–39 Changing constraint types

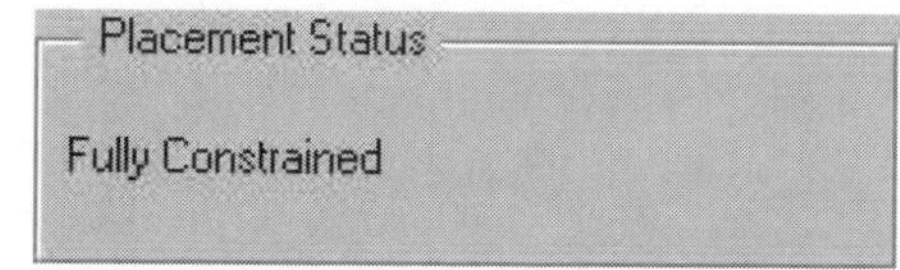

Figure 12–40 Placement status

STEP 12: **Select the NEW CONSTRAINT icon on the dialog box (Figure 12–38).**

STEP 13: **With Automatic as the constraint type, pick the two Align surfaces shown in Figure 12–38.**

The next constraint will align the front surfaces of the parts as shown in the figure.

STEP 14: **If necessary, enter 0.00 as the offset value.**

Aligned and Mated surfaces can have an offset value. When two surfaces are offset, they will remain parallel but will be separated by the user-specified offset value.

When a constraint type is added to the component, the model and work screen change to reflect this constraint. Use dynamic rotation (control key and middle mouse button) to better visualize and select the components. If the components overlap during placement, use options under the Move Tab to adjust the temporary location of the component being placed. When the component is fully constrained, the dialog box will provide the message shown in Figure 12–40.

STEP 15: **With Automatic as the constraint type, pick the two Align axes shown in Figure 12–38.**

The next constraint will align the two hole axes shown in the figure.

STEP 16: **When the component is fully constrained, select OKAY on the dialog box.**

A component that is fully constrained is a parametric model. Pro/ENGINEER does allow components to be placed that have no constraints or that are only partially constrained. This is referred to as a *packaged* component. The Component menu's Package option can be used to directly place a packaged component or to constrain a packaged component.

STEP 17: **Use the ASSEMBLE option to open *Part_Two* (Figure 12–41).**

STEP 18: **Use the Align constraint and the Oriented Offset option to orient the slot surface shown in Figure 12–41.**

Orient the bottom surface of the key slot on Part Two with the bottom of Part One. This constraint combination will orient the surfaces in the same direction. You must manually set the Oriented option as shown in the Figure.

STEP 19: **Select the NEW CONSTRAINT icon.**

STEP 20: **MATE the back of the head of Part Two with the front surface of the boss feature on Part One (use a 0.00 offset value).**

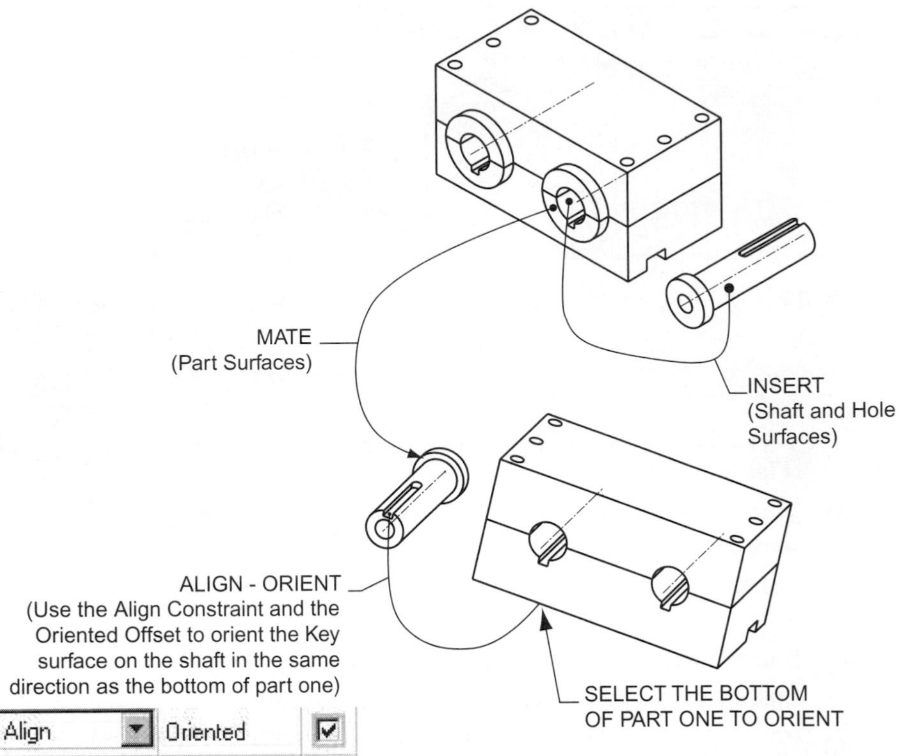

Figure 12-41 Part Two placement

> **MODELING POINT** If your components overlap causing the selection of surfaces and entities to become difficult, you can use the Move tab to temporarily reposition components.

STEP 21: Using the INSERT constraint type, pick the surface of the shaft of Part Two, then the surface of the hole feature of Part One.

The Insert constraint will align the centerlines of each revolved feature.

STEP 22: When your part is fully constrained, select the OKAY option.

STEP 23: Use the same constraint types from the first instances of *Part_Two* to place the second instance of the part (refer to Figure 12–33).

STEP 24: Use the ASSEMBLE option to place *Part_Three* (Figure 12–42).

Place the constraint types in the following order:

- **Mate** Mate the back of the head of Part Three with the top surface of Part Five.
- **Insert** Insert the surface of the shaft of Part Three into the surface of one on the six hole features on Part Five.
- Allow the assumption shown in the Placement Status menu (Figure 12–43). When two revolved features are constrained together, Pro/ENGINEER will assume the constraint around the axis of revolution.

STEP 25: Select the COMPONENT >> PATTERN option.

Six instances of ***Part_Three*** exist in the assembly. Instead of placing each instance of the part, you can pattern the first instance around the reference

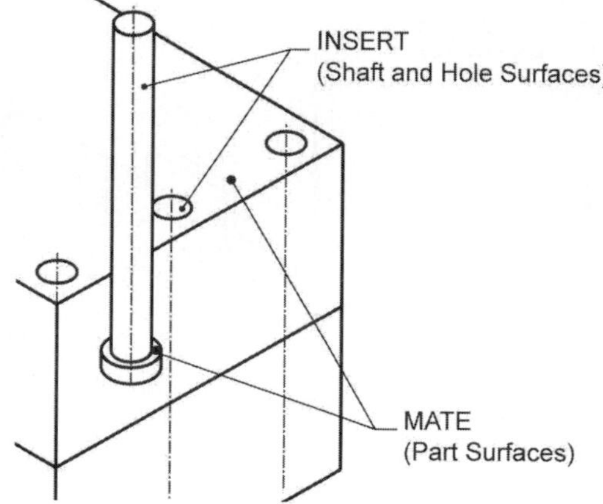

Figure 12–42 Part Three placement

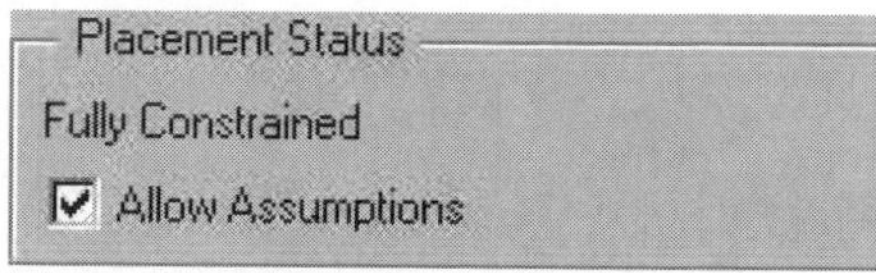

Figure 12–43 Allow assumptions

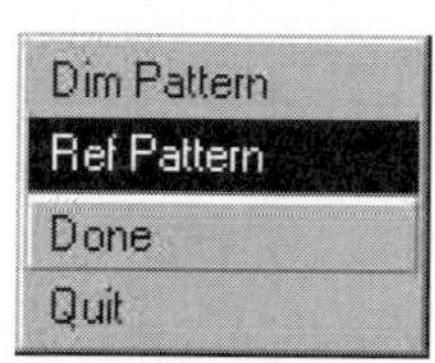

Figure 12–44

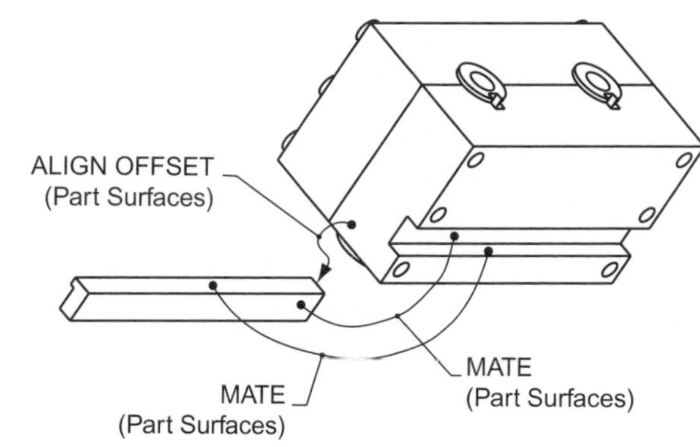

Figure 12–45 Placing Part Four

pattern used to create the holes in Part mode. This technique will only work if you created the six holes using the Pattern option. If you created the holes using a different technique, you must place each ***Part_Three*** instance individually.

STEP 26: On the work screen or on the model tree, select the first instance of *Part_Three.*

STEP 27: On the PRO PAT TYPE menu (Figure 12–44), select the REF PATTERN option, then select DONE.

STEP 28: Use the ASSEMBLE option to place *Part_Four* (Figures 12–45 and 12–46).

Place the constraint types in the following order:

- **Mate** Mate the bottom of Part Four with the bottom of the cut slot.
- **Mate** Mate the back of Part Four with the side of the cut slot.
- **Align offset** Use the Align constraint and the Offset suboption to offset the end of Part Four from the side of Part One (Figure 12–46).

STEP 29: Save your assembly file.

Your complete assembly should appear as shown in Figure 12–47. Observe the components on your Model Tree. Modify, feature creation, and feature manipulation options can be accessed by clicking the right mouse button to pick the component on the Model Tree.

Type	Offset	
Mate	Coincident	☑
Mate	Coincident	☑
Align	0.1250	☑

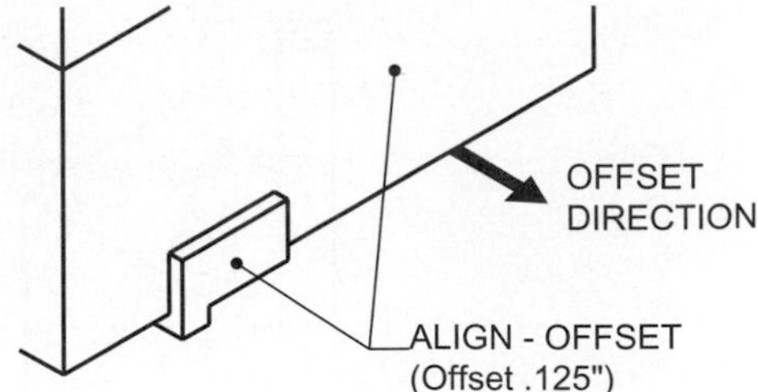

Figure 12-46 Align offset

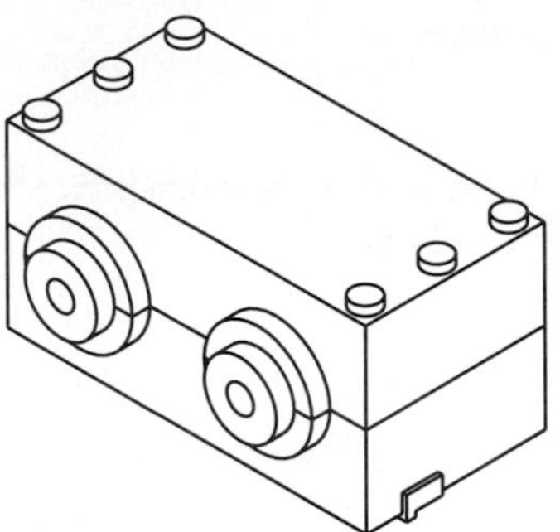

Figure 12-47 Finished assembly

Creating an Exploded Assembly

This segment of the tutorial will create the explode state shown in Figure 12–48. Within any one assembly object, multiple explode states can be created.

Step 1: Select VIEW >> EXPLODE on the Menu bar.

It is rare that an initial exploded view will be adequate for presentation purposes. This tutorial will use the Explode State option to adjust the orientation of components. Once the Explode option is selected, the View menu changes this selection to Unexplode.

Step 2: Select VIEW >> DEFAULT ORIENTATION.

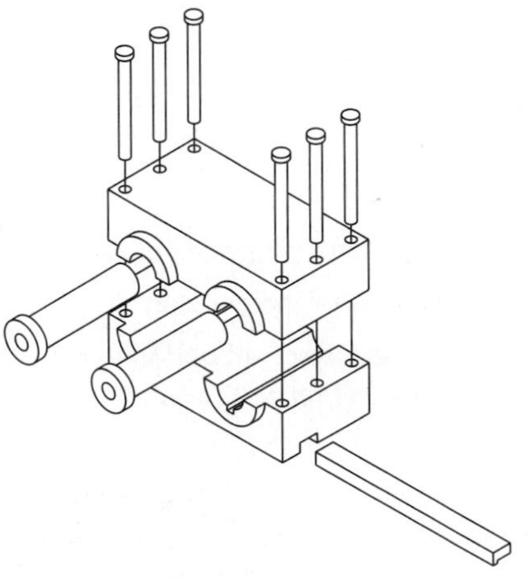

Figure 12-48 Exploded assembly

STEP 3: Select the EXPLODESTATE option on the Assembly menu.

STEP 4: Select CREATE on the Explode State menu.

STEP 5: In Pro/ENGINEER's textbox, enter *EXPLODE1* as the name for the explode state.

After entering the name of the explode state, the model will appear unexploded on the work screen. Despite the display of the model on the work screen, the model is still considered exploded.

STEP 6: On the Explode Position dialog box, select ENTITY/EDGE as the Motion Reference (Figure 12–49).

STEP 7: Select TRANSLATE as the Motion Type (Figure 12–49).

STEP 8: Select the entity edge shown in Figure 12–49.

STEP 9: Select and move the component as shown.

STEP 10: On the Explode Position dialog box, select the Motion Reference pick icon, then pick the entity edge shown in Figure 12–50.

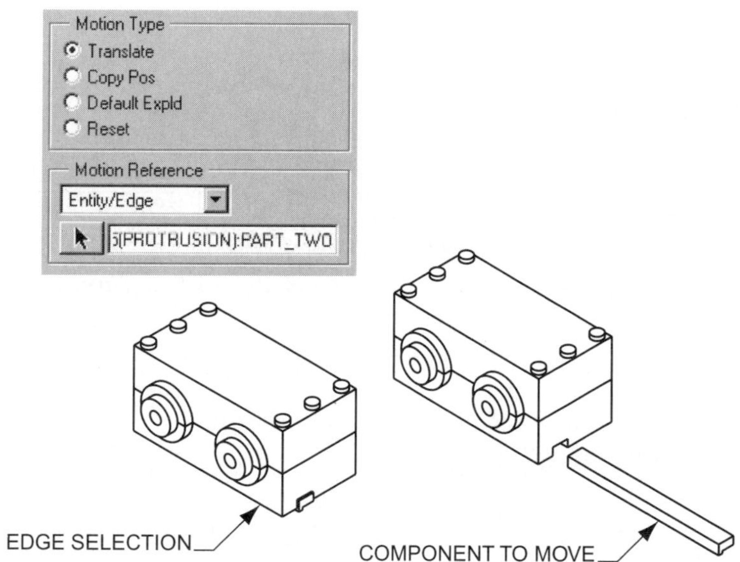

Figure 12–49 Entity/Edge motion

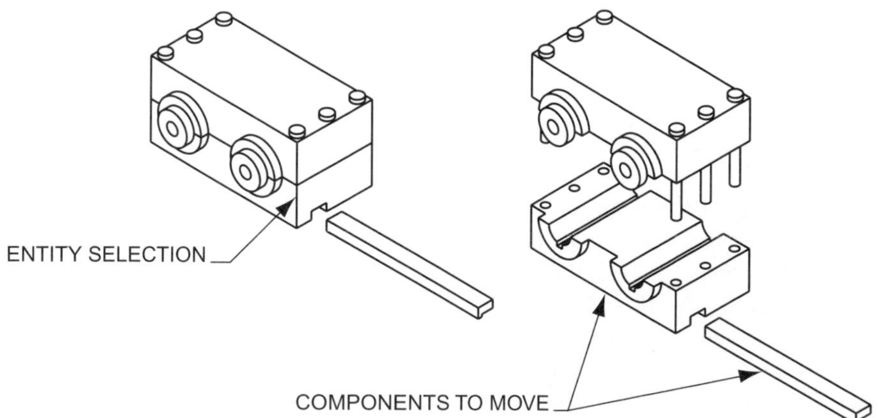

Figure 12–50 Entity/Edge motion

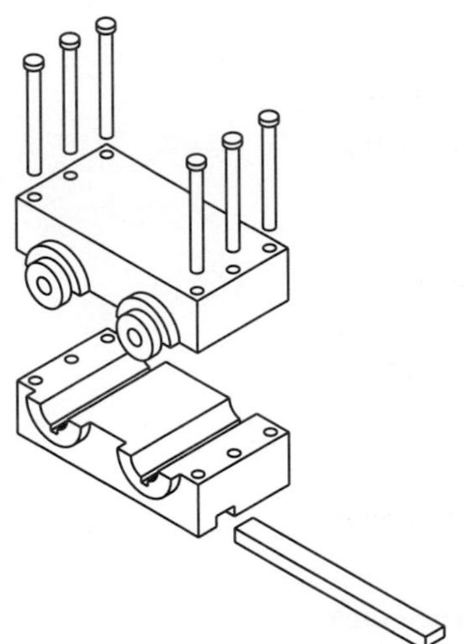

Figure 12-51 Entity/Edge motion

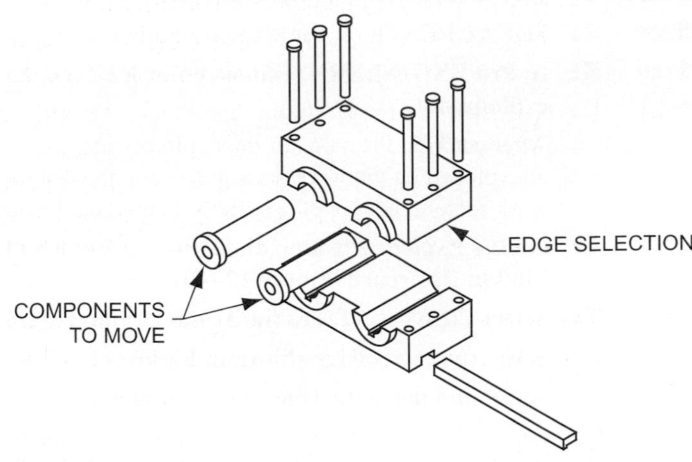

Figure 12-52 Entity/Edge motion

This step will define a needed motion reference. You must first reselect the reference through the pick icon located on the dialog box. As a note, the Entity/Edge selection works with edges and axes. Within this step, you could also pick one of the six vertical axes available within the model.

STEP 11: Move the two components shown in Figure 12–50.

STEP 12: Individually, move each instance of *Part Three* to the locations shown (Figure 12–51).

Once an entity motion type has been selected on the Motion Preference menu, this type remains current until changed. In this case, the previous entity edge selected remains current.

STEP 13: Define the edge shown in Figure 12–52 as the motion reference and move the two components shown.

STEP 14: Select OKAY to exit the Explode Position dialog box.

STEP 15: Select DONE/RETURN to exit the Model Explode menu, then select DONE/RETURN to exit the Explode State menu.

STEP 16: Select VIEW >> UNEXPLODE on the menu bar.

STEP 17: Save your assembly object file.

CREATING AN ASSEMBLY DRAWING (REPORT)

This segment of the tutorial will take the exploded assembly created in the last two segments and create an assembly drawing through the use of the Report module. Assembly views of an object can be placed into a drawing or report using the same procedures for placing a part view. In addition to the assembly view, a bill of material and balloon notes will be created. Pro/ENGINEER, through its Pro/REPORT module, allows reports such as a bill of materials to be placed into a drawing with full associativity. In this tutorial, a bill of materials will be created with a table through the use of a repeat region. A repeat region allows a table to expand to incorporate a list of all the components of an assembly.

STEP 1: Start a New Report object file named *Assembly*.

Use the File >> New option and select Report as the mode to use. Enter *Assembly* as the name of the file, then select OKAY.

STEP 2: On the New Report dialog box, select the options shown in Figure 12–53.

Select the assembly file created in this tutorial as the Default Model. Select the Empty-with-Format option, then Browse to find an A size format. By selecting an A size format, the drawing sheet will be set to the size of the format.

STEP 3: Select OKAY to accept the New Report Dialog Box options.

After selecting OKAY, Pro/ENGINEER will launch a new report session. Notice how the options in Report mode are similar to the options in Drawing mode.

STEP 4: Select TABLE >> CREATE on the Report menu.

The first step in creating an assembly drawing is to define the bill of materials table and repeat region.

STEP 5: Select the ASCENDING >> LEFTWARD >> BY NUM CHARS options on the Table Create menu.

STEP 6: Select the table start point shown in Figure 12–54.

STEP 7: On the work screen, select four Number Characters for the first column (Figure 12–54) followed by 20 Number Characters for the second column and five Number Characters for the third column.

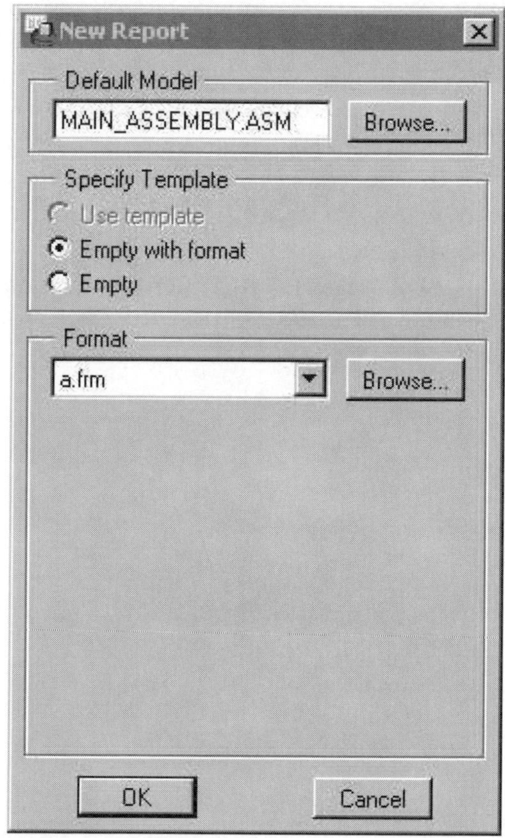

Figure 12-53 New Report dialog box

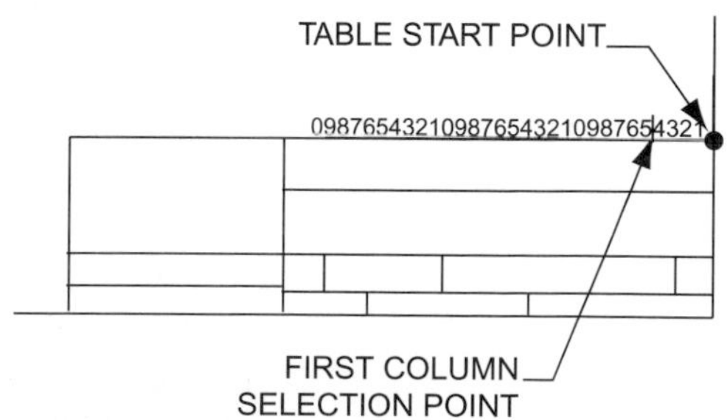

Figure 12-54 Table creation

When the By-Number-Characters (By Num Chars) option is selected, Pro/ENGINEER provides numerical characters on the work screen to define the width of each column. On the work screen, select the spacing for each column.

STEP 8: On the Table Creation menu, select DONE to end the creation of the table's columns.

STEP 9: Using the BY NUM CHARS option, create two rows each one character high.

STEP 10: On the Table Creation menu, select DONE to end the creation of the table.

STEP 11: Select TABLE >> ENTER TEXT.

STEP 12: Create the column headers shown in Figure 12–55.

Within each individual cell, enter the header text shown in the figure (e.g., ITEM, DESCRIPTION, and QTY).

STEP 13: Select the REPEAT REGION option on the Table menu.

STEP 14: From the Table Regions menu, select the ADD option; then select the Repeat Region Start and End cells shown in Figure 12–55.

STEP 15: Select the ENTER TEXT option.

STEP 16: Select the REPORT SYM option on the Enter Cell menu.

Pro/ENGINEER and Pro/REPORT use report parameters to assign associative data to table cells. In this tutorial, you will assign parameters that define each component's item number (&rpt.index), description (&asm.mbr.name), and quantity (&rpt.qty). You can enter each parameter directly into a cell, or you can select each parameter's options from the Report Sym menu.

STEP 17: On the work screen, select the cell above the ITEM header (Figure 12–56).

STEP 18: On the Report Sym menu, select the *RPT . . .*, then *INDEX* options.

These two selections will add the &rpt.indx parameter to the first cell of the first row of the table. Notice on the work screen how the parameter has been added. Do not worry if the parameter crosses into the next cell.

STEP 19: On the Enter Cell menu, select the REPORT SYM option; then on the work screen, select the cell above the DESCRIPTION header (Figure 12–56).

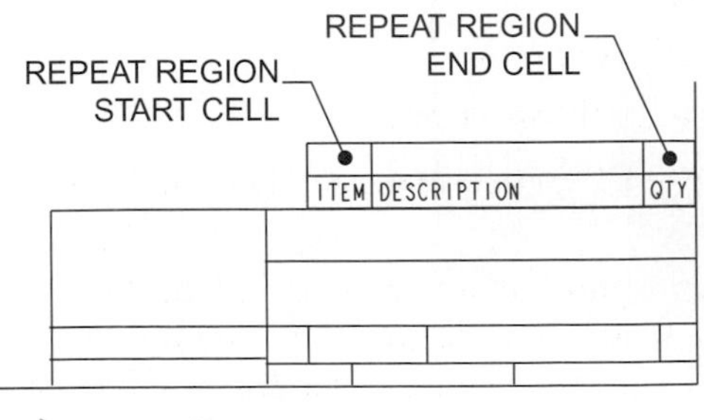

Figure 12-55 Repeat region creation and column headers

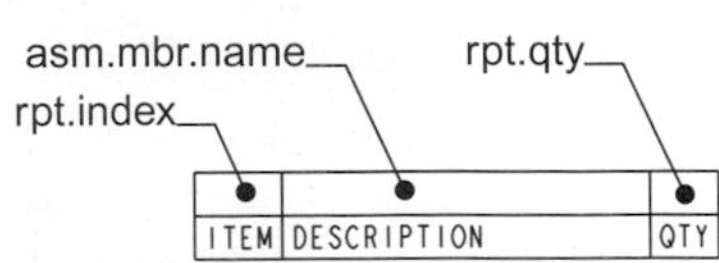

Figure 12-56 Report parameters

STEP 20: On the Report Sym menu, select the *ASM . . .*, then *MBR . . .*, then *NAME* options.

These selections enter the assembly member's name parameter into the second cell.

STEP 21: Enter the *RPT.QTY* (component quantity) parameter into the cell above the QTY header (Figure 12–56).

STEP 22: Select DONE/RETURN to exit the Table menu.

STEP 23: Select VIEWS >> ADD VIEW.

STEP 24: Select the GENERAL >> FULL VIEW >> NoXsec options.

STEP 25: Select the EXPLODED >> SCALE options then select DONE.

The Exploded option will place an exploded view into the drawing.

STEP 26: On the work screen, select the location for the exploded view.

STEP 27: Select the explode state to place, then select DONE.

Any defined explode state can be placed into a drawing.

STEP 28: Enter .20 as the scale for the view.

STEP 29: Select DONE/RETURN to exit the Views menu.

STEP 30: On the Orientation dialog box, select OKAY to placed the view with the default orientation.

After placing the view, notice on the work screen how the table is expanded to include all components of the assembly. Currently, the table is displaying every component on a separate row, even if it is a duplicate. You will change this in the next step.

NOTE: You may not actually see any component names in the repeat region table. They will be revealed later.

STEP 31: Select TABLE >> REPEAT REGION.

STEP 32: Select the ATTRIBUTES option; then on the work screen select the repeat region table.

STEP 33: Select NO DUPLICATES >> DONE/RETURN.

The No Duplicates option will not duplicate a component on the model tree. After selecting the Done/Return option, notice how the table shrinks. The next several steps of the tutorial will have you add the bill of materials balloon notes. Once these balloon notes are added, the bill of materials table will be updated with each component member.

STEP 34: Select the BOM BALLOON option on the Table menu; then on the work screen select the repeat region table.

After selecting the region, verify in the message area that balloon attributes have been added to the repeat region.

STEP 35: Select the SHOW >> SHOW ALL options.

STEP 36: Select DONE/RETURN to exit the Table menu.

STEP 37: Use Pop-Up menu's MOD ATTACH option to modify each balloon's attachment method.

The Pop-Up menu is accessible by preselecting a balloon with the left mouse button, then right-mouse selecting the balloon. The Modify Attachment menu has options for modifying the selected leader's arrowhead type (Arrowhead, Dot, Filled Dot, etc.) and attachment point (On Entity, On Surface, Midpoint, and Intersect).

STEP 38: Save your report object.

The final report drawing is shown in Figure 12–57.

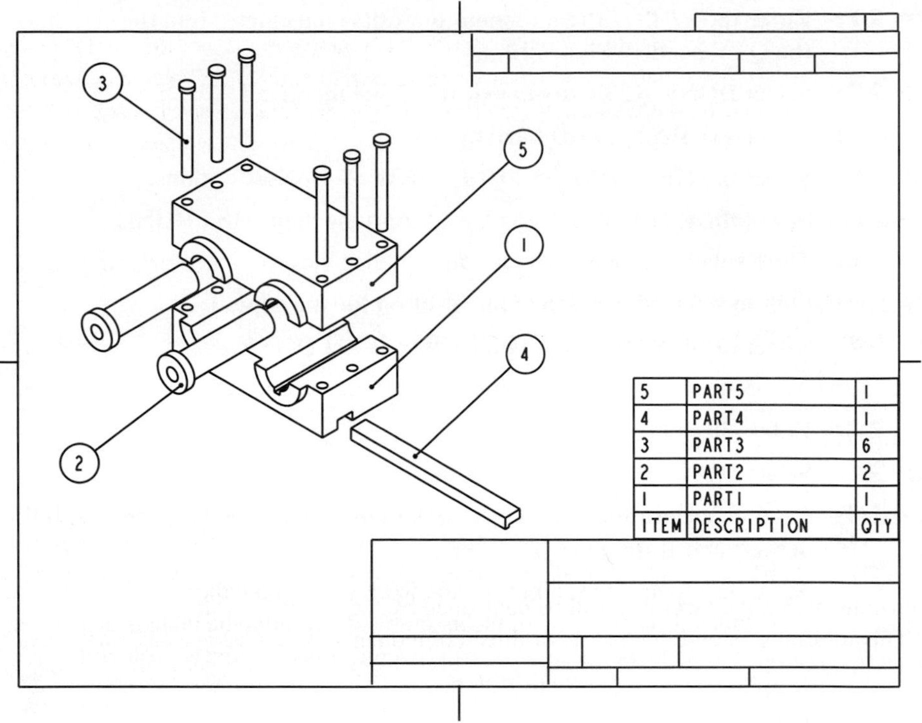

Figure 12–57 Finished assembly drawing

TUTORIAL

Top-Down Assembly Tutorial

This tutorial will cover basic principles of Pro/ENGINEER's top-down assembly design capabilities. Assembly mode allows components (parts, subassemblies, and skeleton models) to be created within the assembly environment. The final assembly for this tutorial is shown in Figure 12–58. This tutorial will cover the following principles of top-down assembly design:

- Creating a layout to capture design intent.
- Creating parts within assembly mode.
- Controlling external references within assembly mode.
- Declaring and using a layout within assembly model.

Creating a Layout

Layouts are similar to an engineer's design notebook. They are useful for capturing design intent and for controlling an assembly from the top down. Within this tutorial, you will create a layout that will help to capture the design intent of the pulley assembly. Key dimensions within the assembly will be controlled by the layout.

This tutorial's assembly consists of three components: a base part, a shaft part, and a pulley part. As shown in Figure 12–59, you will sketch line entities within layout mode to represent these components. When sketching in layout mode, the actual sketched size of the representation is not important. What is important is creating dimensions that will be incorporated within the assembly model. Perform the following steps to create the layout shown in Figure 12–59.

Step 1: Use FILE >> NEW to create a new layout object named *PULLEY*.

Step 2: On the New Layout dialog box, select the EMPTY WITH FORMAT option with an A size standard size sheet; then select OKAY.

Step 3: Use the LINE option to create the horizontal and vertical lines shown in Figure 12–60.

The lines do not have to be a precise length.

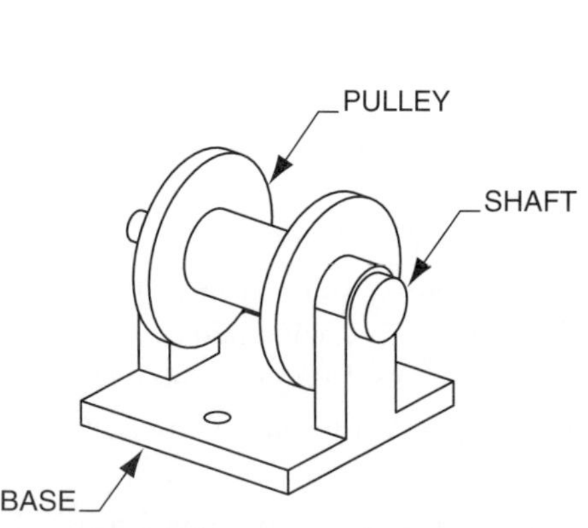

Figure 12–58 Final assembly model

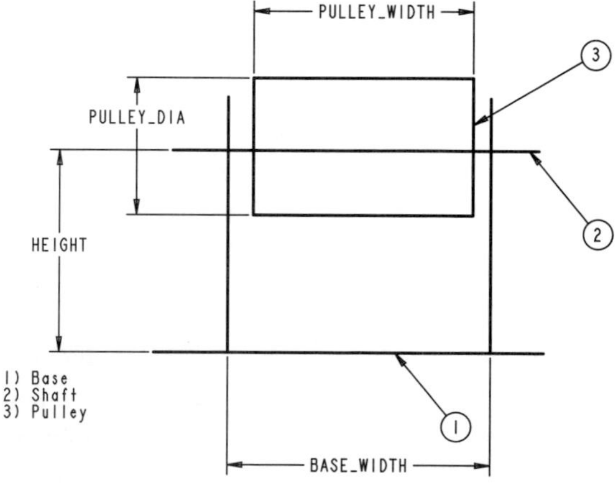

Figure 12–59 Layout of assembly

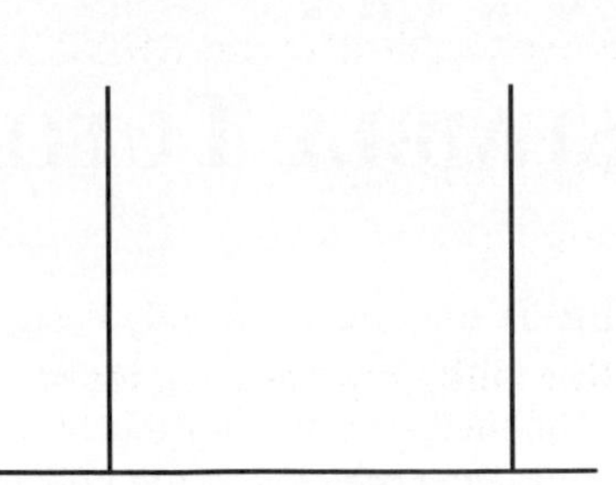
Figure 12–60 Base sketch

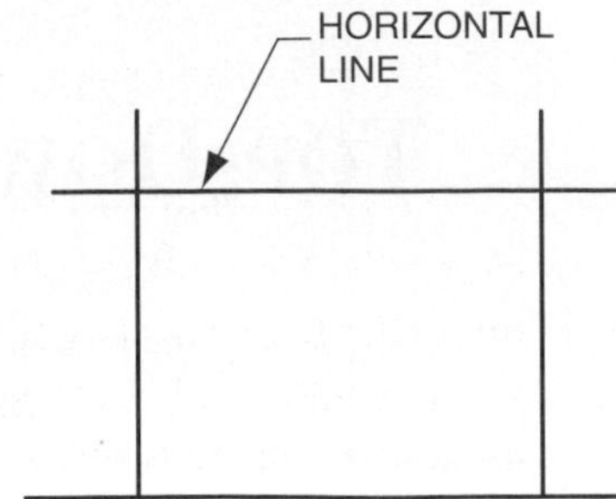

Figure 12–61 Shaft sketch

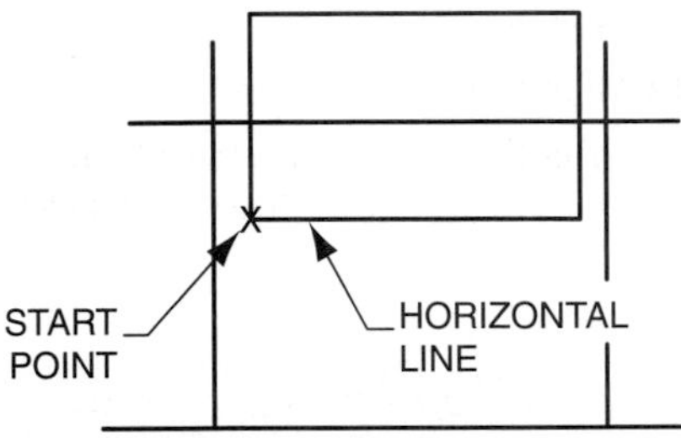

Figure 12–62 Pulley sketch

STEP 4: If necessary, close the Snapping References dialog box.

STEP 5: Select TOOLS >> GROUP >> CREATE.

You will group the three previously created line entities to define them as one component.

STEP 6: On the work screen, pick the three line entities; then select DONE SEL.

STEP 7: Enter *BASE* as the name for the group.

STEP 8: Select DONE/RETURN to exit the Tools menu.

STEP 9: Use the LINE option to create the horizontal line shown in Figure 12–61.

This line represents the shaft part. The line can be constrained horizontal with the Pop-Up menu's Specify Angle option.

STEP 10: Select the LINE icon.

The next sketched entities will represent the Pulley. Within the sketching of this feature, you will create a chain of line entities.

STEP 11: Select the SKETCH CHAIN icon.

The representation of the pulley component will consist of four line entities (see Figure 12–62). The Sketch Chain option will allow you to sketch multiple connected line entities, forming a polyline.

STEP 12: Sketch the four line entities representing the pulley as shown in Figure 12–62.

After sketching the horizontal line, notice how Pro/ENGINEER switches to the Vertical Line option.

STEP 13: When the Pulley representation is complete, deselect the SKETCH CHAIN icon.

STEP 14: If necessary, close the Snapping References dialog box.

STEP 15: Select INSERT >> BALLOON on the menu bar.

STEP 16: Select LEADER >> MAKE NOTE.

STEP 17: On the work screen, pick the horizontal line entity representing the base part (see note 1, Figure 12–63).

Balloon leader note 1 will be attached to this entity.

STEP 18: Select DONE SEL on the Get Select menu.

STEP 19: Select DONE on the Attachment Type men.

STEP 20: On the work screen, pick the attachment point for the balloon note 1 (see Figure 12–63).

STEP 21: Enter BASE as the name for the first balloon note.

If the balloon note is not initially placed correctly, you can use the Move option later to reposition it. Notice on the work screen how Pro/ENGINEER creates a list of named balloon notes.

STEP 22: Create balloon notes for the shaft part and for the pulley part.

STEP 23: When your layout matches Figure 12–63, select DONE/RETURN to exit the Note Types menu.

STEP 24: Individually pick each balloon note and reposition to match Figure 12–63.

STEP 25: Use the INSERT >> DIMENSION >> NEW REFERENCES option to create the four dimensions shown in Figure 12–64.

Dimensions in layout and drawing modes are created similarly to how dimensions are created in a sketcher environment. When creating dimensions in layout mode, the user is required to input a dimension symbol name first, followed by a dimension value. Use the symbol names and values shown in Figure 12–64.

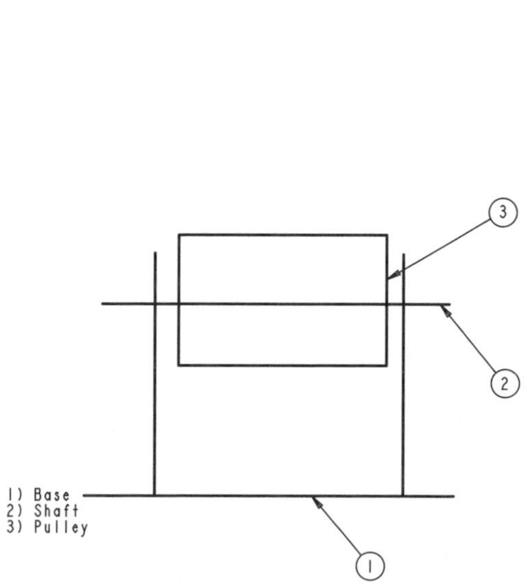

Figure 12–63 Balloon notes

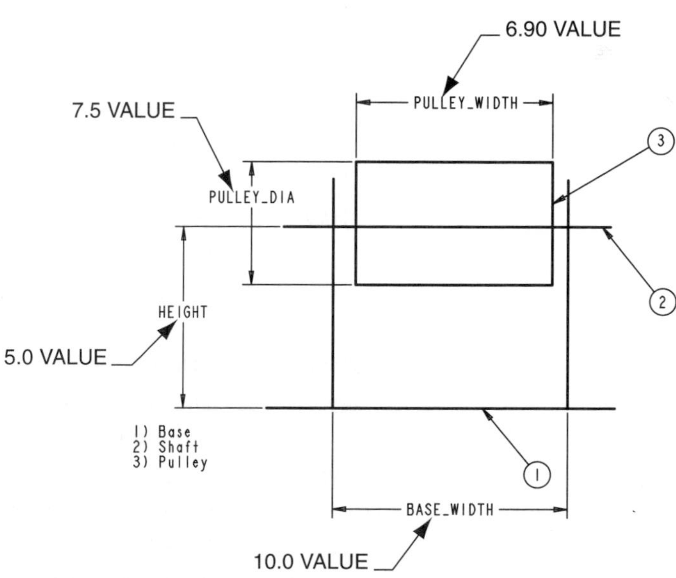

Figure 12–64 Dimension creation

STEP 26: Use the RELATION option to make the *PULLEY_WIDTH* dimension 3.1 inches less than the *BASE_WIDTH* dimension.

This design intent will keep the width of the pulley always 3.1 inches less than the width of the base part. Enter the following equation after selecting the Relation >> Add option: ***PULLEY_WIDTH = BASE_WIDTH − 3.1.***

STEP 27: Use the RELATION option to make the *PULLEY_DIA* dimension 1.25 times the *HEIGHT* dimension.

Enter the following equation: ***PULLEY_DIA = HEIGHT *1.25.***

STEP 28: Save your layout.

CREATING A START PART

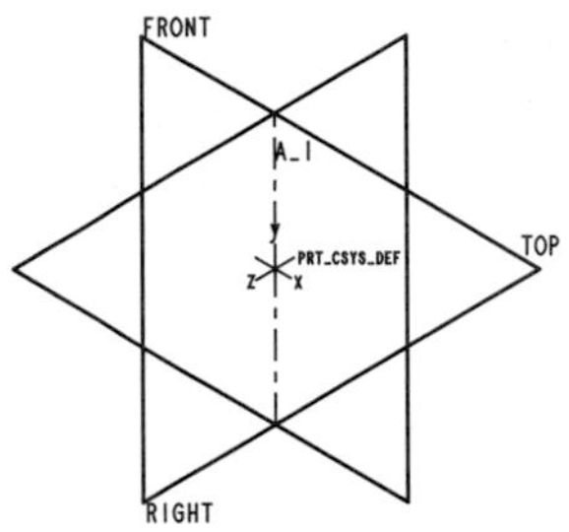

Figure 12-65 Start part

Within top-down assembly design, the first features of a component are often created by copying an existing part. In this tutorial, the first features of the base part will be copied from an existing start part (or template). This start part will consist of Pro/ENGINEER's default datum planes and one datum axis (Figure 12–65). Later in this tutorial, this start part will be used to create components.

STEP 1: If necessary, start Pro/ENGINEER.

STEP 2: Using Pro/ENGINEER's default template (e.g., *inlbs_part_solid*), create a new part named *START*.

The *inlbs_part_solid* template part file includes Pro/ENGINEER's default datum planes and default coordinate system.

STEP 3: Create a datum axis at the intersection of datum planes RIGHT and FRONT.

Select the Datum Axis icon on the toolbar. Use the Two-Planes constraint option to create this axis.

STEP 4: Save the start part.

CREATING THE FIRST COMPONENT (base.prt)

You will create the base part (Figure 12–66) of the assembly by copying the start part. Since it is usually important to control external references within an assembly, copying an existing component is the recommended procedure for creating a new component within assembly mode. In this tutorial, the default datum planes will be copied to start the base part.

STEP 1: Without using a default template (Figure 12–67), create a new assembly model named *PULLEY*.

STEP 2: On the New File Options dialog box, select EMPTY as the template; then select OKAY (see Figure 12–67).

STEP 3: Select COMPONENT >> CREATE on the Assembly menu.

STEP 4: On the Component Create dialog box, select the PART Type and enter *Base* as the Name for the part; then select OKAY (Figure 12–68).

STEP 5: On the Creation Options dialog box, select COPY FROM EXISTING (Figure 12–69).

The Copy-From-Existing option will copy an existing component into the assembly model. Components selected for copying cannot have external references. Once copied into an assembly, the new component is completely independent from the component from which it was copied.

Figure 12-66 Base part

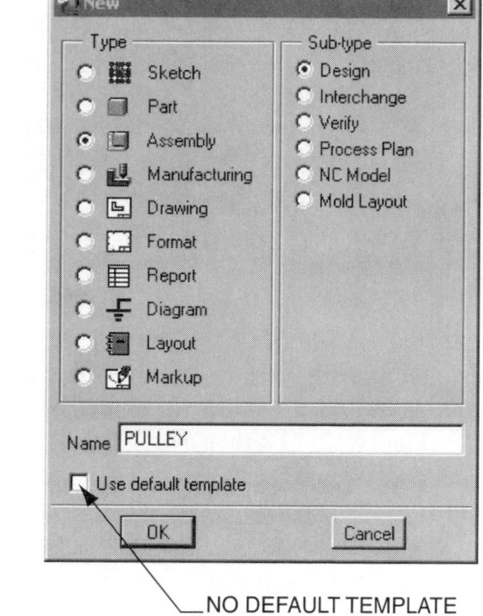

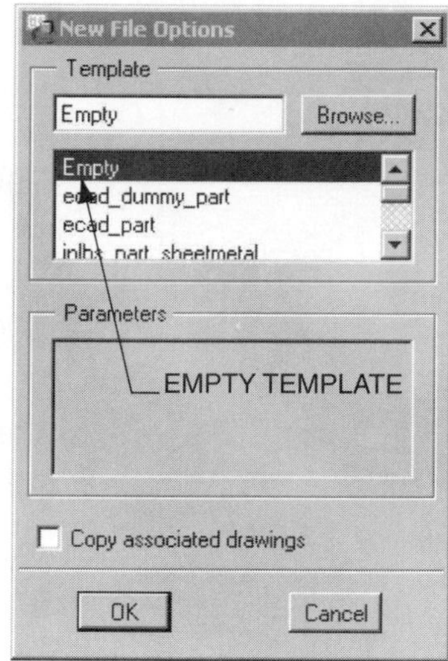

Figure 12-67 New dialog box

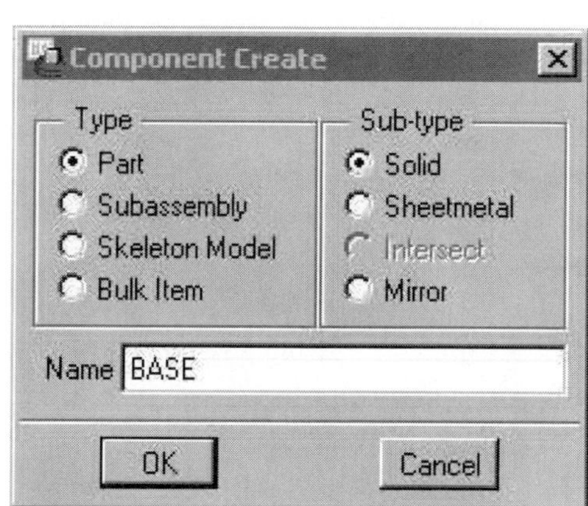

Figure 12-68 Component Create dialog box

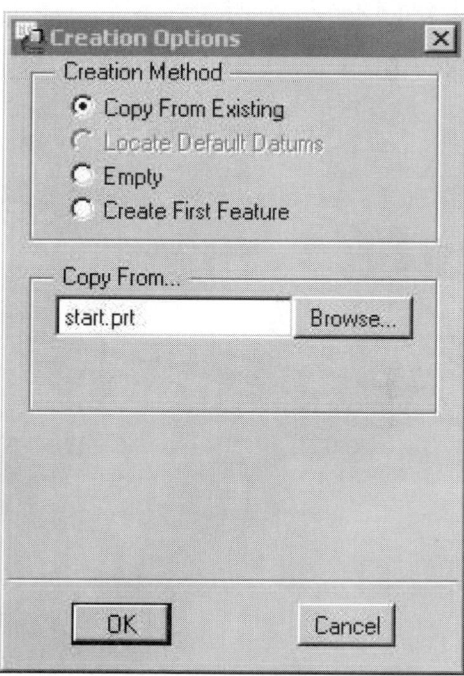

Figure 12-69 Creation Options dialog box

STEP 6: Use the BROWSE option to open the *start* part, then select OKAY on the dialog box.

After selecting OKAY, Pro/ENGINEER will drop the Base part into the assembly. Notice the Base part on the model tree.

STEP 7: On the model tree, select the Base part with the right mouse button; then select the REF CONTROL option.

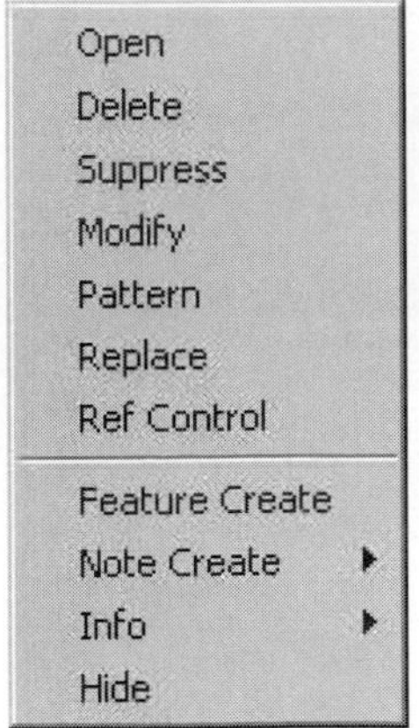

Figure 12-70 Model Tree options

Notice in Figure 12–70 the options that are available when selecting a component on the model tree with the right mouse button. The options observed are identical to the same options found within the menus of assembly mode.

> **MODELING POINT** Using the right mouse button to select a component on the model tree presents many beneficial options. As shown in Figure 12–70, the Open option will open a component into its own window, while the Feature-Create option will allow features to be added to a part.

STEP 8: On the External Reference Control dialog box, select the NONE and deselect the Backup Forbidden References option (as shown in Figure 12–71); then select OKAY.

When modeling components within assembly mode, eternal references can be created to other models within the assembly. Since having external references can limit the future usability of a part, it is usually advisable to avoid the creation of them.

Pro/ENGINEER provides several tools to help control references. In this example, the None option will not allow any external references. The Inside-Subassembly option can be used to only allow external references between components of a subassembly, and the Skeleton-Model option will allow references to skeleton models in a subassembly.

STEP 9: On the model tree, select the Base part with the right mouse button; then select the FEATURE CREATE option.

The Feature Create option allows features to be added to the current base part. Part features created in this manner will be reflected in the part's individual part object file.

STEP 10: Use the PROTRUSION >> EXTRUDE >> SOLID option to create the part feature shown in Figure 12–72.

As shown in the illustration, sketch the feature on datum plane TOP. Orient the remaining datum planes as shown. The holes can be created within the

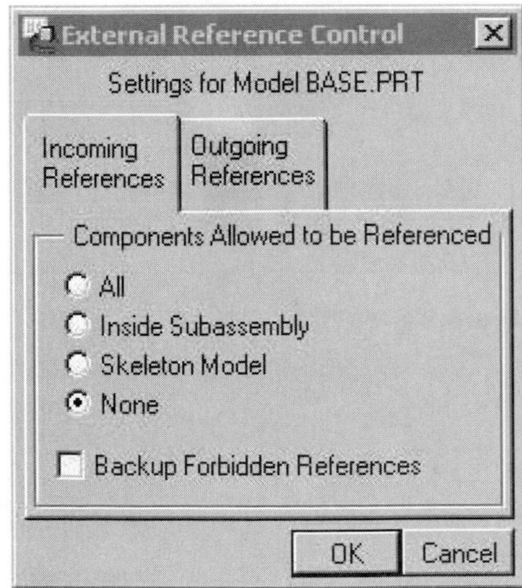

Figure 12-71 Reference Control dialog box

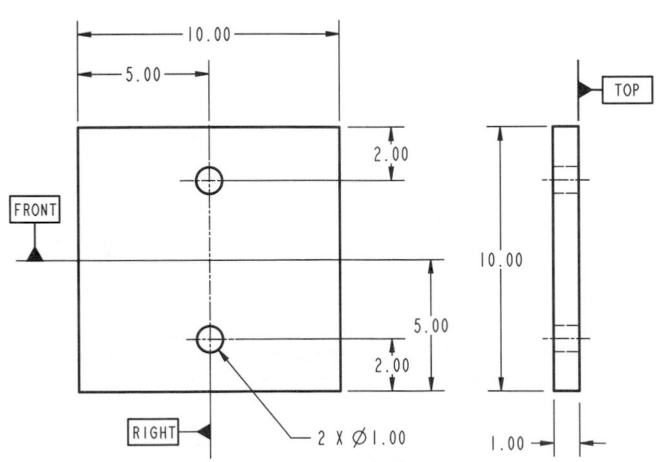

Figure 12-72 Section of first feature

extruded protrusion, or they can be created as separate features. If necessary, create the holes after creating the Protrusion.

STEP 11: Create one of the two protrusion features shown in Figure 12–73.

Use the model tree's *Feature Create* option to modify the part. Create the first feature as a one sided extruded protrusion.

STEP 12: Select MODIFY >> MOD PART on the Assembly menu.

The next several steps will create a dependent copy of the previously created extruded protrusion.

STEP 13: Select the *BASE* part.

STEP 14: Select FEATURE on the Modify Part menu.

STEP 15: Select COPY on the Part Feature menu; then create a Dependent mirrored copy of the previously created extruded protrusion.

Use the Copy >> Mirror >> Dependent option to create the second feature. You should be able to mirror the feature over datum plane RIGHT.

STEP 16: When the mirror is complete, select DONE/RETURN to exit the Part Feature menu; then select DONE to exit the Modify Part menu.

STEP 17: Use the MODIFY >> DIMCOSMETICS >> SYMBOL option to rename the three dimension symbols shown in Figure 12–74.

Modifying a dimension symbol to a more descriptive name makes identifying the dimension easier. After selecting the DimCosmetics >> Symbol option, select the feature holding the dimensions; then pick the dimension to rename. After renaming a dimension symbol in assembly mode, Pro/ENGINEER will add the Session ID to the end of the name.

STEP 18: Select DONE and then DONE/RETURN on the Assembly Modify menu to back out of the Modify option.

STEP 19: Use the RELATIONS >> PART REL option to make the *ASM_BASE_W2* dimension (see Figure 12–74) equal to half of the *ASM_BASE_W* dimension.

You do not have to enter the session ID (e.g., :4) when entering a part relation. Enter the equation: ***ASM_BASE_W2 = ASM_BASE_W/2.***

STEP 20: Select DONE to exit the Model Relations menu, then regenerate the assembly (Regenerate >> Automatic).

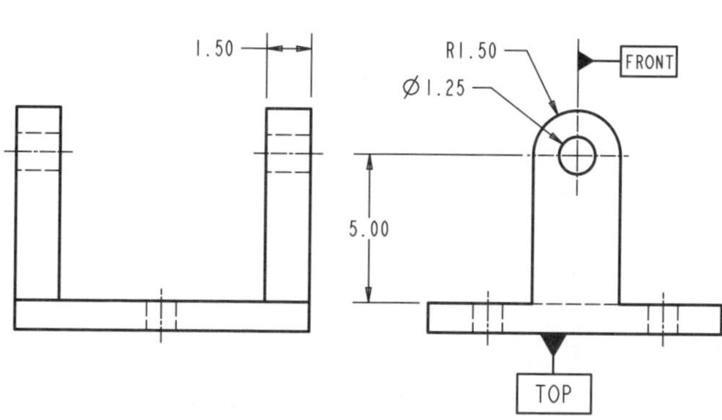

Figure 12–73 Extruded features

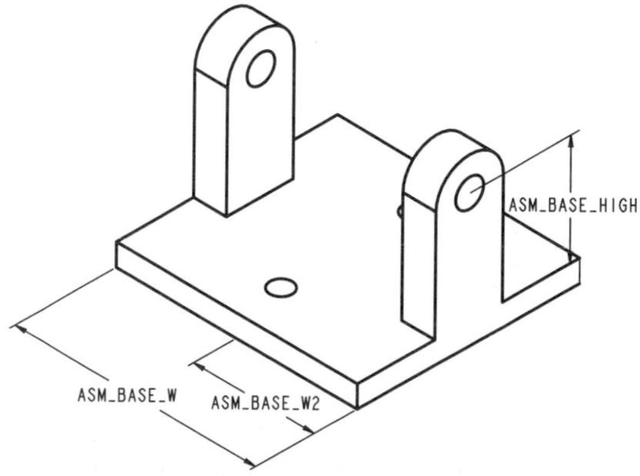

Figure 12–74 Dimension symbols

CREATING THE SECOND COMPONENT (*shaft.prt*)

In this segment of the tutorial, you will create the shaft part shown in Figure 12–75. Similar to the base part, this part will be created with a set of Pro/ENGINEER's default datum planes. Instead of copying the start part, you will use the Locate Default Datums option on the Component Creation dialog box.

STEP 1: Select UTILITIES >> REFERENCE CONTROL on the menu bar.

The Reference Control option located under the Utilities menu will set the default reference control for all assemblies. You could have set the default reference control using this technique before the creation of the base part. This tutorial had you set the reference control individually for the base part for instructional purposes.

STEP 2: On the Reference Control dialog box, select the NONE option and deselect the Backup Forbidden References option (Figure 12–76).

STEP 3: Select OKAY to exit the dialog box.

STEP 4: On the Model Tree, select the *PULLEY.ASM* file name with the right mouse button; then select the COMPONENT >> CREATE option (see Figure 12–77).

A new component can be created within the assembly model using the Component >> Create option on the menu manager, or one can be created using the method described in this step.

STEP 5: On the Component Create dialog box, select PART as the type of component to create (see Figure 12–78).

STEP 6: On the dialog box, enter *shaft* as the name of the new part; then select OKAY.

STEP 7: Select LOCATE DEFAULT DATUMS >> THREE PLANES >> OKAY on the Creation Options dialog box (Figure 12–79).

The Locate-Default-Datums option will place a set of Pro/ENGINEER's default datum planes. Datum planes placed using this option do not create external references. The Three-Planes suboption will allow for the alignment of each plane to existing assembly model planes.

STEP 8: Pick the first plane selection shown in Figure 12–80.

The planes for the default datums of the shaft part will be aligned with the planes picked in the next few steps. The first plane picked in this step will define the sketching plane for the first geometric feature of the shaft part. The second plane will be defined on-the-fly with the Make-Datum option.

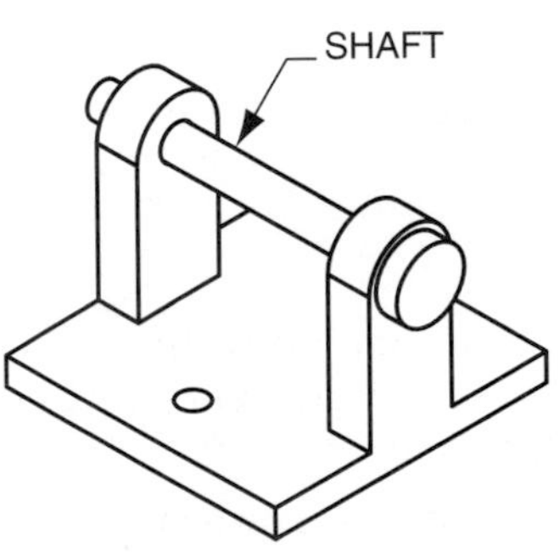

Figure 12–75 Shaft part

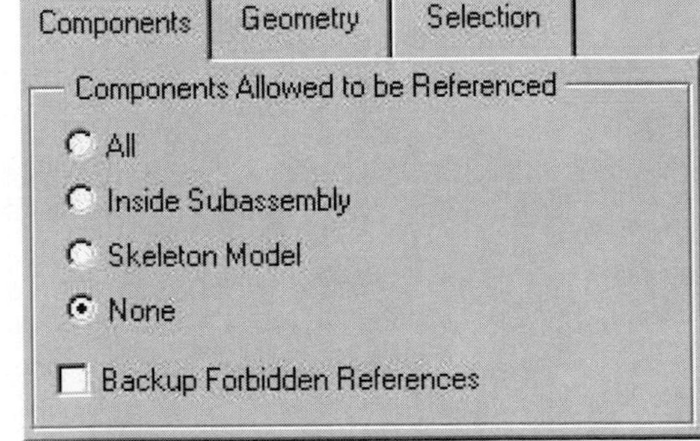

Figure 12–76 Reference Control dialog box

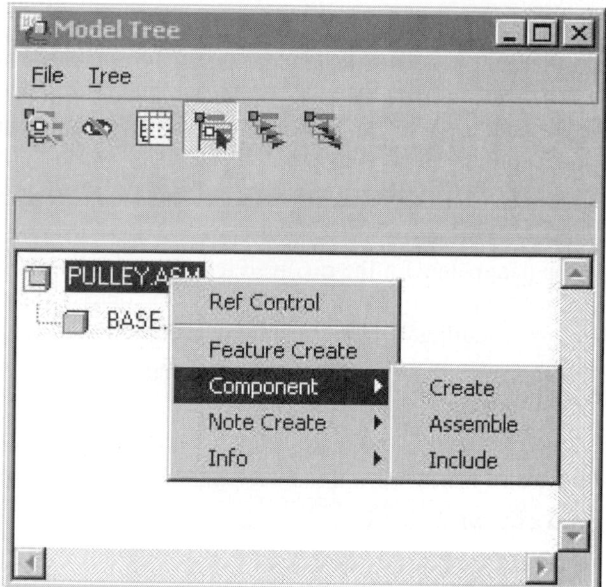

Figure 12–77 Component Create option

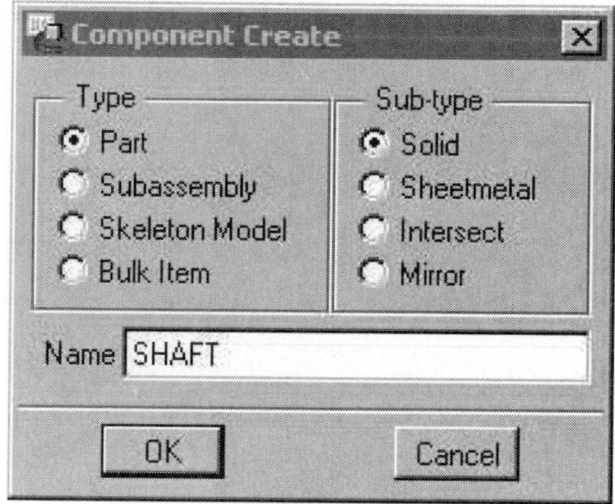

Figure 12–78 Component Create dialog box

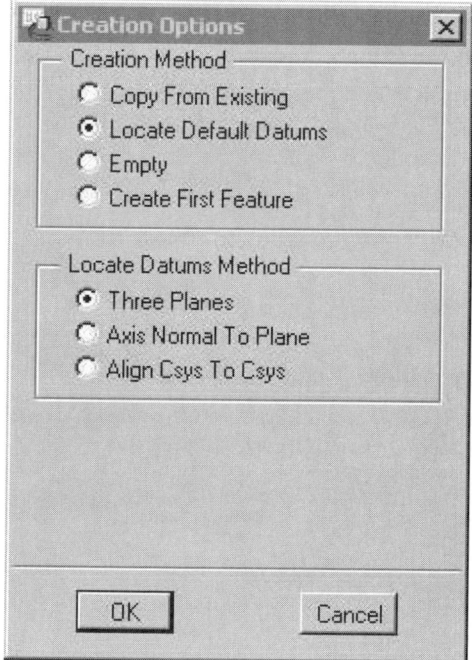

Figure 12–79 Creation Options dialog box

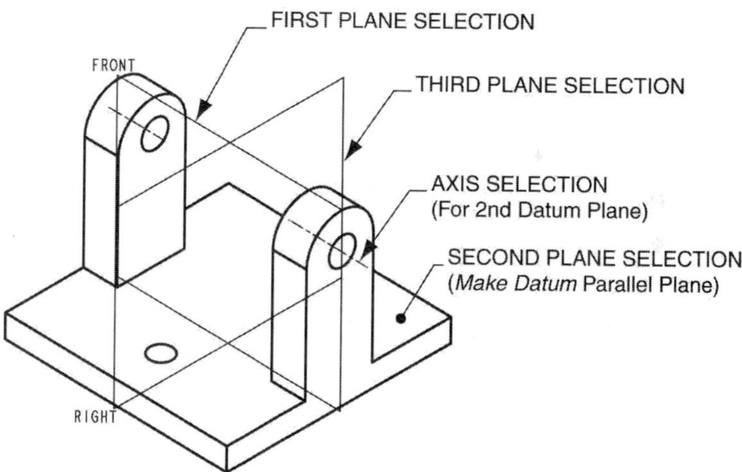

Figure 12–80 Plane selection

STEP 9: Select the MAKE DATUM option to define the second plane; then use the THROUGH >> AXIS and PARALLEL paired constraint options to define an on-the-fly datum plane (see Figure 12–80).

With the Through >> Axis constraint option, pick the axis shown in Figure 12–80. For the Parallel constraint option, pick the part surface shown in the illustration.

STEP 10: Select the third plane selection shown in Figure 12–80.

After defining the three planes notice how the shaft part has been added to the model tree. Also notice how Pro/ENGINEER's Menu Manager automatically provides you with the Component >> Solid menu. This provides you with options for creating the first protrusion feature.

STEP 11: Select PROTRUSION >> REVOLVE >> DONE >> ONE SIDE >> DONE.

After defining the parameters for the revolved feature, Pro/ENGINEER will launch the sketcher environment, without providing the option of selecting a sketching plane. When utilizing default datum planes as the first feature of a new component in assembly mode, the first placed datum plane is defined as the sketching plane.

> **INSTRUCTIONAL POINT** If you make a mistake during the construction of this part's first geometric feature, you can use the Model Tree's Pop-Up menu to access the Feature Create option.

STEP 12: If necessary, use the Reference dialog box (Figure 12–81) to select the two datum planes shown in Figure 12–82 as references.

The shaft part's default datums planes will be utilized as references for this feature.

STEP 13: Sketch the centerline shown in Figure 12–82.

Revolved features require a centerline within the sketcher environment. Use the Centerline icon found on the Sketcher Tools toolbar to create this line. Notice how this line is aligned with datum plane TOP.

STEP 14: Turn off the display of datum planes.

STEP 15: Set Hidden as the model's display mode.

STEP 16: Sketch the geometric entities shown in Figure 12–82.

Use the Line icon on the Sketcher Toolbar to sketch the profile shown in the illustration. The dimensioning scheme shown in the figure matches the

Figure 12–81 Reference dialog box

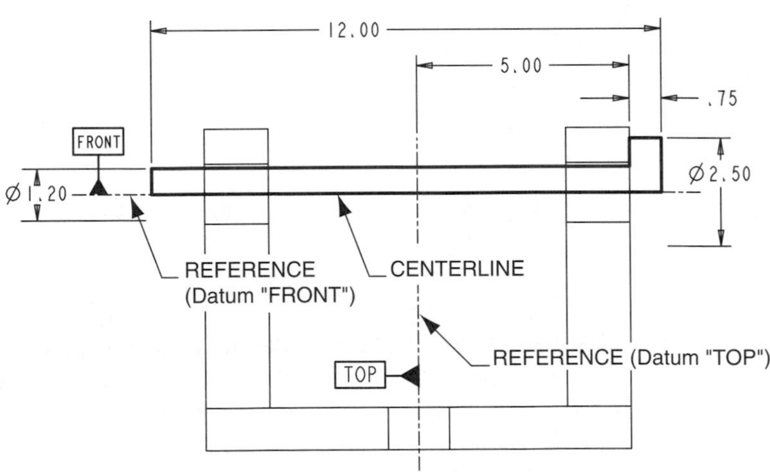

Figure 12–82 Shaft section

design intent for the part and for the assembly. Notice the visualization for the existing base part. You can sketch this section, or any section in top-down design, to fit around or within existing components. Since external references are prohibited for this part, you cannot reference the base part.

STEP 17: When the sketch is complete, exit the sketcher environment.

STEP 18: Revolve the feature 360 degrees.

STEP 19: Preview the feature, then select OKAY on the dialog box.

STEP 20: Save your assembly model.

CREATING THE THIRD COMPONENT (pulley.prt)

In this segment of the tutorial you will create the pulley part (Figure 12–83). You will begin this part by placing the start part (*start.prt*) created earlier in this tutorial.

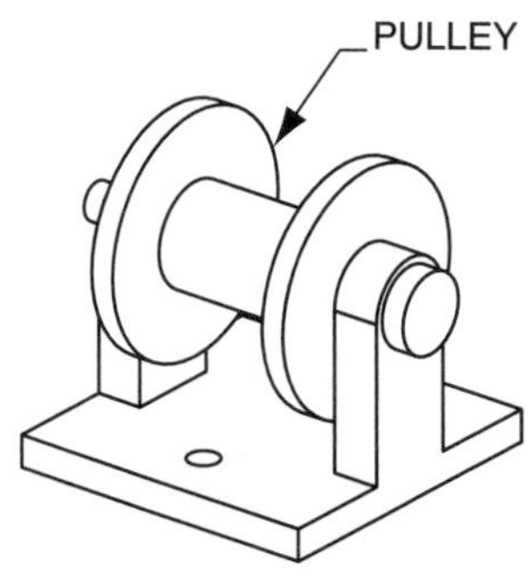

Figure 12-83 Pulley component

STEP 1: Confirm the setting of your reference control.

Use the Utilities >> Reference Control option to set your environment reference settings. You should set your component scope to None, and you should deselect the Backup-Forbidden-References option.

STEP 2: Select COMPONENT >> CREATE on the Assembly menu.

STEP 3: On the Component Create dialog box, enter a Part Type component named *PULLEY;* then select OKAY.

STEP 4: On the Creation Options dialog box, select the COPY FROM EXISTING option, Browse to locate the *START* part, and then select OKAY (Figure 12–84).

Make sure that the display of your datum planes is turned on.

STEP 5: If necessary, use the Move tab's Translate and View Plane suboptions to move the start part to a position that will allow for better entity selection.

STEP 6: Using the Component Placement dialog box, constrain the start part using the constraints shown in Figure 12–85.

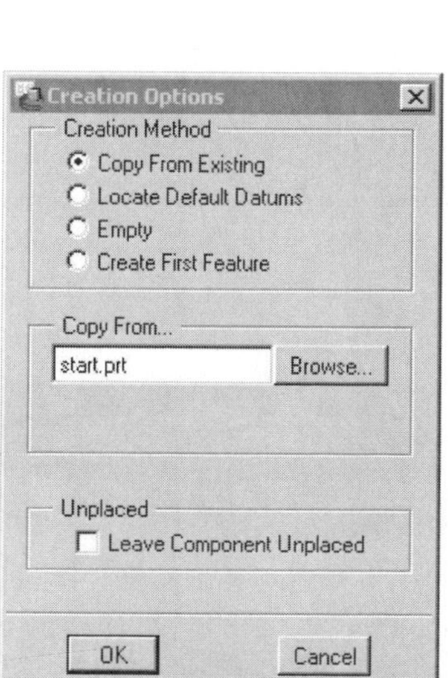

Figure 12-84 Creation Options dialog box

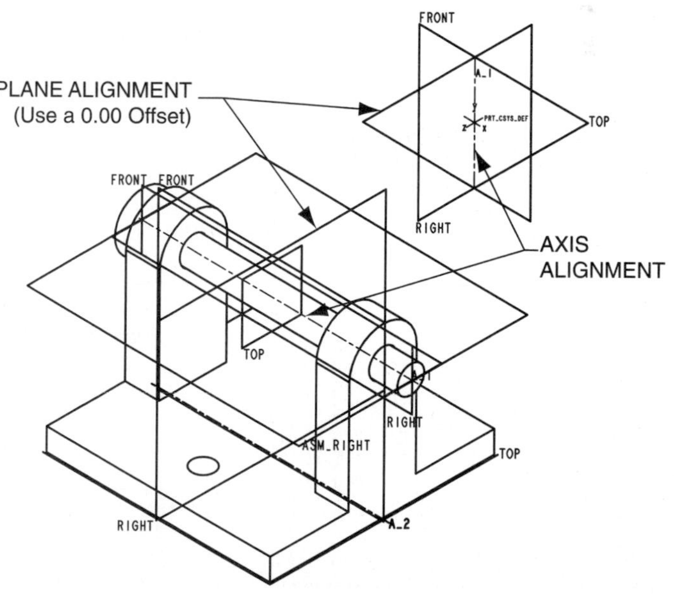

Figure 12-85 Start part placement

Use the Automatic constraint option with the following assigned constraints:

- Align the Axis feature of the start part with the axis of the shaft part. Use Query Select if necessary to pick the shaft's axis.
- Align datum plane TOP from the start part with a vertical datum plane running through the existing assembly.
- Allow for assumptions if necessary.

> **MODELING POINT** While placing a part with the Component Placement dialog box, if the visibility of the part being placed is restricted by existing components, use the Move tab to adjust the temporary location of the new part.

STEP 7: When the component is fully constrained, select OKAY to exit the dialog box; then select DONE/RETURN to exit the Component menu.

STEP 8: On the Model Tree, with the right mouse select the *Pulley* feature; then select the FEATURE CREATE option (see Figure 12–86).

Before proceeding, it is important to ensure that your external reference control has a scope set to None.

STEP 9: Select PROTRUSION >> REVOLVE >> DONE from the Solid menu.

You will create a revolved protrusion sketched on the pulley part's datum plane RIGHT.

STEP 10: Select ONE SIDE >> DONE

STEP 11: For the sketching plane, use QUERY SEL to pick the pulley part's datum plane FRONT (Figure 12–87).

STEP 12: Accept the default feature creation direction, then orient the sketcher environment to match Figure 12–88.

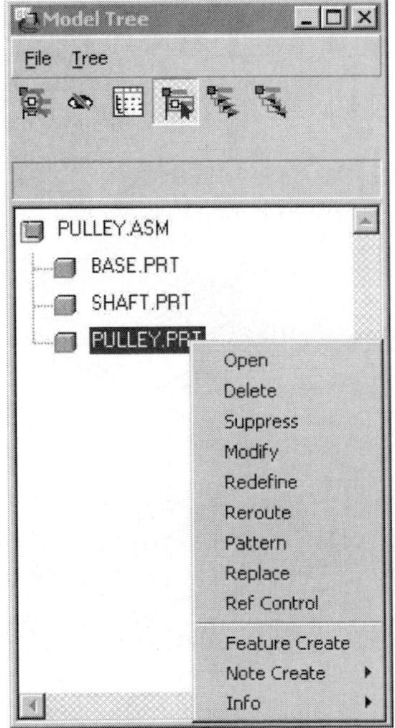

Figure 12-86 Feature create

Figure 12-87 Query bin

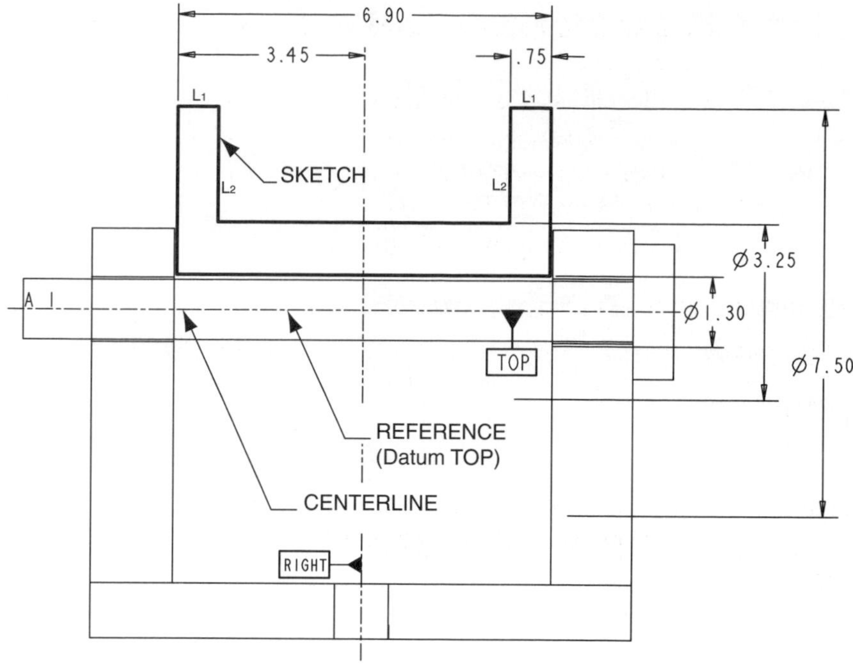

Figure 12–88 Pulley section

STEP 13: **Specify datum planes RIGHT and TOP as references (see Figure 12–88).**

If necessary, use the Query Select option.

STEP 14: **Turn off the display of datum planes.**

STEP 15: **Select HIDDEN as the model's display style.**

STEP 16: **Sketch the centerline shown in Figure 12–88.**

STEP 17: **Sketch the geometry shown in Figure 12–88.**

STEP 18: **Incorporate the dimensional design intent shown in the illustration.**

STEP 19: **Add a dimensional relationship to make the 3.45 dimension in Figure 12–88 equal to half the value of the 6.90 dimension (Sketch >> Relations).**

STEP 20: **Exit the sketching environment.**

STEP 21: **Revolve the feature 360 degrees, then use the Feature Definition dialog box to finish the feature.**

STEP 22: **Save your assembly.**

> **MODELING POINT** When a part is created within assembly mode, the part is saved as a part file and can be opened individually in part mode. If a part created in assembly mode does not have external references, if can be copied to create a new part.

STEP 23: **Select INFO >> GLOBAL REF VIEWER on the Menu Bar.**

STEP 24: **Select the settings shown in Figure 12–89.**

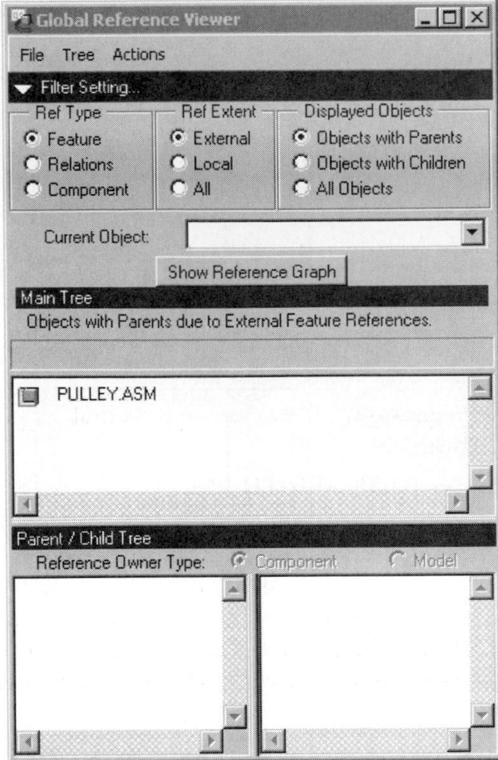

Figure 12–89 Global reference viewer

The Global Reference Viewer is used to view dependencies between features, relations, and components. A component of an assembly has an external reference when it depends upon another model. When utilizing top-down assembly design capabilities, external references should be avoided. The settings shown in Figure 12–89 will display models that have dependencies on other models (External and Objects-with-Parents options). The Local option is used to display references within a model, and the Objects-with-Children option is used to display components that serve as parents to other models. Notice the lack of components displayed under the *PULLEY.ASM* name. This means that no models within the assembly have external references.

DECLARING AND USING A LAYOUT

Layouts can be used to control assembly dimension values. Within this segment of the tutorial, you will utilize the layout created previously in this tutorial to control the assembly model. To perform this function, you have to declare the layout within the assembly.

STEP 1: Open both the assembly model and the layout model; then activate the assembly model's window.

STEP 2: Select SET UP >> DECLARE >> DECLARE LAY within assembly mode.

STEP 3: Select the PULLEY layout on the Declare menu.

> **MODELING POINT** In this tutorial, you will use the dimension values created within the layout to control dimension values within the assembly model. Once a layout has been declared within an assembly, the dimension valves from the layout can be used in relation equations within the assembly model. Notice in Figure 12–90 the dimension symbols illustrated for the layout and for the assembly. These are the dimensions that you will relate within this tutorial.

STEP 4: **Select DONE to exit the Assembly Setup menu.**

STEP 5: **Save the assembly model.**

STEP 6: **Within Assembly mode, select MODIFY >> MOD DIM >> DIMCOSMETICS >> SYMBOL; then on the work screen, select the pulley part.**

STEP 7: **Modify the two pulley dimension symbol names shown in Figure 12–90.**

If necessary, you can use the Move Dimension option on the Dimension Cosmetic menu to move the location of dimensions. Rename the width dimension for the pulley ***ASM_PULLEY*** and the diameter dimension ***ASM_PULLEY_DIA***.

STEP 8: **Select DONE/RETURN to exit the Assembly Modify menu.**

STEP 9: **Select RELATIONS >> ADD and create the following dimensional relationships (refer to Figure 12–90):**

- **ASM_BASE_W:(session ID) = BASE_WIDTH**
- **ASM_BASE_HIGH:(session ID) = HEIGHT**
- **ASM_PULLEY:(session ID) = PULLEY_WIDTH**
- **ASM_PULLEY_DIA:(session ID) = PULLEY_DIA**

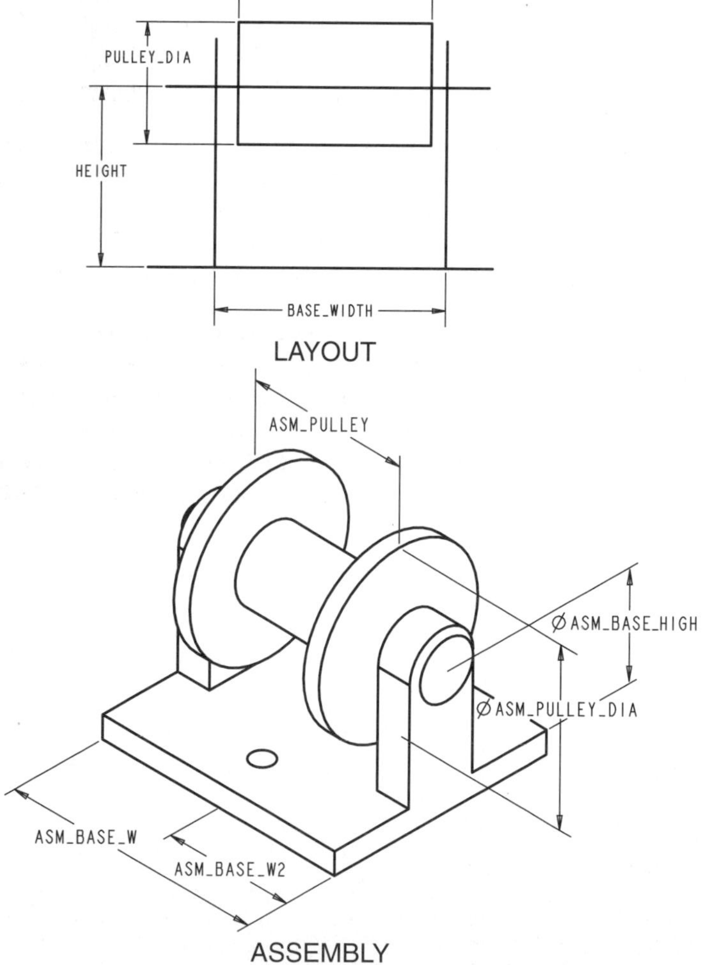

Figure 12–90 Dimension symbols

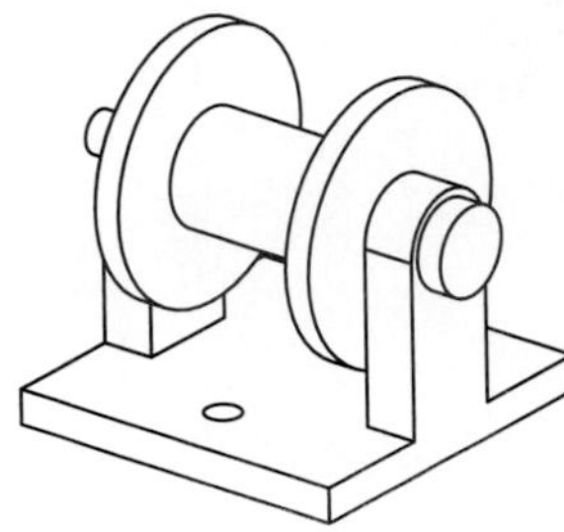

Figure 12-91 Final Model

NOTE: Refer to your model for each equation's session ID. An example of a session ID would be **ASM_BASE_W:0.** You will need to select the features holding the above dimensions to record their dimension symbol names. Each part can have its own unique session ID.

STEP 10: Regenerate the assembly model (Regenerate >> Automatic).

STEP 11: Activate the layout model's window.

STEP 12: Within the layout, use the Pop-Up menu's NOMINAL VAL option and change the *BASE_WIDTH* dimension to a value of 11.

STEP 13: Modify the *HEIGHT* dimension to a value of 6.

STEP 14: Regenerate the layout.

STEP 15: Activate the assembly's window and regenerate the model.

Your assembly model should be enlarged to match the dimensional values of the layout model (see Figure 12–91).

STEP 16: Save your assembly model.

TUTORIAL

MECHANISM TUTORIAL

Within this tutorial, you will explore the assembly and animation capabilities of assembly mode's mechanism option. The nutcracker model shown in Figure 12–92 will be utilized. You will model each component in part mode, then assemble each using mechanism joints.

Assembly mode provides powerful tools for modeling a complete design. Traditional bottom-up constraints necessitate fully constrained components. When a component is not fully constrained, it is considered packaged and presents assembly difficulties. The mechanism option provides tools for assembling components in a manner that replicates a real design. Joints such as pin, cylinder, slider, and ball are available. Notice in Figure 12–92 the connection linkage that exists from the handle through the connection part through to the piston part. The mechanism joints defining these parts only constrain each component within the degrees of freedom required by the design. Within this assembly, the handle can be moved, which in turn will move the piston. This will be demonstrated. The following topics will be covered in this tutorial:

- Modeling assembly parts.
- Assembling a mechanism.
- Manipulating a mechanism.
- Running a mechanism's motion.
- Animating a mechanism.

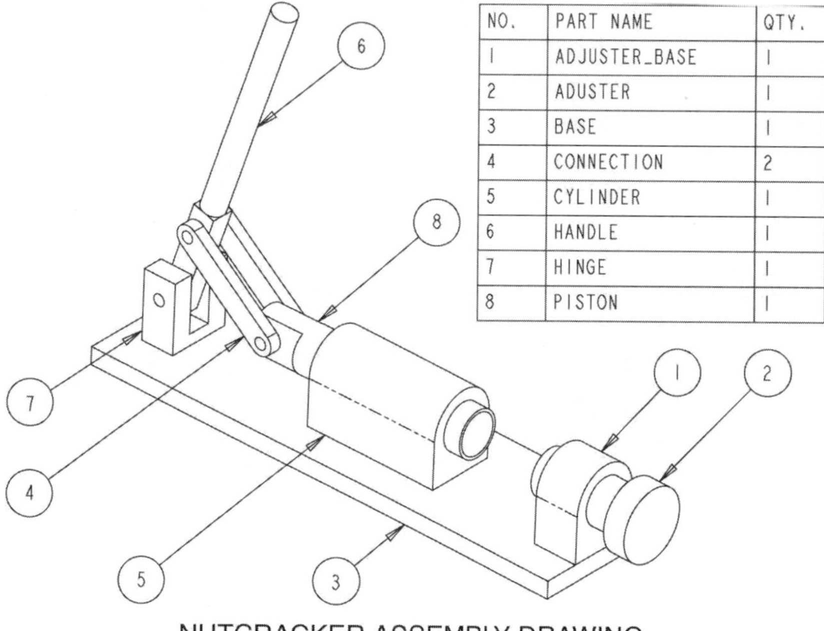

NO.	PART NAME	QTY.
1	ADJUSTER_BASE	1
2	ADUSTER	1
3	BASE	1
4	CONNECTION	2
5	CYLINDER	1
6	HANDLE	1
7	HINGE	1
8	PISTON	1

Figure 12-92 Mechanism assembly

MODELING ASSEMBLY PARTS

The assembly in this tutorial consists of eight different parts: base, cylinder, hinge, piston, adjuster_base, adjuster, connection, and handle. There will be two instances of the connection part. Use part mode to model each of the parts as shown in Figure 12–93. When

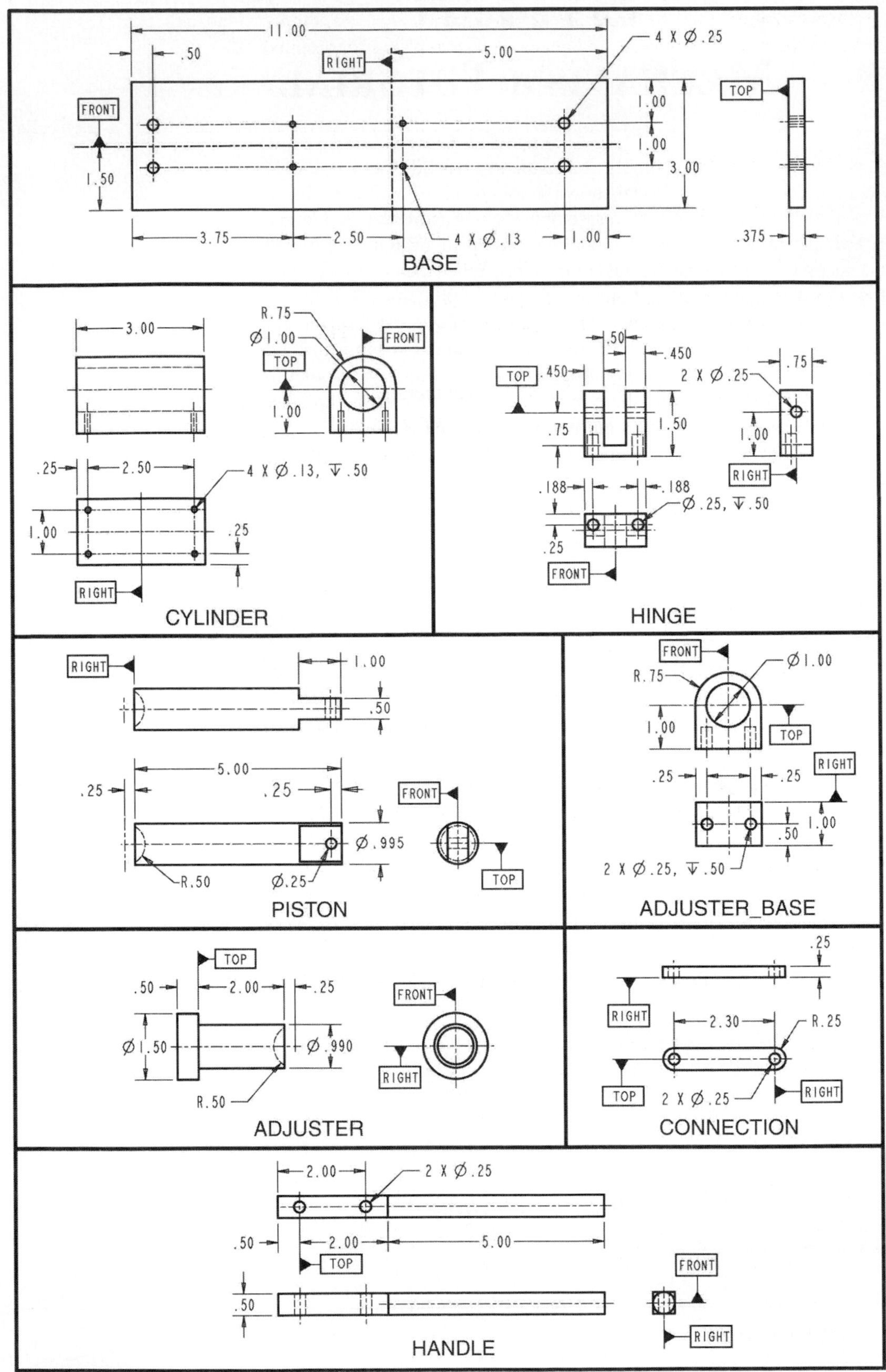

Figure 12-93 Assembly parts

modeling each part, pay careful attention to the locations of your datum planes. For proper assembly of the mechanism, your datum planes should match the datum planes represented in each part's drawing.

ASSEMBLING A MECHANISM

Within this segment of the tutorial you will assemble the parts comprising the design. Within this exercise, you will not use a template file. Do not start this segment of the tutorial until you have modeled all the parts portrayed in Figure 12–93.

STEP 1: Start Pro/ENGINEER, and then select FILE >> NEW.

STEP 2: On the New dialog box, deselect the USE DEFAULT TEMPLATE OPTION.

Within this tutorial, do not use Pro/ENGINEER's default template.

STEP 3: Create a new Assembly object file named *NUTCRACKER.ASM.*

STEP 4: On the New File Options dialog box, select the EMPTY template file; then select OKAY.

STEP 5: Select COMPONENT >> ASSEMBLE on the Assembly menu.

STEP 6: Using the Open dialog box, place the *BASE* part.

Without any existing features or components, Pro/ENGINEER will place the first component without requiring any constraints or joints. If you inadvertently created Pro/ENGINEER's default datum planes, you can mate and/or align the BASE part to these datum planes.

STEP 7: Select COMPONENT >> ASSEMBLE and open the *CYLINDER* part.

STEP 8: Using traditional assembly constraints, assemble the Cylinder part as shown in Figure 12–94.

There are two ways to add components to a mechanism: Fixed and By Connection. The Fixed option is identical to the traditional way of assembling components in Pro/ENGINEER. The By Connection option assembles components through joint definitions. It allows components to move based on the degrees of freedom provided by the selected joint. For the cylinder part, assemble the component with one Mate constraint and two Align constraints as shown in the illustration.

STEP 9: When the Cylinder part is fully constrained, select OKAY to exit the dialog box.

STEP 10: Use the same technique for assembling the cylinder part to constrain the *HINGE* and *ADJUSTER_BASE* parts (Figure 12–95).

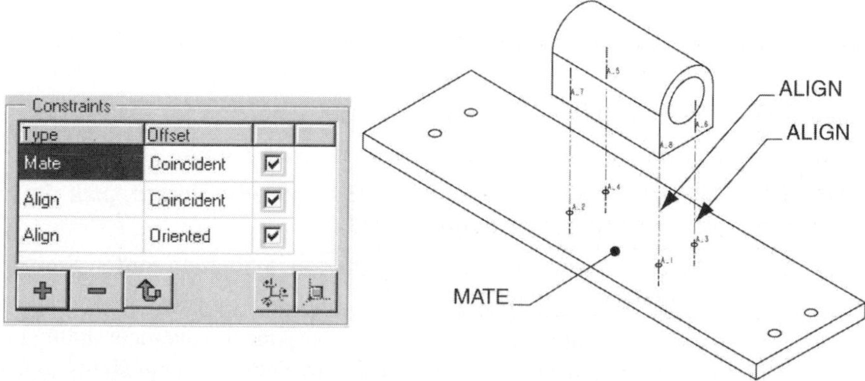

Figure 12-94 Cylinder fixed constraints

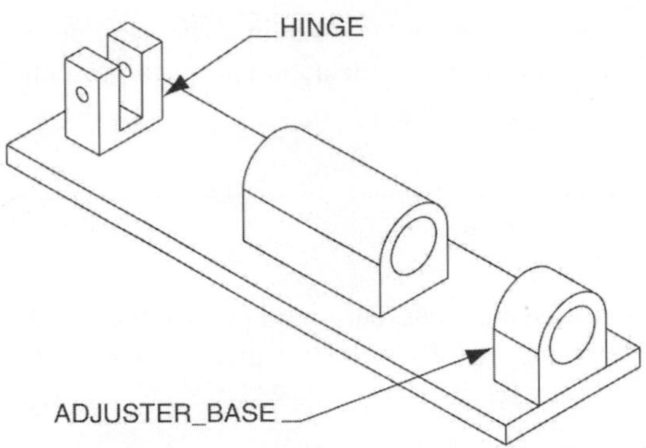

Figure 12-95 *Adjuster_Base* and *hinge* fixed constraints

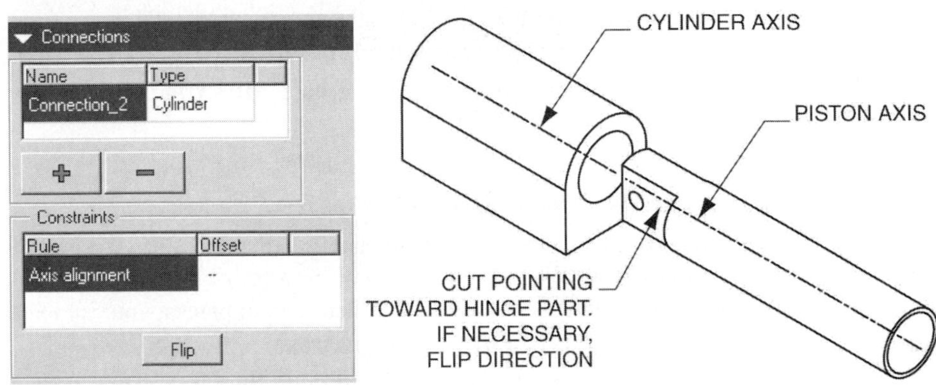

Figure 12-96 Cylinder joint definition

As with the cylinder part, use two align constraints and one mate constraint for each part. Your assembly should appear as shown in the illustration.

Next you will assemble the Piston part using a Cylinder joint. Other available joints include: Pin, Bearing, Slider, Planar, and Ball.

STEP 11: Select ASSEMBLE, then open the *PISTON* part.

STEP 12: On the Component Placement dialog box select CONNECTIONS.

STEP 13: On the Component Placement dialog box, select CYLINDER as the connection type (see Figure 12–96).

STEP 14: Select the two axes shown in Figure 12–96.

A Cylinder joint type is defined through the alignment of two axes. This joint type provides two degrees of freedom: one linear and one rotational.

STEP 15: If necessary, select the FLIP option to point the piston's cut feature toward the hinge part (see Figure 12–96).

STEP 16: On the dialog box, select the MOVE tab.

You will reposition the piston part to match Figure 12–97.

STEP 17: With the Translate and Entity/Edge options selected, pick the axis of the piston part (see Figure 12–97).

STEP 18: Move the Piston part's location to approximately match Figure 12–97; then select the Place tab.

After selecting the Place tab on the component placement dialog box, notice the current placement status of the part. The placement status should state "Connection Definition Complete."

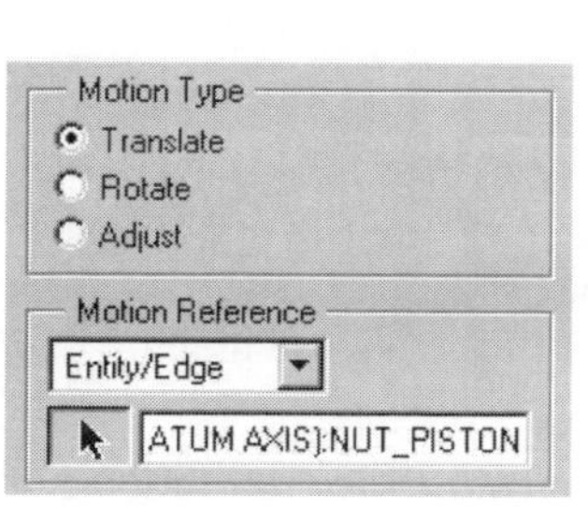

Figure 12–97 Move option

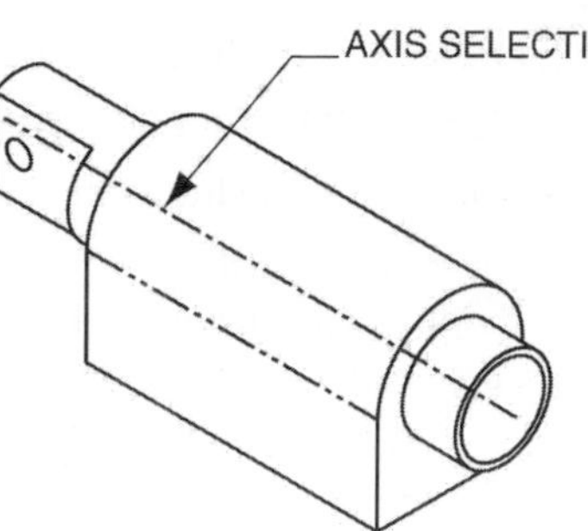

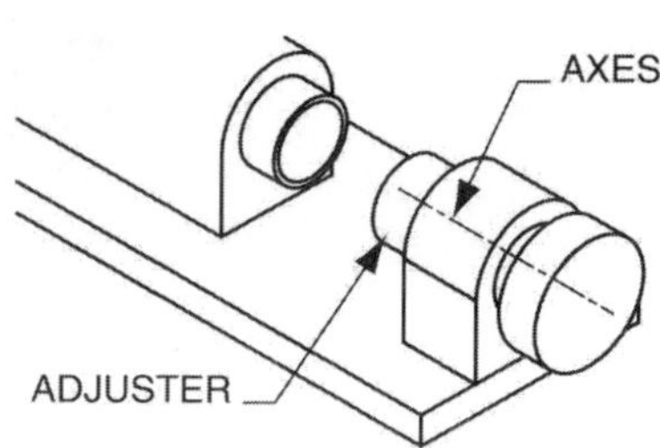

Figure 12–98 Adjuster placement

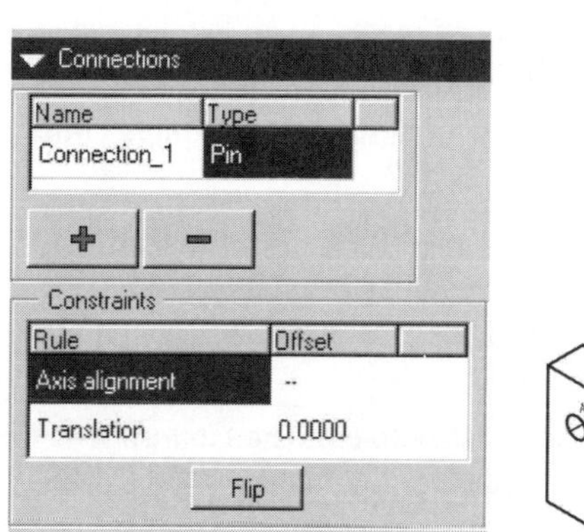

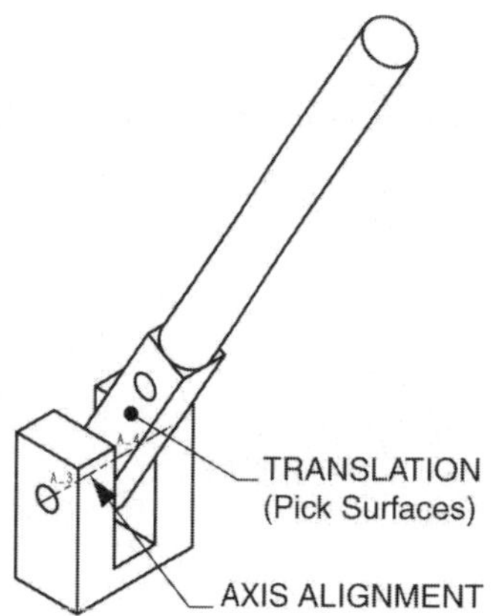

Figure 12–99 Handle Placement

NOTE: You can also use the Mechanism >> Drag option to move components that have mechanism joints.

STEP 19: Select OKAY on the dialog box.

STEP 20: Use the ASSEMBLE option to open the *ADJUSTER* part.

STEP 21: Use the same technique for assembling the piston part to constrain the *ADJUSTER* part (Figure 12–98).

Use Cylinder as the connection for the component. If necessary, use the Flip option and the Move tab to position the component to match the illustration.

STEP 22: Use the ASSEMBLE option to open the *HANDLE* part.

STEP 23: Select the PIN joint type under the Connections option (see Figure 12–99).

Pin connections provide one rotational degree of freedom. It is defined through the alignment of two axes and the aligning or mating of two planes.

STEP 24: Align the hole axes of the handle and hinge parts as shown in Figure 12–99.

STEP 25: With the Translation constraint type selected, mate one side of the handle part with the end side surface of the hinge part.

STEP 26: Use the Move tab to rotate the handle to the approximate location shown in Figure 12–99.

Under the Move tab, use Rotate as the Motion Type and Entity/Edge as the Motion Reference. The handle should be pointing toward the center of the base part.

STEP 27: **Select OKAY to exit the dialog box.**

STEP 28: **Assemble the *CONNECTION* part.**

STEP 29: **Create the PIN joint shown in Figure 12–100.**

The connection part will have two joints: one pin and one cylinder. The pin joint will join the connection part to the handle part. The cylinder joint will join the connection part to the piston. If necessary, use the Flip option to create the mate translation connection.

STEP 30: **Use the Move tab on the dialog box to rotate the connection part to the approximate location shown in Figure 12–100.**

STEP 31: **Create the CYLINDER joint shown in Figure 12–100.**

The Plus icon located under the connection names on the dialog is used to add new joint types. After creating the cylinder joint, your assembly connection will not look like the illustration. This is typical of a Pro/ENGINEER looped mechanism. In a later step, you will execute the Connect option to assume a successful assembly.

STEP 32: **If the placement status signifies a complete connection, select OKAY to exit the dialog box.**

STEP 33: **Use the same technique for assembling the first connection part to place the second instance of the connection part.**

Repeat steps 28 through 32 to place the second instance of the connection part.

STEP 34: **Select DONE/RETURN to exit the Component menu.**

STEP 35: **Select MECHANISM >> CONNECT >> RUN to connect the loop assembly.**

> **INSTRUCTIONAL POINT** If you do not get a successful assembly after selecting the Run option, use the Mechanism >> Settings option to adjust the tolerance of the assembly (see Figure 12–101).

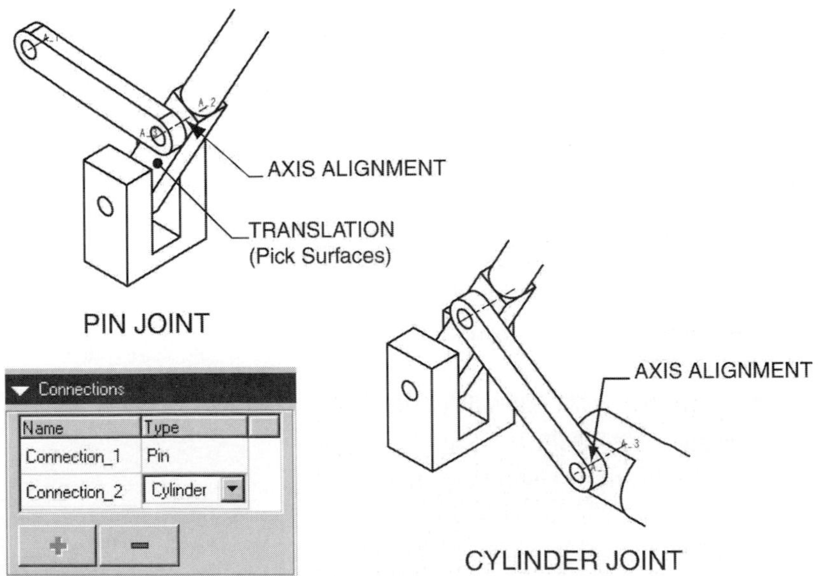

Figure 12–100 Connection part placement

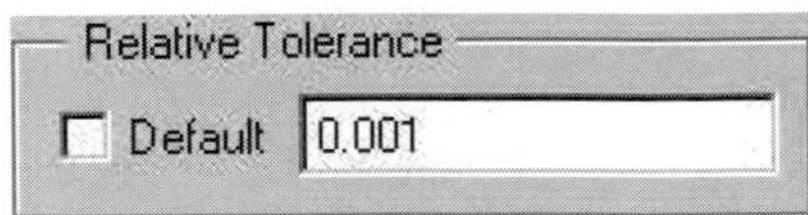

Figure 12–101 Assembly setting

STEP 36: **If you get a positive confirmation message, select YES to accept the successful assembly.**

STEP 37: **Save the assembly.**

MANIPULATING A MECHANISM

This segment of the tutorial will demonstrate how components can be dragged through any defined degrees of freedom. In addition, you will create snapshots of component placements that will be used in the last segment of this tutorial to animate the mechanism.

STEP 1: **Select the DRAG option from under the Mechanism menu.**

STEP 2: **On the Drag dialog box, select the Point Drag icon (Figure 12–102); then select the end of the handle part (see Figure 12–103).**

The Drag dialog box is used to drag components on the screen. Use the Point option to select the end of the handle part. After selecting the handle, you can dynamically drag the component with the mouse. Use the left mouse button to end dragging.

STEP 3: **Drag the handle to the First Position shown in Figure 12–103.**

Figure 12–103 represents a side view of the assembly. You can utilize any orientation to include a user-defined viewpoint.

STEP 4: **On the Drag dialog box, select the SNAPSHOT icon (Figure 12–102).**

Snapshots can be used to restore a mechanism's position and to create animations. You will use the snapshots created in this segment to animate the mechanism in the last segment of this tutorial.

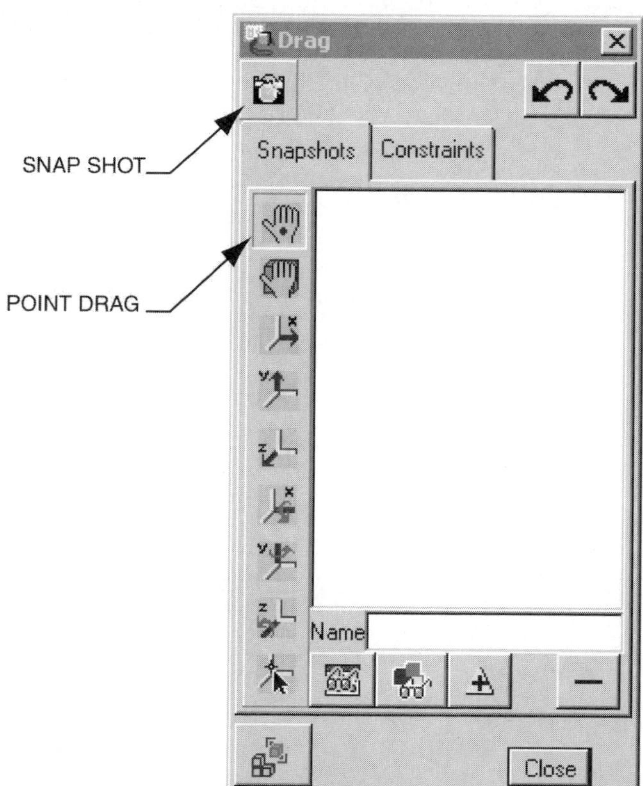

Figure 12–102 Drag entity selection

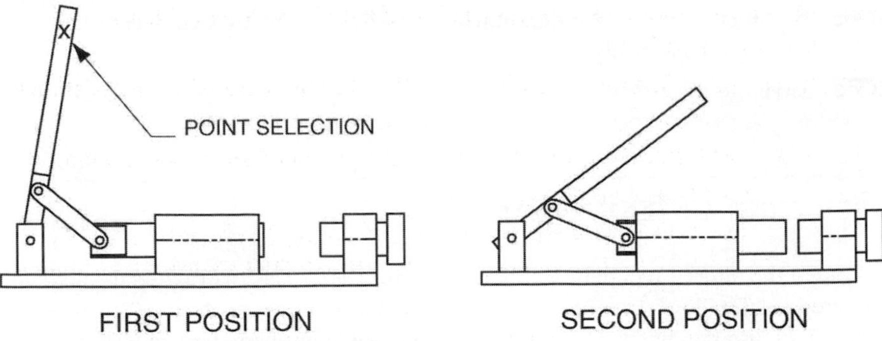

Figure 12–103 Drag positions

STEP 5: Drag the handle to the second position shown in Figure 12–103, then create a second snapshot.

STEP 6: Close the Drag dialog box.

RUNNING A MECHANISM'S MOTION

Motion, as defined by the degrees of freedom within a mechanism, can be animated. Within this segment of the tutorial, you will define the motion of the assembly through the use of a driver.

STEP 1: Select the MODEL >> DRIVERS option on the Mechanism menu.

STEP 2: On the Drivers dialog box, select the ADD option (Figure 12–104).

STEP 3: For the Driven Entity, select the pick icon; then on the work screen pick the pin joint shown in Figure 12–104.

STEP 4: If available, on the Driver Editor dialog box, select ROTATION as the Motion Type (Figure 12–104).

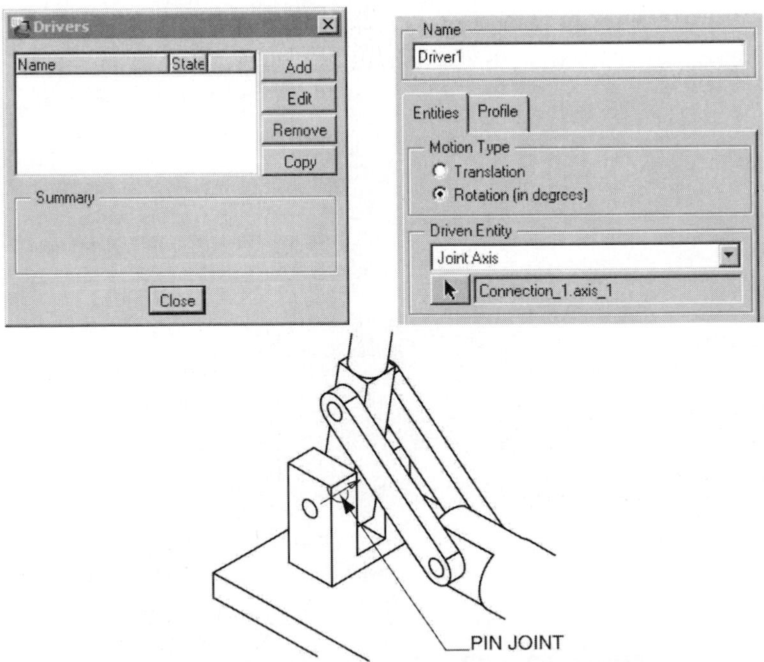

Figure 12–104 Driver creation

STEP 5: **Select the Profile tab, then select VELOCITY as the specification (Figure 12–105).**

STEP 6: **Change the Magnitude option to COSINE, then enter the values shown in Figure 12–105.**

STEP 7: **Select the GRAPH option to observe the graph of your mechanism.**

STEP 8: **Close the Graph windows.**

STEP 9: **Select the SET ZERO option on the dialog box, then set the CYAN BODY REFERENCE and the GREEN BODY REFERENCE as shown in Figure 12–106.**

This will define the starting point for the mechanism animation.

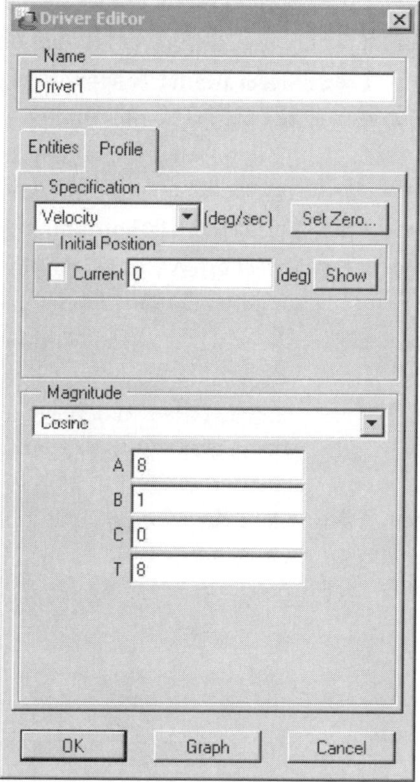

Figure 12–105 Driver Editor dialog box

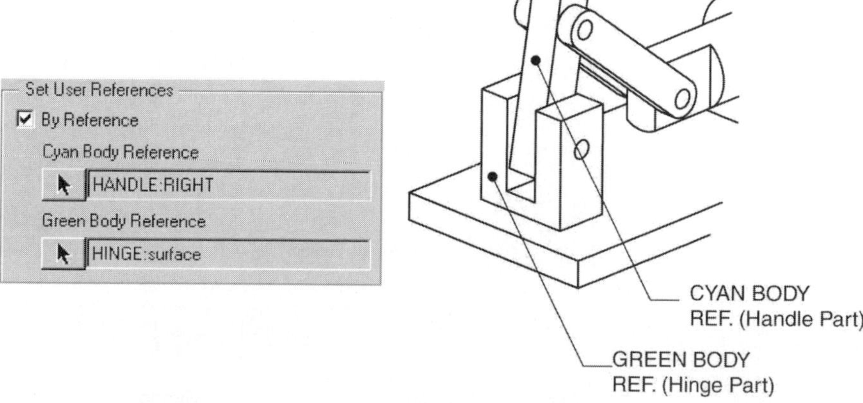

Figure 12–106 Zero reference selection

STEP 10: Select OKAY to exit the Joint Axis Settings dialog box.

STEP 11: On the Driver Editor dialog box, enter 10 as the Initial Angle for the driver (Figure 12–107).

This setting will establish an initial position for the mechanism 10 degrees from the set zero position.

> **INSTRUCTIONAL POINT** Notice on the work screen the arrow representing the joint's joint. Using the right-hand rule with your thumb pointing in the direction of the arrow, your fingers will point in the direction of driver rotation. If necessary, you might have to enter a negative 10 value for the initial angle.

STEP 12: Select OKAY to exit the Driver Editor dialog box.

STEP 13: Close the Drivers dialog box.

STEP 14: Select the RUN MOTION option under the Mechanism menu.

After selecting the Run Motion option, Pro/ENGINEER will launch the Motion Definitions dialog box. This dialog box is used to establish multiple motion definitions for a mechanism.

STEP 15: Select the ADD option on the Motion Definition dialog box.

STEP 16: Select OKAY to accept the default settings on the Motion Definition dialog box (Figure 12–108).

STEP 17: Select RUN on the Motion Definitions dialog box.

With any luck, after selecting the Run option, your mechanism should animate based on defined degrees of freedom and the set driver. If you have unexpected results in your animation, try adjusting the driver's profile values or the driver's initial angle value.

STEP 18: Close the Motion Definitions dialog box.

STEP 19: Select RESULTS >> PLAYBACK on the Mechanism menu.

Figure 12–107 Initial angle setting

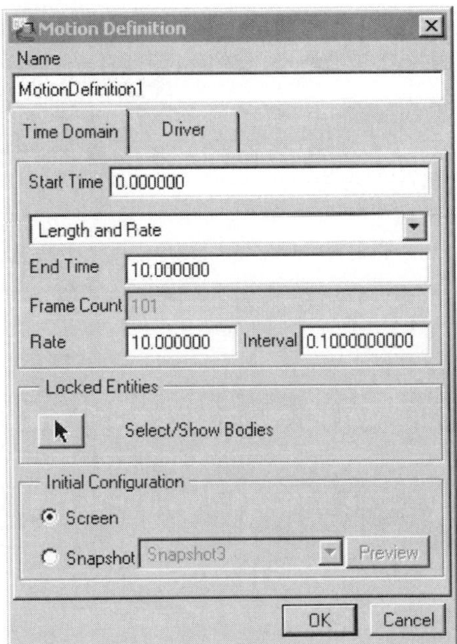

Figure 12–108 Motion Definition dialog box

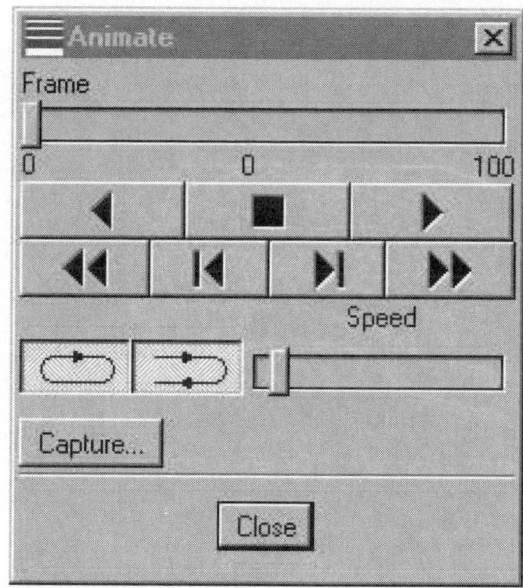

Figure 12–109 Animate dialog box

Step 20: **Select the PLAY option on the Results Playback dialog box.**

Step 21: **Use options on the Animate dialog box to run the results of your motion study (Figure 12–109).**

Step 22: **Close out of the Mechanism option and save your assembly.**

Animating a Mechanism

Mechanisms can be animated using Pro/ENGINEER's Animation mode. Within this segment of the tutorial, you will use the snapshots created previously to animate the nutcracker assembly.

Step 1: **Using Pro/ENGINEER's Menu Bar, select APPLICATIONS >> ANIMATION.**

Upon selecting the Animation option, Pro/ENGINEER will reveal the Animation toolbar (Figure 12–110) and, at the bottom of the screen, a timeline.

Step 2: **Select the ANIMATION icon on the toolbar, then select the NEW icon to create a new animation.**

Step 3: **Close the Animation dialog box.**

Step 4: **Double pick the timeline at the bottom of the work screen and set the time domain values shown in Figure 12–111.**

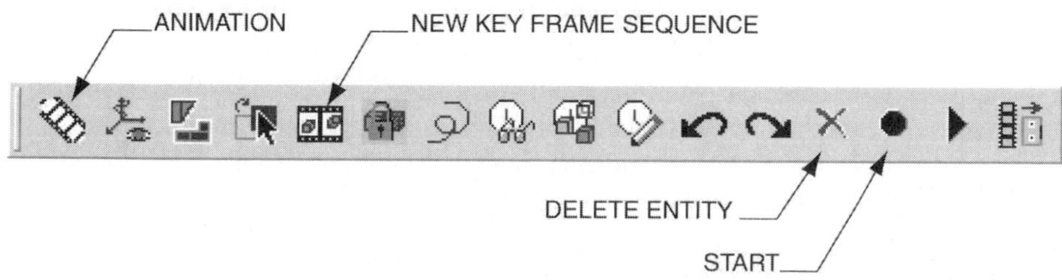

Figure 12–110 Animation dialog box

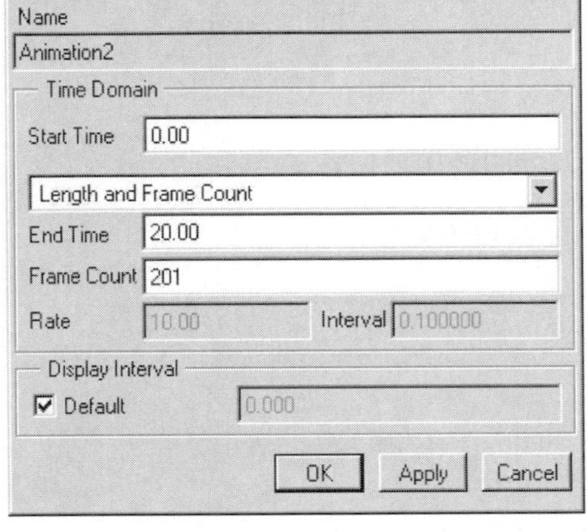

Figure 12–111 Time domain

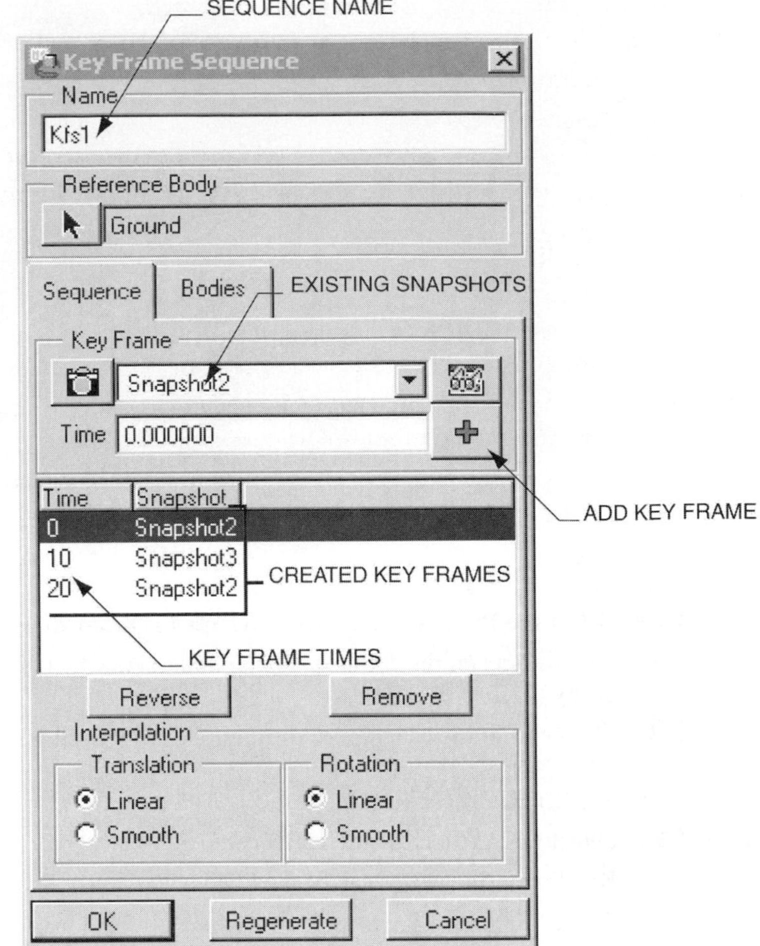

Figure 12–112 Key Frame Sequence dialog box

STEP 5: Select OKAY to create the time domain.

STEP 6: Select the NEW KEY FRAME SEQUENCE icon.

Multiple key frame sequences can be created for an animation. Within this tutorial, two will be used. The first sequence will utilize the two snapshots created previously to animate the handle and piston linkage. The second sequence will animate the adjuster.

STEP 7: On the dialog box, use the ADD KEY FRAME icon to create the three key frames shown on the Key Frame Sequence dialog box (Figure 12–112).

Two snapshots should currently exist. Use the Add Key Frame icon to create the three key frames shown in Figure 12–112 (snapshot1 at 0 sec, snapshot2 at 10 sec, and snapshot1 at 20 second). To perform this, select an existing snapshot, set a specific time value, and then select the Add Key Frame icon.

STEP 8: Select OKAY when your Key Frame Sequence dialog box matches Figure 12–112.

STEP 9: **Select the START icon on the Animation toolbar (Figure 12–111).**

Your handle and piston linkage should animate.

STEP 10: **Select the NEW KEY FRAME SEQUENCE icon.**

The next key frame sequence will animate the adjuster part.

STEP 11: **Select the NEW SNAPSHOT icon on the Key Frame Sequence dialog box.**

The New Snapshot icon will launch the Drag dialog box (Figure 12–113). This dialog box is also accessible directly from the Animation dialog box.

STEP 12: **Select the POINT DRAG icon, then drag the adjuster part to the SNAPSHOT3 position shown in Figure 12–113.**

Your snapshot numbers may be different from those represented in this tutorial. You can approximate the exact location for each shot.

STEP 13: **Select the SNAPSHOT icon to create SNAPSHOT3.**

STEP 14: **Select the POINT DRAG icon, then drag the adjuster part to the SNAPSHOT4 position shown in Figure 12–113.**

STEP 15: **Select the SNAPSHOT icon to create SNAPSHOT4 then close the Drag dialog box.**

STEP 16: **Modify SNAPSHOT4 (see Figure 12–114) to have a value of 10 seconds.**

STEP 17: **Select OKAY to close the Key Frame Sequence dialog box.**

Key Frame Sequences can be modified with the Animation >> Key Frame Sequence option on Pro/ENGINEER's menu bar. Your timeline should look similar to Figure 12–115. Key frames on the timeline are represented by the triangle symbol. They can be manipulated on the timeline by dragging with the mouse.

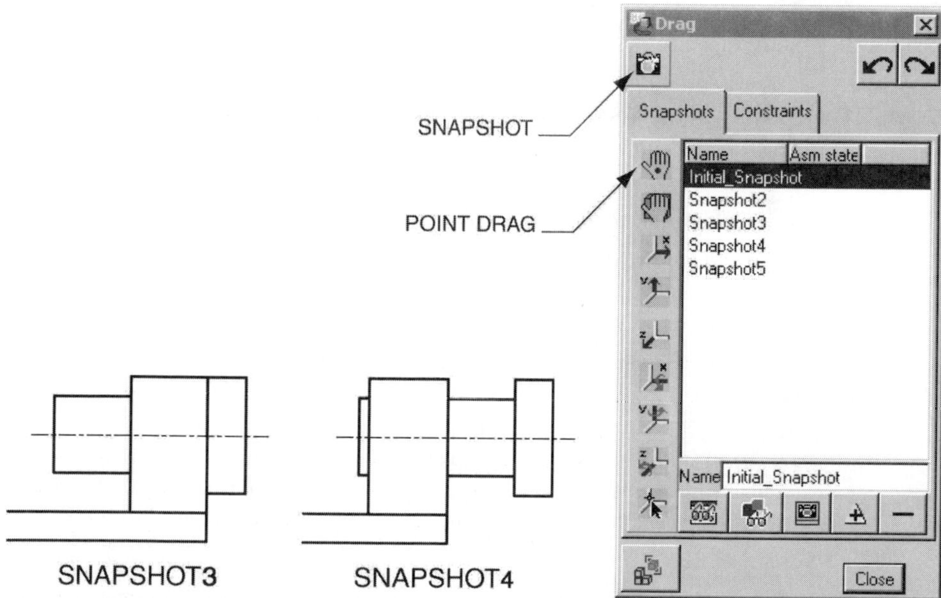

Figure 12-113 Snapshot creation

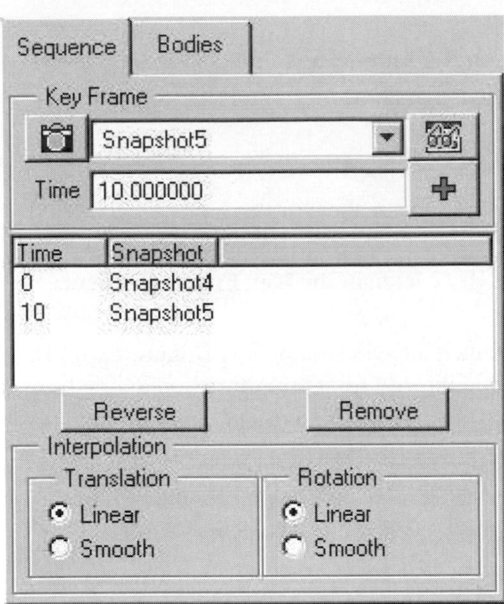

Figure 12–114 Key Frame Sequence dialog box

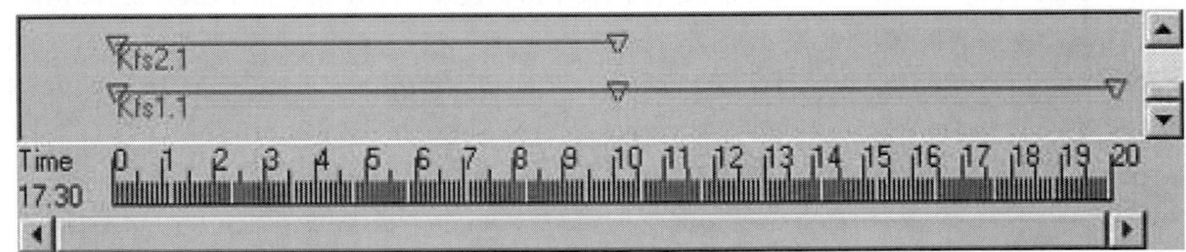

Figure 12–115 Animation timeline

STEP 18: **Use your mouse to drag the second key frame sequence to the position shown in Figure 12–115.**

STEP 19: **Run the animation by selecting the START icon on the Animation toolbar.**

STEP 20: **Playback the created animation by selecting the Playback icon.**

STEP 21: **Save your assembly file.**

PROBLEMS

1. This problem will have you model the assembly shown in Figure 12–116. This assembly consists of five components: one support, one cam, one guide, and two pins. Model each component in Part mode, then place them in the assembly using traditional Pro/ENGINEER constraints.

   Detail drawings of the support and cam parts are shown in Figures 12–117 through 12–118. One of the problems of this exercise is for you to design a guide part that will stay

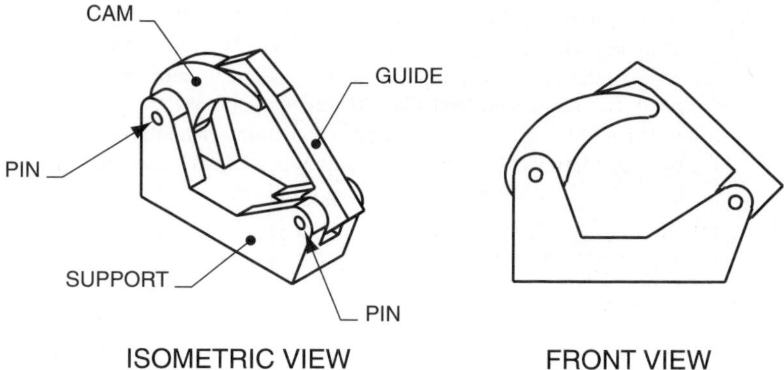

Figure 12-116 Assembly

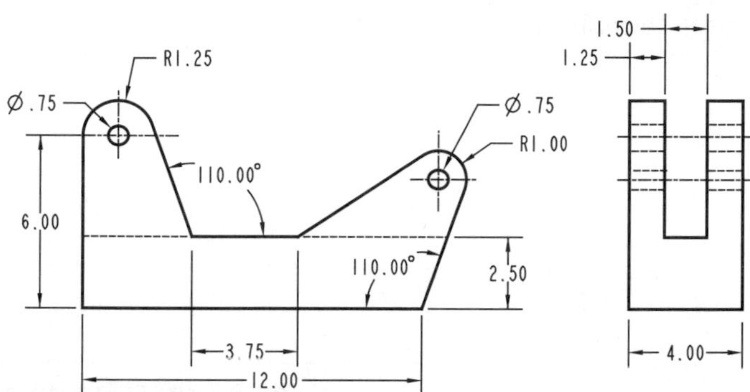

Figure 12-117 Support part

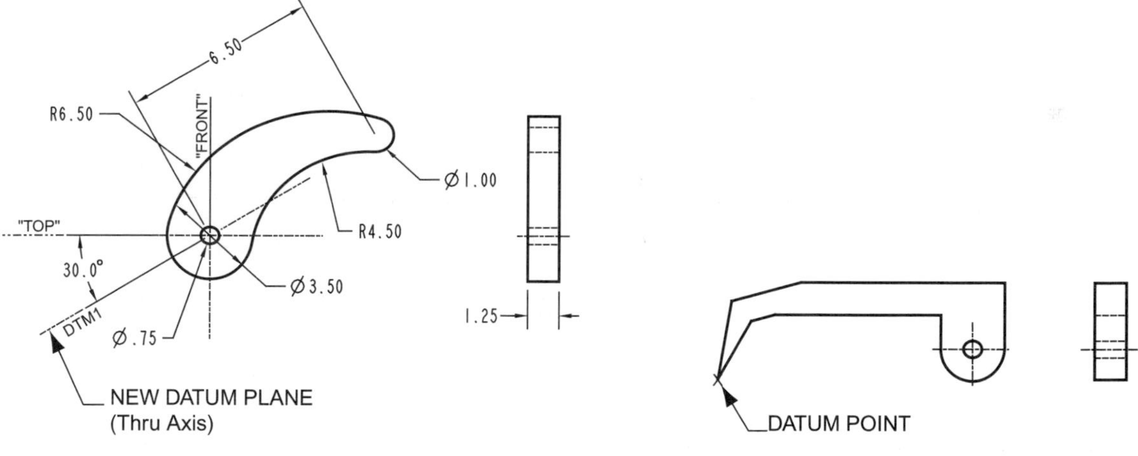

Figure 12-118 Cam part

Figure 12-119 Guide part

mated with the top surface of the cam (see Figures 12–116). As shown in Figure 12–118, when constructing the cam part, create a datum plane (DTM1) that will lie through the central axis of the cam and at an angle to one of Pro/ENGINEER's default datum planes. The angular dimension formed will be used within the assembly to vary the rotational position of the cam. The design intent for the assembly requires the angle shown in the cam drawing to be variable between −10 and 45 degrees (Figure 12–120). Within Figure 12–119, the dimensions for the guide part have been omitted. You must create a guide part that will allow for the −10 through 45 degrees of rotation of the cam, plus any necessary part clearances. Notice in Figures 12–116 and 12–120 how the point of the guide part is constrained to the top of the cam using a Point-On-Surface constraint. In addition, you must design and place the two pin components. After the assembly is complete, use a modify dimension option to change the angular value defining the datum plane.

2. Use Bottom-Up assembly design techniques to create the assembly model shown in Figure 12–121. Model each component in Part mode, then assemble them within Assembly mode. Detail drawings for the *Key* and *Plate* parts are shown in Figure 12–122. The drawing for the *Arm* part can be found in the Problems section of Chapter 5 (Figure 5–85). The drawings for the *Retainer*, *Body*, and *Shaft* parts can be found in the Problems section of Chapter 6 (Figures 6–52, 6–53, and 6–54).

3. Use Top-Down assembly design techniques to create the assembly model shown in Figure 12–121. Model each component in Assembly mode in the order listed in the Bill of Materials shown in the illustration. Detail drawings for the *Key* and *Plate* parts are shown in Figure 12–122. The drawing for the *Arm* part can be found in the Problems section of Chapter 5 (Figure 5–85). The drawings for the *Retainer, Body,* and *Shaft* parts can be found in the Problems section of Chapter 6 (Figures 6–52, 6–53, and 6–54).

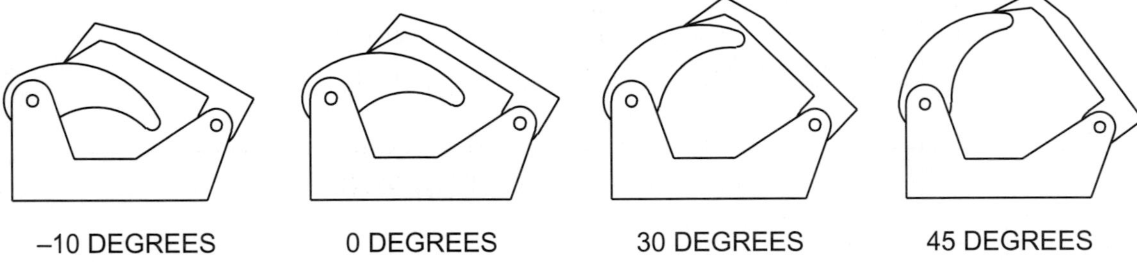

Figure 12–120 Degrees of movement

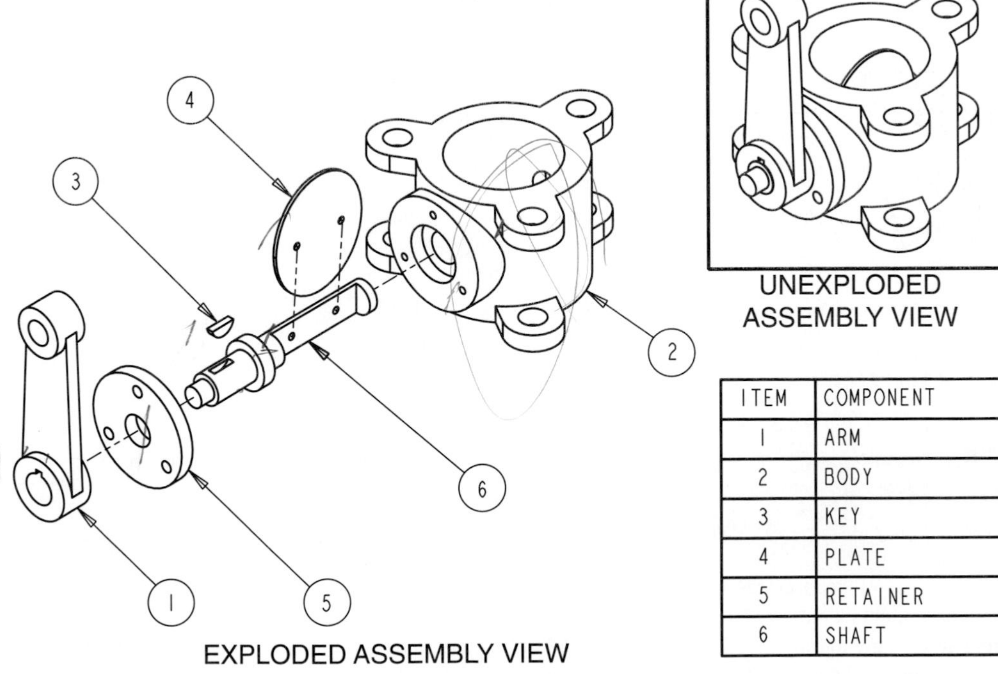

ITEM	COMPONENT
1	ARM
2	BODY
3	KEY
4	PLATE
5	RETAINER
6	SHAFT

Figure 12–121 Exploded and unexploded model

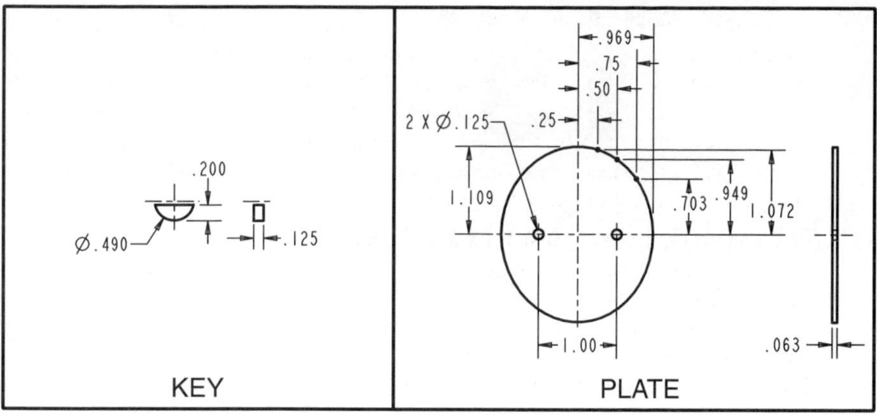

Figure 12-122 Key and plate parts

4. Use Bottom-Up mechanism design techniques to create the assembly model shown in Figure 12–121. Model each component in Part mode, then assemble them as a mechanism within Assembly mode. Detail drawings for the *Key* and *Plate* parts are shown in Figure 12–122. The drawing for the *Arm* part can be found in the Problems section of Chapter 5 (Figure 5–85). The drawings for the *Retainer, Body,* and *Shaft* parts can be found in the Problems section of Chapter 6 (Figures 6–52, 6–53, and 6–54).

Questions and Discussion

1. Describe the basic principles behind Bottom-up assembly design.
2. Describe the basic principles behind Top-down assembly design.
3. Describe the difference between the Align constraint option and the Mate constraint option.
4. How does the Offset Mate constraint differ from the Offset Align constraint?
5. What two constraint options can be used to place the axis of a shaft coaxial with the axis of a hole?
6. Describe four uses of a skeleton model.
7. How do dimension symbols within Assembly mode differ from dimension symbols within Part mode?
8. Describe uses of Pro/ENGINEER's Layout mode.
9. List and describe five mechanism joints.

CHAPTER

13

SURFACE MODELING

Introduction

Pro/ENGINEER's surface creation tools are useful for modeling parts with complex curves and surfaces. While most of Pro/ENGINEER's solid modeling tools are ideal for creating components with planar surfaces and smooth curves, these tools are not always best for complex shapes. Creating a surface model requires a different strategy when compared to a solid model. Even though many of Pro/ENGINEER surface creation tools are similar to solid creation options (e.g., Extrude and Revolve), additional tools such as Boundaries and Merge are available also. The key to creating a surface model is to define the skeleton of the part with appropriate datum options. Once this skeleton has been created, the surface skin can be added. Upon finishing this chapter, you will be able to

- Create extruded, revolved, swept, and blended surface features.
- Create a flat surface feature.
- Create a fillet between two surface features.
- Utilize datum curves to create surface features.
- Merge two quilts.
- Create trimmed features on a surface.
- Create a surface from boundaries.
- Create a solid feature from a merged quilt.

DEFINITIONS

Boundaries A technique for creating a surface feature by defining the surface's boundaries.

Merge The process of joining two surface quilts. Quilts that intersect and quilts that share common boundaries can be joined.

Quilt One or more surface features.

Use Quilt The process of using a quilt to create an additional part feature. The Solid Options menu's Use Quilt option is used to convert a quilt into a solid feature.

INTRODUCTION TO SURFACES

A *surface* is a geometric feature with no defined thickness. Surface features are often confused with thin features. Thin features are actually thin walled solids. The wall of a thin feature has a defined thickness. The wall of a surface feature does not have a defined thickness. Surface tools are used to create geometric shapes with complex contours and undulations. Figure 13–1 shows a comparison of two parts. In the illustration, the solid part's features and geometric shape make an ideal model for using solid creation tools. The surface part, on the other hand, would be better served with Pro/ENGINEER's surface creation tools. On first observation, the Solid part might appear to have surface modeling characteristics. It was actually created as a thin extruded Protrusion. The indentations were created with Tweak menu options. One characteristic of a part created as a surface is the number of features required to constructed the model. Figure 13–1 reveals the model trees for both parts. A typical surface part is composed of multiple datum curves and surfaces.

A surface model is composed of patches of surface features. The surface part shown in Figure 13–1 is composed of multiple surface features. The main body surface and the end features were created with the Boundaries option, while the hole was created with a combination of the Extrude, Merge, and Fillet options. The indented square feature was created with the Draft Offset option. Within Pro/ENGINEER, a **quilt** is a combination of one or more surface features. When surfaces are merged to form a completely enclosed quilt, the solid option **Use Quilt** (found under the Protrusion and Cut commands) can be used to convert the surface feature into a solid feature.

Surface features (quilts) are created in Part mode with the Surface menu option. When no quilts or datum curves currently exist in a model, Pro/ENGINEER will reveal the Surface Options menu (Figure 13–2). This menu is used for creating basic surface features.

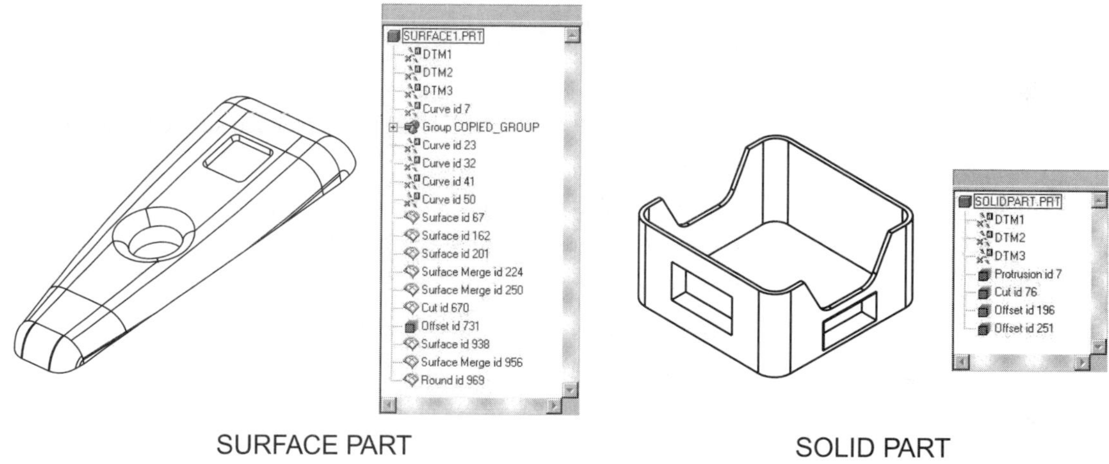

Figure 13–1 Surface versus solid part

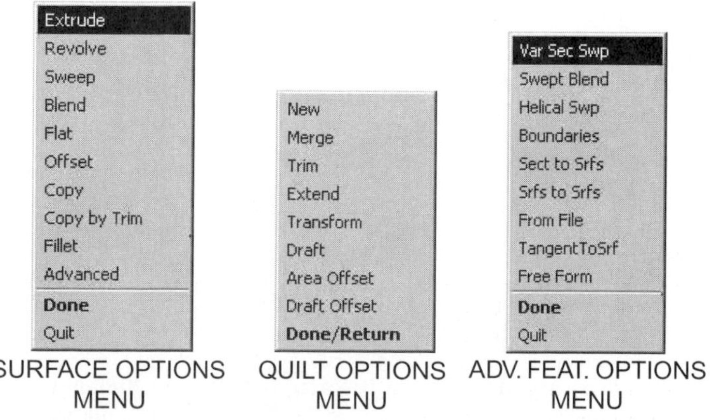

Figure 13–2 Surface menu options

Notice how many of the options available on this menu are the same options found under the Protrusion and Cut commands. Each surface option is similar to the solid option with the same name.

If one or more surface features exist when the Surface option is selected, Pro/ENGINEER will reveal the Quilt Options menu. With the exception of the Draft and Area-Offset options, the selections on this menu are unique to surface features. The New option will reveal the Surface options menu. On the Surface Options menu, the Advanced option will reveal the Advanced Feature Options menu. Notice how some of the options on this menu (Figure 13–2) are the same as options found under the Protrusion and Cut commands Advanced menu.

SURFACE OPTIONS

The Surface Options menu and the Insert >> Surface option are used to create basic surface features. Most of the options available are similar to options found under the Solid Options menu. Figure 13–3 shows four basic surface creation options: Extrude, Revolve, Sweep, and Blend. When a surface feature is created, with the exception of a shaded model, it is displayed in a wireframe format. The following is a description of each option:

EXTRUDE

The Extrude option creates a surface feature by extruding a sketched section. This option functions almost identically to the Extrude option found under the Protrusion and Cut commands. An extra option is provided for either capping the ends of the extrusion or leaving the ends open.

REVOLVE

The Revolve option creates a surface feature by revolving a sketched section around an axis of revolution. Like the various solid revolve options, the axis of revolution is a sketched centerline. This option works virtually identically to the Revolve options found under the Protrusion and Cut commands. Like the Surface-Extrude option, an option exists for either capping the ends of the revolution or leaving the ends option. Ends are available for capping with a less than 360-degree revolution.

SWEEP

The Sweep option creates a surface feature by protruding a section along a sketched or selected trajectory. Like the Extrude and Revolve options, this option functions similarly to options found under the Solid Options menu. Also, like the Extrude and Revolve options, an additional option is available for either capping the ends of the sweep or leaving the ends open.

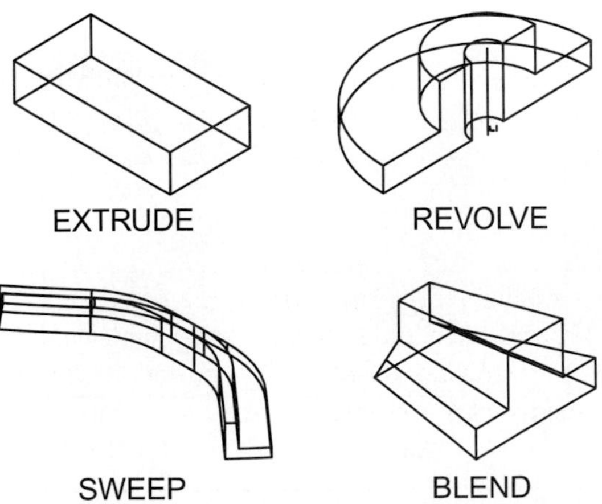

Figure 13–3 Surface creation options

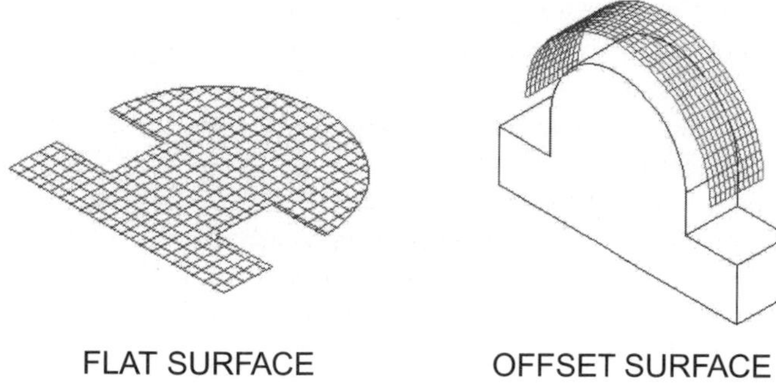

Figure 13–4 Flat and Offset options

BLEND

The Blend option creates a surface feature by protruding between two or more sections. This option functions identically to the Blend option under the Protrusion and Cut commands. Options are available for creating parallel, rotational, and general blends. In addition, the ends of the blend can be capped or left uncapped.

FLAT

The Flat option is used to create a two-dimensional surface feature (Figure 13–4). This option functions similarly to the extrude option, but without the depth parameter. Like the extrude option, the section of the feature is sketched.

OFFSET

The Offset option creates a new surface feature by offsetting from a solid or quilt (Figure 13–4). The user specifies the offset distance and surface to offset from.

COPY

The Copy option creates a surface feature on top of one or more selected surfaces. This option is useful for creating surface features out of existing solid features. The created surface can be exported as an IGES (International Graphics Exchange Standard) file to create a new surface model.

FILLET

The Fillet option functions identically to the Solid menu's Round command. Like the Round command, simple and advanced fillets can be created. Other options such as Variable radius and Full Round are available also. Unlike the Round command, Surf-Surf is the default selection option.

SURFACE OPERATIONS

As defined previously in this chapter, a Quilt is a combination of one or more surface features. Quilt options are available to manipulate and modify existing quilts. These options are accessible by selecting Insert >> Surface-Operations on the menu bar. The following options are available:

MERGE

The **Merge** option is used to join two or more quilts. This option can be used to combine two adjacent surfaces or it can be used to join two intersecting surfaces. Figure 13–5 shows an example of two surfaces joined with the Merge option. Within the figure, three possible solutions to the merge are shown.

TRIM

The Trim option (or Surface Trim) is very similar to the Solid Option menu's Cut command. Trim suboptions available include Extrude, Revolve, Sweep, Blend, Use

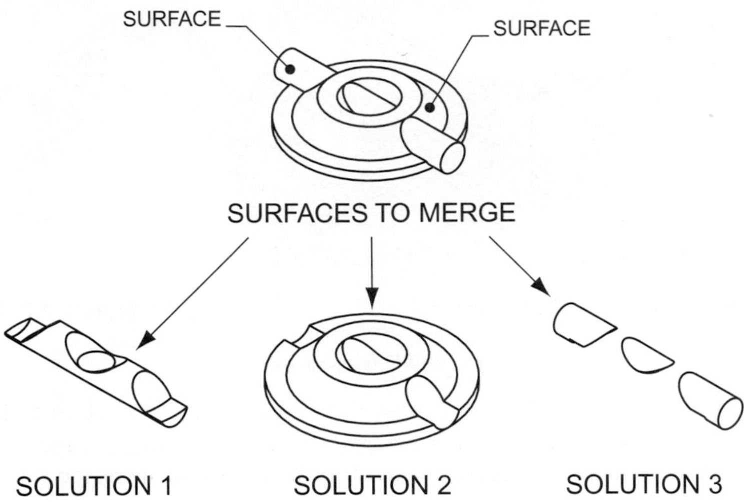

Figure 13-5 Merge solutions

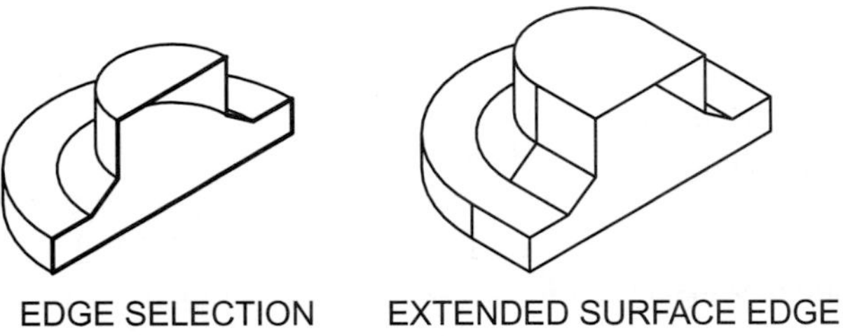

Figure 13-6 Extended edge

Quilt, and Advanced. Notice how many of the options are the same as the options found under Cut. When a cut is created on a surface, surface quilts are removed. As an example, if a circular sketch is extruded through a quilt using the Trim option, surfaces within the extrusion are removed without the addition of surfaces defining the hole's interior boundary.

Extend

The Extend option is used to extend a selected surface edge. Figure 13–6 shows an example of an edge that has been extended. This extended surface was created with the Same Surface option. Other options include Approximate Surface and Along Direction.

Transform

The Transform option is used to move, rotate, and mirror selected surfaces and datum curves. The options available under Transform work similarly to options found under the Copy command.

Draft

The Draft option creates a drafted surface from an existing planar quilt surface (Figure 13–7). This option works identically to the Draft command found under the Tweak menu. Refer to Chapter 5 for more information on creating a draft feature.

Area Offset

The Area Offset option creates a new surface by offsetting from an existing surface (Figure 13–8). A surface can be offset normal to the existing surface or in a given

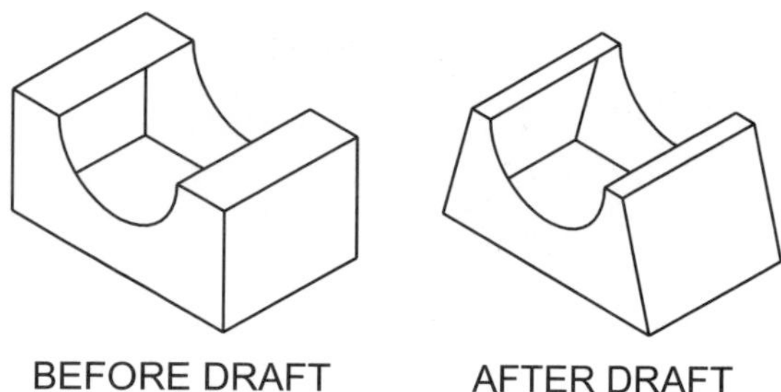

Figure 13–7 Drafted surfaces

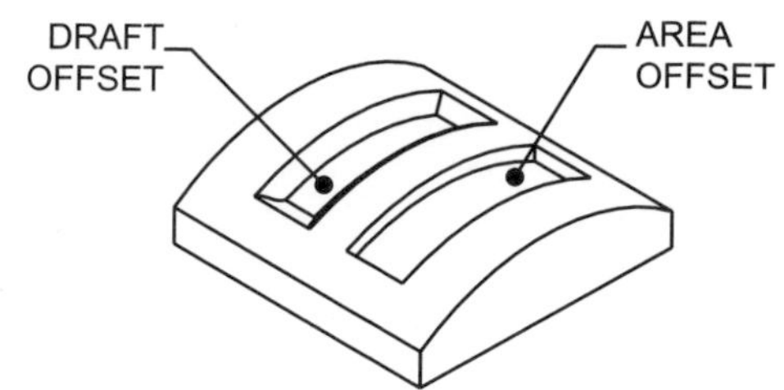

Figure 13–8 Area and draft Offset

direction. An entire surface can be offset, or as shown in the figure, the area to be offset can be sketched. The new surface can protrude away from the parent surface or into the parent feature. A solid Area Offset command exists under the Tweak menu too.

Draft Offset

The Draft Offset option is similar to Area Offset. The Draft Offset allows a beveled side around the offset. In addition, the drafted surface can be either straight or tangent to existing surfaces.

Advanced Surface Options

Advance surface options are available by selecting Insert >> Surface on the menu bar. Options available on this menu are similar to options available for creating solid features. The following is a description of some of the available options:

Variable Section Sweep

The surface Variable Section Sweep option is identical to the same option found for the creation of solids. This option is used to sweep a section along multiple trajectories. Figure 13–9 shows an example of a surface part created with this option. Refer to Chapter 11 for more information on creating a variable-section-sweep feature.

Swept Blend

A Swept Blend is a combination of a sweep and a blend (Figure 13–10). It is created by sweeping one or more sections along a user-defined trajectory. The trajectory can be selected on the work screen or sketched. This option functions identically to the same option for creating solid features.

Helical Sweep

The Helical-Sweep option creates a surface feature by revolving a sketched section around an axis and along a user-defined trajectory (Figure 13–10). Common features

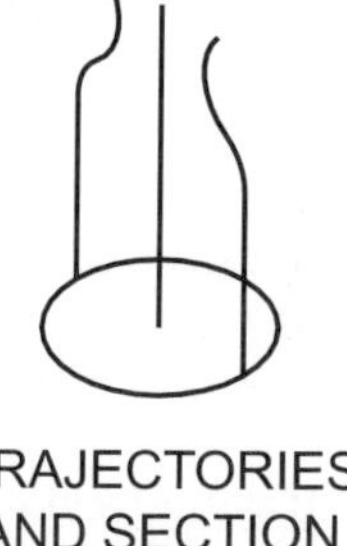

Figure 13–9 Variable section sweep

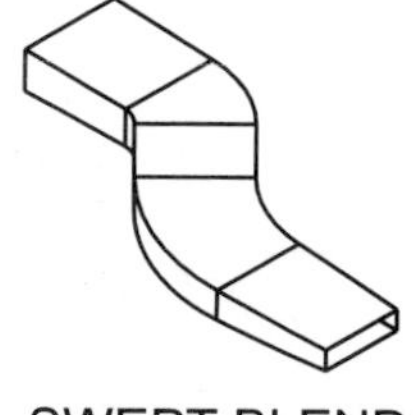

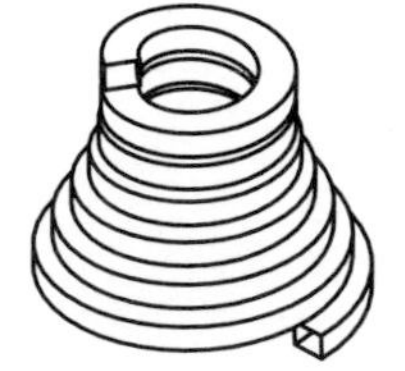

Figure 13–10 Swept blend and helical sweep

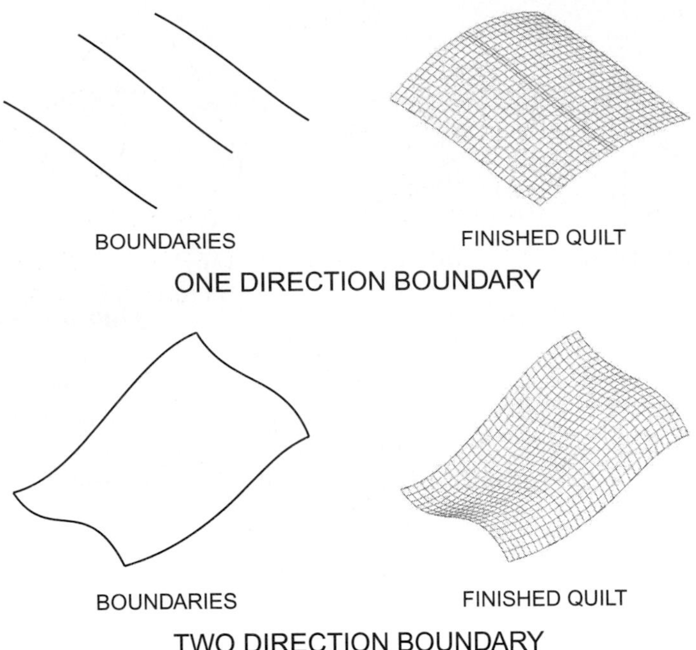

Figure 13–11 Boundaries option

created with this option are threads and springs. This option is identical to the Helical-Sweep option used for the creation of solids. Refer to Chapter 11 for more information.

BOUNDARIES

The **Boundaries** option is used to create a quilt by defining its boundaries. The feature's surface can be defined by selecting reference entities in one or two directions. Figure 13–11 shows examples of quilt surfaces created in one and two directions.

MERGING QUILTS

The Merge option is used to join two or more quilts. Two options are available: Intersect and Join. The Intersect option merges two intersecting quilts (Figure 13–12). The intersection edge of the two quilts is utilized as a knife-edge for trimming each quilt. The Join option merges two adjacent quilts. Each quilt has to be joined on one or more edges. Used in conjunction with the Protrusion >> Use-Quilt option, quilts can be converted to solid features.

This guide will demonstrate the creation of the merged intersecting quilts shown in Figure 13–12. Perform the following steps to merge two intersecting quilts:

STEP 1: Select INSERT >> SURFACE OPERATIONS >> MERGE on the Menu bar.

After selecting the Merge option, Pro/ENGINEER will reveal the Surface Merge dialog box (Figure 13–13). This dialog is used to select quilts to merge and to select the type of merge to execute.

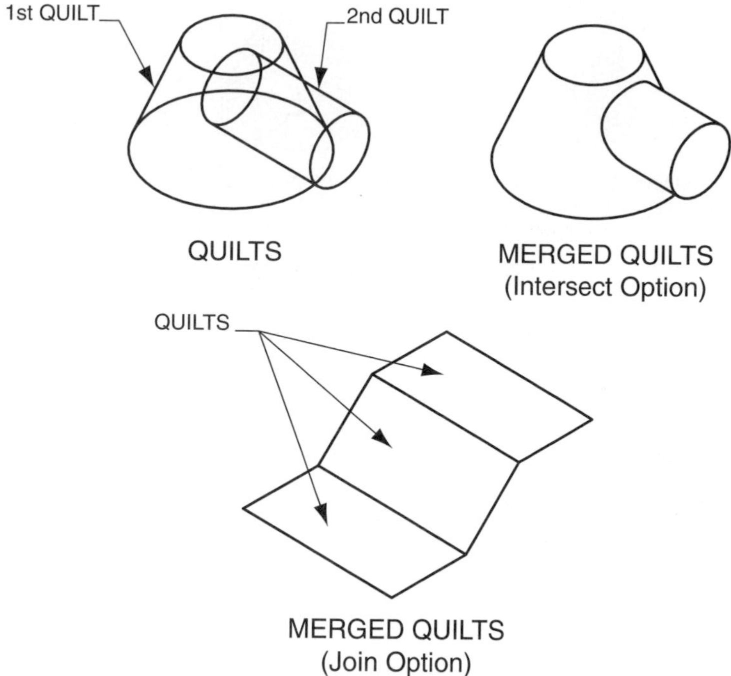

Figure 13-12 Surface merge

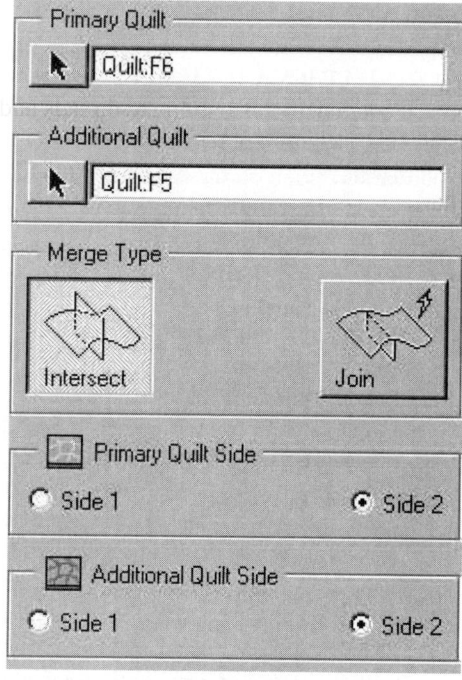

Figure 13-13 Merge Options dialog box

STEP 2: On the work screen, select the Primary Quilt for merging.

The merge option combines two quilts. The first quilt selected is the primary quilt, and the second quilt is the additional quilt.

STEP 3: On the work screen, select the Additional Quilt for merging.

After selecting the Additional Quilt, Pro/ENGINEER will preview the merged quilts based on the Primary and Additional Quilt Side selections (Figure 13–14).

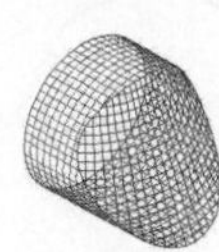

PRIMARY QUILT: Side 1
ADDITIONAL QUILT: Side 1

PRIMARY QUILT: Side 1
ADDITIONAL QUILT: Side 2

PRIMARY QUILT: Side 2
ADDITIONAL QUILT: Side 1

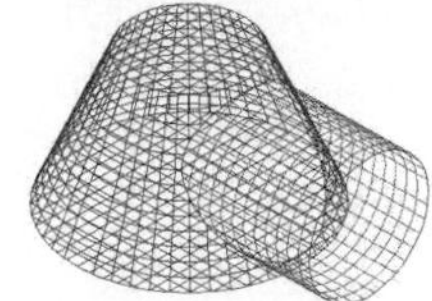

PRIMARY QUILT: Side 2
ADDITIONAL QUILT: Side 2

Figure 13–14 Merged Quilts side options

STEP 4: On the Surface Merge dialog box, select the quilt sides that will create the correctly merged feature (Figure 13–14).

STEP 5: On the Surface Merge dialog box, select the Build Feature icon to create the merged surface.

BOUNDARIES OPTION

A surface feature can be created by selecting bounding edges of the feature through the Boundaries option on the Advanced Features Options menu (or the From Surfaces option on the menu bar). Four suboptions are available for the creation of boundary surfaces. The following is a description of each:

BLENDED SURFACE

The Blended Surface suboption of the Boundaries command creates a quilted surface by defining the surface's external boundaries (Figures 13–15). Selectable entities include datum curves and datum points. Selected entities can lie in either one or two directions. The illustration of the blended surface shown in the figure represents a quilt defined by selecting entities lying in two directions.

CONIC SURFACE

The Conic Surface suboption creates a quilted surface between two selected boundaries. The surface is formed by a third controlling curve (Figure 13–15). Two options are available for this controlling curve: Shoulder Curve and Tangent Curve. With the Shoulder Curve option, the surface feature passes through the control curve. With the Tangent Curve option, the surface feature does not pass through the curve.

APPROXIMATE BLEND

The Blended Surface suboption of the Boundaries menu creates a quilted surface by defining the surface's boundaries (Figure 13–15). Unlike the Blended Surface option, an additional curve can be selected for the manipulation of the surface's form and shape. This additional curve does not form the boundary of the quilt.

N-SIDED SURFACE

The N-Sided Surface suboption is used to create a quilted surface from more than four bounding entities (Figure 13–15).

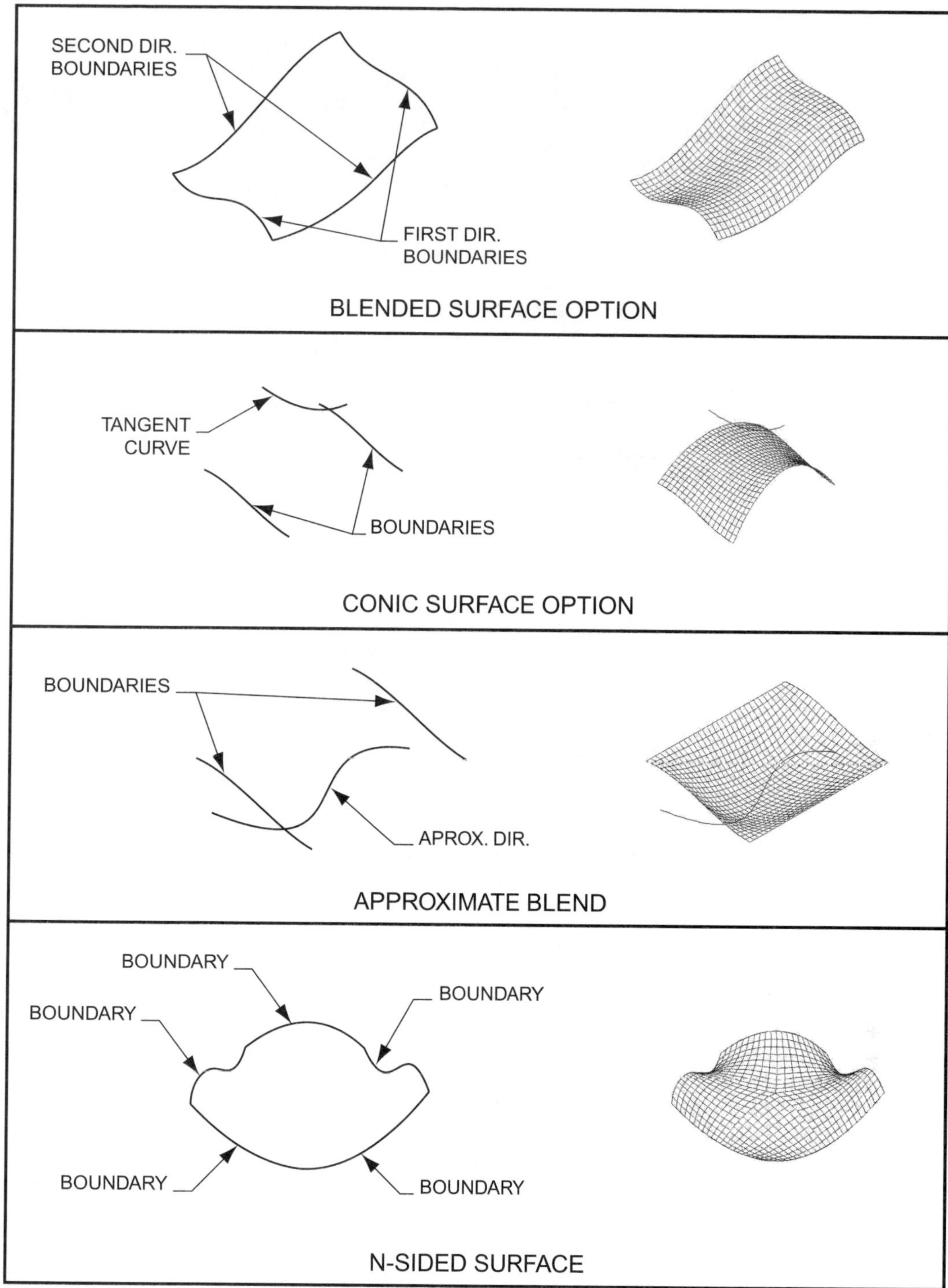

Figure 13–15 Boundaries options

Creating a Blended Surface from Boundaries

This guide will create the quilted surface shown in Figure 13–16. The Boundaries option from the Advanced Feature Options menu will be used with the Blended Surface sub-option. As shown in the figure, the boundaries forming the quilt consist of datum curves running in two directions. Perform the following steps to create this feature.

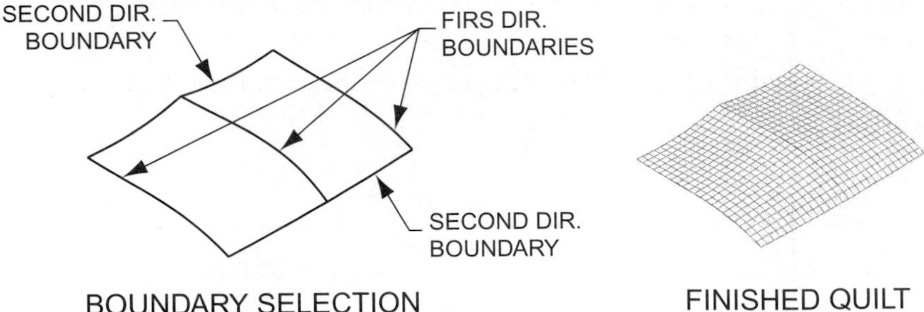

Figure 13–16 Blended surface feature

STEP 1: Select INSERT >> SURFACES >> FROM BOUNDARIES on the Menu bar.

The Boundaries option is also available on the Surface Options Advanced menu (Surface >> New >> Advanced).

STEP 2: Select BLENDED SURF >> DONE on the Boundaries Options menu.

After selecting the Blended Surface option, Pro/ENGINEER will launch the Surface Feature Definition dialog box and the Curve Options menu. The Curve Options menu is used to select bounding curves.

STEP 3: On the Curve Options menu, select the FIRST DIR and ADD ITEM options.

These options should be selected by default. The First-Direction (First Dir) option is used to select bounding curves in the first direction. The Second-Direction (Second Dir) option will be used in a later step to select bounding curves in the second direction. Pro/ENGINEER only requires a boundary definition in one direction.

STEP 4: On the work screen, select curve entities that define the first direction of the surface feature.

Multiple entities can be defined as boundaries for the feature. The following rules apply:

- Curves, edges, datum points, and vertices can be used as bounding entities.
- Entities must be selected in consecutive order.
- For boundaries defined in two directions, the bounding entities must form a closed loop. As shown in Figure 13–16, the vertices of the bounding datum curves are connected to form a closed loop.

> **INSTRUCTIONAL NOTE** Boundaries for a surface can be selected in one direction or two directions. For two directional boundaries, proceed to the next step of this guide. For one directional boundaries, skip to step 7.

STEP 5: On the Curve Options menu, select the SECOND DIR option.

STEP 6: On the work screen, select curve entities that define the second direction of the surface feature.

STEP 7: On the Curve Options menu, select the DONE CURVES option.

The Done-Curves option will end the selection of curves for defining the surface's boundaries.

STEP 8: Preview the feature then select OKAY on the Feature Definition dialog box.

USE QUILT

Merged quilts can be used to create solid features. The Protrusion and Cut commands' **Use Quilt** options can be used to create either positive or negative space features. Quilts defining positive space solids must be completely "water tight." In other words, no holes or openings can exist in the quilt. Quilts defining the boundary of a solid also have to be merged. As with most options under each command, the Thin option is available. A quilt does not have to be completely enclosed when a thin solid is being defined. The Protrusion >> Use Quilt >> Thin option is a handy tool for creating a solid model from complex surface features. Figure 13–17 shows an example of how quilts can be converted to positive and negative space solid features. The primary revolved feature is constructed as a thin protrusion, while the torus-shaped surface feature is constructed as a solid cut feature. Perform the following steps to create a solid from a quilt.

STEP 1: Select FEATURE >> CREATE >> PROTRUSION (or Cut).

Surfaces used to create a solid must be joined with the Merge option before a solid feature can be created from them.

STEP 2: Select USE QUILT on the Solid Options menu.

STEP 3: Select either SOLID or THIN, then select DONE.

The Solid option will create a feature solid throughout its geometry. The Thin option will create a feature with a defined wall thickness. After selecting Done, Pro/ENGINEER will launch the Use Quilt dialog box (Figure 13–18).

STEP 4: On the work screen, select a quilt for use in building the solid feature.

STEP 5: On the Use Quilt dialog box, select an appropriate Material Side (for Thin features only).

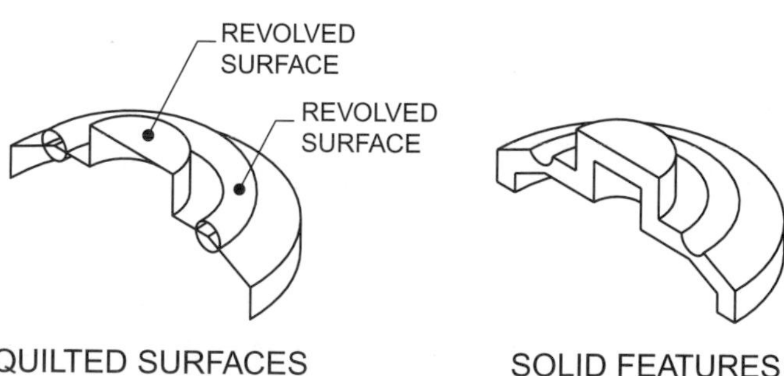

Figure 13–17 Solid features from quilts

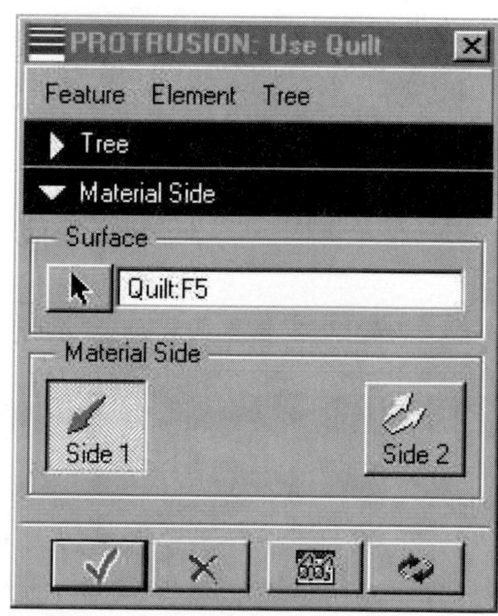

Figure 13–18 Use Quilt option

STEP 6: For thin features, enter a thickness for the wall of the solid (Figure 13–19).

This step is available only with the Thin option.

STEP 7: Select the Build Feature icon on the dialog box to create the feature.

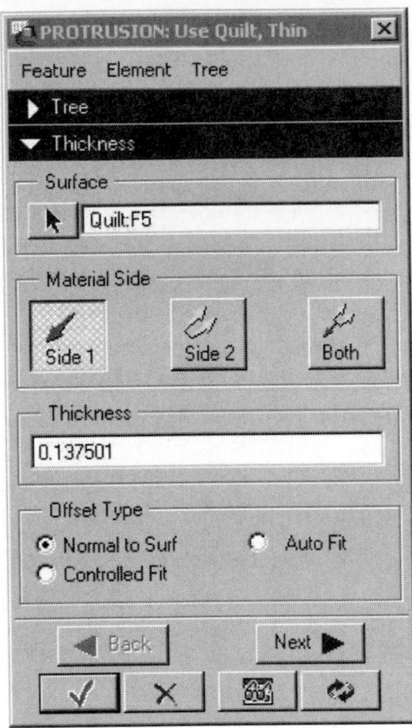

Figure 13–19 Use Quilt Thin dialog box

SUMMARY

Pro/ENGINEER's surface options are useful for the creation of complex geometric shapes. Solid creation tools are useful for creating hard mechanical components, but lack the flexibility to create free forming surfaces. While standard Pro/ENGINEER options such as Extrude, Revolve, Sweep, and Variable Section Sweep are available for creating surfaces and solids, additional surface options like Merge and Boundaries are not available for the creation of solids. Solids can be converted from surface quilts through the Use Quilt option.

TUTORIAL

Surface Tutorial 1

The objective of this tutorial is to familiarize you with the basic techniques behind creating surface features in Pro/ENGINEER.

This tutorial exercise will provide instruction on how to model the part shown in Figure 13–20.

Within this tutorial, the following topics will be covered:

- Creating an extruded surface feature.
- Creating datum curves.
- Trimming surface features.
- Creating an approximate boundary surface.
- Merging quilts.
- Creating a solid from a quilt.

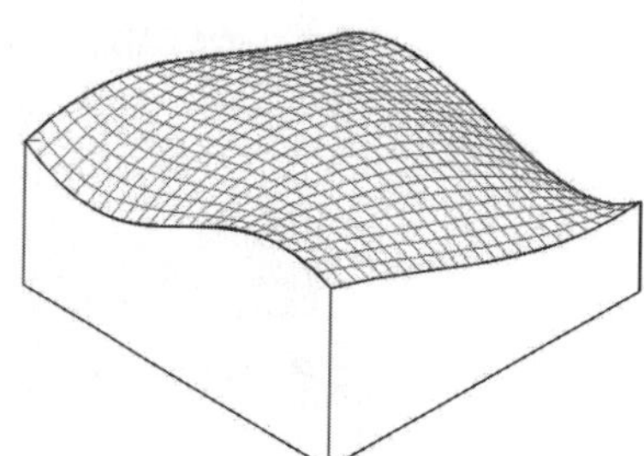

Figure 13-20 Finished model

Create the Base Extruded Surface Feature

The first segment of this tutorial will create an extruded surface feature (Figure 13–21). Extrude options (surface and trim functions) under the Surface menu option work basically the same as the Extrude options under the Protrusion, Cut, and Slot commands. For Extruded surfaces, an option is available for capping the end of the extrusion.

Step 1: Start a New Part file.

Use the File >> New option to create a new Part object file named *SURFACE1* (use the default template file).

Step 2: Select the FEATURE >> CREATE >> SURFACE option.

Step 3: Select EXTRUDE >> DONE on the Surface Options menu.

Do you notice any similarities between the Surface Option menu and options commonly found under the Protrusion and Cut commands? These surface options work similarly to the same options available for the creation of solid features.

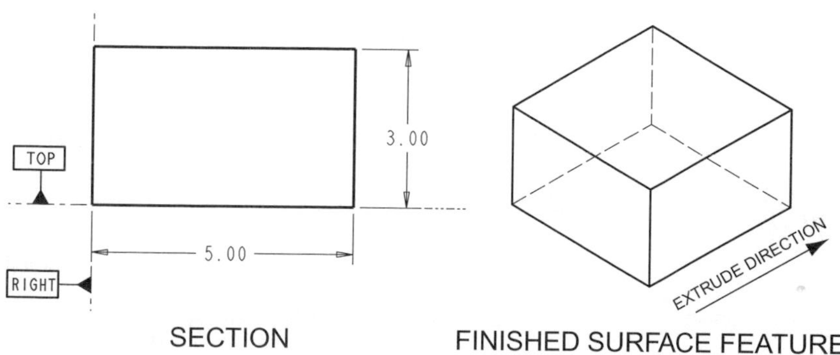

Figure 13-21 Extruded surface feature

STEP 4: On the Attributes menu, select ONE SIDE and CAPPED ENDS as attributes for the surface feature; then select DONE.

As with all extruded features, an option is available for extruding one direction or both directions from the sketching plane. With extruded surface features, the ends of the feature are left open. The Capped Ends option will close these ends.

STEP 5: Select datum plane FRONT as the sketching plane for the feature, accept the default direction of extrusion, and then orient datum plane TOP toward the top of the work screen.

STEP 6: Sketch the Section shown in Figure 13–21.

Use the following options when creating the section:

- Specify datum planes RIGHT and TOP as references.
- Use the Rectangle option to sketch the section.
- Use the dimensional sizes shown in Figure 13–21.

STEP 7: Select the Continue icon to exit the sketching environment, then enter 5.00 as the BLIND extrude distance.

STEP 8: Select the Preview option on the feature definition dialog box and preview the feature in the No-Hidden display mode.

Notice the difference in appearance for a surface model compared to how a solid model would look. In No-Hidden display mode, you cannot see the surface mesh of the surface feature. Due to this, the model can appear ambiguous. Try using the Shade display mode.

STEP 9: Select OKAY to create the feature.

STEP 10: Save your model.

CREATING DATUM CURVES

This segment of the tutorial will create the five datum curves shown in Figure 13–22. Datum curves and points are important features for the creation of surfaces. Within this tutorial, the four connected spline datum curves will be used as boundaries for the creation of a surface quilt. The fifth datum curve will be used as an approximate curve for manipulating the undulation of the surface.

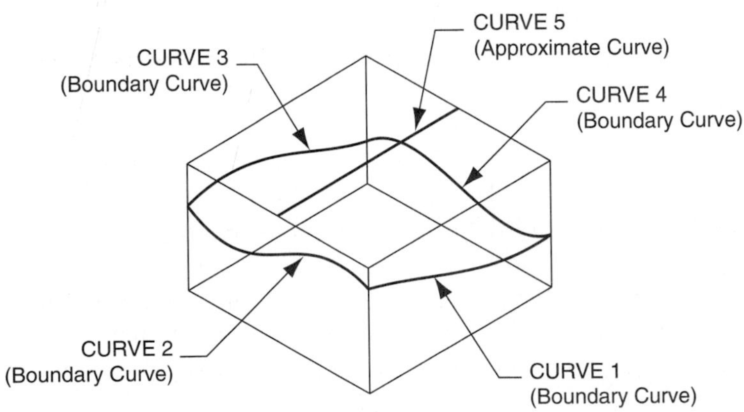

Figure 13–22 Datum curves

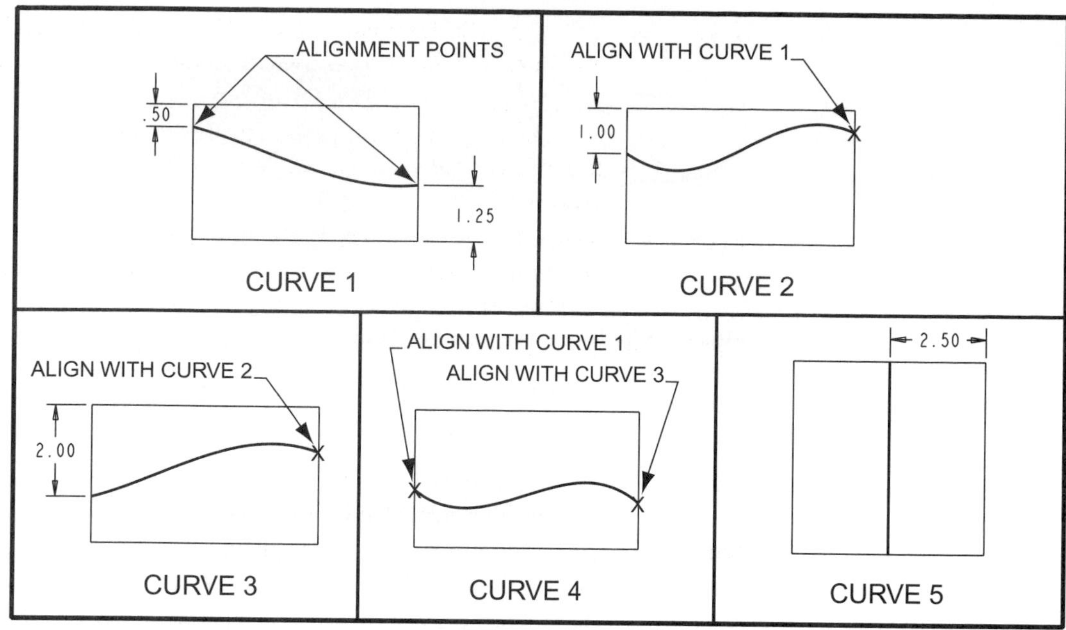

Figure 13-23 Curve regenerated sections

Use the Datum Curve icon and the Sketch suboption to create each datum curve. Use appropriate existing planar surfaces as sketching planes. Each curve is a separate feature. If necessary, use the Query Select option to select hidden sketching planes. While sketching, the vertices of each curve must align. When specifying references for curves 2, 3, and 4, dynamically rotate the model to better select the vertices of existing curves. Within the Sketching environment, use the Spline option to create curves 1 through 4. The sections for each curve are shown in Figure 13–23. Approximate the exact spline shape of each curve.

TRIMMING SURFACE FEATURES

This segment of the tutorial will use the previously created datum curves to trim the base surface feature. The finished trim is shown in Figure 13–24.

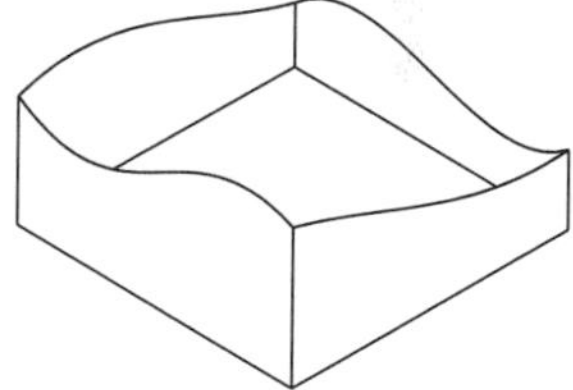

Figure 13-24 Trimmed surface

STEP 1: Select FEATURE >> CREATE >> SURFACE.

STEP 2: Select the TRIM option on the Quilt Surface menu.

STEP 3: USE CURVES >> DONE on the Form menu.

Before selecting Done, did you notice how some of the options on the Form menu resemble options found under the Protrusion and Cut commands. The Extrude, Revolve, Sweep, and Blend options function similarly to their solid counterparts. The Use-Curves option is used to trim a surface quilt along a datum curve.

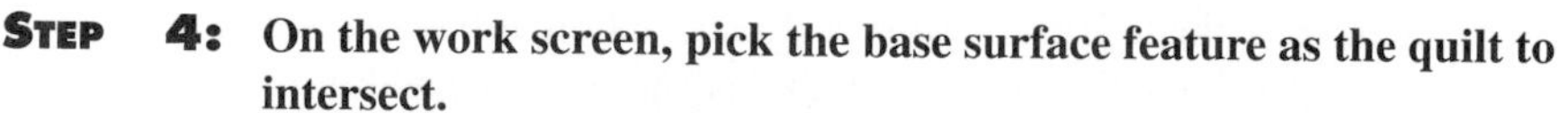

STEP 4: On the work screen, pick the base surface feature as the quilt to intersect.

Pick the first surface feature created in this tutorial. This selection will be the surface that is trimmed. You can trim only one surface at a time.

STEP 5: On the work screen, pick the four spline datum curves (curves 1, 2, 3, and 4) created previously in this tutorial.

STEP 6: On the Chain menu, select DONE to end the selection of curves.

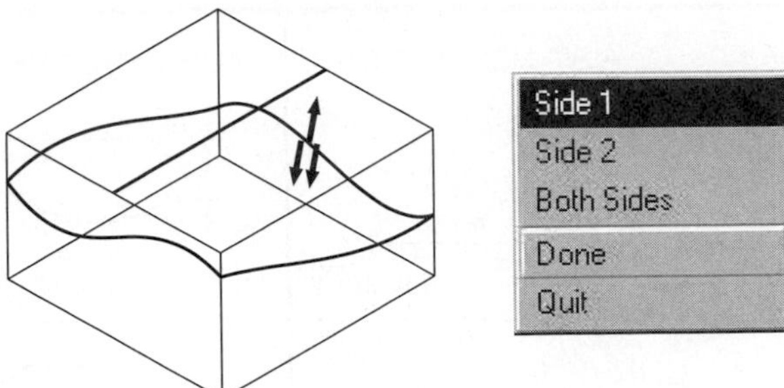

Figure 13–25 Material removal side

STEP 7: Use the Material Side menu to select a material removal side based on information provided in the message area.

Pro/ENGINEER will graphically provide arrows pointing toward possible material removal sides (Figure 13–25). Pro/ENGINEER will also provide the following message: "Arrow(s) point TOWARD area to be KEPT. Pick Side 1 (one red arrow), Side 2 (two yellow arrows), Both Sides, Quit, or Done". Based on the text in the message and the direction of your arrows, select the material removal side.

STEP 8: Select DONE to enter a material removal side (Figure 13–25).

STEP 9: Preview the feature, then select OKAY on the Feature Definition dialog box.

STEP 10: Save your model file.

CREATING AN APPROXIMATE BOUNDARIES SURFACE

This segment of the tutorial will create the Approximate Boundaries Surface shown in Figure 13–26. Surface features created with the Boundaries option require the selection of boundary entities that define the exterior of the surface. With the Blended Surface and Approximate Blend options, these boundaries must form a closed loop. With the Approximate Blend option, an additional curve can be selected that will allow for extra control in the variation of the feature's surface.

STEP 1: Select FEATURE >> CREATE >> SURFACE.

STEP 2: On the Quilt Surface menu, select the NEW option.

STEP 3: Select ADVANCED >> DONE on the Surface Options menu.

STEP 4: Select BOUNDARIES >> DONE on the Advanced Feature Options menu.

STEP 5: On the Boundaries Options menu, select the APPROX BLEND option; then select DONE.

The Approximate Blend (Approx Blend) option will create a surface between defined directional boundaries and with one additional approximate direction. The Blended Surface option produces the same type surface, but without the approximate direction.

With the Approximate Blend and Blended Surface options, directional curves can be selected in one or two directions. When selecting surface boundaries in two directions, the curves must be joined at vertices and exterior curves must close.

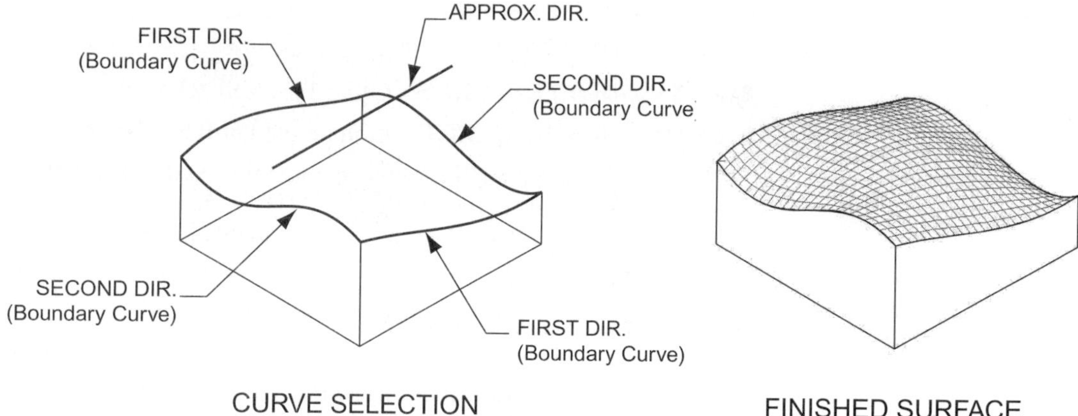

Figure 13–26 Curve selection and finished surface

STEP 6: Select the FIRST DIR option on the Curve Options menu (Figure 13–27).

STEP 7: On the work screen, select the two First Direction curves shown in Figure 13–26.

STEP 8: Select the SECOND DIR option on the Curve Options menu.

STEP 9: On the work screen, select the two Second Direction curves shown in Figure 13–26.

STEP 10: Select the APPROX DIR option.

The Approximate Direction curve option is used to select an additional reference curve. This additional curve will allow for a deviation in the surface of the quilt.

STEP 11: On the work screen, select the Approximate Direction curve shown in Figure 13–26.

STEP 12: Select DONE CURVES.

The Done Curves option is used to end the selection of curves.

STEP 13: Enter .500 as the Smoothness Parameter.

STEP 14: Enter 10 as the number of patches in the First and Second Directions.

The Smoothness Parameter from step 13 must be a value between 0 and 1. The Number of Patches parameter is used to specify the number of patches to create in each direction of the quilt. The more patches, the closer the surface will match the directional curves.

STEP 15: Create the feature by selecting OKAY on the Surface Feature Definition dialog box.

STEP 16: Select DONE/RETURN to exit the Quilt Surface menu.

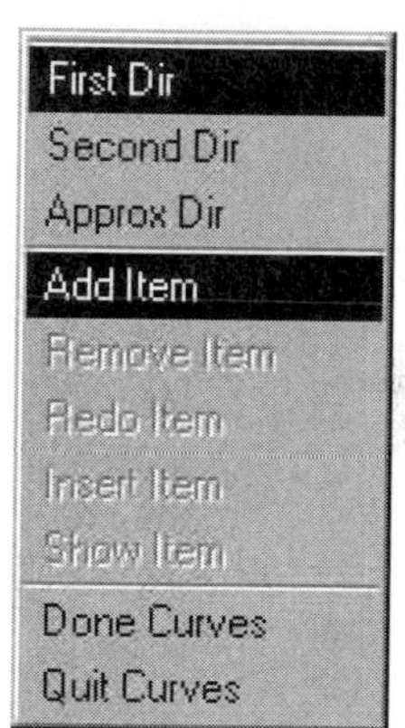

Figure 13–27 Curve Options menu

MERGING QUILTS

Your model currently consists of three surface features: two surfaces and one cut. On the model tree, notice these features and the five datum curves used to create them. Surface models typically have more features than solid models. This segment of the tutorial will merge the two surfaces. The intent of this tutorial is to create a final solid model of the design. To convert a surface model to a solid model, each surface has to be merged.

STEP 1: Select FEATURE >> CREATE >> SURFACE.

STEP 2: Select the MERGE option on the Quilt Surface menu.

After selecting the Merge option, Pro/ENGINEER will launch the Surface Merge dialog box (Figure 13–28). The first two selections of this dialog box require you to select two surfaces to merge. The first surface is the Primary Quilt, and the second surface is the Additional Quilt (see dialog box).

STEP 3: On the work screen, select the Primary Quilt.

Select the Base surface feature created in the first segment of this tutorial.

STEP 4: On the work screen, select the Additional Quilt.

Select the Boundaries surface feature previously created in this tutorial. Notice your Surface Merge dialog box and (in Figure 13–28) how quilt identifications have been entered in the Primary Quilt and Additional Quilt textboxes.

STEP 5: On the Surface Merge dialog box, select JOIN as the Merge Type.

Two Merge Types are available: Intersect and Join. The Intersect option will merge two quilts that overlap. The Join option will merge two quilts that have common boundaries. Since in this model no surfaces are overlapping, the Intersect option is not appropriate.

STEP 6: ✓ On the dialog box, select the Build Feature icon to create the surface feature.

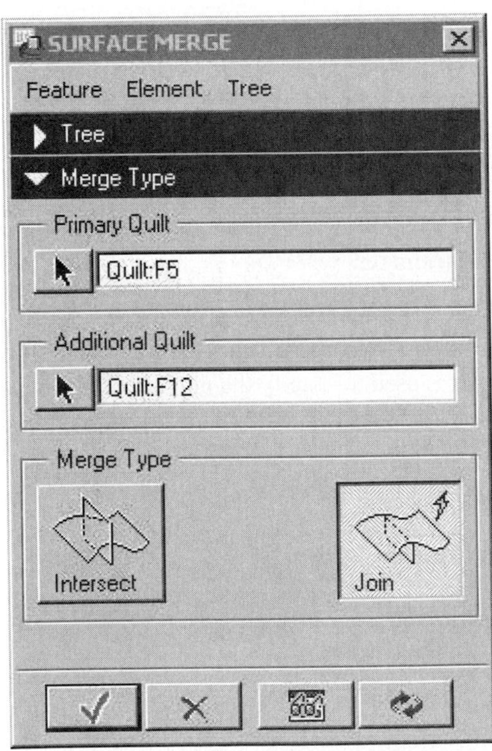

Figure 13–28 Surface Merge dialog box

On your model tree, notice how a Surface Merge feature has been added. With the Merge option, the two parent features of the merge remain separate definable features in the order of regeneration.

STEP 7: Shade your model to better visualize the surface feature.

STEP 8: Save your model file.

CREATING A SOLID FROM A QUILT

Surface features have no defined thickness. How often does a design surface not have at least a minimum surface thickness? Merged surfaces can be converted to solids with the Use Quilt option found under the Protrusion and Cut commands. This segment of the tutorial will create a solid feature from the existing merged quilts.

STEP 1: Select FEATURE >> CREATE >> PROTRUSION.

Since the Use Quilt option creates a solid feature from a surface feature, this option is located under the Solid menu.

STEP 2: Select USE QUILT >> SOLID >> DONE.

When you created the boundary surface earlier in this tutorial, you defined the number of patches to equal 10 in both directions. If you would have accepted the default value of 5, then the model would not have been closed well enough to use the solid option in this step.

After selecting the Done option, Pro/ENGINEER will launch the Use Quilt dialog box.

STEP 3: On the work screen, select the model.

STEP 4: On the Use Quilt dialog box (Figure 13–29), select the Build Feature icon to build the solid model.

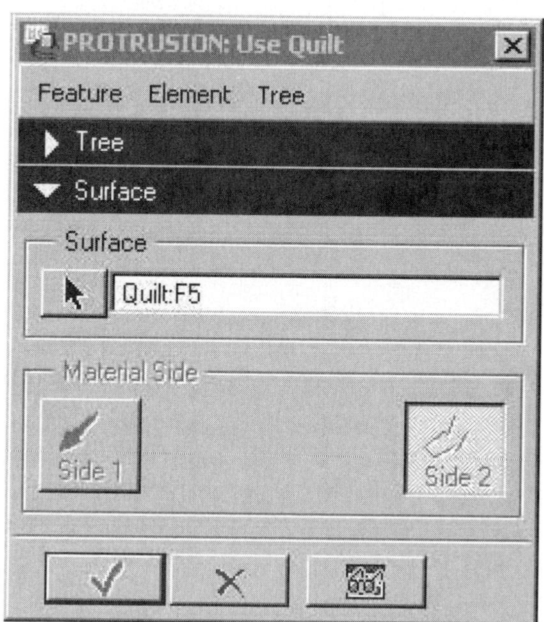

Figure 13–29 Use Quilt dialog box

Figure 13-30 Finished model

You model should appear as shown in Figure 13–30. You can use solid modeling tools (Protrusion, Cut, etc.) to build upon a model created with the Use Quilt option. On your model tree, notice the number of features it took to build this part.

STEP 5: Save your model.

> **MODELING POINT** Notice the absence of datum curves in Figure 13–30. Datum curves are not merged with surfaces and are not affected by the Use Quilt option. The datums in this example were placed on a layer, and the display of the layer was hidden. See Chapter 2 for information on creating layers.

TUTORIAL

SURFACE TUTORIAL 2

This tutorial will expand upon the concepts covered in the first tutorial. This project will focus on modeling the hair dryer shell shown in Figure 13–31. Like most surface models, the first segment of this tutorial will involve the creation of datum curves to control the creation of surface features.

This tutorial exercise will provide instruction on how to model the part shown in Figure 12–31.

Within this tutorial, the following topics will be covered:

- Creating datum curves.
- Creating surfaces from boundaries.
- Defining boundary conditions.
- Merging surfaces.
- Filleting surface features.
- Trimming a surface.
- Creating a draft offset.
- Converting a surface to a solid.

CREATING DATUM CURVES

This segment of the tutorial will establish the object's modeling environment and create the datum curves necessary to define the part.

STEP 1: Start Pro/ENGINEER and create a new Part file named *SURFACE2*.

STEP 2: Use the Datum Curve option and the Sketch suboption to create the first datum curve.

The datum curves used to construct this part are shown in Figure 13–32. With the exception of one mirrored curve, each will be created with the sketch option.

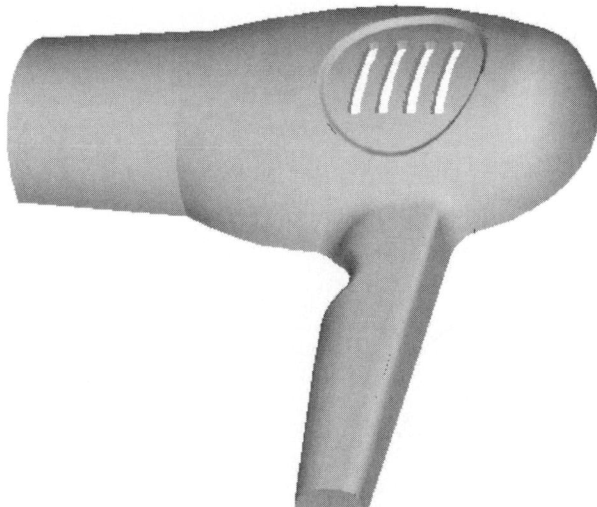

Figure 13–31 Finished model

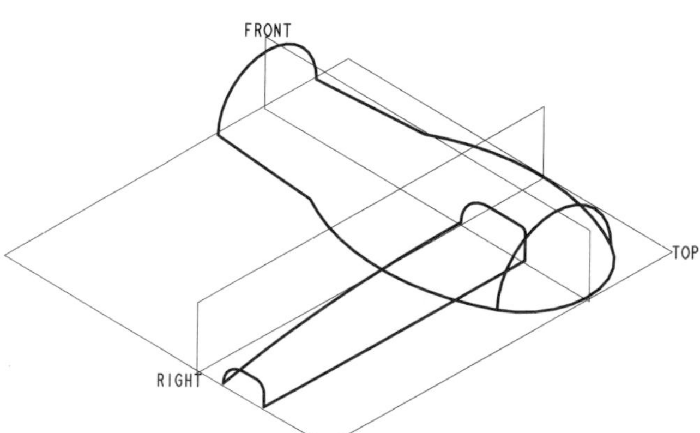

Figure 13–32 Datum curves

STEP 3: Select datum plane TOP as the sketch plane for the datum curve.

STEP 4: Accept the Default Direction and orient datum plane RIGHT toward the right of the work screen (see the orientation of Figure 13–33).

STEP 5: Create the regenerated section shown in Figure 13–33.

Sketch the section using one line and one arc. Create the entities oriented toward the datum planes as shown. Sketch the horizontal centerline aligned with datum plane FRONT and use the dimensioning scheme shown.

NOTE: The center of the arc entity is aligned with datum plane RIGHT, but not with datum plane FRONT.

STEP 6: Exit the sketching environment; then select OKAY on the Feature Definition dialog box.

STEP 7: Use the COPY >> MIRROR option and the DEPENDENT suboption to mirror the first datum curve about datum plane FRONT.

The body of the part is symmetrical about datum plane FRONT. The Dependent option will help to keep the model symmetrical. Your first two datum curves should appear as shown in Figure 13–34.

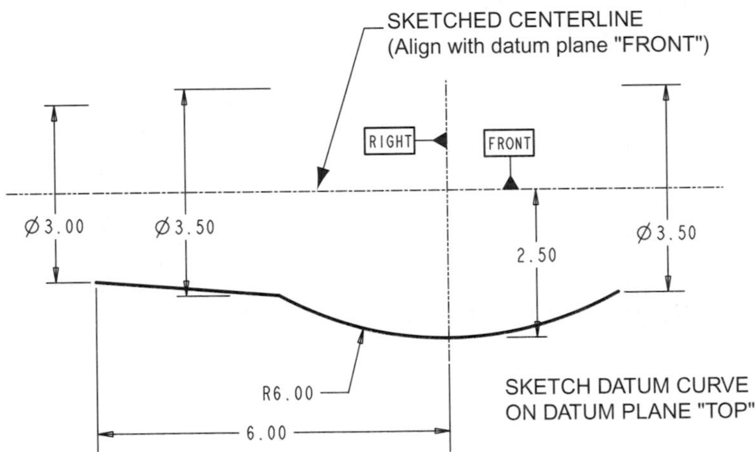

Figure 13–33 The first datum curve

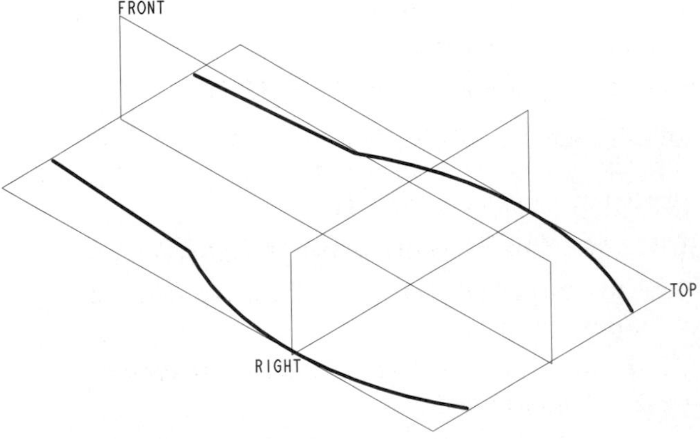

Figure 13–34 Datum curves

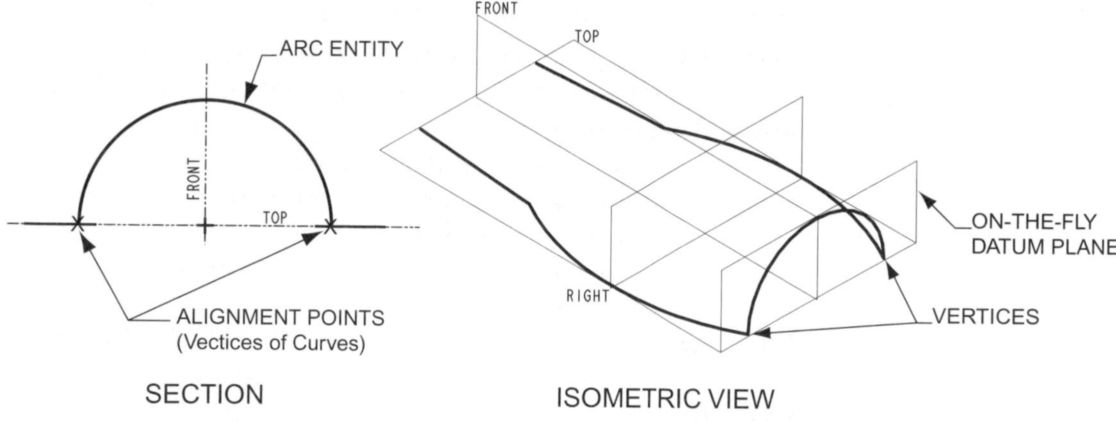

Figure 13–35 Section and isometric view

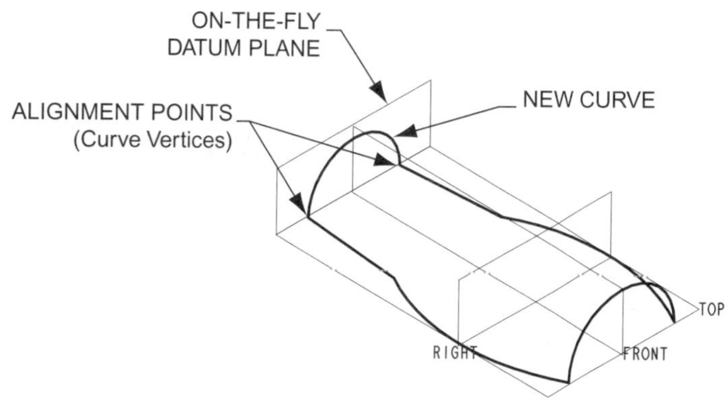

Figure 13–36 New datum curve

STEP 8: Sketch the Datum Curve feature shown in Figure 13–35.

Use the Datum-Curve option and the Sketch suboption to create the curve feature. Sketch the feature on an on-the-fly datum plane that runs through the vertices at the end of the two existing datum curves (see Figure 13–35). In addition, constrain the sketch plane parallel to datum plane RIGHT. When creating the section, use the Arc option and align the ends of the arc with the two vertices shown in the figure.

STEP 9: Sketch the Datum Curve feature shown in Figure 13–36.

Use the same procedure from the previous datum curve to create this datum curve. Sketch the curve on an on-the-fly datum plane running through the two vertices and parallel to datum plane RIGHT.

STEP 10: Sketch the Tangent Datum Curve feature shown in Figure 13–37.

Sketch the curve on datum plane TOP. One end of the curve should be constrained tangent to existing part curves.

STEP 11: Use the Datum Curve option and the Sketch suboption to create CURVE 1 (Figures 13–38 and 13–39).

Sketch CURVE 1 as shown in Figure 13–39. Sketch the entity as a vertical line with one end aligned with datum plane FRONT.

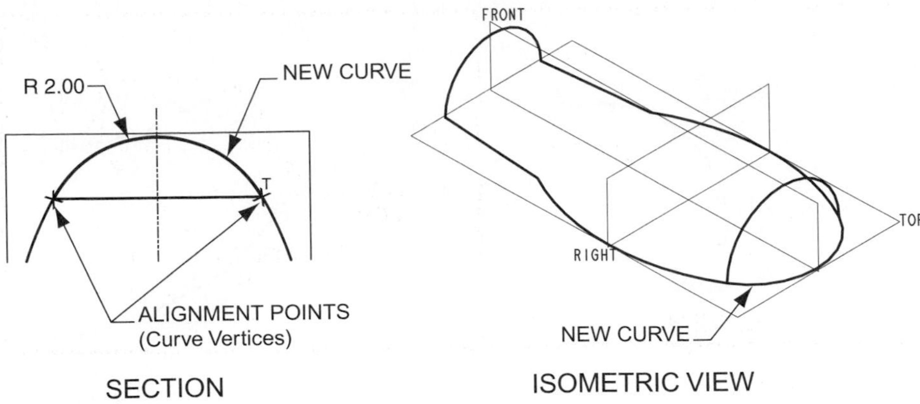

Figure 13-37 New tangent datum curve

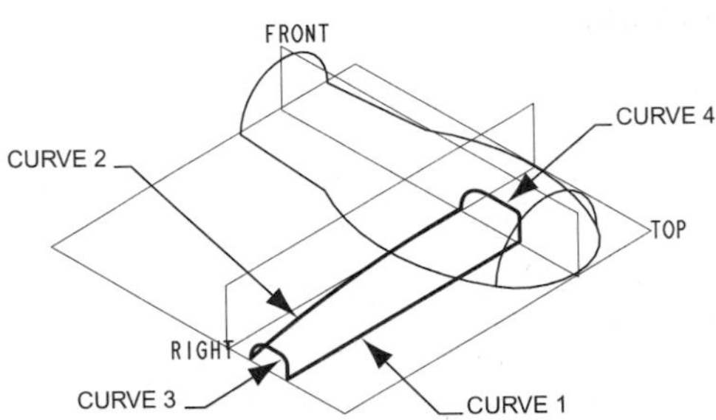

Figure 13-38 Datum curves

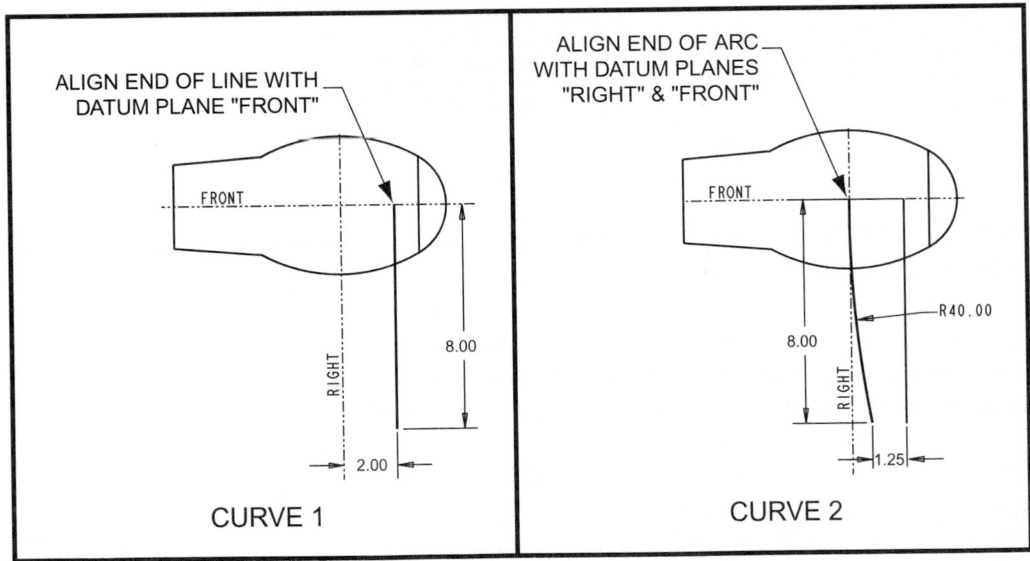

Figure 13-39 Datum curves 1 and 2

STEP 12: Use the Datum Curve option and the Sketch suboption to create CURVE 2 (Figures 13–38 and 13–39).

Sketch CURVE 2 as shown in Figure 13–39. Sketch the entity as an arc with one end aligned at the intersection of datum planes RIGHT and FRONT.

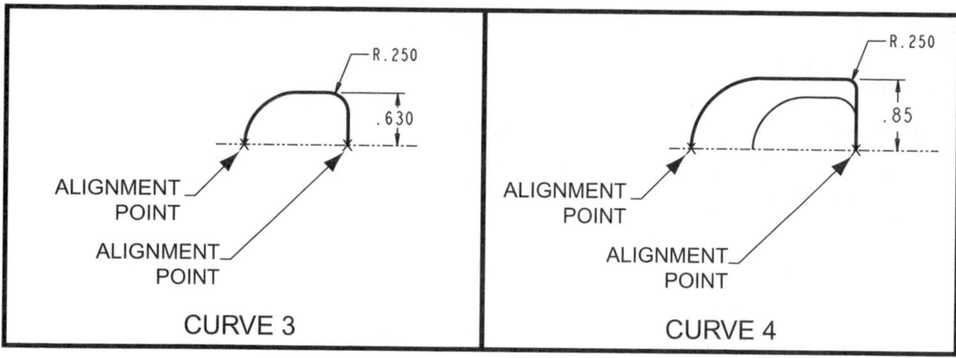

Figure 13-40 Datum curves 3 and 4

STEP 13: Use the Datum Curve option and the Sketch suboption to create CURVE 3 (Figures 13–38 and 13–40).

Sketch CURVE 3 as shown in Figure 13–40. Sketch the section on an on-the-fly datum plane that runs through the ends of CURVE 1 and CURVE 2 and that is Normal to datum plane TOP. Sketch the entities of the section aligned with the ends of CURVE 1 and CURVE 2.

STEP 14: Use the Datum Curve option and the Sketch suboption to create CURVE 4 (Figures 13–38 and 13–40).

Sketch CURVE 4 as shown in Figure 13–40. Sketch the section on an on-the-fly datum plane that runs through the ends of CURVE 1 and CURVE 2. Sketch the entities of the section aligned with the ends of CURVE 1 and CURVE 2.

INSTRUCTIONAL POINT The primary surface quilts of this part will each be constructed with the Boundaries option. Two of the boundaries' surfaces will be constructed in two directions. Datum curves forming the boundaries with this type of surface must be aligned at their vertices. The previous segment of this tutorial specified alignments when necessary. You should double-check to verify that all alignments are correct.

CREATING SURFACES FROM BOUNDARIES

The previous segment of this tutorial created datum curves for use in defining the shape of the part. This segment of the tutorial will use these curves to create surface features by defining each surface quilt's boundary. The four surfaces to be defined are shown in Figure 13–41.

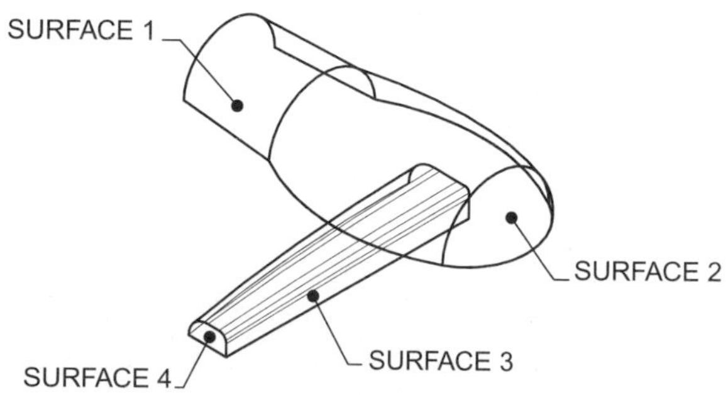

Figure 13-41 Surface creation

STEP 1: On the menu bar, select INSERT >> SURFACE >> FROM BOUNDARIES.

The first surface to be created will be SURFACE 1 shown in Figure 13–41. The Boundaries option is used to define a surface between two or more bounding entities. Boundaries can be defined in one or two directions. When defining a boundary in two directions, the boundary entities must be connected at each entity's vertex.

STEP 2: On the Boundaries Options menu, select the BLENDED SURF option; then select DONE.

STEP 3: With the Curve Options menu's First Direction (First Dir) option selected, pick the two datum curves in the *First Direction* shown in Figure 13–42.

STEP 4: With the Curve Options menu's Second Direction (Second Dir) option selected, pick the two datum curves in the *Second Direction* shown in Figure 13–42.

STEP 5: Select the DONE CURVES option.

STEP 6: Preview the feature, then select OKAY on the Feature Definition dialog box.

STEP 7: Select INSERT >> SURFACE >> FROM BOUNDARIES on the menu bar.

The next surface to be created will be SURFACE 2 shown in Figure 13–42. This surface feature will be created by defining the two boundaries shown in Figure 13–43. In addition, a boundary condition will be defined that makes this surface tangent to SURFACE 1.

STEP 8: On the Boundaries Options menu, select the BLENDED SURF option; then select DONE.

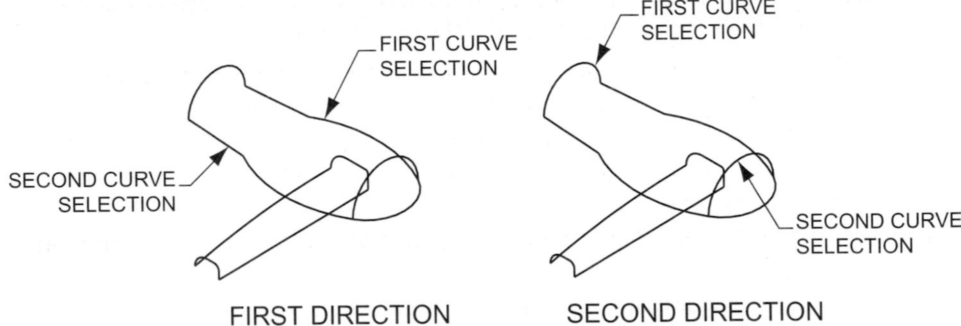

Figure 13–42 First surface creation

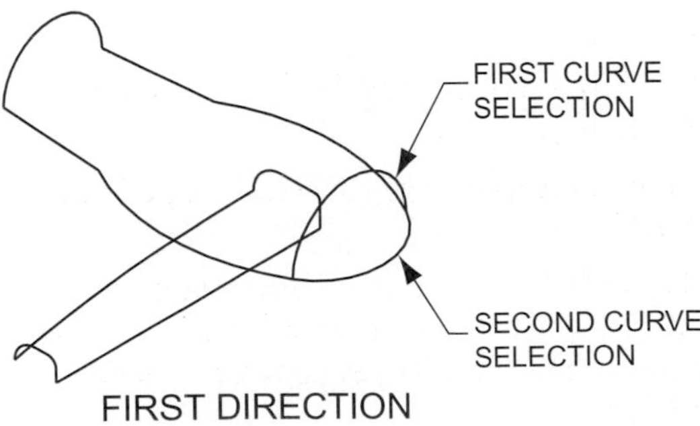

Figure 13–43 Surface 2 creation

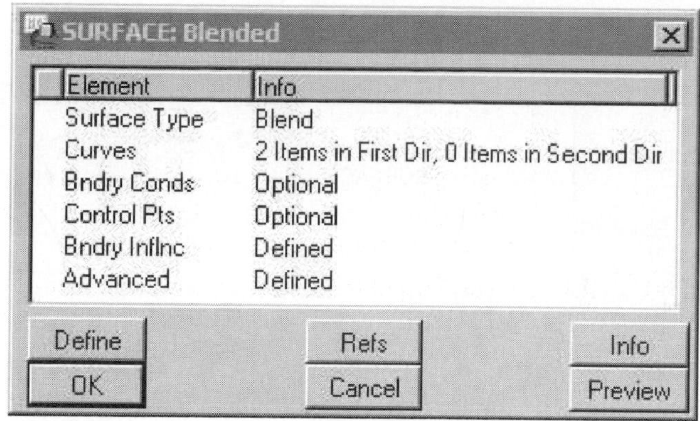

Figure 13–44 Feature Definition dialog box

STEP 9: With the Curve Options menu's First Direction (First Dir) option selected, pick the two datum curves in the *First Direction* shown in Figure 13–43.

STEP 10: Select the DONE CURVES option (do not select OKAY on the dialog box).

STEP 11: Preview the feature on the Feature Definition dialog box.

After selecting the Preview option, shade the model and dynamically rotate it. Notice that the new surface is not tangent with SURFACE 1. The design intent for this part requires tangency between these two surfaces. The Boundary Conditions (Bndry Conds) option on the Feature Definition dialog box (Figure 13–44) is used to establish a specific boundary condition between two surfaces.

STEP 12: On the Feature Definition dialog box, select the BNDRY CONDS element; then select DEFINE.

STEP 13: On the Boundary menu (Figure 13–45), select the Boundary number defining the boundary between the two surfaces

While observing the model, move the cursor over the two available boundary numbers. When the datum curve between the two surfaces is highlighted, this is the required boundary. If you selected the boundary curves in the order shown in Figure 13–43, BOUNDARY #1 should be the correct selection.

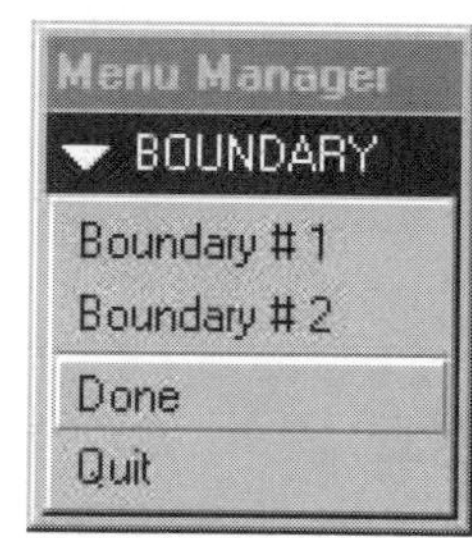

Figure 13–45 Boundary menu

STEP 14: On the Boundary Condition menu, select the TANGENT option; then select DONE.

The following boundary conditions are available:

- **Free** No tangent condition is established between surfaces.
- **Tangent** The surface is created tangent across the boundary.
- **Normal** The surface is created normal across the boundary.
- **Crvtr cont** The surface has curvature continuity across the boundary.

STEP 15: On the Boundary Dialog Box, select the REF TYPE element; then select the DEFINE option (Figure 13–46).

STEP 16: Pick SELECT >> DONE on the Reference Type menu; then pick the surface shown in Figure 13–47.

Notice on the Entity menu, the ENTITY #1 option. When defining the boundary condition, every entity forming the boundary must have a selected conditional surface. In this tutorial, only one entity defines the boundary, so only one surface selection is required.

STEP 17: Select DONE on the Entity menu.

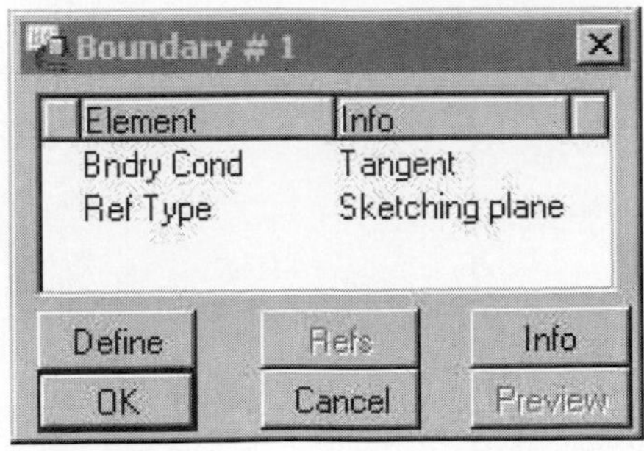
Boundary # 1

Element	Info
Bndry Cond	Tangent
Ref Type	Sketching plane

Define | Refs | Info
OK | Cancel | Preview

Figure 13-46 Boundary 1 Feature Definition dialog box

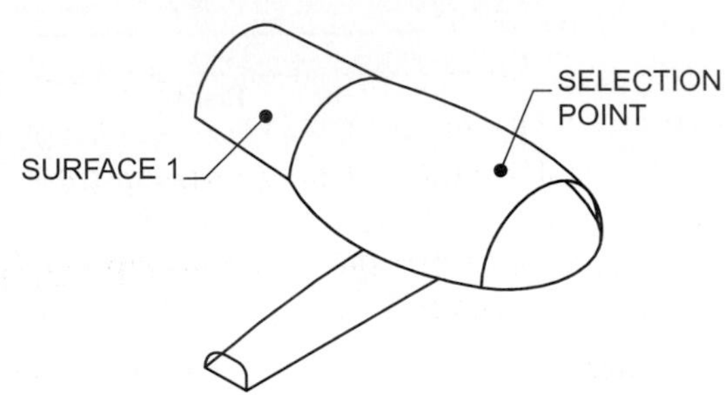

Figure 13-47 Surface selection

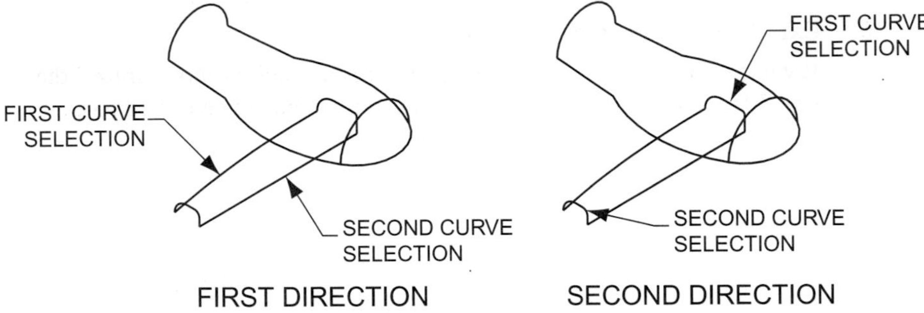

Figure 13-48 Surface creation

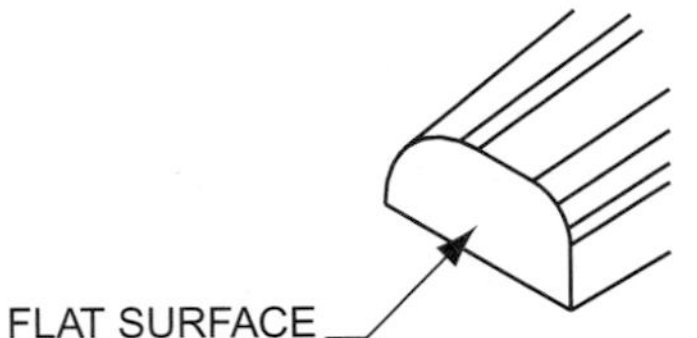

Figure 13-49 Flat Surface option

STEP 18: Select OKAY on the Boundary Dialog Box.

STEP 19: Select DONE on the Boundary menu.

STEP 20: Preview the feature, and then select OKAY on the Feature Definition dialog box.

The next section of this segment will create the surface feature for the handle of the part. You will create the quilt using the Boundaries option.

STEP 21: Use the FROM BOUNDARIES option to create the two–directional boundary feature surface shown in Figure 13–48.

As shown in Figure 13–48, create the surface for the handle using the From-Boundaries surface creation option. Create the Boundaries feature in the two directions shown. In a later segment of this tutorial, this surface will be merged with the surface defining the main body of the part.

STEP 22: Save your object file.

The next section of this tutorial will create the surface feature located at the end of the handle (Figure 13–49). You will create this feature using the New >> Flat option.

STEP 23: Select INSERT >> SURFACE >> FLAT.

The Flat option is used to create a two-dimensional surface feature. The construction technique is similar to the Extrude option, only without the depth definition.

STEP 24: Use the MAKE DATUM option to create the on-the-fly datum plane shown in Figure 13–50.

Use the Through and Normal constraint options to create the three datum plane constraint options shown in Figure 13–50.

STEP 25: Orient and establish the sketching environment.

STEP 26: Use the USE EDGE option and the CHAIN suboption to project the entities shown in Figure 13–51.

The Use Edge option will project existing feature edges onto the sketching plane as sketcher entities. The Chain option will select a chain of entities between two picked entities.

STEP 27: Use the LINE option to create the line shown in Figure 13–52.

STEP 28: Exit the sketching environment.

If you get the error message "*Section must be closed for this feature,*" the Use Edge option probably did not completely define the required loop. Observe the work screen to find the problem area.

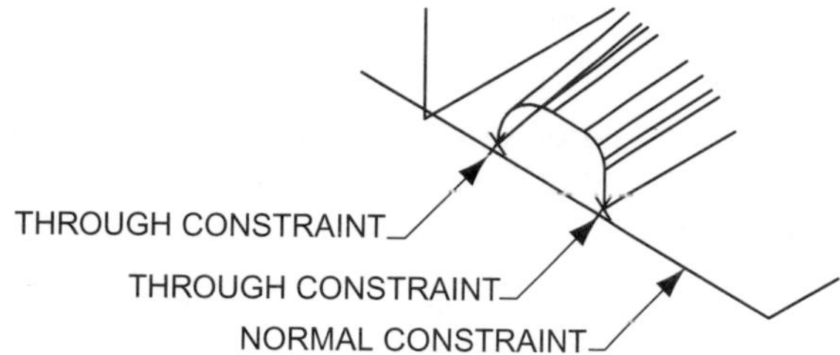

Figure 13–50 Make Datum Constraint options

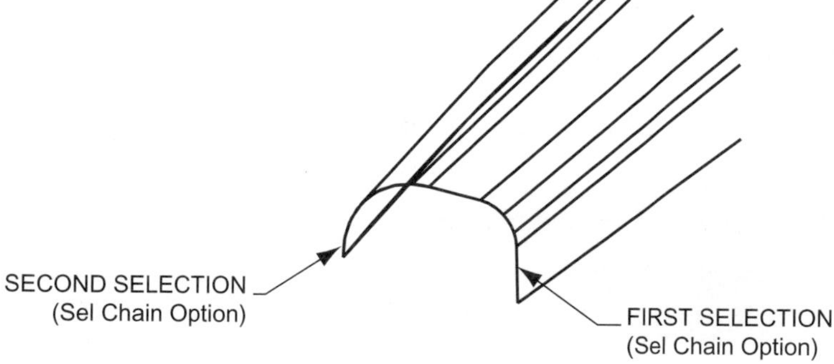

Figure 13–51 Use Edge option

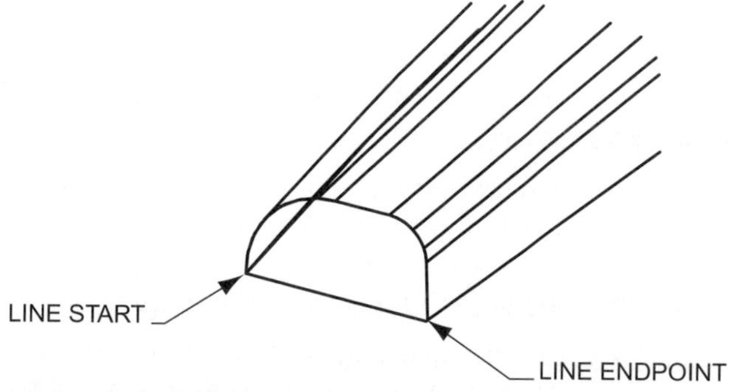

Figure 13–52 Line creation

STEP 29: **Select OKAY on the Feature Definition dialog box to create the surface.**

STEP 30: **Save your part.**

MERGING SURFACES

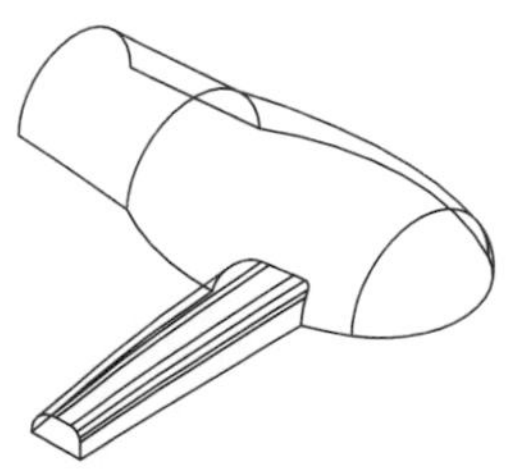

Figure 13-53 Merged surfaces

At this point in the modeling process, your model consists of four surface features. You will use the Merge option to join these surfaces to create one quilted feature. The finished merged feature is shown in Figure 13–53.

STEP 1: **On the menu bar, select INSERT >> SURFACE OPERATION >> MERGE.**

After selecting the Merge option, Pro/ENGINEER will launch the Surface Merge dialog box (Figure 13–54). This dialog box is used to select quilts that either intersect or share a common edge. Quilts that share a common edge can be merged with the Join option. Two quilts that intersect can be merge at the intersection of the quilts.

STEP 2: **With the Primary Quilt option selected on the dialog box, select the Primary Quilt shown in Figure 13–55.**

Select the first surface feature as shown in the illustration.

STEP 3: **With the Additional Quilt option selected on the dialog box, select the Additional Quilt shown in Figure 13–55.**

STEP 4: **On the Surface Merge dialog box, select the JOIN option.**

The Join option is used to merge two quilts that share a common boundary.

STEP 5: **On the Surface Merge dialog box, select the Build Feature option.**

After creating the merged feature, notice the Surface Merge feature on the model tree. The Merge option creates a separate feature.

STEP 6: **Select INSERT >> SURFACE OPERATIONS >> MERGE.**

STEP 7: **With the Primary Quilt option selected on the dialog box, select the Primary Quilt shown in Figure 13–56.**

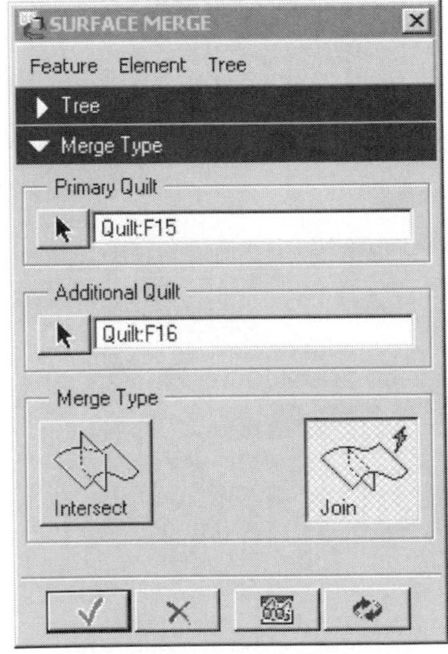

Figure 13-54 Surface Merge dialog box

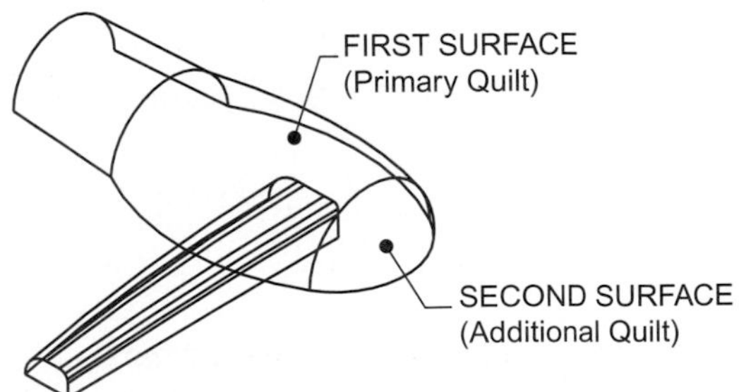

Figure 13-55 Surface selection

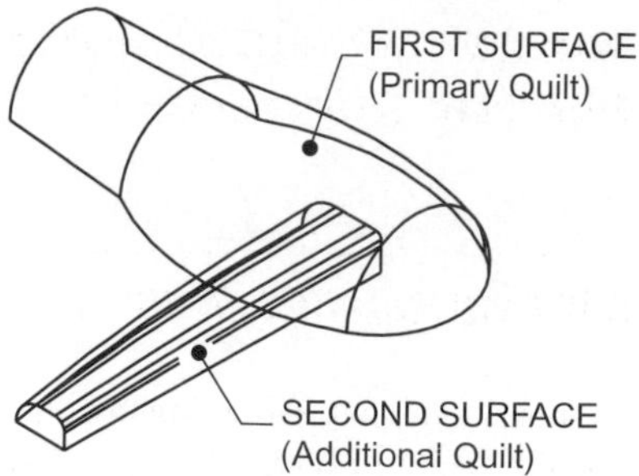

Figure 13-56 Merge selections

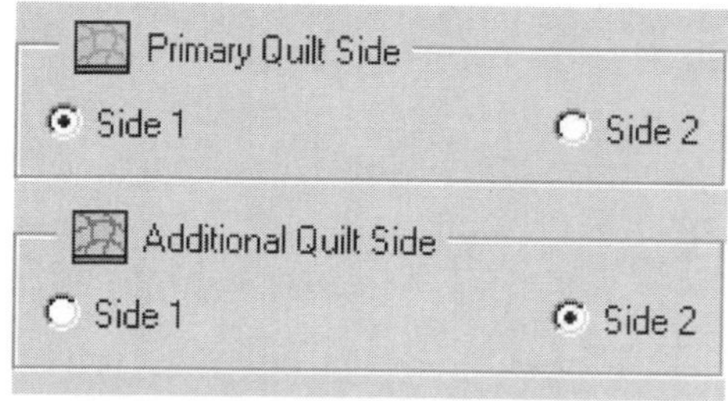

Figure 13-57 Side options

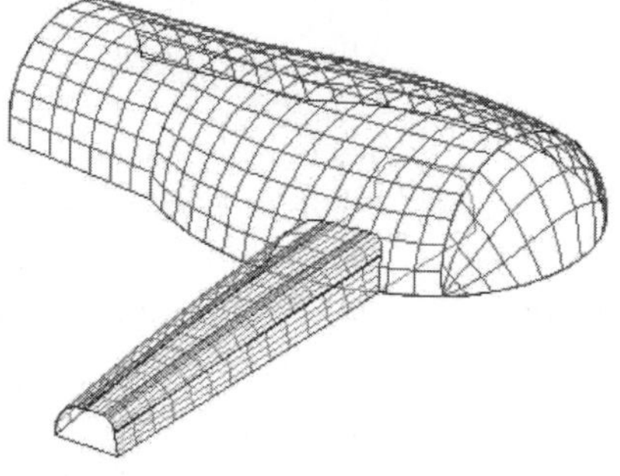

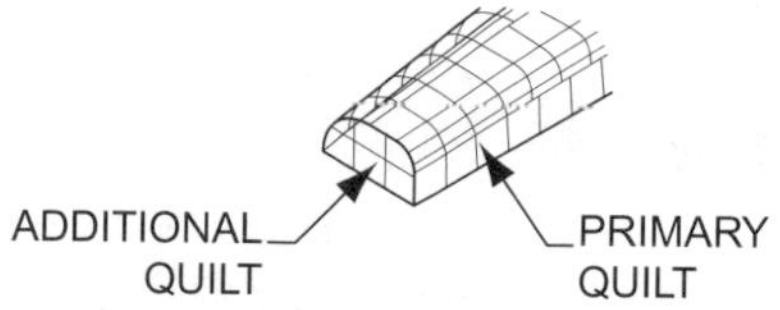

Figure 13-58 Merged quilts

STEP 8: With the Additional Quilt option selected on the dialog box, select the Additional Quilt shown in Figure 13–56.

STEP 9: On the Surface Merge dialog box, select the INTERSECT option.

The Intersect option will merge two quilts that intersect each other. After selecting two quilts, with the Intersect option, there are generally four possible ways to trim the surfaces. The dialog box provides the Primary Quilt Side and Additional Quilt Side options (Figure 13–57) to set appropriate trim sides.

STEP 10: On the dialog box, select the Quilt Side options to create the merged quilt shown in Figure 13–57.

STEP 11: On the dialog box, select the Build Feature option.

STEP 12: Use the MERGE >> JOIN option to join the Flat end quilt with the previously defined merged quilt (Fig. 13–58).

CREATING ADDITIONAL SURFACES

This segment of the tutorial will create additional surface features (Figure 13–59). The first feature created will be a fillet between the body of the part and the handle. The second surface feature will be a revolved quilt forming the trigger finger location.

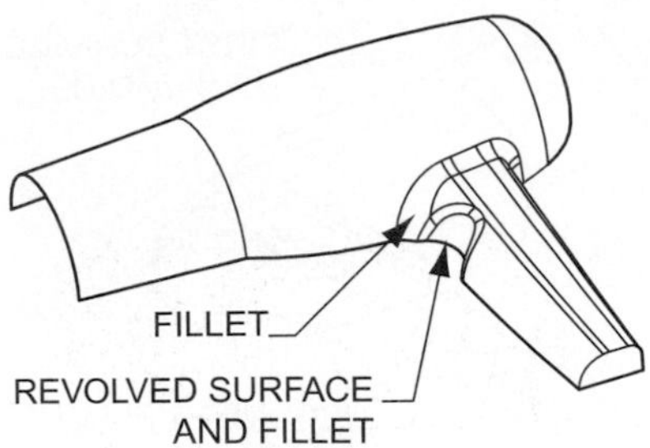

Figure 13–59 Additional surfaces

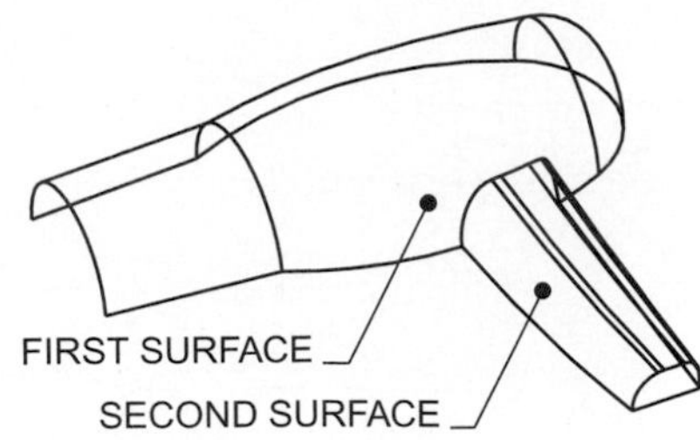

Figure 13–60 Surface selection

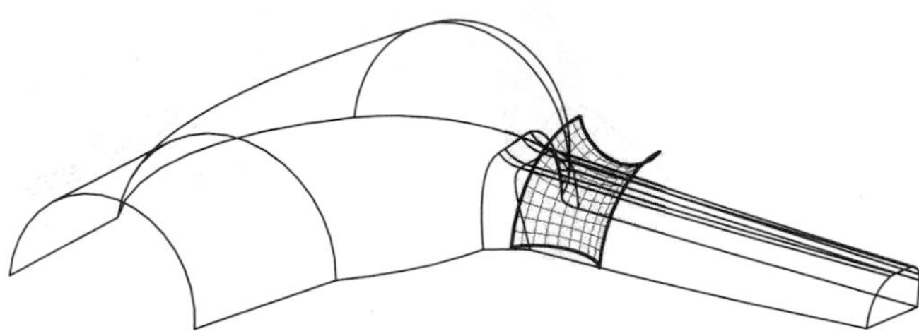

Figure 13–61 Revolved surface

STEP 1: Select INSERT >> SURFACE >> FILLET.

The surface Fillet option works identically to the solid Round command.

STEP 2: Select SIMPLE >> DONE on the Round Type menu.

STEP 3: Select CONSTANT >> SURF-SURF on the Round Set Attribute menu, then select DONE.

The Constant option will create a fillet with a constant radius while the Surface-Surface (Surf-Surf) option defines a fillet between two selected surfaces.

STEP 4: Select the two surfaces shown in Figure 13–60.

STEP 5: Enter .500 as the radius value of the fillet.

STEP 6: Preview the feature, then select OKAY on the Feature Definition dialog box.

STEP 7: Select FEATURE >> CREATE >> SURFACE >> NEW.

Next, you will create the revolved surface shown in Figure 13–61. After creation, you will merge this surface with the remaining part surfaces. The Revolve option is also available under the Insert >> Surface option on the menu bar.

STEP 8: From the Surface Options menu, use the REVOLVE option.

STEP 9: Create the revolved surface shown in Figure 13–61.

Referring to Figure 13–62, use the following options when creating the revolved surface:

- Use One-Side and Open-Ends as attributes.
- Use datum plane TOP as the sketching plane.
- Orient datum plane RIGHT toward the bottom on the sketcher environment.

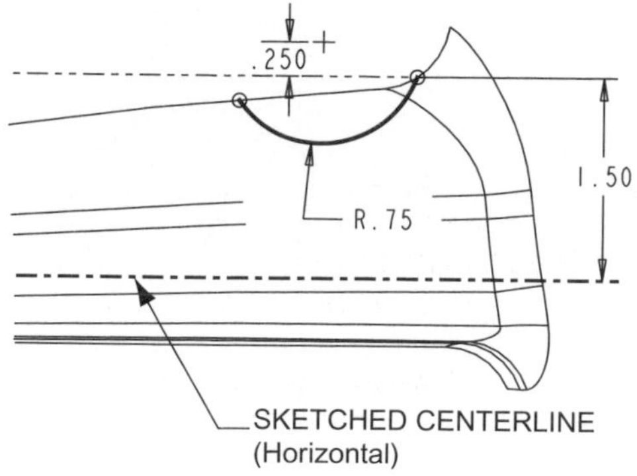

Figure 13–62 Surface section

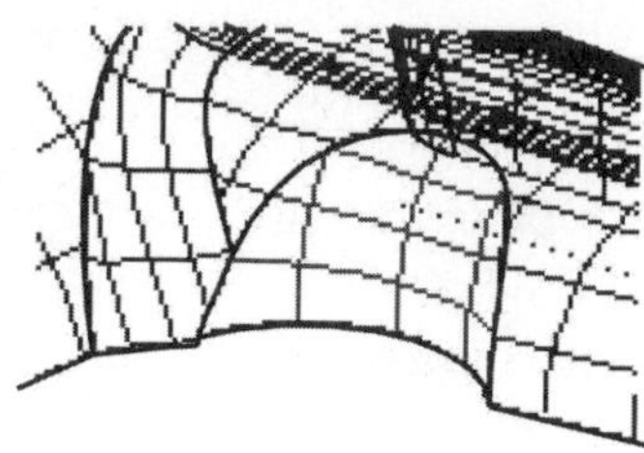

Figure 13–63 Merged surfaces

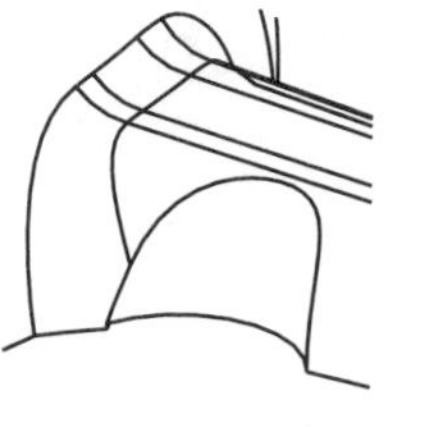

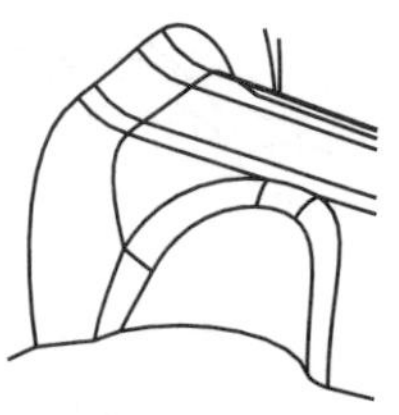

Figure 13–64 Filleted surface

- Sketch the feature's section as shown in Figure 13–62.
- Use the Arc-Center/Ends option to sketch the section.
- Revolved features require a centerline as shown.
- Use a 90-degree rotation angle.

STEP 10: Use the MERGE option to join the previously created revolved surface.

Use the Merge option and the Intersect suboption to join the revolved surface with any intersecting surfaces. The final merge is shown in Figure 13–63.

STEP 11: Use the FILLET option to create the 0.250 radius fillet shown in Figure 13–64.

CREATING A DRAFT OFFSET

This segment of the tutorial will create the Draft Offset feature shown in Figure 13–65. The Draft Offset command is available on the Quilt Surface menu or from the Insert menu on the menu bar. It is also available for solid features under the Tweak menu. The Draft Offset command creates a new surface by offsetting from an existing surface. The sides of the new surface can be beveled similarly to a drafted surface. The new surface can protrude into or away from the selected parent surface. In this tutorial, the surface will protrude into the part.

STEP 1: Select INSERT >> SURFACE OPERATIONS >> DRAFT OFFSET.

STEP 2: Select the NORM-TO-SURF and OFFSET options, and then select DONE.

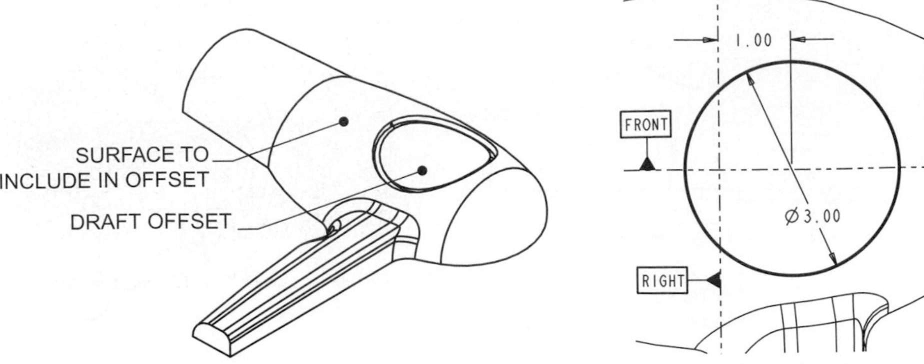

Figure 13–65 Draft offset

Figure 13–66 Feature section

STEP 3: **On the work screen, select the individual surface shown in Figure 13–65 for inclusion in the Draft Offset.**

STEP 4: **On the Surface Selection menu, select the DONE option.**

STEP 5: **Select datum plane TOP as the sketching plane, and orient datum plane FRONT toward the bottom of the sketching environment.**

Draft Offset features are created by projecting a section onto a receiving surface. In this example, the section will be sketched on datum plane TOP.

STEP 6: **Create the section shown in Figure 13–66.**

STEP 7: **When the section is complete, exit the sketching environment.**

STEP 8: **Select TANGENT >> DONE on the Profile Type menu.**

With the Tangent option, the surfaces of the feature will be blended with round, tangent edges. With the Straight option, sharp edges remain.

STEP 9: **Enter an Offset Value that will create a 0.125 offset toward the interior of the surface features (you might have to enter a negative value).**

Observe the work screen and make sure you enter an Offset Value that will protrude the feature into the part. Negative values can be entered if necessary. This value can be modified later.

STEP 10: **Enter a Bevel Value of 30 degrees.**

STEP 11: **Preview the feature, then select OKAY on the dialog box.**

TRIMMING A SURFACE

This segment of the tutorial will create the four trimmed features shown in Figure 13–67. The first feature will be created with the Surface Trim command. You will pattern this feature to create the remaining three features.

STEP 1: **Select INSERT >> SURFACE TRIM >> EXTRUDE.**

STEP 2: **On your work screen, select the Quilt to Intersect shown in Figure 13–68.**

STEP 3: **Select ONE SIDE >> DONE.**

STEP 4: **Select datum plane TOP as the sketching plane for the feature.**

STEP 5: **If necessary, FLIP the extrude direction so the feature will intersect the part.**

Figure 13-67 Trimmed features

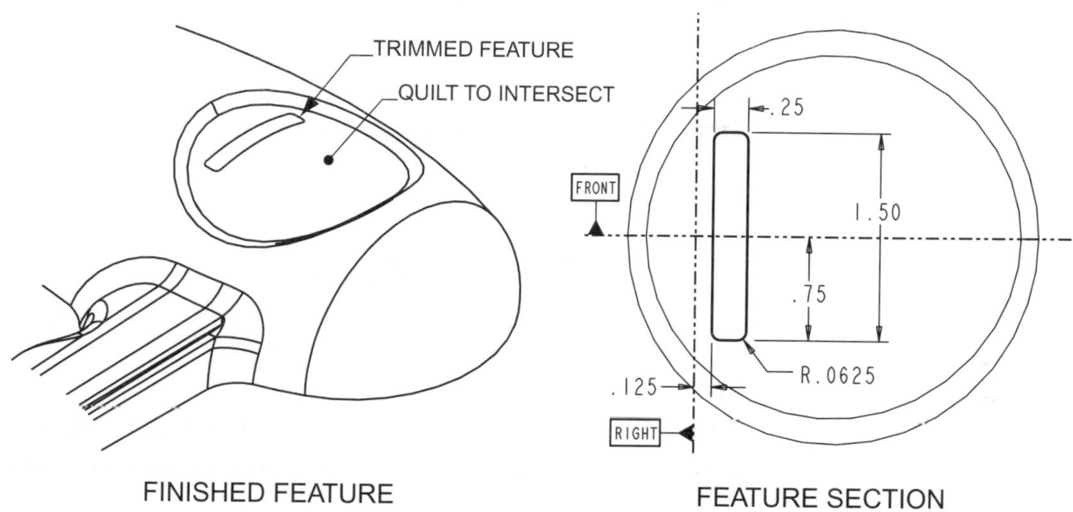

Figure 13-68 Trimmed feature construction

STEP 6: Select OKAY to except the feature creation direction; then accept the Default sketcher orientation.

STEP 7: Sketch the section shown in Figure 13–68.

STEP 8: When the section is complete, exit the sketching environment.

STEP 9: Using the Material Side menu, select a material removal side that will create the appropriate feature (Figure 13–69).

The Material Side menu has options for either removing material from Side 1, Side 2, or Both Sides. The arrows on the work screen point toward the side of the section to be kept. Refer to Pro/ENGINEER's message area when determining a side. This element can be redefined on the Feature Definition dialog box.

STEP 10: Select THRU ALL >> DONE as the Depth option.

STEP 11: Preview the feature, then select OKAY.

Attributes and definitions used to create the trimmed feature can be redefined on the Feature Definition dialog box. The next two steps will pattern this feature.

STEP 12: Use the PATTERN command to create the pattern shown in Figures 13–67 and 13–70.

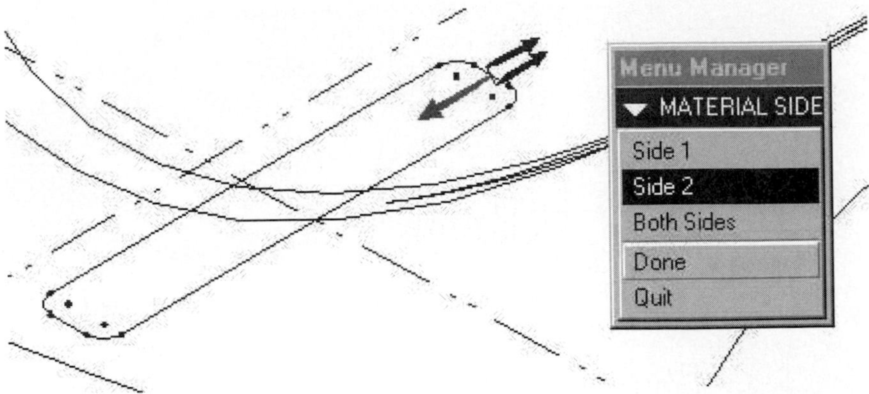

Figure 13-69 Material removal side

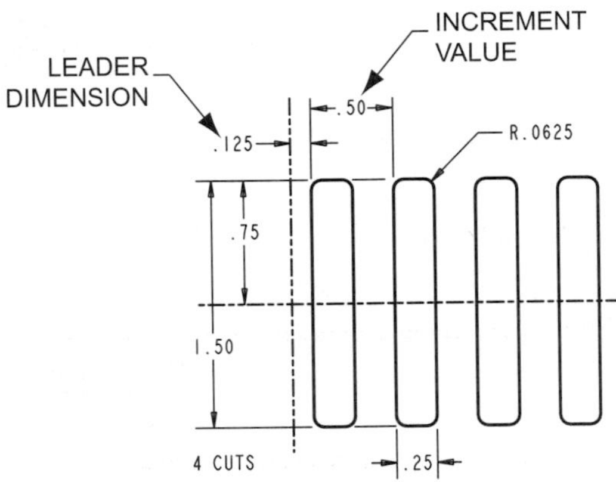

Figure 13-70 Patterned feature

Use the Feature >> Pattern option to create the one-directional linear pattern. Select the Leader dimension shown in Figure 13–70. Increment 4 instances of the pattern a distance of .50 inches.

STEP 13: Save your part file.

CONVERTING A SURFACE TO A SOLID

This segment of the tutorial will convert the part from a surface model to a solid model. In this tutorial, the part will be defined as a thin solid. Once a surface quilt is converted to a solid, normal solid commands can be used to manipulate the feature.

STEP 1: Select FEATURE >> CREATE >> PROTRUSION.

STEP 2: Select USE QUILT >> THIN >> DONE.

STEP 3: Select any surface of the part.

STEP 4: On the Use Quilt dialog box (Figure 13–71), select a Material Side that will create solid material toward the inside of the existing surface features.

STEP 5: Enter .005 as the Thickness for the thin solid feature.

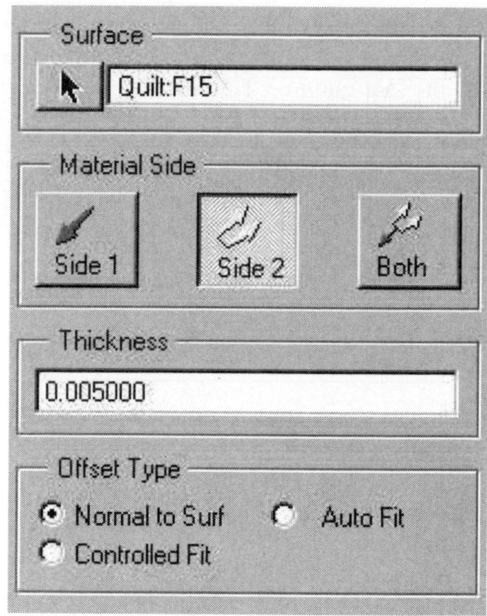

Figure 13–71 Use Quilt dialog box

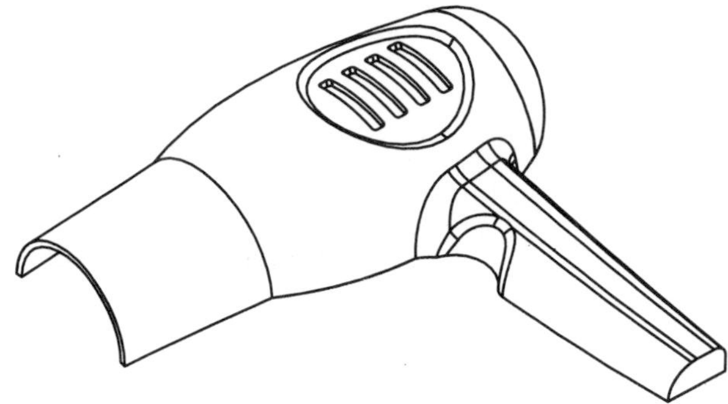

Figure 13–72 Finished part

STEP 6: On the dialog box, select the Build Feature option.

STEP 7: Save your part model.

Your final part should appear as shown in Figure 13–72.

PROBLEMS

1. Model the plastic soap dish shown in Figure 13–73. In this problem, do not model the lid of the dish to be removable from the base.

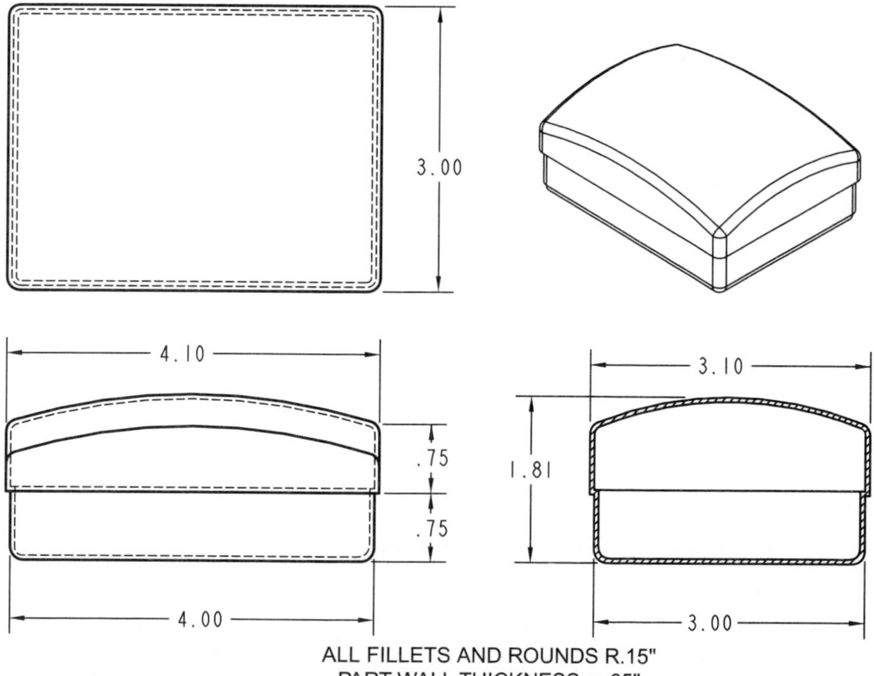

Figure 13–73 Problem one

2. Model the plastic bottle shown in Figure 13–74. The figure includes orthographic views of the part and a view with dimensions for the part's datum curves. Create the part using Pro/ENGINEER's surfacing tools. Use the Use-Quilt >> Thin option under the Protrusion command to convert the surfaces to a solid feature.

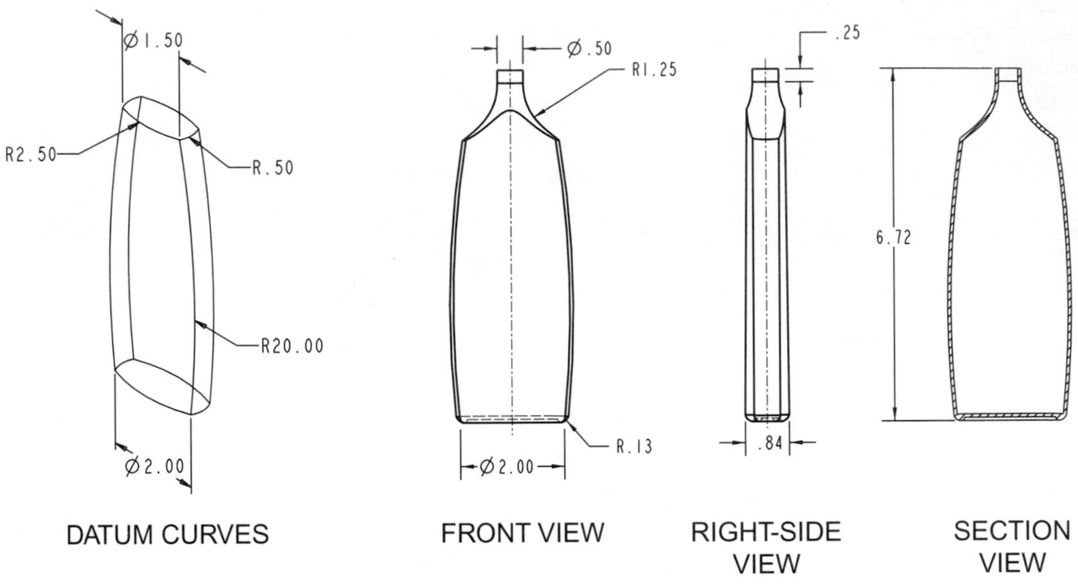

Figure 13–74 Problem two

3. Model the part shown in Figure 13–75.

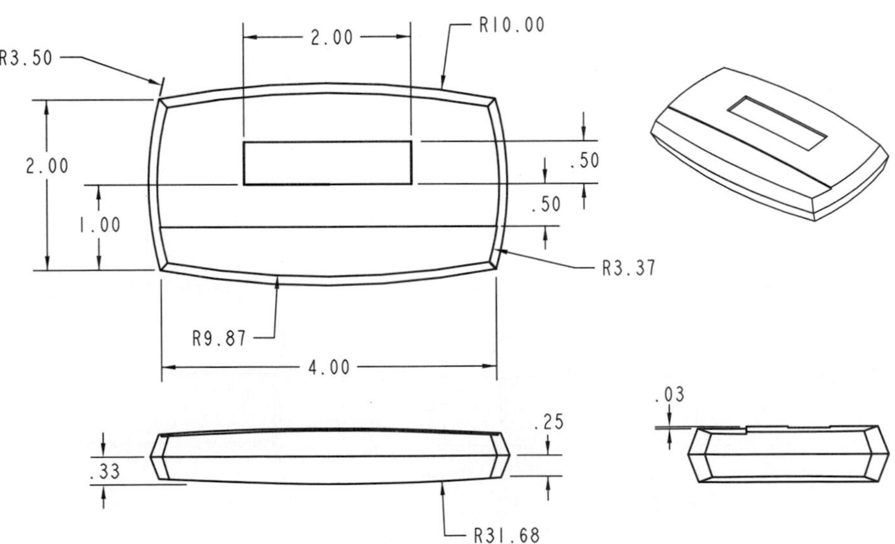

Figure 13–75 Problem three

4. Model the part shown in Figure 13–76.

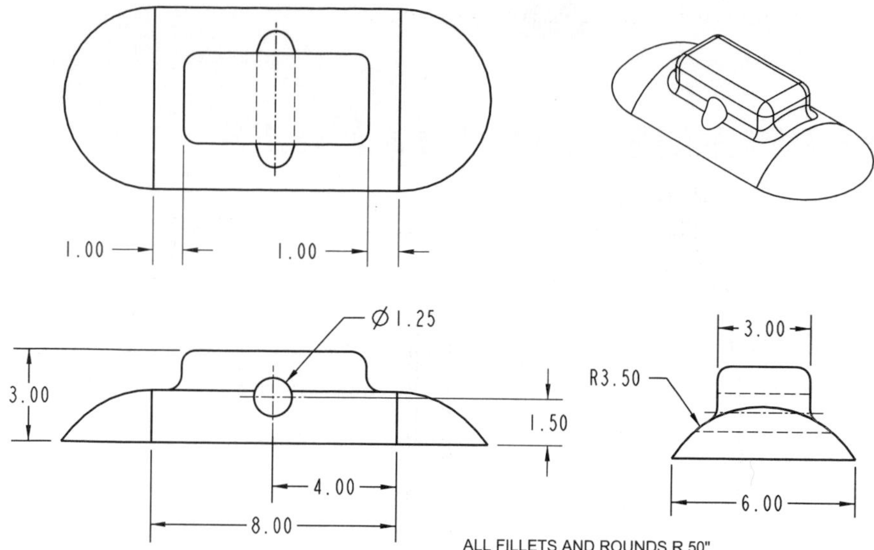

Figure 13-76 Problem four

QUESTIONS AND DISCUSSION

1. How does a surface feature differ from a thin solid feature?
2. How does the Fillet surface option differ from the Round command?
3. Compare the Area Offset option to the Draft Offset command. In addition to the Quilt Options menu, where else within Pro/ENGINEER can these two commands be found?
4. Describe when two surfaces can be merged using the Join suboption.
5. Describe when two surfaces can be merged using the Intersect suboption.

APPENDIX

SUPPLEMENTAL CD FILES

CHAPTER 2	
interface.prt	Used in the Interface Tutorial
CHAPTER 4	
datum1.prt	Used in the Datum Tutorial
CHAPTER 5	
draft1.prt	Used in the “Creating a Neutral Plane No Split Draft” guide
draft2.prt	Used in the “Creating a Neutral Plane Split Draft” guide
draft3.prt	Used in the “Creating a Neutral Curve Draft” guide
pattern1.prt	Used in the “Creating a Linear Pattern” guide
rib.prt	Used in the “Creating a Rib” guide
CHAPTER 7	
copy_mirror.prt	Used to practice the Copy-Mirror option
copy_new_ref.prt	Used to practice the Copy-New Reference option
copy_rotate.prt	Used to practice the Copy-Move-Rotate option
copy_translate.prt	Used to practice the Copy-Move-Translate option
generic_bolt.prt	Used in the “Creating a Family Table” guide
group_pattern.prt	Used in the “Patterning a Group” guide
pattern1.gph	Used in the “Placing a User-Defined Feature” guide
udf_part.prt	Used in the “Creating a User-Defined Feature” guide
CHAPTER 8	
dtl_no_tol.dtl	Drawing Setup File with tolerances not displayed
dtl_yes_tol.dtl	Drawing Setup File with tolerances displayed
detail_view.prt	Part used to practice the establishment of a drawing
CHAPTER 9	
align.drw	Drawing used in the “Aligned Section Views” guide
align.prt	Part used in the “Aligned Section Views” guide
auxiliary.drw	Drawing used in the “Auxiliary Views” guide
auxiliary.prt	Part used in the “Auxiliary Views” guide
broken_out.drw	Drawing used in the “Broken Out Section” guide
broken_out.prt	Part used in the “Broken Out Section” guide
offset.drw	Drawing used in the “Offset Sections” guide

offset.prt	Part used in the "Offset Sections" guide
revolve_sect.drw	Drawing used in the "Revolved Sections" guide
revolve_sect.prt	Part used in the "Revolved Sections" guide
section.drw	Drawing used to practice the creation of section views
section.prt	Part used to practice the creation of section views
section1.prt	Used in "Advanced Drawing Tutorial 1"

CHAPTER 11

var_sec_swp_traj.prt	Used in the "Creating a Variable Section Sweep" guide

CHAPTER 13

surface1.prt	Used in the "Merging Quilts" guide
surface2.prt	Used in the "Creating a Blended Surface from Boundaries" guide
surface3.prt	Used in the "Use Quilt" guide

MISCELLANEOUS

start.prt

config.pro

APPENDIX

B

CONFIGURATION FILE OPTIONS

Option/Possible Values	Description
ANGULAR_TOL *ANGULAR_TOL_0.00*	Used to set angular tolerance values.
ALLOW_ANATOMIC_FEATURES *YES* *NO*	Use to set the display of menu options to include Slot, Shaft, Neck, and Flange.
BELL *YES* *NO*	Used to turn on or off Pro/ENGINEER's message bell.
DEFAULT_DEC_PLACES	Used to set the default number of decimal places.
DEFAULT_DRAW_SCALE	Used to set a default initial drawing scale factor.
DEF_LAY *LAYER_GEOM_FEAT features*	Used to create a default layer and to automatically add items to the layer.
DRAWING_SETUP_FILE C:\dtl\dtl_no_tol.dtl	Used to establish a default drawing setup file.
FONTS_SIZE *SMALL* *MEDIUM* *LARGE*	Used to set the size of menu fonts.
LINEAR_TOL *LINEAR_TOL_0.000*	Used to set linear tolerance values.
ORIENTATION *ISOMETRIC* *TRIMETRIC* *USER-DEFAULT*	Used to set the default orientation of models.
PEN#_LINE_WEIGHT *PEN1_LINE_WEIGHT 2*	Used to set the line weight for a specific pen number. Possible settings can range from 1-16. Each increment equals a width value of .005″.
PRO_CROSSHATCH_DIR	Used to specify the directory where hatch patterns are saved.
PRO_DTL_SETUP_DIR	Used to set a default drawing setup file.
PRO_FORMAT_DIR	Used to set the default directory for locating drawing formats.
PRO_GROUP_DIR	Used to set the default directory for locating User-Defined Features.
PRO_MATERIAL_DIR	Used to set the default directory for locating material files.
PRO_PLOT_CONFIG_DIR	Used to specify the directory where Pro/ENGINEER searches for plotter configuration files (.pcf files).
SAVE_MODEL_DISPLAY	Used to set the shading display in a graphical simplified representation. Possible options include *wireframe, no_display, shading_low, shading medium, shading_high,* and *shading_lod*. The default is *wireframe*.
SKETCHER_DISPLAY_CONSTRAINTS *YES* *NO*	Used to control the display of constraints in the sketcher environment.

(Continued)

Option/Possible Values	Description
SKETCHER_DISPLAY_DIMENSIONS *YES* *NO*	Used to control the display of dimensions in the sketcher environment.
SKETCHER_DISPLAY_GRID *YES* *NO*	Used to control the display of grids in the sketcher environment.
SKETCHER_DISPLAY_VERTICES *YES* *NO*	Used to control the display of vertices in a sketcher environment.
SKETCHER_INTENT_*MANAGER* *YES* *NO*	Used to activate Intent Manager in a sketcher environment.
SPIN_CENTER_DISPLAY *YES* *NO*	Used to control the display of the spin center symbol.
SYSTEM_BACKGROUND_COLOR 100 100 100	Used to set the background color of Pro/ENGINEER's work screen.
SYSTEM_DIMMED_MENU_COLOR	Used to set the color of dimmed entities and menu options.
TANGENT_EDGE_DISPLAY *NO* *SOLID* *CENTERLINE* *PHANTOM* *DIMMED*	Used to set the default tangent edge display style.
TOL_DISPLAY *YES* *NO*	Used to set the default display of tolerances.
TOL_MODE *NOMINAL* *LIMITS* *PLUSMINUS* *PLUSMINUSSYM*	Used to set the default display mode of tolerances.
TOLERANCE_STANDARD *ANSI* *ISO*	Used to select a default tolerance standard.

INDEX

A

Adding relations to a part, 259–260
Additional surfaces, 549–551
Advanced drawing tutorials, 355–376
Advanced rounds, 171–173
Alignment, 69
  section view, 358–360
Angular dimensions, 72–73
ANSI tolerance standard, 32
Approximate boundaries surface, 534–535
Assembly drawing (report), 480–484
Assembly mode menu, 450–451
Assembly modeling, 450–517
  assembly relations, 465–466
  constraints, 451–455
  creating features, 460
  exploded, 468–470
  introduction, 450–451
  layout mode, 466–467
  mechanism design, 456–457
  modifying assemblies and parts, 458–460
  placing components, 451–456
  simplified representation, 467–468
  top-down, 460–465
Assembly tutorial, 471–484
Axis, point, 67

B

Base features, 391, 397–399, 401
  extruded, 531–532
  geometric, 161–163, 175
  protrusions, 216, 253–255, 261
Blend tutorial, 391–396
Blended surface from boundaries, 527–529
Blends, 381–384
  creating, 391–394
  parallel, 382–384
Bolt shaft, 433
Bolt tutorial, 433–438
Boss features, 262–263
Boundaries option, 526–529
Broken front views, 369–370

C

Cartesian coordinates, 388–390
CD files, 558–559
Centerlines
  adding, 371–373
  and dimensions, 362–365
Chamfers, 145–146
  creating, 165–167, 182
Coaxial holes, 179, 255–256, 264
Components
  for an assembly, 471–472
  moving, 455
  packaged, 455
  parametric, 455–456
Conditional relationships, 266–271
Conditional statements, 238–240
Configuration
  file options, 560–561
  printer, 39–40
Constant pitch helical sweep features, 420–421
Constraint modeling, 6–7
Constraint options, 61
  paired, 104
  stand-alone, 103–104
Constraints, 60–61
  assembly, 451–455
  with intent manager, 60–61
  without intent manager, 61
  sketches, 60–61
Construction entities, 68
Construction geometry, 296
Context-sensitive help, 21
Coordinate systems, 387–390
  cartesian, 388–390
Copying features, 229–233
  independent, *versus* dependent, 230
  mirroring, 230–231
  with new references, 233
  options, 229–230
  rotating, 231–232
  selecting a model, 230
  spoke group, 406–409
  translated, 232
Cosmetic features, 154–156
  threads, 155–156
Cross sections, 245–247
  modified, 246
  offset, 247
  planar, 246–247
Cut features, 95–96
  creating, 182–183, 219–220, 395–396
  solid *versus* thin, 96

D

Datum axes, 203–204
Datum curves, 384–387, 402–403, 532–533, 539–543
  projected and formed, 385
Datum display, 21
Datum features, 9–19
Datum planes, 102–104
  creating, 103
  on-the-fly, 104
  paired constraint options, 104
  Pro/ENGINEER's defaults, 103
  stand-alone constraint options, 103–104
Datum points, 386–387
Datum tutorial, 131–138
Default layers, 47
Design intent
  capturing, 56–57
  parametric design, 11–13
Detailed views, 308–310
Dimensional relationships, 7
Dimensional tolerance, 31–37, 288–292
  ANSI tolerance standard, 32
  geometric, 34–37, 292
  ISO tolerance standard, 32–33
  modifying, 33–34
Dimensions, 70–75
  adding, 371–373
  angular, 72–73
  and centerlines, 371–373
  centerlines and, 362–365
  creating, 310–316, 328–330
  linear, 70–71
  manipulating, 289–290
  modifying, 74–75, 290–291
  ordinate, 73
  perimeter, 73
  radial, 71–72
  reference, 73–74
  sketches, 70–75
  variation in, 157
Draft geometry, 295–296
Draft offsets, 551–552
Drafts, 146–151
  creating, 185–186
  neutral curve drafts, 149–151
  neutral plane drafts, 146–149
  neutral planes and curves, 146
Drawing
  starting, 304–305, 355–356, 367–368
  starting with a template, 322–323
Drawing formats, adding, 305
Drawing setup values, 310, 323–324, 357, 368–369
Drawing tutorials, 303–340
Drawing views, 278–281
  menu, 278–279
  types, 279–281
  visibilities, 281
Dynamic viewing, 25–26

E

Editing relations, 240
Elliptic fillets, 68
Explode states, 468–470
Exploded assemblies, 478–480
Extrude tutorial, 109–130
Extruded features, 97–100
  adding, 163–164
  creating, 100–102, 177–179
  depth options, 98–99
  extending, 435
  extrude direction, 97–98
  material side, 100
  open and closed sections in, 99–100
  protrusions, 177–179, 446–449

F

Family tables, 240–245
  adding items to, 241
  creating, 241–245
Feature construction tutorials, 161–190
Feature manipulation, 225–271
  copying, 229–233
  cross sections, 245–247
  family tables, 240–245
  grouping, 225–228
  mirror geometry command, 234
  model tree, 248–250
  regenerating features, 250–251
  relations, 238–240
  simplified representations, 251–252
  user-defined, 234–238
Feature references, 7–8
Feature-based modeling, 5–6, 91–92
Features, 139–190
  blended, 381–384
  chamfers, 145–146
  cosmetic, 154–156
  drafts, 146–151
  holes, 139–143
  patterned, 156–160
  ribs, 152–154
  rounds, 143–145
  shelled parts, 151–152
  suppressing, 248
File management, 20–25
  activating an object, 25
  configuration options, 560–561
  creating a new object, 24
  filenames, 21
  memory, 22
  opening an object, 22–24
  saving an object, 24–25
  working directory, 22
Fillets, elliptic, 68
First component (base.prt), 488–491
First feature, 93, 423
Flange features, 399–400
Formats, 276
Formed datum curves, 385
Fundamentals of sketching, 56–57
  capturing design intent, 56–57
  sketch plane, 57
  sketching elements, 57

G

General views, 306–307, 324–325, 357–358

Geometric dimensioning and tolerancing, 34–37, 292
Geometric tools, for sketches, 76–78
Geometry
  construction, 296
  draft, 295–296
Grid options, 58
Grouping features, 225–228, 264–265, 405
  menu, 226
  patterning, 227–228
  types, 227

H

Helical sweeps, 419–421
Help, context-sensitive, 21
Hole features, 139–143
  depth options, 141–142
  placement options, 140
  straight coaxial, 143
  straight linear, 142–143
  types, 140–141

I

Inserting features, 248
Integration, 8
Intent manager
  constraints with, 60–61
  constraints without, 61
ISO tolerance standard, 32–33
Items
  showing all types, 286–287
  showing and erasing, 285–287

L

Layers, 45–47
  creating, 46
  default, 47
  establishing, 52–54
  setting items to, 46–47
Layout mode, 466–467
Layouts
  creating, 485–488
  declaring and using, 498–500
Left-side views, 370–371
Linear dimensions, 70–71
Linear holes, 169–170, 180
Linear patterns, 157–160, 180

M

Manipulation tutorials, 253–271
Mapkeys, 43–45
Materials, 30–31
  assigning, 31
  writing to disk, 30–31
Mechanism tutorial, 501–517
Mechanisms
  animating, 511–517
  assembling, 503–507
  designing, 456–457
  manipulating, 507–508
  running, 508–511
Memory, 22
Menu bar, 17–19
Menus
  assembly mode, 451
  drawing views, 278–279
  grouping features, 226
  pop-up, 287–288
  user-defined features, 234–235
  views, 278–279
Merging
  quilts, 535–537
  surfaces, 548–549
Mirroring features, 230–231
  extruded, 256–257
Model analysis, 38–39
Model display, 20–21, 26–27
Model tree, 8, 248–250
  inserting features, 248
  reordering features, 249
  rerouting features, 249–250
  suppressing features, 248
Modeling
  advanced techniques, 410–449
  assembly, 450–517
  assembly parts, 501–503
  constraint, 6–7
  dimension values, 373
  feature-based, 5–6
Models, selecting, 230
Modification
  assemblies and parts, 458–460
  component features, 459–460
  cosmetic dimension, 106–107
  cross sections, 246
  dimension, 74–75, 105, 459
  features, 105–107
  formats, 275
  numbers of holes, 214
  tolerance, 290–291
  tolerance table, 33–34
Moving components, 455

N

Naming, views, 28
Neutral curve drafts, 146, 149–151
Neutral plane drafts, 146–147
  creating no split, 147–148
  creating split, 149
New objects
  creating, 24
  creating in sketch mode, 79–80
  starting, 161
New part features, 459
New parts, starting, 175
New sections, 83
Notes
  creating, 292–293, 317–318
  with a standard leader, 293
  without leader, 292–293

O

Objects
  activating, 25
  opening, 22–24, 49–50
  saving, 24–25
  viewing, 50
Offset cross sections, 247
Offset datum planes, 175–177
Opening objects, 22–24, 49–50
Operations, order of, 64, 69–70
Order of operations, 64, 69–70
Ordinate dimensions, 73
Origin trajectory, 439

P

Packaged components, 455
Paired constraint options, 104
Parallel blends, 382–384
Parametric components, 455–456
Parametric design, 1–16
  computer-aided, 1–2
  concurrent, 10–11
  design intent, 11–13
  engineering graphics, 2–3
  Pro/ENGINEER modes, 13–15
Parametric modeling concepts, 3–10
  constraint modeling, 6–7
  datum features, 9–19
  dimensional relationships, 7
  feature references, 7–8
  feature-based modeling, 5–6
  integration, 8
  model trees, 8
  sketching, 6
Parent/child relationships, 38, 92
Partial auxiliary views, 370–371
Partial broken out section views, 361–362
Parts
  adding relations to, 259–260
  creating, 304, 321–322, 355, 367, 461–464
Pattern options, 156–157
Patterned features, 156–160
  creating a linear pattern, 157–160
  dimensions variation, 157
  pattern options, 156–157
Patterning a group, 227–228
  boss, 265–266
Patterns of the cut, 221–224
Perimeter dimensions, 73
Placing components, 451–456
  assembly constraints, 451–455
  into an assembly, 472–477
  moving, 455
  packaged, 455
  parametric, 455–456
Planar cross sections, 246–247
Planes, datum, 102–104
Printing in Pro/ENGINEER, 39–40
Pro/ENGINEER drawing, 272–340
  broken out sections, 348–350
  detailed views, 283–285
  dimensioning and tolerancing, 288–292
  draft cross sections, 298–299
  draft dimensions, 298
  drawing tables, 293–295
  fundamentals, 272–273
  general view, 282–283
  line styles and fonts, 297–298
  manipulating draft geometry, 299–301
  multiple model, 282
  new, 276–277
  notes, 292–293
  pop-up menus, 287–288
  setting a display mode, 283
  setup file, 273–274
  sheet formats, 274–276
  showing and erasing items, 285–287
  two-dimensional, 295–296
  views, 278–281
Pro/ENGINEER interface tutorial, 49–54
  establishing layers, 52–54
  opening an object, 49–50
  setting an object's units, 51
  viewing the object, 50
Pro/ENGINEER's user interface, 17–54
  configuration file, 42–43
  dimensional tolerance set up, 31–37
  environment, 40–42
  file management, 21–25
  layers, 45–47
  mapkeys, 43–45
  materials, 30–31
  menu bar, 17–19
  naming features, 38
  obtaining model properties, 38–39
  printing, 39–40
  selecting features and entities, 47–48
  setting up a model, 28–29
  toolbars, 19–21
  units, 29
  viewing models, 25–28
Projected datum curves, 385
Projection views, 307–308
Protrusions, 95–96
  solid *versus* thin features, 96

R

Radial dimensions, 71–72
Radial hole patterns
  creating, 211–212
  placing, 197
  sketching, 209–211
Redefining, 107–108, 459–460
References
  copying with new, 233
  dimensioning, 73–74
Regenerating, 250–251
Relations, 238–240
  adding and editing, 240
  conditional statements, 238–240
Renaming datum planes, 326–327
Reordering, 249
Rerouting, 249–250

Revolved features, 191–224
creating, 212–213
creating radial pattern, 202
datum axes, 203–204
flange and neck options, 199–200
parameters of, 192–193
revolved cuts, 192–194
revolved feature fundamentals, 191–192
rotational patterns, 201–202
shaft option, 199
sketching and dimensioning, 191–192
Revolved features tutorial, 205–214
Revolved hole options, 195–198
sketching, 195–197
straight-diameter (radial), 197–198
Revolved protrusions, 192–194
creating, 193–194, 205–209
Ribs, 152–154
creating, 153–154, 183–185
Right-side views, 326
Rotation, 231–232
of extruded features, 258–259
Round radii options, 143–144
Round reference options, 144
Rounds, 143–145
creating, 164–165, 187–190, 264, 405
creating a simple round, 144–145
round radii options, 143–144
round reference options, 144

S

Saving
objects, 24–25
views, 28
Second blends, 394–395
Second component (shaft.prt), 492–495
Second feature, 424
Section information, 60
Section tools, 58–60
grid options, 58
placing sections, 58–59
section information, 60
Sections, 341–376
aligned views, 350–351
auxiliary views, 352–354
full, 343–345
fundamentals, 341–342
half, 345–346
offset, 346–348
revolved, 351–352
view types, 342
Settings
configuration options, 215
datum planes, 326–327
dimensional tolerances, 335–337
display modes, 318–320
geometric tolerances, 330–335
items in a layer, 46–47
units, 29, 51
Shaft tutorial, 215–224
Shafts, 217–218
Sheet formats, 274–276
creating, 276
modifying, 275
Shelled parts, 151–152
Shells, inserting, 173–174
Simple rounds, 144–145
Simplified representations, 467–468
creating, 468
Skeleton models, 464–465
Sketch entities, 64–68, 80–82
Sketch plane, 57
Sketcher relations, 75–76
Sketcher tutorials, 79–90
Sketches, 55–90
constraints, 60–61
creating, 83–90
dimensioning, 70–75, 191–192
display options, 61–63
geometric tools, 76–78
section tools, 58–60
Sketching, 6
arcs, 64–66
axis point, 67
circles, 66
elements, 57
elliptic fillets, 68
entities, 64–68
fundamentals of, 56–57
holes, 195
with intent manager, 63–64
without intent manager, 68–70
order of operations, 69–70
lines, 64
rectangles, 67
splines, 67
text, 67
Solids
converting a surface to, 554–557
creating from quilts, 537–538
Splines, 67
Spoke group, copying, 406–409
Spring tutorial, 429–432
Stand-alone constraint options, 103–104
Standard coaxial holes, 167–169
Start parts, 488–491
Straight coaxial holes, 143
Straight linear holes, 142–142
Straight-diameter (radial) holes, 198
Surface modeling, 518–557
boundaries option, 526–529
introduction to surfaces, 518–524
merging quilts, 524–526
options, 520–521, 523–524
use quilt, 529–530
Surface operations, 521–523
Surface tutorials, 531–557
Surfaces
converting to a solid, 554–557
creating from boundaries, 543–548
Sweep tutorials, 397–409
Sweeps
creating, 404–405
helical, 419–421
with a sketched section, 379–381
Swept and blended features, 377–409, 424–428
blended features, 381–384
coordinate systems, 387–390
datum curves, 384–387
fundamentals, 377
Swept blend option, 410–414
Swept blend tutorial, 423–428
Swept features, 378–381, 444–446
Systems of units, 29

T

Third component (pulley.prt), 495–498
Threads
bolt, 433–435
creating cosmetic, 155–156
Through >> axis datum plane, 261–262
Title blocks
creating, 338–340
information in, 374–376
notes in, 366
Tolerance set up
dimensional, 31–37
geometric, 34–35, 292
Tolerance standards
ANSI, 32
ISO, 32–33
Tolerance table, modifying, 33–34
Toolbar, 19–21
context-sensitive help, 21
datum display, 21
file management, 20
model display, 20–21
view display, 20
Tools
feature construction, 139–190
feature manipulation, 225–271
geometric, 76–78
Top-down assembly design, 460–465
creating parts in assembly mode, 461–464
skeleton models, 464–465
Top-down assembly tutorial, 485–500
Trajectory creation, 441
Translated features, 232
Trimming surfaces, 552–554
Tutorials
advanced drawing, 355–376
assembly, 471–484
blend, 391–396
bolt, 433–438
datum, 131–138
drawing, 303–340
extrude, 109–130
feature construction, 161–190
manipulating, 253–271
mechanism, 501–517
Pro/ENGINEER interface, 49–54
revolved features, 205–214
shaft, 215–224
sketcher, 79–90
spring, 429–432
surface, 531–557
sweep, 397–409
swept blend, 423–428
top-down assembly, 485–500
variable section sweep, 439–449
Two-dimensional drafting, 295–296
construction geometry, 296
draft geometry, 295–296

U

Units, 29
creating a system of, 29
setting an object's, 51
User-defined features, 234–238
creating, 235–236
menu, 234–235
placing, 236–238

V

Variable section sweep, 414–418, 442–444
creating, 416–418
Variable section sweep tutorial, 439–449
Variable section sweeps, 416–418
View orientation, 27–28
View visibilities, 281
Viewing models, 25–28
dynamic viewing, 25–26
model display, 26–27
naming and saving views, 28
view orientation, 27–28
Viewing the object, 50
Views
displaying, 20
naming, 28
saving, 28
types, 279–281
Views menu, 278–279

W

Wheel handle, 401
Working directory, 22
Writing a material to disk, 30–31

X

X-trajectory creation, 440

OPTION AND COMMAND GUIDES

FEATURE CREATION

Blend – Parallel 382
Boundaries – Blended Surface 527
Chamfer 145
Coordinate System 388
Cosmetic Thread 155
Datum Axis 203
Datum – Curve 384
Datum Plane – Creation 102
Datum – Point 386
Draft – Neutral Curve 150
Draft – Neutral Plane No Split 147
Draft – Neutral Plane Split 149
Flange 199
Helical Sweep 420
Hole – Coaxial 143
Hole – Linear Placement 142
Hole – Diameter (Radial) Placement 198
Hole – Sketched 196
Neck 199
Protrusion – Extrude 100
Protrusion – Revolve 193
Rib 153
Round – Simple 144
Section – Placing 58
Shell 151
Sweep 379
Swept Blend 412
Variable Section Sweep 416
Use Quilt 529

FEATURE MANIPULATION

Copy – New References 233
Copy – Mirror 230
Copy – Rotate 231
Copy – Translate 232
Cross Section – Offset 247
Cross Section – Planar 246
Family Table – Creating 241
Group – Creating 227
Group – Pattern 227
Insert Mode 248
Mirror Geometry 234
Merging Quilts 524
Pattern – Linear 157
Pattern – Rotational (Radial) 202
Redefining Features 107
Relations – Adding 238
Reordering Features 249
Rerouting Features 249
Simplified Representations (Features) 251
Suppressing Features 248
User-Defined Features – Creating 235
User-Defined Features – Placing 236

DRAWINGS

Aligned Section View – Creating 350
Auxiliary View – Creating 352
Broken Out Section View – Creating 348
Detail View 283
Dimensions (Manipulating) 289
Dimensions (Showing) 285
Display Mode 283
Drawing – Creating 276
Drawing Setup File 273
Drawing Table – Creating 293
Drawing View – Creating (General) 282
Format – Creating 276
Format – Modifying 275
Full Section View – Creating 343
Half Section View – Creating 345
Notes – Creating 292
Offset Section View – Creating 346
Revolved Section View Creating 351
Show/Erase 285

ASSEMBLY OPERATIONS

Creating Parts 461
Explode State 468
Mechanism Assembly 456
Modifying Dimensions 459
Placing a Component 455
Redefining a Component 459
Simplified Representations 467

MISCELLANEOUS OPTIONS

Configuration File 42
Dynamic Viewing 25
Environment Options 40
Geometric Tolerance 35
Layers 45
Mapkeys 43
Materials 30
Opening an Object 22
Printing 39
Saving an Object 24
Tolerance Setup 31
Units 29